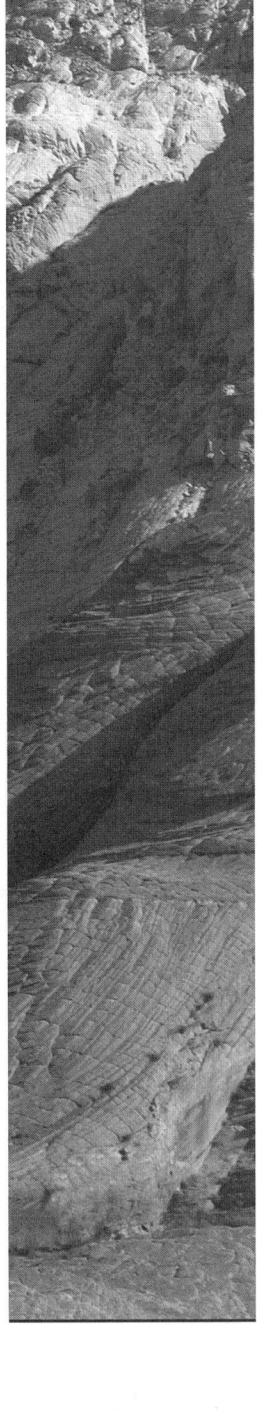

A Dictionary for the Oil and Gas Industry

First Edition

Published by

PETROLEUM EXTENSION SERVICE
THE UNIVERSITY OF TEXAS AT AUSTIN
Continuing & Extended Education
Austin, Texas

2005

Library of Congress Cataloging-in-Publication Data

A dictionary for the oil and gas industry—1st ed.
   p. cm.
   "Based on A dictionary of petroleum terms, third edition revised"—
ISBN 0-88698-213-8
   1. Petroleum engineering—Dictionaries. 2. Gas engineering—
Dictionaries. 3. Petroleum—Dictionaries. 4. Natural gas—Dictionaries.
5. Petroleum industry and trade—Dictionaries. 6. Gas industry—
Dictionaries. I. Dictionary of petroleum terms.
   TN865.D48 2005
   665.5'03—dc22   2005017112

Based on *A Dictionary for the Petroleum Industry,*
third edition revised
©2001 Petroleum Extension Service

©2005 by The University of Texas at Austin
All Rights Reserved
First edition published 1991. Third edition published 1999
Revised edition published 2001. First edition 2005
Second impression 2006
Printed in the United States of America

Catalog No. 1.35010
ISBN 0-88698-213-8

The following publishers have graciously granted permission to use definitions or altered definitions from copyrighted material:

Council of Petroleum Accountants
Societies
P. O. Box 12131
Dallas, TX 75225
    Various accounting definitions.
    Used with permission.

Doubleday Division of Bantam,
Doubleday, Dell Publishing Group,
Inc.
501 Franklin Avenue
Garden City, NY 11530
American Geological Institute,
Dictionary of Geological Terms, rev.
(©1976, 1984 by American Geological
Institute). Used with permission. All
rights reserved.
    body wave; dip-slip fault;
    downthrow; nuclear log; shear
    wave; slip plane; solid solution;
    stratigraphic unit; surface wave;
    ultimate strength

Elsevier Science Publishers Ltd.
Crown House
Linton Road
Barking, Essex IG11 8JU, England
Jenkins Oil Economists' Handbook
(1985). Used with permission. All
rights reserved.
    based on or adapted from
    definitions of arm's length
    bargaining; branded distributor;
    swing producer

Energy Information Administration
Office of Energy Markets & End Use
U.S. Department of Energy
Washington, DC 20585
Monthly Energy Review (August 1989).
Used with permission. All rights
reserved.
    parts of glossary

Gulf Publishing Company
P.O. Box 2608
Houston, TX 77252-2608
David F. Tver, *Gulf Publishing
Dictionary of Business and Science*,
third ed. Houston: Gulf Publishing
Company, 1974. Used with
permission. All rights reserved.
    compression wave; critical
    angle; Joule's law

Gulf Publishing Company
P.O. Box 2608
Houston, TX 77252-2608
Harry Whitehead, ed., *A–Z Dictionary
of Offshore Oil & Gas*, second ed.
Houston: Gulf Publishing, 1983. Used
with permission. All rights reserved.
    visbreaking

McGraw-Hill, Inc.
1221 Avenue of the Americas
New York, NY 10020
Daniel N. Lapedes, editor-in-chief,
*McGraw-Hill Dictionary of Scientific
and Technical Terms*, second ed. New
York: McGraw-Hill, 1978. Used with
permission. All rights reserved.
    caliper; micrometer

PennWell Books, a division of
PennWell Publishing Company
P.O. Box 1260
Tulsa, OK 74101
Langenkamp, *Handbook of Oil Industry
Terms and Phrases*, fourth ed. Tulsa:
PennWell Books, 1985. Used with
permission. All rights reserved.
    middle distillate; straight-run

Langenkamp, *Illustrated Petroleum
Reference Dictionary*, third ed. Tulsa:
PennWell Books, 1984. Used with
permission. All rights reserved.
    carrier bar; charge stock; gas oil;
    process stream; Seven Sisters;
    sling; swag

Preface

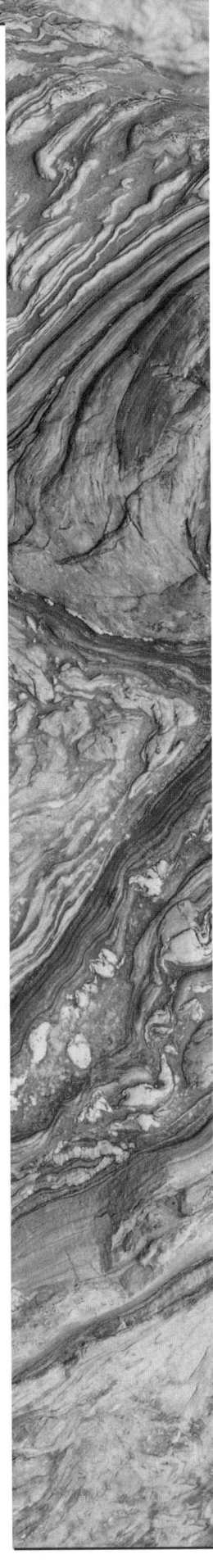

Compiling a dictionary of petroleum terms is a daunting task. PETEX staff first began collecting petroleum terms in the 1950s when they realized the value of a comprehensive dictionary for the petroleum industry. PETEX issued the first edition of *A Dictionary of Petroleum Terms* in the 1970s; it contained about 500 terms. Shortly, however, we realized that this edition had merely scratched the surface. So it was that PETEX released a second edition in the 1980s. The second edition had over twice the number of terms as the first. In 1991, the manual was renamed *A Dictionary for the Petroleum Industry* and we compiled over 5,000 words for it. By 1997, the second edition listed over 7,000 entries. Foolishly, we believed that 7,000 terms would hold us for awhile. Nothing could have been further from the truth, for users of the dictionary quickly informed us of terms we had failed to define. One user in particular, Tom Thomas, who is Transocean's modular training coordinator, sent us hundreds of words that we overlooked. Thus, the third edition grew to over 8,700 terms. The third edition, revised, included almost 9,000 definitions. This first edition has over 11,000 definitions.

We've now reconciled ourselves to the fact that a dictionary of petroleum terms is never going to be finished. Instead, it will continue to grow and change as the industry itself grows, changes, and becomes more technical. Readers are invited and encouraged to send us entries for possible inclusion in future editions.

Definitions in this dictionary come from many sources—writers and editors, industry personnel, PETEX instructors and coordinators, and various published works. Although this dictionary could not have been compiled without these sources, PETEX is solely responsible for its content. Further, while we worked very hard to ensure that our information is accurate and up-to-date, bear in mind that this dictionary is intended for training purposes only. Nothing in it is to be considered approval or disapproval of any product or practice.

Susan Toalson
Director
PETEX

A *abbr*: 1. ampere. 2. cross-sectional area, in.² 3. well spacing, acres.

AAPG *abbr*: American Association of Petroleum Geologists.

AAPL *abbr*: American Association of Petroleum Landmen.

abaft *adv*: 1. toward the stern of a ship or mobile offshore drilling rig. 2. behind. 3. farther aft than. See *aft*.

abandon *v*: to cease producing oil and gas from a well when it becomes unprofitable or to cease further work on a newly drilled well when it proves not to contain profitable quantities of oil or gas. Several steps are involved: part of the casing may be removed and salvaged; one or more cement plugs are placed in the borehole to prevent migration of fluids between the different formations penetrated by the borehole; and the well is abandoned. In most oil-producing states, it is necessary to secure permission from official agencies before a well may be abandoned.

abandoned well *n*: a well not in use because it was a dry hole originally, or because it has ceased to produce. Statutes and regulations in many states require the plugging of abandoned wells to prevent the seepage of oil, gas, or water from one stratum of underlying rock to another.

abandonment *n*: termination of a jurisdictional sale or service. Under Section 7(b) of the Natural Gas Act, the Federal Energy Regulatory Commission must determine in advance that the "present or future public convenience and necessity" or depletion of gas supplies requires termination.

abandonment pressure *n*: the average reservoir pressure at which an amount of gas insufficient to permit continued economic operation of a producing gas well is expelled.

ABC choke *n*: an old-style choke, which is no longer manufactured. A steel rod went through the center of a 2-in. (50.8-mm) pipe, which a rubber boot surrounded. Hydraulic pressure expanded the boot to close the choke. It tended to wear and fail in the full-open position.

abd, abdn *abbr*: abandoned; used in drilling reports.

abeam *adv*: to or at the side of a ship, vessel, or offshore drilling rig and especially at right angles to the ship, vessel, or rig's length.

abnormal pressure *n*: strictly speaking, pressure in a formation that is less than or more than the pressure to be expected at a given depth. However, in the field, abnormal pressure is often considered to be pressure only that is higher than that which is expected at a given depth. Normal pressure increases approximately 0.465 pounds per square inch per foot of depth or 10.5 kilopascals per metre of depth (this value may be slightly more or slightly less than 0.465 psi or 10.5 kilopascals, depending on a particular area). In general, however, normal formation pressure at 1,000 feet is 465 pounds per square inch; at 1,000 metres it is 10,500 kilopascals. See *pressure gradient*.

aboard *adv*: on or in a ship, offshore drilling rig, helicopter, or production platform.

abrasion *n*: wearing away by friction.

ABS *abbr*: American Bureau of Shipping.

ABS certification *n*: a one-time verification that a ship or other marine structure meets an ABS standard.

abscissa *n*: the horizontal coordinate of a point in a plane obtained by measuring parallel to the x-axis. Compare *ordinate*.

ABS classification *n*: a process that occurs over the life of a vessel, ship, offshore rig, or other structure to ensure that such structures are not only built, but also maintained to ABS and industry-accepted standards.

absolute (abs) *adj*: independent or unlimited, such as an absolute condition, or completely unadulterated, such as alcohol.

absolute density *n*: the density of a solid or liquid substance at a specified temperature. Sometimes referred to as true density or density in vacuo. See *density*.

absolute dynamic viscosity *n*: the force in dynes that a stationary flat plate with a surface area of 1 square centimetre exerts on a similar parallel plate 1 centimetre away and moving in its own plane with a velocity of 1 centimetre per second, the space between the plates being filled with the liquid in question. It is a measure of the resistance that the liquid offers to shear.

absolute error *n*: the difference between the result of a measurement and the true value of the measured quantity as determined by means of a suitable standard device.

absolute humidity *n*: the amount of moisture present in the air. It may be expressed in milligrams of water per cubic metre of air. Compare *relative humidity*.

absolute kinematic viscosity *n*: the value obtained when the absolute dynamic viscosity is divided by the density (expressed in grams per cubic centimetre) of the liquid at the temperature concerned.

absolute mass *n*: the expression of a fluid's weight (mass) in terms of its weight in a vacuum.

absolute open flow (AOF) *n*: the maximum flow rate that a well could theoretically deliver with zero pressure at the face of the reservoir.

absolute open flow potential (AOFP) *n*: see *absolute open flow*.

absolute ownership *n*: the theory that minerals such as oil and gas are fully owned in place before they are extracted and reduced to possession. Despite this theory, title to oil and gas may be lost by legitimate drainage and by the rule of capture. Also called ownership in place. See *rule of capture*.

absolute permeability *n*: a measure of the ability of a single fluid (such as water, gas, or oil) to flow through a rock formation when the formation is totally filled (saturated) with that fluid. The permeability measure of a rock filled with a single fluid is different from the permeability measure of the same rock filled with two or more fluids.

Compare *effective permeability, relative permeability.*

absolute porosity *n*: the percentage of the total bulk volume of a rock sample that is composed of pore spaces or voids. See *porosity.*

absolute pressure *n*: total pressure measured from an absolute vacuum. It equals the sum of the gauge pressure and the atmospheric pressure. Expressed in pounds per square inch.

absolute temperature scale *n*: a scale of temperature measurement in which zero degrees is absolute zero. On the Rankine absolute temperature scale, which is based on degrees Fahrenheit, water freezes at 492° and boils at 672°. On the Kelvin absolute temperature scale, which is based on degrees Celsius, water freezes at 273° and boils at 373°. See *absolute zero temperature.*

absolute viscosity *n*: the property by which a fluid in motion offers resistance to shear and flow. Usually expressed as newton-seconds/metre.

absolute zero temperature *n*: a hypothetical temperature at which there is a total absence of heat. Since heat is a result of energy caused by molecular motion, there is no motion of molecules with respect to each other at absolute zero.

absorb *v*: 1. to take in and make part of an existing whole. 2. to recover liquid hydrocarbons from natural or refinery gas in a gas-absorption plant. The wet gas enters the absorber at the bottom and rises to the top, encountering a stream of absorption oil (a light oil) traveling downward over bubble-cap trays, valve trays, or sieve trays. The light oil removes, or absorbs, the heavier liquid hydrocarbons from the wet gas. See *bubble-cap tray, sieve tray, valve tray.*

absorbent *n*: see *absorption oil.*

absorber *n*: 1. a vertical, cylindrical vessel that recovers heavier hydrocarbons from a mixture of predominantly lighter hydrocarbons. Also called absorption tower. 2. a vessel in which gas is dehydrated by being bubbled through glycol. See *absorb.*

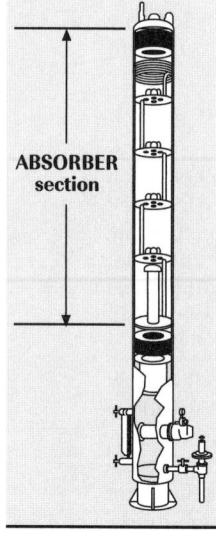

ABSORBER section

absorber capacity *n*: the maximum volume of natural gas that can be processed

through an absorber at a specified absorption oil rate, temperature, and pressure without exceeding pressure drop or any other operating limitation.

absorption *n*: 1. the process of sucking up, taking in and making part of an existing whole. Compare *adsorption.* 2. the process in which shortwave radiation is retained by regions of the earth.

absorption dynamometer *n*: a device that measures mechanical force. The energy measured is absorbed by frictional or electrical resistance.

absorption gasoline *n*: the gasoline extracted from natural gas by putting the gas into contact with oil in a vessel and subsequently distilling the gasoline from the heavier oil.

absorption oil *n*: a hydrocarbon liquid used to absorb and recover components from natural gas being processed. Also called wash oil.

absorption plant *n*: a plant that processes natural gas with absorption oil.

absorption-refrigeration cycle *n*: a mechanical refrigeration system in which the refrigerant is absorbed by a suitable liquid or solid. The most commonly used refrigerant is ammonia; the most commonly used absorbing medium is water. Compare *compression-refrigeration cycle.*

absorption tower *n*: see *absorber.*

abstract-based title opinion *n*: a title opinion based on a complete abstract of title and other relevant documents. Compare *stand-up title opinion.*

abstract company *n*: a private company in the business of preparing abstracts of title and performing related services. Also called abstract plant.

abstract of title *n*: a collection of all of the recorded instruments affecting title to a tract of land. Compare *base abstract.*

abstract plant *n*: see *abstract company.*

abyssal *adj*: of or relating to the bottom waters of the ocean.

Ac *abbr*: altocumulus.

AC *abbr*: alternating current.

AC bus *n*: in a diesel-electric power system, a common set of conductors made up of large, heavy-duty copper cables that carry alternating current generated by the system's alternators (AC generators).

accelerated aging test *n*: a procedure whereby a product may be subjected to intensified but controlled conditions of heat, pressure, radiation, or other variables to produce, in a short time, the effects of long-time storage or use under normal conditions.

acceleration *n*: the rate at which something increases in velocity with respect to time.

acceleration stress *n*: when a crane is hoisting a load, the additional force the load imposes on a wire rope or a sling when the load's speed increases.

accelerator *n*: a chemical additive that reduces the setting time of cement. See *cement, cementing materials.*

accelerometer *n*: an instrument that detects changes in motion or measures acceleration.

accessory *n*: a secondary part or assembly of parts which contributes to the overall function and usefulness of a machine.

accessory equipment *n*: any device that enhances the utility of a measurement system, including readouts, registers, monitors, and liquid- or flow-conditioning equipment.

accrete *v*: to enlarge by the addition of external parts or particles.

accumulate *v*: to amass or collect. When oil and gas migrate into porous formations, the quantity collected is called an accumulation.

accumulator *n*: 1. a vessel or tank that receives and temporarily stores a liquid used in a continuous process in a gas plant. See *drip accumulator.* 2. on a drilling rig, an assembly of devices such as bottles, control valves, pumps, and hydraulic fluid reservoirs that stores hydraulic fluid under pressure and provides a way for personnel to operate (open and close) the blowout preventers. See *blowout preventer, blowout preventer control unit.*

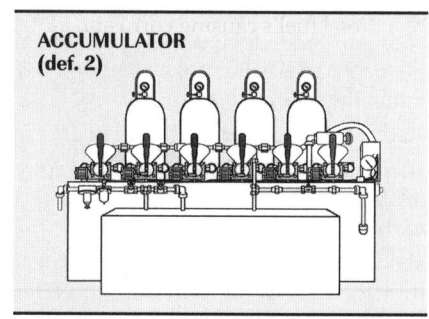

ACCUMULATOR (def. 2)

accumulator bottle *n*: a bottle-shaped steel cylinder located in a blowout preventer control unit to store nitrogen and hydraulic fluid under pressure (usually at 3,000 pounds per square inch). The fluid operates the blowout preventers. See *blowout preventer control unit.*

accuracy *n*: the ability of a measuring instrument to indicate values closely approximating the true value of the quantity measured.

accuracy curve of a volume meter *n*: a plot of meter factor as a function of flow rate used to evaluate the meter's performance. See *flow rate, meter factor.*

acetic acid *n*: an organic acid compound sometimes used to acidize oilwells. It is not as corrosive as other acids used in well treatments. Its chemical formula is $C_2H_4O_2$, or CH_3COOH.

acetylene welding *n*: a method of joining steel components in which acetylene gas and oxygen are mixed in a torch to attain the high temperatures necessary for welding. As an early type of welding (it was also called oxyacetylene welding), its primary disadvantage was the seepage of molten weld material onto the interior surface of the pipe, often leading to corrosion problems.

ACGIH *abbr*: American Conference of Governmental and Industrial Hygienists.

acid *n*: 1. any chemical compound, one element of which is hydrogen, that dissociates in solution to produce free hydrogen ions. For example, hydrochloric acid, HCl, dissociates in water to produce hydrogen ions, H^+, and chloride ions, Cl^-. This reaction is expressed chemically as $HCl + H^+ + Cl^-$. See *ion*. 2. a liquid solution having a pH of less than 7; a liquid acid solution turns blue litmus paper red.

acid brittleness *n*: see *hydrogen embrittlement.*

acid clay *n*: a naturally occurring clay that, after activation, usually with acid, is used mainly as a decolorant or refining agent, and sometimes as a desulfurizer, coagulant, or catalyst.

acid corrosiveness *n*: a characteristic of diesel fuel that indicates the likelihood of a diesel fuel's causing corrosion as the engine burns fuel. In general, a fuel with a high acid content will be more corrosive than a fuel with low acid content.

acid fracture *v*: to part or open fractures in productive hard limestone formations by using a combination of oil and acid or water and acid under high pressure. See *formation fracturing.*

acid gas *n*: a gas that forms an acid when mixed with water. In petroleum production and processing, the most common acid gases are hydrogen sulfide and carbon dioxide. Both cause corrosion, and hydrogen sulfide is very poisonous.

acidity *n*: the quality of being acid. Relative acid strength of a liquid is measured by pH. A liquid with a pH below 7 is acid. See *pH*.

acidize *v*: to treat oil-bearing limestone or other formations with acid for the purpose of increasing production. Hydrochloric or

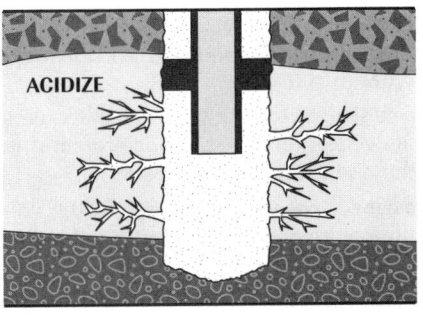

other acid is injected into the formation under pressure. The acid etches the rock, enlarging the pore spaces and passages through which the reservoir fluids flow. Acid also removes formation damage by dissolving material plugging the rock surrounding the wellbore. The acid is held under pressure for a period of time and then pumped out, after which the well is swabbed and put back into production. Chemical inhibitors combined with the acid prevent corrosion of the pipe.

acid recovery plant *n*: plant for the recovery of sulfuric acid from acid sludge.

acid sludge *n*: material of high specific gravity formed during the chemical refining treatment of oils by sulfuric acid and usually separable by settling or centrifuging. Also called acid tar.

acid stimulation *n*: a well stimulation method using acid. See *acidize.*

acid tar *n*: see *acid sludge.*

acid treatment *n*: a method by which petroleum-bearing limestone or other formations are put into contact with an acid to enlarge the pore spaces and passages through which the reservoir fluids flow.

acid-up *v*: to fracture a well using acids (acidizing).

acid wash *n*: an acid treatment in which an acid mixture is circulated through a wellbore to clean the perforations.

acknowledgment *n*: a declaration or an avowal of any act or fact made by a signatory party to a document to a notary public or other public official authorized to take an acknowledgment to give it legal effect.

acoustic backup system *n*: in offshore drilling from a floating drilling rig using a subsea blowout preventer stack, devices that send acoustic signals to a subsea receiver to operate the blowout preventer (BOP) components. The system consists of a miniature control pod with several subsea pilot manipulated (SPM) valves to operate the selected BOP components. The system is used when the hydraulically operated system fails.

acoustic log *n*: a record of the measurement of porosity, done by comparing

depth to the time it takes for a sonic impulse to travel through a given length of formation. The rate of travel of the sound wave through a rock depends on the composition of the formation and the fluids it contains. Because the type of formation can be ascertained by other logs, and because sonic transit time varies with relative amounts of rock and fluid, porosity can usually be determined in this way.

acoustic position reference *n*: a system consisting of a beacon positioned on the seafloor to transmit an acoustic signal, a set of three or four hydrophones mounted on the hull of a floating offshore drilling vessel to receive the signal, and a position display unit to track the relative positions of the rig and the drill site. Monitoring of the display unit aids in accurate positioning of the rig over the site.

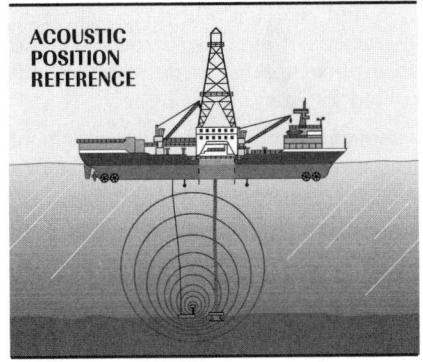

acoustic signatures *n pl*: the characteristic patterns for various degrees of cement bonding between the casing and the borehole that appear on an oscilloscope display when a sonic cement bond log is made.

acoustic survey *n*: a well-logging method in which sound impulses are generated and transmitted into the formations opposite the wellbore. The time it takes for the sound impulses to travel through the rock is measured and recorded. Subsequent interpretation of the record (log) permits estimation of the rock's porosity and fluid content.

acoustic well logging *n*: the process of recording the acoustic characteristics of subsurface formations, based on the time required for a sound wave to travel a specific distance through rock. The rate of travel depends on the composition of the formation, its porosity, and its fluid content. Also called sonic logging.

acquired land *n*: land owned by the United States, acquired by deed or otherwise. Such land has never been in the public domain or was in the public domain at one time and was later reacquired by purchase, condemnation, or donation.

acreage contribution *n*: acreage owned in the vicinity of a test well being drilled by another party and contributed to the driller of the well in return for information obtained from its drilling. The assignment of information is usually made on completion of the well.

acre-foot *n*: a unit of volume often used in oil reservoir analysis, equivalent to the volume (as of oil or water) necessary to cover 1 acre to a depth of 1 foot.

acre-ft *abbr*: acre-foot.

across *prep*: over. The term usually relates conditions of fluid flow on one side of a piece of equipment to conditions on the opposite side (e.g., a pressure drop across a separator).

ACSR *abbr*: aluminum cable steel-reinforced conductor.

ACT *abbr*: automatic custody transfer.

activated charcoal *n*: a form of carbon characterized by a high absorptive and adsorptive capacity for gases, vapors, and colloidal solids.

activation *n*: a reaction in which an element has been changed into an unstable isotope during bombardment by neutrons.

active mud tank *n*: one of usually two, three, or more mud tanks that holds drilling mud that is being circulated into a borehole during drilling. They are called active tanks because they hold mud that is currently being circulated.

actual pressure *n*: the sum of the ideal pressure and the dynamic pressure.

actual residue gas remaining *n*: the volume of gas remaining after processing in a plant.

actual strength *n*: see *breaking strength*.

actuator *n*: a device that activates or puts into motion a process or an action by use of pneumatic, hydraulic, or electronic signals; for example, a valve actuator opens or closes a valve.

adapter kit *n*: a kit comprising a setting sleeve, adapter rod, and adjustment nut for setting drillable or permanent tools on wireline or by using hydraulic setting tools.

adapter spool *n*: a joint to connect blowout preventers of different sizes or pressure ratings to the casinghead.

addend *n*: one of a collection of numbers to be added.

additive *n*: 1. in general, a substance added in small amounts to a larger amount of another substance to change some characteristic of the latter. In the oil industry, additives are used in lubricating oil, fuel, drilling mud, and cement. 2. in cementing, a substance added to cement to change its characteristics to satisfy specific conditions in the well. A cement additive may work as an accelerator, retarder, dispersant, or other reactant.

adhesion *n*: a force of attraction that causes molecules of one substance to cling to those of a different substance.

adiabatic change *n*: a change in the volume, pressure, or temperature of a gas, occurring without a gain or loss of heat.

adiabatic compression *n*: compression of air or gas that exists when no heat is transferred between the air or gas and surrounding bodies (such as the cylinders and pistons in a compressor). It is characterized by an increase in temperature during compression and a decrease in temperature during expansion.

adiabatic expansion *n*: the expansion of a gas, vapor, or liquid stream from a higher pressure to a lower pressure with no change in enthalpy of the stream.

adjustable choke *n*: a choke in which the position of a conical needle, sleeve, or plate may be changed with respect to their seat to vary the rate of flow; may be manual or automatic. See *choke*.

adjustable kickoff tool (AKO) *n*: a part of a downhole motor assembly used to kick off or deflect the hole from vertical in a directional hole. The critical equipment is the adjustable kickoff sub (as opposed to fixed bent sub or bent housing). The kickoff angle can be set in an adjustable sub then pulled and reset without changing tools. It can be used to increase or to decrease hole angle. See *bent housing*.

adjustable spacer sub *n*: a sub run below a dual or triple packer to permit the correct space-out. See *space-out*.

adjustment *n*: the operation of bringing a measuring instrument, such as a meter, into a satisfactory state of performance and accuracy.

administrative penalties *n pl*: penalties or fines handed down by administrative agencies, such as the EPA.

administrator *n*: person appointed by the court to administer the estate of someone who dies without a will (intestate).

ADS *abbr*: atmospheric diving system.

adsorbent *n*: a solid used to remove components from natural gas that is being processed.

adsorption *n*: the adhesion of a thin film of a gas or liquid to the surface of a solid. Liquid hydrocarbons are recovered from natural gas by passing the gas through activated charcoal, silica gel, or other solids, which extract the heavier hydrocarbons. Steam treatment of the solid removes the adsorbed hydrocarbons, which are then collected and recondensed. The adsorption process is also used to remove water vapor from air or natural gas. Compare *absorption*.

adsorption plant *n*: a plant that processes natural gas with an adsorbent.

ad valorem tax *n*: 1. a state or county tax based on the value of a property. 2. tax imposed at a percent of a value.

advance payment agreement *n*: an agreement between a producer and a purchaser to make an advance payment for gas to be delivered at a future date.

advection fog *n*: a fog caused by the movement of warm, moist air over a surface with a temperature less than the dew point of the air. Also called sea fog.

adverse possession *n*: a method of asserting and gaining title to property against other claimants, including the owner of record. The claim through adverse possession must include certain acts (as required by statute) over an uninterrupted interval of time. It must also be, in most states, "open," "notorious," and "hostile."

aeolian *adj*: deposited by wind.

aeolian deposit *n*: a sediment deposited by wind.

aerated drilling *n*: a rotary drilling method in which the circulated drilling fluid is a mixture of liquid and air or gas.

aerated drilling mud *n*: a drilling mud, usually water based, into which air or gas is injected to lighten the mud. Aerated drilling typically uses conventional mud pumps in conjunction with large auxiliary air compressors. The mud is pumped to the standpipe where high-pressure air is injected into the mud stream before it enters the drill stem. Aerated muds may be used in wells where it is not possible to drill with air or gas alone but it is possible to drill with lightweight mud. Using air, gas, or aerated mud increases the rate of penetration because the hydrostatic pressure of such fluids, being less than the hydrostatic pressure of liquid mud, does not tend to hold down the cuttings made by the bit. Thus, the drill cuttings easily move away from the bottom of the hole, which allows the bit to work on fresh, uncut formations. Aerated mud has higher cuttings carrying capacity than air or gas drilling.

aeration *n*: the injection of air or gas into a liquid. In the oil industry a common form of aeration is the injection of natural gas into reservoir liquids standing in a well. Aeration with natural gas reduces the density of the liquids and allows declining reservoir pressure to lift the liquids. See *gas lift*.

aerial cooler *n*: see *air-cooled exchanger*.

aerial river crossing *n*: a river crossing technique in which a pipeline is either suspended by cables over a waterway or attached to the girders of a bridge designed to carry vehicular traffic.

aerobic *adj*: requiring free atmospheric oxygen for normal activity.

aerobic bacteria *n pl*: bacteria that require free oxygen for their life processes. Aerobic bacteria can produce slime or scum, which accumulates on metal surfaces, causing oxygen-concentration cell corrosion.

aerosol *n*: suspension of very small particles in a gas.

AESC *abbr*: Association of Energy Service Companies.

AFCI *abbr*: arc fault circuit interrupter.

AFE *abbr*: authority for expenditure.

affiant *n*: the person who makes a sworn statement.

affidavit *n*: a written affirmation of fact made and sworn to before a notary public or other authorized official. The official signs a certificate called a jurat, which states that the affidavit was signed and sworn to before him or her.

affiliated company *n*: a company that is either directly or indirectly controlled and/or owned by another firm or company.

A-frame *n*: 1. a derrick or crane shaped like the letter A and used to handle heavy loads. 2. an A-shaped openwork structure that is the stationary and supporting component of the mast of a jackknife rig and to which the mast is anchored when it is in an upright or drilling position. 3. the uppermost section of a standard derrick, shaped like the letter A and used as a support in lifting objects such as the crown block to the water table. See *gantry*.

aft *adv*: toward or near the stern of a ship or offshore drilling rig.

aftercooler *n*: on a supercharged engine, a device, cooled by either air or by engine coolant, that reduces the temperature of the engine's exhaust. It is necessary to cool the exhaust's temperature because the exhaust drives the supercharger, which forces air into the engine's intake manifold. The temperature of the supercharged air must be at an acceptable level; otherwise, the engine will run too hot. See *supercharger*.

afternoon tour (pronounced tower) *n*: see *evening tour*.

AGA *abbr*: American Gas Association.

age *v*: to allow cement to mature, or reach a stage harder than that of immediate setting. The process is sometimes called curing. See *cure*.

agency *n*: a federal entity such as a department, a bureau, a corporation, a commission, or a division under the U.S. government that is charged by Congress to administer, implement, and enforce U.S. law.

agent of the state *n*: the surface owner given the right to grant an oil and gas lease, with the owner and state splitting any benefits from that lease.

agglomerate *n*: the groups of larger individual particles usually originating in sieving or drying operations.

agglomeration *n*: the grouping of individual particles.

aggregate *n*: a group of two or more particles held together by strong forces. Aggregates are stable with normal stirring, shaking, or handling; they may be broken by treatment such as ball milling a powder or shearing a suspension.

aggregation *n*: formation of aggregates. In drilling fluids, aggregation results in the stacking of the clay platelets face to face. The viscosity and gel strength decrease in consequence.

aging test *n*: see *accelerated aging test*.

agitator *n*: a motor-driven paddle or blade used to mix the liquids and solids in drilling mud.

A-h *abbr*: ampere-hour.

AHTS *abbr*: anchor-handling tug/supply vessel.

AHV *abbr*: anchor-handling vessel.

AIME *abbr*: American Institute of Mining, Metallurgical, and Petroleum Engineers.

air *n*: a colorless, odorless, tasteless, gaseous mixture, mainly nitrogen (approximately 78 percent) and oxygen (approximately 21 percent) with lesser amounts of argon, carbon dioxide, hydrogen, neon, helium, and other gases.

air-actuated *adj*: equipment activated by compressed air, as are the clutch and the brake system in drilling equipment.

air-balanced pumping unit *n*: a beam pumping unit on which the counterbalancing is done by means of an air cylinder.

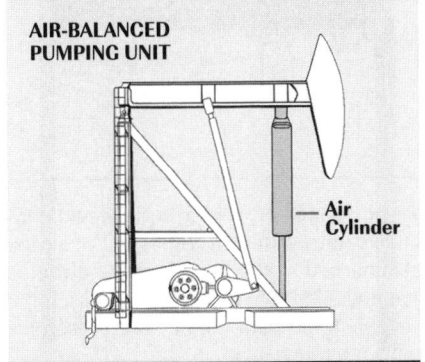

AIR-BALANCED PUMPING UNIT

— Air Cylinder

air bazooka *n*: a special aeration unit that forces air into dry mud material (such as bentonite), and which assists crew members in transferring the dry material from a bulk tank on the rig to a transport truck.

air bit *n*: a roller cone bit that is specially designed for air or gas drilling. It is very similar to a regular bit, but features screens over the bearings to protect them from clogging with cuttings and thicker hardfacing on the shirttail to protect them from abrasive, high-velocity air or gas drilling fluid.

air cleaner *n*: a device installed on an engine's air intake to remove foreign materials, such as dirt or dust, from the air before it enters the engine.

air cleaner element *n*: see *air cleaner filter element*.

air cleaner filter element *n*: the part of an engine's air cleaner that traps dust and dirt particles in the intake air that passes through the element. Filter elements are designed to trap most, but not all, of the foreign material that may be in the air taken in by an engine. (If the filter elements trapped all particles, they would have to be replaced too frequently to be practical.) Some elements, after they become clogged with dust, can be cleaned and reused; others must be discarded and replaced.

air-cooled exchanger *n*: an atmospheric fin-tube exchanger that utilizes air for cooling. Ambient air contacts the external fins by fan-forced or natural draft. Also called air-fin unit or aerial cooler.

air-cored coil *n*: in a coil of wire carrying electric current, the central interior portion of the coil in which only air exists.

air-cut *adj*: having inadvertent mechanical incorporation of air into a liquid system.

air diving *n*: diving in which a diver uses a normal atmospheric mixture of oxygen and nitrogen as a breathing medium. It is limited to depths less than 190 feet (58 metres) because of the dangers of nitrogen narcosis; however, dives with bottom times of 30 minutes or less may be conducted to a maximum of 220 feet (67 metres).

air drilling *n*: a method of rotary drilling that uses compressed air as the circulation medium. The conventional method of removing cuttings from the wellbore is to use a flow of water or drilling mud. Compressed air removes the cuttings with equal or greater efficiency. The rate of penetration is usually increased considerably when air drilling is used; however, a principal problem in air drilling is the penetration of formations containing water, since the entry of water into the system reduces the ability of the air to remove the cuttings.

air eliminator *n*: a part of a LACT unit installed ahead of the metering equipment to separate and remove mass volumes of air and gas from the fluid.

air-fin unit *n*: see *air-cooled exchanger*.

airfoil *n*: a body designed to provide a desired reaction force when in motion relative to the surrounding air.

air gap *n*: the distance from the normal level of the sea surface to the bottom of the hull or base of an offshore drilling platform or rig.

air gun *n*: 1. a hand tool that is powered pneumatically. 2. a chamber filled with compressed air, often used offshore in seismic exploration. As the gun is trailed behind a boat, air is released, making a low-frequency popping noise, which penetrates the subsurface rock layers and is reflected by the layers. Sensitive hydrophones receive the reflections and transmit them to recording equipment on the boat.

air hammer *n*: a drilling tool that, when placed in the drill stem just above a special bit, delivers high frequency percussion blows to the rotating bit. Hammer drilling combines the basic features of rotary and cable-tool drilling (i.e., bit rotation and percussion).

air hoist *n*: a hoist operated by compressed air; a pneumatic hoist. Air hoists are often mounted on the rig floor and are used to lift joints of pipe and other heavy objects.

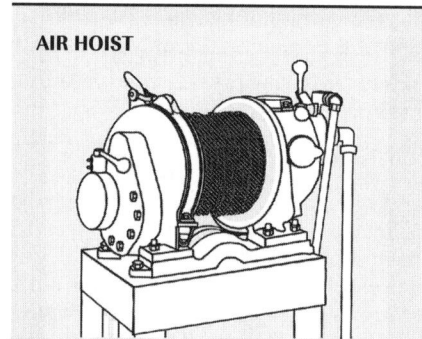

AIR HOIST

air induction system *n*: the network of devices by which air is taken into an engine for combustion in the cylinders.

air injection *n*: 1. the injection of air into a reservoir in a pressure maintenance or an in situ combustion project. 2. the injection of fuel into the combustion chamber of a diesel engine by means of a jet of compressed air.

air intake manifold *n*: on a diesel engine, an arrangement of pipes and passageways through which air is conducted to the engine's combustion chambers.

air-jacketed thermometer *n*: a glass-stem thermometer totally encased in a glass sheath, which provides air space between the thermometer and the liquid in which the unit is immersed.

air knocking *n*: on a diesel engine, a phenomenon that occurs when trapped air in the fuel injection system enters the engine's cylinder with the fuel. The fuel-air mixture ignites but, because of the extra air in the fuel, the engine cylinder misfires and knocks or hammers. The problem should be corrected promptly to prevent damage to the engine.

air mass *n*: a body of air that remains for an extended period of time over a large land or sea area with uniform heating and cooling properties. The air mass will acquire characteristics (such as temperature and moisture content) of the underlying region.

air mass source region *n*: an area over which an air mass rests and develops temperature and moisture characteristics typical of that location.

air motor *n*: a motor powered by compressed air.

air-motor starter *n*: on an engine, a device powered by compressed air that starts the engine. The compressed air, when allowed to enter the starter motor by means of a valve, causes a gear on the starter to engage a gear attached to the outer edge of the engine's flywheel. The rotating starter gear moves the flywheel gear, which causes the engine's pistons to move. If fuel, air, and, on spark-ignition engines, electric spark are present in the engine, the engine will start after a few rotations. As soon as the engine starts, the starter gear disengages from the flywheel gear. Air-motor starters are installed on large industrial engines like those used on a drilling rig.

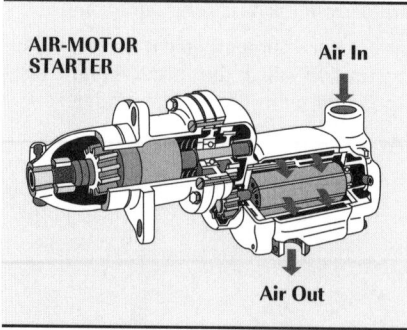

AIR-MOTOR STARTER   Air In

Air Out

air-purge system *n*: in a liquid-level control system, a device consisting of a pipe submerged in the liquid whose outlet is connected above the vessel in which the level is to be maintained. It represents a zero reference level.

air shutoff valve *n*: on a diesel engine, a special valve that, when activated, prevents air from entering the engine's combustion chambers, thereby stopping the engine. Air shutoff valves are a safety feature that may be needed when a well blows out. If natural gas is present in the blowout's fluids, a diesel engine can take in the gas and continue to run even when its normal fuel source is cut off.

air slide *n*: a mechanism using pressurized air through a diaphragm to fluidize powdered materials so that they will flow from a delivery truck to a storage tank on the rig site.

air-trap method *n*: a method, similar to a diaphragm box, that overcomes objections of internal mounting in corrosive liquids and effects of high temperature. A box is placed in a position to trap air under the box and liquid pressure is transmitted to an indicator-recorder.

air tube clutch *n*: a clutch containing an inflatable tube that, when inflated, causes the clutch to engage the driven member. When the tube is deflated, disengagement occurs.

air tugger *n*: see *air hoist*.

air vapor eliminator *n*: a device used to separate and remove gases from a liquid to be measured to prevent an error in liquid measurement.

alarm *n*: a warning device triggered by the presence of abnormal conditions in a machine or system. For example, a low-water alarm automatically signals when the water level in a vessel falls below its preset minimum. Offshore, alarms are used to warn personnel of dangerous or unusual conditions, such as fires and escaping gases.

alidade *n*: a surveying instrument consisting of sighting device, index, and reading or recording device.

aliphatic hydrocarbons *n pl*: hydrocarbons that have a straight chain of carbon atoms. Compare *aromatic hydrocarbons*.

aliphatic series *n*: a series of open-chained hydrocarbons. The two major classes are the series with saturated bonds and the series with unsaturated bonds.

alkali *n*: a substance having marked basic (alkaline) properties, such as a hydroxide of an alkali metal. See *base*.

alkaline *n*: having a pH of more than 7; a liquid alkaline solution turns red litmus paper blue.

alkaline (caustic) flooding *n*: a method of improved recovery in which alkaline chemicals such as sodium hydroxide are injected during a waterflood or combined with polymer flooding. The chemicals react with the natural acid present in certain crude oils to form surfactants within the reservoir. The surfactants enable the water to move additional quantities of oil from the depleted reservoir. Compare *polymer flooding, waterflooding.*

alkalinity *n*: the combining power of a base, or alkali, as measured by the number of equivalents of an acid with which it reacts to form a salt. Any solution that has a pH greater than 7 is alkaline. See *pH.*

alkane *n*: see *paraffin.*

alkanolamine *n*: a chemical family of specific organic compounds, including monoethanolamine (MEA), diethanolamine (DEA), and triethanolamine (TEA). These chemicals, and proprietary mixtures containing them and other amines, are used extensively for the removal of hydrogen sulfide and carbon dioxide from other gases and are particularly adapted for obtaining the low acid gas residuals that are usually specified by pipelines.

alkanolamine process *n*: a continuous-operation liquid process for removing acid gas from natural gas by using chemical absorption with subsequent heat addition to strip the acid gas components from the absorbent solution.

alkyl *n*: a compound derived from an alkane by removing one hydrogen atom.

alkylation *n*: a process for manufacturing components for 100-octane gasoline. An alkyl group is introduced into an organic compound either with or without a catalyst. Now usually used to mean alkylation of isobutane with propene, butenes, or hexenes in the presence of concentrated sulfuric acid or anhydrous hydrofluoric acid.

alligator grab *n*: a fishing device used to pick up relatively small objects like wrenches that have fallen or have been dropped into the wellbore. The alligator grab's jaws are pinned open before the tool is run into the well. The jaws snap shut over the fish when contact is made.

alligatoring *n*: a phenomenon that occurs in paint and other coatings, which is characterized by wide and extensive breaks in the paint or coating that do not penetrate to the substrate.

all-level sample *n*: in tank sampling, a sample that is obtained by submerging a stoppered beaker or bottle to a point as near as possible to the draw-off level, then opening the sampler and raising it at a rate that makes it about three-quarters full (maximum 85%) as it emerges from the liquid.

allocation *n*: the distribution of oil or gas produced from a well per unit of time. In a state using proration, this figure is established monthly by its conservation agency.

allocation meter *n*: see *tail gate.*

allonge *n*: see *rider.*

allotted land *n*: Native American land designated for use by a specific individual, although the title is still held by the United States.

allowable *n*: the amount of oil or gas that can be produced legally from a well. In a state using proration, this figure is established monthly by its conservation agency. See *proration.*

alloy *n*: a substance with metallic properties that comprises two or more elements in solid solution. See *ferrous alloy, nonferrous alloy.*

alluvial fan *n*: a large, sloping sedimentary deposit at the mouth of a canyon, laid down by intermittently flowing water, especially in arid climates, and composed of gravel and sand. The deposit tends to be coarse and unworked, with angular, poorly sorted grains in thin, overlapping sheets. A line of fans may eventually coalesce into an apron that grows broader and higher as the slopes above are eroded away.

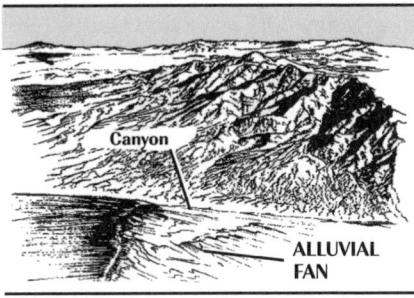

alnico *n*: a metal alloy of aluminum, nickel, and one or more of the elements cobalt, copper, and titanium. This alloy is used to make strong permanent magnets.

alnico magnet *n*: a permanent magnet made from the alloy alnico. See *alnico.*

ALP *abbr*: articulated loading platform.

alpha particle *n*: one of the extremely small particles of an atom that is ejected from a radioactive substance (such as radium or uranium) as it disintegrates. Alpha particles have a positive charge.

alternating current (AC) *n*: current that flows in one direction in a circuit and then reverses flow. Reversal of current flow happens rapidly—for example, in many AC circuits in the United States, reversal happens 60 times in 1 second.

alternation *n*: in an alternating current generator, the rising and falling of induced positive and negative voltage as the coil rotates from 0 to 180 degrees to produce positive voltage and from 180 to 360 degrees to produce negative voltage. The rising and falling of positive voltage from 0 to 180 degrees is one alternation. The rising and falling of negative voltage from 180 to 360 degrees is also one alternation. One positive and 1 negative alternation is 1 cycle.

alternator *n*: an electric generator that produces alternating current.

altocumulus (Ac) *n*: a white or gray mid-level cloud that appears as closely arranged rolls. This type of cloud is composed of either ice crystals or water droplets.

altostratus *n*: a bluish or grayish layer of uniform mid-level clouds that cover large portions of the sky. This type of cloud is composed of either ice crystals or water droplets.

alumina *n*: aluminum oxide (Al_2O_2), used as an abrasive, refractory, catalyst, and adsorbent.

aluminum bronze *n*: an alloy of copper and aluminum that may also include iron, manganese, nickel, or zinc.

aluminum cable steel-reinforced conductor *n*: wireline made of braided aluminum wires, which is strengthened by the addition of steel wires, and which conducts electricity.

aluminum stearate *n*: an aluminum salt of stearic acid used as a defoamer. See *stearate.*

AM *abbr*: amplitude modulation.

Amagat-Leduc rule *n*: states that the volume occupied by a mixture of gases equals the sum of the volumes each gas would occupy at the pressure and temperature of the mixture.

A-mast *n*: an A-shaped arrangement of upright poles, usually steel, used for lifting heavy loads. See *A-frame.*

ambient conditions *n pl*: the conditions (pressure, temperature, humidity, etc.) of the medium surrounding the case of a meter, instrument, transducer, etc.

ambient pressure *n*: the pressure of the medium that surrounds an object.

ambient temperature *n*: the temperature of the medium that surrounds an object.

Amerada bomb *n*: a wireline instrument for measuring bottomhole temperature or pressure. It contains a clock-driven recording section and either a pressure element or a temperature element. It is thin enough to pass downhole through small tubing.

American Association of Petroleum Geologists (AAPG) *n*: a leading national industry organization established to disseminate scientific and technical ideas and data in the field of geology as it relates to oil and natural gas exploration and production. Its official publications are the *AAPG Bulletin* and the *AAPG Explorer*. Address: P.O. Box 979; Tulsa, OK 74101; 918-584-2555.

American Association of Petroleum Landmen (AAPL) *n*: an international trade organization of landmen and related professionals. Its official publication is *The Landman*. Address: 4100 Fossil Creek Blvd.; Fort Worth, TX 76137; 817-847-7700.

American Bureau of Shipping (ABS) *n*: a leading ship classification society, its main purpose is to determine the structural and mechanical fitness of ships and other marine structures through classification procedures. ABS establishes and administers standards, known as rules, for the design, construction, and operational maintenance of marine vessels and structures. They operate as three divisions: ABS Americas, ABS Europe, and ABS Pacific. Corporate headquarters are in Houston, Texas. Address: ABS Plaza, 16855 Northchase Drive, Houston, TX 77060; 281-877-5800; www.eagle. org.

American Conference of Governmental and Industrial Hygienists (ACGIH) *n*: a professional multinational organization of persons employed by governmental units responsible for full-time programs of hygiene and worker health and protection. Address: 1330 Kemper Meadow Drive; Cincinatti, Ohio 45240; 513-742-2020.

American Gas Association (AGA) *n*: a national trade association whose members are U.S. and Canadian distributors of natural, manufactured, and mixed gases. AGA provides information on sales, finances, utilization, and all phases of gas transmission and distribution. Its official publications are *AGA Monthly* and *Operating Section Proceedings*. Address: 400 N. Capitol Street, NW; Washington, DC 20001; 202-824-7000; fax 202-824-7115.

American Institute of Mining, Metallurgical, and Petroleum Engineers (AIME) *n*: parent group of the Society of Petroleum Engineers (SPE). See *Society of Petroleum Engineers*. Its official publication is the *Journal of Petroleum Technology*. Address: P.O. Box 270728, Littleton, CO 80127; 303-948-4255.

American National Standards Institute (ANSI) *n*: serves as clearinghouse for nationally coordinated voluntary standards for fields ranging from information technology to building construction. Address: 11 W. 42nd Street, 13th floor; New York, NY 10036; 212-642-4900; www.ansi.org.

American Petroleum Institute (API) *n*: an oil trade organization (founded in 1920) that is the leading standardizing organization for oilfield drilling and producing equipment. Headquartered in Washington D.C., it publishes materials concerning exploration and production, petroleum measurement, marine transportation, marketing, pipelining, refining, safety and fire protection, storage tanks, valves, training, health and environment, policy, and economic studies. Address: 1220 L Street NW; Washington, DC 20005; 202-682-8000; www.api. org.

American Society for Testing and Materials (ASTM) *n*: organization that sets guidelines for the testing and use of equipment and materials. Its publications are *ASTM Standardization News, Book of Standards, Cement and Concrete Aggregates Journal, Composites Technology & Research, Geotechnical Testing Journal, Journal of Forensic Science,* and *Journal of Testing and Evaluation*. Address: 100 Barr Harbor Dr.; West Conshohocken, PA 19428-2959; 610-832-9500.

American Society of Mechanical Engineers (ASME) *n*: organization whose equipment standards are sometimes used by the oil industry. Its official publication is *Mechanical Engineering*. Address: 3 Park Avenue; New York, NY 10016; 212-591-7000; fax 212-591-7739.

American Society of Safety Engineers (ASSE) *n*: organization that establishes safety practices for several industries. Its publications are *Professional Safety* and *Society Update*. Address: 1800 East Oakton; Des Plaines, IL 60018-2187; 847-699-2929.

American standard code for information interchange (ASCII) *n*: in computer science, a standard for assigning numerical values to the set of letters in the Roman alphabet and to typographic characters. For example, in ASCII code, the binary number 1001011 is assigned the letter K, and the binary number 1011110 is assigned the typographic character ^.

American War Standard (AWS) *n*: outdated specification for supplies and equipment purchased by the military in World War II. It has been replaced by military specification (MIL).

American wire gauge (AWG) *n*: a standard that assigns a number (a gauge) to electrical wires of various diameters, or sizes. The largest AWG wire is listed as 0000, which is often written as 4/0, and has a diameter of 460 mils, or 0.46 inches. (A mil is 0.001 inches.) A wire often used in small household appliances, such as small table lamps, is 16 gauge, which is 51 mils, or 0.051 inches, in diameter. The smallest is 40-gauge wire, which is 3.1 mils, or 0.0031 inches in diameter. As the AWG number increases, the gauge (thickness) of the wire decreases.

AMI *abbr*: area of mutual interest.

amidships *n*: the half-way point in the overall length of a floating offshore drilling rig, especially a drill ship.

amine *n*: any of several compounds employed in treating natural gas. Amines are generally used to remove hydrogen sulfide from water solutions and carbon dioxide from gas or liquid streams.

amine clay *n*: see *organophilic clay*.

amine salt *n*: an organic compound derived from ammonia, in which organic compounds replace one or more of the hydrogen atoms in the ammonia.

ammeter *n*: an instrument for measuring electric current in amperes.

amortization *n*: 1. the return of a debt (principal and interest) in equal annual installments. 2. the return of invested principal in a sinking fund.

amp *abbr*: ampere.

ampere (A) *n*: the fundamental unit of electrical current; 1 ampere = 6.28×10^{18} electrons passing through the circuit per second. One ampere delivers 1 coulomb in 1 second.

ampere-hour *n*: a unit of electricity equal to the amount produced in 1 hour by a flow of 1 ampere. See *ampere*.

ampere-hour rating *n*: a measure of a battery's ability to produce current in terms of ampere-hours. For example, if a battery has an ampere-hour rating of 30, the battery could produce 1 ampere of current for a 30-hour period.

ampere turn *n*: the number of turns in a coil multiplied by the number of amperes of current flowing through the coil.

amplifier *n*: an electronic device for increasing the magnitude of a quantity such as an electrical measurement signal. Amplifiers may be used to increase a transmitted and received measurement signal.

amplitude *n*: 1. in marine architecture, the maximum absolute value of a periodically varying quantity, such as the roll of a floating vessel. 2. in electronics, a measure of the extent of a vibratory movement, or an oscillation. It expresses the maximum deviation that a periodically varying quantity takes from its normal value of zero. 3. in well logging, the shapes and heights of the peaks in a spontaneous potential curve. 4. a measure of the extent of a vibratory movement, or an oscillation. It expresses the maximum deviation that a periodically varying quantity takes from its normal value of zero.

amplitude modulation (AM) *n*: the varying (modulating) of a wave in which the height (amplitude) of the wave is the characteristic varied in accordance with the intelligence to be transmitted. Compare *frequency modulation*.

anaerobic *adj*: active in the absence of free oxygen.

anaerobic bacteria *n pl*: bacteria that do not require free oxygen to live or that are not destroyed by its absence. Under certain conditions, anaerobic bacteria can cause scale to form in water-handling facilities in oilfields or hydrogen sulfide to be produced from sulfates.

analog *adj*: 1. of or pertaining to an instrument or equipment that measures a continuous variable that is proportional to another variable over a given range. For example, temperature can be represented or measured as voltage, its analog. 2. pertaining to devices, data, circuits, or systems that operate with variables that are represented by continuously measured voltages or other quantities. Compare *digital*.

analog data *n*: information indicated by a continuous form, usually a needle or pointer moving across a dial face. Compare *digital readout*.

analog input module *n*: a device in a programmable logic controller (PLC) that accepts an analog signal from a proportional transmitter and converts the analog signal to a digital signal for processing by the system processor.

analog output module *n*: a device in a programmable logic controller (PLC) that typically delivers an analog current signal of 4 to 20 milliamps (mA) to an analog transducer such as a current-to-pressure (I/P) transducer.

analog signal *n*: the representation of the magnitude of a variable in the form of a measurable physical quantity that varies smoothly rather than in discrete steps.

analysis *n*: examination and testing of the drilling fluid to determine its physical and chemical properties and condition.

anchor *n*: a device that secures, or fastens, equipment. In producing wells with sucker rod pumps, a gas anchor is a special section of perforated pipe installed below the pump. It provides a space for gas to break out of the oil. In offshore drilling, floating drilling vessels are often secured over drill sites by large anchors like those used on ships. For pipelines, a device that secures pipe in a ditch. In deck gratings on offshore installations, the device that secures the grating to its supports is an anchor.

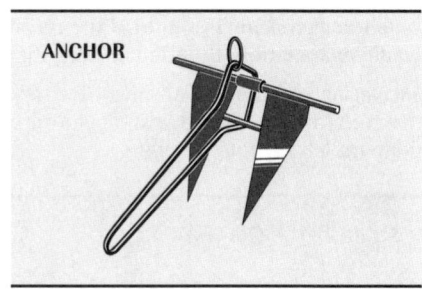

ANCHOR

anchor bolster *n*: a structure (usually tubular) that is mounted on a vessel's hull to allow the racking of anchors.

anchor buoy *n*: a floating marker used in a spread mooring system to position each anchor of a semisubmersible rig or drill ship. See *spread mooring system*.

anchor crown *n*: the top part of an anchor fluke to which the anchor pendant (or pennant) is attached.

anchor fluke *n*: the part of a drag embedment anchor (DEA) that embeds into the seafloor (either wholly or partially) and provides the greatest amount of anchor-holding capacity.

anchor-handling tug/supply vessel AHTS) *n*: a combined supply and anchor-handling ship. The AHTS is an offshore supply vessel specially designed to provide anchor-handling services and to tow offshore rigs, platforms, barges, and production modules/vessels. The vessel is often used as a standby rescue vessel for oilfields in production. The AHTS is often equipped for fire fighting, rescue operations, and oil recovery. The AHTS is also used in general supply service for all kinds of platforms, transporting both wet and dry cargo in addition to deck cargo.

ANCHOR HANDLING TUG/SUPPLY VESSEL

anchor-handling vessel (AHV) *n*: a relatively small but powerful watercraft used to set and retrieve anchors that moor an offshore drilling unit on location.

anchoring system *n*: in pipeline construction, a combination of anchors used to hold a lay barge on station and move it forward along the planned route. Lay-barge anchors may weigh in excess of 20 tons (18 tonnes), and a dozen or more may be needed.

anchor key *n*: a device on the deadline tie-down anchor used to secure the drilling line.

anchor packer *n*: a packer designed for sitting on a pipe that rests on bottom, such as a tail pipe or liner. See *packer*.

anchor pattern *n*: the pattern of minute projections from a metal surface produced by sandblasting, shot blasting, or chemical etching to enhance the adhesiveness of surface coatings.

anchor pendant (pennant) system *n*: a system of wire, chain, connectors, and buoys connected to the crown of an anchor which is used to buoy off the anchor. It is used in anchor deployment or recovery when a permanent chain chaser (PCC) is not used.

anchor seal assembly *n*: a seal assembly run on the production tubing that allows the tubing to be landed properly in the casing's seal bore when tubing weight alone is not sufficient to seat the tubing.

anchor shackle *n*: the connector used to connect wire, chain, or a swivel to an anchor.

anchor shank *n*: the part of an anchor to which the fluke is connected at the base and to which the mooring wire and/or chain is connected at the top.

anchor washpipe spear *n*: a fishing tool installed inside washover pipe to prevent a fish stuck off bottom from falling to bottom during a washover. Slips on the anchor washpipe spear engage the inside of the washover pipe as the pipe travels downhole around the fish.

anchor weight *n*: a weight installed in a tank to which the guide wires or cables for an automatic tank gauge float are attached to hold them taut and plumb.

andesite *n*: finely crystalline, generally light-colored extrusive igneous rock composed largely of plagioclase feldspar with smaller amounts of dark-colored minerals. Compare *diorite*.

AND gate *n*: a circuit that has two or more input-signal ports that deliver an output only if and when every input port is simultaneously energized.

anemometer *n*: an instrument for measuring wind speed in the atmosphere. The most common types are cup, vane, and hot-wire anemometers.

aneroid barometer *n*: a barometer consisting of a flexible, spring-filled metal cell from which air has been removed and a mechanism that registers. See *barograph*.

angle-azimuth indicator *n*: see *riser angle indicator*.

angle-control section *n*: the part of a groove on the drawworks drum that changes direction. Grooves run parallel to each other, except in the angle-control section, where each groove is machined at a slight angle. This angled part of the groove causes the wire rope being spooled onto the drum to change direction so that the next wrap of rope on the drum lies in between the wraps of rope already on the drum. Also called a crossover section.

angle indicator (boom) *n*: an accessory which measures the angle of the boom above horizontal.

angle of declination *n*: the difference, in angular degrees, between magnetic north and true north.

angle of deflection *n*: in directional drilling, the angle at which a well diverts from vertical; usually expressed in degrees, with vertical being 0°.

angle of deviation *n*: see *drift angle*.

angle of dip *n*: the angle at which a formation dips downward from the horizontal.

angle of drift *n*: see *drift angle*.

angle of heel *n*: the angle in degrees that a floating vessel inclines to one side or the other.

angle of loading *n*: in crane operations, when a sling with two legs is attached to each end of a load, the slant or the angle the two legs make as measured from horizontal. A reduction in the strength of a sling must be allowed for angles other than 90 degrees.

angle of loll *n*: see *heel*.

angle of wrap *n*: the distance that the brake band wraps around the brake flange. Drawworks have an angle of wrap of 270° or more.

angle-stem assembly *n*: a type of thermometer used on oil tanks. The graduated part of the thermometer is angled at least 90° from the temperature-sensitive portion. The angle conforms to the shell of the tank and permits easy reading of the instrument.

angle-stem thermometer *n*: a glass-stem thermometer in which the tail is bent at an angle to the stem so that the tail can be mounted in a horizontally disposed thermowell, leaving the scale vertical for greater ease of reading.

angle sub *n*: see *bent sub*.

angular misalignment *n*: one type of misalignment in a chain-and-sprocket drive. The shafts are not parallel to each other (they form an angle) in either the horizontal or the vertical plane. This pulls the link plates on one side tighter than those on the other side; thus, one side of the chain and sprockets wears faster than the other.

Link plates on only one side of the chain break because of fatigue.

angular unconformity *n*: an unconformity in which formations above and below are not parallel. See *unconformity*.

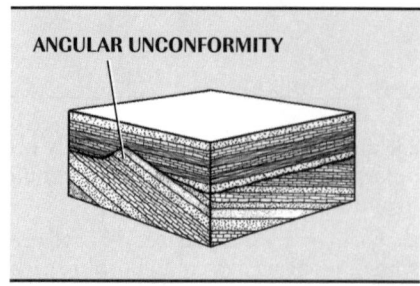

ANGULAR UNCONFORMITY

angular velocity *n*: a measure of the time required for a flowing medium, such as a fluid, to change its angular displacement.

anhydrite *n*: the common name for anhydrous calcium sulfate, $CaSO_4$.

anhydrous *adj*: without water.

aniline *n*: liquid chemical derived from benzene or from benzene and ammonia. It is used as a solvent and in the organic synthesis of other substances.

aniline point *n*: the lowest temperature at which the chemical aniline and a solvent (such as the oil in oil-base muds) will mix completely. In general, the oil of oil-base muds should have an aniline point of at least 150°F (66°C) to obtain maximum service life from the rubber components in the mud system.

anion *n*: 1. a negatively charged ion. 2. the ion in an electrolyzed solution that migrates to the anode. See *ion*. Compare *cation*.

annealed glass *n*: glass that has been treated with a heating and cooling process that toughens it and reduces its brittleness.

anniversary date *n*: the date, usually one year from the effective date of the lease, by which rentals must be paid to maintain the lease in effect in the absence of drilling or production.

annubar *n*: a gas measurement device that consists of a multiple-ported Pitot tube installed inside a pipe through which gas is flowing; it is installed perpendicular to the flow of gas. The length of the annubar is equal to the diameter of the pipe in which it is installed. An annubar senses the difference between total flowing pressure and static pressure; gas volume is calculated from this difference. See *Pitot tube*.

annular *adj*: pertaining to the annulus. The annulus is sometimes referred to as the annular space.

annular blowout preventer *n*: a large valve, usually installed above the ram blowout preventers, that, when closed, forms a seal in the annular space between the pipe and the wellbore or, if no pipe is present, in the wellbore itself. Compare *ram blowout preventer*.

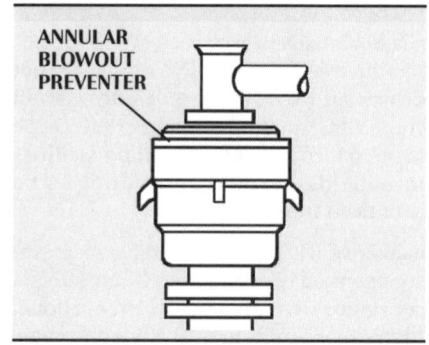

ANNULAR BLOWOUT PREVENTER

annular piston *n*: in drilling from floating offshore drilling rigs using a subsea blowout preventer stack and a Cameron model HC hydraulic connector, which is a device that connects the lower marine riser package (LMRP) to the subsea blowout preventer stack, a solid cylinder (a piston) that completely surrounds the locking segments in the connector and which provides an additional locking force. The additional locking force available through the annular piston increases the pressure rating to 15,000 psi (105,000 kPa). See *hydraulic connector, lower marine riser package*.

annular pressure *n*: fluid pressure in an annular space, as around tubing within casing.

annular pressure loss *n*: a reduction in the pressure of the fluid in the annulus caused by its motion against the wellbore, which may be open or cased. As the fluid moves through the annulus, friction between the fluid and the annular wall and within the fluid itself creates a pressure loss. The faster the fluid moves, the greater are the losses.

annular production *n*: production of formation fluids through the production casing annulus.

annular space *n*: 1. the space that surrounds a cylindrical object within a cylinder. 2. the space around a pipe in a wellbore, the outer wall of which may be the wall of either the borehole or the casing; sometimes termed the annulus.

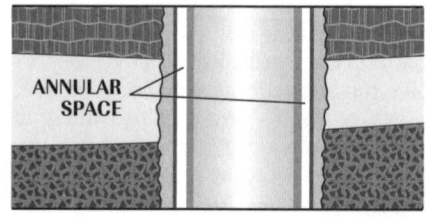

ANNULAR SPACE

annular valve *n*: a valve installed on the surface of a well which, when closed, seals the annular space between the tubing and the casing.

annular velocity *n*: the rate at which mud is traveling in the annular space of a drilling well.

annulus *n*: see *annular space*.

annulus master valve *n*: a valve on a wellhead that, when closed, isolates the annulus if well fluids enter the annulus because a packer or the tubing fails. The valve is normally closed during production operations unless the well is on gas lift, in which case, it is opened.

annulus swab valve *n*: see *production swab valve*.

annulus wing valve (AWV) *n*: if gas lift is required, it is used to control the fluid flow from the gas-lift line into the annulus.

annunciator *n*: a signaling device that operates electromechanically and serves to indicate visually or visually and audibly whether current is flowing, has flowed, or has changed direction of flow in one or more circuits.

anode *n*: 1. one of two electrodes in an electrolytic cell; represented as the negative terminal of the cell, it is the area from which electrons flow. In a primary cell, it is the electrode that is wasted or eaten away. 2. in cathodic protection systems, an electrode to which a positive potential of electricity is applied, or a sacrificial anode, which protects a structure by forming one electrode of an electric cell. 3. in a semiconductor diode, the terminal toward which forward current flows from an external circuit. See *diode*.

anomaly *n*: a deviation from the norm. In geology the term indicates an abnormality such as a fault or dome in a sedimentary bed.

anoxia *n*: an undersupply of oxygen reaching the tissues of the body, possibly causing permanent damage or death. Also called hypoxia.

ANSI *abbr*: American National Standards Institute.

Antarctic front *n*: in meteorology, an area that separates cold air from the Antarctic from the warm air that lies north of the South Pole. Sometimes called a polar front.

anticipated load *n*: 1. in designing or ordering hoisting (lifting) equipment, the maximum weight that the equipment will be able to safely lift. 2. on diesel electric rigs, the maximum amount of electric power that the generators will have to produce to adequately power the rig.

anticipated surface pressure *n*: pressure expected to be measured at the surface after a well is drilled and completed.

anticlinal trap *n*: a hydrocarbon trap in which petroleum accumulates in the top of an anticline. See *anticline*.

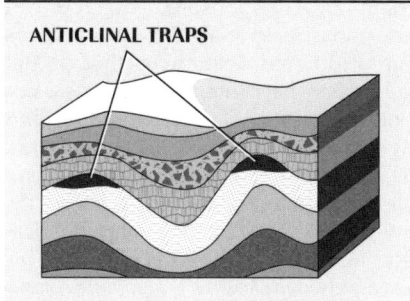

ANTICLINAL TRAPS

anticline *n*: rock layers folded in the shape of an arch. Anticlines sometimes trap oil and gas. Compare *syncline*.

anticyclone *n*: a large system of atmospheric high pressure marked by circulating winds moving clockwise from the center in the Northern Hemisphere and counterclockwise in the Southern Hemisphere. Compare *cyclone*.

anticyclonic wind *n*: the wind associated with a high-pressure area.

antidifferential sticking additive *n*: a chemical added to the drilling fluid to minimize the possibility of the drill stem becoming stuck to the side of the hole.

antiextrusion plate *n*: in a ram blowout preventer, a flat steel piece (a plate) on the ram block that prevents rubber from being forced out of the area between the drill pipe and the ram block when wellbore pressure is applied, and which feeds reserve rubber into contact with the drill pipe as rubber is lost because of wear. See *ram, ram blowout preventer*.

antifoam *n*: a substance used to prevent foam by greatly increasing surface tension. Compare *defoamer*.

antifreeze *n*: a chemical added to liquid that lowers its freezing point. Often used to prevent water in an engine's cooling system from freezing.

antiknock compound *n*: a substance, such as tetraethyl lead or other compounds, added to the fuel of an internal-combustion engine to prevent detonation of the fuel. Antiknock compounds effectively raise the octane rating of a fuel so that it burns properly in the combustion chamber of an engine. See *octane rating, tetraethyl lead*.

antiknock rating *n*: the measurement of how well an automotive gasoline resists detonation or pinging in a spark-ignition engine.

antilog *abbr*: antilogarithm.

antilogarithm *n*: a second number whose logarithm is the first number. See *logarithm*.

antisludge agents *n pl*: common additives which prevent an acid from reacting with certain types of crude and forming an insoluble sludge that blocks channels or reduces permeability.

antiwhirl bit *n*: a drill bit, usually a polycrystalline diamond bit, that is designed to prevent the bit's drilling a spiral-shaped hole because it whirls off-center as it rotates. See *bit whirl*.

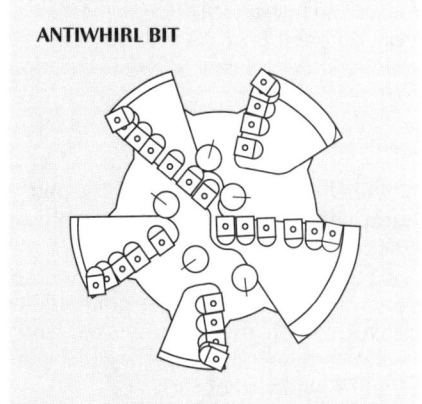

ANTIWHIRL BIT

ANWR *abbr*: Arctic National Wildlife Refuge.

AOF *abbr*: absolute open flow.

AOFP *abbr*: absolute open flow potential.

Ap *abbr*: cross-sectional area through the packer seal, in.^{2}

API *abbr*: American Petroleum Institute.

API cement class *n*: a classification system for oilwell cements, defined in *API Specification 10A*.

API gamma ray unit *n*: the standard unit of gamma ray measurement. Standardization of this unit results from the normalization of the detector-measurement systems of all primary service companies in the API test pits at the University of Houston. The API gamma ray unit is defined as $1/200$ of the difference in log deflection between two zones of different gamma ray intensity. The test pit is constructed so that the average midcontinent shale will record about 100 API gamma ray units.

API gravity *n*: the measure of the density or gravity of liquid petroleum products on the North American continent, derived from relative density in accordance with the following equation:

$$\text{API gravity at } 60°F = 141.5/\text{specific density} - 131.5$$

API gravity is expressed in degrees, a specific gravity of 1.0 being equivalent to 10° API. See *gravity*.

API *MPMS* *abbr*: American Petroleum Institute's *Manual of Petroleum Measurement Standards*.

API neutron unit *n*: the standard unit of measurement for neutron logs. Standardization of this unit results from the calibration of each logging tool model in the API neutron test pit at the University of Houston.

API scale *n*: a scale of liquid gravity measurement units called degrees API, devised and adopted by the American Petroleum Institute. Although the scale is very different from an ordinary specific gravity scale, it bears a definite relation to it as follows:

$$°API = \frac{140}{G} - 130$$

where *G* is specific gravity of the petroleum with reference to water, both at 60°F.

The API scale has particular advantages: it provides finer graduations between whole number units and lends itself to schemes for correcting to a temperature standard of 60°F.

API Spec 7 *abbr*: *API Specification for Rotary Drilling Equipment*, a publication that lists the specifications for drilling equipment.

API standards *n pl*: publications developed and written by knowledgeable and experienced persons—often employees of oil companies that are API members—to establish standard manufacturing, handling, inspection, and other aspects of oilfield equipment, materials, and practices. API standards cover such sectors of the petroleum and natural gas industry as drilling, producing, pipelining, refining, and measuring. Once the documents are written, API publishes and distributes them to the oil industry and other interested parties. An example of a standard is API Specification 6A, *Wellhead and Christmas Tree Equipment*, which provides manufacturing specifications for equipment used in pressure control systems for the production of oil and gas. API also publishes recommended practices (RPs) and documents such as technical reports, bulletins, and forms. RPs give recommended practices that personnel should follow when performing certain operations. For example, RP 13A, *Standard Procedure for Field Testing of Water-Based Drilling Fluids*, shows the equipment procedures that personnel should use to test water-based drilling fluids in the field. Bulletins provide information and guidance about practices that affect the petroleum industry. For example, Bulletin D10, *Procedure for Selecting Rotary Drilling Equipment*, describes a system of analysis to select a rig suitable for drilling a specific well. Technical reports cover a variety of subjects, such as the results of API-sponsored research. API also publishes special items such as forms and additional documents that do not fall under the previous categories. For example, Form 5UO1, *Voluntary Unit Agreement*, provides a common standard of reference for parties involved in forming a cooperative operation to increase the ultimate recovery of oil and gas. Publication 4633, *Barium in Produced Water: Fate and Effects in the Marine Environment*, summarizes information about the physical and chemical behavior of barium in the ocean. For further information, contact API at **www.api.org** or call 800-854-7179 or 303-397-7956.

APO *abbr*: after payout; commonly used in the land department of an oil company.

apparent compressibility *n*: the algebraic sum of the actual compressibility of a liquid and the volume change per unit volume of the confining container caused by a unit change in pressure at constant temperature.

apparent power *n*: the vector sum of the power in watts plus the reactive power in volt-ampere reactive (var) in a circuit. See *reactive power*.

apparent viscosity *n*: the viscosity of a drilling fluid as measured with a direct-indicating, or rotational, viscometer.

appraisal well *n*: a well drilled to confirm and evaluate the presence of hydrocarbons in a reservoir that have been found by a wildcat well.

APR *abbr*: a trademark name for an annular pressure-responsive valve for a DST string.

apron *n*: 1. a body of coarse, poorly sorted sediments formed by the coalescence of alluvial or detrital fans along the flanks of a mountain range. 2. a similar body of turbidite sediments formed by the coalescence of submarine debris fans along the base of the continental slope.

aquadag *n*: the graphite coating on the inside of certain cathode-ray tubes. It collects secondary electrons emitted by the face of the tube.

aquifer *n*: 1. a permeable body of rock capable of yielding groundwater to wells and springs. 2. the part of a water drive reservoir that contains the water.

arc *n*: a section of a circle, an ellipse, or other curved figure.

arc fault circuit interrupter (AFCI) *n*: a special type of circuit breaker that has a coil and special electronic circuitry that can sense the small fault currents created by arcing between components and that trips the circuit to prevent fires caused by arcing.

Archie's equation *n*: the formula for evaluating the quantity of hydrocarbons in a formation. The form of the equation depends on its specific use. The basic equation is—

$$S_w{}^2 = aR_w / 0^m R_t.$$

Archimedes's Principle *n*: the buoyant force exerted on a body suspended in a fluid is proportional to the density of the fluid.

Arctic front *n*: in meteorology, an area that separates cold air from the Arctic from warm air that lies south of the North Pole. Sometimes called a polar front.

Arctic National Wildlife Refuge (ANWR) *n*: nineteen million acres on Alaska's northeast coast that are believed to contain 600 million to 9 billion barrels (954 billion to 143 trillion litres) of oil. Opening of the area to oil exploration and production is the subject of much debate between environmentalists and oil-industry personnel.

arctic submersible rig *n*: a mobile submersible drilling structure used in arctic areas. The rig is towed onto the drilling site and submerged during periods when the water is free of ice. All equipment below the waterline is surrounded by a caisson to protect it from damage by moving ice. The drilling deck has no square corners so that moving ice can better flow around it. See *submersible drilling rig*.

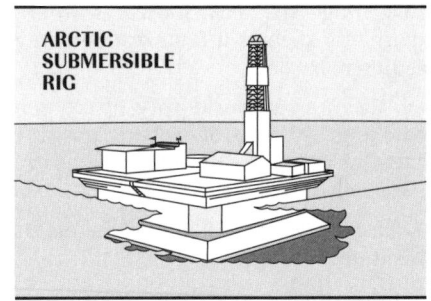

ARCTIC SUBMERSIBLE RIG

arc weld *v*: to join metals by utilizing the arc created between the welding rod, which serves as an electrode, and a metal object. The arc is a discharge of electric current across an air gap. The high temperature generated by the arc melts both the electrode and the metal, which fuse.

area *n*: the extent of a surface enclosed within a boundary; the extent of the surface of all or part of a solid. For two-dimensional plane surfaces, area is usually stated in square units. For example, a rectangle 2 feet long on one side and 3 feet long on the other side has an area of 2×3 feet, which equals 6 square feet (ft^2). Thus, the area of the rectangle is 6 ft^2.

area drilling superintendent *n*: an employee of a drilling contractor whose job is to coordinate and oversee the contractor's drilling projects in a particular region or area.

area of mutual interest *n*: an area, usually outlined on a plat attached to a farmout agreement or described in an exhibit, that allows both parties the first right of refusal on leases acquired by either party after the agreement is executed.

arenaceous *adj*: pertaining to sand or sandy rocks (such as arenaceous shale).

arenite *n*: a sandstone in which less than 15% of the total volume is silt and clay.

argillaceous *adj*: pertaining to a formation that consists of clay or shale (such as argillaceous sand).

arithmetic converter *n*: in a computer, a circuit or component that changes (converts) electronic signals into units that additional circuitry can add, subtract, multiply, divide, or compare.

arkose *n*: sandstone composed largely of feldspar grains and deriving from granitic source rocks.

armature *n*: a part made of coils of wire placed around a metal core, in which electric current is induced in a generator, or in which input current interacts with a magnetic field to produce torque in a motor.

armored case *n*: a corrosion-resistant metal case in which a glass-stem thermometer can be placed to minimize the risk of breakage.

arm's-length bargaining *n*: negotiations between a willing buyer and a willing seller, which should result in a price that truly reflects the market.

aromatic hydrocarbons *n pl*: hydrocarbons derived from or containing a benzene ring. Many have an odor. Single-ring aromatic hydrocarbons are the benzene series (benzene, ethylbenzenes, and toluene). Aromatic hydrocarbons also include naphthalene and anthracene. Compare *aliphatic hydrocarbons*.

aromatization *n*: the conversion of aliphatic or alicyclic compounds to aromatic hydrocarbons.

arpent *n*: a French unit of measurement, equal to 191.833 feet (58.5 metres).

arsenic (As) *n*: a chemical element that occurs as a brittle, steel-gray hexagonal mineral and that is added as an impurity to semiconductors to give them a negative charge.

artesian well *n*: a well in which water flows to the surface under natural pressure.

articulated loading platform (ALP) *n*: offshore platform in which the riser is jointed to allow for changes in water level, current, and so on.

artificial island *n*: an artificial gravel island sometimes used in shallow Arctic waters as a base on which drilling and production equipment is erected.

artificial lift *n*: any method used to raise oil to the surface through a well after reservoir pressure has declined to the point at which the well no longer produces by means of natural energy. Sucker rod pumps, gas lift, hydraulic pumps, and submersible electric pumps are the most common means of artificial lift.

artificial magnet *n*: a bar or other shape of soft iron or hard steel alloy that is magnetized by placing the bar or other shape in a magnetic field. Magnetite (lodestone) is a natural magnet; other magnets are thus artificial.

artificial satellite *n*: an information-collecting instrument that orbits the earth.

As *abbr*: altostratus.

As *sym*: arsenic.

asbestos *n*: term applied to many heat resistant and chemically inert fibrous minerals, some forms of which are used in certain drilling fluids. It finds only limited use because prolonged exposure can cause respiratory illness.

asbestos felt *n*: a wrapping material consisting of asbestos saturated with asphalt; one element of pipeline coatings. It is no longer commonly used because asbestos can cause respiratory illness.

ASCII *abbr*: American standard code for information interchange.

as-delivered Btu *n*: the number of Btus contained in a cubic foot of natural gas adjusted to reflect the actual water content of the gas at delivered pressure, temperature, and gravity conditions.

ash *n*: noncombustible residue from the gasification or burning of coal or a heavy hydrocarbon.

ASME *abbr*: American Society of Mechanical Engineers.

aspect ratio *n*: the ratio of a screen's (or frame's) width to the screen's height. For television sets in the U.S. and Great Britain, for example, the TV screen's aspect ratio is 3 to 4, which means that if the screen's height is 3 inches, its width is 4 inches; or, if its height is 12 inches, its width is 16 inches.

asphalt *n*: a hard brown or black material composed principally of hydrocarbons. It is insoluble in water but soluble in gasoline and can be obtained by heating some petroleums, coal tar, or lignite tar. It is used for paving and roofing and in paints.

asphalt-base oil *n*: see *naphthene-base oil*.

asphalt enamel *n*: an asphalt-base enamel applied as a coating to pipe that is to be buried. The asphalt is combined with finely ground mica, clay, soapstone, or talc and applied while hot. Combined with a subsequent wrapping, this coating protects the buried pipe from corrosion.

asphaltic crude *n*: petroleum with a high proportion of naphthenic compounds, which leave relatively high proportions of asphaltic residue when refined.

asphaltic material *n*: one of a group of solid, liquid, or semisolid materials that are predominantly mixtures of heavy hydrocarbons and their nonmetallic derivatives and are obtained either from natural bituminous deposits or from the residues of petroleum refining.

aspiration *n*: in internal combustion engines, the method which the engine uses to take air into its cylinders. Engines can be naturally aspirated, which means that the pistons draw in air at atmospheric pressure as they move down the cylinder; or they can be blown or supercharged, in which an engine-driven compressor of some type raises the pressure of the air above that of the atmosphere and forces it into the cylinder.

ASSE *abbr*: American Society of Safety Engineers.

assembly *n*: a group of components that make up a mechanism, machine, or similar device.

assignment *n*: a transfer of rights and interests in real or personal property or rights under a contract—for example, the transfer of an oil and gas lease from the original lessee to another party.

assignment clause *n*: a clause in any instrument that allows either party to the instrument to assign all or part of his or her interest to others.

assistant driller *n*: a member of a drilling rig's crew whose job is to aid and assist the driller during rig operations. This person not only controls the drilling operation at certain times, but also keeps records, handles technical details, and, in general, keeps track of all phases of the operation. See *driller*.

assistant rig superintendent *n*: an employee of a drilling contractor whose job includes aiding the rig superintendent; in some cases, the assistant rig superintendent takes over for the rig superintendent during nighttime hours. Consequently, the assistant rig superintendent is sometimes called the night toolpusher. See *rig superintendent*.

associated free gas *n*: see *associated gas*.

associated gas *n*: natural gas that overlies and contacts crude oil in a reservoir. Where reservoir conditions are such that the production of associated gas does not substantially affect the recovery of crude oil in the reservoir, such gas may also be reclassified as nonassociated gas by a regulatory agency. Also called associated free gas. See *gas cap*.

Association of Desk and Derrick Clubs *n*: see *Desk and Derrick Clubs*.

Association of Energy Service Companies (AESC) *n*: organization that sets some of the standards, principles, and policies of oilwell servicing contractors. Its official publication is *Well Servicing*. Address: 10200 Richmond Ave., #253; Houston, TX 77042; 713-781-0758.

astern *adv or adj*: 1. at or toward the stern of a ship or an offshore drilling rig; abaft. 2. behind the ship or rig.

ASTM *abbr*: American Society for Testing and Materials.

ASTM distillation *n*: any distillation made in accordance with ASTM procedure. Generally refers to a distillation test to determine the initial boiling point, the temperature at which percentage fractionations of the sample are distilled, the final boiling point, and quantity of residue.

athwart *prep or adv*: from side to side of a ship or offshore drilling rig.

atm *abbr*: atmosphere.

atmosphere (atm) *n*: a unit of pressure equal to the atmospheric pressure at sea level, 14.7 pounds per square inch (101.325 kilopascals).

atmospheres absolute *n pl*: total pressure at a depth underwater, expressed as multiples of normal atmospheric pressure.

atmospheric diving system (ADS) *n*: a one-person atmospheric suit rated to depths of 2,300 feet (701 metres). Advantages include allowing the operator to work in a safe environment for debris salvage or recovery operations.

atmospheric pressure *n*: the pressure exerted by the weight of the atmosphere. At sea level, the pressure is approximately 14.7 pounds per square inch (101.325 kilopascals), often referred to as 1 atmosphere. Also called barometric pressure.

atmospheric pressure cure *n*: the aging of specimens for test purposes at normal atmospheric pressure for a designated period of time under specified conditions of temperature and humidity.

atoll *n*: a coral island consisting of a reef surrounding a lagoon.

atom *n*: the smallest quantity of an element capable of either entering into a chemical combination or existing alone.

atomic number *n*: a number that expresses the number of protons and an equal number of electrons in each atom of an element.

atomic particle *n*: one of the small bits of matter that makes up an atom, such as an electron, proton, or neutron.

atomic weight *n*: a number that approximates the number of protons plus the number of neutrons in an atom; it expresses the relative weights of atoms.

atomize *v*: to spray a liquid through a restricted opening, causing it to break into tiny droplets and mix thoroughly with the surrounding air.

attachment *n*: an alternate designation for front-end equipment; also, any device that may be added as a unit or assembly.

attachment efficiency *n*: in crane operations, the relative strength of a sling's end attachment, such as a loop eye or a swaged terminal, which attaches the sling to the load to be lifted by the crane. The less efficient the attachment is, the more the sling's strength is reduced.

attack hoses *n pl*: smaller-sized hoses used to put out small fires or a fire that is more or less under control.

attapulgite *n*: a fibrous clay mineral that is a viscosity-building substance; used principally in saltwater-base drilling muds. Also called fuller's earth.

attenuating probe *n*: in electronics, a small device (a probe) that is put into contact with or inserted into a circuit to make measurements of the circuit and that reduces (attenuates) the strength of the value (such as voltage) being measured in the circuit. The amount of attenuation is usually expressed as a ratio—for example, a 10-to-1 probe reduces the values strength by 10.

attest *v*: to verify or witness, by a designated company official, the signature of the signing company officer and the affixing of the official company or corporation seal. The title of the company official executing the instrument and the title of the company official attesting it should appear under the signatures.

audio transformer *n*: an iron-core transformer used to bring together (to electronically couple) circuits that carry audio frequencies.

augend *n*: a quantity to which another quantity is added.

auger *n*: a boring tool that consists of a shaft with spiral channels. An auger may be used to bore the hole for a pipeline that must cross beneath a roadbed.

authority for expenditure *n*: an estimate of costs prepared by a lease operator and sent to each nonoperator with a working interest for approval before work is undertaken. Normally used in connection with well-drilling operations.

automated plant *n*: a plant that contains instruments for the measurement and control of temperatures, pressures, flow rates, and properties of resulting products and thereby makes necessary corrections in the plant operating conditions so as to maintain specification products. Such a plant contains shutdown and other automatic devices for minimizing damage in the absence of operating personnel.

automatic cathead *n*: see *breakout cathead, cathead, makeup cathead*.

automatic choke *n*: an adjustable choke that is power-operated to control pressure of flow. See *adjustable choke*.

automatic control *n*: a device that regulates various factors (such as flow rate, pressure, or temperature) of a system without supervision or operation by personnel. See *instrumentation*.

automatic custody transfer *n*: a system for automatically measuring and sampling oil or products at points of receipt or delivery. See *lease automatic custody transfer*.

automatic driller *n*: a mechanism used to regulate the amount of weight on the bit without the presence of personnel. Automatic drillers free the driller from the sometimes tedious task of manipulating the drawworks brake to maintain correct weight on the bit. Also called an automatic drilling control unit.

automatic drilling control unit *n*: see *automatic driller*.

automatic fill-up shoe *n*: a device usually installed on the first joint of casing to

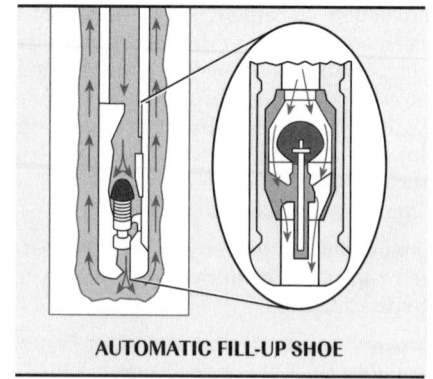

AUTOMATIC FILL-UP SHOE

regulate automatically the amount of mud in the casing. The valve in this shoe keeps mud from entering the casing until mud pressure causes the valve to open, allowing mud to enter.

automatic flooding valve *n*: in a subsea marine riser, a device that prevents collapse of the riser from the surrounding water pressure when the riser is inadvertently emptied of drilling mud. The valve is incorporated into a riser joint and, when it senses an increase in pressure from the seawater surrounding the empty riser, it opens to allow seawater into the riser to prevent its collapse.

automatic gauge *n*: a measuring device installed on the outside of a tank to permit observation of the depth of the liquid inside.

automatic gauging tape *n*: the flexible measuring or connecting element that is used to measure the liquid level in tanks by the automatic gauge method.

automatic J *n*: a mechanism in packers and other tools in which straight pickup or set-down action will set or release the tool. Rotation is not needed to set or release a tool with an automatic J.

automatic pipe racker *n*: a device used on a drilling rig to automatically remove and insert drill stem components from and into the hole. It replaces the need for a person to be in the derrick or mast when tripping pipe into or out of the hole.

automatic positioner *n*: see *hook positioner*.

automatic positioner lock *n*: a device on a drilling rig's hook that allows the hook to rotate freely when hoisting the drill string out of open (uncased) hole but automatically rotates the elevators to face the derrickhand when they reach the monkeyboard. Allowing the elevators to rotate prevents them from damaging the open hole and prevents the drilling line from becoming twisted. The positioner lock can also lock the elevators into one position when tripping in cased hole where it is not necessary to allow the elevators to rotate freely.

automatic pumping station *n*: an automatically operated station installed on a pipeline to provide pressure when a fluid is being transported.

automatic roughneck *n*: a large, self-contained pipe-handling machine used by drilling crew members to make up and break out tubulars. The device combines a spinning wrench, a torque wrench, and backup wrenches. See *Iron Roughneck*.

automatic sampler *n*: a device that, when installed in a pipe or flow channel and actuated by automatic control equipment, enables a representative sample to be obtained of the fluid flowing therein.

automatic sampling *n*: the gathering of small quantities of liquid from a pipeline on a systematic basis, which can be tested to determine the quality of the liquid flowing in the line. An automatic sampling device consists of a probe inserted into the line, an extractor, a timer or other device that controls the sampling rate, and a small tank (sample receiver) that holds the samples.

automatic shutdown *n*: a system in which certain instruments are used to control or to maintain the operating conditions of a process. If conditions become abnormal, this system automatically stops the process and notifies the operator of the problem.

automatic slips *n pl*: a device, operated by air or hydraulic fluid, that fits into the opening in the rotary table when the drill stem must be suspended in the wellbore (as when a connection or trip is being made). Automatic slips, also called power slips, eliminate the need for roughnecks to set and take out slips manually. See *slips*.

automatic tank gauge *n*: an instrument that automatically measures and displays liquid levels or ullages in one or more tanks continuously, periodically, or on demand.

automatic tank gauge tape *n*: a metal tape used to connect the liquid level-detecting element and the gauge-head mechanism.

automatic temperature compensator *n*: a meter accessory that enables a meter that is measuring volume at stress temperature to register the equivalent volume at a reference or base temperature.

automatic vessel tank gauging system *n*: a system that automatically measures and displays liquid levels or ullage in one or more vessel tanks on a continuous, periodic, or on-demand basis.

automatic welding *n*: a welding technique for joining pipe ends. Two general types of automatic welding used in pipeline construction are submerged-arc welding and automatic wire welding.

automatic wire welding *n*: an automatic welding process utilizing a continuous wire feed and a shielding gas. Automatic wire welding is similar to semiautomatic welding except that manual adjustment of the rate of wire feed and amount of shielding gas is unnecessary. See also *semiautomatic welding*.

automation *n*: automatic, self-regulating control of equipment, systems, or processes. See *instrumentation*.

autotransformer *n*: a transformer with only one winding, part of which is used for either the primary or secondary conductor, and part of which is used as both primary and secondary conductors.

auxiliaries *n pl*: equipment on a drilling or workover rig that is not a direct part of the rig's drilling equipment, such as the equipment used to generate electricity for rig lighting or the equipment used to mix drilling fluid.

auxiliary brake *n*: a braking mechanism on the drawworks, supplemental to the mechanical brake, that permits the lowering of heavy hook loads safely at retarded rates without incurring appreciable brake maintenance. There are two types of auxiliary brake—the hydrodynamic and the electrodynamic. In both types, work is converted into heat, which is dissipated through liquid cooling systems. See *electrodynamic brake, hydrodynamic brake*.

auxiliary equipment *n*: see *auxiliaries*.

auxiliary hoist *n*: see *whip line*.

auxiliary meter equipment *n*: equipment such as strainers, air separators, or flow conditioners installed in conjunction with a meter to protect or improve the performance of the meter; does not include instrumentation and accessories driven by the meter's output rotation or pulses.

average (avg) *n*: approximately or resembling an arithmetic mean, specifically, about midway between extremes.

average current rating *n*: in a solid-state diode, the amount of current that the diode is designed to handle over a range of operating current in a circuit, from a minimum to a maximum amount.

average reservoir pressure (p) *n*: the pressure that occurs when a producing well is shut in to prevent any movement of formation fluids within the reservoir.

average sample *n*: in tank sampling, a sample that consists of proportionate parts from several sections of the tank.

average value *n*: the average of all the instantaneous values of current or voltage in one-half a cycle of AC electricity; equal to 0.636 times the maximum value.

avg *abbr*: average.

Avogadro's law *n*: states that under the same conditions of pressure and temperature, equal volumes of all gases contain equal numbers of molecules. Also called Avogadro's hypothesis.

Avogadro's number *n*: 6.024×10^{23}.

AWG *abbr*: American wire gauge.

AWS *abbr*: American War Standard.

axial compression *n*: pressure produced parallel with the cylinder axis when casing hits a deviation in the hole or a sticky spot and stops. The force pushing down on the pipe causes axial compression.

axial flow turbine meter *n*: a velocity-measuring device in which gas flow is parallel to the rotor axis and the speed of rotation is proportional to the rate of flow.

axial lead (pronounced"leed") *n*: in an electronic component such as a composition resistor, a single wire (a lead) that exits each end of the component from the component's center and to which the electrical connections are made to the circuit in which the component is installed. Compare *radial lead*.

axial load *n*: the vertical weight an object carries that extends perpendicularly into the object.

axial tension *n*: vertical upward force (tension) placed on an object.

axis of rotation *n*: the vertical line through the axis around which the crane superstructure rotates.

azeotrope *n*: a mixture of liquids that distills without a change in composition. The azeotropic mixture exhibits a constant maximum or minimum boiling point, which is either higher or lower than that of its components.

azimuth *n*: 1. in directional drilling, the direction of the wellbore or of the face of a deflection tool in degrees (0°–359°) clockwise from true north. 2. an arc of the horizon measured between a fixed point (such as true north) and the vertical circle passing through the center of an object.

azimuth angle indicator *n*: see *riser angle indicator*.

B *abbr*: 1. bottom of; used in drilling reports. 2. center of buoyancy; used mostly in making buoyancy, stability, and trim calculations for offshore drilling rigs. 3. formation volume factor. 4. shear modulus, psi/radian.

babbitt *n*: metal alloy, either tin-based or lead-based, used primarily in friction bearings.

backbarrier complex *n*: the depositional environments associated with a shallow lagoon shoreward from a coastal barrier island. These environments are highly variable and may include tidal channels, salt marshes, shell reefs, and mangrove swamps, among others.

back electromotive force (emf) *n*: 1. voltage induced in the primary winding of a transformer that opposes the incoming line voltage. 2. voltage induced in the armature of a motor by the magnetic field in which it is turning. Also called counter emf and counter-voltage. See *electromotive force*.

back emf *abbr*: back electromotive force.

backfilling *n*: the technique for covering a completed pipeline so that adequate fill material is provided underneath the pipe as well as above it. Backfilling prevents pipe damage caused by loose rock, abrasion, shifting, and washouts.

background gas *n*: in drilling operations, gas that returns to the surface with the drilling mud in measurable quantities but does not cause a kick. Increases in background gas may indicate that the well is about to kick or has kicked.

backhaul *n*: a transport of gas by displacement against the flow on a single pipeline, so that the gas is redelivered upstream of its point of receipt.

backhoe *n*: an excavating machine fitted with a hinged arm to which is rigidly attached a bucket that is drawn toward the machine in operation. The backhoe is used for excavating and for clearing blasted rock out of the ditch during ditching and pipe laying.

back-in *n*: an option right reserved by the granting company of a farmout to convert an overriding royalty to a working interest once the conditions for such back-in have been met. Compare *election at casing point*.

back-in unit *n*: a portable servicing or workover rig that is self-propelled, using the hoisting engines for motive power. Because the driver's cab is mounted on the end opposite the mast support, the unit must be backed up to the wellhead. See *carrier rig, drive-in unit*.

back off *v*: to unscrew one threaded piece (such as a section of pipe) from another.

back-off *n*: the procedure whereby one threaded piece (such as a pipe) is unscrewed from another.

back-off connector *n*: see *washover back-off connector*.

back-off joint *n*: a section of pipe with left-hand threads on one end and conventional right-hand threads on the other. In setting a liner, a back-off joint is attached to it so that the drill pipe may be disengaged from the liner by conventional right-hand rotation.

back-off wheel *n*: see *wheel-type back-off wrench*.

backout *v*: to overcome the positive electrical potentials of anodic areas in cathodic protection systems.

back-pressure *n*: 1. the pressure maintained on equipment or systems through which a fluid flows. 2. in reference to engines, a term used to describe the resistance to the flow of exhaust gas through the exhaust pipe. 3. the operating pressure level measured downstream from a measuring device.

back-pressure control *n*: a pneumatically operated device—a controller—which maintains a fixed pressure by controlling the amount of upstream pressure on a vessel such as a fractionating tower or an oil and gas separator.

back-pressure regulator *n*: a device that reduces or increases upstream, or back-pressure to maintain a given amount of back-pressure in a pneumatic line.

back-pressure valve *n*: 1. a valve used to regulate back-pressure on equipment or systems through which a fluid flows. 2. a valve used to regulate automatically a uniform pressure on the inlet side of the valve.

back rake angle *n*: in a PDC bit, the amount of distance, in degrees, between vertical (90°) and the face of the cutter.

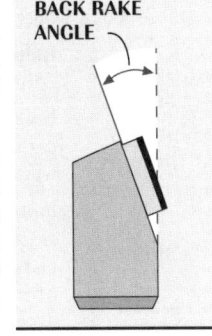

BACK RAKE ANGLE

back ream *v*: to enlarge the wellbore by raising and rotating the drill string that has a reamer made up in it. Back reaming enlarges tight spots in the wellbore. See *reaming*.

backshore *n*: that part of the seashore that lies between high-tide and storm-flood level.

backside *n*: the area above a packer between casing ID and tubing OD.

back up *v*: to hold one section of an object such as pipe while another section is being screwed into or out of it.

backup element *n*: a metal sealing ring on either side of the center packing element of an annular blowout preventer to limit the packer's extrusion.

backup ring *n*: a cylindrical ring employed to back up (or assist) a sealing member against extrusion under temperature and pressure.

backup system *n*: a set of devices in an assembly or in a process that can take over the operation of the assembly or process should the devices normally used to operate the assembly or process fail.

backup tongs *n pl*: the tongs latched on the drill pipe joint hanging in the rotary by the slips, used to keep the pipe from turning as the makeup or breakout tongs (the lead tongs) apply torque to make up or break out the tightened tool joint connection. Compare *lead tongs*.

17

backup wrench *n*: any wrench used to hold a pipe or a bolt to prevent its turning while another length of pipe or a nut is being screwed into or out of it.

backwash *v*: to reverse the flow of fluid from a water injection well to get rid of sediment that has clogged the wellbore.

BACT *abbr*: best-available control technology.

bacteria *n pl*: a large, widely distributed group of typically one-celled microorganisms. See *anaerobic bacteria, sulfate-reducing bacteria*.

bactericide *n*: anything that destroys bacteria.

bad oil *n*: oil that contains a sufficient amount of sediment and water to make it unacceptable to a pipeline or other oil transportation system.

baffle plate *n*: 1. a partial restriction, generally a plate, placed to change the direction, guide the flow, or promote mixing within a tank or a vessel. 2. a device that is seated on the bit pin, in a tool joint, or in a drill pipe float, to centralize the lower end of a go-devil while permitting the bypass of drilling fluid. The go-devil contains a surveying instrument.

bag filter *n*: on an engine, a bag-shaped piece made of cotton or fiber cloth that fits into a special holder in the fuel system piping. Fuel is circulated through the bag, which removes foreign matter from the fuel.

baguio *n*: a tropical cyclone in the Philippines. Also called baruio.

bail *n*: 1. a cylindrical steel bar (similar to the handle or bail of a bucket, only much larger) that supports the swivel and connects it to the hook. Sometimes, the two cylindrical bars that support the elevators and attach them to the hook are called bails or links. 2. the U-shaped handle on a nozzle used to close a valve that shuts off water flow when pushed forward. *v*: to recover bottomhole fluids, samples, or drill cuttings by lowering a cylindrical vessel called a bailer to the bottom of a well, filling it, and retrieving it.

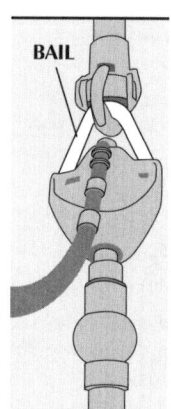

bailer *n*: a long, cylindrical container fitted with a valve at its lower end, used to remove water, sand, mud, drilling cuttings, or oil from a well in cable-tool drilling.

bailing *v*: the operation of cleaning mud cuttings and other material from the bottom of the wellbore with a bailer.

bailing drum *n*: the reel around which the bailing line is wound. See *bailing line*.

bailing line *n*: the cable attached to a bailer, passed over a sheave at the top of the derrick, and spooled on a reel.

bail pin *n*: a large steel dowel (pin) that attaches the swivel's bail to the swivel's body. Normally, swivels have two bail pins, one for each side of the bail where it attaches to the body. See *bail*.

bail throat *n*: the inside curve in the swivel's bail where the bail hangs from the traveling block's hook. See *bail*.

balanced *n*: when the pressure of drilling fluid in the wellbore is the same as the pressure in the formation.

balanced plug placement *n*: the displacement of enough cement out of the drill pipe to achieve a column of cement in both pipe and annulus that are equal in height.

balance point *n*: the point at which enough pipe has been snubbed into the hole to allow the pipe's weight to overcome the upward force of well pressure. See *snubbing, stripping in*.

Baldt connecting link *n*: a type of chain-connecting link manufactured by Baldt.

ball *n*: a spherical object that, when pumped down the hole, operates certain hydraulic tools.

ball-and-seat valve *n*: a device used to restrict fluid flow to one direction. It consists of a polished sphere, or ball, usually of metal, and an annular piece, the seat, ground and polished to form a seal with the surface of the ball. Gravitational force or the force of a spring holds the ball against the seat. Flow in the direction of the force is prevented, while flow in the opposite direction overcomes the force and unseats the ball.

ballast *n*: 1. for ships, water taken on board into specific tanks to permit proper angle of repose of the vessel in the water, and to assure structural stability. 2. for semisubmersible drilling rigs, seawater added to the rig's pontoons to increase its draft, or to submerge it to a desired depth below the sea's surface.

ballast control *n*: the act of maintaining a floating offshore drilling rig on even keel, regardless of weather or load conditions.

ballasted condition *n*: the condition of a floating offshore drilling rig when ballast has been added.

ballast engineer *n*: on a semisubmersible drilling rig, the person responsible for maintaining the rig's stability under all weather and load conditions.

ballast leg *n*: see *ballast movement*.

ballast movement *n*: a voyage or voyage leg made without any paying cargo in the vessel's tanks. To maintain proper stability, trim, or draft, seawater is usually carried during such movements. Also called ballast leg or ballast passage.

ballast passage *n*: see *ballast movement*.

ballast tank *n*: any shipboard tank or tanker compartment normally used for carrying saltwater ballast. When these compartments or tanks are not connected with the cargo system, they are called segregated ballast tanks or systems.

ball bearing *n*: a bearing in which a finely machined shaft (a journal) turns on freely rotating hardened-steel spheres that roll easily within a groove or track (a race) and thus convert sliding friction into rolling friction. See *ball race*.

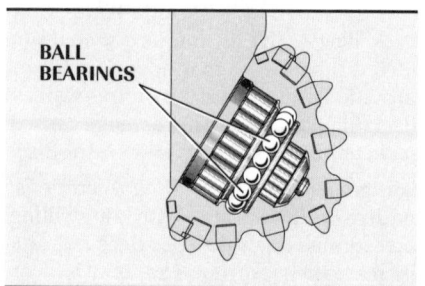

ball catcher *n*: a cylindrical tube placed around the retrieving neck of a retrievable bridge plug to catch debris or frac balls.

ball cock *n*: a device for regulating the level of fluid in a tank. It consists of a valve connected to a hollow floating ball that, by rising or falling, shuts or opens the valve.

balled-up bit *n*: see *ball up*.

ballistic prover *n*: see *small-volume prover*.

ball joint *n*: in drilling from floating offshore drilling rigs that employ a subsea blowout preventer system, a device mounted between the annular preventer and the riser adapter on the lower marine riser package (LMRP) to prevent excessive bending forces from being exerted on the marine riser, the LMRP, and the blowout preventer components. It is a forged-steel ball and socket containing a cylindrical neck extension with a riser adapter attached at the top of the neck. Compare *flex joint*.

ball mill *n*: a hollow drum that contains material to be ground or pulverized, heavy steel or ceramic balls, and a liquid such as water. The drum is rolled or agitated so that the balls crush or polish the material as they roll about.

ball-mill *v*: to use a ball mill to grind, polish, or pulverize metal or stone particles.

balloon *v*: to swell or puff up. In reference to tubing under the effects of temperature changes, sucker rod pumping, or high internal pressure, to increase in diameter while decreasing in length. Compare *reverse-balloon*.

ball-out *v*: to plug open perforations by using ball sealers.

ball race *n*: a track, channel, or groove in which ball bearings turn.

ball sealers *n pl*: balls made of nylon, hard rubber, or both and used to shut off perforations through which excessive fluid is being lost.

ball up *v*: 1. to collect a mass of sticky consolidated material, usually drill cuttings, on drill pipe, drill collars, bits, and so forth. A bit with such material attached to it is called a balled-up bit. Balling up is frequently the result of inadequate pump pressure or insufficient drilling fluid. 2. in reference to an anchor, to fail to hold on a soft bottom, pulling out, instead, with a large ball of mud attached.

ball valve *n*: a flow-control device employing a ball with a rotating mechanism to open or close the tubing.

banana peel *n*: a thin sheet of steel created when a washover pipe's rotary shoe grinds into the casing and mills a portion of the casing. It can be prevented by selecting a shoe diameter that is much smaller than the diameter of the casing but is still of adequate diameter to remove the material that is sticking the fish.

band *n*: on offshore installations that have grated flooring or decks, a length of flat metal welded to the side or end of a panel of grating or to the edges of a cutout section of grating, that does not extend above or below the bearing bars, which are the steel or aluminum bars that support the grating.

band brake *n*: circular type of brake either of external contracting type or internal expanding type, having a strap lined with heat and wear resistant friction material.

band clutch *n*: circular type of clutch either of external contracting type or internal expanding type, having a strap lined with heat and wear resistant friction material.

bank draft *n*: see *draft*.

bar *n*: a unit of pressure equal to 1 million pascals, 1 million newtons per square metre, or 10 million dynes per square centimetre. *v*: to move or turn (as a flywheel) with a bar used as a lever.

bareboat charter *n*: an agreement to lease, or charter, a ship or offshore drilling unit without crew or equipment, usually for a predetermined time.

barefoot completion *n*: see *open-hole completion*.

barge *n*: 1. a flat-decked, shallow-draft vessel, usually towed or pushed by a tugboat. Barges are not self-propelled. They are used to transport oil or products on rivers, lakes, and inland waterways. Also, a complete drilling rig may be assembled on a barge and the vessel used for drilling wells in lakes and in inland waters and marshes. Similarly, well service and workover equipment can be mounted on a barge. 2. a drill barge. See *drill barge*.

barge control operator *n*: an employee on a semisubmersible rig whose main duty is to monitor and control the stability of the rig. From a special work station on board the rig, this person controls the placement of ballast water inside the rig's pontoons to maintain the rig on even keel during all operations.

barge engineer *n*: see *barge control operator*.

barge-engineering department *n*: on offshore rigs, the personnel on the rig who are concerned with its buoyancy and stability characteristics. They are responsible for keeping track of factors that affect the rig's stability and for taking the required corrective measures to ensure that the rig maintains its stability.

barge master *n*: see *barge control operator*.

barge tow *n*: a string of barges laced together with ratchets and steel cable.

baric wind law *n*: see *Buys-Ballot's law*.

barite *n*: barium sulfate, BaSO₄; a mineral frequently used to increase the weight or density of drilling mud. Its relative density is 4.2 (i.e., it is 4.2 times denser than water). See *barium sulfate, mud*.

barite plug *n*: a settled volume of barite particles from a barite slurry placed in the wellbore to seal off a pressured zone.

barite sack *n*: a container (a sack) of powdery material (barite) that weighs about 100 pounds (about 45 kilograms). Barite is a dense mineral with a specific gravity of about 4.2—it is over four times heavier than water. On drilling rigs, barite (barium sulfate, or BaSO₄) can be supplied in heavy-duty paper sacks. In drilling from floating drilling rigs, sacks of barite are sometimes placed on top of the temporary guide base before it is run to

the seafloor. The sacks provide additional weight to the guide base and ensure that it makes firm contact with the seafloor. Also, barite is often used to increase the density, or weight, of drilling mud. See *barite, temporary guide base*.

barite slurry *n*: a mixture of barium sulfate, chemicals, and water of a unit density between 18 and 22 pounds per gallon.

barium sulfate *n*: a chemical compound of barium, sulfur, and oxygen (BaSO₄), which may form a tenacious scale that is very difficult to remove. Also called barite.

bar magnet *n*: a piece of hard steel in the shape of a rectangular-shaped rod (a bar) that has been strongly magnetized and holds its magnetism. It is therefore a type of permanent magnet. See *permanent magnet*.

barograph *n*: an aneroid barometer that continuously records atmospheric pressure values on a graph or chart.

barometer *n*: an instrument for measuring atmospheric pressure. See *aneroid barometer, mercury barometer*.

barometric pressure *n*: see *atmospheric pressure*.

barrel (bbl) *n*: 1. a measure of volume for petroleum products in the United States. One barrel is the equivalent of 42 U.S. gallons or 0.15899 cubic metres (9,702 cubic inches). One cubic metre equals 6.2897 barrels. 2. the cylindrical part of a sucker rod pump in which the pistonlike plunger moves up and down. Operating as a piston inside a cylinder, the plunger and barrel create pressure energy to lift well fluids to the surface. 3. the lagging or body portion of a rope drum.

barrel compressor *n*: a special type of centrifugal compressor with a barrel-shaped housing.

barrel equivalent *n*: a laboratory unit used for evaluating or testing drilling fluids. One gram of material, when added to 350 millilitres of fluid, is equivalent to 1 pound of material added to one 42-gallon barrel of fluid.

barrel-mile *n*: a unit of measure for pipeline shipment of oil or liquid product that signifies 1 barrel moved 1 mile.

barrel reamer *n*: in pipeline construction, a cylindrical device fitted on both ends with hollow cutting teeth, used in directionally drilled river crossings. Used during the pullback portion of a crossing effort, the barrel reamer opens the hole and keeps the pull on course.

barrels per day (bpd) *n*: in the United States, a measure of the rate of flow of a well; the total amount of oil and other fluids produced or processed per day.

barrier *n*: in well control and under-balanced drilling, a means of preventing the uncontrolled flow of fluids from a well. Barriers range from drilling mud whose hydrostatic pressure is greater than formation pressure to mechanical devices such as blowout preventers and drill stem valves, which, when closed, prevent the exit of fluids from the well.

barring *n*: turning over an engine by hand. A solid-steel rod (a bar) is inserted into special holes in the engine's flywheel. By lifting up or pressing down on the bar, the flywheel can be turned, which rotates the pistons in the cylinders. Barring is used as a precautionary measure before starting an engine that has not been run in some time.

bar sand *n*: reservoir rock formed from a mass of sand, gravel, or alluvium deposited on the bed of a stream, sea, or lake by waves and currents.

baruio *n*: see *baguio*.

baryte *n*: variation of barite. See *barite*.

basalt *n*: an extrusive igneous rock that is dense, fine grained, and often dark gray to black. Compare *gabbro*.

base *n*: 1. a substance capable of reacting with an acid to form a salt. A typical base is sodium hydroxide (caustic), with the chemical formula NaOH. For example, sodium hydroxide combines with hydrochloric acid to form sodium chloride (a salt) and water. This reaction is written chemically as $NaOH + HCl \rightarrow NaCl + H_2O$. 2. in a junction transistor, the region that lies between the transistor's emitter and collector and into which minority carriers are injected. The base is the transistor's control terminal and receives current from an external source into its region. See *collector, emitter*.

base abstract *n*: an abstract of title that contains full and complete copies of all recorded instruments from the sovereignty of the soil to the date the same is completed as set forth in the abstractor's certificate.

base exchange *n*: the replacement of cations associated with a clay surface by those of another element, e.g., the conversion of sodium clay to calcium clay.

base gas *n*: the volume of gas needed as permanent inventory to maintain adequate underground storage reservoir pressures and deliverability rates throughout the withdrawal season.

base line *n*: 1. the fore and aft reference line at the upper surface of the flat plate keel at the center line for flush shell plate vessels. 2. the thickness of the garboard strake above that level for vessels having lap seam shell plating.

base load *n*: as applied to gas, a given send-out of gas remaining fairly constant over a period of time and usually not temperature sensitive. For example, residential base load is a given send-out of gas consumed by clothes dryers, water heaters, and in cooking.

base map *n*: horizontal representation of nongeologic surface features such as streams, roads, buildings, survey benchmarks, and property lines.

basement rock *n*: igneous or metamorphic rock, which seldom contains petroleum. Ordinarily, it lies below sedimentary rock. When it is encountered in drilling, the well is usually abandoned.

base metal *n*: 1. any of the reactive metals at the lower end of the electrochemical series. 2. metal to which cladding or plating is applied.

base (mounting) *n*: the pedestal upon which the revolving superstructure is mounted.

base plate *n*: 1. in drilling from floating rigs, a flat steel piece (a plate) that is part of the riser-support spider (the device that supports riser pipe in the rotary table as it is being run or retrieved). The base plate supports the spider and fits into the rig's rotary table. 2. in drilling from floating rigs, a flat steel piece (a plate) at the top of the lower marine riser package (LMRP) on top of which the control pods are attached and which supports the pods on the LMRP. See *lower marine riser package, riser-support spider*.

base pressure *n*: the pressure exerted by a specific number of molecules contained in a specific volume (usually 1 cubic foot) at a specific temperature. Base pressure is a factor used in calculating gas volume. Standard base pressure varies from state to state. In Texas, for example, base pressure is 14.73 psia at 60°F. In Louisiana, base pressure is 15.025 psia at 60°F.

base price *n*: the value of natural gas, usually at the wellhead, before any imposition of taxes, gathering, compression, or other charges, as stated in a gas sales contract.

base quantity *n*: one of a small number of physical quantities in a system of measurement that is defined, independent of other physical quantities, by means of a physical standard and by procedures for comparing the quantity to be measured with the standard. Also known as fundamental quantity.

basicity *n*: pH value above 7 and the ability to neutralize or accept protons from acids.

basic sediment *n*: see sediment.

basic sediment and water (BS&W) *n*: see *sediment and water (S&W)*.

basin *n*: 1. a local depression in the earth's crust in which sediments can accumulate to form thick sequences of sedimentary rock. 2. the area drained by a stream and its tributaries. 3. a geologic structure in which strata are inclined toward a common center.

basket *n*: a device placed in the drill or work string that catches debris when a drillable object is being milled or drilled downhole.

basket grapple *n*: an expandable, cylindrically shaped gripping mechanism that is fitted into an overshot to retrieve fish from the borehole. See *grapple*.

basket hitch *n*: in crane operations, a method of attaching a sling to a load in which both ends of the sling are passed under the load and then attached to the crane's hook.

basket sub *n*: a fishing accessory run above a bit or a mill to recover small, nondrillable pieces of metal or junk in a well.

batch *n*: 1. a specific quantity of material that is processed, treated, or used in one operation. 2. in corrosion control, a quantity of chemical corrosion inhibitors injected into the lines of a production system. 3. in oilwell cementing, a part of the total quantity of cement to be used in a well. 4. in pipelining, a quantity of one weight or type of crude or liquid product pumped next to one of different weight or type.

batch cementing *n*: in oilwell cementing, the pumping of cement in partial amounts, or batches, as contrasted with pumping it all in one operation.

batching *n*: in pipelining, the pumping of a quantity of crude or product of one weight next to one of different weight or type. Usually, a small amount of mixing occurs where the two batches come in contact.

batching sphere *n*: a large rubber ball placed in a pipeline to separate batches and prevent mixing.

batch mixer *n*: a cement-mixing system in which dry cement and water are blended by a stream of air. The primary disadvantages are uneven mixing and volume limitations.

batch treating *n*: the process by which a single quantity of crude oil emulsion is broken into oil and water. The emulsion is gathered and stored in a tank or container prior to treating. Compare *flow-line treating*.

batch treatment *n*: in corrosion control, the injection of a quantity of chemical corrosion inhibitors into the lines of a production system, usually on a regular schedule.

bath *n*: liquid placed in a container and held at a controlled temperature to regulate the temperature of any system placed in it or passing through it.

bathymetric (bathymetry) data *n pl*: seafloor conditions (e.g., contours or anomalies).

battery *n*: 1. an installation of identical or nearly identical pieces of equipment (such as a tank battery or a battery of meters). 2. an electricity storage device consisting of two or more galvanic cells. See *galvanic cell*.

battery charger *n*: an electrical device (a rectifier) that converts alternating current into direct current and is used to reestablish the battery's ability to produce current by reversing the chemical reaction that occurs in a secondary (wet cell) battery when it discharges. See *rectifier*.

baud *n*: in digital communication terminology, a unit of speed in data transmission equal to one bit per second.

Baumé gravity *n*: specific gravity as measured by the Baumé scale. Two arbitrary scales are employed: one for liquids lighter than water and the other for liquids heavier than water. This scale is also used to describe the density of acid solutions.

Baumé scale *n*: either of two arbitrary scales—one for liquids lighter than water and the other for liquids heavier than water—that indicates specific gravity in degrees.

bbl *abbr*: barrel.

bbl/acre-ft *abbr*: barrels per acre-foot.

bbl/d *abbr*: barrels per day.

B$_c$ *abbr*: Bearden units of consistency-dimensionless.

BCD *abbr*: binary coded decimal.

Bcf *abbr*: billion cubic feet.

Bcf/d *abbr*: billion cubic feet per day.

b/d *abbr*: barrels per day; often used in drilling reports.

B/D *abbr*: barrels per day.

BDC *abbr*: bottom dead center.

b$_{dn}$ *abbr*: slope of flowmeter velocity vs rps response curve (down runs).

beam *n*: 1. the extreme width (breadth) of the hull of a ship or mobile offshore drilling rig. 2. a walking beam. See *walking beam*.

beam-balanced pumping unit *n*: a beam pumping unit that has a counterbalance weight on the walking beam.

beam counterbalance *n*: the weights on a beam pumping unit installed on the end of the walking beam, which is opposite the end over the well. The counterbalance offsets, or balances, the weight of sucker rods and other downhole equipment installed in the well.

beam pumping unit *n*: a machine designed specifically for sucker rod pumping. An engine or motor (prime mover) is mounted on the unit to power a rotating crank. The crank moves a horizontal member (walking beam) up and down to produce reciprocating motion. This reciprocating motion operates the pump. Compare *pump jack*.

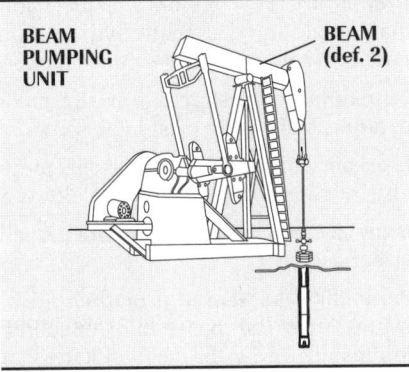

BEAM PUMPING UNIT — BEAM (def. 2)

beam well *n*: a well whose fluid is being lifted by rods and pump actuated by a beam pumping unit.

bean *n*: a nipple or restriction placed in a line (a pipe) to reduce the rate of flow of fluid through the line. Beans are frequently placed in Christmas trees to regulate the flow of fluids coming out of the well. Also called a flow bean. See *Christmas tree*.

Bearden units of consistency (B$_c$) *n*: the thickness, or viscosity, of a slurry (a liquid in which solid particles are suspended), measured in terms of a unit developed by an engineer named Bearden. (Cement particles suspended in water is a common oilfield slurry.) Bearden units of consistency are dimensionless—that is, such units can only be compared to each other when expressing a slurry's consistency. Therefore, Bearden units cannot be converted to other units of viscosity.

bearing *n*: 1. an object, surface, or point that supports. 2. a machine part in which another part (such as a journal or pin) turns or slides.

bearing assembly *n*: in a machine, any device consisting of a set of bearings on or against which two or more surfaces move. The assembly holds the bearings in place as the surfaces move.

bearing bar *n*: on offshore installations with grated floors or decks, load carrying bars made from steel strips or from rolled or extruded aluminum and extending in the direction of the grating span.

bearing cap *n*: a device that is fitted around a bearing to hold or immobilize it.

bearing pin *n*: a machined extension around which are placed bit bearings.

Beaufort scale *n*: a numerical ranking system that provides an estimate of the force of wind and the height of waves. Higher-scale values correlate with higher forces.

bed *n*: a specific layer of earth or rock that presents a contrast to other layers of different material lying above, below, or adjacent to it.

bedding plane *n*: the surface that separates each successive layer of a stratified rock from the preceding layer. It is here that minor changes in sediments or depositional conditions can be observed.

bed load *n*: the gravel and coarse sand that are rolled and bounced along the bottom of a flowing stream. Compare *dissolved load, suspended load*.

bedrock *n*: solid rock just beneath the soil.

belching *n*: slang. See *flow by heads*.

bell *v*: to flare the end of a cylindrical object so that it resembles the bottom of a bell.

belled box *n*: a tool joint box which has been subjected to a torque which has resulted in permanent enlargement of the box diameter. This normally occurs adjacent to the box sealing shoulder.

bell hole *n*: a hole shaped like a bell, larger at the top than at the bottom. A bell hole may be dug beneath a pipeline to allow access for workers and tools.

bell nipple *n*: a short length of pipe (a nipple) installed on top of the blowout preventer. The top end of the nipple is flared, or belled, to guide drill tools into the hole and usually has side connections for the fill line and mud return line.

bellows *n pl*: a pressure-sensing element of cylindrical shape whose walls contain convolutions that cause the length of the bellows to change when pressure is applied.

bellows meter *n*: see *orifice meter*.

belt *n*: a flexible band or cord connecting and wrapping around each of two or more pulleys to transmit power or impart motion.

belt guard *n*: a protective grill or cover for a belt and pulleys.

benchmark *n*: a marking on the pin and the box of a tool joint that gauges the amount of metal that can be safely removed when dressing (facing) the pin or box shoulder.

benchmark price *n*: the price per barrel of certain crude oils such as West Texas intermediate and North Sea Brent, which serves as an indicator of worldwide oil prices.

bending *n*: occurs when tension is increased on one side of the pipe while compression is increased on the other.

bending load pressure *n*: the force created on an object by applying the weight of another object so that the weight bends or tends to bend the first object.

bending mandrel *n*: an attachment that keeps large-diameter pipe from buckling or wrinkling when it is bent.

bending moment *n*: a force that lateral movement (bending) creates on an object.

bending shoe *n*: an attachment that works off the side-boom tractor and winch and bends small-diameter pipe.

bending stress *n*: in offshore drilling from floating rigs, a force placed on a marine riser system and subsea blowout preventer stack that originates from movements of the water and from the weight of the upper components. Wind, waves, and currents act on the equipment to cause it to bend as does the weight of components in the system.

Bendix *n*: the brand name for a type of friction clutch in an electric starter for small engines. When electric current is applied to the starter, the friction clutch (the Bendix) moves forward to engage a pinion gear on the starter with a ring gear on the engine flywheel. As the starter's pinion rotates, it rotates the ring gear, which moves the flywheel to turn the engine over.

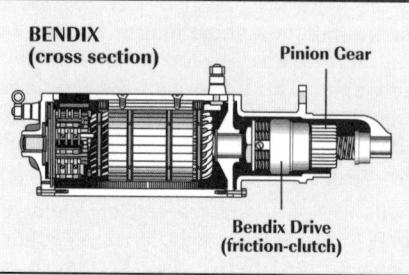

BENDIX (cross section)   Pinion Gear

Bendix Drive (friction-clutch)

bends *n*: a highly painful and potentially fatal condition in which air or other breathable gases come out of solution in the bloodstream and cause distress or death. So named because the bending joints of the body are most often affected. Also called decompression sickness.

bent housing *n*: a special housing for the positive-displacement downhole mud motor, which is manufactured with a bend of 1° to 3° to facilitate directional drilling.

bentonite *n*: a colloidal clay, composed primarily of montmorillonite, that swells when wet. Because of its gel-forming properties, bentonite is a major component of water-base drilling muds. See *gel, mud*.

bentonite extenders *n pl*: a group of polymers that can maintain or increase the viscosity of bentonite while flocculating other clay solids in the mud. With bentonite extenders,

desired viscosity can often be maintained using only half the amount of bentonite that would otherwise be required.

bent sub *n*: a short cylindrical device installed in the drill stem between the bottommost drill collar and a downhole motor. Its purpose is to deflect the downhole motor off vertical to drill a directional hole. See *drill stem*.

benzene *n*: C_6H_6, colorless, volatile, flammable toxic liquid aromatic hydrocarbon used as a solvent and as a motor fuel.

benzene ring *n*: the structural arrangement of atoms believed to exist in benzene.

bequeath *v*: to make a gift of personal property by means of a will. Compare *devise*.

bergy bit *n*: an unscientific term for a small iceberg.

Bernoulli's theorem *n*: a mathematical expression of the conservation of energy in streamline flow; the theorem states that the sum of the ratio of the pressure to the mass density, the product of the gravitational constant and the vertical height, and the square of the velocity divided by 2 are constant.

best-available control technology (BACT) *n*: an air emission standard of technology determined by the states on a case-by-case basis. The BACT is applied in attainment areas that meet or exceed NAAQS, and states can consider cost and economic impact when determining exactly what the BACT is and how it will be applied in a specific case.

beta factor *n*: a factor used in calculating the amount of fluid flowing through a measurement system that employs an orifice plate. It is the ratio of the orifice's diameter to the pipe's diameter.

beta particle *n*: one of the extremely small particles, sometimes called rays, emitted from the nucleus of a radioactive substance such as radium or uranium as it disintegrates. Beta particles have a negative charge.

beta ratio (β) *n*: measure of the relationship of the size of the pipe through which fluid is flowing and the size of a restriction, such as an orifice.

bevel gear *n*: one of a pair of toothed wheels whose working surfaces are inclined to nonparallel axes.

BFPH *abbr*: barrels of fluid per hour; used in drilling reports.

BHA *abbr*: bottomhole assembly.

bhhp *abbr*: bit hydraulic horsepower.

bhp *abbr*: brake horsepower.

BHP *abbr*: bottomhole pressure.

BHT *abbr*: bottomhole temperature.

bias *n*: direct-current voltage applied to a transistor control electrode to establish a desired operating point.

bias drilling *n*: see *directional drilling*.

bicarb *n*: see *sodium bicarbonate*.

bidirectional meter *n*: a meter that can measure flow from either direction.

Big Inch *n*: the first cross-country pipeline with a 24-inch (61-centimetre) diameter. The 1,340-mile (2,157-kilometre) Big Inch was begun in 1942 with government financing as a part of an emergency construction program (War Emergency Pipelines) to meet the demand for petroleum products during World War II.

big-inch pipe *n*: thin-walled pipe of high tensile strength with a diameter of 20 inches (51 centimetres) or more.

bilge radius *n*: the radius of the rounded portion of a vessel's shell that connects the bottom to the sides.

bilging *n*: see *damage stability*.

billet *n*: a solid steel cylinder used to produce seamless casing. The billet is pierced lengthwise to form a hollow tube that is shaped and sized to produce the casing.

bill of lading *n*: a document by which the master of a ship acknowledges having received in good order and condition (or the reverse) certain specified goods consigned by a particular shipper, and binds himself or herself to deliver them in similar condition, unless the perils of the sea, fire, or enemies prevent it, to the consignees of the shippers at the point of destination on their paying the stipulated freight.

bimetallic *n*: two metals with different thermal coefficients of expansion—one with a low coefficient and the other with high coefficient of expansion—held together through brazing, welding, or riveting, for the purpose of bending into an arc. When subjected to temperature variations, the bimetallic elements can be used for actuation, indication, and control.

bimetal thermometer *n*: a temperature sensing and measuring device that has two strips of metal, one with an extremely low coefficient of thermal expansion and the other a rather high coefficient. Temperature is inferred from the different expansion rates of the two metals.

binary *adj*: 1. characterized by or consisting of two parts or components; twofold. 2. of or relating to a system of numeration having 2 as its base. In instrumentation, a binary device or signal has two discrete states or conditions. For example, an on-off switch, or a voltage or current switching on and off over time are binary devices. A binary number uses two symbols, 1 and 0, that represent two discrete states or conditions; in specific combinations, the 1s and 0s can represent equivalent decimal numbers.

binary coded decimal (BCD) code *n*: a system of number representation in which a binary number represents each digit of a decimal number.

binary number *n*: a method of representing numbers using only the digits 0 and 1 in which successive digits are interpreted as coefficients of successive powers to the base 2. For example, the decimal number 3 is 10 in binary numbers and the decimal number 4 is 11 in binary numbers.

binary point *n*: the character, or the location of an implied symbol, that separates the integral part of the numerical expression from its fractional part in binary notation.

binder *n*: a substance that, when added to a collection of loose substances, causes the loose substances to cohere.

binding bar *n*: a steel bar or section fixed to the edges of a panel of deck grating on an offshore installation that is flush with the top of the grating supports.

Bingham plastic *n*: a non-Newtonian fluid that has a given yield stress that must be exceeded before the fluid starts to flow. After a Bingham plastic begins to flow, the rate of shear versus the shear stress curve is linear. See *Newtonian fluid*.

biochemical *adj*: involving chemical reactions in living organisms.

biodegradation *n*: a process similar to composting.

biofacies *n*: a part of a stratigraphic unit that differs in its fossil fauna and flora from the rest of the unit.

biogenic *adj*: produced by living organisms.

bioherm *n*: a reef or mound built by small organisms and their remains, such as coral, plankton, and oysters. Originally a wave-resistant coral structure served as an anchor for calcareous debris that formed limestone. It was tectonically submerged, or the sea level rose faster than the corals could build it, and it was eventually buried beneath marine shales. A bioherm is often porous enough to hold large accumulations of hydrocarbons, especially if it has been dolomitized. A bioherm is a stratigraphic trap.

biomass *n*: the total mass of living organisms per unit volume per unit of time.

biopolymer *n*: a polymer produced by the action of a particular strain of bacteria on carbohydrates, used for increasing apparent viscosity and yield point with moderately good filtration control. Also called X-C polymer.

biopolymer mud *n*: a drilling mud formulated with a biopolymer.

bioremediation *n*: 1. the process of breaking down organic wastes with microbes. Bacteria that are naturally present in the environment use microbial enzymes to break down the materials into a soluble form that passes through the cell walls of the bacteria. The bacteria metabolize the material and convert it into components that are more readily assimilated in the environment, such as water or carbon dioxide (i.e., a gardener's compost pile). 2. the creation of engineered and managed conditions to boost natural bioremediation processes. Nutrients are added to stimulate the bacteria naturally present in the waste. Oxygen and pH levels are adjusted for maximum effectiveness, and, if conditions are right, the bacteria grow in large numbers and break down the hydrocarbons much more rapidly than with natural bioremediation. In cases in which the necessary bacteria are not already present in the waste, suppliers can provide them along with the appropriate nutrients.

biosphere *n*: the thin zone of air, water, and soil where all terrestrial life exists.

biota *n pl*: the animals, plants, fungi, etc., of a region or period.

biotic *adj*: relating to life, biologic; relating to the actions of living organisms.

biotite *n*: a type of mica that is high in magnesium and dark in color.

bipolar *adj*: of a material capable of assuming a positive or negative electric charge, such as a bipolar transistor.

birdcage *n*: a wire rope that has been flattened and the strands of which have been spread. *v*: to flatten and spread the strands of a wire rope.

birdcaged *adj*: see *wickered*.

birdcaged wire *n*: wire rope used for hoisting that has had its wires distorted into the shape of a birdcage by a sudden release of load.

bird-dog *v*: to supervise another too closely or continuously.

bistable *adj*: capable of assuming either of two stable states; (pronounced bi-stable).

bistable mutivibrator *n*: see *flip-flop circuit*.

bit *n*: the cutting or boring element used in drilling oil and gas wells. The bit consists of a cutting element and a circulating element. The cutting element is steel teeth, tungsten carbide buttons, industrial diamonds, or polycrystalline diamonds (PDCs). These teeth, buttons, or diamonds penetrate and gouge or scrape the formation to remove it. The circulating element permits the passage of drilling fluid and

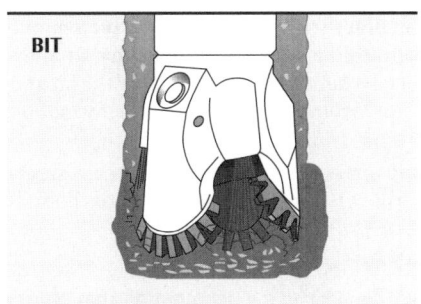

BIT

utilizes the hydraulic force of the fluid stream to improve drilling rates. In rotary drilling, several drill collars are joined to the bottom end of the drill pipe column, and the bit is attached to the end of the drill collars. Drill collars provide weight on the bit to keep it in firm contact with the bottom of the hole. Most bits used in rotary drilling are roller cone bits, but diamond bits are also used extensively.

bit balling *n*: an occurrence in which a mass of sticky consolidated material, usually drill cuttings, collect on the bit. A balled-up bit usually cannot drill efficiently.

bit breaker *n*: a special device that fits into a bit breaker adapter (a plate that goes into the rotary table) and conforms to the shape of the bit. Rig workers place the bit to be made up or broken out of the drill stem into the bit breaker and lock the rotary table to hold the bit breaker and bit stationary so that they can tighten or loosen the bit.

bit breaker adapter *n*: a heavy plate that fits into the rotary table and holds the bit breaker; a device used to hold the bit while it is being made up or broken out of the drill stem.

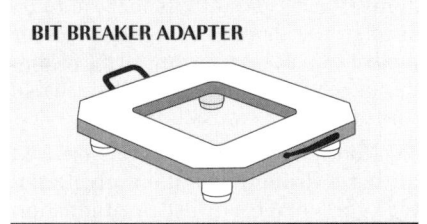

BIT BREAKER ADAPTER

bit cone *n*: on a roller cone bit, a cone-shaped steel device from which the manufacturer either mills or forges steel teeth, or into which the manufacturer inserts tungsten carbide buttons. Most roller cone bits have three cones, which roll, or rotate, on bearings as the bit rotates. As the cones roll over the formation, the cutters on the cone scrape or gouge the formation to remove the rock.

bit cutter *n*: the cutting elements of a bit.

bit dresser *n*: 1. a member of a cable-tool drilling crew who repairs bits. 2. a machine used to repair, sharpen, and gauge bits.

bit drift *n*: the tendency of the bit to move other than vertically, caused by an interaction between the rotation of the bit and the varying resistance of the formation being drilled.

bit energy *n*: a measure of the work a drill bit performs when it is on bottom and drilling.

bit float *n*: see *drill pipe float*.

bit flounder *n*: a phenomenon that occurs when the drilling rate becomes so high that drilling fluid—even if it is air, gas, or foam—cannot remove the cuttings from under the bit; consequently, the cuttings prevent the bit from drilling and it flounders.

bit gauge *n*: a circular ring used to determine whether a bit is of the correct outside diameter. Bit gauges are often used to determine whether the bit has been worn down to a diameter smaller than specifications allow; such a bit is described as undergauge.

bit hydraulic horsepower *n*: the measure of hydraulic power expended through the bit nozzles for cleaning the bit cutters and the hole bottom.

bit matrix *n*: on a diamond bit, the material (usually powdered and fused tungsten carbide) into which the diamonds are set.

bit nozzle *n*: see *nozzle*.

bit pin *n*: the threaded element at the top of a bit that allows it to be made up in a drill collar or other component of the drill stem.

bit plot *n*: on diamond bits, the pattern in which the diamonds are placed in the face of the bit.

bit profile *n*: in fixed-head bits, the shape of the cross section of the body of the bit.

bit program *n*: a plan for the expected number and types of bits that are to be used in the drilling of a well. The bit program takes into account all the factors that affect bit performance so that reliable cost calculations can be made.

bit record *n*: a report that lists each bit used during a drilling operation, giving the type, the footage it drilled, the formation it penetrated, its condition, and so on.

bit run *n*: the placing of a bit on the bottom of the hole, drilling with it until it drills the prescribed amount of hole, or until it wears out, and pulling it out of the hole.

bit shank *n*: the threaded portion of the top of the bit that is screwed into the drill collar. Also called the pin.

bit sub *n*: a sub inserted between the drill collar and the bit. See *sub*.

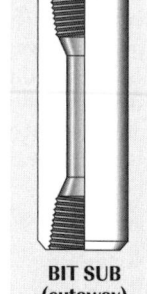

BIT SUB (cutaway)

bitumastic material *n*: a compound of asphalt and filler that is used to coat metals exposed to corrosion or weathering.

bitumen *n*: a substance of dark to black color consisting almost entirely of carbon and hydrogen with very little oxygen, nitrogen, or sulfur. Bitumens occur naturally and can also be obtained by chemical decomposition.

bituminous shale *n*: see *oil shale*.

bit walk *n*: the tendency of the bit to drill in the direction of rotation in an inclined hole; e.g., a right-rotating bit walks to the right.

bit whirl *n*: the motion a bit makes when it does not rotate around its center but instead drills with a spiral motion. It usually occurs to a bit drilling in a soft or medium soft formation when the driller does not apply enough weight or does not rotate the bit fast enough. A whirling bit drills an overgauge hole (a hole larger than the diameter of the bit) and causes the bit to wear abnormally.

bl *abbr*: black; used in drilling reports.

black granite *n*: see *diorite*.

blackout *n*: the total loss of power on a drilling rig or production facility. *v*: to totally lose power on a drilling rig or production facility.

blank casing *n*: casing without perforations.

blanket gas *n*: a gas phase above a liquid phase in a vessel. It is placed there to protect the liquid from contamination, to reduce the hazard of detonation, or to pressure the liquid. The gas has a source outside of the vessel.

blank flange *n*: a solid disk used to dead-end, or close off, a companion flange.

blanking plug *n*: a plug used to cut off flow of liquid.

blanking pulse *n*: in an oscilloscope, a positive or negative square-wave variation in electrical power (a pulse) that switches off a part of the oscilloscope's screen for a predetermined length of time.

blank joint *n*: a heavy wall sub placed in the tubing string opposite flowing perforations.

blank liner *n*: liner with no perforations.

blank off *v*: to close off (as with a blank flange or bull plug).

blank pipe *n*: pipe, usually casing, with no perforations.

blasthole drilling *n*: the drilling of holes into the earth for the purpose of placing a blasting charge (such as dynamite) in them.

blasting *n*: see *shooting rock*.

blasting mats *n pl*: coverings used to contain flying debris and rock caused by the use of explosives during pipeline ditching.

blast joint *n*: a tubing sub made of abrasion-resistant material. It is used in a tubing string where high-velocity flow through perforations may cause external erosion.

bld *abbr*: bailed; used in drilling reports.

bleed *v*: to drain off liquid or gas, generally slowly, through a valve called a bleeder. To bleed down, or bleed off, means to release pressure slowly from a well or from pressurized equipment.

bleeder valve *n*: a special valve used to bleed pressure slowly from a line or vessel. Also called a blow-down valve.

bleed line *n*: a pipe through which pressure is bled, as from a pressurized tank, vessel, or other pipe.

blender *n*: a device used to blend slurries or gels, usually mobile equipment.

blending stock *n*: a petroleum product that is mixed, or blended, with another petroleum product.

blend sample *n*: a representative sample taken from any suitable point in a tank, or from a tank side sample connection after mixing the tank contents and before any significant separation of phases has occurred.

blind *n*: a circular metal disk installed in a pipeline to prevent flow by fastening it between flanges. *v*: to close a line to prevent flow.

blind drilling *n*: a drilling operation in which the drilling fluid is not returned to the surface; rather, it flows into an underground formation. Sometimes blind-drilling techniques are resorted to when lost circulation occurs.

blind end *n*: the end of a device, such as the end of a cylinder used in a riser tensioning system, which is capped and completely closed, or blinded.

blind ram *n*: an integral part of a blowout preventer, which serves as the closing element on an open hole. Its ends do not fit around the drill pipe but seal against each other and shut off the space below completely. See *ram*.

blind ram preventer *n*: a blowout preventer in which blind rams are the closing elements. See *blind ram*.

blind-shear ram preventer *n*: see *shear ram*.

blister *n*: on the pontoons of some of the earliest semisubmersible rigs, a relatively small protuberance on the pontoon's side that widened the pontoon and increased the rig's stability.

blistering *n*: isolated convex deformation of a paint that occurs when one or more of the coats become detached. The deformations resemble blisters that occur under human skin.

blizzard box *n*: a housing built around controls on equipment that is sensitive to cold, such as heater-treaters.

BLM *abbr*: Bureau of Land Management.

block *n*: any assembly of pulleys on a common framework; in mechanics, one or more pulleys, or sheaves, mounted to rotate on a common axis. The crown block is an assembly of sheaves mounted on beams at the top of the derrick or mast. The drilling line is reeved over the sheaves of the crown block alternately with the sheaves of the traveling block, which is hoisted and lowered in the derrick or mast by the drilling line. When elevators are attached to a hook on a conventional traveling block, and when drill pipe is latched in the elevators, the pipe can be raised or lowered in the derrick or mast. See *crown block, traveling block*.

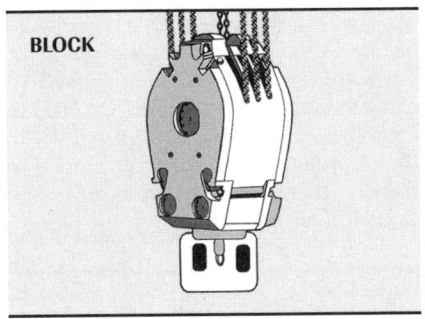

BLOCK

block and bleed valve *n*: a high-integrity valve with double seals and with provision for determining whether either seal leaks.

block billing *n*: FERC proposal calling for separation of old and new gas into blocks, or batches, for interstate sales and shipping.

block diagram *n*: a three-dimensional perspective view of a cube, or block, of earth. It is developed from cross sections of the area.

blocking diode *n*: a diode that restricts (blocks) the flow of current in an undesirable direction in a circuit. See *diode*.

blocks *n pl*: heavy lifting mechanism used on rigs to provide a mechanical pulling and running advantage.

block squeeze *n*: a technique in squeeze cementing in which (1) the zone below the producing interval is perforated and a high-pressure squeeze carried out; (2) the zone above the producing interval is perforated and squeezed off in a similar manner; (3) the hole is drilled out; and (4) the producing interval is perforated. The purpose of block squeezing

is to isolate the producing interval and prevent communication with the sand immediately above and below the producing interval.

block valve *n*: a valve that completely shuts off flow of a fluid. It is usually either completely open or completely closed.

blooey line *n*: the discharge pipe from a well being drilled by air drilling. The blooey line is used to conduct the air or gas used for circulation away from the rig to reduce the fire hazard as well as to transport the cuttings a suitable distance from the well. See *air drilling*.

bloom *n*: the color of fluorescent light exhibited by some oils when viewed by reflected light. This usually differs from the color as seen by transmitted light.

blooming *n*: a thin film on the top of a coat of paint that reduces the luster or veils its depth of color.

blow *v*: 1. to supercharge an engine. See *supercharge*. 2. to depressure. See *depressure*.

blowby *n*: the percentage of gases that escape past the piston rings from the combustion chamber into the crankcase of an engine.

blow case *n*: 1. a pressurized device capable of transferring liquid; sometimes used to transfer crude oil and water mixtures if pump agitation would create unwanted emulsions. 2. a small tank in which liquids are accumulated and drained by applying gas or air pressure above the liquid level. Such a vessel is usually located below a pipeline or other equipment at a location where an outside power source is not convenient for removing the drained liquids. Sometimes referred to as a drip.

blow down *v*: to empty or depressure a vessel. Also called depressure.

blowdown *n*: 1. the emptying or depressurizing of material in a vessel. 2. the material thus discarded.

blowdown period *n*: that period following the completion of a cycling or pressure maintenance operation in a reservoir in which the remaining gas is produced from the reservoir without being replaced by injected gas.

blow-down valve *n*: see *bleeder valve*.

blown *adj*: supercharged. See *supercharge*.

blown oil *n*: fatty oil the viscosity of which has been increased by blowing air through it at an elevated temperature.

blowoff cock *n*: a device that permits or arrests a flow of liquid from a receptacle or through a pipe, faucet, tap, or stop valve.

blowout *n*: an uncontrolled flow of gas, oil, or other well fluids into the atmosphere or into an underground formation. A blowout may occur when formation pressure exceeds the pressure applied to it by the column of drilling fluid and rig crew members fail to take steps to contain the pressure. Before a well blows out, it kicks; thus a kick precedes a blowout. See *kick*.

blowout preventer (BOP) *n*: one of several valves installed at the wellhead to stop (prevent) the escape of pressure either in the annular space between the casing and the drill pipe or in open hole (i.e., hole with no drill pipe) during drilling or completion operations. Blowout preventers on land rigs are normally located beneath the rig at, or slightly below, the land's surface; on jackup or platform rigs, at the water's surface; and on floating offshore rigs, on the seafloor. See *annular blowout preventer, ram blowout preventer*.

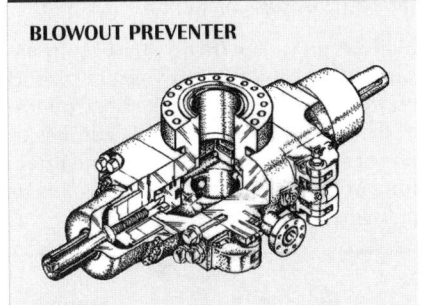

BLOWOUT PREVENTER

blowout preventer control panel *n*: controls, usually located near the driller's position on the rig floor, that are manipulated to open and close the blowout preventers. See *blowout preventer*.

blowout preventer (BOP) control unit *n*: a device that stores hydraulic fluid under pressure in special containers and provides a method to open and close (operate) the blowout preventers quickly and reliably. Usually, compressed air and hydraulic pressure provide the opening and closing force in the unit. See *blowout preventer*. Also called an accumulator.

blowout preventer drill *n*: a training procedure to determine that rig crews are familiar with correct operating practices to be followed in the use of blowout prevention equipment.

blowout preventer (BOP) mandrel *n*: in drilling from offshore floating rigs, a metal fitting connected to the top of the annular preventer in the subsea blowout preventer stack. The mandrel has a mating-and-seal profile to match that of the lower marine riser package's (LMRP's) connector. The connector normally has the same pressure rating as the LMRP's annular preventer.

blowout preventer (BOP) manifold regulator *n*: in drilling from floating drilling rigs using a subsea blowout preventer, a special device that controls (regulates) the pressure to the inlet ports of a subsea blowout preventer's subsea pilot manipulated (SPM) valves, which, in turn, operate the ram preventers, wellhead and lower marine riser package (LMRP) connectors, and subsea choke and kill valves.

blowout preventer operating and control system *n*: see *blowout preventer control unit.*

blowout preventer (BOP) orientation pin *n*: a device that engages the helix of a tubing hanger's running and orientation tool to orient a dual-bore tubing hanger within the wellhead. The pin is a spring retracted hydraulic piston that has the same pressure rating of the blowout preventer because it could be exposed to wellbore fluids.

blowout preventer rams *n pl*: the closing and sealing components of a preventer. Corresponds to the gate in the gate valve.

blowout preventer (BOP) stack *n*: an assembly of blowout preventers placed on top of each other (stacked one on top of the other), which typically consists of one or two annular preventers and three, four, or more ram preventers. See *annular preventer, ram preventer.*

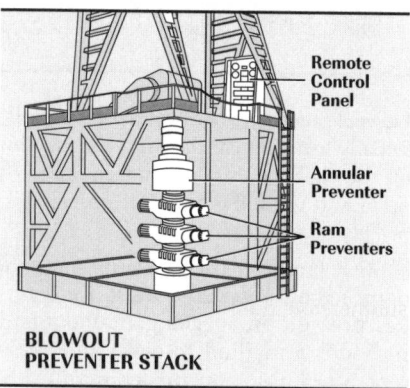

Remote Control Panel

Annular Preventer

Ram Preventers

BLOWOUT PREVENTER STACK

blowout sticking *n*: jamming or wedging of the drill string in the borehole by sand or shale that is driven uphole by formation fluids during a blowout.

BLPD *abbr*: barrels of liquid per day; usually used in reference to total production of oil and water from a well.

blue pod *n*: see *hydraulic control pod.*

blushing *n*: in clear paints, a milky opalescence caused by the deposition of moisture from the atmosphere or the precipitation of one or more of the solid constituents of the paint.

BO *abbr*: barrels of oil; used in drilling reports.

bob *n*: see *gauger's bob.*

bob-tail plant *n*: an extraction plant without fragmentation facilities.

body *n*: used to indicate the consistency of a paint.

body lock ring *n*: an internal mechanism employed in certain tools to lock cones to the mandrel.

body wave *n*: transverse or longitudinal seismic waves transmitted in the interior of an elastic solid or fluid and not related to a boundary surface.

boilaway test *n*: see *weathering test.*

boiler *n*: a closed pressure vessel that has a furnace equipped to burn coal, oil, or gas and that is used to generate steam from water.

boiler feed water *n*: water piped into a boiler from which steam is generated.

boiler fuel *n*: fuel suitable for generating steam or hot water in large industrial or electricity-generating utility applications.

boiler fuel gas *n*: natural gas used as a fuel for the generation of steam or hot water.

boiler house *v*: (slang) to make up a report on a condition as fact without knowledge of its accuracy. Sometimes referred to as "doghouse."

boiling point *n*: the temperature at which the vapor pressure of a liquid becomes equal to the pressure exerted on the liquid by the surrounding atmosphere. The boiling point of water is 212°F or 100°C at atmospheric pressure (14.7 pounds per square inch gauge or 101.325 kilopascals).

boll weevil *n*: (slang) an inexperienced rig or oilfield worker; sometimes shortened to weevil.

boll weevil corner *n*: (slang, obsolete) the work station of an inexperienced rotary helper, on the opposite side of the rotary from the rotary helper who sets pipe back on the rig floor during trips.

boll weevil hanger *n*: a tubing hanger.

bolt *n*: a metal fastener made from a threaded pin or rod, usually with a square or hexagonal head at one end, which is inserted through holes in assembled parts and secured by a mated nut that is tightened by applying torque.

bolted-flange riser connector *n*: in drilling from floating drilling rigs using a subsea blowout preventer and riser system, a riser connector in which bolts are screwed into threaded holes in a flange welded on the end of the riser. The bolts are tightened to strict specifications with special torque wrenches to avoid failure of the connection. See *riser pipe.*

bomb *n*: a thick-walled container, usually steel, used to hold devices that determine and record pressure or temperature in a wellbore. See *bottomhole pressure.*

bomb hanger *n*: a device set in a tubing coupling to facilitate the landing of pressure bombs (recorders).

bond *n*: the adhering or joining together of two materials (as cement to formation). *v*: to adhere or to join to another material.

bonnet *n*: 1. the part of a valve that packs off and encloses the valve stem. 2. on a ram blowout preventer, the housing that surrounds the rams and contains the operating piston and rod.

bonus consideration *n*: a cash payment by the lessee for the execution of an oil and gas lease by the mineral owner; expressed as dollars per acre. Occasionally, an oil payment or overriding royalty may be reserved as a bonus by a lessor in addition to regular royalty.

Boolean logic *n*: a system of symbolic logic based on algebraic symbols representing such logical operations as AND, OR, and NOR.

boom *n*: 1. a movable arm of tubular or bar steel used on some types of cranes or derricks to support the hoisting lines that carry the load. 2. floating device used to contain oil. 3. a period of high activity in the oil industry. 4. a member hinged to the revolving superstructure and used as part of an attachment.

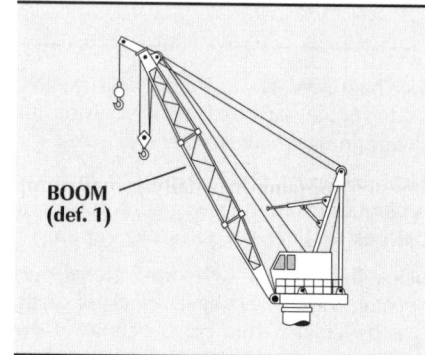

BOOM (def. 1)

boom angle *n*: the angle above horizontal of the longitudinal centerline of the boom base section.

boom chord *n*: a main corner member of a lattice-type boom.

boom dog *n*: a ratchet device on a crane that prevents the boom of the crane from being lowered but still permits it to be raised. Also called a boom ratchet.

boomer *n*: 1. (slang) an oilfield worker who moves from one center of activity to another; a floater or transient. 2. a device used to tighten chains on a load of pipe or other equipment on a truck to make it secure.

boom hoist *n*: rope drum(s) and its drive, or other mechanism for controlling the angle of the boom.

boom lacing *n*: structural truss members at angles to and supporting the boom chords of a lattice-type boom.

boom length *n*: the straight line distance from the centerline of a boom foot pin to the centerline of a boom point load hoist sheave pin.

boom pendant *n*: the large-diameter wire rope or the strands that support the boom of a crane.

boom ratchet *n*: see *boom dog*.

boom sections *n pl*: crane booms are usually in two sections, upper and lower; such booms may be lengthened by insertion of one or more additional sections.

boom splices *n pl*: splicing connections for sections of a basic crane boom and additional sections usually of the splice plate type, pin type, or butt type.

boom stop *n*: the steel projections on a crane struck by the boom if it is raised too high or lowered too far.

booster pump *n*: on a diesel engine, a small manually or electrically operated pump that an engine operator can use to prime the engine's fuel system for starting the engine. When activated, the pump moves fuel from a tank, through the engine's fuel lines, and to the engine's injectors, ensuring that fuel is available for starting the engine.

booster station *n*: an installation on a pipeline that maintains or increases pressure of the fluid coming through the pipeline and being sent on to the next station or terminal.

booster-type pump *n*: usually a small pump mounted upstream of another pump that charges the fluid being pumped prior to the fluid's reaching the main pump.

boot *n*: 1. a tubular device placed in a vertical position, either inside or outside a larger vessel, and through which well fluids are conducted before they enter the larger vessel. A boot aids in the separation of gas from wet oil. Also called a flume or conductor pipe. 2. a large pipe connected to a process tank to provide a static head that can absorb surges of fluid from the process tank. Also called surge column.

boot basket *n*: see *junk sub*.

boot sub *n*: a device made up in the drill stem above the mill to collect bits of junk ground away during a milling operation. During milling, drilling mud under high pressure forces bits of junk up the narrow space between the boot sub and hole wall.

When the junk reaches the wider annulus above the boot sub and pressure drops slightly, the junk falls into the boot sub. A boot sub also can be run above the bit during routine drilling to collect small pieces of junk that may damage the bit or interfere with its operation. Also called a junk sub or junk boot.

BOP *abbr*: blowout preventer.

BOP assembly *n*: see *blowout preventer stack*.

BOP control system *n*: see *blowout preventer control unit*.

BOPD *abbr*: barrels of oil per day.

BOPE *abbr*: blowout preventer equipment.

BOP orientation pin *n*: its purpose is to engage with the tubing hanger running and orientation tool helix and orientate the dual-bore tubing hanger within the wellhead. The pin is simply a spring retract hydraulic piston and must be rated to the operating pressure of the BOP since it could be exposed to wellbore fluids.

BOP stack *n*: the assembly of blowout preventers installed on a well. See *blowout preventer stack*.

bore *n*: 1. the inside diameter of a pipe or a drilled hole. 2. the diameter of the cylinder of an engine. *v*: to penetrate or pierce with a rotary tool. Compare *tunnel*.

bored crossing *n*: a hole under a road, railroad, stream, or other obstacle under which a pipeline must cross. It is created by using an auger or other type of drill.

borehole *n*: a hole made by drilling or boring; a wellbore.

borehole ballooning *n*: an effect that occurs when certain shale zones are penetrated by the wellbore and the pressure exerted by the drilling mud exceeds the pressure exerted by the shale. The pressure differential causes the borehole to flex, or balloon, and the borehole no longer behaves as a fixed, rigid cylinder.

borehole-contact log *n*: any logging device whose operation depends on a portion of the logging tool touching the wellbore.

borehole effect *n*: false influence on well logging measurement caused by the borehole environment, e.g., diameter, shape, rugosity, type of fluid, or mud cake.

borehole pressure *n*: the total pressure exerted in the wellbore by a column of fluid and any back-pressure imposed at the surface.

bore protector *n*: a device suspended inside a wellhead that protects the casing hanger's sealing surfaces inside the wellhead housing before the casing hangers are installed.

borings sample *n pl*: obtained by collecting the chips made by boring holes with a ship auger from the top to the bottom of the material contained in a barrel, case, bag, or cake.

borrowing *n*: in the mathematical operation of subtraction, the act of taking the number 1 from the tens, hundreds, or other column to the left of the ones column, and placing it to the left of the number in the ones column, to raise the number in the ones column to the next higher order digit. This action reduces the number that was borrowed from by 1. Borrowing then continues, if necessary, in each subsequent column to the left until the solution is achieved. For example, to subtract 679 from 831, 1 is borrowed from 3 in the tens column of 831, which leaves 2. Nine is then subtracted from 11 to leave 2. Next, 7 is subtracted from 12, because 1 was borrowed from the hundreds column to make 12. This action leaves 5. Finally, 6 is subtracted from 7 in the hundreds column, which was reduced by 1 when it was borrowed from earlier. This action leaves 1. So, the answer is 152.

bottled gas *n*: liquefied petroleum gas placed in small containers for sale to domestic customers.

bottleneck *n*: an area of reduced diameter in pipe caused by excessive longitudinal strain or by a combination of longitudinal strain and the swaging action of a body. A bottleneck may result if the downward motion of the drill pipe is stopped with the slips instead of the brake.

bottleneck elevator *n*: an elevator that is bored to match the 18-degree taper of a tool joint.

bottles *n pl*: large-diameter steel cylinders used to support submersible and semi-submersible rigs.

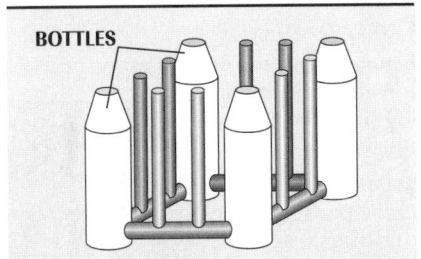

bottle test *n*: a test in which different chemicals are added to bottle samples of an emulsion to determine which chemical is the most effective at breaking the emulsion into oil and water. Once an effective chemical is determined, varying amounts of it are added to bottle samples of the emulsion to determine the minimum amount required to break the emulsion effectively.

bottle-type semisubmersible rig *n*: see *semisubmersible drilling rig*.

bottle-type submersible rig *n*: a mobile submersible drilling structure constructed of several steel cylinders or bottles. When the bottles are flooded, the rig submerges and rests on bottom; when water is removed from the bottles, the rig floats. The latest designs of this type of rig drill in water depths up to 100 feet (30.5 metres). See *submersible drilling rig*.

BOTTLE-TYPE SUBMERSIBLE RIG

Bottles

bottom angle *n*: the angle formed where the bottom and the side of the oil tank meet, usually nearly 90°.

bottom dead center (BDC) *n*: the positioning of the piston at the lowest point possible in the cylinder of an engine; often marked on the engine flywheel.

bottom guide wire anchor *n*: a bar welded to the bottom of a tank to which guide wires or cables for the float of an automatic tank gauge are attached.

bottom hold-down *n*: a mechanism for anchoring a bottomhole pump in a well; located on the lower end of the pump. Compare *top hold-down*.

bottomhole *n*: the lowest or deepest part of a well. *adj*: pertaining to the bottom of the wellbore.

bottomhole assembly *n*: the portion of the drilling assembly below the drill pipe. It can be very simple—composed of only the bit and drill collars—or it can be very complex and made up of several drilling tools.

bottomhole choke *n*: a device with a restricted opening

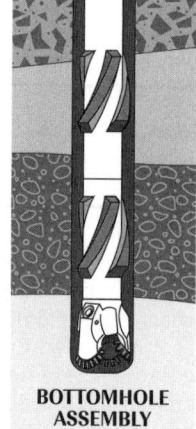

BOTTOMHOLE ASSEMBLY

placed in the lower end of the tubing to control the rate of flow. See *choke*.

bottomhole contract *n*: a contract providing for the payment of money or other considerations upon the completion of a well to a specified depth.

bottomhole flow regulator *n*: see *bottomhole choke*.

bottomhole heater *n*: an electric heater screwed onto lengths of tubing and lowered into the well. It raises the bottomhole temperature of the well to prevent solidification of paraffin in subsurface equipment.

bottomhole letter *n*: a contract providing for the payment of money or other considerations on the completion of a well to a specified depth, regardless of whether the well is a producer of oil or gas or is a dry hole.

bottomhole money *n*: money paid by a contributing company in exchange for the information received from the drilling on the completion of a well to a specified depth, regardless of whether the well is a producer of oil or gas or is a dry hole.

bottomhole packer *n*: a device, installed near the bottom of the hole, that blocks passage through the annular space between two strings of pipe. See *packer*.

bottomhole plug *n*: a bridge plug or cement plug placed near the bottom of the hole to shut off a depleted, water-producing, or unproductive zone.

bottomhole pressure (BHP) *n*: 1. the pressure at the bottom of a borehole. It is caused by the hydrostatic pressure of the wellbore fluid and, sometimes, by any back-pressure held at the surface, as when the well is shut in with blowout preventers. When mud is being circulated, bottomhole pressure is the hydrostatic pressure plus the remaining circulating pressure required to move the mud up the annulus. 2. the pressure in a well at a point opposite the producing formation, as recorded by a bottomhole pressure bomb.

bottomhole pressure bomb *n*: a pressure-tight container (bomb) used to record the pressure in a well at a point opposite the producing formation.

bottomhole pressure gauge *n*: an instrument to measure bottomhole pressure. Also called bottomhole pressure bomb.

bottomhole pressure test *n*: a test that measures the reservoir pressure of the well, obtained at a specific depth or at the midpoint of the producing zone. A flowing bottomhole pressure test measures pressure while the well continues to flow; a shut-in bottomhole pressure test measures

pressure after the well has been shut in for a specified period of time. See *bottomhole pressure, bottomhole pressure gauge*.

bottomhole pump *n*: any of the rod pumps, high-pressure liquid pumps, or centrifugal pumps located at or near the bottom of the well and used to lift the well fluids. See *centrifugal pump, hydraulic pumping, submersible pump, sucker rod pumping*.

bottomhole separator *n*: a device used to separate oil and gas at the bottom of wells to increase the volumetric efficiency of the pumping equipment.

bottomhole temperature *n*: temperature measured in a well at a depth at the midpoint of the thickness of the producing zone.

bottom-intake electric submersible pump *n*: a submersible pump configuration in which the pump and the motor sections are reversed, with the pump intake through a stinger at the bottom of the unit. It is used where casing size prohibits the desired production volume because of tubing friction loss or pump diameter restriction.

bottom loading pressure *n*: the pressure exerted on the bottom hull of a column-stabilized semisubmersible drilling rig when the hull is submerged to drilling water depth.

bottom plug *n*: a cement wiper plug that precedes cement slurry down the casing. The plug wipes drilling mud off the walls of the casing and prevents it from contaminating the cement. See *cementing, wiper plug*.

bottom-pull method *n*: an offshore pipeline construction technique in which the pipe string remains below the surface while it is towed to its final location.

bottoms *n pl*: 1. the liquids and the residue that collect in the bottom of a vessel (such as tank bottoms) or that remain in the bottom of a storage tank after a period of service. 2. the residual fractions remaining at the bottom of a fractionating tower after lighter components have been distilled off as vapors.

bottom sample *n*: in tank sampling, a sample obtained from the material near the bottom surface of the tank, container, or line at a low point.

bottomset bed *n*: the part of a marine delta that lies farthest from shore. It consists of silt and clay extending well out from the toe of the steep delta face. Such beds grow slowly, out of reach of most of the effects of river current and wave action.

bottom sub *n*: a device run at or near the bottom of the tubing string to which production tools can be attached.

bottoms up *n*: a complete trip from the bottom of the wellbore to the top.

bottom-supported offshore drilling rig *n*: a type of mobile offshore drilling unit (MODU) that has a part of its structure in contact with the seafloor when it is on site and drilling a well. The remainder of the rig is supported above the water. The rig can float, however, allowing it to be moved from one drill site to another. Bottom-supported units include submersible rigs and jackup rigs. See *mobile offshore drilling unit*.

bottoms-up time *n*: the time required for mud to travel up the borehole from the bit to the surface.

bottom time *n*: the total amount of time, measured in minutes, from the time a diver leaves the surface until he or she begins the ascent.

bottom water *n*: water found below oil and gas in a producing formation.

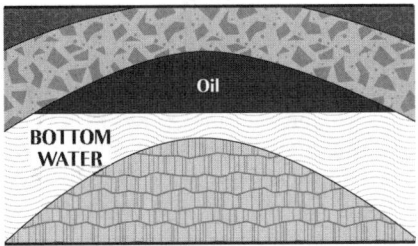

bottom-water drive *n*: see *water drive*.

bottom-water sample *n*: a spot sample of free water taken from beneath the petroleum contained in a ship or barge compartment or a storage tank.

bottom wiper plug *n*: a device placed in the cementing head and run down the casing in front of cement to clean the mud off the walls of the casing and to prevent contamination between the mud and the cement.

bounce dive *n*: a rapid dive with a very short bottom time to minimize decompression time.

Bourdon tube *n*: a pressure-sensing element consisting of a twisted or curved tube of noncircular cross section, which tends to straighten when pressure is applied internally. By the movements of an indicator over a circular scale, a Bourdon tube indicates the pressure applied.

bow *n*: the nautical term for the forward part of a ship or vessel.

bowl *n*: in drilling operations, an insert that fits into the opening of a master bushing and accommodates the slips.

bowline knot *n*: a knot used primarily in lifting heavy equipment with the catline, since it can be readily tied and untied regardless of the weight of the load on it.

bow lines *n pl*: the lines running from the bow of a mobile offshore drilling rig, especially the forward mooring lines.

bow thruster *n*: a control barge used between the towboat and the barge to give more control in steering the barge down the waterway.

box *n*: the female section of a connection. See *tool joint*.

box and pin *n*: see *tool joint*.

box tap *n*: old-style tap with longitudinal grooves across the threads. See *tap*, *taper tap*.

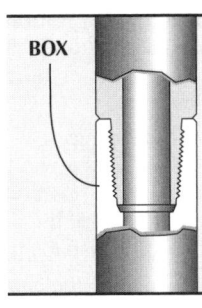

box threads *n pl*: threads on the female section, or box, of a tool joint. See *tool joint*.

Boyle's law *n*: the principle that states that at a fixed temperature, the volume of an ideal gas or gases varies inversely with its absolute pressure. As gas pressure increases, gas volume decreases proportionally.

bpd or **BPD** *abbr*: barrels per day.

BPH *abbr*: barrels per hour; used in drilling reports.

B-P mix *n*: a liquefied hydrocarbon product composed chiefly of butanes and propane. If it originates from a refinery, it may also contain butylenes and propylene.

BPO *abbr*: before payout; commonly used in land departments.

brackish water *n*: water that contains relatively low concentrations of soluble salts. Brackish water is saltier than fresh water but not as salty as salt water.

bradding *n*: a condition in which the weight on a bit tooth has been so great that the tooth has dulled until its softer inner portion caves over the harder case area.

bradenhead *n*: (obsolete) casinghead. Glen T. Braden invented a casinghead in the 1920s that became so popular that all casingheads were called bradenheads.

bradenhead flange *n*: a flanged connection at the top of the oilwell casing.

bradenhead gas *n*: see *casinghead gas*.

bradenhead squeeze *n*: a process used to repair a hole in the casing by pumping cement down tubing or drill pipe. First, the casinghead, or bradenhead, is closed to prevent fluids from moving up the casing. Then the rig's pumps are started. Pump pressure moves the cement out of the tubing or pipe and, since the top of the casing is closed, the cement goes into the hole in the casing. The tubing or pipe

is pulled from the well and the cement allowed to harden. The hardened cement seals the hole in the casing. Although the term "bradenhead squeezing" is still used, the term "bradenhead" is obsolete. See *annular space, casinghead, squeeze*.

braided sling *n*: in crane operations, a sling made from a braided wire rope.

braided wire *n*: see *stranded wire*.

braided wire rope *n*: a rope formed by plaiting (braiding) several strands of wire rope.

brake *n*: a device for arresting the motion of a mechanism, usually by means of friction, as in the drawworks brake. Compare *electrodynamic brake, hydrodynamic brake*.

brake band *n*: a part of the brake mechanism consisting of a flexible steel band lined with a material that grips a drum when tightened. On a drilling rig, the brake band acts on the flanges of the drawworks drum to control the lowering of the traveling block and its load of drill pipe, casing, or tubing.

brake block *n*: a section of the lining of a brake band shaped to conform to the curvature of the band and attached to it with countersunk screws. See *brake band*.

brake flange *n*: the surface on a winch, drum, or reel to which the brake is applied to control the movement of the unit by means of friction.

brake horsepower *n*: the power produced by an engine as measured by the force applied to a friction brake or by an absorption dynamometer applied to the shaft or the flywheel.

brake lining *n*: the part of the brake that presses on the brake drum. On the drawworks, the circular series of brake blocks are bolted to the brake bands with countersunk brass bolts; the lining is the frictional, or gripping, element of a mechanical brake.

brake linkage *n*: everything from the dead ends of the brake bands to the brake lever.

brake rider *n*: (slang) a mostly obsolete term for the driller. See *driller*.

brake rim *n*: see *brake flange*.

brake shoe *n*: that part of a shoe-type brake or clutch which makes contact with a brake drum or clutch drum.

braking capacity *n*: how much weight the brake can stop in a given length of time.

branch line *n*: a line, usually a pipe, joined to and diverging from another line.

branded distributor *n*: gasoline wholesaler who, under the supplier's brand name, provides his or her own financing, transportation, and storage and who may also retail.

brash *n*: a mass of ice fragments that floats on the surface of a sea.

brass running nipple *n*: a device used in the flow cross of the Christmas tree as a thread protector while the rods are being run. Because it is brass, it prevents friction sparks.

breadth *n*: the greatest overall dimension measured perpendicular to the longitudinal centerline of the hull of a mobile offshore drilling rig. Also called the beam.

break *v*: to begin or start (e.g., to break circulation or break tour).

breakaway torque *n*: see *starting torque*.

break circulation *v*: to start the mud pump for restoring circulation of the mud column. Because the stagnant drilling fluid has thickened or gelled during the period of no circulation, high pump pressure is usually required to break circulation.

breakdown *n*: 1. a failure of equipment. 2. in a fracture treatment or a squeeze job, the failure of formation rock to withstand pressure, allowing it to rupture. *adj*: pertaining to the amount of pressure needed at the wellhead to rupture the formation in a fracture treatment or a squeeze job (as in formation breakdown pressure).

breakdown torque *n*: see *pullout torque*.

breaker *n*: in oceanography, a surface wave that has become too steep to be stable.

breaker plate *n*: see *bit breaker adapter*.

breaker points *n*: contacts that interrupt the current in the primary circuit of the ignition system in a spark-ignition engine.

break-in *n*: the drilling of the first few feet of hole with a new bit.

breaking down *v*: unscrewing the drill stem into single joints and placing them on the pipe rack. The operation takes place on completion of the well, or in changing from one size of pipe to another. See *lay down pipe*.

breaking strength *n*: in crane operations, the amount of load that causes a wire rope or a sling to fail. Also called actual strength. Compare *minimum acceptance strength, nominal strength*.

break it *v*: see *break out*.

break out *v*: 1. to unscrew one section of pipe from another section, especially drill pipe while it is being withdrawn from the wellbore. During this operation, the tongs are used to start the unscrewing operation. 2. to separate, as gas from a liquid or water from an emulsion.

breakout block *n*: a heavy plate that fits in the rotary table and holds the drill bit while it is being unscrewed from the drill collar. See *bit breaker*.

breakout cathead *n*: a device attached to the catshaft of the drawworks that is used as a power source for unscrewing drill pipe; usually located opposite the driller's side of the drawworks. See *cathead*.

breakout tanks *n pl*: tanks located at main-line junction stations that hold products temporarily until they can be relayed to other tanks or terminals farther up the line.

breakout tongs *n pl*: tongs that are used to start unscrewing one section of pipe from another section, especially drill pipe coming out of the hole. Compare *makeup tongs*. See also *tongs*.

breakover *n*: the change in the chemistry of a mud from one type to another. Also called conversion.

break tour (pronounced "tower") *v*: to begin operating 24 hours a day. Moving the rig and rigging up are usually carried on during daylight hours only. When the rig is ready for operation at a new location, crews break tour.

breathe *v*: to move with a slight, irregular rhythm. Breathing occurs in tanks of vessels when vapors are expelled and air is taken in. For example, a tank of crude oil expands because of the rise in temperature during the day and contracts as it cools at night, expelling vapors as it expands and taking in air as it contracts. Tubing breathes when it moves up and down in sequence with a sucker rod pump.

breather *n*: a small vent in an otherwise airtight enclosure for maintaining equality of pressure inside and outside.

breathing *n*: see *tubing movement*.

breathing bag *n*: part of the semiclosed circuit breathing apparatus, used to mix gas and ensure low breathing resistance.

breccia *n*: a conglomerate rock composed largely of angular fragments greater than 0.08 inch (2 millimetres) in diameter.

brecciation *n*: the breaking of solid rock into coarse, angular fragments by faulting or crushing.

Brent *n*: a major area of oil production in the British sector of the North Sea.

bridge *n*: 1. an obstruction in the borehole, usually caused by the caving in of the well or the borehole or by the intrusion of a large boulder. 2. a tool placed in the hole to retain cement or other material; it may later be removed, drilled out, or left permanently. See *floating harness*.

bridge over *n*: a phenomenon that sometimes occurs when a well blows out. Rocks, sand, clay, and other debris clog the hole and stop the blowout.

bridge plug *n*: a downhole tool, composed primarily of slips, a plug mandrel, and a rubber sealing element, that is run and set in casing to isolate a lower zone while an upper section is being tested or cemented.

bridge socket *n*: a forged or cast steel fitting for a wire rope or a strand. It has a basket with adjustable bolts that secure rope ends.

bridging materials *n pl*: the fibrous, flaky, or granular material added to a cement slurry or drilling fluid to aid in sealing formations in which lost circulation has occurred. See *lost circulation, lost circulation material*.

bridle *n*: a cable on a pumping unit, looped over the horsehead and connected to the carrier bar to support the polished rod clamp. See *sucker rod pumping*.

bridle sling *n*: in crane operations, a sling composed of several legs, the tops of which are gathered in a fitting that goes over the crane's lifting hook. See *leg*.

bright oil *n*: oil that contains little or no water.

bright spot *n*: a seismic phenomenon that shows up on a seismic, or record, section as a sound reflection that is much stronger than usual. A bright spot sometimes directly indicates natural gas in a trap. See *record section*.

brine *n*: water that has a large quantity of salt, especially sodium chloride, dissolved in it; salt water.

bring in a well *v*: to complete a well and put it on producing status.

bring on a well *v*: to bring a well on-line, that is, to put it on producing status.

British thermal unit (Btu) *n*: a measure of heat energy equivalent to the amount of heat needed to raise 1 pound of water 1°F at or near its point of maximum density (39.1°F). Equivalent to 0.252 kilogram-calories or 1,055 joules.

brittleness *n*: the state of having rigidity but little tensile strength. Compare *toughness*.

brkn *abbr*: broken; used in drilling reports.

broaching *n*: blowing out of formation fluids outside the casing and under the rig.

broadside wind force *n*: the force of the wind that acts on a floating offshore rig or ship when it is perpendicular to the direction of the wind.

broken cone *n*: a cone on a bit that has become cracked.

bromine value *n*: the number of centigrams of bromine that are absorbed by 1 gram of oil under certain conditions. This is a test for the degree of unsaturatedness of a given oil.

Brownian movement *n*: the random movement exhibited by microscopic particles when suspended in liquids or gases. It is caused by the impact of molecules of fluid surrounding the particle.

brush *n*: a carbon block used to make an electrical connection between the rotor of a generator or motor and a circuit.

BS&W *abbr*: basic sediment and water. API Committee on Petroleum Measurement prefers simply S&W.

Bscf/d *abbr*: billion standard cubic feet per day.

Btu *abbr*: British thermal unit.

Btu rating *n*: a measure of the heat energy a fuel, such as natural gas, is capable of generating in terms of British thermal units. The higher the Btu rating is, the more energy contained in the fuel.

bubble bucket *n*: a bucket of water into which air from the drill stem is blown through a hose during the first flow period of a drill stem test. In drill stem testing, the flow into the bubble bucket is an easy way to judge flow and shut-in periods.

bubble cap *n*: a metal cap, mounted on a tray, that has openings allowing vapor bubbles in a gas-processing tower to contact cool liquids, causing some of the vapor to condense to liquid.

bubble-cap tray *n*: a perforated steel tray on which bubble caps are mounted. Bubble caps and trays are arranged in bubble towers, cylindrical vessels set vertically. See *sieve tray, valve tray.*

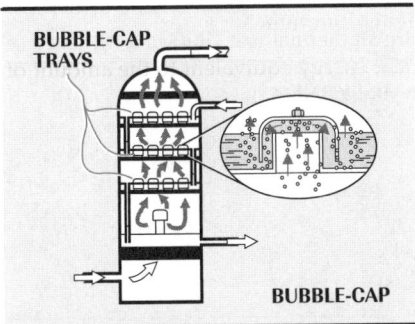

bubble chopper *n*: a special well-control tool made up in the drill pipe usually just above the drill collars. Should a large quantity, or bubble, of gas enter the annulus, the bubble chopper allows mud in the drill string to be diverted into the gas bubble by means of an opening, or port, in the tool. The mud from the drill string mixes with and spreads out the large gas bubble so that the gas cannot expand to a great extent as it nears the surface. Reducing the amount of bubble expansion reduces pressure on the casing near the surface and thus minimizes the chances of formation fracture and an underground blowout.

bubble point *n*: 1. the temperature and pressure at which part of a liquid begins to convert to gas. For example, if a certain volume of liquid is held at constant pressure, but its temperature is increased, a point is reached when bubbles of gas begin to form in the liquid. That is the bubble point. Similarly, if a certain volume of liquid is held at a constant temperature but the pressure is reduced, the point at which gas begins to form is the bubble point. Compare *dew point*. 2. the temperature and pressure at which gas, held in solution in crude oil, breaks out of solution as free gas.

bubble tower *n*: a vertical cylindrical vessel in which bubble caps and bubble-cap trays are arranged.

bubble tray *n*: see *bubble-cap tray.*

buckling stress *n*: bending of pipe that may occur because of hole deviation. Pipe may bend because of the angle of the hole or because of an abrupt deviation such as a dogleg.

buck up *v*: to tighten up a threaded connection (such as two joints of drill pipe).

buddy system *n*: a method of pairing two persons for their mutual aid or protection. The buddy system is used to ensure that each crew member is accounted for, particularly in situations where hydrogen sulfide may be encountered.

buffer *n*: 1. any substance or combination of substances that, when dissolved in water, produces a solution that resists a change in its hydrogen ion concentration with the addition of acid or base. 2. in solid-state electronics, an isolating circuit in a computer that prevents the action of a driven circuit from affecting the corresponding driving circuit. 3. an electric circuit or component that prevents an undesirable electrical interaction between two circuits or components.

bug blower *n*: (slang) any large fan installed on a drilling rig or production facility to move large quantities of air across the rig floor or other area.

builder's monogram *n*: a plate on a tank built to API specifications that records the name of the manufacturer, erection date of the tank, and nominal diameter, height, and capacity.

building assembly *n*: see *fulcrum assembly.*

build up *v*: to increase the rate of inclination, or drift angle, in the hole in order to bring the hole towards the horizontal. Usually expressed as degrees per 100 ft (30 m).

buildup test *n*: a test in which a well is shut in for a prescribed period of time and a bottomhole pressure bomb run in the well to record the pressure. From these data and from knowledge of pressures in nearby wells, the effective drainage radius or the presence of permeability barriers or other production deterrents surrounding the wellbore can be estimated.

bulb *n*: the temperature-sensing (detecting) element of a temperature-measuring device.

bulkhead *n*: an interior wall that subdivides a ship or a mobile offshore drilling rig into compartments.

bulkhead deck *n*: the highest deck to which water bulkheads extend on a ship or a mobile offshore drilling rig.

bulk modulus *n*: the ratio of the intensity of stress to the volume of strain produced by stress.

bulk plant *n*: a wholesale distributing point for products made from natural gas and petroleum.

bulk tank *n*: on a drilling rig, a large metal bin that usually holds a large amount of a certain mud additive, such as bentonite, that is used in large quantities in the makeup of the drilling fluid. Also called a P-tank.

bullet perforator *n*: a tubular device that, when lowered to a selected depth within a well, fires bullets through the casing to provide holes through which the formation fluids may enter the wellbore.

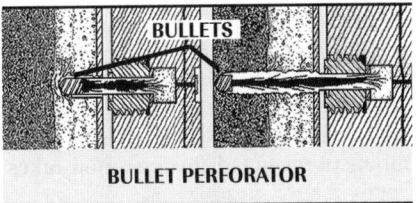

BULLET PERFORATOR

bullets *n pl*: the devices loaded into perforating guns to penetrate casing and cement and for some distance into the formation when the guns are fired. See *gun-perforate.*

bull gear *n*: the large circular gear in a mud pump that is driven by the prime mover and that, in turn, drives the connecting rods. See *swing gear.*

bullheading *n*: 1. forcing gas back into a formation by pumping into the annulus from the surface. 2. any pumping procedure in which fluid is pumped into the well against pressure.

bull plug *n*: a threaded nipple with a rounded, closed end used to stop up a hole or close off the end of a line.

bull shaft *n*: see *crankshaft.*

bullwheel *n*: one of the two large wheels joined by an axle and used to hold the drilling line on a cable-tool rig.

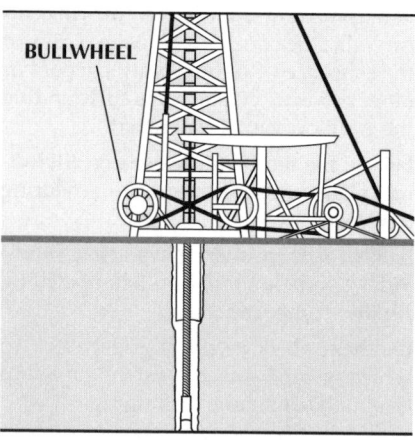

BULLWHEEL

bump a well *v*: to cause the pump on a pumping unit to hit the bottom of the well by having too long a sucker rod string in the unit.

bumped *adj*: in cementing operations, pertaining to a cement plug that comes to rest on the float collar. A cementing operator may say, "I've a bumped plug" when the plug strikes the float collar.

bumper jar *n*: a device made up in the drill string that, when actuated, delivers a heavy downward blow to the string. A bumper jar has a hollow body that moves upward when the drill string is picked up. When the string is dropped quickly, the jar body produces a sharp downward blow on the tubing or pipe made up below the jar. If downward blows can free a fish, a bumper jar can be very effective.

BUMPER JAR

bumper sub *n*: a percussion tool run on a fishing string to jar downward or upward on a stuck fish to knock it free. The bumper sub body moves up and down on a mandrel.

bump off a well *v*: to disconnect a pull-rod line from a central power unit.

bumps *v*: see *bumped*.

Buna-N *n*: a nitrile rubber used in seals and packing units throughout the oilfield.

bunker C oil *n*: see *residuals*.

bunkers *n*: heavy fuel oil (#6 oil) used by ships as fuel to power the vessel.

bunker suit *n*: protective firefighting clothing stored on a rig that consists of a helmet, hood, coat, pants, gloves, boots, and oxygen supply.

buoyancy *n*: the apparent loss of weight of an object immersed in a fluid. If the object is floating, the immersed portion displaces a volume of fluid the weight of which is equal to the weight of the object.

buoyant riser joint *n*: in drilling from offshore floating rigs in deep water, a marine riser joint to which has been added syntactic foam modules. The syntactic foam modules, which contain many hardened spheres of air that are embedded in a rugged foam-like material, are attached to the riser joint. The spheres provide buoyancy. Buoyant riser joints are required when drilling in deep water (depths beyond 4,000 feet or 1,200 metres) to relieve stress on the riser tensioning system. See *riser pipe, syntactic foam*.

buoyant weight *n*: the weight of the drill stem in a mud-filled borehole. Buoyant weight is less than the weight of the drill stem in air because of the buoyant effect of the mud on the drill stem.

b$_{up}$ *abbr*: slope of flowmeter velocity vs rps response curve (up runs).

Bureau of Land Management (BLM) *n*: an agency within the Department of the Interior (DOI) that is responsible for managing the nation's public lands and resources in a combination of ways that best serves the needs of the American people. The resources on public lands under BLM authority include recreation; forestry; fish and wildlife; range; timber; minerals; watershed; wilderness; and natural, scientific, and cultural values. Address: Dept. of the Interior, 1849 C Str. NW; Washington, DC 20240; Director's Office 202-452-5125.

burette *n*: graduated glass tube with small opening and stopcock used for delivering measured quantities of liquid or for measuring liquid or gas received or discharged.

buried hill *n*: an elevation on an ancient land surface that is covered by younger sedimentary rocks.

buried valley *n*: a depression in an ancient land surface that is concealed by younger deposits.

burner tip *n*: an attachment to a natural gas line for a burner head that forms a burner port modified for a specific application. The burner tip represents the end of the transportation of natural gas from the wellhead and includes its consumption. By definition, it also includes its consumption as a feedstock (an ingredient

in manufactured products) and its use in pipeline operations and other overhead activities.

burning point *n*: the lowest temperature at which an oil or fuel will burn when an open flame is held near its surface.

burn over *v*: to use a mill to remove the outside area of a permanent downhole tool.

burn pit *n*: an earthen pit in which waste oil and other materials are burned.

burn shoe *n*: a milling device attached to the bottom of washpipe that mills or drills debris accumulated around the outside of the pipe being washed over. Usually, a burn shoe has pieces of very hard tungsten carbide embedded in it. Also called a rotary shoe. See *washpipe*.

burst pressure *n*: the amount of internal pressure, or stress, required to burst casing or other pipe. When the pipe's internal pressure is greater than its external pressure, the pipe bursts.

burst pressure rating *n*: the pressure at which a manufacturer has determined that a pipe or vessel will burst from internal pressure.

burst strength *n*: a pipe or vessel's ability to withstand rupture from internal pressure.

bury barge *n*: a vessel used to bury a pipeline beneath the seafloor. The barge moves forward by means of anchors. A jet sled is lowered over the pipeline; as the barge pulls it over the pipe, high-pressure jets of water remove soil from beneath the pipe, allowing the pipe to fall into the jetted-out trench.

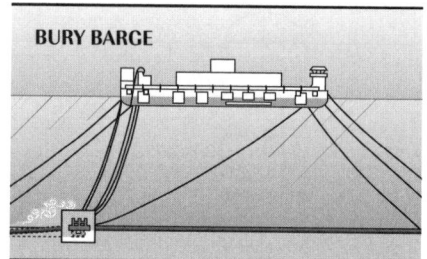

BURY BARGE

bury sled *n*: in pipeline construction, a pipe-straddling device, fitted with nozzles on either side, that is towed by a bury barge. As water is pumped at high pressure through the nozzles, spoil from beneath the pipe is removed and pumped to one side of the trench. The line then sags naturally into position in the trench.

bus *n*: an assembly of electrical conductors for collecting current from several sources and distributing it to feeder lines so that it will be available where needed. Also called bus bar.

bus bar *n*: see *bus*.

bushing *n*: 1. a pipe fitting on which the external thread is larger than the internal thread to allow two pipes of different sizes to be connected. 2. a removable lining or sleeve inserted or screwed into an opening to limit its size, resist wear or corrosion, or serve as a guide.

butadiene *n*: a flammable gaseous hydrocarbon, C_4H_6.

butane *n*: a paraffin hydrocarbon, C_4H_{10}, that is a gas in atmospheric conditions but is easily liquefied under pressure. It is a constituent of liquefied petroleum gas. See *commercial butane, field-grade butane, normal butane.*

butane, commercial *n*: see *commercial butane.*

butanes required *n pl*: the quantity of butane extracted in a processing plant for which lease settlement is included in the settlement made for natural gasoline by virtue of the gasoline settlement's being used on a higher vapor pressure natural gasoline than the vapor pressure of the natural gasoline actually extracted. Sometimes called free butane or excess butanes.

butene *n*: see *butylene.*

Butterworth tank cleaning system *n*: trade name for apparatus for cleaning and freeing oil tanks of gas by means of high-pressure jets of hot water. It consists essentially of opposed double nozzles, which rotate slowly about their horizontal and vertical axes and project two streams of hot water at a pressure of 175 psi (1,206 kilopascals) against all inside surfaces of the deck, bulkheads, and shell plating.

button bit *n*: a drilling bit with tungsten carbide inserts on the cones that resemble plugs or buttons. See *roller cone bit.*

button slip *n*: a slip employing tungsten-carbide "buttons" in lieu of conventional wicker-type teeth to set tools in very hard casing.

button up *v*: to secure the wellhead or other components.

buttress *n*: a special type of heavy-duty threads.

butylene *n*: hydrocarbon member of the olefin series, with the chemical formula C_4H_8. Its official name is butene.

Buys-Ballot's law *n*: a law that clarifies the relationship between the horizontal wind direction in the earth's atmosphere and the distribution of pressure. The law states that if one stands facing the wind, the barometric pressure to the right is lower than to the left in the Northern Hemisphere. This relationship is reversed in the Southern Hemisphere. Also known as the baric wind law.

BW *abbr*: barrels of water; used in drilling reports.

BWPD *abbr*: barrels of water per day.

BWPH *abbr*: barrels of water per hour; used in drilling reports.

by heads *n*: intermittent flow in a flowing well.

bypass *n*: 1. a pipe connection around a valve or other control mechanism that is installed to permit passage of fluid through the line while adjustments or repairs are being made on the control. 2. a delivery of gas to a customer by means of a pipeline other than that customer's traditional supplier. For example, delivery of gas to an end user directly off a transmission pipeline without moving the gas through the end user's traditional local distribution company supplier.

bypass gate *n*: a device that allows the mud flow to be directed around the shale shaker.

bypass valve *n*: a valve that permits flow around a control valve, a piece of equipment, or a system.

byproducts *n pl*: side products of a process to produce a main product.

C *abbr*: 1. capacitance. 2. Celsius (formerly centigrade). See *Celsius scale*.

C *sym*: 1. carbon 2. coulomb.

C' *n*: 1. the rate of flow in cubic feet per hour at base conditions. 2. a factor used in the calculation of gas volume flow through an orifice meter, now largely supplanted by an equation formulated in 1992. Mathematical accounting for variations in pressure, temperature, density, and so on, of a gas as it flows through an orifice of a particular size. Also called orifice flow constant; formerly called flow coefficient.

CAA *abbr*: Clean Air Act.

cab *n*: a housing which covers the revolving superstructure machinery and operator's station.

cable *n*: 1. a rope of wire, hemp, or other strong fibers; often, but not always, personnel in the petroleum industry call it cable wire rope. 2. in electronics, a wire made of a conducting material such as copper or aluminum. Usually, cable is braided from several single strands of wire into a single conductor that is easier to bend than a solid wire of the same gauge. 3. in electronics, a wire made of a conducting material such as copper or aluminum. Usually, cable is braided from several single strands of wire into a single conductor that is easier to bend than a solid wire of the same gauge.

cable-guide assembly *n*: fishing equipment used to recover wireline tools stuck in the hole when the wireline or cable is intact. It consists of a cable hanger, spear point rope socket, spearhead overshot, conventional overshots, drill pipe, and elevators.

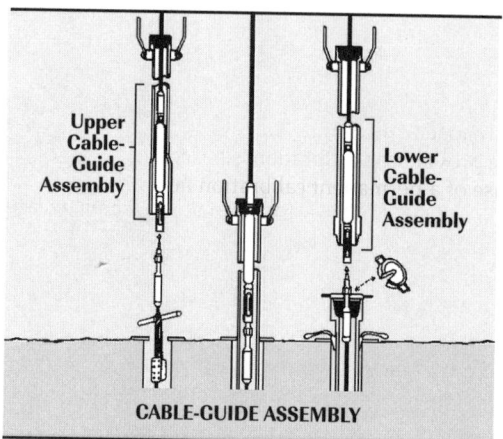

CABLE-GUIDE ASSEMBLY

cable-guide fishing *n*: a method of retrieving wireline tools using a cable-guide assembly.

cable hanger *n*: a fishing device made up on a cable-guide assembly to prevent the cable from falling in the hole. It is used in the recovery of wireline tools.

cable reel *n*: in offshore drilling from floating drilling rigs, a large spool around which is wrapped the electric cable that provides the link between the surface power and electronics and the subsea power and electronics on the lower marine riser

package and the subsea blowout preventer stack. Because most subsea blowout preventers have two operational systems in case one fails, two cable reels are usually provided. A typical cable reel can store over 10,000 feet (3,000 metres) of cable.

cable-suspended submersible pump *n*: a form of electric submersible pumping for artificial lift, in which the pump can be run or pulled on the power cable. This type of pump installation is made possible by power cables designed for tensile loads in excess of 100,000 pounds (45,360 kilograms).

cable-tool drilling *n*: a drilling method in which the hole is drilled by the rig's equipment dropping a sharply pointed and heavily weighted bit on bottom. The bit and weight are attached to a cable, and the rig's equipment repeatedly drops the cable to drill the hole.

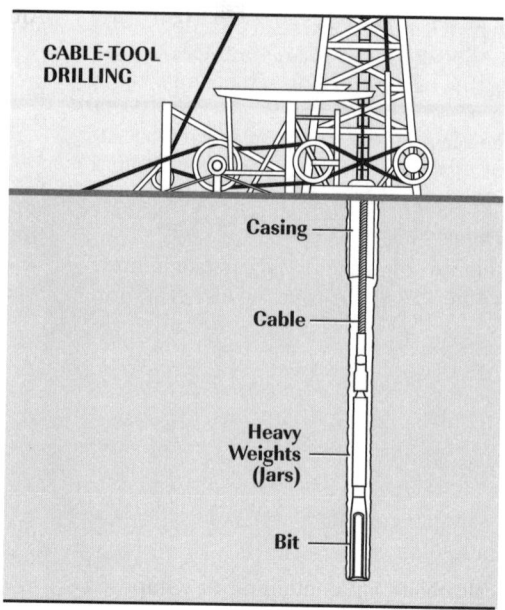

cable-tool rig *n*: a drilling rig that uses wire rope (cable) to suspend a weighted, chisel-shaped bit in the hole. Machinery on the rig repeatedly lifts and drops the cable and bit. Each time the bit strikes the bottom of the hole, it drills deeper. Rotary drilling rigs have virtually replaced all cable-tool rigs.

cable-wind pumping unit *n*: a sucker rod pumping unit, not commonly used, that employs cable wound around a spool, one end going into the hole, the other with a suspended counterweight. This type of unit can achieve 40- to 80-foot (12- to 24-metre) stroke lengths, reducing dynamic loads, rod reversals, gas locking, and power usage and can be applied to deep pump applications.

cage *n*: in a sucker rod pump, the device that contains and confines the valve ball and keeps it the proper operating distance from the valve seats.

cage wrench *n*: a special wrench designed for use in connecting the cage of a sucker rod pump to the sucker rod string.

caisson *n*: 1. one of several columns made of steel or concrete that serve as the foundation for a rigid offshore platform rig, such as the concrete gravity platform rig. 2. a steel or concrete chamber that surrounds equipment below the waterline of an Arctic submersible rig, thereby protecting the equipment from damage by moving ice.

caisson-type drilling platform *n*: a drilling platform that rests on a steel or concrete caisson. Used mainly in the Arctic, caisson-type drilling platforms are mobile offshore drilling units built to withstand the ravages of moving ice in Arctic waters.

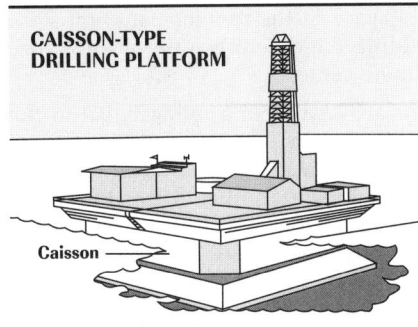

CAISSON-TYPE DRILLING PLATFORM

Caisson

cake *n*: see *filter cake*.

cake consistency *n*: the character or state of the drilling mud filter cake. According to *API RP 13B*, such notations as "hard," "soft," "tough," "rubbery," and "firm" may be used to convey some idea of cake consistency.

cake thickness *n*: the thickness of drilling mud filter cake.

calcareous *adj*: containing or composed largely of calcium carbonate, or calcite ($CaCO_3$).

calcite *n*: calcium carbonate, $CaCO_3$.

calcium *n*: one of the alkaline earth elements with a valence of 2 and an atomic weight of about 40. Calcium compounds are a common cause of water hardness. Calcium is also a component of lime, gypsum, and limestone.

calcium carbonate *n*: a chemical combination of calcium, carbon, and oxygen, $CaCO_3$. It is the main constituent of limestone. It forms a tenacious scale in water-handling facilities and is a cause of water hardness.

calcium chloride *n*: a moisture-absorbing chemical compound, or desiccant, $CaCl_2$, used to accelerate setting times in cement and as a drying agent. It is also added to fresh water to increase the water's weight and to give it desirable properties during well completion or workover.

calcium contamination *n*: dissolved calcium ions in sufficient concentration to impart undesirable properties, such as flocculation, reduction in yield of bentonite, and increased fluid loss, in a drilling fluid. See also *anhydrite, calcium carbonate, calcium sulfate, gypsum, lime*.

calcium hydroxide *n*: the active ingredient of slaked (hydrated) lime, and the main constituent in cement (when wet). Referred to as "lime" in field terminology. Its symbol is $Ca(OH)_2$.

calcium magnesium carbonate *n*: a chemical combination of calcium, magnesium, carbon, and oxygen, $CaMgCO_3$. It is the main constituent of dolomite. Compare *calcium carbonate*.

calcium sulfate *n*: a chemical compound of calcium, sulfur, and oxygen, $CaSO_4$. Although sometimes considered a contaminant of drilling fluids, it may at times be added to them to produce certain properties. Like calcium carbonate it forms scales in water-handling facilities, which may be hard to remove. See *anhydrite, gypsum*.

calcium-treated mud *n*: a freshwater drilling mud using calcium oxide (lime) or calcium sulfate (gypsum) to retard the hydrating qualities of shale and clay formations, thus facilitating drilling. Calcium-treated muds resist salt and anhydrite contamination but may require further treatment to prevent gelation (solidification) under the high temperatures of deep wells.

calibration *n*: the adjustment or standardizing of a measuring instrument or of a standard capacity measure, a tank prover, or a pipe prover. Log calibration is based on the use of a permanent calibration facility of the American Petroleum Institute at the University of Houston to establish standard units for nuclear logs.

calibration coefficient *n*: pulse per unit volume.

calibration constant *n*: in turbine operations, a constant that accounts for the difference between theoretical rotor speed and actual rotor speed.

calibration table *n*: 1. a table that shows the capacities of, or volumes in, a tank for various liquid levels measured from the dip point or from the ullage reference point. Also called gauge or tank capacity table. 2. a table that is developed by recognized industry methods to represent volumes in each tank according to the liquid (innage) or empty space (ullage) measured in the tank. Also called tank capacity table, tank table.

calibration tank *n*: see *prover tank*.

caliper *n*: an instrument with two legs or jaws that can be adjusted for measuring linear dimensions, thicknesses, or diameters.

caliper log *n*: a record showing variations in wellbore diameter by depth, indicating undue enlargement due to caving in, washout, or other causes. The caliper log also reveals corrosion, scaling, or pitting inside tubular goods.

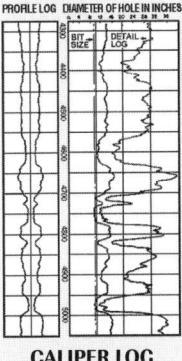

CALIPER LOG

caliper logging tool *n*: a device lowered into a wellbore to measure and record the wellbore's diameter.

call on oil *n*: an option to buy crude oil for an extended period of time.

CALM *abbr*: catenary anchor leg mooring.

calorie (cal) *n*: the amount of heat energy necessary to raise the temperature of 1 gram of water 1° Celsius. It is the metric equivalent of the British thermal unit.

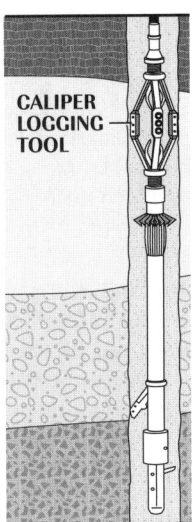

CALIPER LOGGING TOOL

calorimeter *n*: an apparatus used to determine the heating value of a combustible material.

cam *n*: an eccentrically shaped disk, mounted on a camshaft, that varies in distance from its center to various points on its circumference. As the camshaft is rotated, a set amount of motion is imparted to a follower riding on the surface of the cam. In the internal-combustion engine, cams are used to operate the intake and exhaust valves.

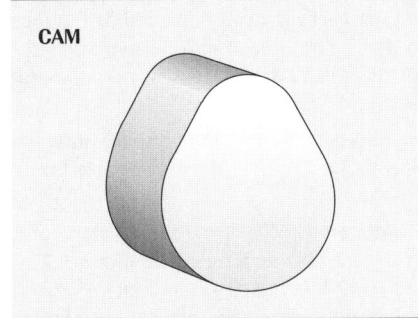

CAM

cam-actuated running tool (CART) *n*: in drilling from floating rigs that employ a subsea blowout preventer system, a tool used to run the permanent guide base. It allows the drill pipe to be disconnected from the permanent guide base after it lands in the temporary guide base. When the drill pipe is rotated, the cams move into a position that allows the pipe to be removed from the base. See *permanent guide base, temporary guide base.*

Cameron gauge *n*: a pressure gauge usually used in lines or manifolds.

cam follower *n*: output link of a cam mechanism.

cam ring *n*: a part of a hydraulic connector that attaches the blowout preventer (BOP) stack to the wellhead and the lower marine riser package (LMRP) to the BOP stack. The cam ring is connected to hydraulic pistons in the connector and the ring's inside diameter has a 4° taper. When hydraulic locking pressure is applied, the pistons pull the cam ring downward over the locking dogs or segments and the 4° taper forces the dogs or segments inward to lock on the wellhead's profile.

camshaft *n*: the cylindrical bar used to support a rotating device called a cam.

Canadian Association of Oilwell Drilling Contractors (CAODC) *n*: a trade association that represents virtually 100% of the rotary drilling contractors and the majority of well service rig operators in western Canada. The organization concerns itself with research, education, accident prevention, government relations, and other matters of interest to members. Address: 540 5th Avenue SW, Suite 800; Calgary, Alberta, Canada T2P 0M2; (403) 264-4311; fax (403) 263-3796.

canal-lay construction *n*: a pipeline construction technique used in swamps and marshes. The first of several barges clears the right-of-way and digs a trench large enough to float itself and the following barges.

candela (cd) *n*: the fundamental unit of luminous intensity in the metric system.

canted leg *adj*: pertaining to an independent-leg jackup rig designed so that the legs may be slanted outward to increase support against lateral stresses when the unit is on the seafloor.

cantilever *n*: beams that project outward from a structure and are supported only at one end.

cantilevered jackup *n*: a jackup drilling unit in which the drilling rig is mounted on two cantilevers that extend outward from the barge hull of the unit. The cantilevers are supported only at the barge end. Compare *keyway*.

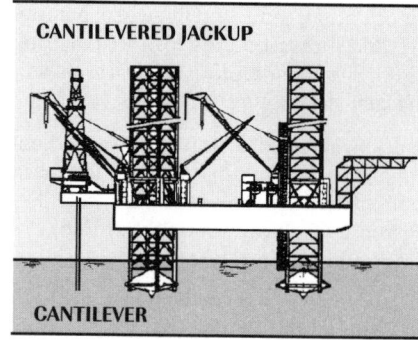

CANTILEVERED JACKUP

CANTILEVER

can-type capacitor *n*: a capacitor whose components are housed inside a small, can-shaped cylinder. See *capacitor.*

canvas packer *n*: (obsolete) a device for sealing the annular space between the top of a liner and the existing casing string.

CAO *abbr*: computer assisted operations.

CAODC *abbr*: Canadian Association of Oilwell Drilling Contractors.

cap *n*: to control a well that is flowing out of control; often accomplished by attaching a valve in the open position on top of the well and then closing it to seal off the flow.

capacitance (C) *n*: the ratio of an impressed electrical charge on a conductor to the change in electrical potential; measured in farads. See *capacitor.*

capacitance probe *n*: a device used in most net-oil computers that senses the different dielectric constants of oil and water in a water-oil emulsion. See *dielectric constant.*

capacitive coupling *n*: in electronics, the transfer of energy from one circuit to another through a capacitor.

capacitive reactance *n*: the impedance of current flow in an alternating current circuit (the reactance) caused by a capacitor in the circuit.

capacitor *n*: an electrical device that, when wired in an electrical circuit, stores a charge of electricity and returns the charge to the line when certain electrical conditions occur. Physically consists of two parallel conductors separated with an insulating material called the dielectric. Its surface area, plate separation, and dielectric constant determine the amount of capacitance in farads. Also called a condenser.

capacity *n*: 1. volume capability of an electrical capacitor in relation to its ability to store electrical charges. 2. volume capability of a container. 3. in instrumentation and process control, the amount of energy contained in a part of the process. For example, a storage tank stores energy in the form of the weight and the hydrostatic pressure of the liquid it contains. Single capacity systems have only one form of stored energy, while multiple capacity systems have two or more. An example of a multiple capacity system is a tank containing liquid that not only stores energy in the form of hydrostatic pressure, but also stores energy in the form of heat, both of which affect the control process. See *capacitance, storage capacity.*

capacity indicator *n*: a device fitted to a proving tank to indicate the position of the liquid surface in relation to the reference mark corresponding to the nominal capacity of the proving tank, thus enabling the determination of the tank's liquid contents.

capacity rating *n*: a rating equal to the maximum number of cubic feet of gas that will pass through a meter when the pressure differential across the meter equals a 0.5-inch water column and the flowing pressure is 0.25 psig.

capacity table *n*: a table that tells the quantity of liquid contained in a tank at any given level.

cap a well *v*: to control a blowout by placing a very strong valve on the wellhead. See *blowout.*

cap bead *n*: the final welding pass made to complete the uniting of two joints of pipe.

cap gas *n*: natural gas trapped in the upper part of a reservoir and remaining separate from any crude oil, salt water, or other liquids in the well.

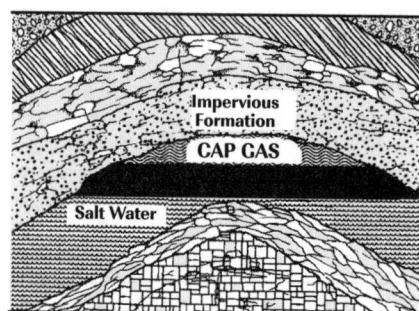

Impervious Formation
CAP GAS
Salt Water

capillaries *n pl*: very small fissures or cracks in a formation through which water or hydrocarbons flow.

capillarity *n*: the rise and fall of liquids in small-diameter tubes or tubelike spaces, caused by the combined action of surface tension (cohesion) and wetting (adhesion). See *capillary pressure.*

capillary meter seal *n*: the liquid seal that reduces slippage between moving parts of a meter.

capillary pressure *n*: a pressure or adhesive force caused by the surface tension of water. This pressure causes the water to adhere more tightly to the surface of small pore spaces than to larger ones. Capillary pressure in a rock formation is comparable to the pressure of water that rises higher in a small glass capillary tube than it does in a larger tube.

capitalize *v*: in accounting, to include expenditures in business accounts as assets instead of expenses.

capitalized *adj*: pertaining to expenditures treated as assets instead of expenses.

capped well *n*: a well capable of production but lacking wellhead installations and a pipeline connection.

caprock *n*: 1. a disklike plate of anhydrite, gypsum, limestone, or sulfur overlying most salt domes in the Gulf Coast region. 2. impermeable rock overlying an oil or gas reservoir that tends to prevent migration of oil or gas out of the reservoir.

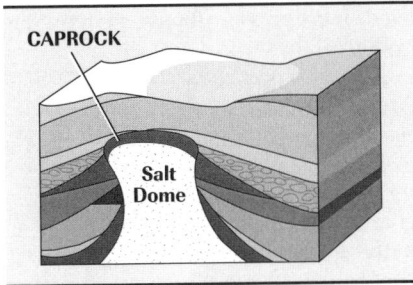

capsule *n*: 1. a small container that holds liquid or gas. 2. a lightweight metal unit to which is attached a single metallic diaphragm. 3. an escape capsule. See *escape capsule*.

captive customers *n pl*: buyers who can purchase natural gas from only one supplier and have no access to alternate fuel sources. This term is usually applicable to residential and small commercial users, but can, under certain conditions of alternative fuel availability, be applied to large industrial and electric utility users as well. Also called core customers.

capture cross section *n*: the tendency of elements in their compounds to reduce energy or number of particles by absorbing them. The more densely populated an area may be, the more certain it is that energy will be absorbed or that the particles will be retained within the atomic structure.

capture gamma ray *n*: a high-energy gamma ray emitted when the nucleus of an atom captures a neutron and becomes intensely excited. Capture gamma rays are counted by the neutron logging detector.

carbon (C) *n*: a naturally abundant nonmetallic element that occurs in many inorganic and in all organic compounds, exists freely as graphite and diamond and as a constituent of coal, limestone, and petroleum, and is capable of chemical self-bonding to form an enormous number of chemically, biologically, and commercially important molecules.

carbonate *n*: 1. a salt of carbonic acid. 2. a compound containing the carbonate radical (CO_3). 3. a carbonate rock.

carbonate mud *n*: a mud that forms on the seafloor by the accumulation of calcite particles. It may eventually become limestone.

carbonate reef *n*: see *reef*.

carbonate rock *n*: a sedimentary rock composed primarily of calcium carbonate (limestone) or calcium magnesium carbonate (dolomite); sometimes makes up petroleum reservoirs.

carbonation *n*: 1. a chemical reaction that produces carbonates. 2. in geology, a form of chemical weathering in which a mineral reacts with carbon dioxide (in solution as carbonic acid) to form a carbonate mineral.

carbon black *n*: very fine particles of almost pure amorphous carbon, usually produced from gaseous or liquid hydrocarbons by thermal decomposition or by controlled combustion with a restricted air supply. It is used in the manufacture of carbon paper, tires, cosmetics, and so on.

carbon dioxide *n*: a colorless, odorless gaseous compound of carbon and oxygen, CO_2. A product of combustion and a filler for fire extinguishers, this heavier-than-air gas can collect in low-lying areas, where it may displace oxygen and present the hazard of anoxia. In contact with steel, it may cause corrosion.

carbon dioxide corrosion *n*: corrosion that occurs when steel is in contact with carbon dioxide and which is characterized by a dark gray scale and pits.

carbon dioxide excess *n*: see *hypercapnia*.

carbonic *adj*: of or relating to carbon, carbonic acid, or carbon dioxide.

carbonize *v*: to convert into carbon or a carbonic residue.

carbon log *n*: a record that indicates the presence of hydrocarbons by measuring carbon atoms and that reveals the presence of water by measuring oxygen atoms. Oil and water saturations can be closely approximated without the requirement of adequate salinity of known concentration to calculate saturations (as in resistivity and pulsed neutron surveys).

carbon monoxide *n*: a colorless, odorless gaseous compound of carbon and oxygen, CO. A product of incomplete combustion, it is extremely poisonous to breathe.

carbon residue *n*: in a diesel fuel, the amount of carbon remaining in a special container after the fuel is burned under controlled conditions. Generally, high-quality fuels have low carbon residue.

carbon tetrachloride *n*: a liquid used for degreasing metals; it is toxic when ingested, breathed, or exposed to the skin and is a known carcinogen.

carbonyl sulfide *n*: a chemical compound of the aldehyde groups containing a carbonyl group and sulfur, COS. A contaminant in gas liquids, it is usually removed to meet sulfur specifications.

carboxymethyl cellulose (CMC) *n*: a nonfermenting cellulose product used in drilling fluids to combat contamination from anhydrite (gypsum) and to lower the water loss of the mud.

carburetion *n*: the mixing of fuel and air in the carburetor of an engine.

carburetor *n*: the device in a spark-ignition internal-combustion engine in which fuel and air are mixed in controlled quantities and proportions.

carburize *v*: to impregnate or combine with carbon.

carcinogen *n*: cancer-causing material or substance.

CARE *abbr*: Conservation Award for Respecting the Environment.

cargo manifest *n*: a document that lists the goods or materials carried by a ship, train, airplane, or truck.

cargo quantity option certificate *n*: a certificate signed by vessel representatives acknowledging the amount of cargo that they intend to load. In general, most product cargoes have a tolerance based on supplier, receiver, or vessel capabilities. Each party involved with the loading agrees to the quantity to be loaded.

carnotite *n*: see *radioactive tracer*.

carriage *n*: the transportation of a third party's natural gas by a pipeline as a separate service for a fee, as opposed to the pipeline's transportation of its own system's supply of natural gas.

carried interest *n*: an interest in oil and gas properties that belongs, for example, to a working interest owner or unleased landowner who agrees to a joint operation without being willing to pay a share of the costs of the operation. An interest may be carried until the well pays out, at which point it may stop or may continue for the life of production. Stated another way, an agreement between two or more working interests whereby one party (the carried party) does not share in lease revenue until a certain amount of money has been recovered by the other party (the carrying party). The carrying party pays costs applicable to the carried party's interests in the property and is reimbursed out of the revenue applicable to the carried party's interest.

carrier *n*: 1. a pipeline, trucking company, or marine transportation company that transports oil or gas. 2. on offshore installations with deck grating, a flat or angled length of steel welded to the leading edge of a grated stair that supports the stair tread.

carrier bar *n*: yoke or clamp fastened to a pumping well's polished rod and to which the bridle of the pumping unit's horsehead is attached.

carrier pipe *n*: term used to refer to a pipeline when other pipe, called casing, is used with it in crossing under roadbeds and railroad rights-of-way. See *casing*.

carrier rig *n*: a large, specially designed, self-propelled workover rig that is driven directly to the well site. Power from a carrier rig's hoist engine or engines also propels the rig on the road. While a carrier rig is primarily intended to perform workovers, it can also be used to drill relatively shallow wells. A carrier rig may be a back-in type or a drive-in type. Compare *back-in unit, drive-in unit*.

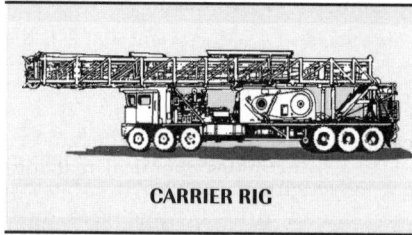

CARRIER RIG

carrier unit *n*: see *carrier rig*.

carry *v*: in the mathematical operation of addition, to perform the necessary actions to obtain a correct answer when the sum of the digits in a given position equals or exceeds the base of the number system. For example, when adding 352 and 49 in the decimal system, 9 is added to 2 in the ones column, which equals 11. In this case, 1 is written below the ones column and the other 1 is added (carried) to the 5 in the tens column of 352, making the 5 a 6. Then, 6 is added to 4 and a 0 placed below the tens column. The 1 is carried to the 3 in the hundreds column, making it 4. The answer is 401.

CART *abbr*: cam-actuated running tool.

carved-out interest *n*: an interest in oil and gas created out of a greater interest and assigned by the owner. Examples are the grant of an overriding royalty interest out of a working interest and the grant of an oil payment out of a working interest.

cascade refrigeration *n*: uses two-stage refrigeration to separate methane from the NGLs.

cascade system *n*: in systems supplying a breathable source of air to workers wearing breathing equipment in a toxic atmosphere, a serial connection of air cylinders in which the output of air from one adds to that of the next.

case *n*: the outer cylinder of a concentric cylinder centrifuge. See *concentric cylinder centrifuge*.

cased *adj*: pertaining to a wellbore in which casing has been run and cemented. See *casing*.

cased hole *n*: a wellbore in which casing has been run.

cased-hole fishing *n*: the procedure of recovering lost or stuck equipment in a wellbore in which casing has been run.

case-hardened *adj*: hardened (as for a ferrous alloy) so that the surface layer is harder than the interior.

case law *n*: see *common law*.

cash flow *n*: the difference between inflow and outflow of funds over a period of time. Cash flow can be positive (profit) or negative (loss).

cash flow analysis *n*: an economic analysis that relates investments to subsequent revenues and also makes possible a comparison between investments. It usually also includes the general plan to be used for the figuring of federal income tax on the investments.

casing *n*: 1. steel pipe placed in an oil or gas well to prevent the wall of the hole from caving in, to prevent movement of fluids from one formation to another, and to improve the efficiency of extracting petroleum if the well is productive.

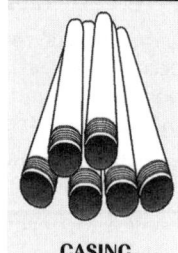

CASING

Most casing joints are manufactured to specifications established by API, although non-API specification casing is available for special situations. Casing manufactured to API specifications is available in three length ranges. A joint of range 1 casing is 16 to 25 feet (4.8 to 7.6 metres) long; a joint of range 2 casing is 25 to 34 feet (7.6 to 10.3 metres) long; and a joint of range 3 casing is 34 to 48

feet (10.3 to 14.6 metres) long. The outside diameter of a joint of API casing ranges from 4½ to 20 inches (114.3 to 508.0 millimetres). Casing is made of many types of steel alloy, which vary in strength, corrosion resistance, and so on. 2. large pipe in which a carrier pipeline is contained. Casing is used when a pipeline passes under railroad rights-of-way and some roads to shield the pipeline from the unusually high load stresses of a particular location. State and local regulations identify specific locations where casing is mandatory.

casing adapter *n*: a swage nipple, usually beveled, installed on the top of a string of pipe that does not extend to the surface. It prevents a smaller string of pipe or tools from hanging up on the top of the column when it is run into the well.

casing burst pressure *n*: the amount of pressure that, when applied inside a string of casing, causes the wall of the casing to fail. This pressure is critically important when a gas kick is being circulated out, because gas on the way to the surface expands and exerts more pressure than it exerted at the bottom of the well.

casing centralizer *n*: a device secured around the casing at regular intervals to center it in the hole. Casing that is centralized allows a more uniform cement sheath to form around the pipe.

casing collar *n*: a coupling between two joints.

casing coupling *n*: a tubular section of pipe that is threaded inside and used to connect two joints of casing.

CASING CENTRALIZER

casing crew *n*: the employees of a company that specializes in preparing and running casing into a well. Usually, the casing crew makes up the casing as it is lowered into the well; however, the regular drilling crew also assists the casing crew in its work.

casing cutter *n*: a heavy cylindrical body, fitted with a set of knives, used to free a section of casing in a well. The cutter is run downhole on a string of tubing or drill pipe, and the knives are rotated against the inner walls of the casing to free the section that is stuck.

casing elevator *n*: see *elevators*.

casing float collar *n*: see *float collar*.

casing float shoe *n*: see *float shoe*.

casing gun *n*: a perforating gun run into the casing string. Most perforating guns are run into the well through the tubing string.

casing hanger *n*: a circular device with a frictional gripping arrangement of slips and packing rings used to suspend casing from a casinghead in a well.

casing hanger pack-off *n*: in a wellhead, a device that isolates the casing hanger from the inside diameter of the wellhead by means of a metal-to-metal seal. The pack-off keeps pressure from escaping the wellhead from around the casing hanger.

casinghead *n*: a heavy, flanged steel fitting connected to the first string of casing. It provides a housing for slips and packing assemblies, allows suspension of intermediate and production strings of casing, and supplies the means for the annulus to be sealed off. Also called a spool.

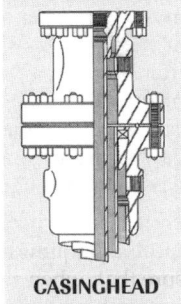

CASINGHEAD

casinghead gas *n*: gas produced with oil.

casinghead gas contract *n*: a contract used by industry for the purchase and sale of casinghead gas.

casinghead gasoline *n*: (obsolete) natural gasoline.

casing overshot *n*: see *casing-patch tool*.

casing pack *n*: a means of cementing casing in a well so that the casing may, if necessary, be retrieved with minimum difficulty. A special mud, usually an oil mud, is placed in the well ahead of the cement after the casing has been set. Nonsolidifying mud is used so that it does not bind or stick to the casing in the hole in the area above the cement. Since the mud does not gel for a long time, the casing can be cut above the cemented section and retrieved. Casing packs are used in wells of doubtful or limited production to permit reuse of valuable lengths of casing.

casing-pack *v*: to cement casing in a well in a way that allows for easy retrieval of sections of casing. See *casing pack*.

casing-patch tool *n*: a special tool with a rubber packer or lead seal that is used to repair casing. When casing is damaged downhole, a cut is made below the damaged casing, the damaged casing and the casing above it are pulled from the well, and the damaged casing is removed from the casing string. The tool is made up and lowered into the well on the casing until it engages the top of the casing that remains in the well, and a rubber packer or lead seal in the tool forms a seal with the casing that is in the well. The casing-patch tool is an overshot-like device and is sometimes called a casing overshot.

casing point *n*: 1. the depth in a well at which casing is set, generally the depth at which the casing shoe rests. 2. the objective depth in a drilling contract, either a specified depth or the depth at which a specific zone is penetrated. When the depth is reached, the operator makes a decision with respect to running and setting a production string of casing. Under some farmout and letter agreements, some owners are carried to casing point.

casing pressure *n*: the pressure in a well that exists between the casing and the tubing or the casing and the drill pipe.

casing protector *n*: a short threaded nipple screwed into the open end of the coupling and over the threaded end of casing to protect the threads from dirt accumulation and damage. It is made of steel or plastic. Also called thread protector.

casing rack *n*: see *pipe rack*.

casing roller *n*: a tool composed of a mandrel on which are mounted several heavy-duty rollers with eccentric roll surfaces. It is used to restore buckled, collapsed, or dented casing in a well to normal diameter and roundness. Made up on tubing or drill pipe and run into the well to the depth of the deformed casing, the tool is rotated slowly, allowing the rollers to contact all sides of the casing and restore it to roughly its original condition.

casing scraper *n*: a bladed tool used to scrape away junk or debris from inside casing; it is usually run into the casing on drill pipe or tubing.

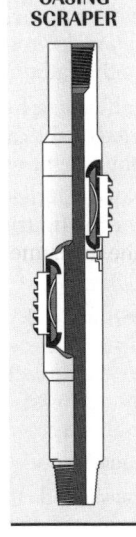

CASING SCRAPER

casing seal receptacle *n*: a casing sub containing a seal bore and a left-handed thread. It is run as a crossover sub between casing sizes and to provide a tubing anchor.

casing seat *n*: the location of the bottom of a string of casing that is cemented in a well. Typically, a casing shoe is made up on the end of the casing at this point.

casing seat test *n*: a procedure whereby the formation immediately below the casing shoe is subjected to a pressure equal to the pressure expected to be exerted later by the drilling fluid or by the drilling fluid and the back-pressure created by a kick. See *leak-off test*.

casing-shear rams *n*: usually high-capacity shear rams capable of shearing drill collars and casing strings that are installed below the blind-shear rams in a BOP stack. Casing-shear rams can be used in addition

to the blind-shear rams to shear pipe. See *shear ram*.

casing shoe *n*: see *guide shoe*.

casing slip *n*: see *spider*.

casing spear *n*: a fishing tool designed to grab casing from the inside so that when the spear is retrieved, the attached casing comes with it.

casing spider *n*: see *spider*.

casing string *n*: the entire length of all the joints of casing run in a well. Most wells have more than one string of casing, which are run one inside the other as drilling progresses. Generally, wells have a length, or string, of surface casing, intermediate casing, and production casing, although a great deal of variation occurs according to a well's depth, the formations it penetrates, and other factors. See *casing, intermediate casing, production casing, surface casing*.

casing sub *n*: a sub used to join two dissimilar joints of casing.

casing swage *n*: a solid cylindrical body pointed at the bottom and equipped with a tool joint at the top for connection with a jar. It is used to make an opening in a collapsed casing and drive it back to its original shape.

casing tongs *n pl*: large wrench used for turning when making up or breaking out casing. See *tongs*.

casing-tubing annulus *n*: in a wellbore, the space between the inside of the casing and the outside of the tubing.

Cat *n*: 1. short for an industrial or gas engine manufactured by Caterpillar, Inc. 2. short for any piece of industrial equipment, such as a bulldozer, manufactured by Caterpillar, Inc.

catalog strength *n*: see *nominal strength*.

catalyst *n*: a substance that alters, accelerates, or instigates chemical reactions without itself being affected.

catalytic cracking *n*: 1. the breaking down of heavier hydrocarbon molecules into lighter hydrocarbons using catalysts and relatively low temperatures and pressures. 2. a motor gasoline refining process in which heavy hydrocarbon components are broken down into light hydrocarbon components using catalysts and relatively low temperatures and pressures. It produces a gasoline that has a higher octane rating and a lower sulfur content than that produced by thermal cracking.

catastrophism *n*: the theory that the earth's landforms assumed their present configuration in a brief episode at the beginning of geologic history—possibly in a single great catastrophic event—and have remained relatively unchanged since that time. Compare *uniformitarianism*.

catcher *n*: a device fitted into a junk basket that retains the junk picked up by the basket.

catch samples *v*: to obtain cuttings for geological information as formations are penetrated by the bit. The samples are obtained from drilling fluid as it emerges from the wellbore or, in cable-tool drilling, from the bailer. Cuttings are carefully washed until they are free of foreign matter, dried, and labeled to indicate the depth at which they were obtained.

cat cracker gas *n*: gas that is a by-product of refining liquid hydrocarbons in a catalytic cracker.

categorical exclusion *n*: a category of actions that do not individually or cumulatively have a significant effect on the human environment and that have been found to have no such effect in procedures adopted by a federal agency when implementing NEPA regulations. Therefore, neither an environmental assessment nor an environmental impact statement is required.

catenary *n*: 1. the curve assumed by a perfectly flexible line hanging under its own weight between two fixed points. A suspension bridge is an example of a catenary structure; an anchor chain is a catenary. 2. the part of a mooring leg suspended between a rig and the seafloor.

catenary anchor leg mooring (CALM) *n*: a type of offshore mooring in which the facility is anchored by at least six anchors; it generally has no storage facility.

catenary riser *n*: on floating offshore production platforms, a flexible metal pipe—a riser—that conducts well fluids from the subsea wellhead to production facilities on the platform. On the platform at the water's surface, buoyant air cans or tensioning devices support the riser. The riser is flexible enough to assume a catenary shape when deployed. That is, the riser runs vertically into the water but turns to lie flat on the seafloor, roughly in the shape of a J. As the platform heaves, the riser flexes. See *catenary*.

catenary spread mooring (CSM) *n*: multileg spread mooring using components in a pure catenary configuration.

cathead *n*: a spool-shaped attachment on the end of the catshaft, around which rope

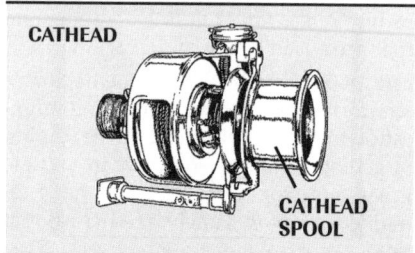

CATHEAD

CATHEAD
SPOOL

for hoisting and moving heavy equipment on or near the rig floor is wound. See *breakout cathead, makeup cathead*.

cathead spool *n*: see *cathead*.

cathode *n*: 1. one of two electrodes in an electrolytic cell, represented as the positive terminal of a cell. 2. in cathodic protection systems, the protected structure that is representative of the cathode and is protected by having a conventional current flow from an anode to the structure through the electrolyte. 3. the terminal of a semiconductor diode that is negative with respect to the other terminal when the diode is biased in the forward direction. See *diode*.

cathode ray *n*: a stream of electrons, such as those emitted by a heated filament in a tube, or those emitted by the cathode of a gas-discharge tube when the cathode is bombarded by positive ions.

cathode-ray tube (CRT) *n*: an electron tube in which a bean of electrons can be focused to a small area and varied in position and intensity on a surface.

cathodic protection *n*: a means of preventing the destructive electrochemical process of corrosion of a metal object by using it as the cathode of a cell with a sacrificial anode. Current at least equal to that caused by the corrosive action is directed toward the object, offsetting its electrical potential.

cation *n*: a positively charged ion; the ion in an electrolyzed solution that migrates to the cathode. See *ion*. Compare *anion*.

catline *n*: a hoisting or pulling line powered by the cathead and used to lift heavy equipment on the rig. See *cathead*.

catshaft *n*: an axle that crosses through the drawworks and contains a revolving spool called a cathead at either end. See *cathead*.

catwalk *n*: 1. the ramp at the side of the drilling rig where pipe is laid to be lifted to the derrick floor by the catline or by an air hoist. See *catline*. 2. any elevated walkway.

caustic *n*: see *caustic soda*.

caustic soda *n*: sodium hydroxide, NaOH. It is used to maintain an alkaline pH in drilling mud and in petroleum fractions.

caustic treater *n*: a vessel holding sodium hydroxide or other alkalis through which a solution flows for removal of sulfides, mercaptans, or acids.

cave-in *n*: the collapse of the walls of the wellbore.

cavern *n*: a natural cavity in the earth's crust that is large enough to permit human entry. Commonly formed in limestone by groundwater leaching. Compare *vug*.

cavernous formation *n*: a rock formation that contains large open spaces, usually resulting from the dissolving of soluble substances by formation waters that may still be present. See *vug*.

caving *n*: collapsing of the walls of the wellbore. Also called sloughing.

cavings *n pl*: particles that fall off (are sloughed from) the wall of the wellbore. Compare *cuttings*.

cavitation *n*: the formation and collapse of vapor- or gas-filled cavities that result from a sudden decrease and increase of pressure. Cavitation can cause mechanical damage to adjacent surfaces in meters, valves, pumps, and pipes at locations where flowing liquid encounters a restriction or change in direction.

cavity *n*: a hollow area or a hole within the body of a piece of equipment, a device, or a rock.

Cb *abbr*: cumulonimbus.

CBHT *abbr*: circulating bottomhole temperature.

CBL *abbr*: cement bond log.

Cc *abbr*: cirrocumulus.

cc *abbr*: cubic centimetre.

CCA *abbr*: cold cranking ampere.

CCL *abbr*: casing collar log.

cd *abbr*: candela.

ceiling *n*: the height above the ground up to the base of the lowest layer of clouds when over half the sky is obscured.

ceiling price *n*: the maximum lawful price that may be charged for the first sale of a specified NGPA category of natural gas.

cell *n*: a device that delivers an electric current as the result of a chemical reaction; a receptacle, such as a jar, tube, or box, that contains two dissimilar metals immersed in or surrounded by a liquid or solid chemical (an electrolyte) that reacts with both metals.

cellar *n*: a pit in the ground, usually lined with concrete, steel pipe, or wood, that provides additional height between the rig floor and the wellhead to accommodate the installation of blowout preventers, rathole, mousehole, and so forth. It also collects drainage water and other fluids for subsequent disposal.

cellar deck *n*: the lower deck of a double-decked semisubmersible drilling rig. See *main deck, Texas deck*.

cellophane *n*: a thin, transparent material made from cellulose and used as a lost circulation material. See *cementing materials*.

Celsius scale *n*: the metric scale of temperature measurement used universally by scientists. On this scale, 0° represents the freezing point of water and 100° its boiling point at a barometric pressure of 760 mm. Degrees Celsius are converted to degrees Fahrenheit by using the following equation:

$$°F = \tfrac{9}{5}\,(°C) + 32.$$

or

$$°F = 1.8\,(°C) + 32.$$

The Celsius scale was formerly called the centigrade scale; now, however, the term "Celsius" is preferred in the International System of Units (SI).

cement *n*: a powder consisting of alumina, silica, lime, and other substances that hardens when mixed with water. Extensively used in the oil industry to bond casing to the walls of the wellbore.

cement additive *n*: a material added to cement to change its properties. Chemical accelerators, chemical retarders, and weight-reduction materials are common additives. See *cementing materials*.

cementation *n*: 1. the crystallization or precipitation of soluble minerals in the pore spaces between clastic particles, causing them to become consolidated into sedimentary rock. 2. precipitation of a binding material around grains or minerals in rocks.

cement bond *n*: the adherence of casing to cement and cement to formation. When casing is run in a well, it is set, or bonded, to the formation by means of cement.

cement bond log *n*: an acoustic logging method based on the fact that sound travels at different speeds through materials of different densities. The fact that sound travels faster through cement than through air can be used to determine whether the cement has bonded properly to the casing.

cement bond survey *n*: an acoustic survey or sonic-logging method that records the quality or hardness of the cement used in the annulus to bond the casing and the formation. Casing that is well bonded to the formation transmits an acoustic signal quickly; poorly bonded casing transmits a signal slowly. See *acoustic survey, acoustic well logging*.

cement casing *v*: to fill the annulus between the casing and wall of the hole with cement to support the casing and prevent fluid migration between permeable zones.

cement channeling *n*: when casing is being cemented in a borehole, the cement slurry can fail to rise uniformly between the casing and the borehole wall, leaving spaces devoid of cement. Ideally, the cement should completely and uniformly surround the casing and form a strong bond to the borehole wall.

cement clinker *n*: a substance formed by melting ground limestone, clay or shale, and iron ore in a kiln. Cement clinker is ground into a powdery mixture and combined with small amounts of gypsum or other materials to form cement.

cement dump bailer *n*: a cylindrical container with a valve used to release small batches of cement in a remedial cementing operation.

cementer *n*: a retrievable cement squeeze tool that is used in remedial cementing. See *remedial cementing, secondary cementing*.

cement hydration *n*: reaction with water that begins when water is added to powdered cement. The cement gradually sets to a solid as hydration continues.

cementing *n*: the application of a liquid slurry of cement and water to various points inside or outside the casing. See *primary cementing, secondary cementing, squeeze cementing*.

cementing barge *n*: a barge containing the cementing pumps and other equipment needed for oilwell cementing in water operations.

cementing basket *n*: a collapsible or folding metal cone that fits against the walls of the wellbore to prevent the passage of cement; sometimes called a metal-petal basket.

cementing company *n*: a company whose specialty is preparing, transporting, and pumping cement into a well. Usually, a cementing company's crew pumps the cement to secure casing in the well.

cementing head *n*: an accessory attached to the top of the casing to facilitate cementing of the casing. It has passages for cement slurry and retainer chambers for cementing wiper plugs. Also called retainer head.

cementing materials *n pl*: a slurry of portland cement and water and sometimes one or more additives that affect either the density of the mixture or its setting time. The portland cement used may be high early strength, common (standard), or slow setting. Additives include accelerators (such as calcium chloride), retarders (such as gypsum), weighting materials (such as barium sulfate), lightweight additives (such as bentonite), or a variety of lost circulation materials (such as mica flakes).

cementing pump *n*: a high-pressure pump used to force cement down the casing and into the annular space between the casing and the wall of the borehole.

cementing time *n*: the total elapsed time needed to complete a cementing operation.

cement plug *n*: 1. a portion of cement placed at some point in the wellbore to seal it. 2. a wiper plug. See *cementing, wiper plug*.

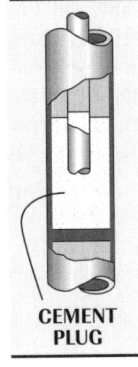

CEMENT PLUG

cement retainer *n*: a tool set temporarily in the casing or well to prevent the passage of cement, thereby forcing it to follow another designated path. It is used in squeeze cementing and other remedial cementing jobs.

cement system *n*: a particular slurry containing cement and water, with or without additives.

Cenozoic era *n*: the time period from 65 million years ago until the present. It is marked by rapid evolution of mammals and birds, flowering plants, grasses, and shrubs, and little change in invertebrates.

center coring *n*: a condition on a bit in which the inside row of cutters on the cones wears or breaks or the nose of one or more of the cones breaks.

center distance *n*: in a chain-and-sprocket drive, the distance between the centers of the two sprockets or the shafts they fit on.

center jet *n*: on roller cone bits, especially those designed for drilling soft, sticky formations, an additional jet nozzle that is added above the cones and in the middle of the bit. In some cases, the center jet, in addition to the three jets on the sides of the bit, can help keep the bit cutters clean.

center-latch elevators *n pl*: elevators that have the handles and latch arranged so that when a crew member opens the latch, they part in the middle. Compare *side-door elevators*. See *elevators*.

centerline *n*: the middle line of the hull of a mobile offshore drilling rig from stem to stern, as shown in a waterline view.

center of buoyancy (B) *n*: the center of gravity of the fluid displaced by a floating body (such as a ship or mobile offshore drilling rig).

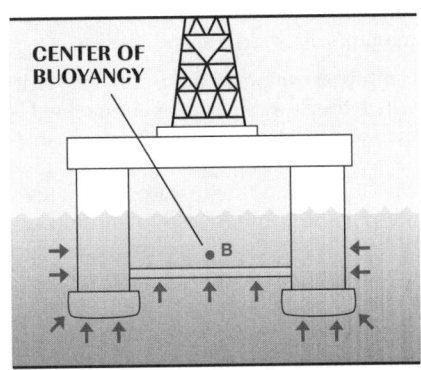

CENTER OF BUOYANCY

center of flotation (F) *n*: the geometric center of the water plane at which a mobile offshore drilling rig floats and about which a rig rotates when acted on by an external force without a change in displacement.

center of gravity (G) *n*: the point at which an object can be supported so that it balances, and at which all gravitational forces on the body and the weight of the body are concentrated; the center of mass.

center of pressure *n*: the point at which all wind pressure forces are concentrated.

center pin *n*: vertical pin or shaft which acts as a rotation centering device and connects the revolving superstructure and base mounting; also called center pintle or center post.

center spear *n*: a barbed fishing tool used to snag and retrieve broken wireline from the wellbore.

centigrade scale *n*: see *Celsius scale.*

centimetre (cm) *n*: a unit of length in the SI metric system equal to one-hundredth of a metre (10^{-2} metre).

centimetre-gram-second (cgs) *n*: a variant of the SI metric system in which the measures are founded on the centimetre, gram, and second rather than directly on the kilogram and metre.

centipoise (cp) *n*: one-hundredth of a poise; a measure of a fluid's viscosity, or resistance to flow.

central facility *n*: an installation serving two or more leases and providing one or more of such functions as separation, compression, dehydration, treating, gathering, or delivery of gas and oil.

centralizer *n*: see *casing centralizer.*

central oil-treating station *n*: a processing network used to treat emulsion produced from several leases, thus eliminating the need for individual treating facilities at each lease site.

central processing unit (CPU) *n*: in computer technology, the electronic device that contains the circuits necessary to receive, interpret, and carry out instructions received from peripherals such as keyboards, monitors, hard drives, and the like.

centrifugal compressor *n*: a compressor in which the flow of gas to be compressed is moved away from the center rapidly, usually by a series of blades or turbines. It is a continuous-flow compressor with a low-pressure ratio and is used to transmit gas through a pipeline. Gas passing through the compressor contacts a rotating impeller, from which it is discharged into a diffuser, where its velocity is slowed and its kinetic energy changed to static pressure.

Centrifugal compressors are nonpositive-displacement machines, often arranged in series on a line to achieve multistage compression.

centrifugal degasser *n*: a device for removing entrained gas from drilling mud in which a vacuum pump draws mud into degassing tubes inside the device. The tubes spin to create centrifugal force, which disperses the mud into thin layers on the wall of the tubes. A vacuum then pulls gas out inside an expansion tank.

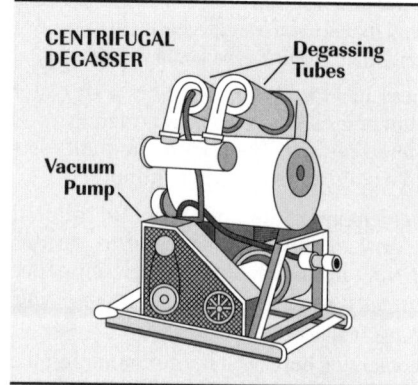

centrifugal force *n*: the force that tends to pull all matter from the center of a rotating mass.

centrifugal pump *n*: a pump with an impeller or rotor, an impeller shaft, and a casing, which discharges fluid by centrifugal force. An electric submersible pump is a centrifugal pump.

centrifugal units *n pl*: see *centrifugal compressor.*

centrifuge *n*: a machine that uses centrifugal force to separate substances of varying densities. A centrifuge is capable of spinning substances at high speeds to obtain high centrifugal forces. Also called the shake-out or grind-out machine.

centrifuge test *n*: a test to determine the amount of S&W in samples of oil or emulsion. The samples are placed in tubes and spun in a centrifuge, which breaks out the S&W.

CEQ *abbr*: Council on Environmental Quality.

CERCLA *abbr*: Comprehensive Environmental Response, Compensation, and Liability Act of 1980.

certificate *n*: a certificate of public convenience and necessity issued under the Natural Gas Act.

certificate natural gas *n*: any natural gas that has a certificate issued and that in effect is transported by an interstate pipeline.

certified copy *n*: a copy made from records in a recorder's or county clerk's office and certified to by the recorder or county clerk

as being an exact copy of the paper on file or of record.

certs *n pl*: (slang) certifications of materials on physical and chemistry properties.

cessation of production clause *n*: a clause in an oil and gas lease that provides the lessee with the right to begin new operations within a stated time period should production cease.

cetane number *n*: a measure of the ignition quality of fuel oil. The higher the cetane number, the more easily the fuel is ignited.

CFG *abbr*: cubic feet of gas; used in drilling reports.

CFR *abbr*: 1. Code of Federal Regulations. 2. Coordinating Fuels and Equipment Research Committee.

chain *n*: in offshore drilling, a heavy line constructed of iron bars looped together and used for a mooring line.

chain and gear drive *n*: see *chain drive.*

chain-and-sprocket drive *n*: a type of drive that consists of a loop of chain that goes around two shafts with sprockets fitted on the shafts. The driving shaft provides power and the driven shaft receives power. When the sprocket on the driving shaft turns, the chain moves and turns the driven sprocket.

chain-and-sprocket transmission *n*: a selective transmission made up primarily of chain-and-sprocket drives. There are three two-strand chain-and-sprocket drives and a pair of gears for reversing, located between the input shaft coming from the compound, and the output shaft on a mechanically driven drawworks. Each of these chain-and-sprocket drives has a different sprocket ratio to provide three speeds to the high-drum drive and three to the low-drum drive. Compare *gear transmission.*

chain case *n*: see *chain guard.*

chain drive *n*: a mechanical drive using a driving chain and chain gears to transmit power. Power transmissions use a roller chain, in which each link is made of side bars, transverse pins, and rollers on the pins. A double roller chain is made of two connected rows of links, a triple roller chain of three, and so forth.

chain guard *n*: in a chain-and-sprocket drive, a case that protects workers from the moving drive. It also protects the drive from dirt and is part of the cooling and lubrication systems. Some enclose only one drive and some enclose several drives. May be made of sheet metal or plate metal. Heavier guards also support the shafts on bearings. Both types have access panels for inspection and maintenance.

chain locker *n*: a compartment on a rig (usually located in a column) used to store mooring chain.

chain of title *n*: recorded transfers (links) in title from patent to present.

chain reducer *n*: a reduction unit between a prime mover and a beam pumping unit that reduces the prime mover's output speed by means of a series of chains.

chain stopper *n*: a device on a rig that locks off a chain mooring and takes the mooring line tension in the survival condition.

chain stud *n*: a steel component inserted into the center of a chain link. The stud may or may not be welded to the link.

chain tongs *n pl*: a hand tool consisting of a handle and chain that resembles the chain on a bicycle. In general, chain tongs are used for turning pipe or fittings of a diameter larger than that which a pipe wrench would fit. The chain is looped and tightened around the pipe or fitting, and the handle is used to turn the tool so that the pipe or fitting can be tightened or loosened.

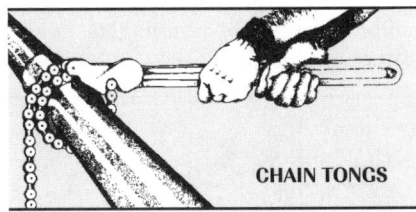

CHAIN TONGS

chain whelp *n*: the part of a chain windlass that holds the weight of the chain during heave in or payout.

chain width *n*: in roller chain, the width of the rollers, which is the distance between the inside faces of the roller link plates.

chain wildcat *n*: see *chain whelp*.

chamber gas lift *n*: a specialized form of intermittent-flow gas lift that functions in the same way as other forms, except that, when the injection gas is off, incoming well fluids accumulate in a chamber having a larger diameter than the tubing. For the same volume of fluid, hydrostatic head and wellbore pressure at the formation are both reduced.

change house *n*: a small building, or doghouse, in which members of a drilling rig or roustabout crew change clothes, store personal belongings, and so on.

change of heel (COH) *n*: the difference between initial and final heel caused by a load change in the port-to-starboard or starboard-to-port direction.

change of trim (COT) *n*: the difference between the initial and final trim caused by a load change in the forward or aft direction. See *trim*.

change rams *v*: to take rams out of a blowout preventer and replace them with rams of a different size or type. When the size of a drill pipe is changed, the size of the pipe rams must be changed to ensure that they seal around the pipe when closed (unless variable-bore pipe rams are in use).

channel *n*: the main current path between the source and drain electrodes in a field-effect transistor.

channeling *n*: when casing is being cemented in a borehole, the cement slurry can fail to rise uniformly between the casing and the borehole wall, leaving spaces, or channels, devoid of cement. Ideally, the cement should completely and uniformly surround the casing and form a strong bond to the borehole wall. See *cement channeling*.

charcoal filter *n*: a vessel filled with charcoal, which absorbs chemical contaminants such as condensates and well treatment chemicals from the glycol stream.

charcoal test *n*: a test standardized by the American Gas Association and the Gas Processors Association for determining the natural gasoline content of a given natural gas. The gasoline is adsorbed from the gas on activated charcoal and then recovered by distillation. The test is described in Testing Code 101–43, a joint AGA and GPA publication.

charge stock *n*: oil that is to be treated in, or charged to, a particular refinery unit.

Charles's law *n*: an ideal gas law that states that at constant pressure the volume of a fixed mass or quantity of gas varies directly with the absolute temperature; that is, at a constant pressure, as temperature rises, gas volume increases proportionately.

charter party *n*: an agreement by which a shipowner agrees to place an entire ship, or part of it, at the disposal of a merchant or other person to carry cargo for an agreed-upon sum.

chart integration department *n*: part of a company that interprets and processes the information obtained from the orifice meter charts. The department's responsibilities include calculating the gas flow from the information on the chart. Today, charts have been largely supplanted by electronic readouts generated by computer controlled sensors at the orifice meter station. These readouts are transmitted to the company's computers in a central, or district, office, where they are recorded and read.

chase pipe *v*: to lower the drill stem rapidly a few feet into the hole and then stop it suddenly with the drawworks brake. A surge of pressure in the mud in the drill stem and annular space results and may help to flush out debris accumulated in or on the pipe. The pressure surge may break down a formation, however, causing lost circulation, or may damage the bit if it is near the bottom.

chase threads *v*: to clean and deburr the threads of a pipe so that it will make up properly.

chassis *n*: the framework to which the components of a radio, television, or other electronic equipment are attached.

chatter *n*: a phenomenon in which the cutters of a bit, while rotating on bottom, do not adequately penetrate the formation; instead, the cutters tend to bounce off the formation. The bounce is sometimes transmitted to the drill stem in the form of a vibration, or chatter.

cheater *n*: a length of pipe fitted over a wrench handle to increase the leverage of the wrench. Use of a larger wrench is usually preferred. Also called a snipe.

checkerboard farmout *n*: an agreement for the acquisition of mineral rights (i.e., oil and gas leases) in a checkerboard pattern of alternate tracts, usually beginning, in the case of farmouts, with the drill site tract.

checking *n*: slight breaks in the surface of a paint film that do not render the underlying surface visible when the film is viewed at a magnification of ten times. Types include irregular pattern checking in which the breaks do not have a definite pattern, line checking in which the breaks are generally parallel, and crow-foot checking in which the breaks assume a three-pronged shape and the prongs radiate from a point at an angle of about 120 degrees.

check meter *n*: a device for measuring gas throughput, usually installed in conjunction with the primary measuring device owned and operated by the party having the obligation to provide measurement services at a specific point on a pipeline system.

check valve *n*: a valve that permits fluid to flow in one direction only. If the gas or liquid starts to reverse, the valve automatically closes, preventing reverse movement. Sometimes referred to as a one-way valve.

Chemelectric treater *n*: a brand name for an electrostatic treater.

chemical *n*: a substance defined under HAZCOM (OSHA) as any element, chemical compound, or mixture of elements and/or compounds.

chemical absorption *n*: in removing contaminants, the liquid absorbent reacts chemically with the acid gases but not with the natural gas.

chemical barrel *n*: a container in which various chemicals are mixed prior to addition to drilling fluid.

chemical consolidation *n*: the procedure by which a quantity of resinous material is squeezed into a sandy formation to consolidate the sand and to prevent its flowing into the well. The resinous material hardens and creates a porous mass that permits oil to flow into the well but holds back the sand at the same time. See *sand consolidation*.

chemical cutoff *n*: a method of severing steel pipe in a well by applying high-pressure jets of a very corrosive substance against the wall of the pipe. The resulting cut is very smooth.

chemical cutter *n*: a fishing tool that uses high-pressure jets of chemicals to sever casing, tubing, or drill pipe stuck in the hole.

chemical fingerprinting *n*: the process of tracking down the source of an illegal waste discharge through sampling the waste and possible sources to determine which source matches the discharge best. Fingerprinting is expensive, but in large damage cases, it can help prove or disprove whether a substance was discharged from a particular source.

chemical flooding *n*: see *alkaline (caustic) flooding, micellar-polymer flooding*.

chemical inhibitor *n*: liquid chemical compounds that are injected into lines carrying fluids that contain H_2S. Most of these inhibitors are designed to coat equipment surfaces to physically isolate them from corrosive substances. Others react with sulfur compounds to form less-destructive compounds.

chemical injection valve (CIV) *n*: a valve on the wellhead of a producing well that is usually installed between the production master valve and production wing valves. A chemical injection line leading into the well exits from the valve and goes into the well. The chemical injection valve isolates the chemical injection line from the surface.

chemical inventory *n*: an inventory required by HAZCOM. Under HAZCOM, all employers must keep a complete list, or inventory, of all hazardous materials on site.

chemical protective clothing *n*: clothing that is designed to protect against a specific chemical hazard (i.e., suits or aprons made of or coated with chemical-resistant materials like butyl rubber, neoprene, or polyvinyl chloride).

chemical pump *n*: an injection pump used to introduce a chemical into a fluid stream or receptacle.

chemical reaction *n*: a change in which a substance or substances is changed into one or more substances.

chemicals *n pl*: in drilling-fluid terminology, a chemical is any material that produces changes in the viscosity, yield point, gel strength, fluid loss, and surface tension.

chemical treatment *n*: any of many processes in the oil industry that involve the use of a chemical to effect an operation. Some chemical treatments are acidizing, crude oil demulsification, corrosion inhibition, paraffin removal, scale removal, drilling fluid control, refinery and plant processes, cleaning and plugging operations, chemical flooding, and water purification.

chert *n*: a rock of precipitated silica whose crystalline structure is not easily discernible and that fractures conchoidally (like glass). Flint, jasper, and chat are forms of chert.

chicken hook *n*: a long steel pole with a hook on one end that allows one of the rotary helpers to release the safety latch on the drilling hook so that the bail of the swivel can be removed (as when the kelly is set back prior to making a trip). Also called a shepherd's stick, or hook.

Chiksan line *n*: a flexible coupling used in high-pressure lines. Sometimes called Chicksan.

chiller *n*: a heat exchanger that cools process fluids with a refrigerant.

chip *n*: 1. the shaped and processed semiconductor die that is mounted on a substrate to form a transistor, diode, or other semiconductor device. 2. an integrated microcircuit performing a significant number of functions and constituting a subsystem.

chip hold-down effect *n*: the holding of formation rock chips in place as a result of high differential pressure in the wellbore (i.e., pressure in the wellbore is greater than pressure in the formation). This effect limits the cutting action of the bit by retarding circulation of bit cuttings out of the hole.

chipping *n*: the removal of paint and surface contaminants from a substrate by means of impact from a sharpened tool.

chips *n pl*: see *cuttings*.

chk *abbr*: choke; used in drilling reports.

chlorine *n*: a greenish-yellow, acrid gas (Cl) that irritates the skin and mucous membranes and causes breathing difficulties.

chlorine log *n*: a record of the presence and concentration of chlorine in oil reservoirs. See *chlorine survey*.

chlorine survey *n*: a special type of radioactivity-logging survey used to measure the relative amount of chlorine in the formation. Rocks with low chlorine content are likely to contain gas or oil; rocks with high chlorine content usually contain salt water only.

choke (chk) 1. *n*: a device with an orifice installed in a line to restrict the flow of fluids. Surface chokes are part of the Christmas tree on a well and contain a choke nipple, or bean, with a small-diameter bore that serves to restrict the flow. Chokes are also used to control the rate of flow of the drilling mud out of the hole when the well is closed in with the blowout preventer and a kick is being circulated out of the hole. See *adjustable choke, bottomhole choke, positive choke*. 2. in electronics, an inductance used in a circuit to present high impedance to frequencies above a specified frequency range without appreciably limiting the flow of direct current.

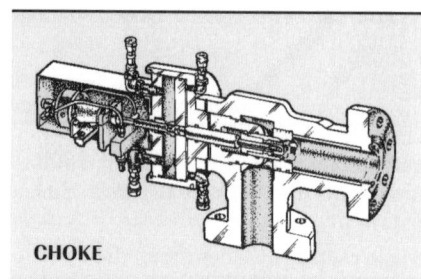

CHOKE

choke and kill system *n*: an assembly of lines (pipes), valves, and operating devices (a system) on a subsea blowout preventer that provides a conduit for drilling mud to be pumped into the well through a kill line and a conduit for fluids exiting from the well to flow through a choke line to the surface. See *choke line, kill line*.

choke and kill valve *n*: a device that, when opened, allows drilling mud to be pumped into the well and drilling and other fluids returning from the well to be directed to the choke manifold on the surface. See *choke line, kill line*.

choke and separator loss *n*: the drop in pressure that occurs when fluid flows through the choke and separator on a drilling or workover rig. The drop is caused by friction as the fluid contacts the sides of the choke and the separator; also, friction within the flowing fluid itself causes a pressure loss.

choke angle *n*: in crane operations, the angle between the vertical part and choked part of a choker sling. As this angle decreases, the load-lifting capacity of the sling also decreases. See *choker sling.*

choke bean *n*: a device placed in a choke line that regulates the flow through the choke. Flow depends on the size of the opening in the bean; the larger the opening, the greater the flow.

choke control panel *n*: a panel or board on the rig floor, usually near the driller's position, that allows a person to remotely adjust the size of the opening of a choke installed in the choke manifold. By adjusting the choke's opening, the operator can increase or decrease the amount of back-pressure being held on the well by the choke. Usually, the panel also has pressure gauges and other instruments to help the operator control the well.

choke flow *n*: see *critical flow.*

choke flow line *n*: see *choke line.*

choke line *n*: a pipe, or conduit, usually installed on outlets below the ram blowout preventers, through which fluids in the annulus can flow to the surface and through the choke manifold when valves to it are open. On floating offshore rigs, flow may be directed either down the annulus or up the annulus; on most land rigs, flow is only upward.

choke manifold *n*: an arrangement of piping and special valves, called chokes. In drilling, mud is circulated through a choke manifold when the blowout preventers are closed. In well testing, a choke manifold attached to the wellhead allows flow and pressure control for test components downstream.

choke nipple *n*: see *bean.*

choke pressure *n*: see *back-pressure.*

choker hitch *n*: in crane operations, a method of attaching a sling to a load in which one end of the sling is passed around the load, through an attachment or an eye on the other end, and then to the crane's hook. As the load is lifted by the crane, the hitch tightens on the load.

choker sling *n*: a wire rope sling that forms a slip noose around an object to be moved or lifted by a crane.

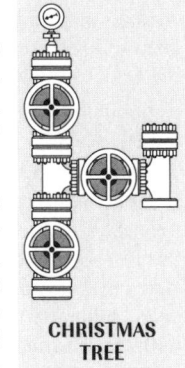

CHRISTMAS TREE

Christmas tree *n*: the control valves, pressure gauges, and chokes assembled at the top of a well to control the flow of oil and gas after the well has been drilled and completed. It is used when reservoir pressure is sufficient to cause reservoir fluids to flow to the surface.

chromate *n*: a compound in which chromium has a valence of 6, e.g., sodium bichromate. Chromate may be added to drilling fluids either directly or as a constituent of chrome lignites or chrome lignosulfonates. In certain areas, chromate is widely used as a corrosion inhibitor, often in conjunction with lime.

chromatograph *n*: an analytical instrument that separates mixtures of substances into identifiable components by means of chromatography.

chromatograph tests *n pl*: a laboratory analysis of a gas sample to determine the composition of the gas. The analysis will determine the percentage of each component of the gas; this percentage can be used to calculate liquid volume percentage, GPMs, Btu, and gravity.

chromatography *n*: a method of separating a solution of closely related compounds by allowing it to seep through an adsorbent so that each compound is adsorbed in a separate layer.

chrome lignite *n*: mined lignite, usually leonardite, to which chromate has been added or has reacted. The lignite can also be causticized with either sodium or potassium hydroxide.

Ci *abbr*: cirrus.

CIDS *abbr*: concrete island drilling system.

circ *abbr*: circulated; used in drilling reports.

circle plot *n*: in diamond bits, a pattern of setting diamonds in which the manufacturer spreads them in concentric circles on the cutting surface (the nose) of the bit. Compare *grid plot.*

circle wrench *n*: see *wheel-type back-off wrench.*

circuit *n*: a complete electrical path from one terminal of a source of electricity to the other, usually connected to a load. When the circuit is closed, electric current flows through it; when it is opened, the current flow stops.

circuit diagram *n*: see *schematic.*

circular mil (cmil) *n*: the cross-sectional area of a wire or cable expressed in thousandths of an inch (a milli-inch).

circular reader *n*: a metering instrumentation system component that counts the revolutions of the index shaft.

circulate *v*: to pass from one point throughout a system and back to the starting point. For example, drilling fluid is circulated out of the suction pit, down the drill pipe and drill collars, out the bit, up the annulus, and back to the pits while drilling proceeds.

circulate-and-weight method *n*: see *concurrent method.*

circulate the short way *v*: see *reverse circulation.*

circulating and releasing overshot *n*: an external gripping tool that allows direct circulation through the tool and the fish to keep the wellbore lubricated during fishing. See *overshot.*

circulating components *n pl*: the equipment included in the drilling fluid circulating system of a rotary rig. Basically, the components consist of the mud pump, the rotary hose, the swivel, the drill stem, the bit, and the mud return line.

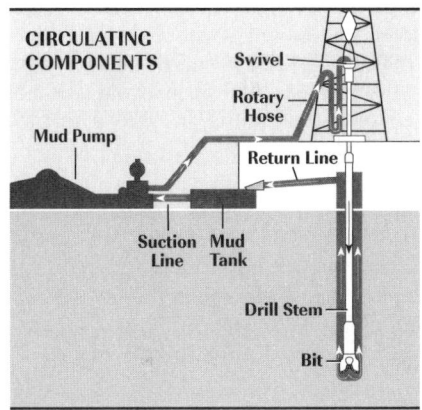

CIRCULATING COMPONENTS

Swivel

Rotary Hose

Return Line

Mud Pump

Suction Line

Mud Tank

Drill Stem

Bit

circulating density *n*: see *equivalent circulating density.*

circulating fluid *n*: see *drilling fluid, mud.*

circulating head *n*: an accessory attached to the top of the drill pipe or tubing to form a connection with the mud system to permit circulation of the drilling mud. In some cases, it is also a rotating head.

circulating pressure *n*: the pressure generated by the mud pumps and exerted on the drill stem.

circulating rate *n*: the volume flow rate of the circulating drilling fluid usually expressed in gallons or barrels per minute in the United States. Elsewhere, it is expressed in cubic metres per minute.

circulating system *n*: responsible for getting mud down the hole through drill pipe and drill collars, jetting mud through the bit so it can pick up cuttings, and carrying cuttings to the surface.

circulation *n*: the movement of drilling fluid out of the mud pits, down the drill stem, up the annulus, and back to the mud pits. See *normal circulation, reverse circulation.*

circulation sleeve *n*: see *sliding sleeve.*

circulation squeeze *n*: a variation of squeeze cementing for wells with two producing zones in which (1) the upper fluid sand is perforated; (2) tubing is run with a packer, and the packer is set between the two perforated intervals; (3) water is circulated between the two zones to remove as much mud as possible from the channel; (4) cement is pumped through the channel and circulated; (5) the packer is released and picked up above the upper perforation, a low squeeze pressure is applied, and the excess cement is circulated out. The process is applicable where there is communication behind the pipe between the two producing zones because of channeling of the primary cement or where there is essentially no cement in the annulus.

circulation valve *n*: an accessory employed above a packer, to permit annulus-to-tubing circulation or vice versa.

cirrocumulus *n*: a white high-level cloud that is composed primarily of ice crystals. This type of cloud forms from cirrus or cirrostratus clouds and has a rippled appearance.

cirrostratus *n*: a very thin, white, high-level cloud that produces halos visible around the sun or moon. This type of cloud is composed of ice crystals and appears as a thin veil in the sky.

cirrus (Ci) *n*: a high-level, white, featherlike cloud that is composed of ice crystals.

cistern barometer *n*: an atmospheric pressure measuring device (a barometer) that is made up of an evacuated glass tube (about 1 yard or 1 metre in length) that is completely filled with mercury and an open dish (exposed to the atmosphere) that is partially filled with mercury. When the open end of the tube is placed in the dish, the mercury in the tube falls to a height equivalent to the atmospheric pressure.

citations *n pl*: notations that appear in government documents for cross-referencing information. If you are reading a regulation that refers to another regulation or to a different section in the same regulation, the citation will be listed so that you may refer to the appropriate material for more information.

city gas station *n*: see *city gate station*.

city gate station *n*: a station to which a transmission company delivers gas to the distribution company. This is the point at which interstate and intrastate pipelines sell and deliver gas to local distribution companies. Usually built adjacent to a natural gas transmission line in a sparsely populated area on the outskirts of a population center. Also called city gas station, town border station.

CIV *abbr*: chemical injection valve.

civil law *n*: see *statute law*.

civil penalties *n pl*: penalties for damages to private individuals or organizations typically determined by a court decision on suits filed against private individuals or corporations.

Cl *form*: chlorine.

clabbered *adj*: (slang) commonly used to describe moderate to severe flocculation of mud due to various contaminants. Also called gelled-up.

cladding *n*: in electronics, braided or solid aluminum or other material that surrounds insulated wires or cables. Cladding protects the wire or cable from being damaged by blows or forces from external sources, such as dropped objects, foot traffic, and the like.

clamp *n*: a mechanical device used to hold an object in place. For example, a leak-repair clamp, or saddle clamp, holds a piece of metal with the same curvature as the pipe over a hole in a line, effecting a temporary seal. A wireline clamp holds the end of a wire rope against the main rope, while a polished-rod clamp attaches the top of the polished rod to the bridle of a pumping unit.

clamp-on meter *n*: an AC voltmeter-ammeter that has a magnetic core in the form of hinged jaws that can be clamped around a current-carrying wire to measure AC voltage or amperage. Also called a snap-on meter.

Class I injection wells *n pl*: a classification of injection wells under the Safe Drinking Water Act that are related to hazardous, industrial nonhazardous, and municipal wastewater disposal below underground sources of drinking water.

Class II injection wells *n pl*: a classification of injection wells under SDWA that is related to oil and gas activity. Typically, these wells are used to dispose of produced waters into depleted oil formations below drinking water sources; to inject produced water from production operations back into the producing zone; to inject fluids for enhanced recovery; and to store hydrocarbons.

clastic rock *n*: a sedimentary rock composed of fragments of preexisting rocks. The principal distinction among clastics is grain size. Conglomerates, sandstones, and shales are clastic rocks.

clastics *n pl*: 1. sediments formed by the breakdown of large rock masses by climatological processes, physical or chemical. 2. the rocks formed from these sediments.

clastic texture *n*: rock texture in which individual rock, mineral, or organic fragments are cemented together by an amorphous or crystalline mineral such as calcite. Compare *crystalline texture*.

Claus process *n*: a process to convert hydrogen sulfide into elemental sulfur by selective oxidation.

clay *n*: 1. a term used for particles smaller than $1/256$ millimetre (4 microns) in size, regardless of mineral composition. 2. a group of hydrous aluminum silicate minerals (clay minerals). 3. a sediment of fine clastics.

clay extender *n*: any of several substances—usually organic compounds of high molecular weight—that, when added in low concentrations to a bentonite or to certain other clay slurries, will increase the viscosity of the system. See *low-solids mud*.

clay hydration *n*: the swelling that occurs when clays in the formation take on water.

clay yield *n*: the number of barrels of a liquid slurry of a given viscosity that can be made from a ton of clay. Clays are often classified as either high- or low-yield. A ton of high-yield clay yields more slurry of a given viscosity than a low-yield clay.

Clean Air Act (CAA) *n*: one of the first environmental laws passed by the federal government. The CAA addresses and regulates airborne pollution that may be potentially hazardous to public health or natural resources.

clean ballast *n*: ballast that is free of any measurable quantity of hydrocarbon.

cleaning and priming machine *n*: a self-propelled machine with a rotating set of brushes and buffers that removes from pipe surface any loose material and applies a thin coat of primer to prepare the pipe for coating.

clean out *v*: to remove sand, scale, and other deposits from the producing section of the well to restore or increase production.

cleanout door *n*: an opening made to permit removal of sediments from the bottom of a tank. Usually a plate near ground level is removed from the side of the tank to make the door.

cleanout tools *n pl*: the tools or instruments, such as bailers and swabs, used to clean out an oilwell.

clean-up operation *n*: (HAZWOPER) an operation during which hazardous substances are removed, contained, incinerated, neutralized, stabilized, cleaned up, or in any other manner processed or handled, with the ultimate goal of making the site safer for people or the environment.

Clean Water Act (CWA) *n*: law that regulates the discharge of toxic and nontoxic pollutants into the surface waters of the United States. Under the jurisdiction of both the EPA and the Army Corps of Engineers, CWA's short-term goal is to make surface waters safe for recreation, fishing, and other uses. Its long-term goal is to eliminate *all* harmful discharges into surface waters.

clear *v*: in electronic instrumentation, to restore a storage device, memory device, or binary stage to a prescribed state, usually that denoting zero. Also known as reset.

clearance *n*: 1. the distance by which one object clears another. 2. the amount of space between two objects.

clearance pocket *n*: a device mounted on the cylinder of a compressor that allows the amount of clearance space to be adjusted. Clearance space may have to be adjusted to compensate for increases or decreases in the load on the compressor.

clearance sample *n*: in tank sampling, a spot sample taken 4 inches (100 millimetres) below the level of the tank outlet.

clearance space *n*: in a reciprocating compressor, the volume between the top of the piston and the cylinder head with the piston at the top of its stroke.

clearance volume *n*: the amount of space between the traveling and standing valves in a sucker rod pump when the pump is at the bottom of its stroke.

clear brine *n*: a drilling fluid made up mainly of chemical salts, such as sodium chloride, calcium chloride, or potassium chloride. Clear brine contains little or no clay or other solid material and is virtually transparent. It is often used when drilling into a producing formation because clear brine minimizes formation damage. See *completion fluid, formation damage.*

clear opening *n*: on a panel of rectangular floor or deck grating on an offshore installation, the distance between the supporting bars of the panel.

clear span *n*: on offshore installations with grated decks or floors, the length of the load bearing bars (the supports) that go between two other supports.

clear the title *n*: to establish the full legal status of the land in question before executing a lease.

clear water drilling *n*: drilling operations in which plain water (usually salt water) is used as the circulating fluid.

cleat *n*: see *coal cleat.*

clevis *n*: see *shackle.*

CLFP *abbr*: choke-line friction pressure.

climate *n*: the average weather or the regular variations in weather in an area over a period of years.

clingage *n*: the amount of oil that adheres to the wall of a measuring or prover tank after draining.

Clinton flake *n*: finely shredded cellophane used as a lost circulation material for cement.

clip *n*: a U-bolt or similar device used to fasten parts of a wire cable together. *v*: in electronics, to prevent the amplitude of a waveform from exceeding a specified level while preserving the shape of the waveform at amplitudes less than the specified level.

clipper *n*: an electronic circuit used to prevent the amplitude of a waveform from exceeding a specified level while preserving the shape of the waveform at amplitudes less than the specified level. Also called a limiter.

clipping *n*: in electronics, the act of preventing the amplitude of a waveform from exceeding a specified level while preserving the shape of the waveform. Also called limiting.

clock *n*: see *timing circuit.*

closed circuit *n*: 1. a life-support system in which the gas is recycled continually while the carbon dioxide is removed and oxygen added periodically. 2. a television installation in which the signal is transmitted by wire to a limited number of receivers.

closed-in pressure *n*: see *shut-in pressure.*

closed isobar *n*: in meteorology, a line drawn on a weather map connecting points of equal atmospheric pressure (an isobar) that forms a circular pattern on the chart and indicates an area of low pressure circulating about a center. Compare *open isobar.*

closed loop *n*: in process control, a system of components that utilizes feedback from its output for comparison with the desired set point; any difference results in automatic correction of the desired output. Closed loop systems are used to automatically control processes and equipment without human intervention.

closed stationary tank prover *n*: see *positive-volume prover.*

closed system *n*: a water-handling system (such as a saltwater-disposal system) in which air is not allowed to enter, thereby preventing corrosion or scale.

close in *v*: 1. to shut in a well temporarily that is capable of producing oil or gas. 2. to close the blowout preventers on a well to control a kick. The blowout preventers

close off the annulus so that pressure from below cannot flow to the surface.

close nipple *n*: a very short piece of pipe threaded its entire length.

closing gauge *n*: the measurement in a tank after a delivery or receipt. Compare *opening gauge.*

closing machine *n*: a machine that braids wires into strands and strands into rope in the manufacture of wire rope. Also called a stranding machine.

closing pressure *n*: in a blowout preventer stack, the amount of pressure supplied by the blowout preventer control unit that is required to close a preventer. See *operating pressure.*

closing ratio *n*: the mathematical relation (the ratio) between the pressure in the hole and the operating-piston pressure needed to close the rams of a blowout preventer. The closing ratio represents the difference in area between the part of the ram preventer's operating piston that is affected by wellbore pressure, and the part of the piston that closing pressure acts on.

closing unit *n*: see *blowout preventer control unit.*

closing-up pump *n*: an electric or hydraulic pump on an accumulator that pumps hydraulic fluid under high pressure to the blowout preventers so that they may be closed or opened.

closure *n*: 1. the vertical distance between the top of an anticline, or dome, and the bottom, an indication of the amount of producing formation that may be expected. 2. the process of closing down a site, such as plugging and abandoning a well or filling in a pit.

cloud *n*: in meteorology, a mass of visible, suspended water droplets or ice crystals that are formed by condensation in the atmosphere.

cloud on a title *n*: a claim or encumbrance that, if upheld by a court, would impair the owner's title to the property.

cloud point *n*: the temperature at which paraffin wax begins to congeal and become cloudy.

clump weights *n pl*: special segmented weights attached to the guy wires of a guyed compliant platform that keep the guy wires taut as the platform jacket moves with the waves and current of the water. As tension on a guy wire is lessened, one or more segments of the weight drop to the seafloor and reestablish tension on the wire. As tension increases on a guy wire, the tension pulls one or more segments off the seafloor so that the wire is not overstressed.

clutch *n*: a coupling used to connect and disconnect a driving and a driven part of a mechanism, especially a coupling that permits the former part to engage the latter gradually and without shock. In the oilfield, a clutch permits gradual engaging and disengaging of the equipment driven by a prime mover. *v*: to engage or disengage a clutch.

cm *abbr*: centimetre.

cm² *abbr*: square centimetre.

cm³ *abbr*: cubic centimetre.

CMC *abbr*: carboxymethyl cellulose.

cmil *abbr*: circular mil.

Co *sym*: cobalt.

CO *form*: carbon monoxide.

CO₂ *form*: carbon dioxide.

coagulation *n*: see *flocculation*.

coal *n*: a carbonaceous, rocklike material that forms from the remains of plants that were subjected to biochemical processes, intense pressure, and high temperatures. It is used as fuel.

coal bed methane *n*: natural gas, primarily methane, that occurs naturally in the fractures and matrix of coal beds.

coal cleat *n*: the cleavage or fracture plane of coal as found in a mine or in a coal deposit. Natural gas often migrates through cleats.

coalesce *v*: to unite as a whole.

coalescence *n*: 1. the change from a liquid to a thickened, curdlike state by chemical reaction. 2. the combining of globules in an emulsion caused by molecular attraction of the surfaces.

coalescer *n*: a device used to cause the separation and removal of one liquid from another. May be used to remove water from a petroleum liquid sample withdrawn from a pipeline.

coal fine *n*: coal crushed sufficiently finely to pass through a screen.

coal gas *n*: gas produced from coal and used for fuel. It contains primarily hydrogen, methane, and carbon monoxide.

coal tar epoxy *n*: a thermosetting resin made from the by-product of the carbonization of bituminous coal. It is used as a coating because of its adhesiveness, flexibility, and resistance to chemicals.

Coastal Zone Management (CZM) Act of 1972 *n*: a congressional act that established policies for the balanced protection and wise development of coastal resources.

coastal zone management (CZM) program *n*: a state-level program that provides for state review of exploration, development, and production plans that affect the coastal zone.

coast ice *n*: see *fast ice*.

coating *n*: any material that forms a continuous film over a metal surface to prevent corrosion damage.

coating flaw *n*: a gap or flaw in pipe coating. Coating flaws, which must be repaired to prevent corrosion problems, are detected through mechanical or visual inspections of the line. Also called holiday.

coating machine *n*: a machine that applies an even layer of coating material to pipe surface. Most coating machines also apply outer wrapping on the pipe.

coax *abbr*: shortened form of *coaxial cable*.

coaxial cable (coax) *n*: electrical conducting wire (cable) that consists of a center conducting wire covered by polyethylene insulation. A second conductor of braided copper (or other conductor such as aluminum) fits over the polyethylene insulation. Finally, a vinyl cover protects the braided conductor. The braided conductor protects, or shields, the center conductor from extraneous and undesirable electrical interference that may affect the signal being carried by the center conductor.

cobalt (Co) *n*: a metallic element used chiefly in alloys.

cobwebbing *n*: a phenomenon that can occur when paint is sprayed onto a surface. Fine filaments of partly dried paint can form a cobweb pattern as the paint is sprayed onto the surface. Cobwebbing usually occurs with the spray application of a fast drying paint.

Code of Federal Regulations (CFR) *n*: an annual paperback set of all the final and permanent agency rules and regulations published in the *Federal Register*. The CFR is divided into 50 titles, which represent broad areas subject to federal regulation.

coefficient *n*: a number that serves as a measure of some property or characteristic and that is commonly used as a factor in computations.

coefficient of cubical expansion *n*: see *coefficient of expansion*.

coefficient of discharge *n*: the ratio of actual flow to theoretical flow.

coefficient of elasticity *n*: see *modulus of elasticity*.

coefficient of expansion *n*: the increment in volume of a unit volume of solid, liquid, or gas for a rise of temperature of 1° at constant pressure. Also called coefficient of cubical expansion, coefficient of thermal expansion, expansion coefficient, expansivity.

coefficient of expansion adjuster *n*: an accessory for a meter that is used with an automatic temperature compensator. It is used to regulate the magnitude of temperature compensation consistent with the coefficient of expansion of the liquid being metered.

coefficient of thermal expansion *n*: see *coefficient of expansion*.

COFCAW *abbr*: combination of forward combustion and waterflooding. Also called wet combustion.

cofferdam *n*: the empty space between two bulkheads that separates two adjacent compartments. It is designed to isolate the two compartments from each other, thereby preventing the liquid contents of one compartment from entering the other in the event of the failure of the bulkhead of one to retain its tightness. In oil tankers, cargo spaces are always isolated from the rest of the ship by cofferdams fitted at both ends of the tank body.

Coflexip hose *n*: a brand name of a flexible, high-strength, steel-braided, high-pressure hose that conducts fluids on a drilling rig and other petroleum installations. Many manufacturers make this type of hose. It is very strong and has high crush, abrasion, and fatigue resistance. It is used for choke and kill lines, cementing lines, surface blowout preventer control lines, injection lines, and other applications.

cogeneration *n*: an energy-producing process involving the simultaneous generation of thermal and electrical energy from a single primary heat source.

COH *abbr*: change of heel.

cohesion *n*: the attractive force between the same kinds of molecules (i.e., the force that holds the molecules of a substance together).

coil *n*: 1. an accessory of tubing or pipe for installation in condensers or heat exchangers. In more complex installations, a tube bundle is used instead of a coil. 2. see *solenoid*.

coiled tubing *n*: a continuous string of flexible steel tubing, often hundreds or thousands of feet long, that is wound onto a reel, often dozens of feet in diameter. The reel is an integral part of the coiled tubing unit, which consists of several devices that ensure the tubing can be safely and efficiently inserted into the well from the surface. Because tubing can be lowered into a well without having to make up joints of tubing, running coiled tubing into the well is faster and less expensive than running conventional tubing. Rapid advances in the use of coiled tubing make it a popular way in which to run tubing into and out of a well. Also called reeled tubing.

coiled-tubing unit *n*: the equipment for transporting and using coiled tubing, including a reel for the coiled tubing, an injector head to push the tubing down the well, a wellhead blowout preventer stack, a power source (usually a diesel engine and hydraulic pumps), and a control console. A unique feature of the unit is that it allows continuous circulation while it is being lowered into the hole. A coiled tubing unit is usually mounted on a trailer or skid.

coiled-tubing workover *n*: a workover performed with a continuous steel tube, normally 0.75 inch to 1 inch (1.9 to 2.54 centimetres) outside diameter, which is run into the well in one piece inside the normal tubing. Lengths of the tubing up to 16,000 feet (4,877 metres) are stored on the surface on a reel in a manner similar to that used for wireline. The unit is rigged up over the wellhead. The tubing is injected through a control head that seals off the tubing and makes a pressure-tight connection.

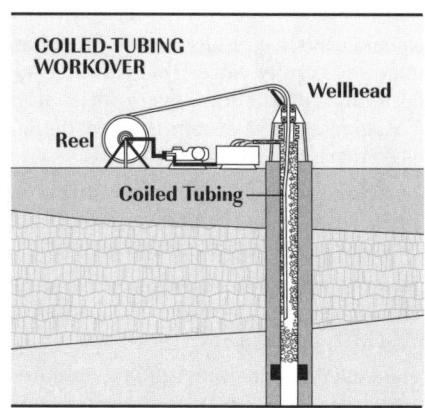

COILED-TUBING WORKOVER

Wellhead

Reel

Coiled Tubing

coke *n*: 1. a solid cellular residue produced from the dry distillation of certain carbonaceous materials that contains carbon as its principal constituent. 2. a residue of heavier hydrocarbons formed by thermal cracking and distillation and deposited in the reservoir during in situ combustion. This residue catches fire and becomes the fuel for continued combustion.

coke breeze *n*: crushed coke used for packing underground anodes in cathodic protection systems to obtain increased anode efficiency at a reduced cost. See *coke*.

coking *n*: a more severe form of thermal cracking.

cold cranking ampere (CCA) *n*: a measure of a battery's ability to provide current to start (crank) an engine when the engine is started cold—that is, when the engine has not been running for several hours, such as overnight.

cold front *n*: the leading edge of a cold air mass that replaces warmer air. The warm air is lifted to ride up over the cool air.

cold occlusion *n*: a weather phenomenon that occurs when a cold front overtakes a warm or stationary front and is characterized by the coldest air being behind the cold front. The warm front is forced aloft. Compare *neutral occlusion, warm occlusion*.

cold wall *n*: the sharp water-temperature gradient between the Gulf Stream and the Labrador Current.

cold-work *n*: plastic deformation of metal at a temperature low enough to insure or cause permanent strain.

coliform *n*: any of several bacilli found in the large intestine of humans and animals; its presence in water indicates fecal pollution.

collapse pressure *n*: the amount of force needed to crush the sides of pipe until it caves in on itself. The pipe collapses when the pressure outside it is greater than the pressure inside it.

collapse resistance *n*: the ability of the wall of a pipe or vessel to resist collapse.

collapse strength *n*: the maximum bearable external pressure that a joint of casing or other tubular can withstand without collapsing.

collar *n*: 1. a coupling device used to join two lengths of pipe, such as casing or tubing. A combination collar has left-hand threads in one end and right-hand threads in the other. 2. a drill collar. See *drill collar*.

collar clamp *n*: a device that wraps around the drill collar and when closed and tightened prevents the drill collar from slipping through the rotary.

collar-lift elevator *n*: an elevator that is bored to match a square shouldered tool joint.

collar locator *n*: a logging device used to determine accurately the depth of a well; the log measures and records the depth of each casing collar, or coupling, in a well. Since the length of each joint of casing is written down, along with the number of joints of casing that were put into the well, knowing the number and depth of the collars allows an accurate measure of well depth.

collar locator log *n*: see *collar locator*.

collar pipe *n*: heavy pipe used between the drill pipe and the bit in the drill stem. See *drill collar*.

collector *n*: in a junction transistor, a semiconductive region through which the primary flow of charge carriers leaves the

base of the transistor. The collector is the transistor's output terminal and is connected to the positive polarity of voltage from the supply through a resistor. See *base* (def. 2), *emitter*.

collet *n*: a fingerlike device used to lock or position certain tool components by manipulating the tubing string or downhole tool.

collision bulkhead *n*: the foremost bulkhead that extends from the bottom to the freeboard deck of a drill ship. It keeps the main hull watertight if a collision occurs.

colloid *n*: 1. a substance whose particles are so fine that they will not settle out of suspension or solution and cannot be seen under an ordinary microscope. 2. the mixture of a colloid and the liquid, gaseous, or solid medium in which it is dispersed.

colloidal *adj*: pertaining to a colloid, i.e., involving particles so minute (less than 2 microns) that they are not visible through optical microscopes. Bentonite is an example of a colloidal clay.

colloidal composition *n*: a colloidal suspension containing one or more colloidal constituents.

colloidal phase *n*: see *reactive phase, solid phase*.

colloidal suspension *n*: finely divided particles of ultramicroscopic size suspended in a liquid.

color code *n*: in electronics, the indication of resistor and capacitor values by the application of 12 different colors and no color. Each color is assigned a numerical value and also indicates a multiplier as well as the power rating. By correctly reading the color sequence, the resistance or capacity, as well as the power rating, of a resistor or a capacitor can be determined.

colorimetric testing *n*: method of testing pH using chemically treated paper strips that change color.

color test *n*: a visual test made against fixed standards to determine the color of a petroleum or other type of product.

column *n*: 1. on a semisubmersible drilling rig, one of several metal cylinders or rectangular-shaped cubes that rise from the pontoons of the rig to a great height above the water's surface. A deck is built on top of the columns and the drilling equipment, crew quarters, and so forth are installed on the deck. 2. a vertical, cylindrical vessel used in chemical and petroleum processing to increase the degree of separation of liquid mixtures by distillation or extraction. Also called a tower.

column racker *n*: an automated crane and rack system that stores pipe vertically and moves it to the well center and back.

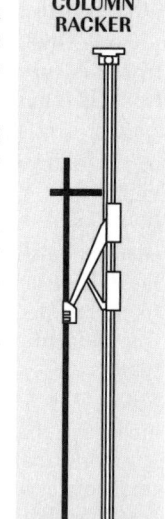

COLUMN RACKER

column-stabilized drilling unit *n*: see *semisubmersible drilling rig*.

combination drive *n*: a combination of two or more natural energies that work together in a reservoir to force fluids into a wellbore. Possible combinations include gas-cap and water drive, solution gas and water drive, and gas-cap drive and gravity drainage.

combination hook-block *n*: a piece of drilling equipment that joins the hook and traveling block into one unit designed to support the swivel and drill stem during drilling and to raise and lower drilling tools in and out of the hole.

combination rig *n*: a light rig that has the essential elements for both rotary and cable-tool drilling. It is sometimes used for reconditioning wells.

combination strand *n*: a wire rope strand that uses two or more strand designs to form a single strand. Compare *Seale strand, Warrington strand*.

combination string *n*: a casing string with joints of various collapse resistance, internal yield strength, and tensile strength, designed for various depths in a specific well to best withstand the conditions of that well. In deep wells, high tensile strength is required in the top casing joints to carry the load, whereas high collapse resistance and internal yield strength are needed in the bottom joints. In the middle of the casing, average quality is usually sufficient. The most suitable combination of types and weights of pipe helps to ensure efficient production at a minimum cost.

combination trap *n*: 1. a subsurface hydrocarbon trap that has the features of both a structural trap and a stratigraphic trap. 2. a combination of two or more structural traps or two or more stratigraphic traps.

combination wire/chain system *n*: a mooring system using a combination of wire/chain mooring components.

combined misalignment *n*: a type of chain misalignment that may result from combined angular and offset misalignment, or from two shafts that are not level with each other.

combining weight *n*: see *equivalent weight*.

combustible gas indicator *n*: device used to measure the percentage of combustible gas present in a sample of atmosphere.

combustion *n*: 1. the process of burning. Chemically, it is a process of rapid oxidation caused by the union of oxygen from the air with the material that is being oxidized or burned. 2. the organized and orderly burning of fuel inside the cylinder of an engine.

combustion cup *n*: on some diesel engines, a relatively small chamber, usually located just on top of the engine cylinder, into which fuel is injected. Since the cup also contains hot, compressed air, ignition occurs in the cup and the expanding gases push the piston down.

come-along *n*: a manually operated device that is used to tighten guy wires or move heavy loads. Usually, a come-along is a gripping tool with two jaws attached to a ring so that, when the ring is pulled, the jaws close.

come in *v*: to begin to produce; to become profitable.

come out of the hole *v*: to pull the drill stem out of the wellbore to change the bit, to change from a core barrel to the bit, to run electric logs, to prepare for a drill stem test, to run casing, and so on. Also called trip out.

come to see you *v*: (slang) to blow out. A well will "come to see you" if it blows out.

commercial amount *n*: see *commercial quantity*.

commercial butane *n*: a liquefied hydrocarbon consisting chiefly of butane or butylenes and conforming to the GPA specification for commercial butane defined in GPA Publication 2140.

commercial gauger *n*: someone licensed by the U.S. Customs Service to perform the duties of independent inspector or independent surveyor.

commercial laboratory *n*: U.S. Customs Service–approved laboratory, including those of commercial gaugers.

commercial production *n*: oil and gas production of sufficient quantity to justify keeping a well in production.

commercial propane *n*: a liquefied hydrocarbon product consisting chiefly of propane and/or propylene and conforming to the GPA specification for commercial propane as defined in GPA Publication 2140.

commercial quantity *n*: an amount of oil and gas production large enough to enable the operator to realize a profit, however small. To keep the lease in force, production must be in quantities sufficient to yield a return in excess of operating costs, even though drilling and equipment costs may never be recovered.

commingle *v*: to mix crude oil or oil products rather than moving them as separate batches. See *batch*.

commingling *n*: the mixing of crude oil products that have similar properties, usually for convenient transportation in a pipeline.

commingling facility *n*: an installation, serving two or more leases, that provides services such as gathering, separating, treating, and storing more economically than several smaller facilities.

commingling gas *n*: gas from two or more sources that is combined in a single stream.

committed reserves *n*: in gas contracts, the amount of gas the seller has available to sell to the buyer or buyers named in the contract.

commodity *n*: any material product as opposed to a service.

common bus *n*: typically, a large conductor made of copper cables that conducts alternating current from several alternators (AC generators) to rectifiers where the AC is converted to direct current (DC).

common carrier *n*: any cargo transportation system available for public use. Nearly all pipelines are common carriers.

common cement *n*: a regular portland cement classified as either API Class A or ASTM Type 1 cement.

common law *n*: a system of law based on court decisions, or judicial precedent, rather than on legislated statutes or executive decrees. Common law began in England and was later used in English colonies. It is still applied in most of the United States; however, Louisiana operates under the Napoleonic Code. Also called case law.

common rail *n*: the line in a certain type of fuel-injection system for a diesel engine that keeps fuel at a given pressure and feeds it through feed lines to each fuel injector.

common-rail injection *n*: a fuel-injection system on a diesel engine in which one

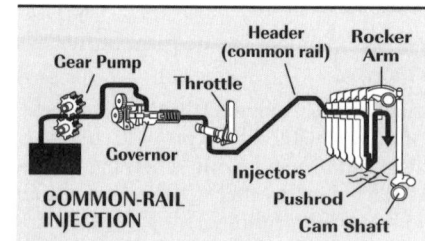

COMMON-RAIL INJECTION

line, or rail, holds fuel at a certain pressure and feed lines run from it to each fuel injector.

community property *n*: property, usually acquired after marriage, held jointly by husband and wife.

Community Right to Know *n*: see *SARA Title III*.

commutator *n*: a series of bars connected to the armature coils of an electric motor or generator. As the commutator rotates in contact with fixed brushes, the direction of flow of current to or from the armature is in one direction only.

comp *abbr*: completed or completion; used in drilling reports.

compact *n*: see *insert*.

compaction *n*: a decrease in the volume of a stratum due to pressure exerted by overlying strata, evaporation of water, or other causes.

compaction anticline *n*: see *draped anticline*.

company hand *n*: see *company representative*.

company man *n*: see *company representative*.

company representative *n*: an employee of an operating company who supervises the operations at a drilling site or well site and coordinates the hiring of logging, testing, service, and workover companies. Also called company hand, company man.

compartment *n*: a subdivision of space on a floating offshore drilling rig, a ship, or a barge.

compensated neutron log *n*: a measure and record of limestone porosity. The log is produced using one source and two detectors instead of just one detector, as an uncompensated neutron log does. A compensated neutron log is less influenced by borehole effects than an uncompensated neutron log. See *borehole effect*.

compensating index *n*: a meter index that has a pressure correction factor built into the gear ratio of the dial.

compensation *n*: provision of a supplemental device, circuit, or special materials to counteract known sources of error.

compensator governor *n*: a type of engine governor that prevents hunting (an engine's speeding up and slowing down as it seeks to run at the speed dictated by the engine governor). A compensator on the governor anticipates the engine's return to its set speed. When an engine's speed goes faster than the set speed, the compensator drops the engine's rpm; when engine speed drops below set speed, the compensator increases the engine's rpm.

Normally, engine operators set the compensator to keep the drop small. With a small speed drop, the governor and compensator quickly make the engine go back to control speed. See *governor, hunting*.

compensatory royalty *n*: payments to royalty owners as compensation for losses in income that they may be suffering because of failure to develop a lease adequately.

competitive field *n*: an oil or gas field comprising wells operated by various operators.

competitive leasing *n*: a procedure, based on competitive bidding, used to acquire oil and gas leases to federal lands within areas designated by USGS as known geologic structures (KGS) or on offshore federal lands.

complete a well *v*: to finish work on a well and bring it to productive status. See *well completion*.

completion *n*: to finish work on a well and bring it to productive status. See *well completion*.

completion fluid *n*: low-solids fluid or drilling mud used when a well is being completed. It is selected not only for its ability to control formation pressure, but also for the properties that minimize formation damage.

compliant piled tower *n*: an offshore platform jacket that flexes with wind, wave, and current forces and is supported by piles driven through guides attached to outside legs of the jacket.

compliant platform rig *n*: an offshore platform jacket that flexes with the wind, wave, and current forces. Two types are the guyed-tower platform rig and the compliant piled tower.

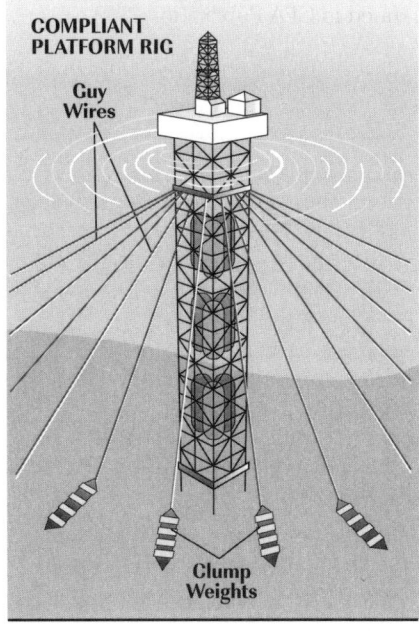

COMPLIANT PLATFORM RIG

Guy Wires

Clump Weights

composite sample *n*: a sample of a substance that is made up of equal portions of two or more spot samples obtained from a tank or pipeline. In a crude oil storage tank, one type of composite sample is taken at the top, at the bottom, and in the middle.

composite spot sample *n*: in tank sampling, a blend of spot samples mixed in equal proportions for testing.

composite stream *n*: 1. a flow of oil and gas in one stream. 2. a flow of two or more different liquid hydrocarbons in one stream.

composition *n*: a special material made of conductive and nonconductive substances, which, when blended and formed into a resistor, give the resistor a specific resistance.

composition resistor *n*: a resistor made of conductive and nonconductive materials, which are combined and formed into a relatively small cylinder. Axial or radial leads in the resistor provide a way to connect the resistor to a circuit.

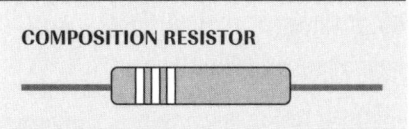

COMPOSITION RESISTOR

compound *n*: 1. a mechanism used to transmit power from the engines to the pump, the drawworks, and other machinery on a drilling rig. It is composed of clutches, chains and sprockets, belts and pulleys, and a number of shafts, both driven and driving. 2. a substance formed by the chemical union of two or more elements in definite proportions; the smallest particle of a chemical compound is a molecule. *v*: to connect two or more power-producing devices, such as engines, to run driven equipment, such as the drawworks.

compounding *n*: 1. the act of connecting two or more power-producing devices to run driven equipment. 2. the act of paying an amount on the accrued interest and the principal. Compare *discounting*.

compounding transmission *n*: on a mechanical-drive rig, the type of transmission that sends power from the engines to the drawworks and the rotary table, and sometimes to the mud pumps. See also *transmission*.

Comprehensive Environmental Response, Compensation, and Liability Act of 1980 (CERCLA) *n*: a congressional act that gives the government the authority to clean up any site where there is an unremedied release of a hazardous substance. Frequently, these sites, which are referred to as "Superfund" sites, are areas where hazardous waste has been disposed of improperly.

compressed air starting *n*: a method of starting a diesel engine in which an air compressor supplies high-pressure air to the engine's intake manifold. Since the air enters the engine under high pressure and temperature, combustion occurs when fuel is introduced and the engine starts.

compressibility *n*: the change in volume per unit of volume of a liquid caused by a unit change in pressure at constant temperature.

compressibility factor *n*: the ratio of the actual volume of gas at a given temperature and pressure to the volume of gas when calculated by the ideal gas law.

compression *n*: the act or process of squeezing a given volume of gas into a smaller space.

compression ignition (CI) *n*: an ignition method used in diesel engines by which the air in the cylinder is compressed to such a degree by the piston that ignition occurs upon the injection of fuel. About a 1-pound (7-kilopascal) rise in pressure causes a 2°-F (1°-C) increase in temperature.

compression-ignition engine *n*: a diesel engine; an engine in which the fuel-air mixture inside the engine cylinders is ignited by the heat that occurs when the fuel-air mixture is highly compressed by the engine pistons.

compression load *n*: the weight (load) placed on a string of pipe, such as riser or casing, that results when force from above presses down on the pipe.

compression packer *n*: a packer that is locked in its set position by tubing weight set down on it.

compression pressure *n*: the pounds-per-square-inch (kilopascal) increase in pressure at the end of the compression stroke in an engine, about 500 pounds per square inch (3,500 kilopascals) in a diesel engine.

compression ratio *n*: 1. the ratio of the absolute discharge pressure from a compressor to the absolute intake pressure. 2. the ratio of the volume of an engine cylinder before compression to its volume after compression. For example, if a cylinder volume of 10 cubic inches (10 cubic centimetres) is compressed into 1 cubic inch (1 cubic centimetre), the compression ratio is 10:1.

compression-refrigeration cycle *n*: the refrigeration cycle in which refrigeration is supplied by the evaporation of a liquid refrigerant such as propane or ammonia. Compare *absorption-refrigeration cycle*.

compression stroke *n*: in a diesel engine, the upward movement of the piston in the cylinder in which the volume of air in the cylinder is reduced by the piston's moving upward in the cylinder. The piston compresses the air so much that it gets hot enough to ignite diesel fuel injected into the cylinder near the top of the piston's travel.

compression test *n*: a type of testing used extensively on casinghead gas that has a propane content of above 0.5 to 1.0 gallon per thousand cubic feet to determine the liquid hydrocarbon content of the gas.

compression wave *n*: wave in an elastic medium that causes an element of the medium to change volume without rotating.

compressive strength *n*: the degree of resistance of a material to a force acting along one of its axes in a manner tending to collapse it; usually expressed in pounds of force per square inch (psi) of surface affected or in kilopascals.

compressive yield strength *n*: the maximum stress a metal, subjected to compression, can withstand without a predefined amount of permanent deformation.

compressor *n*: a device that raises the pressure of a compressible fluid such as air or gas. Compressors create a pressure differential to move or compress a vapor or a gas, consuming power in the process. They may be positive-displacement compressors or nonpositive-displacement compressors. See *centrifugal compressor, jet compressor, reciprocating compressor*.

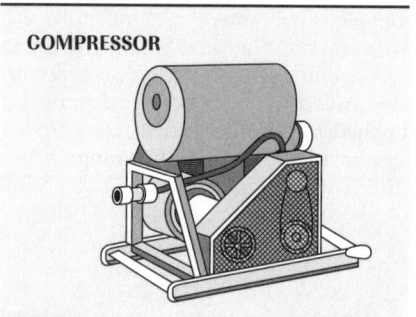

COMPRESSOR

compressor clearance *n*: the ratio of the volume remaining in a compressor cylinder at the end of a compression stroke to the volume displaced by one stroke of the piston. Usually expressed as a percentage.

compressor fuel *n*: natural gas burned as fuel to operate a compressor.

compressor pressure *n*: the force put out by a compressor that is often measured in pounds per square inch (psi) or kilopascals (kPa).

compressor station *n*: a facility consisting of one or more compressors with the necessary auxiliaries for delivering compressed gas.

compressor valve *n*: a device in a compressor that closes while gas is being compressed and opens to allow the compressed gas to discharge and uncompressed gas to be drawn in.

Compton effect *n*: a reaction in which gamma rays with intermediate energy levels (0.6 to 2.5 million electron volts) lose their energy by colliding with orbital electrons.

computer assisted operations *n pl*: a method of data acquisition and control that uses computer technology such as programmable logic controllers or digital controllers.

computerized production control (CPC) *n*: centralized computer control for a completely automated lease or multilease operation.

computing counter-printer *n*: a counter-printer with selective means for setting price per standard unit of measurement. It can compute the total price for a quantity of product delivered.

concentric cylinder centrifuge *n*: a mud centrifuge that consists of two cylinders rotating one within the other, and which, when mud is put into it, separates solid particles, such as barite or other solid particles, from the liquid mud.

concentric piston *n*: tubing pressure acting on the net piston area and causing a force to be exerted on a mandrel.

concentric-tubing workover *n*: a workover performed with a small-diameter tubing work string inside the normal tubing. Equipment needed is essentially the same as that for a conventional workover except that it is smaller and lighter.

concession *n*: a tract of land granted by a government to an individual or company for exploration and exploitation of minerals.

concrete gravity rigid platform rig *n*: a rigid offshore drilling platform built of steel-reinforced concrete and used to drill development wells. The platform is floated to the drilling site in a vertical position. At the site, one or more tall caissons that serve as the foundation of the platform are flooded so that the platform comes to rest on bottom. Because of the enormous weight of the platform, the force of gravity alone keeps it in place. See *platform rig*.

CONCRETE GRAVITY RIGID PLATFORM RIG

concrete island drilling system (CIDS) *n*: a structure made primarily of concrete which is designed for drilling in arctic waters. The structure is very strong and is designed to withstand the great force of moving ice.

CONCRETE ISLAND DRILLING SYSTEM

concurrent method *n*: a method for killing well pressure in which circulation is commenced immediately and mud weight is brought up in steps, or increments, usually a point at a time. Also called circulate-and-weight method.

condensate *n*: a light hydrocarbon liquid obtained by condensation of hydrocarbon vapors. It consists of varying proportions of butane, propane, pentane, and heavier fractions, with little or no methane or ethane.

condensate liquids *n pl*: hydrocarbons that are gaseous in the reservoir but that will separate out in liquid form at the pressures and temperatures at which separators normally operate. Sometimes called distillate.

condensate ratio *n*: the ratio of the volume of liquid produced to the volume of residue gas produced; usually expressed in barrels per million cubic feet.

condensate reservoir *n*: a reservoir in which both condensate and gas exist in one homogeneous phase. When fluid is drawn from such a reservoir and the pressure decreases below the critical level, a liquid phase (condensate) appears.

condensate trap *n*: a device installed on or near the fuel tank of a drilling rig engine into which the engine operator drains condensate water. See *condensate water*.

condensate water *n*: water that condenses out the air inside a fuel tank. When the temperature drops, water vapor in the air that lies above the fuel in a tank condenses into liquid water. This water tends to fall to the bottom of the tank where it can be drained. It is good practice to keep the tank full of fuel to minimize the amount of air out of which condensate water can drop.

condensate well *n*: a gas well producing from a condensate reservoir.

condensation *n*: the process by which vapors are converted into liquids, chiefly accomplished by cooling the vapors, lowering the pressure on the vapors, or both. Condensation is often the cause of the presence of water in fuels.

condensation trail *n*: a visible trail of condensed water vapor in the form of ice crystals, which an aircraft leaves when flying at high altitude. It is caused by the pressure drop created by the wing moving through air that is almost totally saturated with water vapor. The pressure change is sufficient to cause the air to become saturated. Also called contrail, vapor trail.

condenser *n*: 1. a form of heat exchanger in which the heat in vapors is transferred to a flow of cooling water or air, causing the vapors to form a liquid. 2. a capacitor.

condistometer *n*: a thickening-time tester having a stirring apparatus to measure the relative thickening time for mud or cement slurries under predetermined temperatures and pressures. For more information see the API publication *RP10B*.

condition *v*: to treat drilling mud with additives to give it certain properties. Sometimes the term applies to water used in boilers, drilling operations, and so on. To condition and circulate mud is to ensure that additives are distributed evenly throughout a system by circulating the mud while it is being conditioned.

conditioning *n*: see *mud*.

conditions of approval *n pl*: special conditions or additional requirements that are attached to permits to drill, deepen, or plug back. These conditions range from administrative matters, such as the frequency and number of required reports, to technical or environmental conditions, such as requirements for drilling mud disposal. In all cases, these are specific conditions that amplify or explain a requirement in OCSLA, MMS regulations, OCS Orders, or lease stipulations.

conductivity *n*: 1. the ability to transmit or convey (as heat or electricity); the opposite of resistivity. 2. an electrical logging measurement obtained from an induction survey, in which eddy currents produced by an alternating magnetic field induce in a receiver coil a voltage proportionate to the ability of the formation to conduct electricity. See *induction log*.

conductor *n*: a material with a high concentration of free electrons. It readily (carries) conducts an electric current.

conductor casing *n*: generally, but not always, the first string of casing in a well. It may be lowered into a hole drilled into the formations near the surface and cemented in place; it may be driven into the ground by a special pile driver (in such cases, it is sometimes called drive pipe); or it may be jetted into place in offshore locations. Its purpose is to prevent the soft formations near the surface from caving in and to conduct drilling mud from the bottom of the hole to the surface when drilling starts. Also called conductor pipe, drive pipe.

conductor hole *n*: the hole where the crew starts the top of the well in the middle of the cellar.

conductor housing *n*: in offshore drilling from floating rigs, a special fitting welded to the top of the conductor casing that is profiled to fit into the permanent guide base. It connects the conductor casing to the guide base. See *conductor casing*, *permanent guide base*.

conductor line *n*: a small-diameter conductive line used in electric wireline operations, such as electric well logging and perforating, in which the transmission of electrical current is required. Compare *wireline*.

conductor pipe *n*: 1. see *conductor casing*. 2. a boot, or flume.

conduit *n*: 1. a pipe, line, or channel through which fluids flow. 2. solid or flexible metal or other tubing through which insulated electric wires are run.

cone *n*: a conical-shaped metal device into which cutting teeth are formed or mounted on a roller cone bit. See *roller cone bit*. In mud solids control, another name for a hydrocyclone.

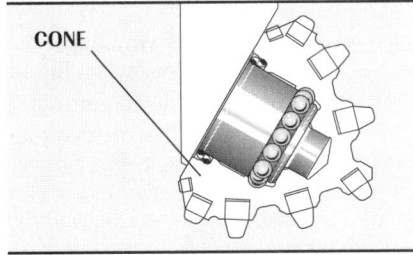

CONE

cone alignment *n*: on a roller cone bit, the way in which the cones of a bit line up with the bit's center axis. If the cones line up with the center axis, they have on-center alignment; if they do not line up with the center axis, they have off-center alignment. Generally, bits with on-center alignment drill hard formations; those with off-center alignment drill soft formations.

cone bit *n*: a roller bit in which the cutters are conical. See *bit*.

cone erosion *n*: a phenomenon in which a bit cone is worn away (eroded) by abrasive materials in the drilling fluid or from other sources.

cone interference *n*: a problem on a bit in which one cone interferes with the action of the other cones. Interference can occur when the bit is jammed into an undergauge (undersize) borehole; the legs of the bit are forced inward, which causes the cones to interfere with each other.

cone offset *n*: the amount by which lines drawn through the center of each cone of a bit fail to meet in the center of the bit. For example, in a roller cone bit with three cones, three lines can be drawn through the center of each cone and extended to the center of the bit. If these cone centerlines do not meet in the bit's center, the cones are said to be offset. In general, bits designed for drilling soft formations have more offset than cones for hard formations, because offset affects the angle at which the bit teeth contact the formation. Since soft formations require a gouging and scraping action by bit teeth, high offset achieves the necessary action.

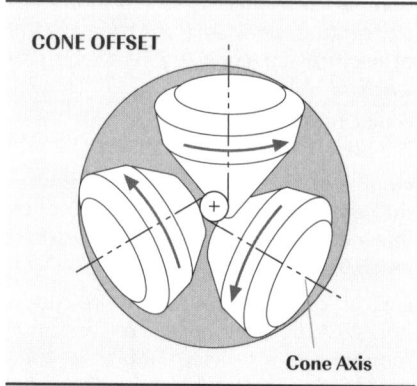

CONE OFFSET

Cone Axis

cone-roof tank *n*: a tank with a fixed conical roof.

cone shake *n*: shaking or vibrating of the cones of a bit that occurs when the bit bearings are worn.

cone shell *n*: that part of the cone of a roller cone bit out of which the teeth are milled or into which tungsten carbide inserts are placed and inside of which the bearings are housed.

cone skidding *n*: locking of a cone on a roller cone bit so that it will not turn when the bit is rotating. Cone skidding results in the flattening of the surface of the cone in contact with the bottom of the hole.

confined space *n*: a space large enough and so configured that an employee can bodily enter and perform work but, 1) has limited entry or exit, 2) has insufficient natural ventilation, 3) could contain hazardous materials, and 4) is not intended for continuous occupancy. Confined space may be either nonpermit or permit required space.

confined space permit *n*: a permit used to ensure that work in a confined space is done safely. The permit sets out conditions to be met before work begins.

confirmation well *n*: the second producer in a new field, following the discovery well.

conformable *adj*: layered in parallel and unbroken rows of rock, indicating that no disturbance occurred during deposition of the rock. Compare *unconformity*.

congl *abbr*: conglomerate; used in drilling reports.

conglomerate *n*: a sedimentary rock composed of pebbles of various sizes held together by a cementing material such as clay. Conglomerates are similar to sandstone but are composed mostly of grains more than 2 millimetres (0.08 inch) in diameter. Most conglomerates are found in discontinuous, thin, isolated layers; they are not very abundant. In common usage, the term "conglomerate" is restricted to coarse sedimentary rock with rounded grains; conglomerates made up of sharp, angular fragments are called breccia.

conical angle *n*: the angle of the cone of a bit. This angle may be steep, in which case the cone has a sharp taper, or it may be shallow, in which case the cone has a flatter taper.

conical drilling unit (CDU) *n*: a drilling structure whose base is shaped like a truncated (cut off) cone. It is designed for drilling in offshore arctic waters, where moving ice can damage the structure. The cone shape helps deflect the ice around the base and keep it intact.

conical funnel *n*: in drilling from floating offshore rigs, one of four cone-shaped devices (funnels) mounted on top of the temporary guide base that guides the blowout preventer stack and lower marine riser package into place on the base. See *blowout preventer stack, lower marine riser package, temporary guide base.*

coning *n*: the encroachment of reservoir water or gas into the oil column and well because of production. The water or gas tends to rise near the wellbore and assume a conical shape.

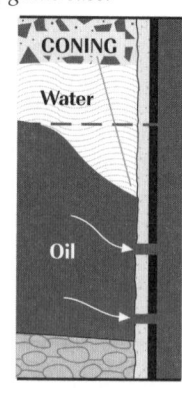

CONING

Water

Oil

connate water *n*: water retained in the pore spaces, or interstices, of a formation from the time the formation was created. Compare *interstitial water.*

connecting rod *n*: 1. a forged-metal shaft that joins the piston of an engine to the crankshaft. 2. the metal shaft that is joined to the bull gear and crosshead of a mud pump.

connecting rod bearing *n*: the bearing between the rod and the crankshaft. Often called the rod bearing.

connection *n*: 1. a section of pipe or fitting used to join pipe to pipe or to a vessel. 2. a place in electrical circuits where wires join. 3. the action of adding a joint of pipe to the drill stem as drilling progresses.

connection gas *n*: the relatively small amount of gas that enters a well when the mud pump is stopped for a connection to be made. Since bottomhole pressure decreases when the pump is stopped, gas may enter the well.

connector *n*: a device used to connect lengths of chain and/or wire.

connector link *n*: in roller chain, a type of link used to make a continuous loop of chain by connecting the two ends of the chain. The connector link is a pin link with either a spring clip or a cotter to hold the pins.

conservation *n*: preservation; economy; avoidance of waste. It is especially important in the petroleum industry, since the amount of oil and gas is finite. Many conservation practices, such as the trapping of condensable vapors, are used in the industry.

Conservation Award for Respecting the Environment (CARE) program *n*: an MMS award program that recognizes and champions exemplary actions and accomplishments by private companies engaged in offshore energy development that support the broader conservation and environmental goals of the nation, the Department of the Interior, and the coastal states.

conservation of energy *n*: see *law of conservation of energy.*

consideration *n*: a promise or an act of legal value bargained for and received in return for a promise; an essential element of a contract. In oil and gas leases, consideration may be payment in money or in kind; it must often be "serious" consideration. Compare *bonus consideration.*

consistency *n*: the cohesion of the individual particles of a given material (i.e., its ability to deform or its resistance to flow).

consistometer *n*: a thickening-time tester with a stirring apparatus to measure the relative thickening time for mud or cement slurries under predetermined temperatures and pressures.

console *n*: see *driller's console.*

consolidated shales *n pl*: compacted beds of clays and silts that are firm.

consolidation *n*: the process by which sand or other loose materials become firm.

constantan *n*: an alloy of 55 percent copper and 45 percent nickel; often used in making precision wire-round resistors.

constant choke-pressure method *n*: a method of killing a well that has kicked, in which the choke size is adjusted to maintain a constant casing pressure. This method does not work unless the kick is all or nearly all salt water. If the kick is gas, this method will not maintain a constant bottomhole pressure because gas expands as it rises in the annulus. In any case, it is not a recommended well-control procedure.

constant pit-level method *n*: a method of killing a well in which the mud level in the pits is held constant while the choke size is reduced and the pump speed slowed. It is not effective, and therefore is not recommended, because casing pressure increases to the point at which the formation fractures or the casing ruptures, and control of the well is lost.

construction *n*: in contract law, the interpretation given by a court of competent jurisdiction—for example, an interpretation of a possibly ambiguous instrument or statute.

consultant *n*: a person who contracts with an oil company to supervise the operations at a drilling site or well site and to coordinate the hiring of logging, testing, service, and workover companies.

contact *n*: 1. in geology, any sharp or well-defined boundary between two different bodies of rock. 2. a bedding plane or unconformity that separates formations. 3. in a petroleum reservoir, a horizontal boundary where different types of fluids meet and mix slightly; for example, a gas-oil or oil-water contact. Also called an interface.

contact angle *n*: the angle formed when two immiscible fluids meet a solid surface. For example, in a reservoir where water contacts a grain of sand or rock, it conforms to the shape of the solid and its angle of contact is at or near zero. Because of surface tension, any oil in the area does not conform to the shape of the water and grain; instead, the oil surrounds the wet grain and produces angles at the point of contact between the oil and the water.

contact area *n*: the gas-oil or oil-water interface in a reservoir.

contact log *n*: any log in which the logging sonde must be put into contact with the walls of the hole or casing to obtain the log.

contact metamorphism *n*: a type of metamorphism that occurs when an intruded body of molten igneous rock changes the rocks immediately around it, primarily by heating and by chemical alteration.

contactor *n*: 1. a vessel or piece of equipment in which two or more substances are brought together. In a glycol dehydration system, the vertical vessel in which wet gas is brought into contact with triethylene glycol. 2. a switch used to open or close an electric circuit.

contaminant *n*: a material, usually a mud component, that becomes mixed with cement slurry during displacement and that affects it adversely.

contamination *n*: the presence in a drilling fluid of any foreign material that may tend to produce detrimental properties of the drilling fluid.

continental crust *n*: rock that is thick—10 to 30 miles (16 to 48 kilometres)—and relatively light. Continents are made of continental crust.

continental drift *n*: according to a theory proposed by Alfred Wegener, a German meteorologist, in 1910, the migration of continents across the ocean floor like rafts drifting at sea. Compare *plate tectonics*.

continental margin *n*: a zone that separates emergent continents from the deep sea bottom.

continental rise *n*: the transition zone between the continental slope and the oceanic abyss.

continental shelf *n*: a zone, adjacent to a continent, that extends from the lower waterline to the continental slope, the point at which the seafloor begins to slope off steeply into the oceanic abyss.

continental slope *n*: a zone of steep, variable topography forming a transition from the continental shelf edge to the ocean basin.

continuous bleed relay *n*: a pneumatic relay whose operating air pressure escapes on a gradual basis.

continuous development clause *n*: in an oil and gas lease, a clause designed to keep drilling operations going steadily after the primary term has expired. In some clauses, designated intervals between completion of one well and commencement of the drilling of another may require the operator to develop the leased land up to its allowable density.

continuous-flow gas lift *n*: see *gas lift*.

continuous flowmeter log *n*: a log used to determine the contribution of each zone to the total production or injection. These surveys are used to indicate changes in the flow pattern versus changes in conditions at the surface, in time, in type of operation, or after stimulation treatments; particularly useful for measuring gas well flow.

continuous flow-responsive sampler *n*: a sampler that automatically adjusts the quantity of sample in proportion to the rate of flow.

continuous line sampling *n*: see *automatic sampling*.

continuous phase *n*: the liquid in which solids are suspended or droplets of another liquid are dispersed; sometimes called the external phase. In a water-in-oil emulsion, oil is the continuous phase. Compare *internal phase*.

continuous sample *n*: a pipeline sample that is withdrawn in a uniform and continuous rate or in one or more increments per minute.

continuous spinner flowmeter *n*: a production log used to determine which of several zones contributes the most to the total production or, in the case of an injection well, which zone is receiving the most injected fluids.

continuous steam injection *n*: see *steam drive*.

continuous time-cycle sampler *n*: a sampler that transfers equal increments of oil from the pipeline to the sample container at a uniform rate.

continuous treatment *n*: in corrosion control, the constant injection of small quantities of chemical corrosion inhibitors into the lines of a production system.

contour map *n*: a map constructed with continuous lines connecting points of equal value, such as elevation, formation thickness, and rock porosity.

contract *n*: a written agreement that can be enforced by law and that lists the terms under which the acts required are to be performed. A drilling contract covers such factors as the cost of drilling the well (whether by the foot or by the day), the distribution of expenses between operator and contractor, and the type of equipment to be used.

contract carrier *n*: a facility that voluntarily provides its services to others on a private contractual basis.

contract demand *n*: the amount of gas a seller agrees to deliver on a periodic (daily, monthly, annually) basis in accordance with a service agreement. The buyer need not take this maximum quantity during the applicable period.

contract depth *n*: the depth of the wellbore at which a drilling contract is fulfilled.

contracted reserves *n pl*: natural gas reserves dedicated to the fulfillment of gas purchase contracts.

contract market *n*: oldest system for trading oil. Suppliers offer long-term contracts to customers with the contracts specifying all terms except price, which is fixed at the time of sale.

contract maximum quantity *n*: the maximum quantity of gas the seller is required to make available to the purchaser during a specified period under terms of a gas sales contract.

contract minimum quantity *n*: the minimum quantity of gas the seller is required to make available to the purchaser during a specified period under terms of a gas sales contract.

contractor *n*: see *drilling contractor*.

contract volume *n*: quantity of gas based on measurement conditions and procedures specified in a gas sales contract.

contrail *n*: short for condensation trail. See *condensation trail*.

control *n*: 1. manual or automatic regulation of a process. A means or device to direct and regulate a process or sequence of events. 2. the section of a digital computer that carries out instructions in proper sequence, interprets each coded instruction, and applies the proper signals to the arithmetic unit and other parts in accordance with this interpretation. 3. a mathematical check used in some computer operations. 4. a test made to determine the extent of error in experimental observations or measurements. 5. a procedure carried out to give a standard of comparison in an experiment.

control agent *n*: in instrumentation, a device, equipment, or fluid used to bring about corrective action to achieve the desired values of level, temperature, or speed.

control board *n*: a panel on which are grouped various control devices such as switches and levers, along with indicating instruments.

control chart *n*: a chart of successive meter factors (or relative meter errors) generally plotted as a function of time. Used to evaluate meter stability and to determine when meter performance has departed from its normal range.

control head *n*: a special blowout preventer used in snubbing.

controlled aggregation *n*: a condition in which clay platelets remain stacked by a polyvalent cation, such as calcium, and are deflocculated by use of a thinner.

controlled directional drilling *n*: see *directional drilling*.

controlled variable (CV) *n*: a process variable, such as temperature, speed, pressure, and level, which is regulated by varying other processes or control agents.

controller *n*: an electric device used for governing the power that goes to an apparatus to which it is connected.

controller pressure *n*: pressure used to control devices in a process; for example,

a current-to-pressure transducer's output pressure controls the position of a proportional valve.

control line *n*: a small hydraulic line used to run fluid from the surface to a downhole tool, such as a subsurface safety valve.

controlling means *n*: elements of a controller that produce a corrective action. Information is sent back from the process, compared to the desired set point, and corrective action of the controlled variable is taken if deviation exists.

control panel *n*: in various types of systems, the switches and devices to start, stop, measure, monitor, or signal what is taking place within the system.

control pod *n*: see *hydraulic control pod*.

control rack *n*: on a diesel engine, the rack-and-pinion gear that allows the engine operator to regulate the amount of fuel the injector forces into the combustion chamber of the engine. See *rack-and-pinion gear*.

control sleeve *n*: in a diesel engine's mechanical governor, a device that transfers the force of a spring and the centrifugal force of flyweights to a control lever that speeds up or slows down the engine. Usually, flyweight force tends to slow down the engine while spring force tends to speed up the engine.

control valve *n*: a valve designed to regulate the flow or pressure of a fluid.

control voltage *n*: in a diesel-electric system using alternators (AC generators), low powered electrical current that flows to a silicon controlled rectifier (SCR). When control voltage is applied to the SCR, the current generated by the AC generators flows through the SCR in one direction, thus converting alternating current to direct current. When control voltage is turned off, no current flows through the SCR.

control well *n*: a well previously drilled in an area of drilling interest, the data from which may be a reliable source of information in the planning of a new well.

convective cloud *n*: a cloud that forms when large volumes of heat in the atmosphere move upward; this upward movement of heat causes water vapor to condense as the heat collides with cool moist air. A thunderhead is a convective cloud.

conventional chart *n*: a circular paper or lightweight plastic chart on which circular graduations are printed and on which a pen or stylus draws a recording indicating the quantities being measured. Conventional charts have a uniform scale in that minor graduations are spaced an equal distance between major graduations.

conventional completion *n*: a method for completing a well in which tubing is set

inside 4.5-inch (1.4-centimetre) or larger casing. Compare *miniaturized completion*.

conventional coupling *n*: piping configuration that uses ten diameters of straight full pipe upstream of a meter and straightening vanes; the downstream contains five or more full pipe diameters.

conventional electric log *n*: an electric log in which current flow in the logging tool was dispersed through the mud in the wellbore prior to entering the formation. As a result, formation resistivity measurements were often inaccurate. These logs are no longer used.

conventional gas-lift mandrel *n*: see *gas-lift mandrel*.

conventional gravel pack *n*: a type of gravel pack in which the well's production packer is removed and a service packer is run in with the gravel pack assembly. After packing, the service tool is retrieved and the production packer rerun.

conventional MODU moorings *n pl*: moorings that are stored on the MODU and are deployed and recovered with the MODU on site.

conventional mud *n*: a drilling fluid containing essentially clay and water; no special or expensive chemicals or conditioners are added.

conventional pollutants *n pl*: those air pollutants that fall under NAAQS: particulates, ozone, carbon monoxide, nitrogen dioxide, lead, and sulfur dioxide. Sometimes called criteria pollutants.

conventional pump *n*: see *fixed pump*.

conventional river crossing *n*: a waterway crossing in which pipeline construction techniques similar to those on land are used. In a conventional crossing, the route is graded and ditched; the pipe is welded, tested, and pulled across the stream; and then the pipe is tied in on each side to the cross-country line.

conventional rotating system *n*: the rotary drilling system that uses a conventional swivel, a kelly, a kelly drive bushing, a master bushing, and a rotary table to turn the drill stem and bit. Compare *top drive*.

conventional system *n*: a system meeting industry standards.

conventional tank *n*: a tank of a shape commonly used in the petroleum industry. It is not constructed to withstand any appreciable pressure or vacuum in the vapor space; it may, therefore, be gauged directly through an open hatch.

convergence pressure *n*: the pressure at a given temperature for a hydrocarbon system of fixed composition at which the vapor-liquid equilibrium values of the various components in the system become or tend to become unified. The convergence pressure is used to adjust vapor-liquid equilibrium values to the particular system under consideration.

conversion *n*: the change in the chemistry of a mud from one type to another. Reasons for making a conversion may be (1) to maintain a stable wellbore, (2) to provide a mud that will tolerate higher weight, or density, (3) to drill soluble formations, or (4) to protect producing zones. Also called a breakover.

converter *n*: a device for changing alternating current to direct current or vice versa; usually, if the device changes AC to DC, it is called a rectifier.

convey *v*: to transfer title to property from one party to another, usually by means of a written instrument.

conveyance *n*: the granting of interest in the petroleum to a person or company for the purposes of exploring, drilling, and producing.

coolant *n*: a cooling agent, usually a fluid, such as the liquid applied to the edge of a cutting tool to carry off frictional heat or a circulating fluid for cooling an engine.

coolant pump *n*: see *water pump.*

coolant temperature gauge *n*: a device on an engine, usually of the dial-and-needle type, that indicates the temperature of the engine's coolant at the point where the gauge's sensor is installed in the coolant system. Often, gauge sensors are installed to indicate coolant temperature at different places in the system; for example, where the coolant enters the engine's radiator or heat exchanger and where the coolant exits the engine after flowing through the cooling system.

cooler *n*: a heat exchanger that reduces the temperature of a fluid by transferring the heat to a nonprocess medium.

cooling tower *n*: a structure in which air contact is used to cool a stream of water that has been heated by circulation through a system. The air flows countercurrently or crosscurrently to the water.

cooling water *n*: treated fresh water that circulates inside a diesel engine to transfer heat.

cooperative *n*: an organization or company that collects money or resources from several companies into a common-use pool, which can be utilized by any cooperative member if the need arises. One of the largest private oil spill response cooperatives is the Marine Spill Response Corporation (MSRC), formed by a group of U.S. oil companies.

Coordinating Research Council, Inc. (CRC) *n*: a nonprofit organization supported jointly by the American Petroleum Institute and the Society of Automotive Engineers, Inc. It administers work of the CFR and other committees that correlate test work and other studies on fuels, lubricants, engines, and engine equipment. Address: 3650 Mansell Rd., #140; Alpharetta, GA 30022; 678-795-0506.

co-owners *n pl*: see *cotenants.*

copolymer *n*: a substance formed when two or more substances polymerize at the same time to yield a product which is not a mixture of separate polymers but a complex having properties different from either polymer alone. See *polymer.* Examples are polyvinyl acetate-maleic anhydride copolymer (clay extender and selective flocculant), acrylamide-carboxylic and copolymer (total flocculant).

copper *n*: a ductile, malleable, reddish-brown metallic element that is an excellent conductor of heat and electricity and is widely used for electrical wiring, water piping, and corrosion-resistant parts, either pure or in alloys such as brass and bronze. Copper's chemical symbol is Cu.

copper loss *n*: see I^2R *loss.*

copper strip test *n*: a test using a small strip of pure copper to determine qualitatively the corrosivity of a product.

copper sulfate electrode *n*: a commonly used nonpolarizing electrode used in corrosion control to measure the electrical potential of a metal structure to a surrounding electrolyte to determine the potential for corrosion damage or to monitor the effectiveness of existing control measures. See *half-cell.*

cord *n*: a small-diameter wire rope or strand.

cordage *n*: all of the rope on a ship or an offshore drilling rig.

cordonazo *n*: a tropical cyclone in Mexico.

core *n*: 1. a cylindrical sample taken from a formation for geological analysis. Usually a conventional core barrel is substituted for the bit and procures a sample as it penetrates the formation. 2. the metallic, partly solid, and partly molten interior of the earth, about 4,400 miles (7,084 kilometres) in diameter. 3. the central, axial member of a wire rope around which the strands are laid. 4. in electricity, the interior and central part of a coil; it may be air or a solid material such as iron. See *solenoid.* *v*: to obtain a solid, cylindrical formation sample for analysis. See *radiator core.*

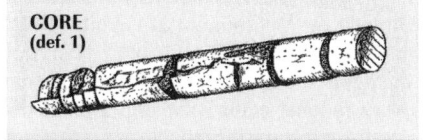

CORE
(def. 1)

core analysis *n*: laboratory analysis of a core sample to determine porosity, permeability, lithology, fluid content, angle of dip, geological age, and probable productivity of the formation.

core barrel *n*: a tubular device, usually from 10 to 60 feet (3 to 18 metres) long, run at the bottom of the drill pipe in place of a bit and used to cut a core sample.

core bit *n*: a bit that does not drill out the center portion of the hole, but allows this center portion (the core) to pass through the round opening in the center of the bit and into the core barrel.

core catcher *n*: the part of the core barrel that holds the formation sample.

core customers *n pl*: see *captive customers.*

core cutterhead *n*: the cutting element of the core barrel assembly. In design it corresponds to one of the three main types of bits: drag bits with blades for cutting soft formations; roller bits with rotating cutters for cutting medium-hard formations; and diamond bits for cutting very hard formations.

core-drill *v*: to obtain a core from the bottom of an existing well by using a core bit. See *core* (defn. 1), *core bit.*

core loss *n*: in a transformer, the reduction (loss) in electrical energy that occurs in the core of the transformer's windings. See *core* (defn. 4).

core marker *n*: a metal device that is placed in the inner core barrel before coring. When all of the core has been removed from the core barrel, the core marker falls out to indicate that the barrel is empty. Also called a rabbit.

corer *n*: a tool used to obtain cylindrical samples of rock and other materials from the wellbore.

core sample *n*: 1. a small portion of a formation obtained by using a core bit in an existing wellbore. See *core bit.* 2. a spot sample of the contents of an oil or oil product storage tank usually obtained with a thief, or core sampler, at a given height in the tank.

core thief *n*: see *thief.*

core-type junk basket *n*: a device made up on a fishing string that drills a formation core surrounding and containing the fish to be retrieved. It has two sets of catchers: one to break off the drilled core, and another to form a basket below the core and the fish.

coring *n*: the process of cutting a vertical, cylindrical sample of the formations encountered as an oilwell is drilled. The purpose of coring is to obtain rock samples, or cores, in such a manner that the rock retains the same properties that it had before it was removed from the formation.

coring reel *n*: see *sand reel*.

Coriolis force *n*: an apparent force caused by the rotation of the earth about a vertical axis. This force causes wind or water to move clockwise in the Northern Hemisphere and counterclockwise in the Southern Hemisphere.

corkscrew *n*: the buckling of tubing in a large-diameter pipe or casing.

Corod *n*: a trade name for a special form of sucker rod. Corod, or continuous rod, normally has no joints between the downhole pump and the surface. It is installed into the well by unwinding it from a reel. See *sucker rod*.

corona effect *n*: a phenomenon that occurs when high voltage is transmitted through uninsulated conductors. At very high voltages, the voltage ionizes the surrounding atmosphere and the atmosphere also begins to conduct electricity. See *ionization*.

correlate *v*: to relate subsurface information obtained from one well to that of others so that the formations may be charted and their depths and thicknesses noted. Correlations are made by comparing electrical well logs, radioactivity logs, and cores from different wells.

correlative rights *n pl*: rights afforded the owner of each property in a pool to produce without waste his or her equitable share of the oil and gas in such pool.

corrodent *n*: a corrosion agent. For example, salt water (oilfield brine) is a corrodent.

corrosion *n*: any of a variety of complex chemical or electrochemical processes, e.g., rust, by which metal is destroyed through reaction with its environment.

corrosion cap *n*: a fitting placed on a subsea wellhead to protect the wellhead not only from corrosion, but also from contamination by debris and marine growth.

corrosion cell *n*: an area on a corrodable substance (usually metal) where corrosion occurs because electrical current is able to flow.

corrosion control *n*: the measures used to prevent or reduce the effects of corrosion. These practices can range from simply painting metal, to isolating it from moisture and chemicals and insulating it from galvanic currents, to cathodic protection, in which a galvanic or impressed direct electric current renders a pipeline cathodic, thus causing it to be a negative element in the circuit. The use of chemical inhibitors and closed systems are other examples of corrosion control.

corrosion-control agent *n*: a chemical added to drilling fluid to minimize corrosion to the drill stem.

corrosion coupon *n*: a metal strip inserted into a system to monitor corrosion rate and to indicate corrosion inhibitor effectiveness.

corrosion fatigue *n*: metal fatigue concentrated in corrosion pits. See *fatigue*.

corrosion inhibitor *n*: a chemical substance that minimizes or prevents corrosion in metal equipment.

corrosion-resisting steel *n*: a steel alloy that contains chromium and nickel and that does not corrode as quickly as normal steel alloys.

corrosion test *n*: one of a number of tests to determine qualitatively or quantitatively the corrosion-inducing compounds in a product.

corrosive *n*: a corrosion agent, e.g., acid.

corrosiveness *n*: the tendency to wear away a metal by chemical attack.

corrosive product *n*: a hydrocarbon product that contains corrosion-inducing compounds in excess of the specification limits for a sweet product.

corrosivity *n*: the quality of being corrosive.

cos *abbr*: cosine.

cosec *abbr*: cosecant.

cosecant (cosec) *n*: in a right triangle, the ratio of the hypotenuse to the side opposite a given angle.

cosine (cos) *n*: in a right triangle, the ratio of the length of the side adjacent to an acute angle to the length of the hypotenuse.

cosurfactant *n*: a surfactant, generally an alcohol, added to a micellar solution to adjust the viscosity of the solution, maintain its stability, and prevent adsorption of the main surfactant by reservoir rock.

COT *abbr*: change of trim.

cotan *abbr*: cotangent.

cotangent *n*: in a right triangle, the ratio of the side adjacent to a given angle to the side opposite.

cotenants *n pl*: persons who hold possessory interests, from title or a lease, in the same piece of land. Also called co-owners or tenants in common.

cotton core *n*: see *fiber core*.

coul *abbr*: coulomb.

coulomb (C, coul) *n*: the amount of electric charge that crosses a surface in 1 second when a steady current of 1 ampere is flowing across the surface.

Council on Environmental Quality (CEQ) *n*: a council established under NEPA as the agency responsible for ensuring that other federal agencies comply with the provisions stated in NEPA.

counterbalance *n*: a force or influence equally counteracting another. For example, drilling fluids of the correct height and weight, or density, should exert enough pressure to counterbalance formation pressure. *v*: to put in balance; to balance, equalize, stabilize, or steady.

counterbalance effect *n*: the effect of counterweights on a beam pumping unit. The approximate ideal counterbalance effect is equal to half the weight of the fluid plus the buoyant weight of the rods.

counterbalance system *n*: see *two-step grooving system*.

counterbalance weight *n*: a weight applied to compensate for existing weight or force. On pumping units in oil production, counterweights are used to offset the weight of the column of sucker rods and fluid on the upstroke of the pump, and the weight of the rods on the downstroke.

counterbalancing *n*: the placing of a weight on a mechanism to restore it to equilibrium, or balance.

counterbore *n*: flat-bottomed enlargement of the mouth of a cylindrical bore. *v*: to enlarge part of a hole by means of a counterbore.

countercurrent *n*: a current that flows adjacent to, but in the opposite direction of, an accompanying current.

counter electromotive force *n*: see *back electromotive force*.

counter emf *n*: see *back electromotive force*.

countershaft *n*: a shaft that gets its movement from a main shaft and transmits it to a working part.

countervoltage *n*: see *back electromotive force*.

counterweight *n*: see *counterbalance weight*.

counting rate meter *n*: an instrument that indicates the time rate of occurrence of input pulses to a radiation counter, averaged over a time interval. Also known as rate meter.

couplant *n*: that which couples.

couple *n*: in mechanics, a pair of forces of equal magnitude acting in parallel but opposite directions, capable of causing rotation but not translation. A couple can therefore cause a body to rotate but not to move in a linear fashion. *v*: to join, connect with a coupling.

coupled design *n*: uses couplings to join two pin-threaded pipe ends.

coupling *n*: 1. in piping, a metal collar with internal threads used to join two sections of threaded pipe. 2. in power transmission, a connection extending longitudinally between a driving shaft and a driven shaft. Most such couplings are flexible and compensate for minor misalignment of the two shafts. 3. in electronics, the relation between two circuits that permits energy transfer from one to the other, usually through a wire, resistor, transformer, capacitor, or other device.

courtesy rights *n*: the rights of a husband to a life interest in all of his wife's inheritable lands. These rights come into effect on her death, provided the couple have children capable of inheriting. Effective in some states.

covalent bond *n*: see *electron-pair bond*.

covalent reaction *n*: a method of forming compounds in which atoms share electrons, thus forming nonpolar, or covalent, unions.

coverall clause *n*: see *Mother Hubbard clause*.

cover depth *n*: the measurement from the top of a pipeline to the top of the soil used to cover the pipeline along a right-of-way. Ditch depth and cover requirements are regulated by the U.S. Department of Transportation.

cover requirements *n pl*: requirements, usually set by a regulatory agency, concerning the characteristics and manner of covering a pipeline.

CP *abbr*: casing pressure or casing point; used in drilling reports.

CPC *abbr*: computerized production control.

C_pl *abbr*: the correction for pressure on liquid; a factor used during a proving run to adjust for pressure on the liquid in the prover and pipe.

C_ps *abbr*: the correction for pressure on steel; a factor used during a proving run to adjust for pressure on the steel of the prover and pipe.

CPU *abbr*: central processing unit.

cracked cone *n*: a bit cone that has been cracked open for any reason.

cracked gas *n*: hydrocarbon gases formed in the catalytic cracking process.

cracked naphtha *n*: the hydrocarbon liquid lighter than kerosene that can be recovered by catalytic or thermal cracking in a refinery.

cracking *n*: in paint technology, the formation of breaks in a paint film that expose the underlying surface. Types of cracking include irregular cracking, in which the breaks have no definite pattern; line cracking, in which the breaks are generally parallel; and sigmoid cracking, in which the breaks are in relatively large curves that come together or cross each other.

crane *n*: a machine for raising, lowering, and revolving heavy pieces of equipment, especially on offshore rigs and platforms.

crane operator *n*: usually, the roustabout foreman.

crank *n*: an arm keyed at right angles to a shaft and used for changing radius of rotation or changing reciprocating motion to circular motion or circular motion to reciprocating motion. On a beam pumping unit, the crank is connected by the pitman to the walking beam, thereby changing circular motion to reciprocating motion.

crank arm *n*: a steel member connected to each end of the shaft extending from each side of the speed reducer on a beam pumping unit.

crank-balanced pumping unit *n*: a beam pumping unit that has the counterbalance weight on the crank arm.

crankcase *n*: the housing that encloses the crankshaft of an engine.

crank pin *n*: a cylindrical projection on a crank that holds a connecting rod and is held by a bearing.

crankshaft *n*: a rotating shaft to which connecting rods are attached. It changes up and down (reciprocating) motion to circular (rotary) motion.

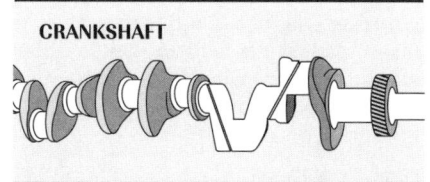

CRANKSHAFT

crank throw *n*: on a crankshaft, the highly polished and accurately machined portion of the crankshaft to which a piston rod is attached.

crank wash *n*: an off-line method of compressor cleaning in which the compressor is washed with water and detergent or solvents. Ineffective for removing baked-on contaminants.

crater *v*: (slang) to cave in; to fail. After a violent blowout, the force of the fluids escaping from the wellbore sometimes blows a large hole in the ground. In this case, the well is said to have cratered. Equipment craters when it fails.

crawler *n*: a self-propelled X-ray machine that rides inside pipe to examine welds for possible defects.

crazing *n*: in paint technology, the formation of tiny crisscross cracks on the surface of a paint film. Crazing resembles checking, but the cracks are deeper and broader and have a polygonal pattern.

CRC *abbr*: Coordinating Research Council, Inc.

crd *abbr*: cored; used in drilling reports.

creaming of emulsions *n*: the settling or rising of the particles of the dispersed phase of an emulsion. Identifiable by a difference in color shading of the layers formed. Creaming can be either upward or downward, depending on the relative densities of the continuous and dispersed phases.

created fracture *n*: fracture induced by means of hydraulic or mechanical pressure exerted on the formation.

crest *n*: the top of a wave.

crest length *n*: in oceanography, the distance between the stillwater level and the beginning and ending of a wave. See *stillwater level*.

Cretaceous *adj*: of or relating to the geologic period from about 135 million to 65 million years ago at the end of the Mesozoic era, or to the rocks formed during this period, including the extensive chalk deposits for which it was named.

crew *n*: 1. the workers on a drilling or workover rig, including the driller, the derrickhand, and the rotary helpers. 2. any group of oilfield workers.

crew chief *n*: the driller or head well puller in charge of operations on a well servicing rig that is used to pull sucker rods or tubing.

crg *abbr*: coring; used in drilling reports.

criminal penalties *n pl*: penalties (fines or imprisonment) for damages against society or government, typically determined by court cases brought against individuals or corporations.

crinkling *n*: the development of wrinkles in a paint film that occurs as the paint is drying.

criteria pollutants *n pl*: the six air pollutants listed under NAAQS in the CAA: ozone, carbon monoxide, sulfur dioxide, lead, nitrogen dioxide, and particulate matter. Sometimes called conventional pollutants.

critical density *n*: the density of a substance at the critical temperature and pressure.

critical flow *n*: the rate of flow of a fluid that is equivalent to the speed of sound in that fluid.

critical-flow prover *n*: a pipe-shaped device with a restriction, usually an orifice or nozzle, that is used to measure the velocity of gas flow during an open-flow test of a gas well.

critical point *n*: 1. the point at which, in terms of temperature and pressure, a fluid cannot be distinguished as being either a gas or a liquid, i.e., the point at which the physical properties of a liquid and a gas are identical. 2. one of the places along the length of drilling line at which strain is exerted as pipe is run into or pulled out of the hole.

critical pressure *n*: the pressure needed to condense a vapor at its critical temperature.

critical speed *n*: the speed reached by an engine or rotating system that corresponds to a resonant frequency of the engine or system. Often, in combination with power impulses, critical speed can cause damaging shock waves.

critical temperature *n*: the highest temperature at which a substance can be separated into two fluid phases—liquid and vapor. Above the critical temperature, a gas cannot be liquefied by pressure alone.

critical value *n*: see *critical point*, defn. 1.

critical velocity *n*: the velocity at the transitional point between laminar and turbulent fluid flow. See *laminar flow*, *turbulent flow*.

critical volume *n*: the specific volume of gas at its critical temperature and pressure.

critical weight *n*: weight placed on the bit that results in tension on the drill string, which causes the drill string to vibrate at the rotary speed being used. A drill stem operating with critical weight and at the critical speed for that weight will have stresses develop that cause very rapid failure.

critical zone *n*: in tanks with floating roofs, the vertical range in which the gauging measurements are not accurate.

crooked hole *n*: a wellbore that has been unintentionally drilled in a direction other than vertical. It usually occurs where there is a section of alternating hard and soft strata steeply inclined from the horizontal.

crooked-hole country *n*: an area in which particular subsurface formations make it difficult to keep a drilled hole straight.

cross assignment *n*: when several producers, either voluntarily or by state regulation, pool acreages to form a "unit." They may cross-assign their leases to one another, creating a common obligation to each royalty owner.

cross bar *n*: on offshore installations with grated decks or floors, a connecting bar, made from strips of steel, sheet steel, or rolled bars, or from rolled or extruded aluminum, which extends across the load bearing supports (bars), usually perpendicular to the supports. A cross bar may be bent into a corrugated or sinuous pattern and, where it intersects a load bearing bar, is welded, forged, or mechanically locked to the bar.

cross-bedding *n*: sedimentation in which laminations are transverse to the main stratification planes.

crosshead *n*: the block in a mud pump that is guided to move in a straight line and serves as a connection between the pony rod and the connecting rod.

crosshead extension rod *n*: the rod connecting the crosshead and the piston rod in a mud pump. Also called pony rod.

crosslinking *n*: a process of molecular bridging of polymers with other chemical substances that alters viscosity and shear rates to enhance lifting of bit cuttings and increase drilling rates.

cross loading *n*: a phenomenon that can occur on a semisubmersible rig when adding ballast (seawater) to the rig pontoons. If significantly more seawater is added to one pontoon than to the other, the rig can experience excessive tilt (heel). Excessive heel can cause structural damage to the rig.

cross-multiply *n*: in mathematics, to multiply the numerator of each of two fractions by the denominator of the other to eliminate the fractions from the equation. For example, in the two fractions $\frac{2}{3} \times \frac{1}{2}$, the 2 in the numerator of the first fraction and the 2 in the denominator of the second fraction are cross-multiplied to cancel each other.

crossover *n*: the place at each end of the drawworks drum where the drilling line being spooled onto or off of the drum stops moving in one direction and starts moving in the other direction. Also called turnback. See *floating harness*.

crossover joint *n*: a length of casing with one thread on the field end and a different thread in the coupling, used to make a changeover from one thread to another in a string of casing.

crossover packer *n*: a type of packer developed for a dual-completion well in which there are both an oil and a gas zone, with the gas zone on the bottom.

crossover section *n*: see *angle-control section*.

crossover sub *n*: a sub that allows different sizes and types of drill pipe to be joined.

crossover valve *n*: a valve employed on producing wells. During workover operations, when a riser or a flow line on the well requires flushing, or when the well is being pressure tested, the crossover valve is opened to allow communication to occur between the production tubing and the annulus between the tubing and casing. Such communication allows fluids to be pumped and removed within the well as required. While the well is producing, the valve is usually closed.

cross-pad flow *n*: see *feeder-collector*.

cross sea *n*: the irregular wave patterns produced when different wave systems cross each other at an angle.

cross section *n*: 1. the property of atomic nuclei of having the probability of collision with a neutron. The nucleus of a lighter element is more likely to collide with a neutron than is the nucleus of a heavier element. Cross section varies with the elements and with the energy of the neutron. 2. a geological or geophysical profile of a vertical section of the earth.

cross-sectional area *n*: the area of an object, such as a pipe, when the object is observed transversally (end on) from one outside edge to the other. Cross-sectional area is usually expressed in square units, such as square inches (in.²).

cross-thread *v*: to screw together two threaded pieces when the threads of the pieces have not been aligned properly.

crown *n*: 1. the crown block or top of a derrick or mast. 2. the top of a piston. 3. a high spot formed on a tool joint shoulder as the result of wobble.

crown block *n*: an assembly of sheaves mounted on beams at the top of the derrick or mast and over which the drilling line is reeved. See *block, reeve the line, sheave*.

crown frame flanges *n pl*: projections on the frame to which the crown block is attached.

CROWN BLOCK

Crown-O-Matic *n*: see *crown saver*.

crown platform *n*: the working platform at the top of the derrick or mast that permits access to the sheaves of the crown block and provides a safe working area for service to the gin pole. See *crown block*.

crown saver *n*: a device mounted near the drawworks drum to keep the driller from inadvertently raising the traveling block into the crown block. A probe senses when too much line has been pulled onto the drum, indicating that the traveling block is about to strike the crown. The probe activates a switch that simultaneously disconnects the drawworks from its power source and engages the drawworks brake.

crown walkaround *n*: the structure at the top of the drilling derrick or mast that supports the crown block. Also called water table.

crow's foot *n*: on a diamond bit, the built-in channel in the bit's nose that helps conduct drilling fluid exiting the center of the bit to additional channels that conduct the fluid over the entire nose.

crow's nest *n*: an elevated walkway where employees work (as on the top of a derrick or a refinery tower).

CRT *abbr*: cathode-ray tube.

crude oil *n*: unrefined liquid petroleum. It ranges in gravity from 9°API to 55°API and in color from yellow to black, and may have a paraffin, asphalt, or mixed base. If a crude oil, or crude, contains a sizable amount of sulfur or sulfur compounds, it is called a sour crude; if it has little or no sulfur, it is called a sweet crude. In addition, crude oils may be referred to as heavy or light according to API gravity, the lighter oils having the higher gravities.

crude oil average domestic first purchase price *n*: the average price at which all domestic crude oil is purchased. Before February 1976, the price represented an estimate of the average of posted prices; since then, it represents an average of actual first purchase prices. Frequently called wellhead price.

crude oil refinery input *n*: total crude oil (including lease condensate) input to crude oil distillation units and other processing units.

crude oil stocks *n*: stocks of crude oil and lease condensate held at refineries, in pipelines, at pipeline terminals, and on leases.

crust *n*: the outer layer of the earth, varying in thickness from 5 to 30 miles (10 to 50 kilometres). It is composed chiefly of oxygen, silicon, and aluminum.

cryogenic nitrogen *n*: nitrogen that is compressed at a very low temperature to liquefy it.

cryogenic plant *n*: a gas-processing plant that is capable of producing natural gas liquid products, including ethane, at very low operating temperatures.

cryogenics *n*: the study of the effects of very low temperatures.

crystal *n*: a substance formed by the solidification of a chemical element, a compound, or a mixture that has a regularly repeating internal arrangement of its atoms. For example, quartz and silicon are crystals.

crystalline texture *n*: rock texture that is the result of progressive and simultaneous interlocking growth of mineral crystals. Compare *clastic texture.*

crystallization *n*: the formation of crystals from solutions or melts.

Cs *abbr*: cirrostratus.

C$_{tl}$ *abbr*: the correction for temperature of liquid; a factor used during proving to convert the temperature of a liquid to a standard temperature.

C$_{ts}$ *abbr*: the correction for temperature of steel; a factor used during proving to adjust the temperature of the prover steel to a standard temperature.

cu *abbr*: cubic.

Cu *abbr*: cumulus.

cubic centimetre (cm^3) *n*: a commonly used unit of volume measurement in the metric system equal to 10^{-6} cubic metre, or 1 millilitre. The volume of a cube whose edge is 1 centimetre.

cubic foot (ft^3) *n*: the volume of a cube, all edges of which measure 1 foot. Natural gas in the United States is usually measured in cubic feet, with the most common standard cubic foot being measured at 60°F and 14.65 pounds per square inch absolute, although base conditions vary from state to state.

cubic metre (m^3) *n*: a unit of volume measurement in the SI metric system, replacing the previous standard unit known as the barrel, which was equivalent to 35 imperial gallons or 42 U.S. gallons. The cubic metre equals approximately 6.2898 barrels.

cumulonimbus *n*: a cloud formed from a cumulus cloud that has reached great vertical development. This type of cloud has a top composed of ice crystals and a bottom composed of water droplets. Also called a thunderhead.

cumulus *n*: a white, puffy cloud with a dark base. Its shape is constantly changing. Prominent in the summer months, cumulus clouds generally cover only 25% of the sky.

cup anemometer *n*: a type of wind-speed measuring device that consists of three or more hemispherical cups extended on horizontal arms from a vertical shaft. The wind rotates the cups, and the rate at which the cups turn is transmitted to instruments that translate this information to wind speed.

cup case thermometer *n*: a holder for a mercury-in-glass thermometer that incorporates a small metal container into which the bulb of the thermometer is inserted and that serves to hold a small sample of liquid the temperature of which is being measured. The liquid in the cup keeps the bulb submerged in liquid until the temperature is recorded.

cup packer *n*: a device made up in the drill stem and lowered into the well to allow the casing and blowout preventers to be pressure-tested. The sealing device is cup-shaped and is therefore called a cup.

cup test *n*: see *packer test.*

cup-type elements *n pl*: rubber seals that are energized by pressure and not by mechanical force.

curb angle *n*: on offshore installations that have grated decks and floors, a member fixed to concrete or supporting steelwork at the perimeter of a flooring area.

cure *v*: to age cement under specified conditions of temperature and pressure.

cure a title *v*: to remedy defects and omissions that, in the opinion of the examining attorney, could make the present owner's claim to property questionable. To cure a title, a title examiner may require additional facts not evident in the material

examined. The curative material is usually obtained in recordable form.

curing mechanism *n*: in paint technology, the means by which the coating goes from a liquid to a dry film.

current (I) *n*: 1. the flow of electric charge or the rate of such flow, measured in amperes. 2. the predominantly horizontal movement of ocean waters.

current gain *n*: that part of the current flowing into the emitter of a transistor that flows through the base region and out the collector.

current meter *n*: an instrument that records an ocean current's speed and direction (along with temperature, salinity, pressure, and other variables). The current meter is moored in position by an anchor or weight on the ocean bottom that is connected to a float at the surface.

current surge rating *n*: a measure of a solid-state diode's ability to operate properly when the circuit in which the diode is installed experiences momentary and intermittent surges in current flowing in the circuit.

current transmitter *n*: electronic circuitry and components that receive and process a signal created by a process variable from the primary element in a process system. After the transmitter processes the signal, it produces a signal output that is usually measured in milliamps (mA). Within the transmitter's circuitry are devices called operational amplifiers (op amps) that amplify, compare, add, subtract, integrate or sum, and so on.

curtailment *n*: reduction in service or purchases below contracted-for levels. Curtailment of gas sales service is a method of balancing a utility's natural gas requirements with its natural gas supply. There is usually a hierarchy of customers for the curtailment plan based on priority of usage according to established regulatory standards of priority. A customer may be required to cut back partially or totally to eliminate this take of gas, depending on the severity of the shortfall between gas supply and demand and the customer's position in the hierarchy. From the customer's standpoint, curtailment may also mean a reduction by the customer of its takes of gas from its supplier. Curtailment of a transportation service occurs when demands for service exceed the capacity of the pipeline. Capacity curtailment is based on contract rather than end-use priority. All interruptible service must be entirely curtailed before any firm service is curtailed. Currently, interruptible transportation service is curtailed based on position in the first-come, first-served queue. Firm transportation service is usually curtailed on a pro rata basis.

cushion *n*: a quantity of water, drilling fluid, or compressed gas placed inside drill pipe or tubing to control both annular and formation pressures. Usually mud in the pipe or tubing supports pressure and prevents the pipe or tubing from collapsing; mud also holds back formation pressure pushing up the pipe or tubing. But sometimes, such as in drill stem testing, it is necessary to have empty pipe or tubing downhole. The cushion protects the pipe or tubing until it is in place. It also allows control of the rising formation pressure as the cushion is removed to prevent formation damage.

cushion gas *n*: see *blanket gas*.

custodian *n*: also called a lease operator or pumper. See *pumper*.

custody transfer *n*: the changing of the ownership of or the responsibility for quantities of gas, petroleum, or petroleum products. See also *lease automatic custody transfer*.

cut *n*: 1. portion or fraction of hydrocarbons that have been separated according to boiling point or gravity. 2. the line of demarcation on the measuring scale made by the material being measured.

cut and fill *v*: to cut down high ground or fill in low ground to achieve a uniform grade for a pipeline.

cut-and-fill boundary *n*: the limits to which a crew may cut and fill when laying a pipeline.

cut drilling fluid *n*: a drilling mud whose density has been decreased by the entrainment of formation fluids or air.

cut fluorescence test *n*: a test involving the observation of a formation sample immersed in a solvent and under ultraviolet light. If any hydrocarbons, which fluoresce under ultraviolet light, are in the sample, they will dissolve and appear as streamers or streaks of color different from the solvent.

cutoff valve *n*: a special valve on an engine that, when activated, blocks the flow of fuel to the engine to make it stop running.

cut oil *n*: an oil that contains water, usually in the form of an emulsion. Also called wet oil.

cutout *n*: an area of deck grating on an offshore installation removed to clear an obstruction or to permit pipes, ducts, columns, and the like to pass through the grating.

cutterhead *n*: in pipeline construction, the lead component in a directional drilling assembly. A circular steel band ringed with conical cutting teeth, the cutterhead does the actual boring of the hole for the pipeline under the waterway being crossed. Also called fly cutter.

cutters *n pl*: 1. on a bit used on a rotary rig, the elements on the end (and sometimes the sides) of the bit that scrapes, gouges, or otherwise removes the formation to make hole. 2. the parts of a reamer that actually contact the wall of the hole and open it to full gauge. A three-point reamer has three cutters; a six-point reamer has six cutters.

cutting in *n*: an action of wire rope during loose drum spooling in which a layer of rope spreads apart so that the next layer travels in the groove produced. Crushing, flattening, or distorting of the rope results.

cuttings *n pl*: the fragments of rock dislodged by the bit and brought to the surface in the drilling mud. Washed and dried cuttings samples are analyzed by geologists to obtain information about the formations drilled.

cuttings-sample log *n*: a record of hydrocarbon content in cuttings gathered at the shale shaker; usually recorded on the mud log.

CWA *abbr*: Clean Water Act.

cycle *n*: 1. the number of strokes a piston makes from one intake stroke to another intake stroke. Diesel engines may have either two strokes or four strokes per cycle. 2. in an alternating current generator (an alternator), one positive and one negative alternation constitutes 1 cycle. As a generator's coil rotates through a magnetic flux to induce voltage in the coil, the magnetic flux induces positive voltage as it rotates from 0 to 180 degrees. It also induces negative voltage as the coil rotates from 180 to 360 degrees. Each positive and negative voltage is one alternation and the two constitute a cycle. See *alternation*.

cycle condensate *n*: condensate produced from cycle gas.

cycle counter *n*: see *frequency meter*.

cycle gas *n*: gas that is compressed and returned to the gas reservoir to minimize the decline of reservoir pressure.

cycles per second *n*: see *hertz (Hz)*.

cycle time *n*: the time it takes drilling mud to go down the drill stem, out of the bit, and up the annulus.

cyclic compound *n*: a compound that contains a ring of atoms.

cyclic steam injection *n*: the injection of steam into the rock surrounding a production well to lower the viscosity of heavy oil and increase its flow into the wellbore. Steam injection may be followed by immediate production or by closing the well (called the soak phase) to allow even heat distribution before production is begun. The cycle of injection, soak, and production is repeated as long as the oil yield is profitable. Also called steam soak and huff 'n' puff.

cyclic stressing *n*: stress that occurs on a pipe, vessel, or machine in cycles, such as the sucker rod string.

cycling *n*: the process by which effluent gas from a gas reservoir is passed through a gas-processing plant or separation system and the remaining residue gas returned to the reservoir. The word *recycling* is sometimes used for this function, but it is not the preferred term.

cycling plant *n*: a plant that cycles residue gas back into the reservoir.

cyclogenesis *n*: the development of a low-pressure system, or cyclone.

cyclohexane *n*: a hydrocarbon often found in natural gasoline, C_6H_{12}.

cyclone *n*: 1. a low-pressure area, around which wind flow is counterclockwise in the Northern Hemisphere and clockwise in the Southern Hemisphere. The term is sometimes used to describe storms occurring in the atmosphere; in the Indian Ocean it is used to designate a tropical cyclone. 2. a device for the separation of various particles from a drilling fluid, most commonly used as a desander. The fluid is pumped tangentially into a cone, and the fluid rotation provides enough centrifugal force to separate particles by mass weight.

cyclonic *adj*: in meteorology, of or pertaining to the clockwise rotation of winds in the Northern Hemisphere or the counterclockwise rotation of winds in the Southern Hemisphere.

cyclonic storm *n*: a hurricane or a typhoon. See *hurricane*.

cyclonic wind *n*: the wind associated with a low-pressure area.

cyclonite *n*: a powerful, highly explosive material (cyclo-trimethylene-trinitramine) used as the main charge in jet perforating guns. Also called RDX. See *jet-perforate*.

cycloparaffin *n*: a saturated nonaromatic hydrocarbon compound with ring-shaped molecules, of the general chemical formula C_nH_{2n}. Also called naphthene.

cylinder *n*: 1. the unit of an internal-combustion engine in which combustion and compression take place. 2. a chamber in a pump from which the piston expels fluid.

cylinder block *n*: a housing that has one or more cylinders in it. See *engine block*.

cylinder gas *n*: liquefied petroleum gas, oxygen, acetylene, or any other gas that is compressed and confined in a pressure cylinder.

cylinder head *n*: the device used to seal the top of a cylinder. In modern drilling rig engines, it also houses the valves and has exhaust passages. In four-cycle operation, the cylinder head also has intake passages. Also called head.

cylinder liner *n*: a removable, replaceable sleeve that fits into a cylinder. When the sliding of the piston and rings wears out the liner, it can be replaced without the block's having to be replaced.

CZM *abbr*: coastal zone management.

D&A *abbr*: dry and abandoned; used in drilling reports.

daily contract quantity (DCQ) *n*: that daily quantity of gas, formulated on some basis (usually a combination of reserves and delivery capacity or delivery capacity only), that the pipeline company will attempt to take; may or may not equal the take-or-pay quantity.

daily drilling report *n*: a record made each day of the operations on a working drilling rig and, traditionally, phoned or radioed in to the office of the drilling company every morning. Also called morning report. See *driller's report*.

Dall tube *n*: a tube-shaped device that creates a pressure drop in a fluid flow system. Flow volume can be inferred from the pressure drop. Similar to a venturi, a Dall tube has a short uniform section that is followed by an abrupt shoulder. The shoulder occurs at the front of an inlet cone. A low-pressure tap at the inlet cone's throat leads to a groove that encircles the throat. Special equipment attached to the tap senses the pressure drop and records it. Named after Horace Dall, a British master optician, who invented the device in the 1950s.

Dalton's law *n*: the law that states that the pressure of a mixture of gases is equal to the sum of the partial pressures of the gases of which it is composed. Also called law of partial pressures.

damage clause *n*: the clause in an oil and gas lease that specifies that the lessee will be liable to the surface owner for damage to growing crops and other listed items.

damage ratio (DR) *n*: in reservoir engineering, a measure of the extent of the formation damage in proportion to the formation's original permeability. Damage ratio expresses the amount of formation damage in terms of a proportion, or ratio. See *formation damage*.

damage stability *n*: in ships or offshore rigs, the condition of a vessel's equilibrium (its stability) when the vessel has suffered enough damage to its hull or structure to allow water to enter (flood) the hull.

dampener *n*: an air- or inert gas-filled device that minimizes pressure surges in the output line of a mud pump. Sometimes called a surge dampener.

dampening *n*: the dissipation of energy in motion of any type, especially oscillatory motion, and the consequent reduction of decay of the motion.

damper *n*: a valve or plate used to regulate the flow of air or other gas.

damping sub *n*: see *Shock Sub*.

daN *sym*: decanewton.

D&P platform *n*: a drilling and production platform. Such an offshore platform is a large structure with room to drill and complete a number of wells.

darcy (*pl*, **darcys**) *n*: a unit of measure of permeability. A porous medium has a permeability of 1 darcy when differential pressure of 1 atmosphere across a sample 1 centimetre long and 1 square centimetre in cross section will force a liquid of 1 centipoise of viscosity through the sample at the rate of 1 cubic centimetre per second. The permeability of reservoir rocks is usually so low that it is measured in millidarcys.

Darcy's law *n*: a law stating that the rate of flow of a fluid through a rock varies directly with the amount of interconnected pore space (permeability) and the applied pressure and varies inversely with the viscosity, or flow resistance, of the fluid.

d'Arsonval movement *n*: an electronic assembly used in a meter that consists of a magnet, a moving coil, a jewel pivot bearing, and a pointer. The pointer sweeps across a dial that can be marked with numerical values expressing current, voltage, resistance, or whatever is appropriate to the application of the meter. It is named after one of its inventors, the French physiologist Jacque-Arsène d'Arsonval.

dart *n*: a device, similar to a ball, used to manipulate hydraulically operated downhole tools. See *ball*.

dart-type inside blowout preventer *n*: a drill pipe inside blowout preventer that is installed on top of the drill stem when the well is kicking through the drill stem. It is stabbed in the open position and then closed against the pressure. The valve that closes is dart-shaped, thus the name.

database *n*: a complete collection of information, such as contained on magnetic disks or in the memory of an electronic computer.

datum *n*: any numerical or geometrical quantity or quantities that serve as a reference or base for other quantities.

datum elevation *n*: a reference elevation, or height, used in mapping, usually sea level.

datum level *n*: a height, or elevation, usually sea level, from which altitudes are measured in surveys.

datum plate *n*: a level metal plate attached to the tank shell or bottom and located directly under the gauging reference point to provide a fixed contact surface from which liquid depth measurement can be made.

datum point *n*: the point to which all measurements for the calibration of a tank are related.

davit *n*: an A-frame that booms out 5 to 10 feet over the side of a barge. It has a cable or chain that is lowered to pick up pipeline. A diver attaches the pipeline to the line and it is reeled in on the davit.

daylight tour (pronounced "tower") *n*: in areas where three 8-hour tours are worked, the shift of duty on a drilling rig that starts at or about daylight. Compare *evening tour, graveyard tour*.

day rate *n*: an hourly or daily contract price the operator agrees to pay for use of rig, crew, and specified equipment. A day rate contract allows the operator to directly supervise the daily drilling operations.

day tank *n*: a fuel tank in the fuel supply system for a diesel engine between the main supply tank and the engine that holds a limited amount of fuel.

day tour (pronounced "tower") *n*: in areas where two 12-hour tours are worked, a period of 12 daylight hours worked by a drilling or workover crew when equipment is being run around the clock.

daywork *adj*: descriptive of work done on daywork rates.

daywork rates *n pl*: the basis for payment on drilling contracts when footage rates are suspended (as when the drilling rig is used in taking extra cores, logging, or other activities that delay actual drilling) or when the contract calls for the entire well to be drilled at daywork rates. In effect, daywork rates pay the drilling contractor by the hour rather than by the foot.

DC *abbr*: 1. direct current. 2. drill collar; used in drilling reports. 3. delivery capacity.

DC inductor *n*: see *choke*.

DCFR *abbr*: discounted cash flow rate of return.

DDC *abbr*: deck decompression chamber.

DEA *abbr*: drag embedment anchor.

dead band *n*: in a process control system, the difference in the input of a device for a fixed output when operated in ascending and descending directions. For example, a level switch on rising liquid may operate at one level (level *x*), but also operate at another level (level *y*) when the liquid level decreases. The difference between level *x* and *y* is the dead band. Also called dead zone.

dead end *n*: the end of a brake band that is anchored to the drawworks frame and does not move.

deadline *n*: the drilling line from the crown block sheave to the anchor, so called because it does not move. Compare *fastline*.

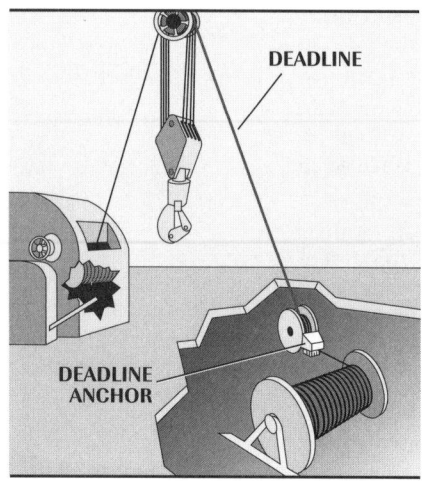

deadline anchor *n*: see *deadline tie-down anchor*.

deadline sheave *n*: the sheave on the crown block over which the deadline is reeved.

deadline tie-down anchor *n*: a device to which the deadline is attached, securely fastened to the mast or derrick substructure. Also called a deadline anchor.

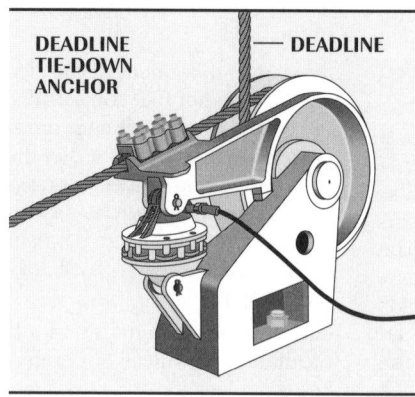

deadman *n*: 1. a buried anchor to which guy wires are tied to steady the derrick, mast, stacks, and so on. 2. an anchoring point against which the winch on a boring machine for pipelining can pull.

dead oil *n*: oil in which little or no gas is dissolved.

dead sheave (pronounced "shiv") *n*: the sheave on a crown block over which the deadline is reeved.

dead time *n*: 1. in a process control system, the time between a change in the input signal and the control system's response to the signal. 2. in a process, the time that elapses after a response to one signal or event during which a system is unable to respond to another. Also known as insensitive time.

deadweight *n*: see *variable load*.

deadweight ton (dwt) *n*: the total carrying capacity of a ship or offshore floating rig expressed in long tons (2,240 pounds); the ship or rig's deadweight is the displacement of the fully loaded ship or rig less the weight of the ship or rig itself.

dead well *n*: 1. a well that has ceased to produce oil or gas, either temporarily or permanently. 2. a well that has kicked and been killed.

deadwood *n*: any fitting, appurtenance, or structural member that affects the capacity of an oil storage tank. It is positive if it increases tank capacity and negative if it decreases capacity.

dead wraps *n*: the first of several wraps of wire rope around the drawworks drum that will never be played out (unspooled) as the rope moves off the drum.

dead zone *n*: see *dead band*.

deaerator *n*: a device used for removing air or other noncondensable gases from a process stream or steam condensate and boiler feed water.

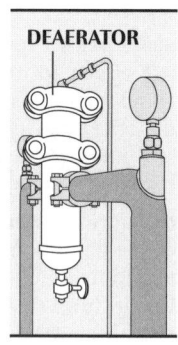

deals *n pl*: a program that may simulate buying and selling transactions.

Dean-Stark apparatus *n*: a device that uses a solvent to reflux, or flow back over, a sample to clean it. Such a device extracts oil or salt-laden water efficiently without harming the core samples.

deasphalting *n*: the removal of asphaltic substances that tend to form carbon deposits when lubricating oils are heated.

DEA unit *n*: a treating system using diethanolamine (DEA) for reduction of hydrogen sulfide, carbon dioxide, carbonyl sulfide, and other acid gases from sour process streams.

debug *v*: to detect, locate, and correct malfunctions in a computer, instrumentation, or other type of system.

deburr *v*: to remove small projections (burrs) from metal pieces or threads.

debutanized liquid *n*: hydrocarbon mixture remaining after the removal of butane and lighter hydrocarbons.

debutanizer *n*: a unit of equipment for separating butane, with or without lighter components, from a mixture of hydrocarbons and leaving a bottoms product that is essentially butane free.

decade *n*: a group or set of ten, as a period of ten years or a set of ten resistor values.

decane *n*: $C_{10}H_{22}$, a liquid hydrocarbon of the paraffin series.

decanewton (daN) *n*: ten times the newton, which is the unit of force in the metric system.

decanting centrifuge *n*: a type of mud centrifuge in which a cone-shaped steel bowl rotates very fast. The action of the rotating bowl creates centrifugal force, which causes the mud solids to separate from the liquid.

decimal number *n*: a base 10 numbering system that includes 10 numbers ranging from 0 to 9; numbers in a system wherein place values are read in powers of 10.

decimal point *n*: a dot written either on or slightly above the line that is used as the point at which place values change from positive to negative powers of 10 in the decimal numbering system.

decision tree *n*: a graphic representation of predicted financial gains or losses for the outcomes of several courses of action.

deck *n*: (nautical) floor.

deck decompression chamber *n*: a chamber in which excessive pressure can gradually be reduced to atmospheric pressure. It is especially equipped to help divers complete their decompression schedules and may also be used to treat diving casualties. One or more of the compartments may be installed on the deck of a work boat or barge.

declination *n*: 1. the angle, variable with geographic position, between the direction in which a magnetic needle points and the true meridian. 2. a turning aside or swerving.

decoder *n*: 1. an electronic device that converts binary numbers to decimal numbers. Compare *encoder*. 2. a demodulator. See *demodulate*.

decompression *n*: the process of gradually lowering elevated ambient pressure to eliminate dissolved gases from a diver's bloodstream and tissues.

decompression sickness *n*: a condition resulting from the formation of gas bubbles in a diver's blood or tissues during ascent. Failure to rid tissues of the inert gas may cause a wide variety of symptoms, including pain, nausea, paralysis, unconsciousness, temporary blindness, and even death. Also called the bends.

decompression table *n*: a profile of ascent rates and breathing mixtures that safely reduce the pressure on a diver to atmospheric after a dive. The table shows depths, bottom times, decompression stops, and total decompression times.

decontaminants *n pl*: materials added to cements or cement slurries to counteract the effects of contamination.

decontamination *n*: (under HAZWOPER) the removal of hazardous substances from employees and their equipment to the extent necessary to preclude the occurrence of foreseeable adverse health effects.

decontrol *n*: the act of ending federal government control of the wellhead price of new natural gas sold in interstate commerce. Also called deregulation.

dedication *n*: the commitment of reserves of natural gas to a buyer or lessor.

deed of trust *n*: an instrument used to transfer legal title to property as security for the repayment of a loan or the fulfillment of some other obligation. Compare *mortgage*.

deep drilling *n*: any drilling project that is deeper than average for a given area or time period.

deepen *v*: to increase the depth of a well. Deepening is generally a workover operation carried out to produce from a deeper formation or to control excessive gas found in the upper levels of a reservoir.

deep-sea dress *n*: see *standard dress*.

deep tank *n*: one of the bulk liquid tanks on freighters, usually under the cargo holds.

deep test well *n*: see *exploration well*.

deep water *n*: in offshore operations, water depths greater than normal for the time and current technology.

deepwater *adj*: of or pertaining to operations in deep water.

deepwater drilling *n*: a relative term that refers to offshore drilling operations in deep oceans or seas—currently, about 5,000 feet (1,500 metres) and deeper. It presents a number of special problems related to water depth.

deepwater riser system *n*: a marine riser used in deepwater drilling operations and that may be equipped with special devices such as automatic flooding valves, which open to admit seawater into the riser and prevent collapse of the riser when accidental evacuation of mud in a riser occurs. Also, some of the riser joints may be equipped with buoyant riser modules. See *buoyant riser joint, riser pipe*.

deep-well pump *n*: a production pump designed for service in a deep well.

deethanized liquid *n*: hydrocarbon mixture remaining after the removal of ethane and lighter hydrocarbons.

defensive hoses *n pl*: large-size hoses used to throw large amounts of water from a distance.

deferred production agreement *n*: an agreement between working-interest owners of a lease under which an owner's share of the gas reserves under the lease is considered to remain in the reservoir while the other owner's share of the gas is being produced.

deficiency gas *n*: the difference between the quantity of gas a purchaser is obligated by a gas sales contract to take or to pay for if not taken and the quantity of gas actually taken.

deficiency payment *n*: the amount paid by the purchaser for the quantity of deficiency gas as required by a gas sales contract.

deflect *n*: see *deflection*.

deflection *n*: a change in the angle of a wellbore. In directional drilling, it is measured in degrees from the vertical.

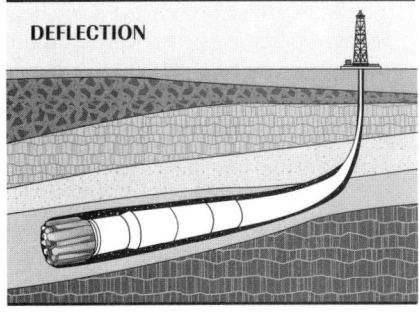

deflection tool *n*: a device made up in the drill string that causes the bit to drill at an angle to the existing hole. It is often called a kickoff tool, because it is used at the kickoff point to start building angle.

deflect-to-connect connection *n*: an underwater pipe-joining technique in which the pipe is pulled to a target area in line with the platform but to one side of it. The connection is made by winding or otherwise deflecting the pipe laterally until it mates with the riser connection. Compare *direct pull-in connection*.

deflocculation *n*: the dispersion of solids that have stuck together in drilling fluid, usually by means of chemical thinners. See *flocculation*.

defoamer *n*: any chemical that prevents or lessens frothing or foaming in another agent.

deformation *n*: the action of earth stresses that results in folding, faulting, shearing, or compression of rocks.

degasser *n*: the device used to remove unwanted gas from a liquid, especially from drilling fluid.

degradation product *n*: an unwanted substance produced because of some reaction such as cracking, dehydrogenation, or polymerization. The term implies the formation of a contaminant or low-value product.

degrade *v*: to break down a compound; to deteriorate.

degree API (°API) *n*: a unit of measurement of the American Petroleum Institute that indicates the weight, or density, of oil. See *API gravity*.

degree-day *n*: a unit of temperature and time, computed per day, equivalent to the difference between a 65°F (18.3°C) base and a daily mean temperature (when the latter is less than 65°F). The total of degree-days for a given period of time can be used to estimate energy requirements such as the amount of fuel oil needed to heat a building.

dehydrate *v*: to remove water from a substance. Dehydration of crude oil is normally accomplished by treating with emulsion breakers. The water vapor in natural gas must be removed to meet pipeline requirements; a typical maximum allowable water vapor content is 7 pounds per million cubic feet (3.2 kilograms per million cubic metres) per day.

dehydration *n*: the removal of water or water vapor from gas or oil.

dehydration plant *n*: a plant having equipment or apparatus for effecting dehydration.

dehydrator *n*: equipment or apparatus for effecting dehydration.

dehydrogenation *n*: the process of removing hydrogen from a compound.

delamination *n*: a phenomenon in which layers of surface metal peel away from the rest of the metal body. Journal (plain) bearings sometimes fail by delamination.

delayed (chronic) health hazard *n*: carcinogens and other hazardous chemicals that cause an adverse effect on a target organ (as defined by the Code of Federal Regulations) that manifests itself after a long period of time following or during repeated contact with the substance.

delay rental *n*: a sum of money payable to the lessor by the lessee for the privilege of deferring the commencement of drilling operations and keeping the lease valid. May be paid monthly, quarterly, or annually.

delineation well *n*: a well drilled in an existing field to determine, or delineate, the extent of the reservoir.

deliquescence *n*: the liquefaction of a solid substance due to the solution of the solid by adsorption of moisture from the air.

deliverability *n*: the ability of a gas to be delivered.

deliverability capacity *n*: the maximum amount of gas a producer can deliver to the purchaser at a specified delivery point.

deliverability plot *n*: a graph that compares flowing bottomhole pressure of a well with production in barrels of oil per day to show the relationship between drawdown and the producing rate. Its main purpose is to find the most efficient flow rate for the well.

deliverability test *n*: a type of test for either an oilwell or a gas well to determine the actual flow rate.

delivery *n*: 1. the actual volume delivered through a meter during a proving or metering operation. 2. a volume of delivered liquid that is measured by a meter.

A batch or tender may also be called a delivery. See *batch*.

delivery capacity (DC) *n*: the total daily rate that seller maintains and can deliver on demand over an extended period to buyer.

delivery gas *n*: natural gas delivered to a distribution point, such as a gas plant or a city gate. Compare *redelivery gas*.

deliveryman *n*: a shipper's representative who takes delivery of oil from a pipeline company at a terminal or junction.

delivery ticket *n*: see *measurement ticket*.

delivery tube *n*: a small device used inside a bottle when tap sampling. The tube allows for the delivery of liquid from the tank into the bottle with little or no splashing.

delta *n*: see *lacustrine delta, marine delta*.

delta connection *n*: in electronics, a combination of three components connected in series to form a triangle similar to the Greek letter delta (Δ). Also known as a mesh connection.

deluge valve *n*: valve that opens quickly and allows a full volume of water to flow immediately.

demand *n*: the quantity of oil, gas, or other petroleum products, or commodities (such as electricity) wanted at a specified time and price.

demand meter *n*: a quantity recorder that measures and integrates the instantaneous demands on a circuit during a given period, usually 15 minutes.

demand rate *n*: the highest rate of power consumption for a minimum period of 15 minutes during a billing period, usually one month. This rate is registered on a demand meter.

demand regulator *n*: the part of the open-circuit diving system that allows a diver to expel all used air directly into the water and avoid rebreathing exhaled carbon dioxide. The regulator reduces the air pressure in the tanks to a diver's ambient pressure so that he or she can breathe the air with no resistance. Also called open-circuit regulator.

demethanizer *n*: see *rich-oil demethanizer*.

demister *n*: in an evaporator, a separator that removes droplets of liquid from water vapor.

demodulate *v*: in electronics, to recover the information transmitted by a modulated wave. See *modulate*.

demodulator probe *n*: in electronics, a small device (a probe) that is placed on or inserted in a circuit and that detects (demodulates) a modulated carrier wave and allows the demodulated signal to be

displayed on an oscilloscope's screen for analysis.

demulsifier *n*: a chemical with properties that cause the water droplets in a water-in-oil emulsion to merge and settle out of the oil, or oil droplets in an oil-in-water emulsion to coalesce, when the chemical is added to the emulsion. Also called emulsion breaker. See *demulsify, emulsion breaker*.

demulsifier-solvent stock *n*: a combination of a demulsifier and solvent that is used during testing for the suspended S&W content of oil and during other operations.

demulsify *v*: to resolve an emulsion, especially of water and oil, into its components. See *emulsion treating*.

DENLA *abbr*: drag embedment normal load anchor.

denominator *n*: in a mathematical fraction, the term or number that divides the other term or number (called the numerator), and is written below the line. For example, in the fraction $7/15$, 15 is the denominator.

densification *n*: the process of making a substance heavier, or denser. For example, in a gas plant, a product may be densified.

Densilog *n*: a logging company's name for a density log.

densimeter *n*: a device that measures the specific gravity or relative density of a gas, liquid, or solid.

densitometer *n*: see *densimeter*.

density *n*: the mass or weight of a substance per unit volume. For instance, the density of a drilling mud may be 10 pounds per gallon, 74.8 pounds per cubic foot, or 1,198.2 kilograms per cubic metre. Specific gravity, relative density, and API gravity are other units of density.

density log *n*: a special radioactivity log for open-hole surveying that responds to variations in the specific gravity of formations. It is a contact log (i.e., the logging tool is held against the wall of the hole). It emits neutrons and then measures the secondary gamma radiation that is scattered back to the detector in the instrument. The density log is an excellent porosity-measure device, especially for shaley sands. Some trade names are Formation Density Log, Gamma-Gamma Density Log, and Densilog.

dent *n*: a small depression made by striking or pressing. Sometimes called a ding.

deoil *v*: to remove, from wax, traces of oil and compounds with a low melting point.

Department of Energy (DOE) *n*: a federal department set up to provide a framework for a complete and balanced national energy plan through coordination and administration of the energy functions of the federal government. It is responsible for long-term, high-risk research and development of energy technology, the marketing of federal power, energy conservation, the nuclear weapons program, energy regulatory programs, and a central energy data collection and analysis program. Address: 1000 Independence Ave. SW; Washington, DC 20585; 202-586-5000.

Department of the Interior (DOI) *n*: a federal department set up as the nation's principal conservation agency. The DOI is responsible for nationally owned lands and natural resources. It fosters wise use of land and water resources, protects fish and wildlife, preserves the environmental and cultural values of national parks and historic places, and provides for the enjoyment of life through outdoor recreation. Address: 1849 C Street NW; Washington, DC 20240; 202-208-3100.

Department of Transportation (DOT) *n*: a federal department that develops regulations governing transportation. Each mode of transportation has a different administration. The administrations under DOT include the U.S. Coast Guard, the Federal Aviation Administration, the Federal Highway Administration, the Federal Railroad Administration, the National Highway Traffic Safety Administration, the Urban Mass Transportation Administration, the Saint Lawrence Seaway Development Corporation, and the Research and Special Programs Administration. Address: 400 7th Street SW; Washington, DC 20590; 202-366-4000.

departure *n*: see *deviation*.

deplete *v*: to exhaust a supply. An oil and gas reservoir is depleted when most or all economically recoverable hydrocarbons have been produced.

depletion *n*: 1. the exhaustion of a resource. 2. a reduction in income reflecting the exhaustion of a resource. The concept of depletion recognizes that a natural resource such as oil is used up over several accounting periods and permits the value of this resource to be expensed periodically as the resource is exhausted.

depletion allowance *n*: a reduction in U.S. taxes for owners of an economic interest in minerals in place to compensate for the exhaustion of an irreplaceable capital asset. This economic interest includes mineral interest, working interest in a lease, royalty, overriding royalty, production payment interest, and net profits interest.

depletion drive *n*: see *gas drive*.

deployment valve *n*: see *drilling deployment valve*.

deposition *n*: the laying down of sediments or other potential rock-forming material.

depositional environment *n*: the set of physical, chemical, and geological conditions (such as climate, stream flow, and sediment source) under which a rock layer was laid down.

depot *n*: a center for storing goods.

depreciation *n*: 1. decrease in value of an asset such as a plant or equipment due to normal wear or passing of time; real property (land) does not depreciate. 2. an annual reduction of income reflecting the loss in useful value of capitalized investments by reason of wear and tear. The concept of depreciation recognizes that the purchase of an asset other than land will benefit several accounting cycles (periods) and should be expensed periodically over its useful life.

depression *n*: in meteorology, an area of low barometric pressure. Depressions often bring rain to the area in which they occur.

depressure *v*: to release the pressure contained in a tank, vessel, or pipe to a predetermined value. See *bleed*.

depropanizer *n*: a unit of equipment for separating propane, with or without lighter components, from a mixture of hydrocarbons and leaving a bottoms product that is essentially propane-free.

depth *n*: 1. the distance to which a well is drilled, stipulated in a drilling contract as contract depth. Total depth is the depth after drilling is finished. 2. on offshore drilling rigs, the distance from the baseline of a rig or a ship to the uppermost continuous deck. 3. the maximum pressure that a diver attains during a dive, expressed in feet (metres) of seawater.

depth in *n*: the depth of the wellbore when a new bit or other tool is run in. Compare *depth out*.

depthometer *n*: a device used to measure the depth of a well or the depth at a specific point in a well (such as to the top of a liner or to a fish) by counting the turns of a calibrated wheel rolling on a wireline as it is lowered into or pulled out of the well.

depth out *n*: the depth of the wellbore when a bit or other tool is pulled out of the hole. Compare *depth in*.

derating factor *n*: the percentage of the rated power output that a given motor can deliver without being thermally overloaded.

deregulation *n*: see *decontrol*.

deregulation clause *n*: a clause in a contract currently governed by regulatory price ceilings to provide a process to reset the price of gas should price controls and regulatory authority cease.

derivative *adj*: in closed-loop control systems, of or pertaining to circuitry or components that change the response of a system to changes in its output or input. Derivative circuits or components are often referred to as rate control circuits or components.

derivative control *n*: a method of control that adjusts an output based on the rate of change of the process under control.

DERRICKS

derrick *n*: a large load-bearing structure, usually of bolted construction. In drilling, the standard derrick has four legs standing at the corners of the substructure and reaching to the crown block. The substructure is an assembly of heavy beams used to elevate the derrick and provide space to install blowout preventers, casingheads, and so forth. Because the standard derrick must be assembled piece by piece, it has largely been replaced by the mast, which can be lowered and raised without disassembly. Compare *mast*.

derrick floor *n*: also called the rig floor or the drill floor. See *rig floor*.

derrickhand *n*: the crew member who handles the upper end of the drill string as it is being hoisted out of or lowered into the hole. On a drilling rig, he or she is also responsible for the circulating machinery and the conditioning of the drilling or workover fluid.

derricking *n*: operation of changing boom angle in a vertical plane.

derrickman *n*: see *derrickhand*.

desalt *v*: to remove dissolved salt from crude oil. Sometimes fresh water is injected into the crude stream to dissolve salt for removal by electrostatic treaters.

desander *n*: a centrifugal device for removing sand from drilling fluid to prevent abrasion of the pumps. It may be operated mechanically or by a fast-moving stream of fluid inside a special cone-shaped vessel, in which case it is sometimes called a hydrocyclone. Compare *desilter*.

descent and distribution laws *n*: the laws in a state that determine the disposition of property among heirs in the absence of a will.

desiccant *n*: a substance able to remove water from another substance with which it is in contact. It may be liquid (e.g., triethylene glycol) or solid (e.g., silica gel).

desiccator *n*: a device used for removing moisture from substances. For drying core samples, the desiccator is a container with a bottom section holding moisture-absorbing chemicals and a top section in which the sample is placed.

design criteria *n pl*: criteria used in the design of a mooring system.

design factor *n*: the ratio of the ultimate load a vessel or structure will sustain to the permissibly safe load placed on it. Such safety factors are incorporated into the design of casing, for example, to allow for unusual burst, tension, or collapse stresses.

design factor of wire rope *n*: see *safety factor of wire rope*.

design load *n*: the most stressful combination of weight or other forces a mechanical system or device is built to sustain.

design water depth *n*: 1. the vertical distance from the ocean bottom to the nominal water level plus the height of astronomical and storm tides. 2. the deepest water in which an offshore drilling rig can operate.

desilter *n*: a centrifugal device for removing very fine particles, or silt, from drilling fluid to keep the amount of solids in the fluid at the lowest possible point. Usually, the lower the solids content of mud, the faster is the rate of penetration. The desilter works on the same principle as a desander. Compare *desander*.

Desk and Derrick Clubs *n*: an association of employees in the petroleum and allied industries. The principal function of the group is to provide informational and educational programs for the enlightenment of its members about the industry they serve. Membership ranges from secretaries through managers and directors of companies. Address: 5153 E. 51st St., Suite 107; Tulsa, OK 74135; 918-622-1749; FAX 918-622-1675.

desorb *v*: to remove an absorbate or adsorbate from an absorbent or adsorbent.

desorption *n*: the process of removing an absorbate or adsorbate from an absorbent or adsorbent.

destructive testing *n*: a procedure in which a weld is torn apart so that its structure can be examined. Destructive testing is used primarily during the qualification procedures required of all welders who work on pipelines.

desulfurize *v*: to remove sulfur or sulfur compounds from oil or gas.

detector *n*: 1. a device used to sense or ascertain the presence of such items as objects, radiation, poisonous gases, or chemicals. 2. in an electronic receiver, the stage in the receiver that recovers (demodulates) the modulated signal the receiver originally gets (receives). A modulated signal is a signal that is varied by its amplitude, frequency, or phase. See *modulate*.

detector probe *n*: see *demodulator probe*.

detector signal *n*: a contact closure change, or other signal, that starts or stops a prover counter/timer and defines the calibrated volume of the prover.

detector switch *n*: in a pipe prover, an electrical switch activated by the displacement sphere as it moves through the U-portion of the prover. Activating the switch sends an electric current through a cable to a pulse generator.

detent *n*: a mechanism that keeps one part in a certain position relative to that of another; it can be released by applying force to one of the parts.

detergent *n*: in lubricating oils and in some engine fuels, a chemical that is added to the oil or to the fuel that suspends dirt, carbon, and other foreign matter in the oil or fuel. As a result of the detergents in motor oil, the oil will very quickly appear dirty because it is suspending the particles.

determinable fee *n*: an interest in property that will end at the happening or non-happening of a particular event. In some states, an oil and gas lease is considered a determinable fee in real estate.

detonation *n*: 1. an explosion. 2. the knock or ping produced when fuel of too-low octane rating is used in the engine. Compare *preignition*.

deuterium *n*: the isotope of the element hydrogen that has one neutron and one proton in the nucleus; atomic weight is 2.0144.

development and production plan *n*: a plan required by the MMS before development and production can take place in the OCS. The MMS approves or disapproves the plan based on environmental, technical, and economic considerations.

development drilling *n*: drilling that occurs after the initial discovery of hydrocarbons in a reservoir. Usually, several wells are required to adequately develop a reservoir.

development well *n*: 1. a well drilled in proven territory in a field to complete a pattern of production. 2. an exploitation well.

deviation *n*: departure of the wellbore from the vertical, measured by the horizontal distance from the rotary table to the target. The amount of deviation is a function of the drift angle and hole depth. The term is sometimes used to indicate the angle from which a bit has deviated from the vertical during drilling. See *drift angle*.

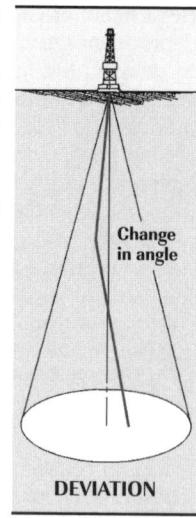

deviation and azimuth indicator *n*: see *double recorder*.

deviation effect *n*: when a gas is confined within a pressure-tight container, the tendency of the gas pressure to be slightly above or below that which is normally expected because of the attraction or repulsion of individual gas molecules.

deviation survey *n*: an operation made to determine the angle from which a bit has deviated from the vertical during drilling. There are two basic deviation-survey, or drift-survey, instruments: one reveals the drift angle; the other indicates both the angle and the direction of deviation.

devise *v*: to make a gift of real property (e.g., land) by means of a will. Compare *bequeath*.

Devonian *adj*: of or relating to the geologic period from about 400 million to 350 million years ago in the Paleozoic era, or to the rocks formed during this period, including those of Devonshire, England, where outcrops of such rock were first identified.

dew point *n*: the temperature and pressure at which a liquid begins to condense out of a gas. For example, if a constant pressure is held on a certain volume of gas but the temperature is reduced, a point is reached at which droplets of liquid condense out of the gas. That point is the dew point of the gas at that pressure. Similarly, if a constant temperature is maintained on a volume of gas but the pressure is increased, the point at which liquid begins to condense out is the dew point at that temperature. Compare *bubble point*.

dew-point recorder *n*: a device used by gas transmission companies to determine and to record continuously the dew point of the gas.

dew-point spread *n*: the difference between the actual air temperature and the dew-point temperature.

dew-point temperature *n*: the temperature at which the rate at which water vapor leaves a gas equals the rate at which water vapor enters the gas at a given pressure.

dew-point tester *n*: a high-pressure stainless steel or nickel-plated brass chamber with a magnifying window, a sample inlet, a pressure gauge, a tripod socket, a gas outlet, and an angle-mounted mirror. It ascertains dew point by observing the temperature and the pressure at which condensation occurs on and disappears from the mirror.

DF *abbr*: derrick floor; used in drilling reports.

diagenesis *n*: the chemical and physical changes that sedimentary deposits undergo (compaction, cementation, recrystallization, and sometimes replacement) during and after lithification.

dial thermometer *n*: a thermometer on which the temperature is indicated by a moving needle or pointer on a circular face or disk rather than with liquid in a glass or a liquid crystal display (LCD).

dial-type assembly *n*: see *dial thermometer*.

diamagnetic *adj*: antimagnetic. Diamagnetic materials resist flux by aligning themselves at right angles to the lines of force in a magnetic field. See *ferromagnetic, paramagnetic*.

diameter *n*: the distance across a circle, measured through its center. In the measurement of pipe diameters, the inside diameter is that of the interior circle and the outside diameter that of the exterior circle.

diamond bit *n*: a drill bit that has small industrial diamonds embedded in its cutting surface. Cutting is performed by the rotation of the very hard diamonds over the rock surface.

diaphragm *n*: a sensing element consisting of a thin, usually circular, plate that is deformed by pressure applied across the plate.

diaphragm actuator *n*: in process control instrumentation, a thin pancake-shaped device in a housing, which is part of a control system. Pneumatic pressure from another part of the system travels through tubing into the actuator housing and contacts the diaphragm to move it. The diaphragm then moves a spring or other mechanical device to vary the size of an orifice (opening) through which the control agent flows.

diaphragm box *n*: in process control instrumentation, the housing that contains a flexible diaphragm.

diaphragm meter tangent *n*: part of a diaphragm meter that regulates the length of the diaphragm stroke; used to adjust the volumetric displacement.

diapir *n*: a dome or anticlinal fold in which a mobile plastic core has ruptured the more brittle overlying rock. Also called piercement dome.

DIAPIR

diapirism *n*: the penetration of overlying layers by a rising column of salt or other easily deformed mineral caused by differences in density.

diastrophism *n*: the process or processes of deformation of the earth's crust that produce oceans, continents, mountains, folds, and faults.

diatom *n*: any of the algae of the class Bacillariophyceae, noted for symmetrical and sculptured siliceous cell walls. After death, the cell wall persists and forms diatomite. Diatoms first appeared in the Cretaceous period and still live today.

diatomaceous earth *n*: an earthy deposit made up of the siliceous cell walls of one-celled marine algae called diatoms. It is used as an admixture for cement to produce a low-density slurry.

diatomite *n*: a rock of biochemical origin, which is composed of the siliceous (glassy) shells of microscopic algae called diatoms.

die *n*: a tool used to shape, form, or finish other tools or pieces of metal. For example, a threading die is used to cut threads on pipe.

die collar *n*: a collar or coupling of tool steel, threaded internally, that can be used to retrieve pipe from the well on fishing jobs; the female counterpart of a taper tap. The die collar is made up on the drill pipe and lowered into the hole until it contacts the lost pipe. If the lost pipe is stuck so that it cannot rotate, rotation of the die collar on top of the pipe cuts threads on the outside of the pipe, providing a firm attachment. The pipe is then retrieved from the hole. It is not often used because it is difficult to release it from the fish should it become necessary. Compare *taper tap*.

dielectric *n*: a substance that is an insulator, or nonconductor, of electricity.

dielectric constant *n*: the value of dielectricity assigned to a substance. A substance that is a good insulator has a high dielectric constant, whereas a poor insulator has a low one. The dielectric constant of oil is lower than that of water, and on this principle a net-oil computer operates.

dielectric strength *n*: the ability of a dielectric material to withstand the effects of voltage applied to it. It is usually measured in terms of the amount of voltage per mil, or thousandths of an inch, required to break down the material.

die nipple *n*: a device similar to a die collar but with external threads.

dies *n pl*: see *insert*.

diesel-based mud *n*: a liquid drilling fluid in which diesel oil is the base fluid. An emulsifier and brine (salt water) is also added to allow the brine to disperse within the oil. See *oil mud*.

diesel-electric power *n*: the power supplied to a drilling rig by diesel engines driving electric generators; used widely.

diesel-electric rig *n*: see *electric rig*.

diesel engine *n*: a high-compression, internal-combustion engine used extensively for powering drilling rigs. In a diesel engine, air is drawn into the cylinders and compressed to very high pressures; ignition occurs as fuel is injected into the compressed and heated air. Combustion takes place within the cylinder above the piston, and expansion of the combustion products imparts power to the piston.

diesel fuel *n*: a light hydrocarbon mixture for diesel engines, similar to furnace fuel oil; it has a boiling range just above that of kerosene.

diesel-oil plug *n*: see *gunk plug*.

diethylene glycol *n*: a liquid chemical used in gas processing to remove water from the gas. See *glycol, glycol dehydrator*.

differential *n*: 1. the difference in quantity or degree between two measurements or units. For example, the pressure differential across a choke is the variation between the pressure on one side and that on the other. 2. the value or volume payment accompanying an exchange of oil for oil. The payment serves as compensation for quality, location, or gravity differences between the oils being exchanged.

differential displacing valve *n*: a special-purpose valve used to facilitate spacing out and flanging up the well, run in on the tubing string.

differential drop *n*: the reduction in pressure that occurs as gas flows through a restriction in a line, such as through an orifice.

differential fill-up collar *n*: a device installed near the bottom of the casing. It has a valve that keeps mud out of the casing so that the casing can be floated into the hole to lessen the weight on the rig's equipment. This valve will, however, automatically open to let mud into the casing when the pressure of the mud outside the casing becomes high enough to collapse the joints of casing near the bottom of the string. Thus, this special collar, with its pressure-sensitive valve, not only allows casing to float, but also keeps it from collapsing.

differential flowmeter *n*: an electronic instrument that measures the difference between the rate of flow of the mud going into the hole and the rate of flow of the mud out of the hole. If the rate of flow coming out of the hole increases, it may be a sign that the well has kicked.

differential head *n*: see *differential pressure.*

differential heating *n*: a phenomenon that causes winds to occur on the earth. Because the sun's light and heat do not strike the earth at the same angle and because some areas of the earth absorb more heat than others, unequal heating takes place. Consequently, some areas receive more light and heat than others. For example, the sun's light and heat strike the equator at a more direct angle than at the poles. Thus, it is hotter at the equator than it is at the poles. Also, because ice at the poles reflects more light than the vegetation at the equator, the equatorial regions are hotter than polar regions.

differential pen *n*: on a flow recorder, the pen used to record differential pressure across the orifice in the meter; usually recorded in red ink on the flow recorder chart.

differential pressure *n*: the difference between two fluid pressures; for example, the difference between the pressure in a reservoir and in a wellbore drilled in the reservoir, or between atmospheric pressure at sea level and at 10,000 feet (3,048 metres). Also called pressure differential.

differential-pressure gauge *n*: a pressure-measuring device actuated by two or more pressure-sensitive elements that act in opposition to produce an indication of the difference between two pressures.

differential-pressure sticking *n*: see *differential sticking.*

differential-pressure transducer *n*: an electrical device that senses very small bellows or diaphragm movement in a bellows meter. Allows a computer to calculate flow rates directly by producing electrical output.

differential-pressure transmitter *n*: 1. a pressure measuring device that senses the difference between two pressures and produces a proportional electrical output, usually 4 to 20 milliamps (mA), which covers the calibrated range of the transmitter. 2. an electrical device that senses very small bellows or diaphragm movement in a bellows meter. 3. an electrical device that allows a computer to calculate flow rates directly by producing an electrical output that is proportional to

the square root of the differential pressure measured by the transmitter.

differential-pressure valve *n*: a valve used to regulate automatically a uniform difference in pressure between two locations in a pipeline.

differential range *n*: the extent of variation between the lowest and highest differential pressures that an orifice meter can accurately measure.

differential signal *n*: in a circuit, an indicator (a signal) that is the voltage difference between two junctions (nodes), neither of which is at ground potential.

differential sticking *n*: a condition in which the drill stem becomes stuck against the wall of the wellbore because part of the drill stem (usually the drill collars) has become embedded in the filter cake. Necessary conditions for differential-pressure sticking, or wall sticking, are a permeable formation and a pressure differential across a nearly impermeable filter cake and drill stem. Also called wall sticking. See *differential pressure, filter cake.*

differential test *n*: a test taken at a particular point in the flow of a meter, which is compared with the same flow rate on a differential curve that was established at the time the meter was installed. If the differential pressure reading obtained during the test is 50% higher than the original test value (at the same flow rate), with pressure and temperature conditions being approximately the same, the meter is removed and the cause of the increased operating resistance determined.

differentiating circuit *n*: a circuit whose output voltage is proportional to the rate of change of the input voltage.

diffuser *n*: a device that uses part of the kinetic energy of a fluid passing through a machine by gradually increasing the cross section of the channel or chamber through which it flows so as to decrease its speed and increase its pressure.

diffusion *n*: 1. the spontaneous movement and scattering of particles of liquids, gases, or solids. 2. the migration of dissolved substances from an area of high concentration to an area of low concentration.

digit *n*: a character that represents a numeric character. See *number.*

digital *adj*: pertaining to data in the form of digits, especially electronic data stored in the form of a binary code. Compare *analog.*

digital logic *n*: a system that describes a statement or condition in a true-false, yes-no, either-or, or on-off manner. Digital logic is used in computer operating systems and in automatic switching and control systems.

adj: of or pertaining to the various types of solid-state devices such as gates, flip flops, and other on-off circuits used to perform problem-solving functions in the many types of digital computers.

digital multimeter *n*: in electronics, a multimeter that is constructed such that its measurements are displayed digitally rather than as analog readouts. See *multimeter.*

digital readout *n*: a type of register on which the information is indicated by directly readable characters, particularly numerals. Compare *analog data.*

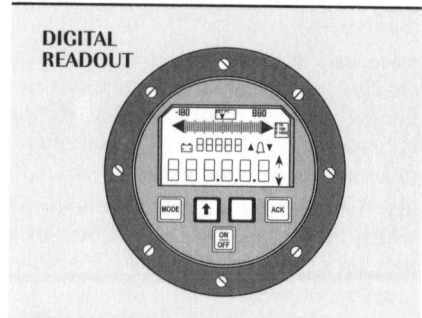

digital signal *n*: the representation of the magnitude of a variable in the form of discrete values or pulses of a measurable physical quantity.

dilatant fluid *n*: a dilatant, or inverted plastic, fluid is usually made up of a high concentration of well-dispersed solids that exhibits a nonlinear consistency curve passing through the origin. The apparent viscosity increases instantaneously with increasing rate of shear. The yield point, as determined by conventional calculations from the direct-indicating viscometer readings, is negative; however, the true yield point is zero.

diluent *n*: liquid added to dilute or thin a solution.

dimensional analysis *n*: a technique that involves the study of the dimensions of physical quantities, which is used primarily as a tool for obtaining information about physical systems too complicated for full mathematical solutions to be feasible.

dimensionless number *n*: a number that does not have a unit of measurement associated with it.

dimple connector *n*: a device used to attach a tool string to the end of coiled tubing. It absorbs vibration created by rotating or reciprocating the tubing and provides a high-pressure seal between the tool string and the coiled tubing.

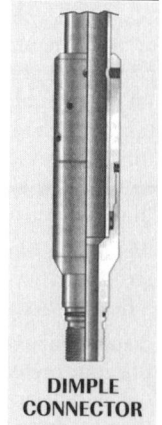

DIMPLE CONNECTOR

dimple-grapple drilling connector *n*: a device used in coiled tubing operations that absorbs forces created when drilling with downhole motors or when using impact hammers. The grapple connector provides high tensile strength and high torque capacity.

diode *n*: 1. a solid-state electronic device that restricts current flow chiefly in one direction. 2. a radio tube that contains an anode and a cathode.

diode junction voltage *n*: in a circuit in which a diode is used, the voltage generated at the point where (the junction) the diode is wired into the circuit. Junction voltage is 0.7 volts and when multiplied by the current flowing in the circuit results in wattage, which takes the form of heat.

diorite *n*: intrusive, or platonic, generally coarse-grained igneous rock composed largely of plagioclase feldspar with smaller amounts of dark-colored minerals. Also known as black granite. Compare *andesite*.

dip *n*: 1. a European term for the depth of liquid in a storage tank. 2. formation dip. See *formation dip, innage, ullage*.

dip gauge point *n*: a European term for the point on the bottom of a container that the dip weight touches during gauging and from which the measurement of the oil and water depths is taken. The dip point usually corresponds with the datum point, but any difference in levels must be designated in the calibration table.

dip hatch *n*: a European term for the opening in the top of a container through which dipping (gauging) and sampling operations are carried out.

dip log *n*: see *dipmeter survey*.

dipmeter log *n*: see *dipmeter survey*.

dipmeter survey *n*: an oilwell-surveying method that determines the direction and angle of formation dip in relation to the borehole. It records data that permit computation of both the amount and direction of formation dip relative to the axis of the hole and thus provides information about the geologic structure of the formation. Also called dipmeter log or dip log.

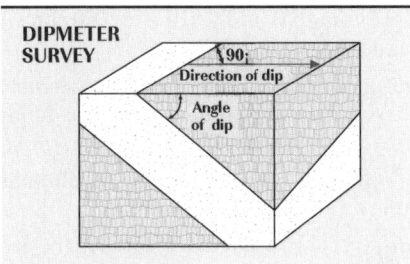

dipping reference point *n*: a European term for a point marked on the dip hatch of a tank directly above the dip point to indicate the position at which dipping should be carried out.

dip plate *n*: a European term for datum plate.

dip rod *n*: a rigid length of wood or metal that is provided with a scale for measurement and usually graduated in units of volume. It is used for manual measurement of liquid in a container. Also called dipstick.

dip slip *n*: upward or downward displacement along a fault plane.

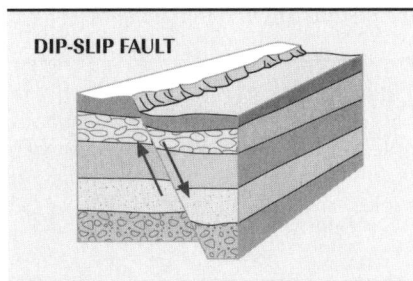

dip-slip fault *n*: a fault in which the slip is practically in the line of the fault dip.

dipstick *n*: see *dip rod*.

dip tape *n*: a European term for gauging tape.

dip tube *n*: the small-diameter tube extending below the pump suction of a sucker rod pump, usually inside the mud anchor, as in a "poor boy" gas anchor.

dip weight *n*: a European term for a plumb bob that is attached to a metal dip tape and that is of sufficient weight to keep the tape taut and of such shape as to facilitate the penetration of any sludge that might be present on the datum plate of a tank. Also called outage bob.

direct-acting electric actuator *n*: a device on a diesel engine's governor that increases the engine's speed by increasing positive voltage going to the actuator and governor. As the engine needs more fuel to go faster, the direct-acting actuator increases positive voltage to make the governor increase the fuel flowing to the fuel injectors. See *electrically-actuated governor*.

direct connection *n*: a straightforward connection that makes the speeds of a prime mover and a driven machine identical.

direct coupling *n*: in electronics, the joining (coupling) of two circuits with a nonfrequency-sensitive device, such as a wire, resistor, or battery, so that both direct and alternating current can flow through the coupling path. For an oscilloscope to measure direct current, its circuits must be direct coupled.

direct current (DC) *n*: in an electric circuit, current flow in one direction only. An example of a simple DC circuit is a battery, a wire, and a light bulb, wherein current flows through the wire from one terminal of the battery, through the light bulb, and through another wire back to the battery's other terminal. Compare *alternating current*.

direct heater *n*: used in emulsion treating to heat noncorrosive emulsions that are under comparatively low pressure. Types are tubular, fluid-jacket, internal firebox, and volume, or jug-type. Compare *indirect heater*.

direct-indicating viscometer *n*: see *direct-reading viscometer*.

directional drilling *n*: 1. intentional deviation of a wellbore from the vertical. Although wellbores are normally drilled vertically, it is sometimes necessary or advantageous to drill at an angle from the vertical. Controlled directional drilling makes it possible to reach subsurface areas laterally remote from the point where the bit enters the earth. It often involves the use of deflection tools. 2. a technique of river crossing in pipeline construction in which the pipe is buried under the riverbed at depths much greater than those of conventional crossings. With this technique, a hole in the form of an inverted arc is drilled beneath the river, and the made-up pipeline is pulled through it.

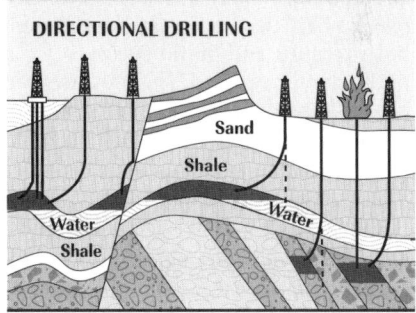

directional drilling service company *n*: a business that provides directional hole planning, sophisticated directional tools, and on-site assistance to the oil company operator of a drilling rig.

directional drilling supervisor *n*: an employee of a directional drilling service company whose main job is to help the driller at a well site keep the wellbore as close as possible to its planned course. Also called directional operator.

directional hole *n*: a wellbore intentionally drilled at an angle from the vertical. See *directional drilling*.

directional operator *n*: see *directional drilling supervisor*.

directional perforating *n*: see *oriented perforating*.

directional survey *n*: a logging method that records drift angle, or deflection from the vertical, and direction of the drift. A single-shot directional-survey instrument makes a single photograph of a compass reading of the drift direction and the number of degrees the hole is off vertical. A multishot survey instrument obtains numerous readings in the hole as the device is pulled out of the well. See *directional drilling*.

directional survey instrument *n*: a tool that, when lowered into the wellbore, makes a photographic record of the angle and drift of the wellbore; that is, it records the number of degrees the hole is off vertical and the direction in which it is off vertical. Several types of instrument are available; some are capable of photographing only a single record—single-shot survey instruments—whereas others are capable of making several records in one run—multishot survey instruments.

direct measurement *n*: a measurement that produces a final result directly from the scale on an instrument.

direct pull-in connection *n*: an underwater pipe-joining technique in which the pipe with its connector sled is steered straight onto the matching receiver at the base of the platform. Compare *deflect-to-connect connection*.

direct reading chart *n*: a type of orifice meter chart that indicates the differential pressure and static pressure for a particular time period. The pressures can be read directly from the chart instead of being calculated from information given on the chart.

direct-reading viscometer *n*: commonly called a "V-G meter." The instrument is a rotational-type device powered by means of an electric motor or handcrank, and is used to determine the apparent viscosity, plastic viscosity, yield point, and gel strengths of drilling fluids. The usual speeds are 600 and 300 rpm. See *API RP13B* for operational procedures.

direct sale *n*: a natural gas sales transaction in which at least one of the intermediary parties in the natural gas delivery system (i.e., pipeline transmission company or local distributing company) does not take title to the natural gas but only transports it. Historically, a sale of natural gas to an end user, as opposed to a "sale for resale." More recently, the term has also been applied to a sale by a producer directly to a local distribution company.

disappearing filament pyrometer *n*: see *optical pyrometer*.

disbonding *n*: a common coating failure in which the coating separates from the pipe.

discharge coefficient *n*: a measure of the efficiency with which gas flows through an actual orifice. Compares the actual orifice's discharge rate with an ideal orifice's discharge rate.

discharge line *n*: a line through which drilling mud travels from the mud pump to the standpipe on its way to the wellbore.

discharge of dredged material *n*: any addition of dredged material into U.S. waters.

discharge valve *n*: on a mud pump, the valve that opens to allow mud to be pushed out of the pump (discharged) by the pistons moving in the liners.

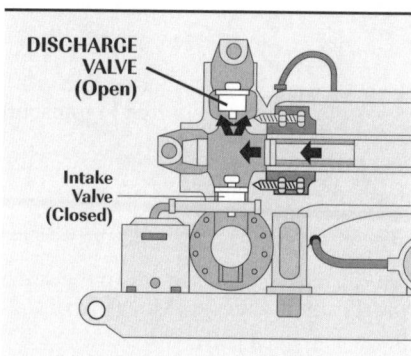

discharge vessel ratio *n*: the total calculated volume (TCV) by vessel measurement on arrival, less remaining on board (ROB), divided by the TCV by shore measurement at discharge.

disclaimer *n*: complete denial and renunciation of any claim to title to property. Surface tenants, for example, often sign disclaimers (in the form of tenant consent agreements) relating to the mineral estate and title of land leased for oil and gas exploration.

disconformity *n*: an interruption in sedimentation caused by erosion; the strata above and below the disconformity are parallel. See *unconformity*. Compare *nonconformity*.

disconnect *n*: a device such as a switch or circuit breaker used to open a circuit.

discontinuous DC current pulse *n*: the intermittent and irregular fluctuation (a pulse) of direct current flowing in a circuit. DC current pulses create heat in the circuit and can affect the operation of solid-state diodes whose temperature rating is not sufficient to withstand the heat created by such pulses.

discounted cash flow rate of return *n*: the rate that causes the sum of the discounted outflows and inflows of funds to equal the net cash outlay in year zero of a project. It is used in evaluating exploration investments.

discounting *n*: taking account of future value in present calculations, i.e., determining present value of future dollars. Discounting is compounding in reverse.

discovery well *n*: the first oil or gas well drilled in a new field that reveals the presence of a hydrocarbon-bearing reservoir. Subsequent wells are development wells.

discrete *adj*: in electronic process control systems, of or pertaining to a device that is either on or off.

discrimination *n*: the ability to sense and record the actual temperature of a liquid to the specified temperature increments.

disk brake *n*: a brake on a drawworks in which two large metal disks are affixed to each end of the drawworks drum. When the brake is actuated, large gripping devices called calipers contact the disks to stop the drawworks from turning. Disk brakes are gradually replacing band brakes on drawworks.

disk lubrication *n*: see *slinger disk lubrication*.

disk-spring compressor valve *n*: a type of compressor valve in which springs and disks are used to open and close the valve.

dispatcher *n*: an employee responsible for scheduling movement of oil through pipelines.

dispersant *n*: a substance added to cement that chemically wets the cement particles in the slurry, allowing the slurry to flow easily without much water.

dispersed mud *n*: a drilling mud to which a chemical—a dispersant—is added to cancel the attractive forces between the particles in the mud that create viscosity. As depth and mud weight increase, these attractive forces may become so high that the mud's viscosity cannot be adequately controlled without dispersants.

dispersed phase *n*: that part of a drilling mud—clay, shale, barite, and other solids—that is dispersed throughout a liquid or gaseous medium, forming the mud.

disperser *n*: see *emulsifying agent*.

dispersible inhibitor *n*: an inhibitor substance that can be dispersed evenly in another liquid with only moderate agitation.

dispersion *n*: a suspension of extremely fine particles in a liquid (such as colloids in a colloidal solution).

dispersion medium *n*: see *continuous phase*.

dispersoid *n*: a colloid or finely divided substance. See also *colloid*.

displacement *n*: 1. the weight of a fluid (such as water) displaced by a freely floating or submerged body (such as an offshore drilling rig). If the body floats, the displacement equals the weight of the body. 2. replacement of one fluid by another in the pore space of a reservoir. For example, oil may be displaced by water.

displacement efficiency *n*: the proportion by volume of oil swept out of the pore space of a reservoir by the encroachment of another fluid. The displacing fluid may be reservoir water or gas or an injected fluid.

displacement engine *n*: see *reciprocating engine*.

displacement fluid *n*: in oilwell cementing, the fluid, usually drilling mud or salt water, that is pumped into the well after the cement is pumped into it to force the cement out of the casing and into the annulus.

displacement meter *n*: 1. a meter in which a piston is actuated by the pressure of a measured volume of liquid, and the volume swept by the piston is equal to the volume of the liquid recorded. 2. a meter in which the measuring element measures a volume of liquid by mechanically separating the liquid into discrete quantities of fixed volume and counting the quantities in volume units.

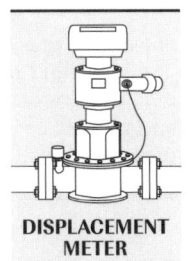

DISPLACEMENT METER

displacement plunger *n*: a device used to pump liquids, usually at high pressures, with an action similar to that of a piston.

displacement rate *n*: a measurement of the speed with which a volume of cement slurry or mud is pumped down the hole.

displacement ton *n*: the weight in long tons (a long ton equals 2,240 pounds) of the water a ship or floating offshore rig displaces. A ship or rig displaces an amount of water whose weight is equal to the ship or rig's weight and its contents.

displacer *n*: a spherical or cylindrical object that is a component part of a pipe prover that moves through the prover pipe. The displacer has an elastic seal that contacts the inner pipe wall of a prover to prevent leakage. The displacer is made to move through the prover pipe by the flowing fluid and displaces a known measured volume of fluid between two fixed detecting devices.

disposal well *n*: a well into which salt water or spent chemical is pumped, most commonly part of a saltwater-disposal system.

dissociation *n*: the separation of a molecule into two or more fragments (atoms, ions) by interaction with another body or by the absorption of electromagnetic radiation.

dissolved gas *n*: natural gas that is in solution with crude oil in the reservoir.

dissolved-gas drive *n*: a source of natural reservoir energy in which the dissolved gas coming out of the oil expands to force the oil into the wellbore. Also called solution-gas drive. See *reservoir drive mechanism*.

dissolved load *n*: in a flowing stream of water, those products of weathering that are carried along in solution. Compare *bed load, suspended load*.

dissolved water *n*: water in solution in oil at a defined temperature and pressure.

distance-velocity lag *n*: the delay caused by the amount of time required to transport material or propagate a signal or condition from one point to another. Also known as transportation lag, transport lag.

distillate *n*: 1. a product of distillation, i.e., the liquid condensed from the vapor produced in a still. Sometimes called condensate. 2. heavy gasoline or light kerosenes used as fuels.

distillate fuel oil *n*: light fuel oils distilled during the refining process and used primarily for space heating, on- and off-highway diesel engine fuel, and electric power generation.

distillation *n*: the process of driving off gas or vapor from liquids or solids, usually by heating, and condensing the vapor back to liquid to purify, fractionate, or form new products.

distilling column *n*: a tall cylindrical steel tower where, in atmospheric distillation, the process takes place at atmospheric (normal) pressure.

distribution *n*: the apportioning of daily production rates to wells on a lease. Because there may be many wells on a lease, such production is apportioned on the basis of periodic tests rather than on the individual receiving and gauging of oil at each well.

distribution main *n*: in a gas distribution system, such as in a city or town, a relatively large pipeline through which gas is supplied to several smaller pipelines. The smaller lines provide gas to homes, schools, and other buildings.

distribution system *n*: a system of pipelines and other equipment by which natural gas or other products are distributed to customers, to lease operations, or to other points of consumption, that is, the mains, services, and equipment that carry or

control the supply of gas from the point of local supply to and including the sales meters.

distribution transformer *n*: an electrical device designed to reduce voltage from primary distribution levels, usually 7,200 or 12,400 volts, to utilization voltages of 480, 240, or 120 volts. See *power transformer*, def. 1.

distributor *n*: a device that directs the proper flow of fuel or electrical current to the proper place at the proper time in the proper amount.

distributor timing *n*: on a diesel engine with a distributor-type fuel injection system, the sequence in which the distributor sends fuel to each engine cylinder.

district regulator station *n*: station to which odorized gas is piped from the city gate station. This station uses regulators and controllers to maintain consistent pressure and to ensure a constant supply of natural gas to consumers. See *city gate station*.

disulfides *n pl*: chemical compounds containing an -S-S linkage. They are colorless liquids completely miscible with hydrocarbons, insoluble in water, and sweet to the doctor test. Mercaptans are converted to disulfides in treating processes employing oxidation reactions.

ditch *n*: a trench or channel made in the earth, usually to bury pipeline, cable, and so on. On a drilling rig, the mud flow channel from the conductor-pipe outlet is often called a ditch. See *mud return line*. *v*: to excavate a trench in which to lay pipe or cable. Ditching equipment and methods for pipe laying vary according to terrain and weather.

ditch breaker *n*: a device that divides a ditch into sections to form internal barriers to water movement. Ditch breakers are used for pipelines in areas where washouts are a threat.

ditch magnet *n*: a powerful magnet placed in the return line between the wellbore and the mud pit to remove metal particles from the drilling mud. A ditch magnet is often used during milling operations.

diurnal tide *n*: a tide having one high water level and one low water level a day.

diverter *n*: in offshore drilling, an assembly of devices used to direct fluids flowing from a well away from the drilling rig. When a kick is encountered at shallow depths, the well often cannot be shut in safely because shutting it in on a shallow formation may create pressures high enough to fracture (break down) the formation. Therefore, a diverter is used. When activated, it allows well fluids to flow through a side outlet to a line (pipe) that carries the well fluids a safe distance away from the rig. A diverter contains a packing element, flow-line seals, and lock-down mechanisms and is run and retrieved with a special handling tool made up on riser pipe.

diverter assembly *n*: see *diverter*.

diverter line *n*: a side outlet on a rig that directs flow away from the rig.

diverter system *n*: see *diverter*.

diverter valve *n*: a type of valve that changes the direction of fluid flow from the intended path to an alternate route.

divided agreement *n*: type of operating agreement that provides for sharing of costs and benefits based on participating areas that may change.

divided interest *n*: a fractional interest in minerals that, when conveyed, gives the new owner a 100% interest in the designated fraction of the described tract. For example, a one-quarter divided interest in an 80-acre (32-hectare) tract results in a 100% interest in 20 specific, describable acres (8 hectares) out of that tract. Compare *undivided interest*.

dividend *n*: in mathematical division, the quantity that is divided by another quantity. For example, in 9 ÷ 3 = 3, 9 is the dividend.

diving bell *n*: a cylindrical or spherical compartment used to transport a diver or dive team to and from an underwater work site.

division order *n*: a contract of sale of oil or gas to a purchaser who is directed to pay for the oil or gas products according to the proportions set out in the division order. The purchaser may require execution thereof by all owners of interest in the property.

division order opinion *n*: a statement of opinion by a title examiner on the state of the title to land, mineral, royalty, or working interests in a producing tract of land. This opinion, usually in letter form, is the basis of payment to all affected owners and must recite all the owners' interests. Compare *drill site opinion, title opinion*.

divisor *n*: in mathematical division, the quantity by which another quantity is divided. For example, in 12 ÷ 3 = 4, 3 is the divisor.

dizzy nut *n*: a mechanism used in packers to lock components together.

dk *abbr*: dark; used in drilling reports.

doctor test *n*: a qualitative method for detecting hydrogen sulfide and mercaptans in petroleum distillates. The test distinguishes between sour and sweet products.

DOE *abbr*: Department of Energy.

dog *n*: 1. a spring-loaded finger in a tubing end locator. 2. any of various simple devices for holding, gripping, or fastening, such as a hook, rod, or a spike with a ring, claw, or lug at the end.

dog collar clamp *n*: see *safety clamp*. Also called collar clamp or drill collar clamp.

doghouse *n*: 1. a small enclosure on the rig floor used as an office for the driller and as a storehouse for small objects. 2. any small building used as an office, a change house, or a place for storage.

dogleg *n*: 1. an abrupt change in direction in the wellbore, frequently resulting in the formation of a keyseat. 2. a sharp bend permanently put in an object such as a pipe, wire rope, or a wire rope sling.

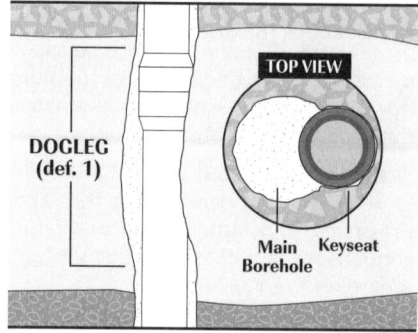

dognut *n*: an upper kelly cock that is enclosed in a ball-shaped housing.

dog-type riser connector *n*: a riser connector that has actuator screw assemblies in the box end. Tightening the screws with pneumatic torque wrenches causes tapered locking dogs to engage mating tapers in the pin end. The application of the correct torque to the actuator screw, combined with the taper geometry of the pin and locking dogs, preloads the connection to the design load rating. Spring-loaded locks prevent vibration from backing out the actuator screws. The actuator screws do not support any of the riser connection's load. See *hydraulic connector*.

DOI *abbr*: Department of the Interior.

dol *abbr*: dolomite; used in drilling reports.

dolly *n*: see *pipe dolly*.

dolly assembly *n*: on a top-drive unit, the device that attaches the unit to the guide rails. See *guide rails, top drive*.

dolo *abbr*: dolomite; used in drilling reports.

dolomite *n*: a type of sedimentary rock similar to limestone but containing more than 50 percent magnesium carbonate; sometimes a reservoir rock for petroleum.

dolomitization *n*: the shrinking of the solid volume of rock as limestone turns to dolomite, i.e., the conversion of limestone to dolomite rock by replacement of a portion of the calcium carbonate with magnesium carbonate.

dolostone *n*: rock composed of dolomite.

dome *n*: a geologic structure resembling an inverted bowl, i.e., a short anticline that dips or plunges on all sides.

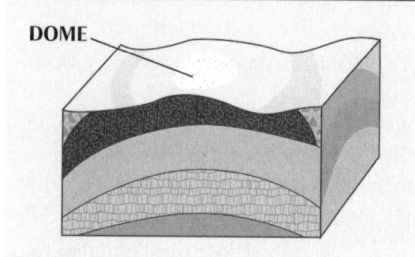

dome plug trap *n*: a reservoir formation in which fluid or plastic masses of rock material originated at unknown depths and pierced or lifted the overlying sedimentary strata.

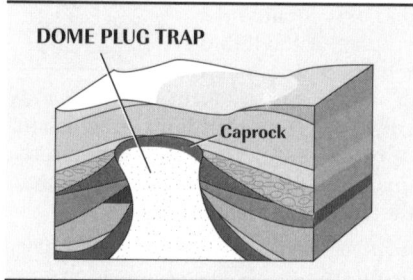

dome-roof tank *n*: a storage tank with a dome-shaped roof affixed to the shell.

doodlebug *n*: (slang) a person who prospects for oil, especially by using seismology. Also called doodlebugger. *v*: (slang) to explore for oil, especially by using seismic techniques in which explosive charges are detonated in shot holes to create shock waves (name taken from the resemblance of these explosions to the puffs of loose dirt thrown up by the doodlebug, or ant lion, when digging its funnel-shaped hole).

door sheet *n*: a plate at the base of a tank shell or wall that is removed so the tank can be cleaned.

dope *n*: a lubricant for the threads of oilfield tubular goods. *v*: to apply thread lubricant.

dope bucket *n*: the container in which dope is stored on the rig floor.

dope pot *n*: a portable container used to melt coal tar enamel and maintain it at the temperature required by a pipe-coating operation.

doping *n*: 1. the addition of impurities to a semiconductor to achieve a desired characteristic, such as producing n- or p-type material. See *n-type structure, p-type structure*. 2. applying lubricant (dope) to a tool joint or other threaded connection.

DOT *abbr*: Department of Transportation.

double *n*: a length of drill pipe, casing, or tubing consisting of two joints screwed together. Compare *fourble, single, thribble*.

double-acting *adj*: in reference to a mud pump, the action of the piston moving mud through the cylinder on both its forward and backward stroke. Compare *single-acting*.

double board *n*: the name used for the working platform of the derrickhand (the monkeyboard) when it is located at a height in the derrick or mast equal to two lengths of pipe joined together. Compare *fourble board, thribble board*.

double-box coupling *n*: a special coupling that has a female connection (box) on each end and that is used to connect sucker rods that have a male connection (pin) on each end.

double-drum hoist *n*: a device consisting of two reels on which wire rope is wound that provides two separate hoisting drums in the assembly. The main drum is used for pulling tubing or drill pipe; the second drum is used for coring and swabbing. See *hoist*.

double extra-strong pipe *n*: a pipe incorporating twice as many safety design factors as normally used.

double grip *n*: a tool employing gripping devices (slips) that limit tool movement from pressure either above or below the tool.

double hull tanker *n*: tanker with two hulls. Theoretically, the first hull absorbs all or most of any impact, leaving the second hull intact or less damaged and thus able to contain the cargo.

double jointing *n*: the process of welding two pipe joints together—usually on a double-joint rack—to form a single piece of pipe. The usual length of a double joint for pipeline construction is 80 feet (24 metres).

double-plane roller assembly *n*: in a kelly bushing, a double set of roller assemblies that are installed on two levels, or planes, within the bushing. Compare *single-plane roller assembly*. See also *kelly bushing, roller assembly*.

double-pole, double-throw (DPDT) switch *n*: a six terminal electrical switch that simultaneously connects one pair of terminals to either of other pairs of terminals.

double-pole, single-throw (DPST) switch *n*: an electrical switch with four terminals that simultaneously opens or closes two separate circuits or both sides of the same circuit.

double-post mast *n*: a well-servicing unit whose mast consists of two steel tubes. Double-pole masts provide racking platforms for handling rods and tubing in stands and extend from 65 to 67 feet (20 to 20.4 metres) so that rods can be suspended as 50-foot (15-metre) doubles and tubing set back as 30-foot (9-metre) singles. See *pole mast*.

double recorder *n*: a device containing an angle indicator, a timing element, and a replaceable chart that is used to measure the direction and the degree of deviation as a hole is being drilled. The timing element goes off when the tool has come to rest just above the drill bit and descends on the sharply pointed angle indicator, which then punches a hole in the chart. The chart rotates 180°, and a second hole is punched to verify a correct reading of the resting angle of the tool.

doughnut *n*: a ring of wedges or a threaded, tapered ring that supports a string of pipe.

dovetail *n*: a cutout section in a cone enabling positive slip movement without the aid of conventional slip return springs.

dowel *n*: a pin fitting into a hole in an abutting piece to prevent motion or slipping.

dower property *n*: in some states, that part of an estate to which a wife is entitled (for her lifetime) on the death of her husband.

downcomer *n*: a tube that conducts liquids downward in a vessel such as an absorber.

downcutting *n*: the direct erosive action of flowing water on a streambed.

downdip *adj*: lower on the formation dip angle than a particular point.

down-draft retort *n*: a device that measures the saturation, or the amount, of each fluid in a core sample by distilling the fluid from the core with heat. The core sample is heated to about 350°F (176.7°C). The fluids are vaporized and cooled until they condense in receiving tubes, where they are measured.

down-flooding angle *n*: in marine architecture, the angle of a vessel at which nonwatertight openings on the main deck start to submerge.

downhole *adj, adv*: pertaining to the wellbore.

downhole motor *n*: a drilling tool made up in the drill string directly above the bit. It causes the bit to turn while the drill string remains fixed. It is used most often as a deflection tool in directional drilling, where it is made up between the bit and a bent sub (or, sometimes, the housing of the motor itself is bent). Two principal types of downhole motor are the positive-displacement motor and the downhole turbine motor. Also called mud motor.

downhole preventer valve *n*: a special valve used in underbalanced drilling; it is run in the drill string and is similar to a drill pipe float. In the downhole preventer valve, however, raising the drill string causes the bit to open the valve. Once open, a keeper, or sleeve, in the upper part of the valve holds it open. Continued upward movement causes the bit to contact the keeper, which closes the valve. With the valve closed against pressure from below, the well can be depressurized and the drill string tripped without stripping or snubbing. When going in the hole, pump pressure opens the valve. See *underbalanced drilling*.

downhole production equipment *n pl*: the tools and devices placed in or on the tubing or casing to produce fluids from a reservoir.

downhole telemetry *n*: signals transmitted from an instrument located downhole to a receiving monitor on the surface—a surface-readout instrument. Downhole telemetry may be transmitted via a special wireline or via mud pulse (much as radio signals are transmitted through the air). Frequently, downhole telemetry is employed in determining the drift angle and direction of a deviated wellbore. See *measurement while drilling*.

downstream *adv, adj*: in the direction of flow in a stream of fluid moving in a line.

downstream market *n*: the sale of products after they are refined or processed.

downstream pipeline *n*: a pipeline receiving natural gas from another pipeline at an interconnection point.

downthrow *n*: the wall of a fault that has moved relatively downward.

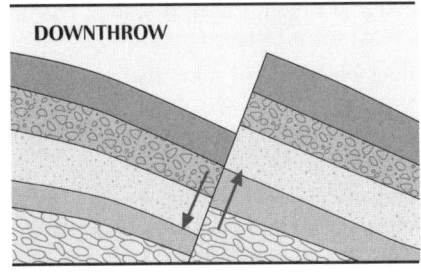

DOWNTHROW

downtime *n*: 1. time during which rig operations are temporarily suspended because of repairs or maintenance. 2. time during which a well is off production.

dozer *n*: a powered machine for earthwork excavation; a bulldozer.

DP *abbr*: drill pipe; used in drilling reports.

DPDT *abbr*: double-pole, double-throw.

DPST *abbr*: double-pole, single-throw.

DR *abbr*: damage ratio.

draft *n*: 1. the vertical distance between the bottom of a vessel floating in water and the waterline. 2. a written order drawn on a solvent bank that authorizes payment or a specified sum of money for a specific purpose to a named person. A draft may or may not be negotiable, depending on how it is drawn up. Also called bank draft.

draft gauge *n*: a modified manometer used to measure very low pressures, such as draft pressure in a furnace, or small differential pressures of less than 2 inches of water (0.07 psi).

draft mark *n*: numbers that show the distance from the bottom of the keel or the lowest projection on a rig to the waterline. Draft marks provide a visual method for a rig operator to keep track of reserve buoyancy and to determine whether the vessel is sitting level in the water.

drag *n*: 1. an excess of draft at the stern of a ship or drill ship, as compared to the bow. 2. friction between a moving device (such as a bit) and another moving or nonmoving part (such as the formation). 3. the bending of rock strata adjacent to a fault.

drag bit *n*: any of a variety of drilling bits that have no moving parts. As they are rotated on bottom, elements of the bit make hole by being pressed into the formation and being dragged across it. See *fishtail bit*.

drag blocks *n pl*: spring-loaded buttons on a packer that provide friction with casing to retard movement of one section of a packer while another section rotates for setting.

drag embedment anchor (DEA) *n*: an anchor type that obtains its holding capacity by being dragged into the seafloor.

drag embedment normal load anchor (DENLA) *n*: a type of vertical loaded anchor (VLA) manufactured by Bruce.

drag fold *n*: minor folds that form in an incompetent bed when competent beds parallel to it and on each side of it move in opposite directions.

drag springs *n pl*: spring-loaded curved metal bands on a packer that operate like drag blocks. See *drag blocks*.

drain *n*: the region into which majority carriers flow in a field-effect transistor. It is comparable to the collector in a bipolar transistor. See *field-effect transistor, majority carrier*.

drainage *n*: the migration of oil or gas in a reservoir toward a wellbore due to pressure reduction caused by the well's penetration of the reservoir. A drainage point is a wellbore (or, in some cases, several wellbores) that drains the reservoir.

drainage clause *n*: clause in a gas sales contract that states that the pipeline company will attempt to take sufficient volume to prevent the seller from being drained.

drainage radius *n*: the area of a reservoir in which a single well serves as a point for drainage of reservoir fluids.

drainage relative permeability *n*: displacement of hydrocarbons or fluids in a reservoir by increasing the nonwetting phase saturation. See *nonwetting phase*.

drainage time *n*: a fixed time period for draining a standard capacity measure calibrated "to deliver." The drain period starts when flow ceases and dripping begins and has the same duration as the drain period used when calibrating the standard.

drain sample *n*: in tank sampling, a sample obtained from the draw-off or discharge valve.

Drake well *n*: the first well drilled in the United States in search of oil. Some 69 feet (21 metres) deep, it was drilled near Titusville, Pennsylvania, and was completed in 1859. It was named after Edwin L. Drake, who was hired by the well owners to oversee the drilling.

draped anticline *n*: an anticline composed of sedimentary deposits atop a reef or atoll, along whose flanks greater thicknesses of sediments have been deposited and compacted than atop the reef itself. Also called compaction anticline.

drawdown *n*: 1. the difference between static and flowing bottomhole pressures. 2. the distance between the static level and the pumping level of the fluid in the annulus of a pumping well.

draw-off level *n*: the height of the valve on a liquid storage tank from which liquid exits the tank.

drawworks *n*: the hoisting mechanism on a drilling rig. It is essentially a large winch

DRAWWORKS

that spools off or takes in the drilling line and thus raises or lowers the drill stem and bit.

drawworks brake *n*: the mechanical brake on the drawworks that can prevent the drawworks drum from moving.

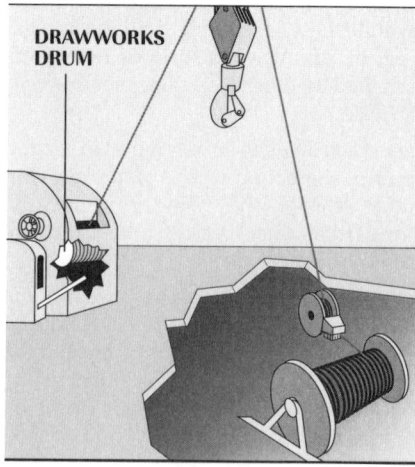

drawworks drum *n*: the spool-shaped cylinder in the drawworks around which drilling line is wound, or spooled.

drawworks-drum socket *n*: a receptacle on the drawworks drum to which the drilling line is attached.

dredged material *n*: material that is excavated or dredged from water.

dress *v*: to sharpen, repair, or add accessories to items of equipment (such as drilling bits and tool joints).

drier *n*: a compound, usually an organic metallic compound that is soluble in organic solvents and binders, which is added to paint to accelerate drying by catalytic oxidation.

drift *n*: 1. an ocean current's speed of motion. 2. an observed change, usually uncontrolled, in meter performance, meter factor, etc., that occurs over a period of time. *v*: 1. to move slowly out of alignment, off center, or out of register. 2. to gauge or measure pipe by means of a mandrel passed through it to ensure the passage of tools, pumps, and so on.

drift angle *n*: the angle at which a wellbore deviates from the vertical, expressed in degrees, as revealed by a directional survey. Also called angle of deviation, angle of drift, and inclination. See *directional survey*.

drift bottle *n*: a bottle released at sea to measure currents. It contains a card stating the date and location of release. The person who finds the bottle writes on the card the date and location of the bottle's recovery and returns the card. Also called a floater.

drift card *n*: a card encased in a buoyant, waterproof envelope that is used to measure currents. The card states the date and location of release. The person who finds it writes on it the date and place that it was found and returns it.

drift diameter *n*: 1. in drilling, the effective hole size. 2. in casing, the guaranteed minimum diameter of the casing. The drift diameter is important because it indicates whether the casing is large enough for a specified size of bit to pass through.

drift indicator *n*: a device dropped or run down the drill stem on a wireline to a point just above the bit to measure the inclination of the well off vertical at that point. It does not measure the direction of the inclination.

drift survey *n*: see *deviation survey*.

drill *v*: to bore a hole in the earth, usually to find and remove subsurface formation fluids such as oil and gas.

drillable *adj*: pertaining to packers and other tools left in the wellbore to be broken up later by the drill bit. Drillable equipment is made of cast iron, aluminum, plastic, or other soft, brittle material.

drillable squeeze packer *n*: a permanent packer, drillable in nature, capable of withstanding extreme working pressures, for remedial work. It has a positive flow-control valve built in.

drill ahead *v*: to continue drilling operations.

drill around *v*: 1. to deflect the wellbore away from an obstruction in the hole. 2. (slang) to get the better of someone (e.g., "He drilled around me and got the promotion").

drill barge *n*: a floating offshore drilling unit with the configuration of a ship but not self-propelled. Compare *drill ship*.

drill bit *n*: the cutting or boring element used for drilling. See *bit*.

drill collar *n*: a heavy, thick-walled tube, usually steel, placed between the drill pipe and the bit in the drill stem. Several drill collars are used to provide weight on the bit and to provide a pendulum effect to the drill stem. When manufactured to

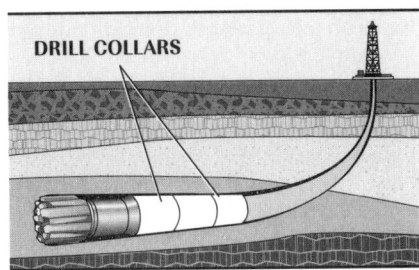

DRILL COLLARS

API specifications, a drill collar joint is 30 or 31 feet (9.14 to 9.45 metres) long. The outside diameter of drill collars made to API specifications ranges from 3⅛ inches to 11 inches (79.38 millimetres to 275 millimetres).

drill collar clamp *n*: see *safety clamp*.

drill collar slips *n pl*: see *slips*.

drill collar sub *n*: a sub made up between the drill string and the drill collars that is used to ensure that the drill pipe and the collar can be joined properly.

drill column *n*: see *drill stem*.

drilled show *n*: oil or gas in the mud circulated to the surface.

drilled solids *n pl*: the fine particles in drilling mud drilled by the bit.

driller *n*: the employee directly in charge of a drilling or workover rig and crew. The driller's main duty is operation of the drilling and hoisting equipment, but this person is also responsible for downhole condition of the well, operation of downhole tools, and pipe measurements.

driller's BOP control panel *n*: a series of controls on the rig floor that the driller manipulates to open and close the blowout preventers.

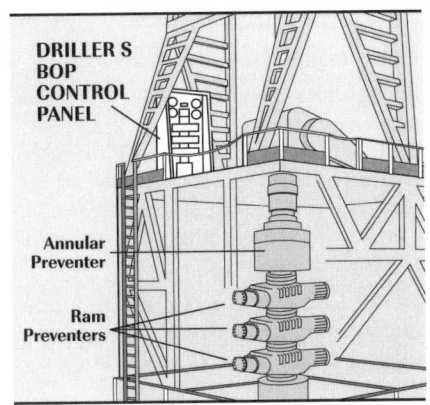

DRILLER'S BOP CONTROL PANEL

Annular Preventer

Ram Preventers

driller's console *n*: a metal cabinet on the rig floor containing the controls that the driller manipulates to operate various components of the drilling rig.

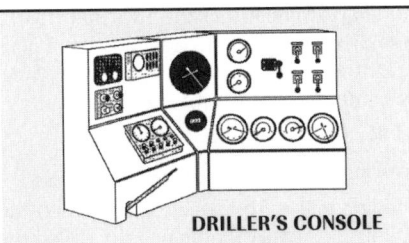

DRILLER'S CONSOLE

driller's control panel *n*: see *driller's console*.

driller's log *n*: a record that describes each formation encountered and lists the drilling time relative to depth, usually in 5- to 10-foot (1.5- to 3-metre) intervals.

driller's method *n*: a well-killing method involving two complete and separate circulations. The first circulates the kick out of the well; the second circulates heavier mud through the wellbore.

driller's panel *n*: see *driller's console*.

driller's position *n*: the area immediately surrounding the driller's console.

driller's report *n*: a record kept on the rig for each tour to show the footage drilled, tests made on drilling fluid, bit record, and other noteworthy items. It is usually telephoned or radioed to the drilling contractor's main office every morning and is therefore sometimes called the morning report.

driller's side *n*: a panel, or console, on the left side of the drawworks (looking at it from the front).

drill floor *n*: also called rig floor or derrick floor. See *rig floor*.

drill in *v*: to penetrate the productive formation after the casing is set and cemented on top of the pay zone.

drill-in fluid *n*: a drilling fluid specially formulated to minimize formation damage as the borehole penetrates the producing zone. See *formation damage*. Compare *completion fluid*.

drilling and delay rental clause *n*: the clause in an oil and gas lease that allows the lease to expire after a given period of time (often 1 year from the date of the lease) unless drilling begins or delay rental is paid. Also called "unless" clause.

drilling and spacing unit *n*: a parcel of land of the size required or permitted by statutory law or by regulations of a state conservation body for drilling an oil or gas well. Also called a proration unit.

drilling blind *n*: drilling without mud returns, as in the case of severe lost circulation.

drilling block *n*: a lease or a number of leases of adjoining tracts of land that constitute a unit of acreage sufficient to justify the expense of drilling a wildcat.

drilling break *n*: a sudden increase in the drill bit's rate of penetration. It sometimes indicates that the bit has penetrated a high-pressure zone and thus warns of the possibility of a kick.

drilling contract *n*: an agreement made between a drilling company and an operating company to drill and complete a well. It sets forth the obligation of each party, compensation, identification, method of drilling, depth to be drilled, and so on.

drilling contractor *n*: an individual or group of individuals who own a drilling rig and contract their services for drilling wells to a certain depth.

drilling control *n*: a device that controls the rate of penetration by maintaining a predetermined constant weight on the bit. Also called an automatic driller or automatic drilling control unit.

drilling crew *n*: a driller, a derrickhand, and two or more helpers who operate a drilling or workover rig for one tour each day.

drilling deployment valve *n*: a downhole valve that closes off the wellbore at a point in the casing. Once the valve is closed, the bottom of the hole opposite the reservoir comes into balance, or equilibrium. Above the valve, no excess pressure occurs; thus, the drill string can be run without stripping or snubbing. During completions, the pipe can be tripped without having to kill the well with mud.

drilling draft *n*: the depth to which the pontoons on a semisubmersible rig are submerged below the water's surface when the rig is drilling. Compare *survival draft*.

drilling engine *n*: an internal-combustion engine used to power a drilling rig. From two to six engines are used on a rotary rig and are usually fueled by diesel fuel, although liquefied petroleum gas, natural gas, and, very rarely, gasoline can also be used.

drilling engineer *n*: an engineer who specializes in the technical aspects of drilling.

drilling fluid *n*: circulating fluid, one function of which is to lift cuttings out of the wellbore and to the surface. It also serves to cool the bit and to counteract downhole formation pressure. Although a mixture of barite, clay, water, and other chemical additives is the most common drilling fluid, wells can also be drilled by using air, gas, water, or oil-base mud as the drilling mud. Also called circulating fluid, drilling mud. See *mud*.

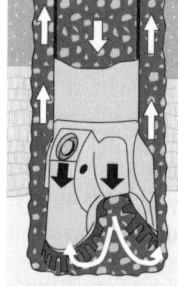

DRILLING FLUID

drilling fluid analysis *n*: see *mud analysis*.

drilling-fluid cycle time *n*: the time required for the pump to move drilling fluid down the hole and back to the surface. The cycle in minutes equals the barrels of mud in the hole divided by barrels per minute.

drilling foreman *n*: the supervisor of drilling or workover operations on a rig. Also called a rig manager, rig supervisor, rig superintendent, or toolpusher.

drilling guide base *n*: see *temporary guide base*.

drilling head *n*: a special rotating head that has a gear-and-pinion drive arrangement to allow turning of the kelly and simultaneous sealing of the kelly against well pressure.

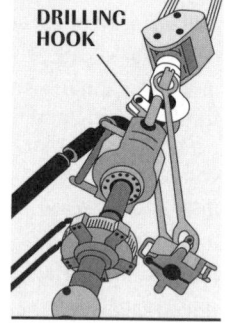

DRILLING HOOK

drilling hook *n*: the large hook mounted on the bottom of the traveling block and from which the swivel is suspended. When drilling, the entire weight of the drill stem is suspended from the hook.

drilling in *n*: the operation during the drilling procedure at the point of drilling into the pay formation.

drilling jar *n*: a special device on top of the drill stem which allows the driller to strike very heavy upward or downward blows on stuck pipe.

drilling line *n*: a wire rope used to support the drilling tools. Also called the rotary line.

drilling mode *n*: the state a drilling rig, especially an offshore drilling rig, is in when it is drilling a well.

drilling motor *n*: see *downhole motor*.

drilling motor head assembly *n*: in coiled tubing operations, a device installed in the tubing string to which a drilling motor is attached. The motor head assembly is designed to withstand vibration and torsional loading associated with downhole motors and hammers.

drilling mud *n*: a specially compounded liquid circulated through the wellbore during rotary drilling operations. See *drilling fluid, mud*.

drilling out *n*: the operation during the drilling procedure when the cement is drilled out of the casing before further hole is made or completion is attempted.

drilling parameters *n pl*: factors that affect a drilling operation, such as the rate of penetration, pump rate, rotary revolutions per minute (rpm), and weight on bit.

drilling pattern *n*: see *well spacing*.

drilling platform rig *n*: see *platform rig*.

drilling rate *n*: the speed with which the bit drills the formation; usually called the rate of penetration (ROP).

drilling recorder *n*: an instrument that records hook load, rate of penetration, rotary speed and torque, pump rate and pressure, mud flow, and so forth during drilling.

drilling rig *n*: see *rig*.

drilling slot *n*: see *keyway*.

drilling spool *n*: a fitting placed in the blowout preventer stack to provide space between preventers for facilitating stripping operations, to permit attachment of choke and kill lines, and for localizing possible erosion by fluid flow to the spool instead of to the more expensive pieces of equipment.

drilling superintendent *n*: an employee, usually of a drilling contractor, who is in charge of all drilling operations that the contractor is engaged in. Also called a drilling foreman, rig manager, rig supervisor, or toolpusher.

drilling template *n*: see *temporary guide base*.

drilling tender *n*: a combination barge-and-platform design that can be towed to a new location after a well is drilled.

drilling time log *n*: a record of the time required for a hole to be drilled; it is a record of the rate of penetration of the bit through various formations encountered during the drilling of a well.

drilling under pressure *n*: the continuation of drilling operations while maintaining a seal at the top of the wellbore to prevent the well fluids from blowing out.

drilling unit *n*: the acreage allocated to a well when a regulatory agency grants a well permit.

drilling with casing (DWC) *n*: the drilling of a well using a special cutting tool on the bottom of a casing string, which, when rotated, drills the hole. By eliminating the need for drill pipe, drilling with casing speeds the drilling process because, once the desired depth is reached, it is not necessary to pull a drill string and run casing. The casing is already in place and can be cemented in place immediately.

drill-off test *n*: a method of determining optimum weight on bit and overall bit performance. A given amount of weight is put on the bit and the drawworks brake is tied off so that no more weight is applied to the bit as it drills. The time it takes for the bit to stop drilling ahead with the given amount of weight is measured. Different weights are applied and the times are compared to determine which amount is best.

drill out *v*: 1. to remove with the bit the residual cement that normally remains in the lower section of casing and the wellbore after the casing has been cemented. 2. to remove the settlings and cavings that are plugged inside a hollow fish (such as drill pipe) during a fishing operation.

drillout tool *n*: a device used to remove settlings and cavings that are plugged inside a hollow fish (such as drill pipe) during a fishing operation. It consists of a body inside of which is a hexagonal kelly tipped with a junk mill. It is run on drill pipe to the top of the fish.

drill pipe *n*: seamless steel or aluminum pipe made up in the drill stem between the kelly or top drive on the surface and the drill collars on the bottom. During drilling, it is usually rotated while drilling fluid is circulated through it. Drill pipe joints are available in three ranges of length: 18 to 22 feet (5.49 to 6.71 metres), 27 to 30 feet (8.23 to 9.14 metres), 38 to 45 feet (11.58 to 13.72 metres). The most popular length is 27 to 30 feet.

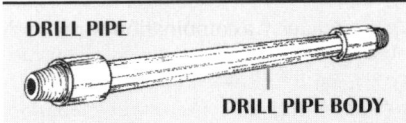

DRILL PIPE   DRILL PIPE BODY

drill pipe body *n*: the tubular part of the drill pipe between the tool joints, which are on each end of the pipe.

drill pipe cutter *n*: a tool to cut drill pipe stuck in the hole. Tools that cut the pipe either internally or externally, permitting some of it to be withdrawn, are available. Jet cutoff or chemical cutoff is also used to free stuck pipe.

drill pipe float *n*: a check valve installed in the drill stem that allows mud to be pumped down the drill stem but prevents flow back up the drill stem. Also called a float.

drill pipe pressure *n*: the amount of pressure exerted inside the drill pipe as a result of circulating pressure, entry of formation pressure into the well, or both.

drill pipe pressure gauge *n*: an indicator, mounted in the mud circulating system, that measures and indicates the amount of pressure in the drill stem. See *drill stem*.

drill pipe pressure loss *n*: the drop in pressure caused by friction as a fluid flows through the drill pipe. The walls of the pipe and internal forces in the fluid itself create friction and this friction reduces the pressure. See *friction loss*.

drill pipe protector *n*: an antifriction device of rubber or steel attached to each joint of drill pipe to minimize wear.

drill pipe rubber *n*: a rubber or elastomer disk that is placed around a joint of drill pipe and is held stationary below the rotary table. As pipe is removed from the well, the rubber wipes mud off the outside of the pipe to minimize corrosion.

drill pipe safety valve *n*: see *drill stem safety valve*.

drill pipe slips *n pl*: see *slips*.

drillship *n*: a self-propelled floating offshore drilling unit that is a ship constructed to permit a well to be drilled from it. Although not as stable as semisubmersibles, drillships are capable of drilling exploratory wells in deep, remote waters. See *floating offshore drilling rig*.

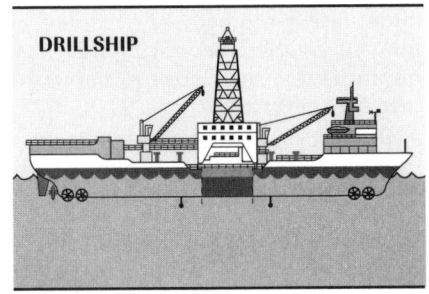

DRILLSHIP

drill site *n*: the location of a drilling rig.

drill site opinion *n*: the written statement of opinion for a title examiner on the status of the title to a drill site, usually in letter form. Compare *division order opinion, title opinion*.

drill stem *n*: all members in the assembly used for rotary drilling from the swivel to the bit, including the kelly, the drill pipe and tool joints, the drill collars, the stabilizers, and various specialty items. Compare *drill string*.

drill stem safety valve *n*: a special valve normally installed below the kelly. Usually, the valve is open so that drilling fluid can flow out of the kelly and down the drill stem. It can, however, be manually closed with a special wrench when necessary. In one case, the valve is closed and broken out, still attached to the kelly to prevent drilling mud in the kelly from draining onto the rig floor. In another case, when kick pressure inside the drill stem exists, the drill stem safety valve is closed to prevent the pressure from escaping up the drill stem. Also called lower kelly cock, mud saver valve.

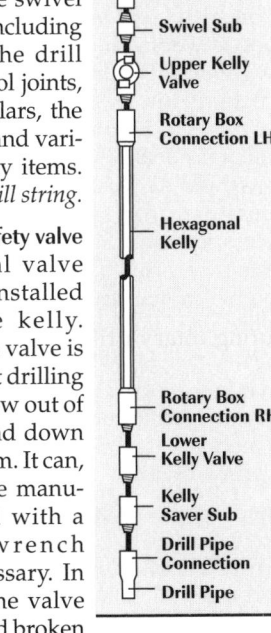

DRILL STEM
- Swivel
- Swivel Sub
- Upper Kelly Valve
- Rotary Box Connection LH
- Hexagonal Kelly
- Rotary Box Connection RH
- Lower Kelly Valve
- Kelly Saver Sub
- Drill Pipe Connection
- Drill Pipe

drill stem test (DST) *n*: the conventional method of formation testing. The basic drill stem test tool consists of a packer or packers, valves or ports that may be opened and closed from the surface, and two or more pressure-recording devices. The tool is lowered on the drill string to the zone to be tested. The packer or packers are set to isolate the zone from the drilling fluid column. The valves or ports are then opened to allow for formation flow while the recorders chart static pressures. A sampling chamber traps clean formation fluids at the end of the test. Analysis of the pressure charts is an important part of formation testing.

drill stem test gases *n pl*: formation gases that are produced during a drill stem test, which is the conventional method of testing formations. Although these gases typically are produced only in very small quantities and consist mainly of natural gas, they become an environmental concern when the gases fall into the extremely hazardous substance category under SARA and are produced in sufficient quantities so that they exceed the TPQ limit set for that substance. Typically, wells are not set up for production operations when the drill stem test is run; consequently, gas-containment equipment is not installed or operational during the running of the test and such gases must be released or flared.

drill string *n*: the column, or string, of drill pipe with attached tool joints that transmits fluid and rotational power from the kelly to the drill collars and the bit. Often, especially in the oil patch, the term is loosely applied to include both drill pipe and drill collars. Compare *drill stem*.

drill string float *n*: a check valve in the drill string that will allow fluid to be pumped into the well but will prevent flow from entering the string.

drill to granite *v*: to drill a hole until basement rock is encountered, usually in a wildcat well. If no hydrocarbon-bearing formations are found above the basement, the well is assumed to be dry. The term comes from the fact that basement rock is sometimes granite.

drill under pressure *v*: to carry on drilling operations with a mud whose density is such that it exerts less pressure on bottom than the pressure in the formation while maintaining a seal (usually with a rotating head) to prevent the well fluids from blowing out under the rig. Drilling under pressure is advantageous in that the rate of penetration is relatively fast; however, the technique requires extreme caution.

drip *n*: 1. the water and hydrocarbon liquids that have condensed from the vapor state in the natural gas flow line and accumulated in the low points of the line. 2. the receiving vessel that accumulates such liquids.

drip accumulator *n*: the device used to collect liquid hydrocarbons that condense out of a wet gas traveling through a pipeline.

drip gasoline *n*: hydrocarbon liquid that separates in a pipeline transporting gas from the well casing, lease separation, or other facilities and drains into equipment from which the liquid can be removed.

drip lubrication *n*: a method of lubricating a chain-and-sprocket drive in which a reservoir drips oil onto the chain at 4 to 20 drops per minute, depending on the speed of the drive. A drip lubricator for a multistrand drive has a pipe to distribute the oil to all strands.

drip pot *n*: a device installed in a flow recorder's manifold to collect liquid that may condense out the gas in the manifold and to minimize the chance of inaccurate differential and static readings caused by liquid in the manifold lines.

drip time *n*: the amount of time it takes for the liquid clinging to the sides of a tank prover to fall and drip out of the vessel.

drive *n*: 1. the means by which a machine is given motion or power, or by which power is transferred from one part of a machine to another. 2. the energy of expanding gas, inflowing water, or other natural or artificial mechanisms that forces crude oil out of the reservoir formation and into the wellbore. *v*: to give motion or power.

drive bushing *n*: see *kelly bushing*.

drive chain *n*: a chain by means of which a machine is propelled.

drive dog *n*: in a gas meter, part of the mechanism that drives the meter.

drive-in unit *n*: a type of portable service or workover rig that is self-propelled, using power from the hoisting engines. The driver's cab and steering wheel are mounted on the same end as the mast support; thus the unit can be driven straight ahead to reach the wellhead. See *carrier rig*.

driven and grouted pile anchor *n*: similar to a driven pile anchor except it is installed by first drilling a hole in the seafloor and then being cemented into this hole after insertion.

driven pile anchor *n*: usually a long, slender tube open at the top and bottom which achieves its holding capacity through skin friction of the tube's surface and the surrounding seafloor. It is installed by being driven into the seafloor.

driven shaft *n*: in a chain-and-sprocket drive, the shaft that receives the power. Compare *driving shaft*.

drive pipe *n*: see *conductor casing*.

drive rollers *n pl*: see *kelly bushing rollers*.

drive shaft *n*: in a top drive, the shaft that comes out of the drive's drilling motor. Usually, crew members attach a saver sub (a short pipe fitting) to the drive shaft and attach the top joint of drill pipe to the sub. The drive shaft transmits the motor's rotating force to the drill string and bit. See *top drive*.

driving fluid *n*: in a hydraulic coupling or torque converter, the moving fluid that impinges against blades to move the member being driven.

driving shaft *n*: in a chain-and-sprocket drive, the shaft that provides the power. Compare *driven shaft*.

drlg *abbr*: drilling; used in drilling reports.

drogue *n*: a current-measuring instrument that is suspended by a buoy and lowered to a specified ocean depth. The effect of the current on the drogue is shown by the movement of the buoy at the surface.

drop off *v*: to reduce the rate of inclination, or drift angle, in the hole, the objective being to bring the hole back towards vertical. Usually expressed as degrees per 100 feet (30 metres).

drum *n*: 1. a cylinder around which wire rope is wound in the drawworks. The drawworks drum is that part of the hoist on which the drilling line is wound. 2. a steel container of general cylindrical form. Some refined products are shipped in steel drums with capacities of about 50 to 55 U.S. gallons, or about 200 litres.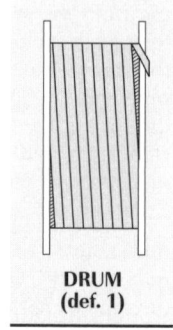

DRUM (def. 1)

drum brake *n*: a device for arresting the motion of a machine or mechanism by means of mechanical friction; in this case, a shoe is pressed against a turning drum.

drum clutch *n*: a type of friction clutch that has a metal housing lined with an expandable diaphragm and pads of friction material called friction shoes. The pressure of the expanded diaphragm squeezes the friction shoes against a drum attached to the driving shaft. Friction between the shoes and the drum allows the clutch to engage. See *clutch*. Compare *plate clutch*.

drum (rope) *n*: a rotating cylinder with side flanges on which rope used in machine operation is wrapped.

drumshaft *n*: in the drawworks, the axle on which the drawworks drum rotates. See *drawworks*.

dry *n*: a hole is dry when the reservoir it penetrates is not capable of producing hydrocarbons in commercial amounts.

dry air cleaner *n*: on an engine, a device that contains an air filter element that does not depend on oil to effectively filter the air entering the engine. Instead, the filter element has a number of folds and chambers that trap dust and dirt going into the engine's air intake. Some dry elements may be cleaned and reinstalled; others are discarded and replaced. Compare *wet air cleaner*.

dry bed *n*: the solid adsorption materials such as molecular sieves, charcoal, or other materials used for purifying or for recovering liquid from a gas. See *adsorption*.

dry-bed dehydrator *n*: a device for removing water from gas in which two or more packages, or beds, of solid desiccant are used. The desiccant bed is usually contained in a tower and wet gas is sent through one bed for drying while the other is prepared for later use.

dry-bottom prover *n*: a type of open tank prover that has a gauge glass on the neck of the prover but not on the bottom of the prover. Compare *wet-bottom prover*.

dry Btu *n*: a measure of heating value for natural gas that is free of moisture. Contractually, natural gas may be defined as "dry" even though it does contain some water vapor, typically less than 7 pounds per million cubic feet (3.2 kilograms per million cubic metres). This standard of measurement reflects the conditions under which natural gas is usually actually delivered for first sales.

dry cell *n*: a primary cell, such as a flashlight battery, in which the electrolyte is a paste. The term "dry" is misleading, because moisture is necessary for the electrolyte to function.

dry combustion *n*: the use of air as the only injected heat carrier during an in situ combustion operation. See *in situ combustion*. The main difficulty with dry combustion is that about 80 percent of the injected heat is lost to the formation. Compare *wet combustion*.

dry drilling *n*: a drilling operation in which no fluid is circulated back up to the surface (often as a result of lost circulation). However, fluid is usually circulated into the well to cool the bit. See *blind drilling*.

dry foam *n*: in underbalanced drilling with foam as the circulating fluid, a foam in which a minimum amount of water is contained in its makeup. Compare *wet foam*. See *foam*.

dry gas *n*: 1. gas whose water content has been reduced by a dehydration process.

2. gas containing few or no hydrocarbons commercially recoverable as liquid product. Also called lean gas.

dry glycol *n*: glycol that has not absorbed water.

dry hole *n*: any well that does not produce oil or gas in commercial quantities. A dry hole may flow water, gas, or even oil, but not in amounts large enough to justify production.

dry hole clause *n*: a clause in an oil and gas lease that allows the operator to keep the lease if he or she drills a dry hole. The operator has a specified period of time in which to drill a subsequent well or begin paying delay rentals again.

dry hole letter *n*: a form of support agreement in which the contributing company agrees to pay so much per foot drilled by another company in return for information gained from the drilling. The contribution is paid only if the well is a dry hole in all formations encountered in drilling.

dry hole money *n*: money paid by a contributing company on the basis of so much per foot drilled by the primary company in return for information gained from the drilling. The contribution is paid only if the well is a dry hole in all formations encountered in drilling.

drying oil *n*: an oil that readily takes oxygen from the air and in so doing changes into a relatively hard, tough, and elastic substance. It is often added to paints to speed drying.

dry monsoon *n*: a winter monsoon that exhibits little or no rainfall.

dry oil *n*: oil that has been treated so that only small quantities of water and other extraneous materials remain in it.

dry string *n*: the drill pipe from which drilling mud has been emptied as it is pulled out of the wellbore.

dry suit *n*: a protective diving garment that is completely sealed to prevent water entry. It is designed to accommodate a layer of insulation between the diver and the suit. This insulation gives a diver maximum warmth and protection from the water.

dry welding *n*: arc, gas, or plasma welding performed in an underwater habitat with a gas environment at ambient pressure.

dry well *n*: see *dry hole*.

D shackle *n*: shackle type shaped like a D. It is used to connect wire and/or chain sections.

DST *abbr*: drill stem test.

DST tool *n*: drill stem test tool; used for formation evaluation.

DTL *abbr*: dynamic tension limit.

dual-bore packer *n*: a packer used in multiple completions that allows the formations opposite to it to be independently produced through separate tubing strings. A dual-bore packer can also permit gas or water and gas to be injected into one formation while producing from the other.

dual casing string *n*: a casing system sometimes used in underbalanced drilling that consists of a special wellhead from which a string of centralized and streamlined casing is suspended inside the intermediate casing. Gas is injected into the annulus between the two strings. When a production string is set, or when the well is finished drilling, the streamlined casing is pulled and saved for the next job.

dual-chamber fitting *n*: an orifice meter fitting having two chambers. The orifice plate is rolled from one chamber to another; the lower chamber is isolated from the upper chamber so that the gas flow can continue unimpeded while the plate is removed. Compare *single-chamber fitting*.

dual completion *n*: a single well that produces from two separate formations at the same time. Production from each zone is segregated by running two tubing strings with packers inside the single string of production casing, or by running one tubing string with a packer through one zone while the other is produced through the annulus. In a miniaturized dual completion, two separate 4.5-inch (11.4-centimetre) or smaller casing strings are run and cemented in the same wellbore.

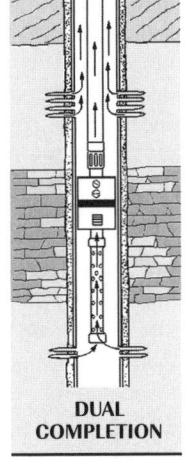

DUAL COMPLETION

dual gradient drilling *n*: in deepwater offshore drilling, the utilization of a subsea mudlift system or a special riser system that allows the riser to be full of seawater (or other relatively light-density fluid) and the borehole to be full of drilling mud of relatively high density. Dual gradient drilling can be advantageous in deepwater drilling because it can reduce the number of casing strings needed to drill the well. See *subsea mudlift drilling*.

dual gradient riser *n*: a special marine riser pipe that incorporates a large-diameter drilling fluid return line and a riser isolation tool. This special riser allows seawater or other relatively light-density fluid to be maintained in the riser while the borehole below the seabed is simultaneously being

circulated with drilling mud of relatively high density. See *dual gradient drilling, subsea mudlift drilling*.

dual induction focused log *n*: a log designed to provide the resistivity measurements necessary to estimate the effects of mud filtrate invasion into the formation surrounding the wellbore so that more reliable values for the true formation resistivity may be obtained. The resistivity curves on this log are made by deep-, medium-, and shallow-investigation induction. Visual observation of the dual induction focused log can give valuable information regarding invasion, porosity, and hydrocarbon content. The three curves on the log can be used to correct for deep invasion and obtain a better value for formation resistivity. Compare *induction survey*.

dual meter counter shifter *n*: an arrangement for connecting two or more meter counters to enable shifting the registration from one counter to another.

dual plug placement *n*: the act of setting two cement plugs in a well. The depth of the first plug is carefully noted so that the second plug can be accurately placed at a given height above the first plug.

dual sheave *n*: in a riser tensioner system, a pair of pulleys (sheaves) that is mounted on the blind end of the system's cylinder and over which the tension lines are reeved.

dual-trace oscilloscope *n*: an oscilloscope that can compare two waveforms on the face of its single cathode-ray tube.

duck's nest *n*: a relatively small, excavated earthen pit into which are channeled quantities of drilling mud that overflow the usual pits.

ductile *adj*: capable of being permanently drawn out without breaking (e.g., wire may be ductile).

ductility *n*: measure of steel's ability to withstand a crack or flaw without fracturing. It can be altered by changing chemical composition, microstructure, and heat treatment. Compare *brittleness*.

dummy pipe *n*: the pipe used to slickbore a road. The dummy pipe is bored through first to create the hole, and the carrier pipe is welded to it. The dummy is pulled through, thereby positioning the carrier pipe.

dummy valve *n*: a blanking valve placed in a gas-lift mandrel to block off annular communication to the tubing.

dump *n*: the volume of oil delivered to a pipeline in a complete cycle of a measuring tank in a LACT installation. A series of such dumps covered by a single scheduling ticket is called a run.

dump bailer *n*: a bailing device with a release valve, usually of the disk or flapper type, used to place, or spot, material (such as cement slurry) at the bottom of the well.

dump meter *n*: a liquid-measuring device consisting of a small tank with narrowed sections at top and bottom that automatically fills and empties itself and records the number of dumps.

dump tank *n*: a calibrated metering tank designed to release a specific volume of liquid automatically. Also called a measuring tank.

dump valve *n*: the valve through which oil and water are discharged from separators, treaters, and so on. It is usually a motor valve, but may be a liquid-level controller as well.

dunefield *n*: an accumulation of wind-borne sand on that part of the seashore that lies above storm-flood level.

duplex pump *n*: a reciprocating pump with two pistons or plungers and used extensively as a mud pump on drilling rigs.

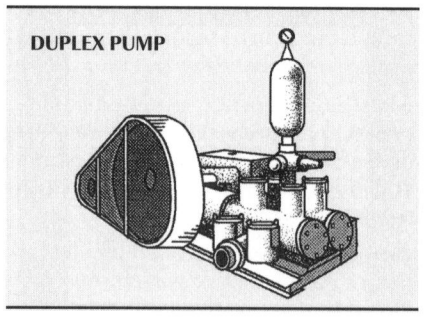

DUPLEX PUMP

duster *n*: a dry hole.

dusting *n*: drilling with dry air or gas.

dutchman *n*: 1. the portion of a stud or screw that remains in place after the head has inadvertently been twisted off. 2. a tool joint pin broken off in the drill pipe box or drill collar box.

DV tool *n*: originally a trademarked name. It is a stage cementing tool, used in selective zone primary cementing.

DWC *abbr*: drilling with casing.

dwell *n*: in conventional automobile ignition systems (those that are not solid-state), the number of degrees through which a distributor cam rotates from the time that the contact points close to the time that they open again. Also called dwell angle.

dwell angle *n*: see *dwell*.

dwell meter *n*: an electronic measuring device that, when properly connected to an automobile's ignition system, measures the number of degrees through which a distributor cam rotates from the time that the contact points close to the time that they open again (the dwell).

dwt *abbr*: deadweight ton.

Dyna-Drill™ *n*: trade name for a downhole motor driven by drilling fluid that imparts rotary motion to a drilling bit connected to the tool, thus eliminating the need to turn the entire drill stem to make hole. Used in straight and directional drilling.

Dynaflex™ tool *n*: the trade name for a directional drilling tool that deflects the drilling assembly off vertical without having to be pulled from the hole. The device that causes the tool to be deflected can be caught and retrieved with a wireline.

dynamic *adj*: moving, active. Compare *static*.

dynamic analysis *n*: the effects of vessel motions and mooring line dynamics on line tensions. These effects can be analyzed using either frequency domain or time domain analysis methods.

dynamic equilibrium *n*: a condition in which several processes act simultaneously to maintain a system in an overall state that does not change with time.

dynamic fluid level *n*: the point to which the static fluid level drops in the casing or tubing of a well under producing conditions.

dynamic load *n*: in drilling from an offshore floating rig, a force created by such environmental factors as wind, waves, and currents. Dynamic loads affect such rig movements as roll, pitch, yaw, sway, heave, and surge.

dynamic loading *n*: the exerting of a force, such as weight, on an object with continuous movement; cyclic stressing.

dynamic miscibility *n*: the ability of substances to mix while moving.

dynamic positioning *n*: a method by which a floating offshore drilling rig is maintained in position over an offshore well location without the use of mooring anchors. Generally, several propulsion units, called thrusters, are located on the hulls of the structure and are actuated by a sensing system. A computer to which the system

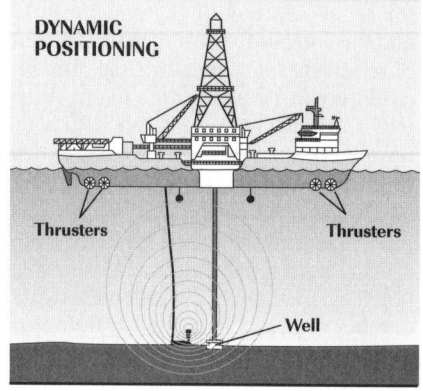

DYNAMIC POSITIONING

Thrusters   Thrusters

Well

feeds signals directs the thrusters to maintain the rig on location.

dynamic positioning operator *n*: an employee on a drill ship or semisubmersible drilling rig whose primary duty is to monitor, operate, and maintain the equipment that maintains the rig on station while drilling.

dynamic pressure *n*: the sum of two deviation forces, one caused by an attractive force between molecules, the other, by a repulsive force between the same molecules.

dynamic stability *n*: in a ship or offshore drilling rig, the characteristic of the vessel that causes it, when disturbed from an original state of steady motion in an upright position, to damp the oscillations set up by restoring moments and to gradually return to its original state.

dynamic tensioning *n*: a sophisticated monitoring system for laying pipe offshore. It is used to control pipe release off a stinger. This system compensates immediately for horizontal and vertical wave motion by either paying out the pipe string or hauling it back. Dynamic tensioning maintains a constant level of tension on the pipe and prevents excessive stress on the line.

dynamic tension limit (DTL) *n*: in a riser tensioning system, the maximum allowable pressure, multiplied by the effective hydraulic area, divided by the number of line parts. It is determined from the equation $D_{tl} = P_a \times A_{cyl} \div N_{lp}$ where D_{tl} = dynamic tension limit, P_a = maximum allowable system operating pressure, A_{cyl} = effective hydraulic area, and N_{lp} = number of line parts.

dynamometer *n*: 1. an instrument or assembly of instruments used to measure torque and other force-related properties of rotating or reciprocating machinery. 2. in sucker rod pumping, a device used to indicate a variation in load on the polished rod as the rod string reciprocates. A continuous record of the result of forces acting along the axis of the polished rod is provided on a dynamometer card, from which an analysis is made of the performance of the well pumping equipment.

dynamometer card *n*: a continuous record made by the dynamometer of the result of forces acting along the axis of the polished rod in sucker rod pumping.

dyne *n*: the unit of force in the centimetre-gram-second system of units, equal to the force that gives an acceleration of 2 centimetres per second squared to a 1 gram mass.

dyneema *n*: a type of high modulus polyethylene (HMPE) fiber made by Axo Chemicals used in synthetic rope construction.

E *abbr*: volt.

E&P *abbr*: 1. exploration and production. 2. those activities that include subsurface studies, seismic and geophysical activities, locating underground hydrocarbon deposits, drilling for hydrocarbon deposits and bringing hydrocarbons to the surface, well completion, and field processing of hydrocarbons prior to entering the pipeline. 3. the upstream end of the petroleum industry.

ears *n pl*: protrusions added to a traveling block to hold the elevators.

earth lifting *n*: see *pressure parting*.

Earth Resource Technology Satellite (ERTS) *n*: see *Landsat*.

easement *n*: a right that one individual or company has on the surface of another's land. In the petroleum industry, it usually refers to the permission given by a landowner for a pipeline or access road to be laid across his or her property.

ebb current *n*: the movement of the tidal current away from the shore.

ebullition *n*: boiling, especially as applied to a system to remove heat from engine jacket water, wherein the water is permitted to boil and the evolved vapors are condensed in air-fin coolers.

eccentric bit *n*: a bit that does not have a uniformly round cross section; instead, the bit has a protuberance that projects from one side. An eccentric bit drills a hole slightly overgauge to compensate for certain formations, such as shale or salt, that deform and enlarge after being drilled. Eccentric bits can also ream undergauge holes.

eccentric fairlead *n*: a wire or chain fairlead where a wire or chain section does not pass through bearings, but where a sheave or wildcat is mounted eccentrically from the bearings.

ECD *abbr*: equivalent circulating density.

ecology *n*: science of the relationships between organisms and their environment.

economic out (eco-out) *n*: provision in gas purchase contracts that permits the purchaser to rescind contracts involving deregulated natural gas if prices of competitive fuels drop to the extent that distributors cannot compete in selling the gas.

Economic Regulatory Administration (ERA) *n*: federal agency that administers federal energy programs other than those regulated by the Federal Energy Regulatory Commission. See *Department of Energy, Federal Energy Regulatory Commission*.

eco-out *abbr*: economic out.

eddy current *n*: an electric current induced within the body of a conductor when the conductor either moves through a magnetic field or is in a region where a change occurs in the magnetic lines of force.

eddy-current loss *n*: in the core of a transformer, losses in power caused by eddy currents in the core. Eddy current is induced in the core by the same magnetic lines of force (flux) that induce voltage in the windings.

edgewater *n*: the water that touches the edge of the oil in the lower horizon of a formation.

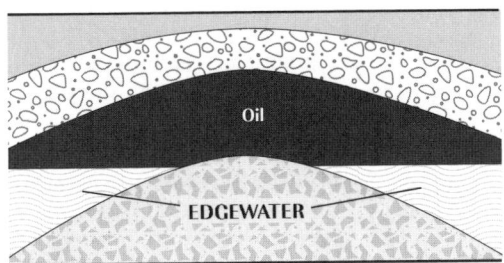

edgewater drive *n*: see *water drive*.

edge well *n*: a well on the outer fringes of a productive subsurface formation.

eductor *n*: a nozzlelike device that increases the velocity of a fluid. It is often used to mix two fluids or a fluid and solids.

EEEIPS *abbr*: extra, extra, extra improved plough (plow) steel.

EEIPS *abbr*: extra, extra improved plough (plow) steel.

EEPROM *n*: see *electrically erasable programmable read-only memory*.

effective inside tank height *n*: the distance from the gauge point on the tank floor or datum plate to the top of the top angle or where the tank contents would begin to overflow.

effective permeability *n*: a measure of the ability of a single fluid to flow through a rock when another fluid is also present in the pore spaces. Compare *absolute permeability, relative permeability*.

effective porosity *n*: the percentage of the bulk volume of a rock sample that is composed of interconnected pore spaces that allow the passage of fluids through the sample. See *porosity*.

effective value *n*: see *RMS value*.

efficiency *n*: the ratio of useful energy produced by an engine to the energy put into it.

effluent *n*: in a hydrocyclone, the liquid that exits the cone through the vortex finder. Also called overflow.

effluent limitation *n*: any restriction imposed on quantities, discharge rates, and concentrations of pollutants that are discharged from point sources into U.S. waters, the waters of the contiguous zone, or the ocean.

effluent water *n*: water that is discharged from a refinery or petrochemical plant.

efm *abbr*: electronic flow measurement.

egress unit *n*: see *emergency escape unit*.

Eh *sym*: oxidation-reduction potential.

EHS *abbr*: extremely hazardous substance.

EIA *abbr*: Electronic Industries Alliance.

EIA *abbr*: Energy Information Administration.

eight round *n*: a tapered connection with 8 threads per inch (centimetre). One turn equals 0.125 inch (0.318 centimetre) of travel. A very common oilfield connection.

EIPS *abbr*: extra improved plough (plow) steel.

ejector *n*: a nozzlelike device that increases the velocity of a fluid. It is often used to mix two fluids or a fluid and solids.

elastic collision *n*: a collision between a neutron and the nucleus of an atom of an element such as hydrogen. In such a collision, the neutron's energy is reduced by exactly the amount transferred to the nucleus with which it collided. The angle of collision and the relative mass of the nucleus determine energy loss.

elastic deformation *n*: temporary changes in dimensions caused by stress. The material returns to the original dimensions after removal of the stress.

elasticity *n*: the capability of an object that is put under stress (e.g., is stretched) to recover its size and shape when the stress is released. Compare *plasticity*.

elasticity modulus *n*: see *modulus of elasticity*.

elastic limit *n*: the point at which a material permanently deforms when stressed.

elastic modulus *n*: see *modulus of elasticity*.

elastomer *n*: a synthetic rubber made from polymers that has the elastic properties of natural rubber; packers (sealing elements) in blowout preventers and downhole packers are often made of elastomer. The term is formed by combining part of the word elastic and part of the word polymer—i.e., elast(ic) and poly(mer).

elbow *n*: a fitting that allows two pipes to be joined at an angle of less than 180°, usually 90° or 45°.

elec log *abbr*: electric log; used in drilling reports.

election at casing point *n*: a decision taken to exercise or not to exercise an option to participate in the completion attempt on a well, including the costs of running completion production casing and all related completion costs. The decision is made when the operator is ready to run casing and complete the well and so ratifies the party owning the election.

electrical filter *n*: a circuit used to eliminate or reduce certain waves or frequencies while leaving others relatively unchanged.

electrical junction box *n*: 1. a housing (box) into which electrical wires from two or more sources enter, join, and exit to other destinations. 2. a housing (box) mounted on an accumulator unit to provide an interface for remote electrical panels and an accumulator unit's pod select valve, pilot valves, and various system pressures; normally two electrical junction boxes are provided in case one fails.

electrically-actuated governor *n*: a hydraulic governor on an engine that has a reversible electric motor (it runs both clockwise and counterclockwise). By manipulating a remote control, the engine operator can adjust the electric motor to closely control the engine's speed. See *governor, hydraulic governor*.

electrically-driven triplex pump *n*: in a blowout preventer's accumulator system, a reciprocating pump with three pistons that electric motors operate.

electrically erasable programmable read-only memory (EEPROM) *n*: an integrated-circuit memory chip that has an internal switch to permit a user to erase the contents of the chip and write new contents into it by means of electrical signals.

electrically neutral *n*: a condition in an electrical circuit or component in which no electrical charge exists. The circuit or component is uncharged—that is, it has neither a positive nor a negative charge.

electrical potential *n*: voltage.

electrical resistance *n*: see *resistance*.

electrical stability test *n*: a test for oil muds that determines the stability of the emulsion.

electric circuit *n*: see *circuit*.

electric coil *n*: see *inductor*.

electric dehydration *n*: see *emulsion breaker*.

electric drive *n*: see *electric rig*.

electric-drive rig *n*: see *electric rig*.

electric generator *n*: a machine by which mechanical energy is changed into electrical energy, such as an electric generator on a drilling rig in which a diesel engine (mechanical power) turns a generator to make electricity (electrical energy).

electrician *n*: the rig crew member who maintains and repairs the electrical generation and distribution system on the rig.

electricity *n*: the physical phenomena arising from the behavior of electrons and protons that is caused by the attraction of particles with opposite charges and the repulsion of particles with the same charge.

electric line *n*: see *conductor line*.

electric log *n*: see *electric well log*.

electric logging *n*: the process of running an electric log. See *electric well log*.

electric motor *n*: an electric device that converts electric energy to mechanical energy. Compare *generator*.

electric potential *n*: work that must be done against electric forces to bring a unit charge from a reference point to another point. It is also called electromotive force and voltage.

electric relay *n*: see *relay*.

electric rig *n*: a drilling rig on which the energy from the power source—usually several diesel engines—is changed to electricity by generators mounted on the engines. The electrical power is then distributed through electrical conductors to electric motors. The motors power the various rig components. Compare *mechanical rig*.

electric starter *n*: a device that uses a battery, an electric motor, gears, and cables to provide a way of starting an engine. An electrically actuated motor turns the engine over by means of a pinion gear in the starter that engages a ring gear on the engine flywheel.

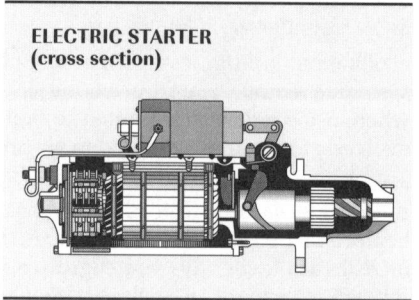

ELECTRIC STARTER
(cross section)

electric submersible pumping *n*: a form of artificial lift that utilizes an electric submersible multistage centrifugal pump. Electric power is conducted to the pump by a cable attached to the tubing.

electric survey *n*: see *electric well log*.

electric well log *n*: a record of certain electrical characteristics (such as resistivity and conductivity) of formations traversed by the borehole. It is made to identify the formations, determine the nature and amount of fluids they contain, and estimate their depth. Also called an electric log or electric survey.

ELECTRIC WELL LOG

electrochemical cell *n*: see *galvanic cell*.

electrochemical series *n*: see *electromotive series*.

electrochemical treater *n*: see *electrostatic treater*.

electrode *n*: a conductor of electric current as it leaves or enters a medium such as an electrolyte, a gas, or a vacuum.

electrodeposition *n*: an electrochemical process by which metal settles out of an electrolyte that contains the metal's ions and is then deposited at the cathode of the cell.

electrodynamic brake *n*: a device mounted on the end of the drawworks shaft of a drilling rig. The electrodynamic brake (sometimes called a magnetic brake) serves as an auxiliary to the mechanical brake when pipe is lowered into a well. The braking effect in an electrodynamic brake is achieved by means of the interaction of electric currents with magnets, with other currents, or with themselves.

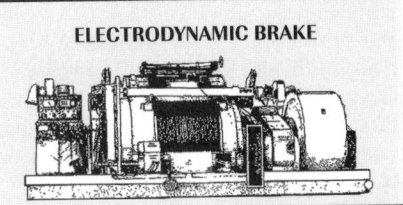

ELECTRODYNAMIC BRAKE

electrodynamometer *n*: a quantity indicator for AC current or voltage.

electrolysis *n*: the decomposition of a chemical compound brought about by the passage of an electrical current through the compound or through the solution containing the compound. Corroding action of stray currents is caused by electrolysis.

electrolyte *n*: 1. a chemical that, when dissolved in water, dissociates into positive and negative ions, thus increasing its electrical conductivity. See *dissociation*. 2. the electrically conductive solution that must be present for a corrosion cell to exist.

electrolytic capacitor *n*: a capacitor that consists of two electrodes separated by an electrolyte. A dielectric film is formed on the surface of one electrode.

electrolytic moisture analyzer *n*: an instrument that uses the principle that moisture is absorbed on a phosphorous pentoxide film between two electrodes to measure the amount of water in a fluid such as gas.

electrolytic property *n*: the ability of a substance, usually in solution, to conduct an electric current.

electrolyze *v*: to decompose by electrolysis.

electromagnet *n*: a magnet consisting of a coil of conducting wire wound around a soft iron or steel core; the core is strongly magnetized when current flows through the coil, and is almost completely demagnetized when current flow ceases.

electromagnetic induction *n*: voltage created when a conductor cuts across, or is cut by, a magnetic field.

electromagnetic measurement while drilling (EMWD) *n*: a special measurement while drilling (MWD) system that uses a low-frequency signal through the earth that is not affected by gas in the drill pipe. It uses equipment similar to conventional MWD tools to measure angle and direction, but an electromagnetic system also transmits signals to receivers on the surface.

electromagnetic propagation tool (EPT) *n*: device that emits microwaves into a formation and measures the way in which these microwaves are affected by the formation. By interpreting the effects, dielectric constants of fluids can be determined. Dielectric constants, in turn, can tell one whether fluid is water or hydrocarbons.

electromagnetism *n*: magnetism produced by the action of a current flowing through a conductor. The electromagnetic field contracts and disappears when the current stops flowing.

electromechanical *adj*: refers to equipment comprising both mechanical and electrical components, such as electromechanical valves and electromechanical counters.

electromechanical relay *n*: a device used to open or close electrical circuits (a relay) that is made up of mechanical parts that are actuated by electricity.

electromotive force (emf) *n*: 1. the force that drives electrons and thus produces an electric current. 2. the voltage or electric pressure that causes an electric current to flow along a conductor. 3. a difference of potential, or electrical, flow through a circuit against a resistance.

electromotive series *n*: a list of elements arranged in order of activity (tendency to lose electrons). The following metals are so arranged: magnesium, beryllium, aluminum, zinc, chromium, iron, cadmium, nickel, tin, copper, silver, and gold. If two metals widely separated in the list (e.g., magnesium and iron) are placed in an electrolyte and connected by a metallic conductor, an electromotive force is produced. See *corrosion*.

electron *n*: a particle in an atom that has a negative charge. An atom contains the same number of electrons and protons (which have a positive charge). Electrons orbit the nucleus of the atom.

electronic flow measurement (efm) *n*: the measurement of natural gas flow in a pipeline by the use of electronic devices, including computers.

Electronic Industries Alliance (EIA) *n*: an organization that writes, promulgates, and publishes *Recommended Standards* (*RSs*) for electronic and physical devices and their means of interfacing. For example, EIA-232 is a standard that defines a computer's serial port, connector pins, and electrical signaling process. Address: 2500 Wilson Blvd., Arlington, VA 22201; 703-907-7500; www.eia.org.

electronic transfer *n*: the exchange of data for business transactions by computer. Often used for nominations, dispatching, billing, and payment.

electron pair *n*: a pair of valence electrons that form a nonpolar bond between two neighboring atoms. See *valance electron*.

electron-pair bond *n*: the combining of atoms (the bonding) in which each atom of a bond pair contributes one electron to form a pair of electrons. Also called *covalent bond*. See *electron pair*.

electron tube *n*: an almost obsolete electronic device in which electrons move through a vacuum or gaseous medium within an airtight glass (or other material) container. Also called radio tube; tube; vacuum tube; or, in Great Britain, valve.

electroscanner *n*: device to scan and analyze paper charts electronically. Greatly increases the number of charts that can be processed daily.

electrostatic meter *n*: a meter that measures voltage directly, rather than in terms of its relation to current.

electrostatic treater *n*: a vessel that receives an emulsion and resolves the emulsion to oil, water, and usually gas, by using heat, chemicals, and a high-voltage electric field. This field, produced by grids placed perpendicular to the flow of fluids in the treater, aids in breaking the emulsion. Also called an electrochemical treater. See *emulsion treating*.

electrovalent reaction *n*: the most common method of forming compounds, in which a positively charged ion and a negatively charged ion attract each other. Compare *covalent reaction*.

element *n*: one of more than a hundred simple substances that consist of atoms of only one kind and that either singly or in combination make up all matter. For example, the simplest element is hydrogen, and one of the most abundant elements is carbon. Some elements, such as radium and uranium, are radioactive.

elev *abbr*: elevation; used in drilling reports.

elevated tank *n*: a vessel above a datum line (usually ground level).

elevate zero *v*: to calibrate a device that measures liquid level by providing a zero starting point when the instrument reading is physically below the actual level.

elevator bails *n pl*: see *elevator links*.

elevator links *n pl*: cylindrical bars that support the elevators and attach them to the hook. Also called elevator bails.

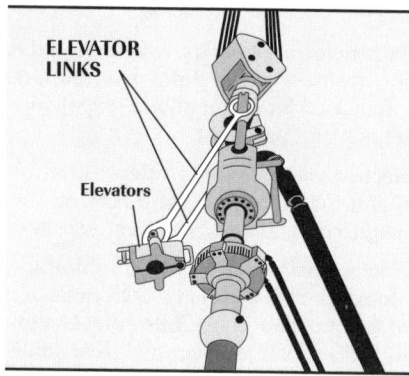

elevators *n pl*: on conventional rotary rigs and top-drive rigs, hinged steel devices with manual operating handles that crew members latch onto a tool joint (or a sub). Since the elevators are directly connected to the traveling block, or to the integrated traveling block in the top drive, when the driller raises or lowers the block or the top-drive unit, the drill pipe is also raised or lowered.

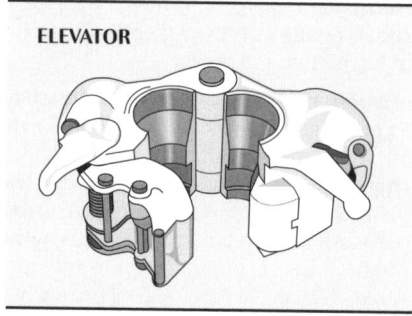

eliminator *n*: see *separator*.

elliptical tank *n*: a tank that has an elliptical cross section.

elongation *n*: see *stretch*.

EMD *abbr*: 1. Electromotive Division of General Motors Corporation. 2. a two-stroke-cycle diesel engine manufactured by the General Motors Corp.

emergency escape unit *n*: a self-contained breathing apparatus that is used only for escape to a safe area. This self-contained breathing apparatus, when properly serviced, furnishes a 5- to 15-minute supply of breathing air. Emergency escape units may have masks or hoods. Also called egress unit. Compare *work unit*.

emergency pack-off recovery tool *n*: a device run on drill pipe or tubing to recover the seal (the pack-off device) between the tubing and the tubing hanger in the wellhead. The recovery tool releases the

tubing hanger's lock-down ring, thus making it possible to retrieve the pack-off device.

Emergency Planning and Community Right-to-Know Act of 1986 (EPCRA) *n*: see *SARA Title Iii*.

emergency response *n*: (under HAZWOPER) a response to an uncontrolled release of a hazardous substance that requires action by employees or other responders from outside the immediate release area. In other words, it is an emergency response if the release poses a potential safety or health hazard, and the release cannot be controlled by employees in the immediate area or by maintenance personnel.

emergency response plan *n*: a plan for emergency response situations that is available to employees and OSHA personnel for inspection and that contains (at a minimum) pre-emergency planning and coordination with outside parties; personnel roles, lines of authority, training, and communication; emergency recognition and prevention; safe distances and places of refuge; site security and control; evacuation routes and procedures; decontamination; emergency alerting and response procedures; critique of response and follow-up; and personal protective equipment (PPE) and emergency equipment.

emf *abbr*: electromotive force.

emitter *n*: in a junction transistor, the region from which charge carriers that are minority carriers in the transistor's base are injected into the base and control the current flowing through the transistor's collector. Base and collector currents are combined in the emitter region and it serves as the outlet for total current in the transistor. See *base* (def. 2), *collector*.

"Employee Right to Know" Standard *n*: see *Hazard Communication Standard*.

emulsified water *n*: water so thoroughly combined with oil that special treating methods must be applied to separate it from the oil. Compare *free water*.

emulsifier *n*: see *emulsifying agent*.

emulsify *v*: to convert into an emulsion.

emulsifying agent *n*: a material that causes water and oil to form an emulsion. Water normally occurs separately from oil; if, however, an emulsifying agent is present, the water becomes dispersed in the oil as tiny droplets. Or, rarely, the oil may be dispersed in the water. In either case, the emulsion must be treated to separate the water and the oil.

emulsion *n*: a mixture in which one liquid, termed the dispersed phase, is uniformly

distributed (usually as minute globules) in another liquid, called the continuous phase or dispersion medium. In an oil-in-water emulsion, the oil is the dispersed phase and the water the dispersion medium; in a water-in-oil emulsion, the reverse holds.

emulsion breaker *n*: a system, chemical, device, or process used for breaking down an emulsion and producing two or more easily separated compounds (such as water and oil). Emulsion breakers may be (1) devices to heat the emulsion, thus achieving separation by lowering the viscosity of the emulsion and allowing the water to settle out; (2) chemical compounds, which destroy or weaken the film around each globule of water, thus uniting all the drops; (3) mechanical devices, such as settling tanks and wash tanks; or (4) electrostatic treaters, which use an electric field to cause coalescence of the water globules.

emulsion mud *n*: see *oil-in-water emulsion mud*.

emulsion test *n*: a procedure carried out to determine the proportions of sediment and dispersed compounds, such as water, in an emulsion. Such tests may range from elaborate distillation conducted in laboratories to simple and expedient practices used in the field.

emulsion treating *n*: the process of breaking down emulsions to separate oil from water or other contaminants. Treating plants may use a single process or a combination of processes to effect demulsification, depending on what emulsion is being treated.

emulsoid *n*: colloidal particles that take up water. See *colloid*.

EMW *abbr*: equivalent mud weight.

EMWD *abbr*: electromagnetic measurement while drilling.

enamel coating *n*: a collective term for a variety of petroleum-based derivatives such as asphalts, coal tars, grease and wax, mastics, and asphalt mastics that are used to coat pipe.

encapsulating agent *n*: a chemical in drilling mud that surrounds (encapsulates) drilled solids.

encoder *n*: an electronic device that converts decimal numbers to binary numbers. Compare *decoder*.

encroachment *n*: see *water encroachment*.

encumbrance *n*: a claim or charge on property; for example, a mortgage or lien for unpaid taxes.

endangered species *n*: a species that is in danger of extinction.

Endangered Species Act (ESA) *n*: an act that declares the intentions of Congress to conserve threatened and endangered species and the ecosystems on which those species depend. ESA provides for federal designation of species determined to be endangered or threatened. It prohibits the taking of an endangered or threatened species of fish, wildlife, or vegetation.

end device *n*: for automated lease operation, a sensor on field equipment to transmit status of operations, report data, or take action as needed.

endpoint *n*: the point marking the end of one stage of a process. In filtrate analysis, the endpoint is the point at which a particular result is achieved through titration.

end-to-end plot *n*: in data acquisition, a printout of data obtained from the start of a process to its completion.

endurance limit *n*: the maximum stress that can be applied to a structure over the structure's expected lifetime.

end user *n*: one who actually consumes or burns a product, as opposed to the one who sells or resells it.

energy *n*: the capability of a body for doing work. Potential energy is this capability due to the position or state of the body. Kinetic energy is the capability due to the motion of the body.

Energy Information Administration (EIA) *n*: a part of the Department of Energy (DOE) the main purpose of which is to disseminate information such as statistics to Congress, the administration, and the general public. Address: 1000 Independence Avenue SW; Washington, DC 20585; 202-586-2363.

Energy Telecommunications and Electrical Association (ENTELEC) *n*: a nonprofit, tax-exempt organization that aids the energy industry in dealing with technological advances by advising and educating members on new equipment and safety. Address: P. O. Box 639; Tomball, TX 77377-0639; 281-357-8700 or 888-503-8700; fax 281-357-8777.

engine *n*: a machine for converting the heat content of fuel into rotary motion that can be used to power other machines. Compare *motor*.

engine block *n*: the cast steel or aluminum body into which the engine manufacturer bores and machines the cylinders; and onto and into which additional engine parts are installed.

engineering analysis *n*: in offshore drilling, the examination and study of such factors as tensile loads, bending stresses, maximum operational water depth, buoyancy

requirements, surface tension, and vessel response for the purpose of designing the best subsea marine riser and blowout prevention system.

engine fan *n*: a multi-bladed propeller usually installed behind an engine's radiator, which moves air over an engine. It helps cool the engine and equalizes the temperature of the air as it flows over the engine, thus preventing hot spots. Engine power usually operates the fan, although some fans have auxiliary electric motors that power them when the engine is stopped.

engine jacket water *n*: coolant, such as water, that circulates through spaces in the engine block.

engine temperature switch (ETS) *n*: a device on an engine that senses overheating and shuts down the engine if overheating occurs.

Engler distillation *n*: a test that determines the volatility of a gasoline by measuring the percentage of the gasoline that can be distilled at various temperatures.

enhanced oil recovery (EOR) *n*: 1. the introduction of artificial drive and displacement mechanisms into a reservoir to produce a portion of the oil unrecoverable by primary recovery methods. To restore formation pressure and fluid flow to a substantial portion of a reservoir, fluid or heat is introduced through injection wells located in rock that has fluid communication with production wells. See *alkaline (caustic) flooding, gas injection, micellar-polymer flooding, primary recovery, secondary recovery, tertiary recovery, thermal recovery, waterflooding.* 2. the use of certain recovery methods that not only restore formation pressure but also improve oil displacement or fluid flow in the reservoir. These methods may include chemical flooding, gas injection, and thermal recovery.

ENTELEC *abbr*: Energy Telecommunications and Electrical Association.

enthalpy *n*: the heat content of fuel. A thermodynamic property, it is the sum of the internal energy of a body and the product of its pressure multiplied by its volume.

entitlement *n*: working-interest owner's share of production. This volume may not equal actual sales because of contractual or market conditions.

entrained *adj*: drawn in and transported by the flow of a fluid.

entrained gas *n*: formation gas that enters the drilling fluid in the annulus. See *gas-cut mud.*

entrained liquid *n*: liquid particles that may be carried out of the top of a distillation

or absorber column with the vapors or residue gas.

entrained water *n*: water suspended in oil. Includes emulsions but does not include dissolved water.

entrapment *n*: 1. the underground accumulation of oil and gas in geological traps. 2. the accumulation in rock pores of large polymer or surfactant molecules unable to move because of small exit openings. The coiled-up molecules reduce the permeability of pores to water but permit oil to pass through the pores.

entrepôt *n*: a center for the collection and distribution of goods.

entropy *n*: the internal energy of a substance that is attributed to the internal motion of the molecules. This energy is within the molecules and cannot be utilized for external work.

environment *n*: 1. the sum of the physical, chemical, and biological factors that surround an organism. 2. the water, air, and land and the interrelationship that exists among and between water, air, and land and all living things. 3. as defined by the U.S. government, the navigable waters, the waters of the contiguous zone, the ocean waters, and any other surface water, groundwater, drinking water supply, land surface, subsurface strata, or ambient air within the United States.

environmental assessment *n*: a detailed environmental review document required under NEPA.

environmental conditions *n pl*: conditions imposed on a vessel due to the effect of wind, waves, and current.

environmental impact statement *n*: as defined in the Environmental Quality Improvement Act of 1970, a statement designed to "serve as an action-forcing device to ensure that the policies and goals of the act are infused into the ongoing policies of the government." Its purpose is to "avoid or minimize adverse impacts" on the environment by providing the environmental analyses (used with other relevant materials) to make sound decisions regarding the environment.

Environmental Protection Agency (EPA) *n*: a federal agency that was created in 1970 from a variety of existing agencies. The EPA administers air pollution, water pollution, pesticide, solid waste, noise control, drinking water, and toxic substances acts. It also has major research responsibilities. Address: 1200 Pennsylvania Avenue NW; Washington, DC 20460; 202-272-0167.

environment of deposition *n*: see *depositional environment.*

EOR *abbr*: enhanced oil recovery.

EPA *abbr*: Environmental Protection Agency.

EP additive *n*: see *extreme-pressure lubricant*.

EPA generator identification number (GIN) *n*: a number issued to any person, by site, whose act or process produces hazardous waste. Such a number is required by the EPA under the Resource Conservation and Liability Act of 1976 for the purpose of tracking hazardous waste.

EPCRA *abbr*: Emergency Planning and Community Right-to-Know Act.

epeiric sea *n*: a shallow arm of the ocean that extends from the continental shelf deep into the interior of the continent. Also called epicontinental sea.

ephemeral streams *n pl*: streams that last a short time.

ephemeris second *n*: the fundamental unit of time of the SI system of measurement, which is equal to $\frac{1}{31,556,925.974}$ of a year.

epicontinental sea *n*: see *epeiric sea*.

epithermal neutron *n*: a neutron having an energy level of 0.02 to 100 electron volts.

epithermal neutron log *n*: a special neutron log used mainly in air- or gas-drilled holes to help determine formation porosity. An epithermal neutron log detects epithermal neutrons, which are neutrons released at energy levels higher than those released in normal neutron logging.

epm/equivalents per million *n*: unit chemical weight of solute per million unit weights of solution. The epm of a solute in solution is equal to the ppm (parts per million) divided by the equivalent weight.

EP mix *abbr*: ethane-propane mix.

epoch *n*: a division of geologic time; a subdivision of a geologic period.

epoxy *n*: 1. any compound characterized by the presence of a reactive chemical structure that has an oxygen atom joined to each of two carbon atoms that are already bonded. 2. any of various resins capable of forming tight crosslinked polymer structures characterized by toughness, strong adhesion, and high corrosion and chemical resistance. Epoxys are used extensively as coatings and adhesives.

EPROM *abbr*: erasable programmable read-only memory.

equalizer *n*: 1. a device used with mechanical drawworks brakes to ensure that, when the brakes are used, each brake band will receive an equal amount of tension and also that, in case one brake fails, the other will carry the load. A mechanical brake equalizer is a dead anchor attached to the drawworks frame in the form of a yoke attached to each brake band. Some drawworks are equipped with automatic equalizers. 2. any device used to distribute force equally on two pieces of equipment—for example, the pitmans on a beam pumping unit.

equalizing slings *n pl*: in crane operations, multiple-leg slings of wire rope and fittings that equally distribute the load among the slings.

equation *n*: in mathematics, a statement that each of two expressions is the same as (is equal to) the other. For example, $a = b$ is an equation as is $2 + 5 = 3 + 4$.

equation of state *n*: a mathematical expression that defines the physical state of a homogeneous substance by relating volume to pressure and absolute temperature for a given mass of the material.

equatorial circumference *n*: the circumference of the horizontal great circle at the equator of a spherical tank. Compare *great circle*.

equilibrium *n*: a state of balance between opposing forces or actions that is either static or dynamic.

equilibrium constant *n*: see *vapor-liquid equilibrium ratio*.

equilibrium vapor pressure *n*: the pressure at which a liquid and its vapor are in equilibrium at a given temperature.

equity *n*: those maxims and general principles that developed in England to moderate the common law and allow remedy for injury. In the broadest sense, justice.

equity line *n*: a company owns part of a pipeline, but does not operate it.

equivalent circulating density (ECD) *n*: the increase in bottomhole pressure expressed as an increase in pressure that occurs only when mud is being circulated. Because of friction in the annulus as the mud is pumped, bottomhole pressure is slightly, but significantly, higher than when the mud is not being pumped. ECD is calculated by dividing the annular pressure loss by 0.052, dividing that by true vertical depth, and adding the result to the mud weight. Also called circulating density, mud-weight equivalent.

equivalent dip *n*: a European term for the depth of liquid in a tank corresponding to a given ullage. It is obtained by subtracting the observed ullage from the height of the ullage reference point above the dip point on the bottom of the tank.

equivalent mud weight (EMW) *n*: pressure expressed as the density of a mud that exerts that pressure. To determine it, the pressure gradient of the mud in use and a value for the constant that is required in the measuring system is used. In the English system, for example, the pressure gradient of a fluid that is 0.6344 pounds per square inch per foot (psi/ft) can be expressed as a mud density of 12.2 pounds per gallon (ppg). This density is derived from the equation $P = MW \times 0.052$, where P is the pressure gradient in psi/ft and MW is mud weight in ppg. When solved for MW, the equation is $MW = P \div 0.052$. Thus, $MW = 0.6344\,\text{psi/ft} \div 0.052 = 12.2$ ppg.

equivalents per million (epm) *n*: the unit chemical weight of solute per million unit weights of solution. The epm of a solute in solution is equal to the ppm (parts per million) divided by the equivalent weight.

equivalent weight *n*: the atomic or formula weight of an element, compound, or ion divided by its valence. Elements entering into combination always do so in quantities proportional to their equivalent weights.

ERA *abbr*: Economic Regulatory Administration.

era *n*: one of the major divisions of geologic time.

erasable programmable read-only memory (EPROM) *n*: in a computer, a type of read-only memory (ROM) that can be erased and thus be reprogrammed for continuous use.

ergonomics *n pl*: the science of designing equipment for the workplace, wherein productivity is increased by reducing operator fatigue and discomfort. Also called biotechnology, human engineering, and human factors engineering.

erosion *n*: the process by which material (such as rock or soil) is worn away or removed (as by wind or water).

erosion drilling *n*: the high-velocity ejection of a stream of drilling fluid from the nozzles of a jet bit to remove rock encountered during drilling. Sometimes sand or steel shot is added to the drilling fluid to increase its erosive capabilities.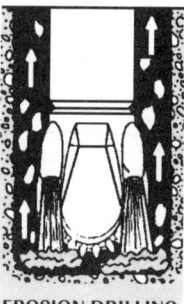

EROSION DRILLING

erosion surface *n*: the upper surface of the most recent sediment layer, formerly smooth and horizontal, modified by running water or other agents.

error and repeatability of measurements *n*: the closeness of the agreement between the results of successive measurements of the same quantity carried out by the same method, under the same environment, by the same observer, with the same measuring instruments, in the same laboratory, at quite short intervals.

error curve *n*: a curve or graph that represents the error of a measuring device such as a pressure gauge as a function either of the quantity measured or of any other quantity that has an influence on the error.

ERTS *abbr*: Earth Resource Technology Satellite.

ES *abbr*: electric survey. See *electric well log*.

ESA *abbr*: Endangered Species Act.

Esaki tunnel diode *n*: see *tunnel diode*.

escape capsule *n*: on offshore drilling rigs and other offshore oil and gas facilities, a special lifeboat, which is usually suspended at deck level to provide easy entry for crewmembers, which is lowered by a single-line launch system, and which is self-propelled. Typically, escape capsules are not boat shaped; rather, they are disk shaped with watertight hatches on the top side of the capsule.

escarpment *n*: a cliff or relatively steep slope that separates level or gently sloping areas of land.

escheat *n*: the reversion of property to the state in the event that the owner thereof dies without leaving a will (intestate) and has no heirs to whom the property may pass by lawful descent. Also called unclaimed property statute.

ESD *abbr*: emergency shut down, an automated platform system to shut in an SCSSV and/or SSV.

est *abbr*: estimated; used in drilling reports.

estate *n*: the nature and extent of a person's ownership or right or interest in land or other property.

ester *n*: an organic compound formed from the reaction between an organic acid and an alcohol.

estimated ROB (remaining on board) *n*: estimated material remaining on board a vessel after a discharge. Includes residue or sediment clingage, which builds up on the interior surfaces of the vessel's cargo compartments.

estoppel *n*: a legal restraint on a person to prevent him or her from contradicting a previous statement.

estuary *n*: a coastal indentation or bay into which a river empties and where fresh water mixes with seawater. Compare *marine delta*.

et al *abbr*: and others (Lat. *et alii*). Commonly used in oil and gas leases.

ETBE *abbr*: ethyl tertiary butyl ether.

ethane *n*: a paraffin hydrocarbon, C_2H_6; under atmospheric conditions, a gas. One of the components of natural gas.

ethene *n*: see *ethylene*.

ethylene *n*: a chemical compound of the olefin series with the formula C_2H_4. Official name is ethene.

ethylene glycol *n*: a colorless liquid used as an antifreeze and as a dehydration medium in removing water from gas. See *glycol, glycol dehydrator*.

ethyl tertiary butyl ether (ETBE) *n*: a compound of ethanol and isobutylene. Compare *methyl tertiary butyl ether*.

ETS *abbr*: engine temperature switch.

et ux *abbr*: and wife (Lat. *et uxor*, and woman). Commonly used in oil and gas leases.

et vir *abbr*: and husband (Lat. *et vir*, and man). Commonly used in oil and gas leases.

EUE *abbr*: external upset end.

evaporate *v*: 1. to convert or change into a vapor. 2. to produce vapor. 3. to draw moisture from, as by heating, leaving only the dry solid portion. 4. to convert a liquid into vapor.

evaporation loss *n*: a loss to the atmosphere of petroleum fractions through evaporation, usually while the fractions are in storage or in process. See *vaporization*.

evaporator *n*: a vessel used to convert a liquid into its vapor phase.

evaporite *n*: a sedimentary rock formed by precipitation of dissolved solids from water evaporating in enclosed basins. Examples are gypsum and salt.

evening tour (pronounced "tower") *n*: the shift of duty on a drilling rig that starts in the afternoon and runs through the evening. Sometimes called afternoon tour. Compare *daylight tour, graveyard tour*.

even keel *n*: on a ship or floating offshore drilling rig, the balance when the plane of flotation is parallel to the keel.

event *n*: a term some regulatory agencies use to describe an occurrence in the oilfield that the agency does not consider normal or usual. For example, a kick (the entry of formation fluids into the wellbore) may be described as a well-control event by such agencies.

examination of title *n*: a thorough inspection of the recorded documents pertaining to a tract's history of ownership. Title examinations are performed by attorneys who look for gaps in the chain of title, ambiguities, or any doubtful points that would cloud the present owner's claim to the property. The examiner then sets forth in a written opinion the facts and instruments that, in his or her judgment, are necessary to make the title merchantable or legally defensible.

excelsior *n*: a fibrous material used as a filtering element in heaters or heater-treaters.

excess butanes *n pl*: see *butanes required*.

excess capacity *n*: the difference in the amount of gas between the producer's deliverability capacity and the purchaser's nominations.

excess flow valve *n*: a valve used automatically to prevent the liquid flow rate in a pipeline from exceeding a high limit.

excess gas used *n*: gas purchased for lease use, surface or subsurface, when the amount the producer is entitled to is insufficient.

exchange agreement *n*: agreement between producers that provides for the exchange of gas produced from one property or interest for gas produced from another property or interest.

exchange gas *n*: gas exchanged under provisions of an exchange agreement.

exchanger *n*: a piping arrangement that permits heat from one fluid to be transferred to another fluid as they travel countercurrently to one another. In the heat exchanger of an emulsion-treating unit, heat from the outgoing clean oil is transferred to the incoming well fluid, cooling the oil and heating the well fluid. In the heat exchanger of a glycol dehydration unit, heat from the hot lean glycol flows through the inner flow tube in the opposite direction of the cool rich glycol, which flows through a shell built around the tube.

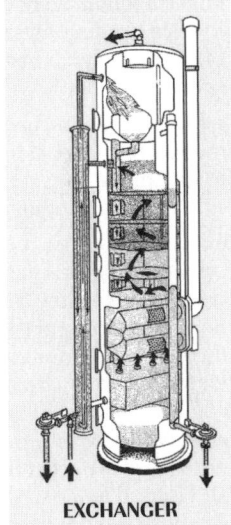

EXCHANGER

excited state of nucleus *n*: the increased energy condition of the nucleus of an atom after it has captured a neutron. On capture, the nucleus emits a high-energy gamma ray called a capture gamma ray.

exciter *n*: a small DC generator mounted on top of a main generator to produce the field for the main generator.

exciting current *n*: the back electromotive force that flows through the primary winding of a transformer when no loads are connected to the secondary winding. See *back electromotive force.*

exclusive or (XOR) *n*: in computer science, an instruction that performs its operation on a bit-by-bit basis for its two operand words, usually storing the result in one of its operand locations. Also called XOR. See *operand.*

execution *n*: the completion of a legal instrument by the required actions—for example, by signing and delivering the instrument. Execution includes the actual delivery of the signed document to the named grantee, lessee, or assignee.

executive rights *n pl*: in regard to mineral rights and interests, the right to execute oil and gas leases. Executive rights may not include the right to bonus or rentals.

executor *n*: the person named in a will to carry out its provisions.

exempt *adj*: free or released from liability or requirement to which others are subject.

exempt interest *n*: an interest owned in a property, usually by a charitable or a governmental agency, that is not subject to state production taxes as provided in the applicable tax regulations. Compare *tax-free interest.*

exhaust *n*: the burned gases that are removed from the cylinder of an engine. *v*: to remove the burned gases from the cylinder of an engine.

exhaust gas *n*: the mixture of gases that comes out of the cylinders of an internal combustion engine as a result of the ignition and burning of the fuel-air mixture in the cylinders. In underbalanced drilling with nitrogen as a circulating fluid, a diesel engine runs compressors and produces exhaust gas, which is high in nitrogen. The nitrogen is extracted from the exhaust gas and is used in the well. Because the nitrogen must be compressed to inject into the well, the cost of gas compression is shared with the cost of producing nitrogen, which makes both less expensive.

exhaust manifold *n*: a piping arrangement, immediately adjacent to the engine, that collects burned gases from the engine and channels them to the exhaust pipe.

exhaust pipe *n*: on an engine, flexible steel tubing that connects the engine exhaust manifold outlet to the muffler. See *muffler.*

exhaust port *n*: an opening in a cylinder wall through which exhaust gas is expelled when the exhaust port is uncovered by the piston.

exhaust silencer *n*: see *muffler.*

exhaust stack *n*: see *tail pipe.*

exhaust stroke *n*: in an engine, the movement of the piston during which time it pushes burned fuel gases out of the cylinder.

exhaust system *n*: valves (or ports), manifolds (passageways and connections), piping, noise silencers (mufflers), and other devices, such as exhaust stacks, that serve to remove burned fuel gases from an engine.

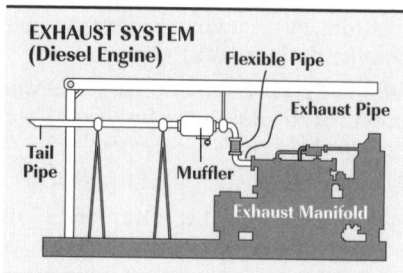

exhaust valve *n*: the cam-operated mechanism through which burned gases are ejected from an engine cylinder.

exhibit *n*: see *rider.*

expanded perlite *n*: a siliceous volcanic rock that is finely ground and subjected to extreme heat. The resulting release of water leaves the rock particles considerably expanded and thus more porous. Expanded perlite is sometimes used in cement to increase its yield and decrease its density without an appreciable effect on its other properties.

expander process *n*: the most commonly used cryogenic method of gas processing.

expanding cement *n*: cement that expands as it sets to form a tighter fit around casing and formation.

expansibility factor *n*: see *expansion factor.*

expansion coefficient *n*: see *coefficient of expansion.*

expansion dome *n*: a cylindrical projection on top of a tank, tank car, or truck into which liquids may expand without overflowing. Gauge point is often in the expansion dome. Shell thickness is part of dome capacity.

expansion factor *n*: a multiplying factor used when calculating gas flow rate. This factor corrects for the reduction in fluid density that a compressible fluid experiences when it passes through an orifice as a result of the increased fluid velocity and the decreased static pressure. Also called expansibility factor.

expansion joint *n*: a device used to connect long lines of pipe to allow the pipe joints to expand or contract as the temperature rises or falls.

expansion loop *n*: a loop built into a pipeline to allow for expansion and contraction of the line.

expansion refrigeration *n*: cooling obtained from the evaporation of a liquid refrigerant or the expansion of a gas.

expansion thermometer *n*: a type of thermometer that uses a known cubical coefficient of expansion of a solid, a liquid, or a gas to provide indication in terms of degrees of temperature.

expansion turbine *n*: a device that converts the energy of a gas or vapor stream into mechanical work by expanding the gas or vapor through a turbine.

expansivity *n*: see *coefficient of expansion.*

expected value concept *n*: a risk analysis process that multiplies expected gain or loss of a decision by its probability of occurrence and averages all possible outcomes to choose the action with the highest expected benefit.

expendable gun *n*: a perforating gun that consists of a metal strip on which are mounted shaped charges in special capsules. After firing, nothing remains of the gun but debris. See *gun-perforate.*

expendable plug *n*: a temporary plug set on a pressure-setting assembly and landed inside a production packer. The plug temporarily converts the packer into a bridge plug.

expendable-retrievable gun *n*: a perforating gun that consists of a hollow, cylindrical carrier into which are placed shaped charges. On detonation, debris created by the exploded charges falls into the carrier and is retrieved when the gun is pulled out of the hole; however, the gun cannot be reused. See *gun-perforate.*

expensed *adj*: deducted from income in the year in which the expenditure is incurred.

exploitation *n*: the development of a reservoir to extract its oil.

exploitation well *n*: a well drilled to permit more effective extraction of oil from a reservoir. Sometimes called a development well.

exploration *n*: the search for reservoirs of oil and gas, including aerial and geophysical surveys, geological studies, core testing, and drilling of wildcats.

exploration and production *n*: see *E&P.*

exploration plan *n*: a plan required by the MMS before exploration can take place in the OCS. The plan must include measures to protect the environment. The MMS reviews the plan, analyzes the environmental effects, and determines any appropriate mitigating measures before approving the plan.

exploration well *n*: a well drilled either in search of an as-yet-undiscovered pool of oil or gas (a wildcat well) or to extend greatly the limits of a known pool. It involves a relatively high degree of risk. Exploratory wells may be classified as (1) wildcat, drilled in an unproven area; (2) field extension or step-out, drilled in an unproven area to extend the proved limits of a field; or (3) deep test, drilled within a field area but to unproven deeper zones.

exploratory drilling *n*: drilling involved with the initial discovery of hydrocarbons. Also called wildcatting.

explosimeter *n*: an instrument used to measure the concentration of combustible gases in the air. Also called a gas sniffer.

explosion cover *n*: see *explosion door*.

explosion door *n*: on an engine, one of usually several spring-loaded, lightweight metal plates placed over openings in the engine's crankcase. Oxygen in the air entering the base of an engine mixes with the oil there. A hot spot could cause the oil and oxygen to explode and damage the crankcase. To prevent such damage, the spring-operated explosion covers (doors) open to release the pressure from the explosion. They then slam closed to prevent more air from entering. Also called explosion cover.

explosion-proof motor *n*: a motor with an enclosure designed to contain an internal explosion and to prevent ignition of surrounding gases or vapors by sparks that may occur in the motor.

explosive fracturing *n*: when explosives are used to fracture a formation. At the moment of detonation, the explosion furnishes a source of high-pressure gas to force fluid into the formation. The rubble prevents fracture healing, making the use of proppants unnecessary. Compare *hydraulic fracturing*.

exponent *n*: in mathematics, a number or symbol, such as the 3 in x^3, placed to the right of and above another number, symbol, or expression, denoting the power to which that number, symbol, or expression is raised to. Also called power.

exposure limits *n pl*: the limits (based on time of exposure, concentration of material, means of contact, and material toxicity) of worker exposure to hazardous materials without ill effect. Exceeding exposure limits set for various materials can result in temporary health problems, chronic illness, acute illness, or death.

expressed covenants *n pl*: the lease contains specific covenants, or obligations, for development. These replace implied covenants.

extended bentonites *n pl*: selected bentonites treated with chemical polymers. See *extender*.

extended nozzle *n*: a special bit nozzle, often used on large bits, that lengthens the nozzle and therefore places the jet of drilling fluid exiting the nozzle close to the bottom of the hole. With large bits, where regular nozzles can be relatively distant from the bottom of the hole, the cleaning power of the jet of drilling fluid may be lost because the velocity, or speed, of the jet diminishes rapidly after it exits the nozzle. By extending the length of the nozzles, the jets are placed closer to bottom for maximum cleaning.

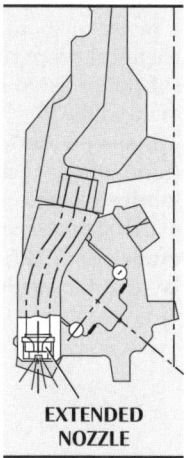

EXTENDED NOZZLE

extended-reach well *n*: a directionally drilled well that has a high degree of deflection.

extender *n*: 1. a substance added to drilling mud to increase viscosity without adding clay or other thickening material. 2. an additive that assists in getting greater yield from a sack of cement. The extender acts by requiring more water than required by neat cement.

extension *n*: a tubular component attached to the bottom of a packer to extend its bore.

extension rod *n*: see *crosshead extension rod*.

extension sub *n*: a device made up on an overshot to lengthen it so that it can pass over the damaged top of a fish and securely engage an undamaged area of the fish. See *overshot*.

external cutter *n*: a fishing tool containing metal-cutting knives that is lowered into the hole and over the outside of a length of pipe to cut it. The severed part of the pipe can then be brought to the surface. Also called an outside cutter. Compare *internal cutter*.

external line-up clamp *n*: a clamp used on the outside of pipe to align two lengths of pipe. External line-up clamps are usually used on pipe with a diameter of 8 inches (20 centimetres) or less. Compare *internal line-up clamp*.

EXTERNAL CUTTER

externally actuated sampler *n*: a device that is operated by a power source other than

the fluid being sampled; for example, an electric or pneumatic motor.

external phase *n*: see *continuous phase*.

external upset end (EUE) *n*: on tubing, casing, or drill pipe, the thickening at each end of the joint such that the internal diameter of the joint is not affected; i.e., it remains uniform throughout the joint's length. Only the outside diameter is enlarged at each end. Pipe is thickened, or upset, at each end to increase its strength so that threads, couplings, or tool joints may be attached. Compare *internal-external upset end*, *internal upset end*.

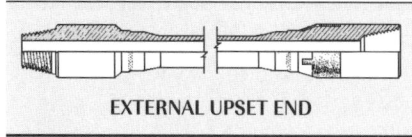

EXTERNAL UPSET END

extract *n*: the substance that is to be separated from the raffinate.

extraction *n*: the process of separating one material from another by means of a solvent. The term can be applied to absorption, liquid-liquid extraction, or any other process using a solvent.

extraction loss *n*: the reduction in volume of wet natural gas due to the removal of natural gas liquids, hydrogen sulfide, carbon dioxide, water vapor, and other impurities from the natural gas stream. See *shrinkage*.

extraction plant *n*: a plant equipped to remove liquid constituents from casinghead gas or wet gas.

extractor *n*: in automatic sampling, a device for removing small amounts of liquid from a flowing stream and diverting these to a storage container.

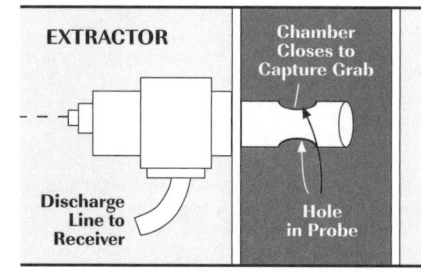

EXTRACTOR · Chamber Closes to Capture Grab · Discharge Line to Receiver · Hole in Probe

extra, extra improved plough (plow) steel (EEIPS) *n*: used in wire rope construction definitions. It is a grade used where a high breaking strength is required. This grade typically provides a breaking strength a minimum of 10 percent higher than EIPS.

extra, extra, extra improved plough (plow) steel (EEEIPS) *n*: it is a very high strength steel that provides a breaking strength 35 percent higher than EIP steel.

extra improved plough (plow) steel (EIPS) *n*: used in wire rope construction definitions. It is a strong, tough, durable steel that combines great strength with high resistance to fatigue.

extra-long rotary slips *n pl*: slips for drill pipe that fit into the tapered insert bowl of a four-pin master bushing and whose taper length is 12¾ inches (324 millimetres) long. Compare *long rotary slips*. See also *master bushing, slips.*

extratropical cyclone *n*: a cyclone that develops in the middle and upper latitudes.

extremely hazardous substance (EHS) *n*: 1. chemical determined by the EPA to be extremely hazardous to a community during an emergency spill or release as a result of its toxicity and physical or chemical properties. 2. (under SARA) a substance listed in appendixes A and B of 40 CFR 355.

extreme-pressure lubricant *n*: additives that, when added to drilling fluid, lubricate bearing surfaces subjected to extreme pressure.

extrusion *n*: 1. the emission of magma (as lava) at the earth's surface. 2. the body of igneous rock produced by the process of extrusion.

extrusive *adj*: volcanic; derived from magnetic materials poured out on the earth's surface, as distinct from intrusive rocks formed from magma that has cooled and solidified beneath the surface.

extrusive rock *n*: igneous rock formed from lava poured out on the earth's surface.

eye splice *n*: a loop, with or without a thimble, formed at the end of a wire rope. Also called an eye. See *thimble.*

f *abbr*: Fanning friction factor.

F *abbr*: Fahrenheit. See *Fahrenheit scale*.

F *sym*: farad.

FAA *abbr*: Federal Aviation Administration.

fabrication *n*: a collective term for the specialized connections and fittings on a pipeline. Fabrication assemblies control product flow, direct products to the proper location, aid in product separation, and facilitate maintenance operations.

fabrication crew *n*: pipeline construction workers responsible for welding fabrication assemblies into the line. The fabrication crew works independently of the rest of the spread.

face mask *n*: a mask made of a rubber frame surrounding a clear, flat lens, used to seal all or a portion of a diver's face from the underwater environment.

face seal *n*: a type of seal in which deformation of the seal is accomplished by a plate or flat surface.

facies *n*: part of a bed of sedimentary rock that differs significantly from other parts of the bed.

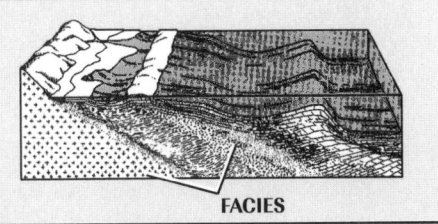

FACIES

facility *n*: an installation serving two or more leases, providing one or more functions such as separation, compression, dehydration, treating, gathering, or delivery.

Faciolog *n*: a log generated at a computer center that identifies various sediments, or facies, raw log data, and a litho-analysis log.

factoring counter *n*: an electronic register capable of expressing true throughput volume. It includes selective means for automatically applying meter factor.

Fahrenheit scale *n*: a temperature scale devised by Gabriel Fahrenheit, in which 32° represents the freezing point and 212° the boiling point of water at standard sea-level pressure. Fahrenheit degrees may be converted to Celsius degrees by using the following formula:

$$°C = \tfrac{5}{9}\,(°F{-}32).$$

fail-safe *adj*: capable of compensating automatically and safely for a failure, as of a mechanism or power source.

fairlead *n*: a device on a rig used to direct either wire or chain from a winch or windlass away from the rig.

fairleader *n*: a device used to maintain or alter the direction of a rope or a chain so that it leads directly to a sheave or drum without undue friction.

fair weather cumulus *n*: a cumulus cloud that lies horizontally and shows little vertical development, indicating good weather conditions.

falls *n pl*: see *parts of line*.

fan-driven heater *n*: a portable electric heater that circulates air over an electric heating element into the space to be heated.

fanglomerate *n*: coarse-grained, poorly sorted sedimentary rock derived from sediments deposited in alluvial fans; a type of conglomerate.

Fanning friction factor (*f*) *n*: a dimensionless number used to study fluid friction in pipes. The factor is equal to the pipe's diameter times the pressure drop in the fluid caused by friction as the fluid flows through the pipe, divided by the product of the pipe's length and the kinetic energy of the fluid per unit volume.

Fann V-G™ meter *n*: trade name of a device used to record and measure at different speeds the flow properties of plastic fluids (such as the viscosity and gel strength of drilling fluids).

FAR *abbr*: federal aviation requirement.

farad (F) *n*: the unit of electrical capacitance. One farad is equal to the capacitance of a capacitor that has a potential difference of 1 volt between its plates when the charge of one of its plates is 1 coulomb, there being an equal and opposite charge on the other plate. See *coulomb*.

Faraday's First Law *n*: the amount of any substance dissolved or deposited in electrolysis is proportional to the total electric charge passed.

farm boss *n*: an oil company supervisor who controls production activities within a limited area.

farmee *n*: the individual or company that has a lease farmed out to it.

farm in *v*: to accept, as an operator, a farm-out. See *farmout*.

farm-in *n*: an agreement identical to a farm-out, with the operator as the third party. The operator takes the farm-in. See *farmout*.

farmor *n*: the individual or company that farms out a lease.

farmout *n*: an agreement whereby the owner of a lease who does not wish to drill at the time agrees to assign the leasehold interest, or some part of it, to a third party who does wish to drill, conditional on the third party's drilling a well within the expiration date of the primary term of the lease. The assignment may include the entire interest together with dry hole money, or partial interest or entire interest with or without an override. If an override is retained, the owner of the lease may retain an option to convert such overriding royalty retained to an agreed-upon working interest. A farmout is distinguished from a joint operating agreement by the fact that the partner farming out does not incur any of the drilling costs. The primary characteristic of a farmout is the obligation of the third party to drill one or more wells on the farmout acreage as a condition prerequisite to completion of the transfer of title to such third party.

farm out *v*: for a lessee, to agree to assign a leasehold interest to a third party, subject to stipulated conditions. See *farmout*.

farm tap regulator *n*: a device in a gas distribution system that decreases the pressure of the gas flowing through a high-pressure line as the gas passes through the regulator. The regulator decreases and maintains the pressure to a low enough value to make it suitable for use in appliances in rural homes or farms.

fast ice *n*: sea ice attached to the shore. Also called coast ice.

fastline *n*: the end of the drilling line that is affixed to the drum or reel of the drawworks, so called because it travels with greater velocity than any other portion of the line. Compare *deadline*.

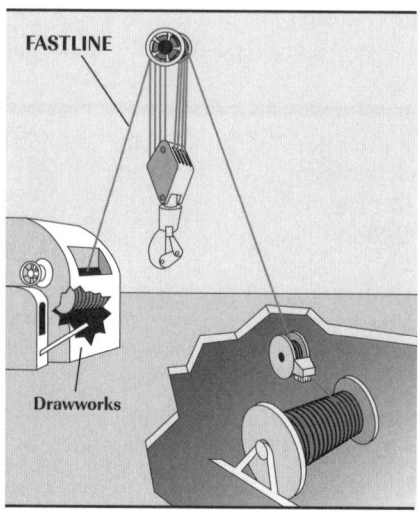

fast sample loop *n*: a secondary pipeline circuit designed for quickly transferring a product sample. This arrangement enables a representative sample passing through the main pipeline to be brought quickly to the sampling equipment, which may be located some distance from the main pipeline.

fast sheave *n*: that sheave on the crown block over which the fastline is reeved.

fathom *n*: a measure of ocean depth equal to 6 feet, or 1.83 metres.

fatigue *n*: the tendency of material such as a metal to break under repeated cyclic loading at a stress considerably less than the tensile strength shown in a static test.

fatigue characteristic *n*: in metallurgy, a measure of the trait of a metal, such as steel, to withstand repeated, or cyclic, stress without failure.

fatigue crack *n*: a fracture starting from a nucleus where there is an abnormal concentration of cyclic stress and propagating through the metal.

fault *n*: a break in the earth's crust along which rocks on one side have been displaced (upward, downward, or laterally) relative to those on the other side.

fault dip *n*: the vertical inclination of a fault's surface, or plane, measured from a horizontal plane.

fault plane *n*: a surface along which faulting has occurred.

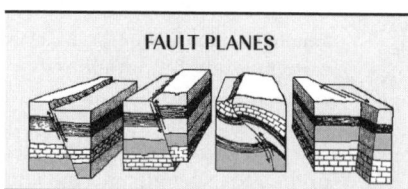

FAULT PLANES

fault trap *n*: a subsurface hydrocarbon trap created by faulting, in which an impermeable rock layer has moved opposite the reservoir bed or where impermeable gouge has sealed the fault and stopped fluid migration.

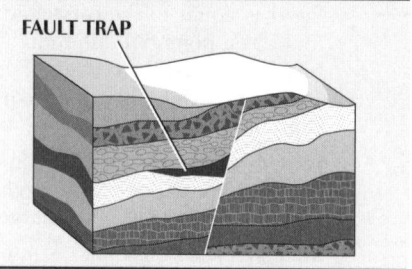

FAULT TRAP

fauna *n pl*: the animals of a given region or period considered as a whole.

faunal succession *n*: the principle that fossils in a stratigraphic sequence succeed one another in a definite, recognizable order.

"favored nation" clause *n*: a provision in a gas sales contract under which the seller is guaranteed that, if the buyer purchases gas within a certain area close to his or her well, that same price will be paid to the seller.

FCC *abbr*: Federal Communications Commission.

fcp *abbr*: final circulating pressure.

Fe *sym*: iron.

Federal Aviation Administration (FAA) *n*: an agency of the U.S. Department of Transportation that sets and enforces federal aviation requirements. Address: 800 Independence Avenue SW; Washington, DC 20591; 202-267-3883.

federal aviation requirement (FAR) *n*: an order, or requirement, issued by the U.S. Federal Aviation Administration that gives rules and regulations that must be followed by aircraft manufacturers, airports, pilots, and anyone involved in civil aviation in the U.S.

Federal Communications Commission (FCC) *n*: the U.S. agency that sets standards and licenses radiotelephone and radiotelegraph stations. Address: 445 12th Street SW, Room CY-C314; Washington, DC 20554; 202-418-0500.

Federal Energy Regulatory Commission (FERC) *n*: an independent agency (created in 1977) of the Department of Energy that has jurisdiction over oil pipelines engaged in interstate commerce. With respect to the natural gas industry, the general regulatory principles of the FERC are defined in the Natural Gas Act (NGA), and the Natural Gas Policy Act (NGPA). The FERC also has jurisdiction over wholesale interstate electric rates, hydroelectric licensing, and oil pipeline rates. Its predecessor was the Federal Power Commission (FPC). Address: 888 1st Street NE; Washington, DC 20426.

Federal Insecticide, Fungicide, and Rodenticide Act (FIFRA) *n*: congressional act that regulates the manufacture, use, and application of pesticides.

Federal Land Policy and Management Act of 1976 (FLPMA) *n*: congressional act that establishes comprehensive land use guidelines for the Bureau of Land Management (BLM) on how to manage lands under its jurisdiction.

federal lease *n*: an oil and gas lease on federal land issued under the Mineral Leasing Act. Federal leases usually provide step-scale or sliding-scale royalty; a flat discovery royalty of one-eighth may also be specified.

Federal Oil and Gas Royalty Management Act of 1982 (FOGRMA) *n*: congressional act that was designed primarily to assure proper and timely revenue accountability from production and leasing of federal lands.

Federal Power Commission (FPC) *n*: before the establishment of the U.S. Department of Energy, gas pipelines were regarded as a utility and operated under the regulation of the FPC.

Federal Register *n*: a daily government publication that publishes proposed changes in agency rules and regulations. It also includes other government actions such as presidential proclamations and a monthly notice of changes in existing rules and regulations.

fee *n*: an estate in real property, completely owned, which the owner can sell or devise to his or her heirs. Often a term for distinguishing private lands (fee lands) from federal or state lands.

feedback *n*: 1. in automatic control of processes, electronic information sent back from the process and compared to the desired set point. Corrective action of the controlled variable is taken if there is any deviation. If a difference exists between the set point (reference) and the actual condition (feedback), the controlling means (such as a valve) is activated to respond for correction. This arrangement of components is commonly called a closed-loop control system. 2. in an electric circuit, the return of a portion of the output current of the circuit to its input.

feedback control *n*: an automatic process by which an operation is monitored and corrective action is taken to eliminate deviations from the operating norm.

feeder-collector *n*: in diamond bits, the type of flow channel that directs drilling fluid across the cutting surface of the bit in which small channels called feeders direct the fluid to flow into other channels called collectors. Compare *radial flow*.

feed in *n*: in drilling, the entrance of formation fluids into the wellbore because hydrostatic pressure is less than formation pressure.

feed off *v*: to lower the bit continuously or intermittently by allowing the brake to disengage and the drum to turn. The feed-off rate is the speed with which the cable is unwound from the drum.

feedstock *n*: material other than a catalyst introduced into a plant for processing. With regard to natural gas, used as an essential component of a process for the production of a product. Use of gas as a feedstock may be required because of the chemical reaction involved, or because of the physical burning characteristics of gas compared with other fuels, such as temperature and by-products. Examples of feedstock use include fertilizer manufacture, glass manufacture, and white brick manufacture.

feed tank *n*: a vessel containing a charge stock or a vessel from which a stream is continuously fed for further processing.

fee in surface *n*: an estate on the surface of land created when the owner separates or severs his or her mineral interests from the surface of the land.

fee simple *n*: a freehold estate on which there are no restrictions or limitations as to who may inherit.

fee simple absolute *n*: an estate limited absolutely to a person and to his or her heirs and assigns forever, without limitation or condition.

fee tail *n*: a freehold estate in which there is a fixed line of inheritable succession limited to the issue of the body of the grantee or devisee and in which the regular and general succession of heirs at law is cut off.

fee tail female *n*: an estate limited by a deed or will that conveys ownership to a person and the female heirs of his or her body. Male heirs cannot inherit the estate.

fee tail male *n*: an estate limited by a deed or will that conveys ownership to a person and the male heirs of his or her body. Female heirs cannot inherit the estate.

feldspar *n*: a group of silicate minerals that includes a wide variety of potassium, sodium, and aluminum silicates. Feldspar makes up about 60 percent of the outer 9.3 miles (15 kilometres) of the earth's crust.

feldspathic *adj*: containing or largely composed of feldspar or feldspar grains.

female connection *n*: a pipe, a coupling, or a tool threaded on the inside so that only a male connection can be joined to it. Compare *male connection*.

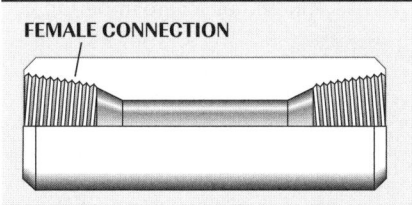

FEMALE CONNECTION

fence diagram *n*: see *panel diagram*.

fencing crew *n*: pipeline construction workers responsible for constructing temporary gates at points where a right-of-way crosses fence lines.

FeO *form*: ferrous oxide.

FERC *abbr*: Federal Energy Regulatory Commission.

FERC/FPC decontrol clause *n*: clause in a gas sales contract that covers what buyer and seller agree to do about price in the event of decontrol.

FERC out *n*: provision in gas purchase contracts that allows the purchaser to reduce the price paid to the producer by an amount that FERC did not allow in the purchaser's rate base for ultimate pass-through to the purchaser's customer.

fermentation *n*: a decomposition process of certain organic substances, such as starch, in which a chemical change is brought about by enzymes, bacteria, or other microorganisms. Often referred to as "souring."

ferromagnetic *adj*: highly magnetic; ferromagnetic materials react to a magnet and can themselves become magnets. See *diamagnetic, paramagnetic*.

ferrous alloy *n*: a metal alloy in which iron is a major component.

ferrous oxide *n*: a black, powdery compound of iron and oxygen (FeO) produced in petroleum operations from

the oxidation of ferrous sulfide (FeS), a reaction that releases a great deal of heat. Also called iron monoxide.

ferrous sulfide *n*: a black crystalline compound of iron and sulfur (FeS) produced in petroleum operations from the reaction of hydrogen sulfide (H_2S) and the iron (Fe) in steel. Also called iron sulfide.

ferrule *n*: a metallic button, usually cylindrical in shape, normally fastened to a wire rope by swaging, but sometimes by brazing.

FeS *form*: ferrous sulfide.

FET *abbr*: field-effect transistor.

fetch *n*: the area of wind wave development where a certain wind force prevails from the same direction for a period of time. Also the distance over which the wind blows to generate the observed waves at a given position or point.

fiber core *n*: cord or rope of vegetable or synthetic fiber used as the axial member of a rope.

fibrous material *n*: any tough, stringy material of threadlike structure used to prevent loss of circulation or to restore circulation in porous or fractured formations.

fiduciary *n*: a person who serves, with or without bond, to act for the benefit of another in all matters connected with a specified undertaking. Fiduciary obligations exist, for example, between trustees and the beneficiaries of the trust.

field *n*: 1. a geographical area in which a number of oil or gas wells produce from a continuous reservoir. A field may refer to surface area only or to underground productive formations as well. A single field may have several separate reservoirs at varying depths. 2. the magnetic field in a motor or generator, or that part of a motor or generator that produces a magnetic field; the magnetic field about any current-carrying electrical conductor.

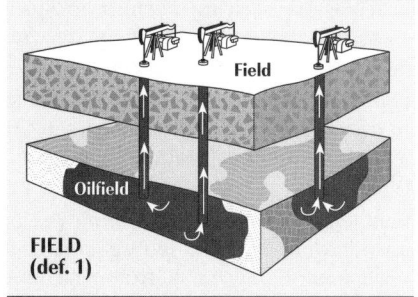

Field

Oilfield

FIELD (def. 1)

field administration *n*: in pipeline construction, the assistant superintendent and crew foremen.

field bevel *n*: a rebeveling of pipe ends in the field, usually required because of damage sustained by the pipe during transport or because a defective weld must be cut out.

field coefficient *n*: a value used to quickly determine gas volume flow; it is derived from several factors, such as the basic orifice factor (F_b), and the pressure base factor (F_{pb}).

field effect *n*: the local change from the normal value that an electric field produces in a semiconductor's charge-carrier region. The field effect is used in field-effect transistors.

field-effect transistor (FET) *n*: a transistor in which the resistance of the current path from source to drain is varied (modulated) by applying a transverse electric field between grid or gate electrodes. The electric field varies the thickness of the depletion layer between the gates to reduce conductance.

field-extension well *n*: see *exploration well*.

field facility *n*: an installation designed for one or more specific field processing units—scrubbers, absorbers, drip points, compressors, single- or multiple-stage separation units, low-temperature separators, and other types of separation and recovery equipment. See *battery*.

field gas facility *n*: see *central facility*.

field-grade butane *n*: a product consisting chiefly of normal butane and isobutane, produced at a gas processing plant. Also called mixed butane.

field office manager *n*: the individual responsible for the contractor's financial affairs on a pipeline spread. The field office manager oversees billing arrangements, payroll, and other money-related matters.

field processing *n*: the processing of oil and gas in the field before delivery to a major refinery or gas plant, including separation of oil from gas, separation of water from oil and from gas, and removal of liquid hydrocarbons.

field processing unit *n*: a unit through which a well stream passes before the gas reaches a processing plant or sales point.

field pump *n*: a pump installed in a field to transfer oil from a production tank to a central gathering station near a main pipeline.

field superintendent *n*: an employee of an oil company who is in charge of a particular oil or gas field from which the company is producing.

field support personnel *n*: in pipeline construction, the mechanics, parts and warehouse workers, truck drivers, and others who service the machinery that actually lays pipe.

FIFRA *abbr*: Federal Insecticide, Fungicide, and Rodenticide Act.

fifty-gallon drum *n*: see *drum*.

FIH *abbr*: fluid in hole; used in drilling reports.

filament *n*: in an incandescent light bulb, a metallic wire or ribbon through which electric current is passed to make it produce light.

file for record *v*: to send a legal document to the county clerk for recording.

filled system *n*: in temperature measurement in process control, a device filled with a fluid that responds to temperature changes. A typical filled system consists of a metallic bulb containing a fluid such as liquid, gas, vapor, or mercury, whose volume or pressure responds to temperature changes; a capillary tube that transmits the changes from the bulb to an indicating device; an indicating device with a spiral Bourdon tube, which drives a pointer or recording pen; and compensating elements to offset the effects of varying ambient temperatures.

filled-system thermometers *n*: a direct-reading device or a device with compensating elements to offset the effects of ambient temperature; used in process control to provide remote indication of temperature or the recording of temperatures; and, are simple in design and inexpensive to manufacture. Gas-filled thermometers have the greatest measurement capabilities, operate over wide temperature ranges, and have fast response to temperature changes.

filler *n*: in surface preparation prior to painting, a composition used for filling fine cracks and indentations to obtain a smooth finish.

filler material *n*: a material added to cement or cement slurry to increase its yield.

filler pass *n*: fourth of five passes or beads in pipeline welding.

filler wire *n*: 1. small spacer wire within a wire-rope strand that positions and supports other wires in the strand. 2. a wire-rope strand pattern that uses filler wires. See *filler-wire strand, strand pattern*.

filler-wire strand *n*: a wire rope strand in which two layers of wire of the same size are laid around a center wire, with the inner layer having half the number of wires of the outer layer. Small wires—the filler wires—are laid in the valleys of the inner layer.

filling density *n*: the percent ratio of the weight of gas in a tank to the weight of water that the tank will hold.

fill line *n*: see *fill-up line*.

fill material *n*: any material used for the primary purpose of replacing an aquatic area with dry land or of changing the bottom elevation of a water body. The term does not include any pollutant discharged into the water.

fill the hole *v*: to pump drilling fluid into the wellbore while the pipe is being withdrawn to ensure that the wellbore remains full of fluid even though the pipe is withdrawn. Filling the hole lessens the danger of a kick or caving the wall of the wellbore.

fill-up line *n*: the smaller of the side fittings on a bell nipple, used to fill the hole when drill pipe is being removed from the well.

fill-up rate *n*: the frequency with which the hole is filled with drilling fluid to replace the pipe removed from the hole.

film *n*: a thin skin or membrane. In electronics, it is used in the manufacture of one type of resistor.

film resistor *n*: a resistor made from a thin skin or membrane (a film). Carbon film, metal film, or metallic oxide film may be used to make resistors whose resistances are very accurate, usually within 5 percent of the desired resistance.

filter *n*: 1. a porous medium through which a fluid is passed to separate particles of suspended solids from it. 2. in electronics, a device in a circuit that senses and allows desired electronic constituents to pass through it to the remainder of the circuit and at the same time senses and removes or weakens (attenuates) electronic constituents not desired in the circuit beyond the point of the filter's placement in the circuit. *v*: to remove or attenuate (weaken) electronic constituents not desired in the circuit beyond the point in the circuit where a filter is placed.

filter cake *n*: 1. compacted solid or semi-solid material remaining on a filter after pressure filtration of mud with a standard filter press. Thickness of the cake is reported in thirty-seconds of an inch or in millimetres. 2. the layer of concentrated solids from the drilling mud or cement

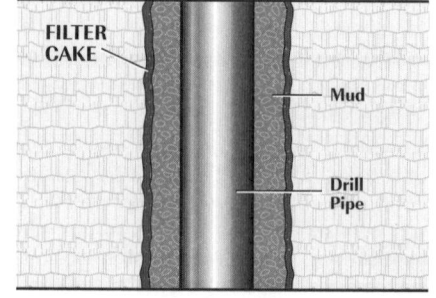

slurry that forms on the walls of the borehole opposite permeable formations; also called wall cake or mud cake.

filter cake thickness *n*: a measurement of the solids deposited on filter paper in thirty-seconds of an inch during a standard 30-minute API filter test. See *cake thickness*. In certain areas the filter cake thickness is a measurement of the solids deposited on filter paper for 7.5 minutes.

filter choke coil *n*: a winding (coil) with a laminated iron core used in a power supply's filter system to pass direct current and simultaneously impede the passage of alternating current.

filter element *n*: see *air cleaner filter element*.

filter loss *n*: the amount of fluid that can be delivered through a permeable filter medium after being subjected to a set differential pressure for a set length of time.

filter paper *n*: porous unsized paper for filtering liquids. The API filtration test specifies one thickness of 9-centimetre Whatman No. 50, S&S No. 576, or the equivalent filter paper.

filter press *n*: a device used in the testing of filtration properties of drilling mud. See *mud*.

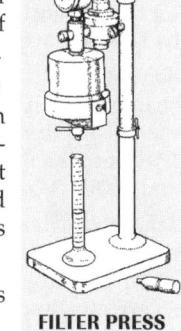

FILTER PRESS

filtrate *n*: 1. a fluid that has been passed through a filter. 2. the liquid portion of drilling mud that is forced into porous and permeable formations next to the borehole.

filtration *n*: the process of filtering a fluid.

filtration loss *n*: the escape of the liquid part of a drilling mud into permeable formations.

filtration qualities *n pl*: the filtration characteristics of a drilling mud. In general, these qualities are inverse to the thickness of the filter cake deposited on the face of a porous medium and the amount of filtrate allowed to escape from the drilling fluid into or through the medium.

filtration rate *n*: see *fluid loss*.

fin *n*: a thin, sharp ridge around the box or the pin shoulder of a tool joint, caused by the use of boxes and pins with different-sized shoulders. See *radiator fin*.

final circulating pressure *n*: the pressure at which a well is circulated during well-killing procedures after kill-weight mud has filled the drill stem. This pressure is maintained until the well is completely filled with kill-weight mud.

final control element *n*: in process control systems, that portion of the controlling

means that directly changes the value of the manipulated variable.

final squeeze pressure *n*: the fluid pressure at the completion of a squeeze-cementing operation.

finding of no significant impact (FONSI) *n*: an NEPA document that is required when agencies determine that there are no significant environmental impacts from a proposed federal action. The FONSI states the reasons why the agency came to that conclusion.

fine *n*: a fragment or particle of rock or mineral that is too minute to be treated as ordinary coarse material.

fingerboard *n*: a rack that supports the tops of the stands of pipe being stacked in the derrick or mast. It has several steel fingerlike projections that form a series of slots into which the derrickman can place a stand of drill pipe after it is pulled out of the hole and removed from the drill string.

FINGERBOARD

fingering *n*: 1. a phenomenon that often occurs in an injection project in which the fluid being injected does not contact the entire reservoir but bypasses sections of the reservoir fluids in a fingerlike manner. Fingering is not desirable, because portions of the reservoir are not contacted by the injection fluid. 2. the same phenomenon in which water bypasses oil in the reservoir on its way to the well.

finger-type junk basket *n*: a fishing tool that uses fingerlike catchers to gather and trap junk at the bottom of the hole for retrieval. Also called a poor-boy junk basket.

finish *n*: the coating, usually paint or galvanizing material, that is applied to offshore deck grating.

Finite Element Analysis (FEA) *n*: can show the reaction of a pipe body or the connection to forces estimated to be in a well environment.

fin-tube *n*: a tube or pipe having an extended surface in the form of fins used in heat exchangers or other heat-transfer equipment.

fire *v*: to start and maintain the fire in a boiler or heater.

fire bending *n*: one of the earliest methods for bending pipe. The joint was first placed over a small bonfire and, when the heat had rendered it sufficiently malleable, it was placed against a tree and pressure was applied until the desired bend was achieved. Fire bends significantly weakened the pipe. A cold-work process is less damaging.

fired heater *n*: a furnace in which natural gas or other fuel is burned to heat the gas or liquid passing through the furnace tubes.

fire flooding *n*: see *in situ combustion*.

fire hazard *n*: flammable, combustible pyrophoric and oxidizer (as defined by the Code of Federal Regulations).

fireman *n*: the member of the crew on a steam-powered rig who is responsible for care and operation of the boilers. Compare *motorman*.

fire point *n*: the temperature at which a petroleum product burns continuously after being ignited. See *flash point*.

fire-stop float *n*: a special drill string float that is used in air drilling when a possibility exists of starting a downhole fire because of the presence of volatile hydrocarbons in the borehole. Basically, a fire-stop float is an upside down drill pipe float that a special zinc ring holds open. Fire melts the ring to close the float and stop the air. With no air, the fire goes out.

fire tube *n*: a pipe, or set of pipes, within a tank through which steam or hot gases are passed to warm a liquid or gas in the tank. See *steam coil*.

fire wall *n*: a structure erected to contain petroleum or a petroleum-fed fire in case a storage vessel ruptures or collapses. Usually a dike is built around the petroleum storage tank and a steel or stone wall is put up between the prime movers and the oil pumps in a pipeline pumping station.

firing line *n*: in pipeline construction, the welding crew that takes over after the root pass and the hot pass have been made. The firing line is responsible for the filler pass and the cap bead, which complete the joint.

firing order *n*: the sequence in which combustion occurs in a multicylinder engine. For example, in an eight cylinder engine, the firing order could be 1–8–4–3–6–5–7–2, which means that combustion occurs first in the first cylinder, then in the eighth cylinder, and so on, until combustion occurs in the second cylinder; then, the sequence starts over.

first responder awareness level *n*: a training level achieved by any employee who has been HAZWOPER trained to witness the release, make proper notifications, and *take no further action*. In other words, if the employee has been trained to recognize an emergency and to notify the proper people, this person has attained the first responder awareness level.

first responder operational level *n*: a training level achieved by any employee who has been HAZWOPER trained to take a defensive role in emergency response. In other words, the employee is not trained to stop the release, but rather to help contain the release from a safe distance and to prevent exposure.

first sale of natural gas *n*: defined by the FERC as any sale of natural gas to any interstate pipeline or intrastate pipeline, to any local distribution company, or to any person for use by such person.

first sale of NGL/NGLP *n*: defined by FERC as the first transfer for value to a class of purchaser for which a fixed price per unit of volume is determined.

fish *n*: an object that is left in the wellbore during drilling or workover operations and that must be recovered before work can proceed. It can be anything from a piece of scrap metal to a part of the drill stem. *v*: 1. to recover from a well any equipment left there during drilling operations, such as a lost bit or drill collar or part of the drill string. 2. to remove from an older well certain pieces of equipment (such as packers, liners, or screen liner) to allow reconditioning of the well.

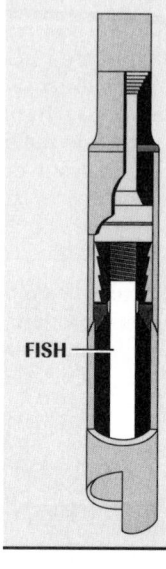

FISH

fish eyes *n pl*: the appearance of the surface of the mud in a mud pit when dry polymer fails to dissolve fully in the mud.

fishing *n*: the procedure of recovering lost or stuck equipment in the wellbore. See *fish*.

fishing assembly *n*: see *fishing string*.

fishing head *n*: a specialized fixture on a downhole tool that will allow the tool to be fished out after its use downhole. See *fish*.

fishing magnet *n*: a powerful permanent magnet designed to recover metallic objects lost in a well.

fishing neck *n*: a device placed on a piece of equipment that is lowered into a wellbore so that the equipment may be retrieved by wireline.

fishing string *n*: an assembly of tools made up on drill pipe that is lowered into the hole to retrieve lost or stuck equipment. Also called a fishing assembly.

fishing tap *n*: a tool that goes inside pipe lost in a well to provide a firm grip and permit recovery of the fish. Sometimes used in place of a spear.

fishing tool *n*: a tool designed to recover equipment lost in a well.

fishing-tool operator *n*: the person (usually a service company employee) in charge of directing fishing operations.

fishtail bit *n*: a drilling bit with cutting edges of hard alloys. Developed about 1900, and first used with the rotary system of drilling, it is still useful in drilling very soft formations. Also called a drag bit.

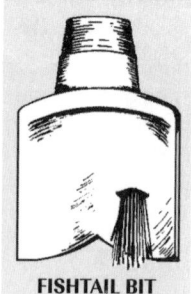

FISHTAIL BIT

fissure *n*: a crack or fracture in a subsurface formation.

fitting *n*: a small, often standardized, part (such as a coupling, valve, or gauge) installed in a larger apparatus.

five-spot *n*: four input or injection wells located in a square pattern with the production well in the center.

fixed-bore ram block *n*: in a ram blowout preventer, a steel block with elastomer surfaces that seal around drill pipe of a specific size. Besides forming a seal around drill pipe to confine well pressure in the annulus, they also support the load of the drill string—that is, the drill string can be hung off on them. Compare *variable bore ram block*. See *hang off, pipe ram*.

fixed choke *n*: a choke whose opening is one size only, that is, not adjustable. Compare *adjustable choke*.

fixed clearance pocket *n*: a clearance pocket that, when open, adds a specific volume to a compressor cylinder.

fixed costs *n pl*: costs that do not vary with the level of operation or units produced or sold. Examples are land costs, depreciation, and license fees.

fixed-cutter bit *n*: see *fixed-head bit*.

fixed-head bit *n*: any bit, such as a diamond bit, whose cutting elements do not move on the face, or head, of the bit. Compare *roller cone bit*.

fixed platform *n*: a structure made of steel or concrete, firmly fixed to the bottom of the body of water in which it rests.

fixed pump *n*: a type of downhole hydraulic pump that is attached to the end

of tubing; the tubing must be pulled to service the pump. See *hydraulic pump*. Compare *free pump*.

fixed-rate contract *n*: a contract that enjoys a known rate of dollars per million Btus during every month of sale; the rate paid may change from month to month, however.

fixed-rate royalty *n*: royalty calculated on the basis of a fixed rate per unit of production, without regard to the actual proceeds received from the sale of such production.

fix-factored correction *n*: a meter correction to which a constant temperature or pressure correction is applied.

fl *abbr*: flowed or flowing; used in drilling reports.

flag *n*: 1. a piece of cloth, rope, or nylon strand used to mark the wireline when swabbing or bailing. 2. an indicator of wind direction used during drilling or workover operations where hydrogen sulfide (sour) gas may be encountered. *v*: 1. to signal or attract attention. 2. in swabbing or bailing, to attach a piece of cloth to the wireline to enable the operator to estimate the position of the swab or bailer in the well.

flag arm *n*: in a two-diaphragm, four-chamber diaphragm displacement meter, the linkage that transfers oscillating motion from the flag rods to turn a tangent and crank assembly.

flammable *adj*: capable of being easily ignited. Sometimes the term "inflammable" is used, but flammable is preferred because it correctly describes the condition.

flange *n*: a projecting rim or edge (as on pipe fittings and openings in pumps and vessels), usually drilled with holes to allow bolting to other flanged fittings.

flanged fitting *n*: a device that holds an orifice plate centered in the line in which a fluid is flowing. It consists of two pieces that are joined by placing them together and tightening bolts and nuts. The orifice plate fits between the two pieces. To remove or inspect the orifice plate in a flanged fitting, the line must be bled and the flow of fluid rerouted so that no pressure exists on the fitting.

flanged orifice fitting *n*: a two-piece orifice fitting with flanged faces that are bolted together.

flange tap *n*: in a flanged orifice fitting, a threaded hole on each side of the orifice plate into which are screwed lines that connect the fitting with the flow recorder.

flange union *n*: a device in which two matching flanges are used to join the ends of two sections of pipe.

flange up *v*: 1. to use flanges to make final connections on a piping system. 2. (slang) to complete any operation, as in, "They flanged up the meeting and went home."

flank *n*: the sides of a bit.

flapper valve *n*: a type of check valve in a pipe or a line that has a hinged closure mechanism (a flapper) and that allows fluid flow in one direction but shuts it off in the other direction.

flare *n*: an arrangement of piping and burners used to dispose (by burning) of surplus combustible vapors, usually situated near a gasoline plant, refinery, or producing well. *v*: 1. to dispose of surplus combustible vapors by igniting them in the atmosphere. Flaring is rarely used, because of the high value of gas and the stringent air pollution controls. 2. in underbalanced drilling, the flame that results when the gaseous fluids exiting the well are ignited.

flare gas *n*: gas or vapor that is flared.

flare line *n*: a line (pipe) that comes out of a mud-gas separator and carries the separated gas a safe distance away from the rig. Usually, the gas is disposed of by burning, or flaring.

flare pit *n*: on land rigs, an earthen pit dug at the end of the flare line. Gas is flared over the flare pit to protect the surrounding area from heat and fire.

flare stack *n*: see *flare line*.

flash *n*: the sudden vaporization of a liquid caused by a rapid decrease in pressure and/or an increase in temperature.

flash gas *n*: gas released from liquid hydrocarbons as a result of an increase in temperature or a decrease in pressure.

flashing *n*: the continuing process by which a liquid is caused to flash.

flash point *n*: the temperature at which a petroleum product ignites momentarily but does not burn continuously. Compare *fire point*.

flash point check *n*: test made to verify that each product vaporizes within the specified proper temperature range.

flash set *n*: a premature thickening or setting of cement slurry, which makes it unpumpable.

flash tank *n*: a vessel used for separating the liquid phase from the gaseous phase formed from a rise in temperature and/or a reduction of pressure on the flowing stream.

flash welding *n*: 1. a form of resistance butt welding used to weld wide, thick members or members with irregular faces together, and tubing to tubing. 2. in pipeline construction, a welding technique in which low voltage is applied to each pipe joint while the ends are in light contact. This contact produces a rapid arcing, called flashing. After the pipe ends have been adequately heated, the current is abruptly increased, and the pipe joints are brought together rapidly and forcefully. The current is then reduced, excess flash material in the pipe is cleared, and the weld is completed.

flat *n*: see *kelly flat*.

flat gel *n*: a condition wherein the 10-minute gel strength is substantially equal to the initial gel strength.

flat pricing *n*: pricing directly from the posting, without reference to gravity.

flattening agent *n*: a material added to paints to reduce the gloss of the film.

fleet angle *n*: the angle created by drilling line between the drawworks drum and the fast sheave. The line is parallel to the sheave groove at only one point on the drum. As the rope moves from this point either way, the fleet angle is created. The fleet angle should be held to a minimum—less than 1.5 degrees for grooved drums.

Fleet Numerical Meteorology and Oceanography (METOC) Detachment *n*: part of the United States Navy, this office mainly provides climatological information for the United States Navy, Marine Corps, and other Department of Defense agencies. However, it also provides surface marine gridded climatology (SMGC) data for any interested party. Address: Officer in Charge, 151 Patton Avenue, Room 563, Asheville, NC 28801-5014; http://navy.ncdc.noaa.gov. See *surface marine gridded climatology data*.

flexible drill pipe *n*: specially manufactured drill pipe that has several pressure-tight joints over the length of the pipe. These joints allow the pipe to bend considerably more than regular drill pipe and are used in directional wells (especially horizontal ones) where the angle of deflection from vertical is relatively abrupt.

flexible hose *n*: a type of tube or pipe that is bendable (flexible) so that repeated movements do not cause it to break; frequently used in subsea blowout preventer systems to conduct operating fluid from the accumulator on the surface to operating devices on the marine riser pipe.

flexible joint (flex joint) *n*: on floating offshore drilling rigs using a subsea blowout preventer, a device mounted between the annular preventer and the riser adapter on the lower marine riser package (LMRP). Flex joints bend laterally to prevent excessive bending forces from being exerted on the marine riser and the LMRP and BOP components. (A bending moment is a force lateral movement creates on an object.) A flex joint typically allows 10 degrees of offset from vertical. The riser adapter can be connected to the top of the flex joint's neck by a flange, a hub, or by welding. Compare *ball joint*.

flexible riser *n*: in offshore production using floating surface facilities, a pipe (riser) that connects flow lines from the subsea well to the facilities on the surface. Because the surface vessel moves with wind, waves, and currents, the riser must be able to bend (flex) to compensate for such movements.

flexure *n*: a phenomenon that occurs in some wells when the formation the well penetrates is affected by the weight of overlying formations. The weight of the overlying formations—the overburden stress—causes the formation opposite the wellbore to expand, or flex, into the wellbore.

flip-flop circuit *n*: a type of electronic circuit in which either of two active devices may remain conducting, with the other nonconducting, until the application of of an external pulse.

flipped *adj*: when the opposite occurs of what is intended in a drilling fluid. In an invert water-in-oil emulsion, the emulsion is said to be flipped when the continuous and dispersed phases reverse.

float *n*: 1. an element of a level-control assembly designed to operate while partially or completely submerged in a liquid the level of which is controlled by the assembly. The buoyancy of the liquid activates the float and the control valve to which it is linked and modifies the rate of the inflow or the outflow of the vessel to maintain a preset level. 2. a drill pipe float. 3. a long flat-bed semitrailer. *v*: 1. to move or rest on the surface of a liquid without sinking 2. to place something or make something move on the surface of a liquid.

float collar *n*: a special coupling device inserted one or two joints above the bottom of the casing string that contains a check valve to permit fluid to pass downward but not upward through the casing. The float collar prevents drilling mud from entering the casing while it is being lowered, allowing the casing to float during its descent and thus decreasing the load on the derrick or mast. A float collar also prevents backflow of cement during a cementing operation.

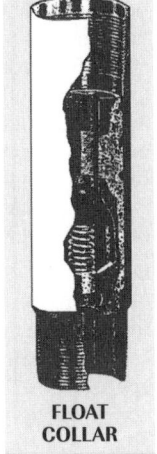

FLOAT COLLAR

floater *n*: see *drift bottle, floating offshore drilling rig*.

float guide wires *n pl*: solid wires or flexible cables used to guide the travel of an automatic gauge float.

floating control *n*: in process control systems, a device in which the speed of correction of the control element (such as a piston in a hydraulic relay) is proportional to the error signal. Also known as proportional-speed control.

floating cover *n*: a lightweight covering of either metal or plastic material designed to float on the surface of the liquid in a tank. Alternatively, it may be supported by a float system so that it is just above the free liquid surface. Used to minimize the evaporation of volatile products in a tank.

floating harness *n*: a frame equipped with sheaves and connected to the boom by stationary ropes usually called pendants.

floating mud cap *n*: see *mud cap*.

floating offshore drilling rig *n*: a type of mobile offshore drilling unit that floats and is not in contact with the seafloor (except possibly with anchors) when it is in the drilling mode. Floating units include barge rigs, drill ships, and semisubmersibles. See *mobile offshore drilling unit*.

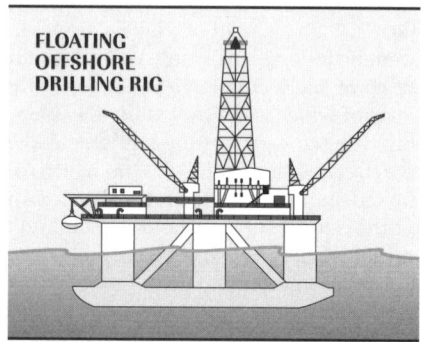

floating production and system off-loader *n*: a floating offshore oil production vessel that has facilities for producing, treating, and storing oil from several producing wells and which puts (off-loads) the treated oil into a tanker ship for transport to refineries on land. Some FPSOs are also capable of drilling, in which case they are termed floating production, drilling, and system off-loaders (FPDSOs).

floating roof *n*: a tank covering that rests on the surface of a hydrocarbon liquid in the tank and rises and falls with the liquid level. A floating roof eliminates vapor space above the liquid in the tank and conserves light fractions of the liquid.

floating roof tank *n*: a tank in which the roof floats freely on the surface of the liquid contents except at low levels, when

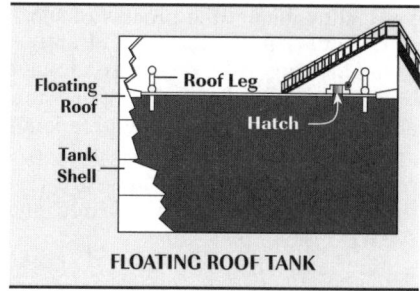

FLOATING ROOF TANK

the weight of the roof is transmitted by its supporting legs to the tank bottom.

floating screen *n*: a lightweight metal or plastic covering that is arranged to float on the surface of a liquid in a container to retard evaporation.

floating tank *n*: a tank with its main gate valve open to the main line at a station. Oil may thus enter or leave the tank as pumping rates in the main line vary.

floating tap system *n*: a type of tap sampling system found on liquid storage tanks. The permanently installed taps on the outside of the tank are attached to pipes and a floating boom inside the tank that rises and falls with the liquid level.

float shoe *n*: a short, heavy, cylindrical steel section with a rounded bottom that is attached to the bottom of the casing string. It contains a check valve and functions similarly to the float collar but also serves as a guide shoe for the casing.

float switch *n*: a switch in a circuit that is opened or closed by the action of a float and that maintains a predetermined level of liquid in a vessel.

float valve *n*: see *drill stem safety valve*.

float well *n*: an enclosure built into the roof of a floating roof tank to contain and guide the float of an automatic tank gauge.

flocculant *n*: substance added to improve the properties of the drilling mud. See *flocculating agent*.

flocculates *n pl*: chemicals that, when added to a liquid, cause it to coagulate.

flocculating agent *n*: material or chemical agent that enhances flocculation.

flocculation *n*: the coagulation of solids in a drilling fluid, produced by special additives or by contaminants.

flocs *abbr*: flocculates.

floe *n*: a floating ice field of any size.

flood *v*: 1. to drive oil from a reservoir into a well by injecting water under pressure into the reservoir formation. See *waterflooding*. 2. to drown out a well with water.

floodable length *n*: the length of a ship or mobile offshore drilling rig that may be flooded without its sinking below its safety

or margin line, usually a few inches below the freeboard deck.

flood current *n*: the movement of the tidal current toward the shore.

flood district office *n*: state or local office that is concerned with areas that are subject to flooding. This office can help determine if a site is in the floodplain and what locations may be involved in flooding. In the event of a hazardous materials incident, these offices can help determine the route of runoff from the incident.

flooded suction *n*: the condition of keeping enough liquid available to a pump's suction so that no air is drawn in with the liquid.

floor crew *n*: those workers on a drilling or workover rig who work primarily on the rig floor. See *rotary helper*.

floorhand *n*: see *rotary helper*.

floorman *n*: see *rotary helper*.

flora *n pl*: the plants of a given region or period considered as a whole.

flotation cell *n*: a large, cylindrical tank in which water that is slightly oil-contaminated is circulated to be cleaned before it is disposed of overboard or into a disposal well. Since oil droplets cling to rapidly rising gas, a device such as a bubble tower is usually installed in the cell to permit the introduction of gas into the water.

flotation level *n*: the depth of submergence of a buoyant automatic gauge float in a liquid of known density or weight.

flotation vest *n*: most commonly worn by sport divers to overcome the buoyancy effect of water and keep them afloat in the proper position. Carbon dioxide cartridges inside the vest are fired when inflation is necessary.

flounder *v*: see *bit flounder*.

flow *n*: a current or stream of fluid.

flowback *n*: see *flow head*.

flow bean *n*: a plug with a small hole drilled through it, placed in the flow line at a wellhead to restrict flow if it is too high. Compare *choke*.

flow by heads *v*: to produce intermittently.

flow chart *n*: a record made by a recording meter that shows the rate of production.

flow check *n*: a method of determining whether a kick has occurred. The mud pumps are stopped for a short period to see whether mud continues to flow out of the hole; if it does, a kick may be occurring.

flow coefficient *n*: see *C'*.

flow computer *n*: computer that handles all calculations, analysis, and processing of electronically transmitted data from a pipeline.

flow-control connection *n*: a device that controls product flow and directs it to the proper location. Mainline valves and side taps are examples of flow-control connections.

flow controller *n*: in control systems, a device used to manage the flow of gases, liquids, slurries, pastes, or solid particles in a line or flow system.

flow coupling *n*: a tubing sub made of abrasion-resistant material and used in a tubing string where turbulent flow may cause internal erosion.

flow drilling *n*: a form of underbalanced drilling in which a liquid drilling fluid (mud) is used, but density of which does not develop sufficient hydrostatic pressure to balance or overbalance formation pressure; consequently, formation fluids flow into the wellbore. Special equipment controls the flow of fluids from the well as it is drilled. Also called underbalanced liquid drilling.

flow efficiency *n*: a measure of the ability of the fluids in a formation to flow to a wellbore.

flow gate *n*: a device on the possum belly that can be moved to regulate mud flow to the shale shakers.

flow head *n*: fluids moving up the well.

flowing bottomhole pressure *n*: pressure at the bottom of the wellbore during normal oil production.

flowing bottomhole pressure test *n*: a bottomhole pressure test that measures pressure while the well is flowing. See *bottomhole pressure test*.

flowing pressure *n*: pressure registered at the wellhead of a flowing well.

flowing temperature *n*: recorded temperature of gas or liquid flowing through a pipe.

flowing well *n*: a well that produces oil or gas by its own reservoir pressure rather than by use of artificial means (such as pumps).

flow line *n*: the surface pipe through which oil travels from a well to processing equipment or to storage.

flow-line manifold *n*: a place in the flow line from a well, heater-treater, or other device where valves, meters, inlets, outlets, and various gauges may be installed.

flow-line sensor *n*: a device to monitor the rate of fluid flow from the annulus.

flow-line temperature *n*: the temperature of the drilling mud as it flows out of the borehole.

flow-line treater *n*: a cylindrical vessel into which an emulsion is piped to be broken down into its components. See *electrostatic treater, heater-treater*.

flow-line treating *n*: the process of separating, or breaking down, an emulsion into oil and water in a vessel or tank on a continuous basis (i.e., without an interruption in the flow of emulsion into the tank or vessel). Compare *batch treating*.

flowmeter *n*: an instrument that monitors, measures, or records the amount of fluid moving through a pipe or other container.

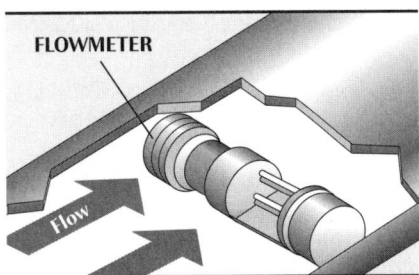

FLOWMETER

flowmeter discrimination *n*: a measure of the smallest increment of change in the pulse output of a flowmeter as it relates to the actual volume being measured.

flow monitor *n*: any device that senses the state of fluid flowing in a pipeline or other container.

flow nozzle *n*: a restriction installed in a line in which fluid is flowing that produces a pressure differential. The volume of fluid can be determined by measurement of the differential. Flow nozzles can handle dirty and abrasive gases better than orifices.

flow period *n*: in formation testing, an interval during which a well is allowed to flow while flow characteristics are being measured.

flow pressure *n*: see *flowing pressure*.

flow properties *n pl*: the properties of a drilling mud that have to do with its flow—viscosity, gel strength, and yield point.

flow proportional sample *n*: a sample taken from a pipeline during the whole period of transfer of a batch at a rate that is proportional to the rate of flow of the liquid in the pipeline at any instant.

flow range *n*: the range between the maximum and the minimum flow rates of a meter, generally determined by the limits of acceptable error.

flow rate (q) *n*: 1. time required for a given quantity of material to move a measured distance. 2. weight or volume of material flowing per unit time. Also known as rate of flow.

flow-rate sensor *n*: a device mounted in the mud return flow line that detects the speed (flow rate) of the mud flow and sends a signal to an instrument on the

rig floor and other rig locations to warn of a change in the return flow rate. An increase in flow may indicate that the well has kicked, while a decrease may indicate a loss of returns.

flow recorder *n*: a device with a chart and pens used to record static and differential pressures and sometimes temperature in an orifice meter installation.

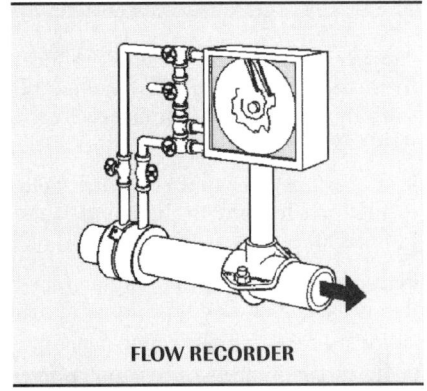

FLOW RECORDER

flow recorder clock *n*: a clock in a flow recorder that moves an orifice chart during a specific time period, such as 24 hours.

flow regulator *n*: a device used to control the flow of a fluid in a system; typically used in a closed-loop system consisting of a flow set point, flow-rate feedback, and flow valve actuator.

flow-sensing device *n*: see *flow-rate sensor*.

flow sensor *n*: see *flow-rate sensor*.

flow straightener *n*: a length of straight pipe containing straightening vanes or the equivalent and installed at the inlet of a flowmeter to eliminate swirl from the liquid from entering the meter and causing measurement errors.

flowstream *n*: the flow of fluids within a pipe.

flowstream samples *n pl*: small quantities of fluid taken at the wellhead or from the flow line and analyzed to determine composition of the flow.

flow string *n*: the string of casing or tubing through which fluids from a well flow to the surface.

flow tank *n*: see *production tank*.

flow test *n*: a preliminary test to confirm flow rate through a tool prior to going downhole.

flow-through meter *n*: any of several types of meter in which the fluid to be measured travels through the meter.

flow treater *n*: a single unit that acts as an oil and gas separator, an oil heater, and an oil- and water-treating vessel. See *heater-treater*.

flow tube *n*: a restriction installed in a line of flowing fluid that produces relatively high differential pressures with relatively low permanent pressure losses as the fluid flows through the device. By measuring the differential, the volume of fluid flowing through the tube can be inferred.

flow velocity *n*: see *flow rate*.

FLPMA *abbr*: Federal Land Policy and Management Act of 1976.

flue gas *n*: the gases that are produced from the combustion of a flammable substance in a special chamber or firebox. Also called stack gas.

fluid *n*: a substance that flows and yields to any force tending to change its shape. Liquids and gases are fluids.

fluid catalytic cracker *n*: a catalytic cracker that is used to crack gas or liquid hydrocarbons.

fluid contact *n*: the approximate point in a reservoir where the gas-oil contact or oil-water contact is located.

fluid coupling *n*: see *hydraulic coupling*.

fluid cutting *n*: see *washout*.

fluid density *n*: the unit weight of a fluid, e.g., pounds per gallon.

fluid end *n*: the portion or end of a fluid pump that contains the parts involved in moving the fluid (such as liners and rods) as opposed to the end that produces the power for movement.

fluid-end body *n*: the steel body in a reciprocating pump that has machined cylinders, openings for the valves, and fluid passageways.

fluid flow *n*: in fluid dynamics, the state of a fluid in motion is determined by the type of fluid (e.g., Newtonian, plastic, pseudo-plastic, dilatant); properties of the fluid such as viscosity and density; geometry of the system; the velocity. Thus, under a given set of conditions and fluid properties, the fluid flow can be described as plug flow, laminar (called also Newtonian, streamline, parallel, or viscous) flow, or turbulent flow.

fluid injection *n*: injection of gases or liquids into a reservoir to force oil toward and into producing wells.

fluidity *n*: the reciprocal of viscosity. The measure of rate with which a fluid is continuously deformed by a shearing stress; ease of flowing.

fluid knock *n*: a pressure concussion caused by suddenly stopping the flow of liquids in a closed container. Also called water hammer or hydraulic hammer.

fluid level *n*: the distance from the earth's surface to the top of the liquid in the tubing or the casing in a well. The static fluid level is taken when the well is not producing and has stabilized. The dynamic, or pumping, level is the point to which the static level drops under producing conditions.

fluid loss *n*: the unwanted migration of the liquid part of the drilling mud or cement slurry into a formation, often minimized or prevented by the blending of additives with the mud or cement.

fluid-loss additive *n*: a compound added to cement slurry or drilling mud to prevent or minimize fluid loss.

fluid pound *n*: the erratic impact of a pump plunger against a fluid when the pump is operating with a partial vacuum in the cylinder, with gas trapped in the cylinder, or with the well pump off.

fluid pressure *n*: a force exerted on a pipe, borehole, or vessel in which a fluid (a gas or a liquid) is confined. The fluid may be at rest or moving. Fluid pressure may be measured in force per unit area, such as pounds per square inch (psi) or as a derived unit from force per unit area, such as pascals (Pa), which is derived from newtons per square meter (N/m^2).

fluid reservoir *n*: 1. a geological formation that contains hydrocarbon fluids (oil and gas); see *reservoir*. 2. a container (a reservoir) on a blowout preventer control unit (an accumulator) that holds hydraulic fluid used to operate the blowout preventers. On offshore floating rigs, where subsea blowout preventers are often employed, two reservoirs are usually mounted on the accumulator unit's skid. One reservoir is used to store the fluid concentrate and one is used to store the mixed fluid. See *blowout preventer control unit*.

fluid sampler *n*: an automatic device that periodically takes a sample of a fluid flowing in a pipe.

fluid saturation *n*: the amount of the pore volume of a reservoir rock that is filled by water, oil, or gas and measured in routine core analysis.

flume *n*: see *boot*.

flume pipe *n*: large pipe used in creek and stream ditching in pipeline construction to allow the water to flow normally and to provide a passage for equipment over the water.

fluor *abbr*: fluorescence; used in drilling reports.

fluorescence *n*: instantaneous reemission of light of a greater wave length than that light originally absorbed.

fluoroelastomer *n*: an elastomer (a material such as synthetic rubber) in which the hydrogen atoms in the hydrocarbons are replaced by fluorine atoms.

flushed zone *n*: the area near the wellbore where invading mud filtrate forces out the movable formation fluids. Often abbreviated R_{xo}.

flush fluids *n pl*: thin fluids that work through a combination of turbulent and surfactant action to separate drilling mud from the cement being pumped downhole, while simultaneously removing coatings of mud left on the formation. Flush removes wall cake and flushes mud ahead of the cement, thereby lessening contamination and ensuring a good bond between the cement and the wall.

flushing case thermometer *n*: an assembly including a mercury-in-glass thermometer affixed to a cylindrical chamber with closures at the top and bottom of the chamber. When lowered into a tank in the open position, liquid flushes through the chamber. The chamber can be closed at any level in the tank, trapping a sample of liquid around the base of the thermometer and allowing for more accurate temperature reading when the thermometer is raised to the surface.

flush-joint casing *n*: a casing in which the outside diameter of the joint is the same as the outside diameter of the casing itself.

flush-joint pipe *n*: pipe in which the outside diameter of the joint is the same as the outside diameter of the tube. Pipe may also be internally flush-joint.

flush production *n*: a high rate of flow from a newly drilled well.

fluted drill collar *n*: see *spirally-grooved drill collar*.

fluvial deposit *n*: sediment deposited by flowing water.

flux *n*: the lines of force in a magnetic field.

flux density *n*: the number of lines of force in a square inch of the cross-sectional area of the core in a coil, or solenoid.

flux field *n*: the area of magnetic or electric lines of force.

flux gate *n*: a detector that produces an electrical signal whose magnitude and phase are proportional to the magnitude and direction of the external magnetic field acting along its axis. It is used to indicate the direction of the earth's magnetic field.

flux linkage *n*: the product of the number of turns in a coil and the magnetic flux passing through the coil.

fly cutter *n*: see *cutterhead*.

flysch *n*: a type of rock consisting of thinly bedded sandstone and shale, thought to be the result of the action of turbidity currents; a succession of turbidites originating in marine depositional basins, usually near the base of the continental slope. Flysch deposits are especially common in the Alpine region of Europe.

flyweight *n*: on a mechanical engine governor, one of usually two small metal weights that spin as the engine runs. When the engine speeds up, centrifugal force on the spinning flyweights increases, which causes a spring to compress and slow the engine down. Conversely, when the engine slows down, centrifugal force on the flyweights decreases which causes the spring to expand and speed the engine up.

flywheel *n*: a large, circular disk, connected to and revolving with an engine crankshaft. It stores energy and disburses it as the engine runs.

fm *abbr*: formation; used in drilling reports.

FM *abbr*: frequency modulation.

foam *n*: a two-phase system, similar to an emulsion, in which the dispersed phase is a gas or air.

foam drilling *n*: a form of underbalanced drilling in which a lightweight emulsion of water and a special foaming agent are circulated to drill the well. The water used to make foam is tightly bound to the foaming chemical and usually makes up a relatively small amount of the total fluid; thus, the water does not easily separate from the fluid. See *mist drilling*.

foamed mist *n*: an underbalanced drilling fluid that has a gas-to-liquid ratio higher than 250 to 1 and lower than 500 to 1. (A fluid with a gas-to-liquid ratio of 250 to 1 is foam; a fluid with a gas-to-liquid ratio of 500 to 1 and higher is mist.) Because pressure at the bottom of the hole is higher than pressure near the surface, foamed mist is foam at the bit and turns into mist uphole.

foam half-life *n*: in underbalanced drilling with foam as a circulating medium, a measure of the effect of foaming agents and a term used to compare foams. The longer a foam's half life is, the better.

foaming agent *n*: a chemical that is added to a liquid such as water to create an emulsion that contains bubbles and liquid, or foam. Foam is sometimes used in underbalanced drilling. See *foam drilling*.

foam quality *n*: in underbalanced drilling with foam as a circulating medium, the percentage of gas in foam at a particular depth or pressure. For example, foam with a quality of 75 percent means it is 75 percent gas and 25 percent water. The same foam deeper in the hole might have a quality of only 60 percent, because increased pressure compresses the gas. Therefore, foam quality varies with hole depth.

foam ratio *n*: in underbalanced drilling with foam as a circulating medium, a measure of the amount of liquid to gas in foam under standard conditions. It is usually expressed as so many units of gas to 1 unit of water or liquid. For example, if a foam has a 100-to-1 ratio, it has 100 cubic feet of gas in 1 cubic foot of liquid.

foam texture *n*: in underbalanced drilling with foam as a circulating medium, a measure of the properties of foam, which are similar to viscosity and gel strength of drilling mud. However, no generally accepted measurements for texture in foam are currently available.

FOB price *n*: see *free on board price*.

fog *n*: 1. a cloud of minute water droplets or ice crystals suspended in the air so that the cloud bottom rests on the earth's surface, either ground or water. See *advection fog, ice fog, steam fog*. 2. a wide, fine spray from a nozzle.

fog bank *n*: a well-defined mass of fog.

FOGRMA *abbr*: Federal Oil and Gas Royalty Management Act of 1982.

fold *n*: a flexure of rock strata (e.g., an arch or a trough) produced by horizontal compression of the earth's crust. See *anticline, syncline*.

foliated metamorphic rock *n*: metamorphic rock that has a layered look not necessarily associated with the original layering in sedimentary rock.

follower *n*: see *cam follower*.

FONSI *abbr*: finding of no significant impact.

fool's gold *n*: see *pyrite*.

footage rates *n pl*: a fee basis in drilling contracts stipulating that payment to the drilling contractor is made according to the number of feet or metres of hole drilled.

foot-pound (ft-lb) *n*: a unit of measure of twisting force, or torque. The amount of energy required to move 1 pound 1 foot vertically. The metric equivalent is the centimetre-kilogram, or, in SI units, the joule.

foot valve *n*: a check valve at the inlet end of the suction pipe of a pump that enables the pump to remain full of liquid when it is not in operation.

footwall *n*: the rock surface forming the underside of a fault when the fault plane is not vertical—that is, if the dip is less than 90°. Compare *hanging wall*.

foraminifera *n pl*: single-celled, mostly microscopic animals with calcareous exoskeletons; mostly marine.

force *n*: that which causes, changes, or stops the motion of a body.

forced draft *n*: air blown into a furnace or other equipment by a fan or blower.

forced pooling *n*: pooling of leased tracts undertaken without the willing cooperation of all the parties. Forced pooling may occur as the result of an order from a state regulatory agency, or an order sought by one or more of the parties affected.

forced unitization *n*: see *statutory unitization*.

force majeure clause *n*: in an oil and gas lease, the clause that usually contains a statement that the lease is subject to state and federal laws. It also excuses the lessee from timely performance of obligations should certain events beyond the lessee's power to control occur. Force majeure means a force or event that cannot be anticipated or controlled.

force per unit area *n*: an expression of pressure measurement in which a force, such as a pound, bears down on a given area, such as a square inch.

fore and aft *n*: the lengthwise measurement of a mobile offshore drilling rig or ship.

foreset bed *n*: a depositional layer on the steep seaward face of a marine delta that lies beyond the topset beds and is composed of finer sedimentary materials than the topset beds.

foreshore *n*: that part of the seashore that lies between low- and high-tide levels.

Forest Service *n*: a service under the Department of Agriculture that has three major program areas: national forest administration, state and private forestry, and research. The Forest Service manages 154 national forests and 19 national grasslands, comprising 188 million acres (75.2 million hectares) in 41 states and Puerto Rico. Address: 201 14th Street SW; Washington, DC 20250; 202-205-1760.

forfeiture *n*: failure to comply with conditions set in lease clauses can result in the loss of property rights by the lessee.

forge *v*: to use hard blows to form and shape metallic ingots into useful items.

formate fluid *n*: a special drilling fluid that contains a salt or an ester of formic acid (formate), which is chemically combined with another element such as potassium or cesium. Formate fluids are very stable at high temperatures and can be made very dense (heavy) without adding weighting materials such as barite. Formate fluids are used as completion fluids because they minimize formation damage. See *completion fluid, formation damage.*

formation *n*: a bed or deposit composed throughout of substantially the same kind of rock; often a lithologic unit. Each formation is given a name, frequently as a result of the study of the formation outcrop at the surface and sometimes based on fossils found in the formation.

formation boundary *n*: the horizontal limits of a formation.

formation breakdown *n*: the fracturing of a formation from excessive borehole pressure.

formation breakdown pressure *n*: the pressure at which a formation will fracture.

formation competency *n*: the ability of the formation to withstand applied pressure. Also called formation integrity.

formation competency test *n*: a test used to determine the amount of pressure required to cause a formation to fracture.

formation damage *n*: the reduction of permeability in a reservoir rock caused by the invasion of drilling fluid and treating fluids to the section adjacent to the wellbore. It is often called skin damage.

Formation Density Log™ *n*: trade name for a density log.

formation dip *n*: the angle at which a formation bed inclines away from the horizontal. Dip is also used to describe the orientation of a fault.

formation evaluation *n*: the analysis of subsurface formation characteristics, such as lithology, porosity, permeability, and saturation, by indirect methods such as wireline well logging or by direct methods such as mud logging and core analysis.

formation face *n*: that part of a formation exposed to the wellbore.

formation fluid *n*: fluid (such as gas, oil, or water) that exists in a subsurface rock formation.

formation fracture gradient *n*: a plot of pressure versus depth that reveals the pressure at which a formation will fracture at a given depth.

formation fracture pressure *n*: the point at which a formation will crack from pressure in the wellbore.

formation fracturing *n*: a method of stimulating production by opening new flow channels in the rock surrounding a production well. Often called a frac job. Under extremely high hydraulic pressure, a fluid (such as distillate, diesel fuel, crude oil, dilute hydrochloric acid, water, or kerosene) is pumped downward through production tubing or drill pipe and forced out below a packer or between two packers. The pressure causes cracks to open in the formation, and the fluid penetrates the formation through the cracks. Sand grains, aluminum pellets, walnut shells, or similar materials (propping agents) are carried in suspension by the fluid into the cracks. When the pressure is released at the surface, the fracturing fluid returns to the well. The cracks partially close on the pellets, leaving channels for oil to flow around them to the well. See *explosive fracturing, hydraulic fracturing.*

formation gas *n*: gas initially produced from an underground reservoir.

formation integrity *n*: see *formation competency.*

formation pressure *n*: the force exerted by fluids in a formation, recorded in the hole at the level of the formation with the well shut in. Also called reservoir pressure or shut-in bottomhole pressure.

formation resistivity *n*: a measure of the electrical resistance of fluids in a formation.

formation sensitivity *n*: the tendency of certain producing formations to react adversely to invading filtrates.

formation strength *n*: the ability of a formation to resist fracture from pressures created by fluids in a borehole.

formation strike *n*: see *strike.*

formation tester *n*: see *wireline formation tester.*

formation testing *n*: the gathering of pressure data and fluid samples from a formation to determine its production potential before choosing a completion method. Formation testing tools include formation testers and drill stem test tools.

formation thickness *n*: the dimension between two surfaces of a rock formation, the surfaces being the top of the formation and the bottom of the formation; in well testing, formation thickness may be abbreviated as h.

formation volume factor *n*: the factor that is used to convert stock tank barrels of oil to reservoir barrels. It is the ratio between the space occupied by a barrel of oil containing solution gas at reservoir conditions and a barrel of dead oil at surface conditions. Also called reservoir volume factor.

formation water *n*: 1. the water originally in place in a formation. See *connate water.* 2. any water that resides in the pore spaces of a formation.

formic acid *n*: a simple organic acid, HCOOH, used for acidizing oilwells. It is stronger than acetic acid but much less corrosive than hydrofluoric or hydrochloric acid and is usually used for high-temperature wells.

formonitrile *n*: see *hydrogen cyanide.*

forward *adv*: in the direction of the bow on a ship or an offshore drilling rig.

forward bias *n*: where a p-type and n-type semiconductor are joined, the condition that occurs when the positive terminal of a battery is connected to the lead on the p-type semiconductor and the negative terminal of the battery is connected to the lead on the n-type semiconductor. Forward biasing attracts free electrons in the n-type material across the junction toward the positive terminal, and attracts holes in the p-type across the junction toward the negative terminal. A forward-biased p-n junction offers low resistance and current flows freely across it. Compare *reverse bias.* See *n-type semiconductor, p-type semiconductor.*

forward combustion *n*: a common type of in situ combustion in which the combustion front moves in the same direction as the injected air. Burning is started at an injection well and moves toward production wells as air is continuously injected into the injection well. Compare *reverse combustion.*

forward voltage drop *n*: the voltage across a semiconducting diode or special transistor that carries current in a forward direction.

FoR$_{xo}$ Log™ *n*: trade name for focused electric log that investigates the flushed zone.

fossil *n*: the remains or impressions of a plant or animal of past geological ages that have been preserved in or as rock.

fossiliferous *adj*: containing fossils.

fossilize *v*: to become changed into a fossil.

fouling factor *n*: a factor used in heat-transfer calculations to represent the resistance to the flow of heat caused by dirt, scale, or other contaminants in the flowing fluids.

foundation pile *n*: the first casing or conductor string (generally with a diameter of 30 to 36 inches—76 to 91 centimetres) set when drilling a well from a floating offshore drilling rig. It prevents sloughing of the ocean-floor formations and is a structural support for the permanent guide base and the blowout preventers.

fourble *n*: a section of drill pipe, casing, or tubing consisting of four joints screwed together. Compare *double, single, thribble.*

fourble board *n*: the name used for the working platform of the derrickman, or the monkeyboard, when it is located at a height in the derrick equal to approximately four lengths of pipe joined together. Compare *double board, thribble board.*

four corner rule *n*: a rule of interpretation holding that an instrument such as an oil and gas lease must be interpreted from within the four corners of the instrument. Interpretation is made without any aid from knowledge of the circumstances under which the instrument came into being; the instrument is construed as a whole, without reference to any one part more than another.

four-pin kelly bushing *n*: a kelly bushing that has four steel dowels, or pins, that fit into corresponding holes in the master bushing. When the pins are engaged with the holes, the rotating master bushing also turns the kelly bushing, which then turns the kelly and the drill stem. See *kelly, kelly bushing, master bushing.*

four-pin master bushing *n*: a master bushing that has four holes symmetrically positioned on its outside perimeter and into which fit four corresponding steel dowels, or pins, on the kelly bushing. When the pins are engaged into the holes and the master bushing turns, the kelly bushing also turns. See *kelly bushing, master bushing.*

four-stroke/cycle engine *n*: an engine in which the piston moves from top dead center to bottom dead center two times to complete a cycle of events. The crankshaft must make two complete revolutions, or 720°.

four-way drag bit *n*: a drag bit with four blades. See *bit, fishtail bit.*

four-wire voltage transmitter *n*: a device that sends (transmits) signals to various components in an electronic process control system. Such a transmitter gets its name from the number of electrical wires used to power and operate it. Two of the wires to the transmitter provide the direct current for operation (typically 24 volts) and the other two wires transmit the signal or voltage.

FP *abbr*: flowing pressure; used in drilling reports.

FPDSO *abbr*: floating production, drilling, and system offloader.

FPSO *abbr*: floating production and system offloader.

frac *abbr*: fractured or fracturing; used in drilling reports.

frac fluid *n*: a fluid used in the fracturing process (i.e., a method of stimulating production by opening new flow channels in the rock surrounding a production well). Under extremely high hydraulic pressure, frac fluids (such as distillate, diesel fuel, crude oil, dilute hydrochloric acid, water, or kerosene) are pumped downward through production tubing or drill pipe and forced out below a packer or between two packers. The pressure causes cracks to open in the formation, and the fluid penetrates the formation through the cracks. Sand grains, aluminum pellets, walnut shells, or similar materials (propping agents) are carried in suspension by the fluid into the cracks. When the pressure is released at the surface, the fracturing fluid returns to the well but leaves behind the propping agents to hold open the formation cracks.

frac gradient *n*: see *fracture gradient.*

frac job *n*: see *formation fracturing.*

fraction *n*: a part of a mixture of hydrocarbons, usually defined by boiling range—for example, naphtha, gas oil, or kerosene.

fractional analysis *n*: a test for the composition of gas or two-phase gas-condensate streams. The analysis generally shows not only the composition in percentage of each hydrocarbon present through hexanes or heptanes but also the gallons per thousand cubic feet of liquids by component and the heating value of the gas.

fractional distillation *n*: the separation of crude oil into different cuts of fractions by distillation.

fractionate *v*: to separate single fractions from a mixture of hydrocarbon fluids, usually by distillation.

fractionating column *n*: the vessel or tower in a gas plant in which fractionation occurs. See *fractionate.*

fractionating tower *n*: see *fractionating column.*

fractionation *n*: see *fractional distillation.*

fracture *n*: a crack or crevice in a formation, either natural or induced. See *explosive fracturing, hydraulic fracturing.*

fracture acidizing *n*: a procedure by which acid is forced into a formation under pressure high enough to cause the for-

mation to crack. The acid acts on certain kinds of rocks, usually carbonates, to increase the permeability of the formation. Also called acid fracturing. Compare *matrix acidizing.*

fracture gradient *n*: the pressure gradient (psi/foot) at which a formation accepts whole fluid from the wellbore. Also called frac gradient.

fracture pressure *n*: the pressure at which a formation will break down, or fracture.

fracture zone *n*: zone of naturally occurring fissures or fractures that can pose problems with lost circulation.

fracturing *n*: shortened form of formation fracturing. See *formation fracturing.*

fracturing fluid *n*: a fluid, such as water, oil, or acid, used in hydraulic fracturing. The fluid carries propping agents that hold open the formation cracks after hydraulic pressure dissipates. See *acid fracturing, hydraulic fracturing, propping agents.*

fractus *n*: a cloud that has a ragged appearance, as if torn. Such clouds are torn from a main cloud bank by strong winds. Also called scud.

free air space *n*: any of the cavities in the human body that contain air and are normally connected to the atmosphere, including lungs, sinuses, and middle ear.

freeboard *n*: the vertical distance between the waterline and the freeboard deck on a ship, boat, or floating offshore drilling rig. Draft plus freeboard equal total height of vessel.

freeboard deck *n*: the uppermost continuous deck on a ship or floating rig that has a permanent means of closing all openings to the sea.

free butane *n*: see *butanes required.*

free electron *n*: an electron on the outer shell of an atom that moves readily from one atom to another.

free gas *n*: a hydrocarbon that exists in the gaseous phase at reservoir pressure and temperature and remains a gas when produced under normal conditions.

free hole *n*: a space (hole) in an atom that is not bound to an impurity in a semiconductor. A free hole is the opposite of a free electron and is a positive charge. See *free electron.*

free on board (FOB) price *n*: the price actually charged at the producing country's port of loading.

free oxygen *n*: oxygen that exists in molecular form (O_2) without being bound in a compound.

free point *n*: an area or point above the point at which a tubular, such as drill pipe, is stuck in the wellbore.

free-point indicator *n*: a device run on wireline into the wellbore and inside the fishing string and fish to locate the area where a fish is stuck. When the drill string is pulled and turned, the electromagnetic fields of free pipe and stuck pipe differ. The free-point indicator is able to distinguish these differences, which are registered on a metering device at the surface.

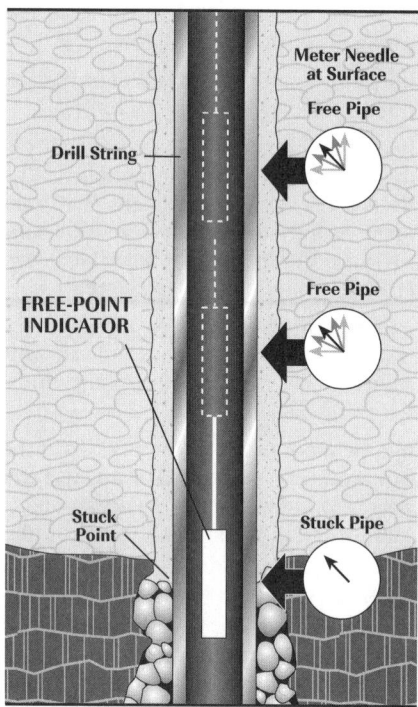

free pump *n*: a type of downhole hydraulic pump that moves in and out of the well by means of circulating fluids. See *hydraulic pump*. Compare *fixed pump*.

freestanding *adj*: of or pertaining to objects in the wellbore that do not contact the sides of the hole or the riser pipe, but which may be suspended from equipment on the surface.

Freestone rider *n*: see *Pugh clause*.

free surface *n*: in a container partially filled with liquid, the area of the top of the liquid. A completely filled container does not have a free surface because the liquid is not free to move about in the container as the container moves. Free surface is of particular concern on floating offshore rigs because any large and partially filled containers, such as the mud pits, or tanks, on board the rig cause the weight of the liquid in the containers to shift as the rig rolls. This shift in weight, if great enough, can create stability problems on the rig.

free water *n*: 1. water produced with oil. It usually settles out within five minutes when the well fluids become stationary in a settling space within a vessel. Compare *emulsified water*. 2. the measured volume of water that is present in a container and that is not in suspension in the contained liquid at observed temperature.

free-water knockout (FWKO) *n*: a vertical or horizontal vessel into which oil or emulsion is run to allow any water not emulsified with the oil (free water) to drop out.

freewheeling *n*: the action in which the clutch used between an auxiliary hydrodynamic brake and the drawworks drum shaft automatically disengages and runs freely while the empty block is being hoisted.

freeze pipe *n*: a device fitted on the vertical support members of the Trans-Alaska Pipeline System (TAPS) to circulate a refrigerant continuously between the subsoil and the top of the pipe. The refrigerant keeps the ground beneath the pipeline frozen to prevent frost heaving.

freeze point *n*: the depth in the hole at which the tubing, casing, or drill pipe is stuck. See *free-point indicator*.

freezing point *n*: the temperature at which a liquid becomes a solid.

frequency *n*: the number of cycles completed by a periodic quantity in a unit time—for example, the number of complete alternations (cycles) per second of alternating electric current.

frequency converter *n*: an electronic instrument for converting frequency (pulse train) to a proportionate analog signal.

frequency counter *n*: an electronic device used to measure the frequency of a signal by counting the number of cycles in the signal during a predetermined time interval.

frequency modulation (FM) *n*: the varying (modulating) of a wave in which the instantaneous frequency of the modulated wave differs from the carrier frequency by an amount proportional to the instantaneous value of the modulating wave. Compare *amplitude modulation*.

frequency output *n*: an output in the form of frequency that varies as a function of the applied measurand (e.g., angular speed and flow rate).

frequency range *n*: the measured values over which a meter or other measuring instrument is intended to measure, specified by their upper and lower limits.

fresh gale *n*: a Beaufort 8 wind blows from 34 to 40 knots.

fresh water *n*: 1. water that has little or no salt dissolved in it. 2. underground water, generally located near the surface, that does not contain a large amount of salt and from which most underground drinking water supplies are drawn. 3. inland surface water, such as lakes, streams, and ponds, that is not salty.

freshwater mud *n*: see *mud*.

friction *n*: resistance to movement created when two surfaces are in contact. When friction is present, movement between the surfaces produces heat.

frictional drag *n*: a force that slows the speed of an object as it moves. Drag occurs as an object moves through a fluid because of the friction that occurs between the object and the fluid; it also occurs as an object moves against another object.

frictional resistance *n*: the opposition to flow created by a fluid when it flows through a line or other container. Frictional resistance occurs within the fluid itself and it is created by the walls of the pipe or container as the fluid flows past them.

friction bearing *n*: see *journal bearing*.

friction cathead *n*: a spool on the side of the drawworks cathead that reels in a soft line, called the catline; used for hoisting light equipment on the rig floor.

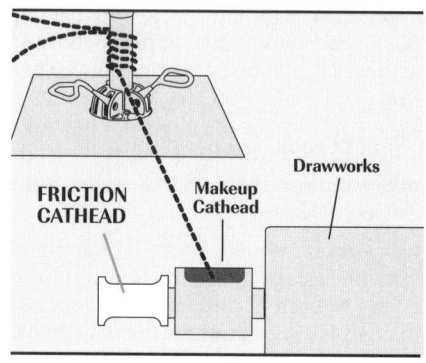

friction clutch *n*: a clutch that makes connection by sliding friction.

friction factor *n*: a dimensionless number used in the study of the flow of fluids in pipes. It is equal to the Fanning friction factor times a dimensionless constant. See *Fanning friction factor*.

friction loss *n*: a reduction in the pressure of a fluid caused by its motion against an enclosed surface (such as a pipe). As the fluid moves through the pipe, friction between the fluid and the pipe wall and within the fluid itself creates a pressure loss. The faster the fluid moves, the greater are the losses.

friction shoe *n*: in a drum clutch, pads of friction material are squeezed against a drum attached to the driving shaft. Friction between the shoes and the drum allows the clutch to engage.

front *n*: an interface between two air masses.

frontal zone *n*: in meteorology, a three-dimensional layer of large, horizontal pressure gradients.

front edging bar *n*: on a grated stair tread or a grated floor panel on an offshore installation, a steel bar that is attached to the front of the tread or floor panel.

frost heaving *n*: movement of the soil resulting from alternate thawing and freezing. Frost heaving generates stress on vertical support members of pipelines in the Arctic and, by extension, on the pipe itself.

frost wedging *n*: the phenomenon resulting when water invades rock, freezes, and, by its expansion, wedges apart the rock. Repeated freeze-thaw cycles can quickly break up any rock that has even the tiniest cracks.

frozen up *adj*: said of equipment the components of which do not operate freely.

ft *abbr*: foot.

ft-lb *abbr*: foot-pound.

ft/min *abbr*: feet per minute.

ft/s *abbr*: feet per second.

ft^2 *abbr*: square foot.

ft^3 *abbr*: cubic foot.

ft^3/bbl *abbr*: cubic feet per barrel.

ft^3/d *abbr*: cubic feet per day.

ft^3/min *abbr*: cubic feet per minute.

ft^3/s *abbr*: cubic feet per second.

fuel centrifuge *n*: a device an engine operator uses to separate water and solid materials from fuel. Centrifugal force created by the rapidly spinning centrifuge causes dirt and water, which are heavier (denser) than fuel, to move to the outside of the centrifuge where they are removed.

fuel content *n*: in improved recovery, the amount of coke available for in situ combustion, measured in pounds per cubic foot of burned area. Coke is formed by thermal cracking and distillation in the combustion zone. The amount of available coke depends on the composition of the reservoir crude oil.

fuel-injection nozzle *n*: see *nozzle*.

fuel injector *n*: a mechanical device that sprays fuel into a cylinder of an engine at the end of the compression stroke.

fuel knock *n*: a hammerlike noise produced when fuel is not burned properly in a cylinder.

fuel modulator *n*: a device installed on a diesel engine to reduce the amount of smoke coming out of the engine's exhaust. If the engine's governor delivers more fuel than air to the engine, the engine smokes too much. A fuel modulator makes the governor increase the fuel supply only at the same rate as the air increase. Such rate control holds down the black smoke from the engine exhaust during acceleration or sudden loading. See *governor*.

fuel oil *n*: 1. diesel fuel. 2. heavier liquid hydrocarbons, such as bunker oil, that is used to fuel turbines on a ship or other vessel.

fuel pump *n*: the pump that pressurizes fuel to the pressure used for injection. In a diesel engine the term is used to identify several different pumps: it is loosely used to describe the pump that transfers fuel from the main storage tank to the day tank; it is also used to describe the pump that supplies pressure to the fuel-injection pumps, although this is actually a booster-type pump.

fuel transfer pump *n*: any relatively small pump in an engine's fuel system that moves fuel from one fuel tank to another or from a tank to another location in the fuel system.

fulcrum *n*: the support about which a lever turns.

fulcrum assembly *n*: a bottomhole assembly, usually made up of drill collars and a stabilizer just above the bit, that can be used to increase hole angle. In holes

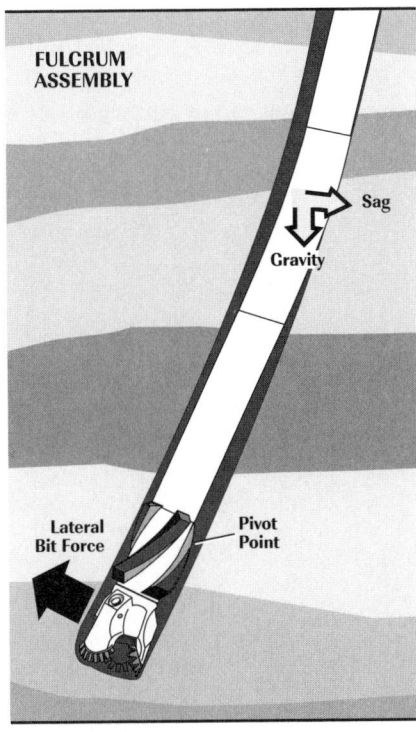

FULCRUM ASSEMBLY

Sag

Gravity

Lateral Bit Force

Pivot Point

inclined 3 degrees or more off vertical, the stabilizer acts as a fulcrum; the drill collars above the fulcrum sag toward the low side of the hole, forcing the bit toward the high side and increasing hole angle.

fulcrum effect *n*: the tendency of a fulcrum assembly to increase hole angle. See *fulcrum assembly*.

full adder *n*: a logic element in an electrical circuit that operates on two binary digits and a carry digit from a preceding stage to produce as output a sum digit and a new carry digit.

full-bore *n*: of a valve, fitting, or other object with an opening placed in a pipe, or line, which, when fully open, matches (or nearly matches) the diameter of the pipe in which the valve, fitting, or object is mounted.

full-bore isolation valve *n*: a completion tool used in deepwater wells. When run into the hole, the valve isolates the tubing from the annulus and allows packers to be set. It has a solid metal barrier that holds pressure from above and below. This barrier may be opened without destroying it, which ensures that it will open reliably.

fuller's earth *n*: see *attapulgite*.

full-gauge bit *n*: a bit that has maintained its original diameter.

full-gauge hole *n*: a wellbore drilled with a full-gauge bit. Also called a true-to-gauge hole.

full load *n*: in electronics, the greatest amount of electric power (the greatest load) that a circuit or device is designed to carry under specified conditions.

full-load displacement *n*: the displacement of a mobile offshore drilling rig or ship when floating at its deepest design draft.

full-wave rectification *n*: the conversion to direct current of both the negative and the positive pulsations of alternating current.

full-well stream *n*: the production from a well as it emerges from the mouth of the well and prior to any separation of the stream's components.

function *n*: a mathematical rule between two sets that assigns to each member of the first set exactly one member of the second.

fundamental quantity *n*: see *base quantity*.

fungible *adj*: relating or pertaining to petroleum products with characteristics so similar that they can be mixed together, or commingled.

funnel viscosity *n*: viscosity as measured by the Marsh funnel, based on the number of seconds it takes for 61 cubic inches (1,000 cubic centimetres) of drilling fluid to flow through the funnel.

FUNNEL VISCOSITY

fuse *n*: a device used to protect electrical equipment from overload. It has a wire that melts at high temperature to open the circuit.

fusible plugs *n pl*: a thermal device employed on surface flow lines as part of an ESD.

fusion-bonded epoxy coating *n*: a powdered resin coating that forms a skin over pipe when applied to its heated steel surface. Usually applied at the mill.

futures market *n*: market in which prices are determined by open bidding for contracts on the trading floor of a commodities market such as the New York Mercantile Exchange. A futures contract is a commitment to deliver or receive a specific quantity and grade of oil or product during a designated future month at a preset price. Delivery of the physical product occurs in only a small number of cases, with most participants liquidating their positions before the end of trading.

FWKO *abbr*: free-water knockout.

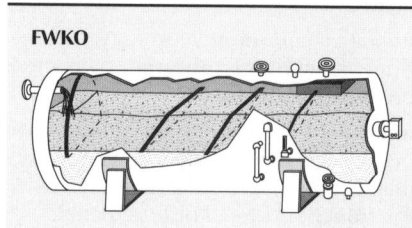

FWKO

g *abbr*: gram.

G *abbr*: center of gravity; this abbreviation is used mostly in buoyancy, stability, and trim calculations for offshore drillng rigs.

Ga *sym*: gallium.

G&OCM *abbr*: gas- and oil-cut mud; used in drilling reports.

gabbro *n*: an intrusive igneous rock with the same composition as basalt.

gage *n, v*: variation of gauge.

gal *abbr*: gallon.

gale *n*: a wind that is blowing at 28 to 55 knots. Gales are classified as moderate, fresh, strong, or whole.

galena (PbS) *n*: lead sulfide. Technical grades (specific gravity about 7) are used for increasing the density of drilling fluids to points impractical or impossible with barite.

gall *n*: damage to steel surfaces caused by friction and improper lubrication.

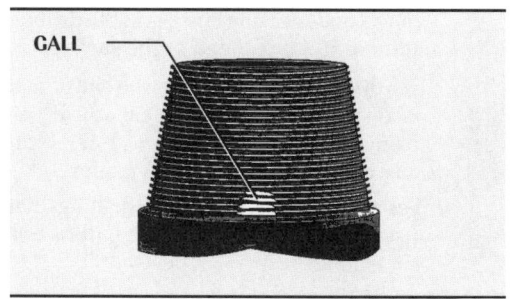

GALL

galling *adj*: the result of the sticking or adhesion of two mating surfaces of metal, not protected by a film of lubricant, and tearing due to lateral displacement.

galling limit *n*: one of the limitations on chain-and-sprocket life. This limitation on the strength of the metal the chains and sprockets are made of may cause the metal to wear even though it is lubricated because of a heavy load or a high speed.

gallium *n*: a silvery-white metallic element often added as an impurity to semiconductors to create a positive charge in the semiconductor.

gallon (gal) *n*: a unit of measure of liquid capacity that equals 3.785 litres and has a volume of 231 cubic inches (0.00379 cubic metres). A gallon of water weighs 8.34 pounds (3.8 kilograms) at 60°F (16°C). The imperial gallon, formerly used in Great Britain, equals approximately 1.2 U.S. gallons.

gallonage *n*: the amount of liquid a fire-fighting device delivers in U.S. gallons.

gal/min *abbr*: gallons per minute.

galvanic *adj*: of, relating to, or producing a direct current of electricity.

galvanic action *n*: the production of current flow when two dissimilar metals are placed in an electrolyte.

galvanic anode *n*: in cathodic protection, a sacrificial anode that produces current flow through galvanic action. See *sacrificial anode*.

galvanic cell *n*: an electrical device that contains two dissimilar metals suspended in or surrounded by an electrolyte, which is capable of producing voltage and current. Two or more galvanic cells constitute a battery. Often, and erroneously, a galvanic cell is called a battery. A battery consists of two or more galvanic cells.

galvanic corrosion *n*: a type of corrosion that occurs when a small electric current flows from one piece of metal equipment to another. It is particularly prevalent when two dissimilar metals are present in an environment in which electricity can flow (as two dissimilar joints of tubing in an oil or gas well).

galvanic protection *n*: corrosion prevention employed at the point where dissimilar metals come into contact and at which point corrosion could occur because of the small current created when dissimilar metals touch. The protection interrupts the current flow and thus prevents corrosion. See *galvanic cell, galvanic corrosion*.

galvanize *v*: to immerse an iron or steel object in molten zinc to produce a coating of zinc-iron alloy on the object. This zinc-iron coating resists corrosion.

galvanized or **galvanization** *n*: zinc coating for corrosion resistance used to extend the life of wire rope.

galvanized steel *n*: steel that is coated with zinc to protect it from rusting.

galvanometer *n*: an instrument that detects or measures a small electric current by movements of a magnetic needle or of a coil in a magnetic field. It can be adapted with shunts or resistors to measure larger currents or voltage. See *resistor, shunt*.

Gamma-Gamma Density Log™ *n*: trade name for a density log.

gamma particle *n*: a short, highly penetrating X-ray emitted by radioactive substances during their spontaneous disintegration. The measurement of gamma particles (sometimes called gamma rays) is the basis for a number of radioactivity well logging methods.

gamma ray *n*: see *gamma particle*.

gamma ray curve *n*: a plot of gamma radiation detected in a wellbore.

gamma ray detector *n*: a device that is capable of sensing and measuring the number of gamma particles emitted by certain radioactive substances.

gamma ray log *n*: a type of radioactivity well log that records natural radioactivity around the wellbore. Shales generally produce higher levels of gamma radiation and can be detected and studied with the gamma ray tool. In holes where salty drilling fluids are used, electric logging tools are less effective than gamma ray tools. See *radioactivity well logging*.

gamma ray spectroscopy log *n*: a special gamma ray log that reveals the relative amounts of radioactive uranium, thorium, and potassium in a formation. This information can help identify shales and clays.

gang pusher *n*: the supervisor of a roustabout crew or the person in charge of a pipeline crew.

gantry *n*: a platform made to carry a traveling crane and supported by towers or side frames running on parallel tracks.

garbet, garbot, or **garbutt rod** *n*: a short rod on the lower end of the traveling valve of a rod pump. It is attached to the standing valve and is used to pull the valve out of its seat when repairs are needed.

gas *n*: a compressible fluid that completely fills any container in which it is confined. Technically, a gas will not condense when it is compressed and cooled, because a gas can exist only above the critical temperature for its particular composition. Below the critical temperature, this form of matter is known as a vapor, because liquid can exist and condensation can occur. Sometimes the terms "gas" and "vapor" are used interchangeably. The latter, however, should be used for those streams in which condensation can occur and that originate from, or are in equilibrium with, a liquid phase.

gas anchor *n*: a tubular, perforated device attached to the bottom of a sucker-rod pump that helps to prevent gas lock. The device works on the principle that gas, being lighter than oil, rises. As well fluids enter the anchor, gas breaks out of the fluid and exits from the anchor through perforations near the top. Remaining fluids enter the pump through a mosquito bill (a tube within the anchor), which has an opening near the bottom. In this way, all or most of the gas escapes before the fluids enter the pump.

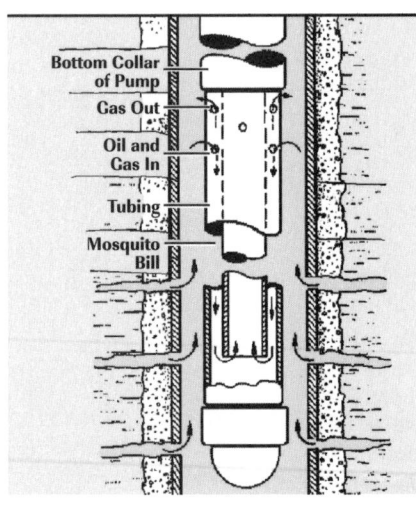

GAS ANCHOR

gas balance *n*: comparison of the sum of the volumes of gas production or receipts with the sum of the volumes of the dispositions of the gas.

gas balancing agreement *n*: an agreement between partners in a joint-interest well to determine the share due each partner if the gas is sold to two or more markets. Stated another way, an agreement covering the manner in which volumes of deferred gas production or exchange gas will be balanced between the parties to the agreement.

gas bubble *n*: an excess supply of natural gas, as when the available supply of natural gas exceeds the demand.

gas buster *n*: see *mud-gas separator*.

gas cap *n*: a free-gas phase overlying an oil zone and occurring within the same producing formation as the oil. See *associated gas, reservoir*.

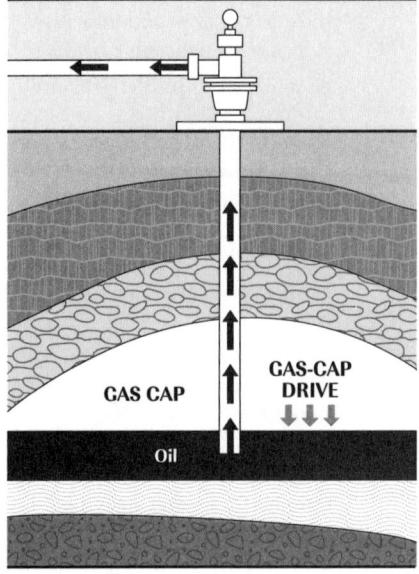

gas-cap drive *n*: drive energy supplied naturally (as a reservoir is produced) by the expansion of the gas cap. In such a drive, the gas cap expands to force oil into the well and to the surface. See *reservoir drive mechanism*.

gas-cap gas *n*: see *associated gas*.

gas chromatograph *n*: a device used to separate and identify gas compounds by their adhesion to different layers of a filtering medium such as clay or paper, sometimes indicated by color changes in the medium.

gas collection cells *n pl*: a device in which cores are thawed and samples of water and gas in the cores are collected. Neon is added to the system to ensure that a full charge of gas is obtained.

gas-condensate-glycol separator *n*: a device used in a glycol dehydration system installed where water is being removed from gas condensate; the separator removes glycol from the gas condensate.

gas constant *n*: a constant number, mathematically the product of the total volume and the total pressure divided by the absolute temperature for 1 mole of any ideal gas mixture of ideal gases at any temperature.

gas contract *n*: a mutually negotiated set of rules governing conduct between the parties thereto relating to all matters of common interest.

gas-cut mud *n*: a drilling mud that contains entrained formation gas, giving the mud a characteristically fluffy texture. When entrained gas is not released before the fluid returns to the well, the weight or density of the fluid column is reduced. Because a large amount of gas in mud lowers its density, gas-cut mud must be treated to reduce the chance of a kick.

gas cutting *n*: a process in which gas becomes entrained in a liquid.

gas detection analyzer *n*: a device used to detect and measure any gas in the drilling mud as it is circulated to the surface.

gas distribution company *n*: the entity that is responsible for moving natural gas from the pipeline to the consumer. Usually called the local gas company.

gas drilling *n*: a method of drilling that uses natural gas as the drilling fluid. See *air drilling*.

gas drive *n*: the use of the energy that arises from the expansion of compressed gas in a reservoir to move crude oil to a wellbore. Also called depletion drive. See *dissolved-gas drive, gas-cap drive, reservoir drive mechanism*.

gaseated fluid *n*: a liquid drilling fluid used in underbalanced drilling into which air or gas is added from the surface.

gas eliminator *n*: see *air eliminator*.

gas expansion *n*: when oil and gas are found in the same reservoir under pressure, the drilling of a well into the reservoir releases the pressure, causing the gas to expand. The expanding gas drives the oil toward and up the wellbore. The expansive energy of the gas can be harnessed whether the gas is in solution or forming a cap above the oil.

gas field *n*: a district or area from which natural gas is produced.

gas flow *n*: in well control, the streaming (the flow) of gas from the well when a formation containing gas is penetrated and is not contained by the drilling mud or well-control equipment.

gas flow recorder clock *n*: see *flow recorder clock*.

gasification *n*: gas cleaning.

gasified liquid drilling *n*: a form of under-balanced drilling in which air or gas is injected into the drill stem from the surface to lower the hydrostatic pressure of a liquid drilling fluid.

gas imbalance *n*: a discrepancy between a transporter's receipts and deliveries of natural gas for a shipper. Most pipelines require that shipper's deliveries to the pipeline and receipts from the pipeline remain essentially in balance within certain tolerances, or the pipeline may assess charges or penalties until the imbalance is rectified.

gasing-up *n*: a condition in a producing oilwell that occurs when lighter hydrocarbons come out of the oil and reach the surface in gaseous form.

gas injection *n*: the injection of gas into a reservoir to maintain formation pressure by gas drive and to reduce the rate of decline of the original reservoir drive. One type of gas injection uses gas that does not mix (i.e., that is not miscible) with the oil. Examples of these gases include natural gas, nitrogen, and flue gas. Another type uses gas that does mix (i.e., that is miscible) with the oil. The gas may be naturally miscible or become miscible under high pressure. Examples of miscible gases include propane, methane enriched with other light hydrocarbons, methane under high pressure, and carbon dioxide under pressure. Frequently, water is also injected in alternating steps with the gas.

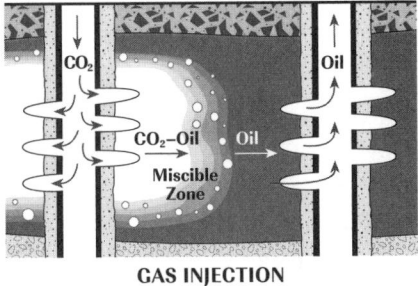

GAS INJECTION

gas injection well *n*: see *gas input well*.

gas input well *n*: a well into which gas is injected for the purpose of maintaining or supplementing pressure in an oil reservoir. More commonly called a gas injection well.

gas into the system *n*: gas that is purchased, taken on exchange, or injected from storage.

gasket *n*: any material (such as paper, cork, asbestos, or rubber) used to seal two essentially stationary surfaces.

gas laws *n pl*: laws such as Charles's, Boyle's, and Avogadro's that describe the behavior of gas volumes under pressure and temperature.

gas leak detector *n*: device used to detect combustible hydrocarbons.

gas lift *n*: the process of raising or lifting fluid from a well by injecting gas down the well through tubing or through the tubing-casing annulus. Injected gas aerates the fluid to make it exert less pressure than the formation does; the resulting higher formation pressure forces the fluid out of the wellbore. Gas may be injected continuously or intermittently, depending on the producing characteristics of the well and the arrangement of the gas-lift equipment.

gas-lift mandrel *n*: a device installed in the tubing string of a gas-lift well onto which or into which a gas-lift valve is fitted. There are two common types of mandrel. In the conventional gas-lift mandrel, the gas-lift valve is installed as the tubing is placed in the well. Thus, to replace or repair the valve, the tubing string must be pulled. In the side-pocket mandrel, however, the valve is installed and removed by wireline while the mandrel is still in the well, eliminating the need to pull the tubing to repair or replace the valve.

gas-lift valve *n*: a device installed on a gas-lift mandrel, which in turn is put on the tubing string of a gas-lift well. Tubing and casing pressures cause the valve to open and close, thus allowing gas to be injected into the fluid in the tubing to cause the fluid to rise to the surface. See *gas-lift mandrel.*

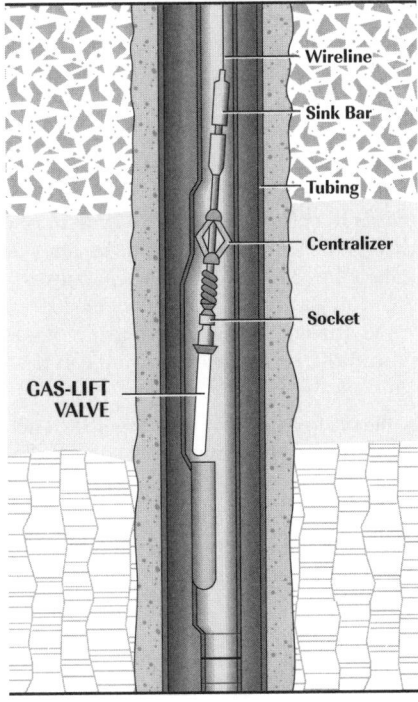

Wireline

Sink Bar

Tubing

Centralizer

Socket

GAS-LIFT VALVE

gas-lift well *n*: a well in which reservoir fluids are artificially lifted by the injection of gas.

gas liquids *n pl*: see *liquefied petroleum gas*.

gas lock *n*: 1. a condition sometimes encountered in a pumping well when dissolved gas, released from solution during the upstroke of the plunger, appears as free gas between the valves. If the gas pressure is sufficient, the standing valve is locked shut, and no fluid enters the tubing. 2. a device fitted to the gauging hatch on a pressure tank that enables manual dipping and sampling without loss of vapor. 3. a condition that can occur when gas-cut mud is circulated by the mud pump. The gas breaks out of the mud, expands, and works against the operation of the piston and valves.

gas meter *n*: the gas meter is probably most evident to consumers of natural gas who use it to monitor their gas usage and form the basis for monthly gas bills. It is an accurate instrument that measures volumetric quantities over a wide range of flow values and does not require any electrical power to operate. Where the gas pressure is not unreasonably low this meter is quite adequate to perform its function.

gas-miscible flooding *n*: see *gas injection*.

gas oil *n*: a refined fraction of crude oil somewhat heavier than kerosene and often used as diesel fuel.

gas-oil contact *n*: the point or plane in a reservoir at which the bottom of a gas sand is in contact with the top of an oil sand.

gas-oil ratio *n*: a measure of the volume of gas produced with oil, expressed in cubic feet per barrel or cubic metres per tonne. Also called solution gas-oil ratio.

gasoline *n*: a volatile, flammable liquid hydrocarbon refined from crude oils and used universally as a fuel for internal-combustion, spark-ignition engines.

gasoline engine starter motor *n*: on a diesel engine, a relatively small engine that runs on spark plugs and gasoline and whose power is used to turn over (move) the pistons in the diesel. As the diesel's pistons move, pressure and heat build in the diesel's cylinders and the diesel starts. As the diesel begins running, the gasoline engine disengages from the diesel.

gasoline plant *n*: also called a natural gas processing plant, a term that is preferred, because it distinguishes the plant from a unit that makes gasoline within an oil refinery. See *natural gas processing plant.*

gas pipeline *n*: a transmission system for natural gas or other gaseous material. The total system comprises pipes and compressors needed to maintain the flowing pressure of the system.

gas plant products *n pl*: liquids recovered from natural gas in a gas processing plant and, in some situations, from field facilities. See *natural gas liquids*.

gas processing *n*: the separation of constituents from natural gas for the purpose of making salable products and also for treating the residue gas to meet required specifications.

gas processing agreement *n*: an agreement under which gas not sold under a percentage of proceeds contract belongs to the producer as it passes through the processing plant. Normal payment is a percentage of the products recovered; the seller retains title to the residue gas.

gas processing plant *n*: see *natural gas processing plant*.

Gas Processors Association (GPA) *n*: an organization of companies that engages in gas processing. Its official publication is *Proceedings*. Address: 6526 E. 60th St.; Tulsa, OK 74145; 918-493-3872.

gas regulator *n*: an automatically operated valve that, by opening and closing in response to pressure, permits more or less gas to flow through a pipeline and thus controls the pressure.

gas reservoir *n*: a geological formation containing a single gaseous phase. When produced, the surface equipment may or may not contain condensed liquid, depending on the temperature, pressure, and composition of the single reservoir phase.

gas sand *n*: a stratum of sand or porous sandstone from which natural gas is obtained.

gas saturation (S_g) *n*: the amount of gas contained within the porosity of a reservoir.

gas separator *n*: see *separator*.

gasser *n*: a well that produces natural gas.

gas show *n*: the gas that appears in drilling fluid returns, indicating the presence of a gas zone.

gas sniffer *n*: see *explosimeter*.

gas transmission system *n*: the central or trunk pipeline system by means of which dry natural gas is transported from field gathering stations or processing plants to the industrial or domestic fuel market. Well pressure is supplemented at intervals along the transmission line by compressors to maintain a flow strong enough to move the gas to its destination.

gas treating *n*: the removal of hydrogen sulfide and carbon dioxide from raw gas or the green gas stream.

gas turbine *n*: an engine in which gas, under pressure or formed by combustion, is directed against a series of turbine blades. The energy in the expanding gas is converted into rotary motion.

gas well *n*: a well that primarily produces gas. Legal definitions vary among the states.

gas zone *n*: an area in a reservoir that is occupied by natural gas.

gate *n*: 1. in a logging tool, a shutterlike device that opens and closes very quickly to permit a measurement of the time a formation is exposed to sound, radioactivity, or other energy; usually timed in millionths of a second (microseconds). 2. in solid-state electronics, one of the electrodes in a field-effect transistor. See *field-effect transistor*.

gate-controlled turnoff *n*: in a semiconductor, the action of switching the semiconductor from its nonconducting, or off, state to its conducting, or on state, by applying a negative pulse to its gate terminal.

gate-turnoff silicon-controlled rectifier (GTO-SCR) *n*: a silicon-controlled rectifier that can be turned off by applying current to its gate. GTO-SCRs are mainly used for direct-current switching because turnoff can be achieved in a fraction of a microsecond. See *silicon-controlled rectifier*.

gate valve *n*: an opening and closing device (a valve) that employs a slab of metal (a gate) with a hole in it that is moved up or down within the valve's body. When the hole in the gate is positioned opposite the opening in the valve, fluid flows through the valve. When the solid part of the gate is positioned opposite the valve's opening, flow stops. Gate valves are not used to regulate the flow of fluids through them—that is, they are either fully open or fully closed.

gate voltage *n*: in an AC-DC diesel-electric system, the small amount of electricity applied to a silicon-controlled rectifier (SCR) to cause electricity to flow in one direction through the SCR. As long as gate voltage is applied, the SCR converts AC voltage to DC voltage as electricity flows through the SCR. If gate voltage is shut off, no electrical current can flow through the SCR.

gather *v*: to cause to come together, such as oil from several wells.

gathering *n*: the process of bringing oil, gas, or both from a well or wells in a field to a point for delivery to a pipeline or other transporting system.

gathering line *n*: a pipeline, usually of small diameter, used to move crude oil or gas from the field to a main pipeline.

gathering station *n*: a central point where there is the accessory equipment for delivering a clean and salable product to the market or to another pipeline.

gathering system *n*: the pipelines and other equipment needed to transport oil, gas, or both from wells to the gathering station. An oil gathering system includes oil and gas separators, emulsion treaters, gathering tanks, and similar equipment. A gas gathering system includes regulators, compressors, dehydrators, and associated equipment.

gauge *n*: 1. the diameter of a bit or the hole drilled by the bit. 2. a device (such as a pressure gauge) used to measure some physical property. 3. in electrical circuits, a measure of the diameter (thickness) of the wire or other device that conducts electricity through a circuit or from one point to another. In general, gauges for wire conductors range from 24 to 4/0 (0000). Some common sizes are 6, 12, and 16 with 6-gauge wire being larger in diameter than 16-gauge wire. See *American wire gauge*. *v*: to measure size, volume, depth, or other measurable property.

gauge area *n*: the outside edges of a bit; those portions of a bit that contact the wall of the hole or parallel the wall of the hole.

gauge cock *n*: a valve that activates or isolates a pressure measuring gauge.

gauge cutters *n pl*: the teeth or tungsten carbide inserts in the outermost row on the cones of a bit, so called because they cut the outside edge of the hole and determine the hole's gauge or size. Also called heel teeth.

gauge glass *n*: a glass tube or metal housing with a glass window that is connected to a vessel to indicate the level of the liquid contents.

gauge hatch *n*: the lidded opening in the top of an oil or oil product storage tank through which gauging and sampling operations are carried out.

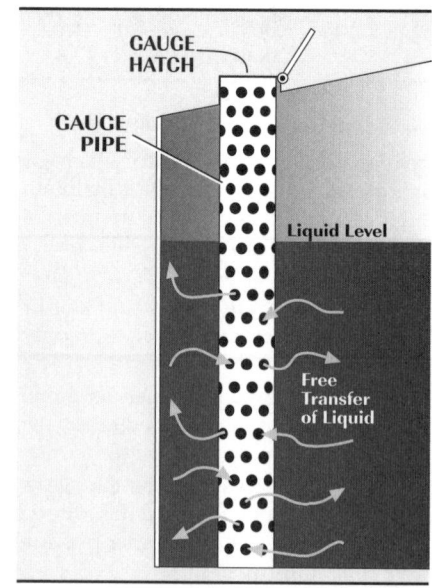

gauge head *n*: the housing of an automatic tank gauge. It may include the indicator and the transmitter.

gauge height *n*: the distance from the gauge, or reference, point from the bottom of a tank to a gauge, or reference, point at the top of the tank, usually on the hatch. On marine vessels, the measurement must be made when the vessel is on even keel.

gauge joint *n*: the heaviest wall casing section of the string, usually located just below the preventers or tree.

gauge lines *n pl*: small-diameter pipes leading from an orifice fitting to a bellows in a flow recorder. They allow gas pressure to be exerted on the bellows so that the pressure drop across the orifice plate can be recorded.

gauge path *n*: the vertical distance from the reference point on the gauge hatch to the bottom of the tank or to the datum plate.

gauge pipe *n*: a vertical pipe that extends from the gauge hatch to the bottom of the tank or to the datum plate.

gauge point *n*: 1. the point at which a tape is lowered and read on a tank, usually at the rim of the hatch, manway, or expansion dome. 2. a point to which all subsequent measurements are related. 3. the point from which the reference height is determined and from which the ullages/innages are taken.

gauge pressure *n*: 1. the amount of pressure exerted on the interior walls of a vessel by the fluid contained in it (as indicated by a pressure gauge). It is expressed in pounds per square inch gauge or in kilopascals. Gauge pressure plus atmospheric pressure equals absolute pressure. 2. pressure measured relative to atmospheric pressure considered as zero.

gauger *n*: a pipeline representative for the sale or transfer of crude oil from the producer to the pipeline. He or she samples or tests the crude oil to determine quantity and quality and uses a calibrated, flexible-steel tape with a plumb bob at the end to measure the oil in the tank.

gauge ring *n*: a cylindrical metal ring used to guide, and centralize, packers or tools inside casing.

gauge rounding *n*: a phenomenon in which the outside areas (the gauge areas) of a bit become worn and rounded by an abrasive formation.

gauge-row patterns *n pl*: the configuration or shape of the teeth in the outermost row of teeth on a bit.

gauger's bob *n*: a graduated weight used with a graduated tape to measure the

height of the liquid in a storage tank. The weight of the bob keeps the tape taut in order to ensure a correct reading of the liquid level.

gauger's tape *n*: a graduated, metal, nonsparking, noncorrosive measuring tape used with a gauger's bob to measure the height of the liquid in a storage tank.

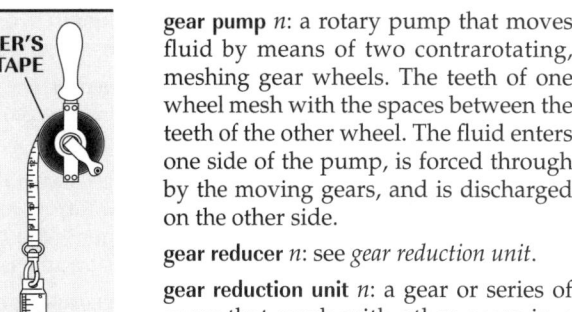
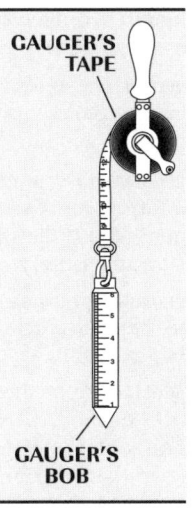

GAUGER'S TAPE

GAUGER'S BOB

gauge surface *n*: the outside surfaces of the outermost rows of teeth on a bit. They determine the diameter or gauge of the hole to be drilled.

gauge trip *n*: running of a gauge on tubing or slickline to verify casing dimensions.

gauging *n*: determining the liquid level of a tank so that its volume can be calculated. Usually done by lowering a weighted graduated steel tape through the tank roof and noting the level at which the oil surface cuts the tape when the weight gently touches the tank bottom. It can also be done by measuring the distance between liquid height and a reference point and subtracting the distance from the gauge height to determine liquid height.

gauging hatch *n*: the opening in a tank or other vessel through which measuring and sampling are performed.

gauging nipple *n*: a small section of pipe in the top of a tank through which a tank may be gauged.

gauging tables *n pl*: tables prepared by computers to show the calculated number of barrels or cubic metres for any given depth of liquid in a tank. Sometimes called strapping tables.

gauging tape *n*: a metal tape used to measure the depth of liquid in a tank.

GC *abbr*: gas-cut; used in drilling reports.

GCC *abbr*: Gulf Cooperation Council.

GCM *abbr*: gas-cut mud; used in drilling reports.

GDC survey *n*: a density log in which a gamma ray log, a Densilog, and a caliper log are recorded simultaneously.

Ge *sym*: germanium.

gear *n*: a toothed wheel made to mesh with another toothed wheel.

gear-and-pinion *adj*: see *rack-and-pinion gear*.

gear pump *n*: a rotary pump that moves fluid by means of two contrarotating, meshing gear wheels. The teeth of one wheel mesh with the spaces between the teeth of the other wheel. The fluid enters one side of the pump, is forced through by the moving gears, and is discharged on the other side.

gear reducer *n*: see *gear reduction unit*.

gear reduction unit *n*: a gear or series of gears that mesh with other gears in a machine and the purpose of which is to reduce the speed of the machine's output shaft or wheel.

gear transmission *n*: a system consisting of several toothed wheels (gears) that, when manipulated (shifted) by an operator, reduces or increases the speed of a member being driven by the gears.

gel *n*: a semisolid, jellylike state assumed by some colloidal dispersions at rest. When agitated, the gel converts to a fluid state. Also a nickname for bentonite. *v*: to take the form of a gel; to set.

gel cement *n*: cement or cement slurry that has been modified by the addition of bentonite.

gelled up *adj*: see *clabbered*.

gel strength *n*: a measure of the ability of a colloidal dispersion to develop and retain a gel form, based on its resistance to shear. The gel, or shear, strength of a drilling mud determines its ability to hold solids in suspension. Sometimes bentonite and other colloidal clays are added to drilling fluid to increase its gel strength.

general gas law *n*: any law relating to the pressure, temperature, or volume of a gas.

generalized viscosity curve *n*: a plot of calibration coefficient versus a modified expression of Reynolds number.

generator *n*: a machine that changes mechanical energy into electrical energy in the form of direct current. Compare *electric motor*.

generator identification number (GIN) *n*: an identification number required for producers of hazardous or acute hazardous waste that exceed EPA minimums under RCRA.

geochemistry *n*: study of the relative and absolute abundances of the elements of the earth and the physical and chemical processes that have produced their observed distributions.

geographic pole *n*: either of two points that lie at either end of an imaginary line drawn through the rotational axis of the earth; one is the North Pole and the other is the South Pole. Compare *magnetic pole*.

geoid *n*: the hypothetical surface of the earth that coincides everywhere with mean sea level. The geoid effectively smoothes out the irregularities in the earth's surface and is useful for accurate determination of the earth's gravity field at various locations.

geological correlation *n*: the relating of subsurface information obtained from one well to that of others.

geologic time scale *n*: the long periods of time dealt with and identified by geology. Geologic time is divided into eras (usually Cenozoic, Mesozoic, Paleozoic, and Precambrian), which are subdivided into periods and epochs. When the age of a type of rock is determined, it is assigned a place in the scale and thereafter referred to as, for example, Mesozoic rock of the Triassic period.

geologist *n*: a scientist who gathers and interprets data pertaining to the rocks of the earth's crust.

Geolograph™ *n*: trade name for a patented device that automatically records the rate of penetration and depth during drilling.

geology *n*: the science of the physical history of the earth and its life, especially as recorded in the rocks of the crust.

geomorphic unit *n*: one of the features that, taken together, make up the form of the surface of the earth.

geophone *n*: an instrument placed on the surface that detects vibrations passing through the earth's crust. It is used in conjunction with seismography. Geophones are often called jugs. See *seismograph*.

geophysical exploration *n*: measurement of the physical properties of the earth to locate subsurface formations that may contain commercial accumulations of oil, gas, or other minerals; to obtain information for the design of surface structures; or to make other practical applications. The properties most often studied in the oil industry are seismic characteristics, magnetism, and gravity.

geophysicist *n*: one who studies geophysics.

geophysics *n*: the physics of the earth, including meteorology, hydrology, oceanography, seismology, vulcanology, magnetism, and radioactivity.

geopressure *n*: abnormally high pressure exerted by some subsurface formations. The deeper the formation, the higher the pressure it exerts on a wellbore drilled into it.

geopressured shales *n pl*: impermeable shales, highly compressed by overburden pressure, that are characterized by large amounts of formation fluids and abnormally high pore pressure.

geoscience *n*: a science dealing with the earth—geology, physical geography, geophysics, geomorphology, geochemistry.

geostatic pressure *n*: the pressure to which a formation is subjected by its overburden. Also called ground pressure, lithostatic pressure, rock pressure.

geostatic pressure gradient *n*: the change in geostatic pressure per unit of depth in the earth.

geosteering *n*: the act of drilling a hole in a desired direction; directional and horizontal wells require geosteering to keep them on course and at the desired angle.

geosyncline *n*: a great downward folding in the earth's crust. See *syncline*.

geothermal *adj*: pertaining to heat within the earth.

geothermal drilling *n*: the boring of a well into a subsurface layer of rock that is very hot or into a rock layer that contains steam or hot water. Once a geothermal well is completed, the heat or the hot water or steam is used to generate electricity. See *geothermal reservoir*.

geothermal gradient *n*: the increase in the temperature of the earth with increasing depth. It averages about 1°F per 60 feet (1°C per 18.3 metres), but may be considerably higher or lower.

geothermal reservoir *n*: 1. a subsurface layer of rock containing steam or hot water that is trapped in the layer by overlying impermeable rock. 2. a subsurface layer of rock that is hot but contains little or no water. Geothermal reservoirs are a potential source of energy.

germanium (Ge) *n*: a brittle, silvery-gray metallic element in the carbon family. It is a rare metal used in semiconductors, alloys, and glass.

Geronimo *n*: see *safety slide*.

get a bite *v*: to set tools in casing.

GFCI *abbr*: ground fault circuit interrupter.

gigohm (Gohm, GΩ) *n*: 1 billion (1,000,000,000) ohms. Sometimes expressed as 1,000 megohms or 10^9 ohms.

GΩ *abbr*: gigohm, or 1 billion (1,000,000,000) ohms.

Gilsonite™ *n*: trade name for asphaltum mined, manufactured, or marketed by or for American Gilsonite Company.

gimbal *n*: a mechanical frame that permits an object mounted in it to remain in a stationary or near-stationary position regardless of movement of the frame.

Gimbals are often used offshore to counteract undesirable wave motion.

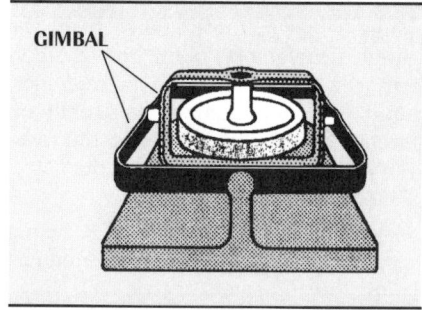

GIN *abbr*: generator identification number.

gin pole *n*: a pole (usually single) with guy wires and used with block and tackle to hoist equipment. On a drilling rig, the gin pole is typically secured to the mast or derrick above the monkeyboard.

gin pole truck *n*: a truck equipped with hoisting equipment and a pole or arrangement of poles for use in lifting heavy machinery.

girt *n*: one of the horizontal braces between the legs of a derrick.

GL *abbr*: ground level; used in drilling reports.

gland *n*: a device used to form a seal around a reciprocating or rotating rod (as in a pump) to prevent fluid leakage. Specifically, the movable part of a stuffing box by which the packing is compressed. See *stuffing box*.

gland packing *n*: material placed around a gland to effect a seal around a reciprocating or rotating rod.

gland-packing nut *n*: a threaded device the sides of which are arranged so that a wrench can be fitted onto them and used to retain the gland packing in place around a rod. See *gland packing*.

glass disk *n*: a sub with a glass blockage in the bore, used to isolate a surge chamber in gravel packing or perforation cleaning operations.

globe valve *n*: see *valve*.

gloss *n*: in paint technology, the visual impression created by the reflecting properties of a painted surface.

glow plug *n*: a small electric heating element placed inside a diesel engine cylinder to heat the air and to make starting easier.

GLR *abbr*: gas-liquid-ratio.

glycol *n*: a group of compounds used to dehydrate gaseous or liquid hydrocarbons or to inhibit the formation of hydrates. Glycol is also used in engine radiators as an antifreeze. Commonly used glycols are ethylene glycol, diethylene glycol, and triethylene glycol.

glycol absorber *n*: a cylinder consisting of several perforated trays with bubble caps mounted over the perforations and functioning as a part of a glycol dehydration system.

glycol/amine process *n*: a process that uses a solution comprising 10 to 30 percent monoethanolamine, 45 to 85 percent glycol, and 5 to 25 percent water for the simultaneous removal of water vapor, H_2S, and CO_2 from gas streams.

glycol dehydrator *n*: a processing unit used to remove all or most of the water from gas. A glycol unit usually includes an absorber, in which the wet gas is put into contact with glycol to remove the water, and a reboiler, which heats the wet glycol to remove the water from it so that it can be recycled.

glycol knockout *n*: a component in the contactor of a glycol dehydration unit that recovers any glycol that has gotten past the mist extractor.

glycol-powered pump *n*: a pump that uses the pressure of glycol and entrained gas from the contactor to circulate regenerated glycol from the storage tank to the contactor.

GM *abbr*: General Motors Corporation; a manufacturer of two-stroke-cycle diesel engines used in the petroleum industry.

GM *abbr*: metacentric height; the abbreviation is used mainly in buoyancy, stability, and trim calculations for offshore floating rigs.

GM Hydrostarter™ *n*: a hydraulic starter motor manufactured by General Motors (GM) that uses hydraulic fluid (light oil) to power a relatively small motor attached to the diesel engine. To start the diesel, the operator activates the Hydrostarter which turns over (moves) the diesel's pistons. As the pistons move, pressure and heat build up in the diesel's cylinders and the diesel starts.

go-devil *n*: 1. a device that is inserted into a pipeline for the purpose of cleaning; a line scraper. Also called a pig. 2. a device that is lowered into the borehole of a well for various purposes such as enclosing surveying instruments, detonating downhole explosive devices, and the like. *v*: to drop or pump a device down the borehole, usually through drill pipe or tubing.

go in the hole *v*: to lower the drill stem, the tubing, the casing, or the sucker rods into the wellbore.

Gondwanaland *n*: the southern part of the supercontinent Pangaea, comprising the future land masses of South America, Africa, Antarctica, Australia, and India.

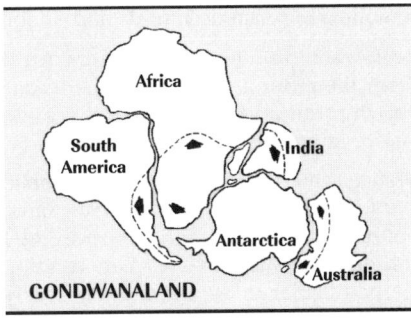

GONDWANALAND

gone to water *adj*: pertaining to a well in which production of oil has decreased and production of water has increased (e.g., "the well has gone to water").

good title *n*: see *merchantable title*.

gooseneck *n*: 1. the curved connection between the rotary hose and the swivel or top drive. 2. any curved length of pipe that serves as a connection between one conduit to another. See *swivel, top drive*.

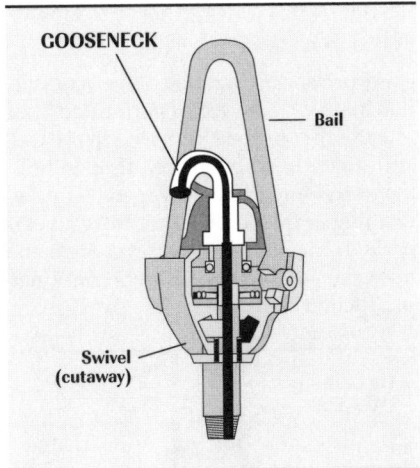

GOOSENECK

GOR *abbr*: gas-oil ratio.

gouge *n*: finely abraded material occurring between the walls of a fault as the result of grinding movement.

governor *n*: any device that limits or controls the speed of an engine.

GPA *abbr*: Gas Processors Association.

GPG *abbr*: grains per gallon; parts per million equals grains per gallon × 17.1.

gpm *abbr*: 1. gallons per minute when referring to rate of flow. 2. gallons per thousand cubic feet when referring to natural gas in terms of chromatograph analysis or theoretical gallons.

gr *abbr*: gray; used in drilling reports.

grab *n*: in sampling petroleum and petroleum products, a small sample, usually 1.5 millilitres (0.05 ounce), obtained by the probe and sampler in an automatic sampling system. Several grabs constitute a sample.

graben *n*: a block of the earth's crust that has slid downward between two faults. Compare *horst*.

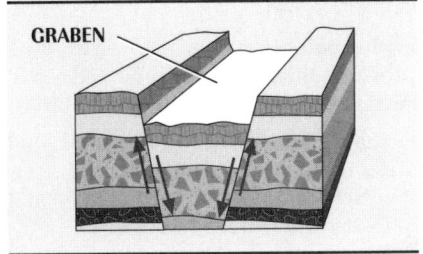

GRABEN

grab sample *n*: sample obtained by collecting loose solids in equal quantities from each part or package of a shipment and in sufficient quantity to be characteristic of all sizes and components.

graded stream *n*: a flowing stream that is stable, or in balance with its average load. It is just steep enough to carry out of its basin the amount of sediment brought in during an average-flow year.

graded string *n*: a casing string made up of several weights or grades of casing and designed to take into account well depth, expected pressures, and weight of the fluid in the well. Also called mixed string.

grades of drill pipe *n*: most present-day seamless drill pipe falls into one of four American Petroleum Institute (API) grades. (1) E-75, with a minimum yield strength of 75,000 psi (517,125 kilopascals-kPa), a maximum yield strength of 105,000 psi (723,975 kPa), and a minimum tensile strength of 100,000 psi (689,500 kPa); (2) X-95, with a minimum yield strength of 95,000 psi (655,025 kPa), a maximum yield strength of 125,000 psi (861,875 kPa), and a minimum tensile strength of 105,000 psi (723,975 kPa); (3) G-105, with a minimum yield strength of 105,000 psi (723,975 kPa), a maximum yield strength of 135,000 psi (930,825 kPa), and a minimum tensile strength of 115,000 psi (792,925 kPa); and (4) S-135, with a minimum yield strength of 135,000 psi (930,825 kPa), a maximum yield strength of 165,000 psi (1,137,675 kPa), and a minimum tensile strength of 145,000 psi (999,775 kPa). V-150 is a non-API, but higher strength, grade. It has a minimum yield strength and a tensile strength of 150,000 psi (1,034,250 kPa).

gradient *n*: ascending or descending with a uniform slope. The rate of change of temperature or pressure.

grading *n*: the process of providing a smooth and even work area to facilitate the movement of equipment onto and along a right-of-way. Grading entails leveling, cutting, and filling.

gradualism *n*: see *uniformitarianism*.

graduated neck *n*: the section of reduced cross-sectional area at the top and/or bottom of an open tank prover. It has visible graduations to enable measuring small volumes.

graduated neck prover *n*: an open tank prover with one or two graduated necks. See *graduated neck*.

Graham's Law of Diffusion *n*: law that states that the rate of diffusion of a gas is inversely proportional to the square root of its density.

grains per gallon (GPG) *n*: a unit of measure for the strength, or concentration, of a solution. It is based on a unit of weight called the grain, which equals 0.002285 ounce, or 0.0648 gram. One grain per gallon equals 0.058479 part per million.

gram *n*: a unit of metric measure of mass and weight equal to 1/1,000 kilogram and nearly equal to 1 cubic centimetre of water at its maximum density.

gram molecular weight *n*: see *molecular weight*.

granite *n*: an igneous rock composed primarily of feldspar, quartz, and mica. It is the most common intrusive rock—that is, it originally solidified below the surface of the earth. Its rock crystals are easily seen by the eye.

grantee *n*: a person to whom property is conveyed. Compare *grantor*.

granting clause *n*: clause in an oil and gas lease that specifies the rights and interests granted by the lessor to the lessee. Such rights usually involve searching and drilling for, then producing, oil and gas.

grantor *n*: a person who conveys property. Compare *grantee*.

graphite *n*: a soft, black, shiny mineral of pure carbon produced when hydrocarbons are subjected to high temperatures and pressures. Used in pencils and crucibles, as a lubricant, and in atomic-energy plants to control the release of radiation from uranium fuel.

grapple *n*: a mechanism that is fitted into an overshot to grasp and retrieve fish from the borehole. The interior of a grapple is wickered to engage the fish. See *basket grapple, spiral grapple*.

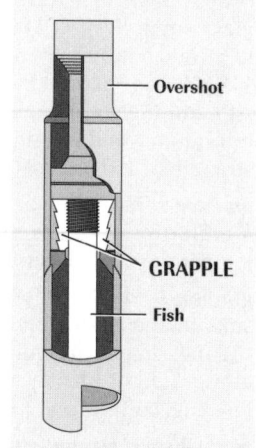

grasshopper *n*: see *water outlet*.

grass-roots refinery *n*: a refinery built from the ground up, as opposed to one to which an addition or a modification has been made.

grating *n*: an open grid assembly of metal bars, in which the load bearing bars, running in one direction, are spaced by a rigid attachment to cross bars running perpendicular to the bearing bars or by bent connecting bars extending between the bearing bars. Grating often serves as flooring or decking to provide access to various areas on the offshore installation.

gravel *n*: sand or glass beads of uniform size and roundness used in gravel packing.

gravel pack *n*: a mass of very fine gravel placed around a slotted liner in a well. See *gravel packing*.

gravel-pack *v*: to place a slotted or perforated liner in a well and surround it with gravel. See *gravel packing*.

gravel packing *n*: a method of well completion in which a slotted or perforated liner, often wire-wrapped, is placed in the well and surrounded by gravel. If open hole, the well is sometimes enlarged by underreaming at the point where the gravel is packed. The mass of gravel excludes sand from the wellbore but allows continued production.

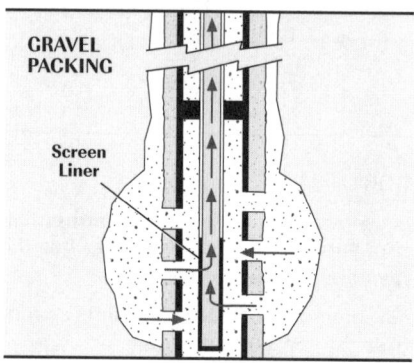

gravel-pack packer *n*: a packer used for the well completion method of gravel packing.

graveyard tour (pronounced "tower") *n*: the shift of duty on a drilling rig that starts at midnight. Sometimes called the morning tour.

gravimeter *n*: 1. an instrument used to detect and measure minute differences in the earth's gravitational pull at different locations to obtain data about subsurface formations. 2. a device for measuring and recording the density or specific gravity of a gas or liquid passing a point of measurement.

gravimetric survey *n*: the survey made with a gravimeter.

gravitational force *n*: the force acting upon objects on earth, where the force is the product of the object's mass, and the acceleration of gravity is at 32 ft/sec/sec. Measured in units of pounds, grams, etc.

gravitometer *n*: also called a densimeter. See *gravimeter*.

gravity *n*: 1. the attraction exerted by the earth's mass on objects at its surface. 2. the weight of a body. See *API gravity, relative density, specific gravity*.

gravity anchor *n*: anchor type that achieves its holding capacity mainly through the weight and friction of its lower surface with the seafloor.

gravity check *n*: in marketing, the checking of the gravity of the fuel at the terminal.

gravity compensator *n*: a double index scale against which a fixed reference pointer may be moved to correct for variations in gravity from a base point computed for water at 60°F. The compensator is marked in both specific gravity and API gravity units.

gravity differential *n*: the difference in density between the water and the oil in an oilfield emulsion. The greater the gravity differential, the easier it is to break the emulsion.

gravity drainage *n*: the movement of fluids in a reservoir resulting from the force of gravity. In the absence of an effective water or gas drive, gravity drainage is an important source of energy to produce oil, and it may also supplement other types of natural drive. Also called segregation drive.

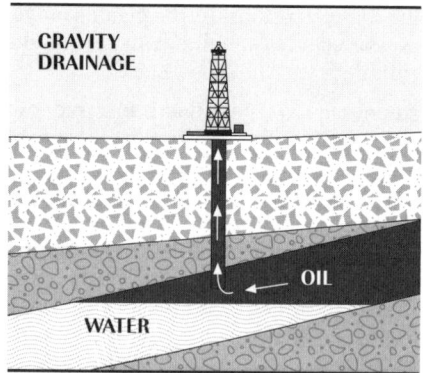

gravity segregation *n*: the tendency of reservoir fluids to separate into distinct layers according to their respective densities. For example, water is heavier than oil; therefore, water injected during waterflooding will tend to move along the bottom portion of a reservoir.

gravity survey *n*: an exploration method in which an instrument that measures the intensity of the earth's gravity is passed over the surface or through the water. In places where the instrument detects stronger- or weaker-than-normal gravity forces, a geologic structure containing hydrocarbons may exist.

Gray code *n*: a modified binary code in which sequential numbers are represented by expressions that differ only in one bit. Because the expressions differ by only one bit, errors are minimized.

Gray valve *n*: see *inside blowout preventer.*

graywacke *n*: a sandstone that contains more than 15 percent silt and clay, and whose grains tend to be angular and poorly sorted.

grease fitting *n*: a device on a machine that is designed to accept the hose of a grease gun so that grease can be added to the part in need of lubrication.

grease injector *n*: a surface device used in pressure control for slickline. See *lubricator.*

grease sample *n*: sample obtained by scooping or dipping a quantity of soft or semiliquid material, such as grease, from a package so that the material on the scoop or dipper is representative of the material in the package.

greasing out *n*: when essentially water-insoluble greasy materials (e.g., emulsifiers, lubricants) separate out of drilling fluids.

great circle *n*: the path on the surface of a sphere-type tank defined by the intersection of a plane surface and including the center of the sphere. Compare *equatorial circumference.*

green gas *n*: raw, untreated gas; gas as it leaves the well and before it enters any field or other treating facilities.

greenhouse effect *n*: a process whereby the earth's atmosphere retains long-wave radiation. The atmosphere works in much the same way as the glass panes of a greenhouse.

greensand *n*: a sand that contains considerable quantities of glauconite, a greenish mineral composed of potassium, iron, and silicate, which gives the sand its color and name.

Grem vane wheel (GVW) *n*: a free-running propeller mounted on or near the shaft-driven main propeller of a ship or tanker ship. The Grem vane wheel turns like a windmill because of the high-speed flow of water created by the main propeller. A GVW improves efficiency and thus can increase a ship's speed without additional power.

grid plot *n*: in diamond drilling bits, a pattern of setting diamonds in which the manufacturer spreads them evenly over the cutting surface (the nose) of the bit. Compare *circle plot.*

grief stem *n*: (obsolete) kelly; kelly joint.

grinding and buffing *n*: in pipeline construction, the process of cleaning pipe ends of dirt, rust, mill scale, or solvent to prepare them for welding. Grinding and buffing tasks are accomplished with power hand tools such as wire brushes and buffers.

grind out *v*: to test for the presence of water in oil by use of a centrifuge.

grind-out machine *n*: see *centrifuge.*

grin through *n*: term that expresses the effect of a paint's not totally obscuring the underlying surface.

grip *n*: see *wire rope grip.*

grn *abbr*: green; used in drilling reports.

GRN *abbr*: gamma-ray-neutron (a well log).

groove *n*: 1. the depression in a sheave (pulley) into which wire rope seats as the wire rope moves over the sheave. 2. on a drawworks drum, one of several depressions on the surface of the drum around which the wire rope drilling line is wrapped. The grooves keep the wire rope from wrapping unevenly across the drum.

groove radius *n*: in measuring grooves of a sheave, the radius of a circle that fits inside a groove. Used to determine wear on wire rope.

gross area *n*: a term applied to grated deck flooring on offshore installations that refers to the total area of the flooring; gross area includes areas of grating that are removed to clear obstructions or to allow pipe, ducts, columns, and the like to pass through the grating. Compare *net area.*

gross energy consumption *n*: total energy use, including electrical system energy losses.

gross heating value *n*: the number of kilojoules (Btus) evolved by the complete combustion, at constant pressure, of 1 standard cubic metre of gas, with all of the water formed by the combustion reaction being condensed to the liquid state.

gross meter throughput *n*: the indicated throughput corrected only for meter performance (i.e., by multiplying by the meter factor).

gross observed volume (GOV) *n*: the total volume of all petroleum liquids and sediment and water, excluding free water, at observed temperature and pressure.

gross production *n*: the total production of oil from a well or lease during a specified period of time.

gross registered tonnage (GRT) *n*: the total volume of a ship's interior, measured in tons (units of 100 ft^3 or 2.83 m^3).

gross standard volume (GSV) *n*: the total volume of all petroleum liquids and sediment and water, excluding free water, corrected by the appropriate temperature correction factor (C_{tl}) for the observed temperature and API gravity, relative density, or density to a standard temperature such as 60°F or 15°C, and also corrected by the applicable pressure correction factor (C_{pl}) and meter factor.

gross standard weight (GSW) *n*: the total weight of all petroleum liquids and sediment and water (if any), excluding free water. It is determined by applying the appropriate weight conversion factors to the gross standard volume.

gross tonnage *n*: the interior capacity of a ship or a mobile offshore drilling unit. The capacity is expressed in tons, although the actual measurement is in volume, 1 ton being equivalent to 100 cubic feet of volume. This rule holds for measuring ship capacity for U.S. maritime purposes. All principal maritime governments have their own rules describing how tonnage is to be measured.

gross volume *n*: the total amount of liquid in a storage tank excluding any adjustments for S&W, temperature, or density. Compare *net volume.*

gross wet gas withdrawal *n*: full well stream volume, including all natural gas plant liquid and nonhydrocarbon gases, but excluding lease condensate. Also includes amounts delivered as royalty payments or consumed in field operations.

gross working interest *n*: a working-interest owner's total ownership in production, before deduction of related royalty, overriding royalty, or production payment interests.

ground *n*: a conducting path, intentional or accidental, between an electric circuit or equipment and the earth, or a conducting body serving in place of the earth. *v*: to connect electrical equipment to earth or to a conducting body, which serves in place of the earth.

ground anchor *n*: see *deadman.*

ground bed *n*: in cathodic protection, an interconnected group of impressed-current anodes that absorbs the damage caused by generated electric current.

ground block *n*: a wireline sheave, or pulley, that is fastened to the ground anchor and that changes a horizontal pull on a wireline to a vertical pull (as in swabbing with a derrick over a well). See *block*.

GROUND BLOCK

Wellhead

ground chain *n*: the mooring chain part that is laid on the seafloor under normal conditions. It may lift off the seafloor in extreme conditions.

ground fault circuit interrupter (GFCI) *n*: a fast-acting circuit breaker that senses very small ground fault currents and which breaks the circuit when it detects such faults.

ground fault detector *n*: in an electric circuit, a device that senses very small ground fault currents that could flow through the body of a person standing on damp ground while touching a live conductor.

ground pressure *n*: see *geostatic pressure*.

groundwater *n*: water that seeps through soil and fills pores of underground rock formations; the source of water in springs and wells.

grout *n*: a mortar-like material used to seal between objects. *v*: to force sealing material into a soil, sand, or rock formation to stabilize it.

growth fault *n*: an active fault that continues to slip while sediments are being deposited, causing the strata on the downthrust side to be thicker than those on the other side. Also called rollover fault.

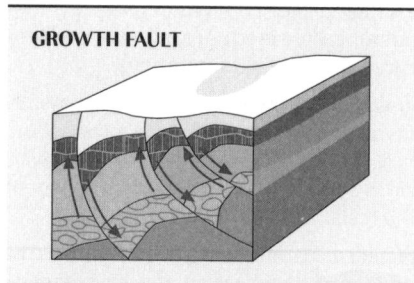

GROWTH FAULT

GRT *abbr*: gross registered tonnage.

GTO-SCR *abbr*: gate-controlled silicon-controlled rectifier.

guaranteed royalty *n*: the minimum amount of royalty income a royalty owner is to receive under the terms of the lease agreement, regardless of the royalty owner's share of actual proceeds from the sale of production.

guard *n*: a metal shield placed around moving parts of machinery to lessen or avoid the chance of injury to personnel. In the oilfield, guards are used on equipment such as belts, power transmission chains, drums, flywheels, and drive shafts.

guard-electrode log *n*: a focused system designed to measure the true formation resistivity in wellbores filled with salty mud. Current is forced by guard electrodes to flow into the formation. Also called Laterolog™.

guardian *n*: person appointed by a court of competent jurisdiction for the purpose of managing property and rights for another person who is considered incapable of managing for himself or herself—for example, a minor child, a mentally ill person, or someone judged mentally incompetent.

guardrail *n*: a railing for guarding against danger or trespass. On a drilling or workover rig, for example, guardrails are used on the rig floor to prevent persons from falling; guardrails are also installed on the mud pits and other high areas where there is any danger of falling.

guar gum *n*: a naturally occurring hydrophilic polysaccharide derived from the seed of the guar plant. The gum is chemically classified as a galactomannan. Guar gum slurries made up in clear fresh or brine water possess pseudoplastic flow properties.

guidance system *n*: in pipeline construction, the means by which a river crossing stays on course. Frequently computerized, guidance systems may be based on information gathered by pendulums, which determine inclination; gyroprobes, which are sensitive to drift and bearing; and sonar. Lasers are also used to guide crossing.

guide base *n*: see *permanent guide base*, *temporary guide base*.

guide fossil *n*: the petrified remains of plants or animals, useful for correlation and age determination of the rock in which they were found.

guide funnel *n*: see *conical funnel*.

guideline *n*: 1. one of usually four lines attached to the temporary guide base and permanent guide base to help position equipment (such as blowout preventers) accurately on the seafloor when a well is drilled offshore from a floating vessel. Guidelines are normally used to drill in water depths of 5,000 feet (1,500 metres) or less. 2. on a spinning chain, a piece of fiber rope (soft line) that is attached to the end of the chain that is wrapped around the tool joint. When the makeup cathead pulls the chain off the drill pipe, a rotary

helper uses the guideline to control the chain's movement. Note that spinning wrenches have replaced spinning chains on most modern drilling rigs.

guidelines *n pl*: lines, usually four, attached to the temporary guide base and permanent guide base to help position equipment (such as blowout preventers) accurately on the seafloor when a well is drilled offshore from a floating vessel.

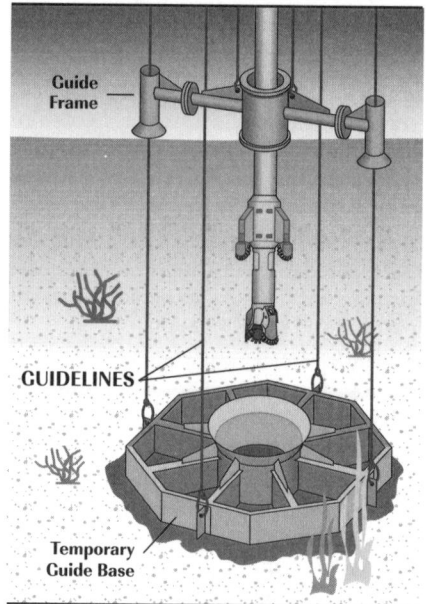

Guide Frame

GUIDELINES

Temporary Guide Base

guideline tensioner *n*: a system of cables (wire ropes), pulleys (sheaves), and pressurized piston-and-cylinder assemblies that maintain an upward tension on the guidelines to keep them taut and to compensate for rig heave (up-and-down movement). See *guideline*.

guide pole *n*: a device, usually in the form of a cylindrical vertical tube, used in floating roof tanks to prevent rotation of the roof.

guide rails *n pl*: on some top drives, the steel tracks on which the top-drive unit travels up and down when the driller raises or lowers the traveling block. The rails also keep the unit from turning when its motor rotates the drill stem. Compare *torque track*, *torque tube*. See also *top drive*.

guide ring *n*: a cylindrical metal ring used to guide packers past casing obstructions.

guide shoe *n*: 1. a short, heavy, cylindrical section of steel filled with concrete and rounded at the bottom, which is placed at the end

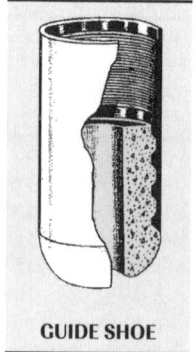

GUIDE SHOE

of the casing string. It prevents the casing from snagging on irregularities in the borehole as it is lowered. A passage through the center of the shoe allows drilling fluid to pass up into the casing while it is being lowered and allows cement to pass out during cementing operations. Also called casing shoe. 2. a device, similar to a casing shoe, placed at the end of other tubular goods.

guide wire *n*: see *guideline*.

Gulf Cooperation Council (GCC) *n*: association of Persian Gulf nations—Bahrain, Kuwait, Qatar, Oman, Saudi Arabia, United Arab Emirates—formed to defend themselves collectively against aggression.

Gulf Stream *n*: warm ocean current that flows north from the Gulf of Mexico along the eastern U.S. coast to an area off the southeastern coast of Newfoundland.

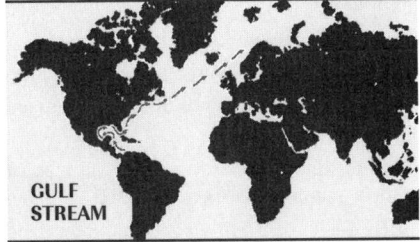
GULF STREAM

gum *n*: any hydrophilic plant polysaccharides or their derivatives that, when dispersed in water, swell to produce a viscous dispersion or solution. Unlike resins, they are soluble in water and insoluble in alcohol.

gumbo *n*: any relatively sticky formation (such as clay) encountered in drilling.

gun barrel *n*: a settling tank used to separate oil and water in the field. After emulsified oil is heated and treated with chemicals, it is pumped into the gun barrel, where the water settles out and is drawn off, and the clean oil flows out to storage. Gun barrels have largely been replaced by unified heater-treater equipment, but are still found, especially in older or marginal fields. Also called a wash tank.

Gunite™ *n*: trade name for a cement-sand mixture used to seal pipe against air, moisture, and corrosion damage.

gunk *n*: the collection of dirt, paraffin, oil, mill scale, rust, and other debris that is cleaned out of a pipeline when a scraper or a pig is put through the line.

gunk plug *n*: a slurry in crude or diesel oil containing any of the following materials or combinations: bentonite, cement,

attapulgite, and guar gum (never with cement). Used primarily in combating lost circulation.

gunk slurry *n*: a mixture of diesel oil and bentonite that is sometimes used to seal a lost circulation zone.

gunk squeeze *n*: a bentonite and diesel oil mixture that is pumped down the drill pipe and into the annulus to mix with drilling mud. The stiff, puttylike material is squeezed into lost circulation zones to seal them.

gun-perforate *v*: to create holes in casing and cement set through a productive formation. A common method of completing a well is to set casing through the oil-bearing formation and cement it. A perforating gun is then lowered into the hole and fired to detonate high-powered jets or shoot steel projectiles (bullets) through the casing and cement and into the pay zone. The formation fluids flow out of

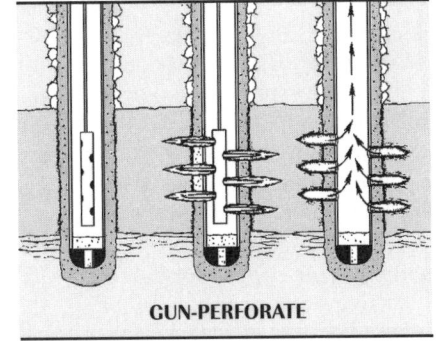
GUN-PERFORATE

the reservoir through the perforations and into the wellbore. See *jet-perforate, perforating gun.*

gun the pits *v*: to agitate the drilling fluid in a pit by means of a mud gun, electric mixer, or agitator.

gunwale *n*: the intersection of the deck plating with the side shell plating on a ship.

gusher *n*: an oilwell that has come in with such great pressure that the oil jets out of the well like a geyser. In reality, a gusher is a blowout and is extremely wasteful of reservoir fluids and drive energy. In the early

GUSHER

days of the oil industry, gushers were common and many times were the only indication that a large reservoir of oil and gas had been struck. See *blowout.*

guyed-tower platform rig *n*: a compliant offshore drilling platform used to drill development wells. The foundation of the platform is a relatively lightweight jacket on which all equipment is placed. A system of guy wires anchored by clump weights helps secure the jacket to the seafloor and allows it to move with wind and wave forces. See *platform rig.*

guying system *n*: the system of guy lines and anchors used to brace a rig.

guy line *n*: a wireline attached to a mast, derrick, or offshore platform to stabilize it. See *load guy line, wind guy line.*

guy line anchor *n*: a buried weight or anchor to which a guy line is attached. See *deadman.*

guy rope *n*: a supporting rope which maintains a constant distance between the points of attachment to the two components connected by the rope.

guy wire *n*: a rope or cable used to steady a mast or pole.

GVW *abbr*: Grem vane wheel.

gyp *n*: (slang) gypsum.

gyp mud *n*: drilling mud that is treated with gypsum to provide a source of soluble calcium in the filtrate to obtain desirable mud properties for drilling in shale or clay formations.

gypsum *n*: a naturally occurring crystalline form of calcium sulfate in which each molecule of calcium sulfate is combined with two molecules of water. See *anhydrite, calcium sulfate.*

gyroscope *n*: a wheel or disk mounted to spin rapidly about one axis but free to rotate about one or both of two axes perpendicular to each other. The inertia of the spinning wheel tends to keep its axis pointed in one direction regardless of how the other axes are rotated.

gyroscopic surveying instrument *n*: a device used to determine direction and angle at which a wellbore is drifting off the vertical. Unlike magnetic surveying instruments, a gyroscopic instrument reads true direction and is not affected by magnetic irregularities that may be caused by casing or other ferrous metals. See *directional drilling, directional survey.*

gyroscopic survey tool *n*: see *gyroscopic surveying instrument.*

GZ *abbr*: the righting arm (or lever).

H *sym*: 1. hydrogen. 2. henry. 3. magnetizing force.

h *abbr*: 1. formation thickness; used in engineering reports during well testing. 2. hour.

HCN *form*: hydrogen cyanide.

H₂S *form*: hydrogen sulfide.

H_2S *form*: hydrogen sulfide.

H₂S trim *n*: see *trim*.

H₂SO₃ *form*: sulfurous acid.

H₂SO₄ *form*: sulfuric acid.

H4 connector *n*: a brand of hydraulic connector that allows the blowout preventer stack to be connected by remote control to the wellhead and to the lower marine riser package. The H4 connector has both a primary and secondary piston connected to a cam ring to provide locking and unlocking force. See *hydraulic connector*.

habendum clause *n*: the clause in an oil and gas lease that fixes the duration of the lessee's interest in both a primary and a secondary term. Also called term clause.

hairspring *n*: a thin spiraled recoil spring that regulates the motion of a pointer in a meter or a clock.

half adder *n*: a logic element in an electronic circuit that operates on two binary digits, but no carry digits, from a preceding stage to produce as output a sum digit and a carry digit.

half-cell *n*: a single electrode immersed in an electrolyte for the purpose of measuring metal-to-electrolyte potentials and, therefore, the corrosion tendency of a particular system.

half-life *n*: the amount of time needed for half of a quantity of radioactive substance to decay or transmute into a nonradioactive substance. Half-lives range from fractions of seconds to millions of years.

half mule shoe *n*: a cutoff pup joint below a packer that is used as a fluid entry device and a seal assemblies guide.

half siding *n*: the flat, horizontal section of the bottom shell plating measured from the centerline of the vessel to the edge of the flat keel plate.

halide ion *n*: in chemistry, a molecule with a negative charge (a negative ion) that is made up of fluorine, chlorine, bromine, or iodine, and oxygen. Light sensitive silver halides, such as silver chloride and silver bromide, are used in conventional photographic processes. In the oilfield, halide ions in contact with steel downhole equipment cause localized pitting or crevice corrosion; halides also cause stress corrosion cracking.

halite *n*: rock salt (NaCl).

hammer *n*: tool used to drive the conductor pipe into the ground. The single-acting diesel is the most popular, as it has the most foot-pounds of energy for the weight of the hammer. Other types of hammer include impact and vibrating.

hammer drill *n*: a drilling tool that, when placed in the drill stem just above a roller cone bit, delivers high-frequency percussion blows to the rotating bit. Hammer drilling combines the basic features of rotary and cable-tool drilling (i.e., bit rotation and percussion).

hammer-drill *v*: see *hammer drill*.

hammer drill bit *n*: a special flat-bottomed steel bit with carbide inserts used in straight hole drilling with air or gas. Normal tricone bits tend to split where the leg is welded to the body under the impact of the hammer. Several designs are available.

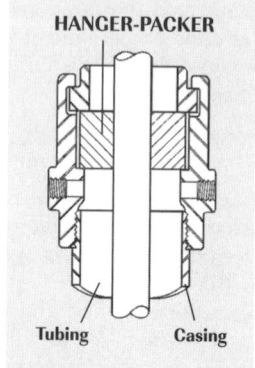

HAMMER BIT

hammering-up *n*: connection of treating line during well servicing, from pump trucks to the tree or wellhead.

hammer test *n*: a method of locating corroded sections of pipe by striking the pipe with a hammer. When struck, a corroded section resounds differently from a noncorroded section.

hand *n*: a worker in the oil industry, especially one in the field.

handling costs *n pl*: see *marketing costs*.

handling-tight coupling *n*: a coupling screwed onto casing tight enough so that a wrench must be used to remove it.

handrail *n*: a railing or pipe along a passageway or stair that serves as a support or a guard.

handy *n*: a connection that can be unscrewed by hand.

hanger *n*: see *casing hanger, tubing hanger*.

hanger-packer *n*: a device that fits around the top of a string of casing or a string of tubing and supports part of the weight of the casing or tubing while providing a pressure-tight seal. See *hanger, packer*.

HANGER-PACKER

Tubing   Casing

hanger plug *n*: a device placed or hung in the casing below the blowout preventer stack to form a pressure-tight seal. Pressure is then applied to the blowout preventer stack to test it for leaks.

hanging load *n*: the amount of weight transferred to the casinghead.

hanging wall *n*: the rock surface forming the upper side of a fault when the fault plane is not vertical—that is, if the dip is less than 90°. Compare *footwall*.

hangline *n*: a single length of wire rope attached to the crown block by which the traveling block is suspended when not in use. Also called a hang-off line.

hang off *v*: to close a ram blowout preventer around the drill pipe when the annular preventer has previously been closed to offset the effect of heave on floating offshore rigs during well-control procedures.

hang-off line *n*: see *hangline*.

hang rods *v*: to suspend sucker rods in a derrick or mast on rod hangers rather than to place them horizontally on a rack.

HAP *abbr*: hazardous airborne pollutant.

hard banding *n*: a special wear-resistant material often applied to tool joints to prevent abrasive wear to the area when the pipe is being rotated downhole.

hardener *n*: in paint technology, a cross-linking agent used to cure a resin or a paint.

hard eye splice *n*: either hand- or mechanically-made splice in either wire or synthetic rope with insertion of steel or synthetic material thimble.

hardfacing *n*: an extremely hard material, usually crushed tungsten carbide, that is applied to the outside surfaces of tool joints, drill collars, stabilizers, and other rotary drilling tools to minimize wear when they are in contact with the wall of the hole.

hard hat *n*: a hard plastic helmet worn by oilfield workers to minimize the danger of being injured by falling objects.

hardness *n*: the resistance of a substance to scratching or to indentation.

hard shut-in *n*: in a well-control operation, closing the BOP without first opening an alternate flow path up the choke line. When the BOP is closed, pressure in the annulus cannot be read on the casing pressure gauge.

hard water *n*: water that contains dissolved compounds of calcium, magnesium, or both. Compare *soft water*.

harmful quantity *n*: any quantity that produces a sheen, sludge, film, or discoloration of the surface water in navigable U.S. waters.

harmonic motion *n*: a periodic motion with a single frequency or amplitude or one that is composed of two or more such motions.

HART *abbr*: highway addressable remote transducer.

HART protocol *n*: a form of digital communication produced by a transducer that uses frequency bursts and lapses to simulate the logic 1 or 0. The frequency can be superimposed on a DC bus or circuit by the transducer or transmitter for data transmission to a remote device for the purposes of programming or modification.

HART protocol programming language *n*: a definite arrangement of digital data produced by the HART transducer that is an established protocol recognized by the receiving unit.

hatch *n*: 1. an opening in the roof of a tank through which a gauging line may be lowered to measure its contents. 2. the opening from the deck into the cargo space of ships.

hawser *n*: a rope, usually of large diameter, used to moor or tow marine vessels.

hay *n*: excelsior, used in emulsion treating to encourage coalescence.

hay pulley *n*: a pulley that is normally attached to the wellhead at a convenient place for the wireline to pass through as it comes from the stuffing box sheave before being spooled onto the wireline reel. The hay pulley prevents any lateral force from being exerted on the lubricator and the wellhead.

hayrack *n*: (obsolete) a rack used to hold pipe on a derrick; a fingerboard.

hayrake *n*: see *hayrack*.

hay section *n*: a section of a heater or a heater-treater that is filled with fibrous material through which oil and water emulsions are filtered.

Hazard Communication Standard (HAZCOM, HCS, or the "Employee Right to Know") *n*: an OSHA standard that guarantees employees the right to know about chemical hazards on the job and how to protect themselves from those hazards. Under HAZCOM, all manufacturers and employers must prepare a written hazard communication program; prepare a list of all hazardous materials in the workplace; label all containers of hazardous materials in the workplace; collect and maintain a material safety data sheet (MSDS) for each hazardous material present in the workplace; and provide employee training on specific topics related to hazardous substances.

hazardous *adj*: involving or exposing one to risk. The lists of material or waste that are considered hazardous vary from agency to agency and from regulation to regulation: hazardous materials in transport are regulated by DOT; hazardous substances in the workplace are regulated by OSHA; hazardous waste is regulated under EPA's RCRA; toxic substances are regulated under EPA's TSCA; hazardous air pollutants are regulated under EPA's CAA; and so on. The hazardous list for that regulation is tailored to that purpose.

hazardous airborne pollutant (HAP) *n*: (CAA) air emissions that are immediately hazardous to human health or that cause cancer, gene mutation, or reproductive harm. Due to its harmful nature, the allowable emission of a hazardous airborne pollutant is much lower than that for a conventional, or criteria, pollutant.

HAZARDOUS AIRBORNE POLLUTANT

hazardous chemical *n*: 1. (OSHA) (HAZCOM) any chemical that is a physical hazard or a health hazard. 2. (SARA) any hazardous chemical as defined under 29 CFR 1910.1200(c), except those that are regulated by other agencies or laws.

hazardous materials (HAZMAT) *n pl*: (DOT) substances or materials in quantities or forms that may pose an unreasonable risk to health, safety, or property when stored, transported, or used in commerce.

hazardous materials specialist level *n*: a training level achieved by any employee who has been HAZWOPER trained to assist and support a hazardous materials technician in making certain emergency action decisions. The duties of hazardous materials specialists parallel those of the technician, but require a more directed or specific knowledge of the various substances they may be called on to contain. The hazardous materials specialist also acts as the site liaison with federal, state, local, and other government authorities in regard to site activities.

hazardous materials technician level *n*: a training level achieved by any employee who has been HAZWOPER trained to take an offensive role in emergency response. Technicians are trained to take certain actions that deal directly with stopping a release, such as approaching the point of release to plug, patch, or otherwise stop the release.

hazardous substance *n*: 1. any substance designated under CWA or CERCLA as posing a threat to waterways and the environment when released. 2. (CERCLA) any substance designated in 40 CFR 302.

hazardous waste *n*: 1. any solid waste ("solid" includes any solid, liquid, semisolid, or contained gaseous material) resulting from industrial, commercial, mining, or agricultural operations, or from community activities, that meets certain characteristics of hazard (i.e., ignitable, corrosive, reactive, toxic) or that is listed as a waste from specific or nonspecific sources or that is a listed commercial chemical product or manufacturing intermediate that is sometimes discarded. 2. (RCRA) discarded materials regulated by the EPA because of public health and safety concerns. 3. (HAZWOPER) a waste or combination of wastes as defined in 40 CFR 261.3, or those substances defined as hazardous wastes in 49 CFR 171.8.3. (CERCLA) those wastes listed in 40 CFR 261.3.

hazardous waste generator *n*: an operator that produces hazardous or acute hazardous waste. If the quantity of waste exceeds EPA minimums under RCRA, the operator must obtain a generator identification number and must meet other RCRA requirements. The hazardous waste generator must place hazardous wastes in proper containers; label the containers; ensure safe handling of the material; manifest shipments to licensed disposal sites; and report discrepancies in waste shipments to the EPA.

hazardous waste manifest *n*: a document that identifies the waste and all parties responsible for it while it is being shipped.

Hazardous Waste Operations and Emergency Response Standard (HAZWOPER) *n*: an OSHA standard that is concerned primarily with worker safety in emergency response situations. HAZWOPER requires employers to protect the safety and health of three specific groups of workers: those involved in emergency response or cleanup at hazardous waste sites; those involved in emergency response at treatment, storage, and disposal (TSD) sites; and those involved in emergency response to incidents involving hazardous substances.

hazard warning *n*: (OSHA) any words, pictures, symbols, or combination thereof appearing on a label or other appropriate form of warning that convey the hazard(s) of the chemical(s) in the container(s).

HAZARD WARNINGS

HAZCOM *abbr*: Hazard Communication Standard.

HAZMAT *abbr*: hazardous material.

HAZMAT team *n*: the designated and trained personnel who respond to hazardous material incidents.

HAZWOPER *abbr*: Hazardous Waste Operations and Emergency Response Standard.

HBP *abbr*: held by production; commonly used in land departments.

HC connector *n*: a brand of hydraulic connector in subsea blowout preventer systems used to connect by remote control the blowout preventer stack to the wellhead and the lower marine riser package to the blowout preventer stack. It uses a large annular piston that completely surrounds locking segments, instead of the nine individual pistons. The additional locking force available through the annular piston increases the pressure rating to 15,000 psi (105,000 kPa) and improves the resistance to bending moments. See *hydraulic connector*.

HCN *form*: hydrogen cyanide.

H-crossover *n*: circulating member with integral landing nipples.

HCR valve™ *n*: trade name for a remote-controlled valve on the choke line that controls the flow of fluids through the line from the well to the choke manifold.

HCS *abbr*: Hazard Communication Standard.

head *n*: 1. the height of a column of liquid required to produce a specific pressure. See *hydraulic head*. 2. for centrifugal pumps, the velocity of flowing fluid converted into pressure expressed in feet or metres of flowing fluid. Also called velocity head. 3. that part of a machine (such as a pump or an engine) that is on the end of the cylinder opposite the crankshaft. Also called cylinder head.

headache *n*: (slang) the position in which the mast on a mobile rig is resting horizontally over the driver's cab.

headache post *n*: the post on cable-tool rigs that supports the end of the walking beam when the rig is not operating.

header *n*: a chamber from which fluid is distributed to smaller pipes or conduits, e.g., a manifold.

head gasket *n*: a thin piece of material made of cork or other similar material placed between an engine's cylinder head and engine block. It seals between the head and block, and flexes to maintain the seal as the head and block expand at different rates with changes in engine temperature.

headgate *n*: the gate valve nearest the pump or compressor on oil or gas lines.

heading *n*: 1. intermittent flow of fluid from a well. 2. direction in which a vessel is pointed.

headlog *n*: in river craft of rectangular shape, the structural member at the extreme end between the rake shell plating and the deck.

head meter *n*: 1. a constriction placed in a closed conduit of flowing fluid. 2. a flowmeter that depends on pressure head change to operate.

head pressure *n*: see *head*.

head room *n*: in crane operations, the vertical distance the crane can hoist a load before it strikes an obstruction.

head well puller *n*: crew chief.

health hazard *n*: (OSHA) the potential human health hazard associated with contact with materials or substances. A chemical that is listed as a health hazard is a chemical for which there is statistically significant evidence based on at least one study conducted in accordance with established scientific principles that acute or chronic health effects may occur in exposed employees.

health, safety, and environment *n*: a phrase covering a company's concern for the health and safety of its employees, as well as the environment in which they work.

heat a connection *v*: to loosen a collar or other threaded connection by striking it with a hammer. Also called warm a connection or whip a connection.

heat checking *n*: a condition that occurs when the cutters of a bit drag on a formation and become very hot because of friction and then are cooled by the drilling fluid; the rapid cooling causes small cracks (heat checks) to develop.

heater *n*: container or vessel enclosing an arrangement of tubes and a firebox in which an emulsion is heated before further treating, or in which natural gas is heated in the field to prevent the formation of hydrates.

heater-treater *n*: a vessel that heats an emulsion and removes water and gas from the oil to raise it to a quality acceptable for

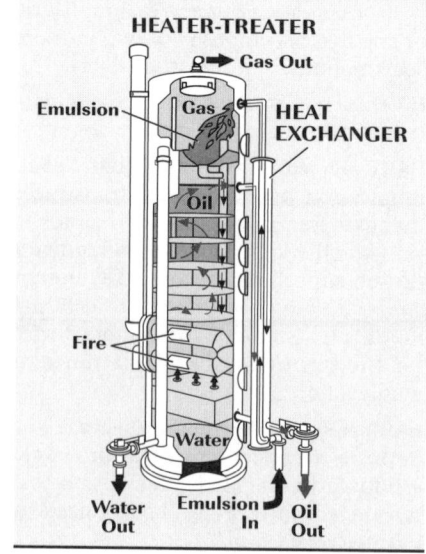

a pipeline or other means of transport. A heater-treater is a combination of a heater, free-water knockout, and oil and gas separator.

heat exchanger *n*: see *exchanger*.

heating cable *n*: flexible, insulated wire of known resistance, which can be wrapped around or strapped to a pipeline or valves for heating.

heating coils *n pl*: (marine) a system of piping in tank bottoms in which steam is carried as required to heat high pour-point liquid cargoes to pumpable viscosity level.

heating medium *n*: a material, whether flowing or static, used to transport heat from a primary source such as combustion of fuel to another material. Heating oil and steam are examples of heating mediums. Also called heat medium.

heating tube *n*: see *fire tube*.

heating value *n*: the amount of heat developed by the complete combustion of a unit quantity of a material. Also called heat of combustion.

heat medium *n*: see *heating medium*.

heat of combustion *n*: see *heating value*.

heat of hydration *n*: heat generated when cement is mixed with water to form a slurry. Also called heat of reaction or heat of solution.

heat of reaction *n*: see *heat of hydration*.

heat of solution *n*: see *heat of hydration*.

heat of vaporization *n*: the quantity of energy required to evaporate 1 mole of a liquid at constant pressure and temperature.

heat pipes *n pl*: pipes, 2 inches (5 centimetres) in diameter, which keep the soil stable in permafrost areas by cooling the ground.

heat pump *n*: an electric heater that is similar to an air conditioner. It heats and cools from a single system by reversing the flow of gas from the compressor to coils.

heave *n*: the vertical motion of a ship or a floating offshore drilling rig.

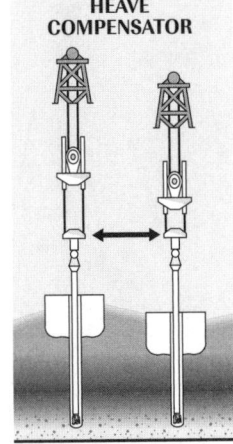

HEAVE COMPENSATOR

heave compensator *n*: a device that moves with the heave of a floating offshore drilling rig to prevent the bit from being lifted off the bottom of the hole and then dropped back down (i.e., to maintain constant weight on the bit). It is used with devices such

as bumper subs. See *motion compensator*.

heaving *n*: the partial or complete collapse of the walls of a hole resulting from internal pressures due primarily to swelling from hydration or formation gas pressures. See *caving*.

heavy ends *n pl*: 1. the parts of a hydrocarbon mixture that have the highest boiling point and the highest viscosity (such as fuel oils and waxes). 2. hexanes and heptanes in a natural gas stream.

heavy hydrocarbons *n pl*: see *heavy ends*.

heavy metals *n pl*: any metal with a density of 5.0 or greater, such as lead, mercury, copper, and cadmium, especially one that is toxic to organisms.

heavy oil *n*: oil composed mainly of heavy ends.

heavy-walled drill pipe *n*: drill pipe that is manufactured with walls that are thicker than those in standard drill pipe. Special extra-length tool joints are attached to the heavier-walled tube. Typically, 5-inch heavy wall drill pipe weighs 50 pounds per foot (75 kilograms per metre); 5-inch conventional drill pipe weighs 19.5 pounds per foot (29 kilograms per metre). Also attached to the tube is a center wear pad that protects the outside diameter of the pipe from abrasion. Heavy wall pipe is often used in the drill stem just above the drill collars, in the transition zone between the stiffer collars and the more limber drill string, and in place of some drill collars to apply weight on the drill bit in small-diameter and horizontal holes. Sometimes erroneously called heavyweight drill pipe. Compare *heavyweight drill pipe*.

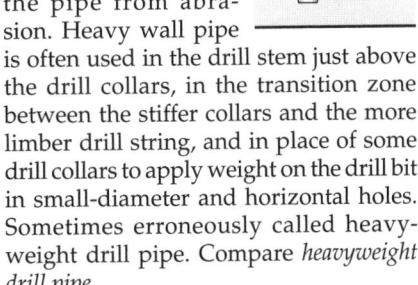

HEAVY-WALLED DRILL PIPE

Box Tool Joint

Center Wear Pad

Pin Tool Joint

heavyweight additive *n*: a substance or material added to cement to make it dense enough for use in high-pressure zones. Sand, barite, and hematite are some of the substances used as heavyweight additives.

heavyweight drill pipe *n*: drill pipe that is like conventional drill pipe except that it is manufactured with a thicker wall, which increases its weight and tensile strength. For example, 5-inch IEU grade E75 standard-weight drill pipe weighs 19.5 pounds per foot (29 kilograms per metre), while 5-inch IEU grade E75

heavyweight drill pipe weighs 25.60 pounds per foot (38.4 kilograms per metre). Heavyweight drill pipe is placed in the top of the string to support a very long string of pipe. Heavyweight drill pipe may be used in the drill stem when high-tensile strength drill pipe is required but high-grade steel cannot be used because of the high-grade steel's susceptibility to hydrogen embrittlement, which is a form of corrosion caused by hydrogen sulfide. Compare *heavy-walled drill pipe*.

hectare *n*: an internationally used unit of measurement equal to 2.47 acres.

hedging *n*: a financial insurance device for minimizing losses and protecting profits during the production, storage, processing, and marketing of commodities. A marketer sells a product on the futures market and locks in prices for later delivery. That contract may then continue to be exchanged on the futures market among other marketers with similar concerns and expectations.

heel *n*: the inclination of a ship or a floating offshore drilling rig to one side, caused by wind, waves, or shifting weights on board. Also called list.

heel teeth *n*: see *gauge cutters*.

HF FAX *n*: see *radiofax*.

helical *adj*: spiral-shaped.

helical groove *n*: spooling of wire rope onto a drum on which the grooves take the form of a helix—that is, like the threads on a pipe end. Also called helical spooling.

helipad *n*: a landing pad for helicopters.

helium unscrambler *n*: an electronic device that lowers the voice of divers breathing helium as part of their breathing-gas mixture so that the surface crew may understand them.

helmet *n*: a protective enclosure for a diver's entire head. It is part of the life-support system and also contains a communications system.

henry (H or L) *n*: a unit of inductance in which an induced voltage (electromotive force) of 1 volt occurs in a circuit when the current changes at the rate of 1 ampere per second.

hepatic gas *n*: see *hydrogen sulfide*.

heptane *n*: a saturated hydrocarbon of the paraffin series; one of the heavy ends in a gaseous hydrocarbon mixture.

hereditaments *n pl*: whatever can be inherited. Of the two kinds of hereditaments, corporeal and incorporeal, the first usually includes tangible things and the second, rights connected with land. Land itself would be a corporeal hereditament; the right to rent would be an incorporeal hereditament.

herringbone gear *n*: see *bull gear*.

Herschel tube *n*: a venturi tube used to measure fluid flow when a low permanent pressure drop is required for measurement.

hertz (Hz) *n*: a unit in the metric system used to measure frequency in cycles per second.

hesitation squeeze *n*: a method of squeeze cementing in which cement is pumped in and the pumps are stopped for a few minutes. Pumping is started and stopped until the desired pressure is obtained.

hexadecimal number *n*: a number based on powers of 16 that uses the decimal digits 0 to 9 and six more digits represented by the letters A, B, C, D, E, and F. A hexadecimal number is usually written with the subscript 16 to indicate that it is a hexadecimal number. For example, $84D1F_{16}$ is a hexadecimal number.

hexagonal kelly *n*: a kelly with a six-sided (hexagonal) cross section. Compare *square kelly*. See also *kelly*.

hexane *n*: C_6H_{14}, a liquid hydrocarbon of the paraffin series.

hex kelly *n*: see *kelly*.

HGOR *abbr*: high gas-oil ratio; used in drilling reports.

hhp *abbr*: hydraulic horsepower.

hiding power *n*: the ability of a paint to obliterate the color difference of a substrate; opacity.

high-capacity torque wrench *n*: a heavy-duty torque wrench used to apply the required torque to flange bolts in the makeup of riser pipe. Correct torque must be used to ensure the correct preload on a riser connection; otherwise, separation at the flanges could lead to failure. See *riser pipe*.

high drum drive *n*: the drive for the drawworks drum used when hoisting loads are light.

high-frequency *adj*: in electronics, involving the rapid cycling of alternating current, radio waves, or sound waves.

high modulus polyethylene (HMPE) *n*: type of synthetic fiber used in rope construction.

high pH mud *n*: a drilling fluid with a pH range above 10.5, i.e., a high-alkalinity mud.

high-pressure area *n*: an area of high atmospheric pressure.

high-pressure distribution system *n*: a system that operates at a pressure higher than the standard service pressure delivered to the customer; thus, a pressure regulator is required on each service to control pressure delivered to the customer. Sometimes called medium pressure.

high-pressure nervous syndrome *n*: a term used to describe symptoms caused by high partial pressures of helium.

high-pressure pipeline *n*: pipeline in which gas is often compressed up to or in excess of 100 times the normal atmospheric pressure.

high-pressure squeeze cementing *n*: the forcing of cement slurry into a well at the point to be sealed with a final pressure equal to or greater than the formation breakdown pressure. See *squeeze cementing*.

high-pressure wellhead *n*: in drilling from floating drilling rigs, a subsea device that is installed on top of the casing to provide a foundation for the blowout preventer stack, to support and house subsequent casing strings, and to provide a seal and a locking arrangement between the surface casing and blowout preventer stack.

high-purity water *n*: water that has little or no ionic content and is therefore a poor conductor of electricity.

high-salinity *adj*: having high salt content.

high tide *n*: the tide at its highest level or the time of day at which it occurs. Compare *low tide*. See *tide*.

high-vacuum range *n*: in vacuum pressure measurement, pressures that exert from 1 micrometre to 1 millimetre of mercury.

high vapor pressure liquid *n*: a liquid that, at the measurement or proving temperature of the meter, has a vapor pressure that is equal to or higher than atmospheric pressure. Compare *low vapor pressure liquid*.

highway addressable remote transducer (HART) *n*: an electronic device that converts one form of energy into another—in this case, frequency bursts and lapses into direct current—so that the direct current can transmit the data to a remote receiver for interpretation, modification, or programming.

high-yield drilling clay *n*: a classification given to a group of commercial drilling-clay preparations having a yield of 35 to 50 barrels (5,565 to 7,950 litres) per ton and intermediate between bentonite and low-yield clays. Usually prepared by peptizing low-yield calcium montmorillonite clays or, in a few cases, by blending some bentonite with the peptized low-yield clay.

hi-lo cam *n*: a mechanism in some packers to set and release the tool with a minimum of rotation.

hinged link *n*: a connector used in joining chain and/or wire where side loading might be experienced.

hinged master bushing *n*: a two-piece master bushing that has a jointed, swinging device (a hinge) on each half into which large pins fit to hold the bushing together. A two-piece insert bowl to hold the slips fits inside this type of master bushing. Compare *solid master bushing, split master bushing*. See also *insert bowl, master bushing, slips*.

hinged panel *n*: a section of steel grating on an offshore installation that has hinges to allow personnel to raise and lower the section of grating.

HLB *abbr*: hydrophilic-lipophilic balance.

HMI *abbr*: human machine interface.

HMPE *abbr*: high modulus polyethylene.

HOCM *abbr*: heavily oil-cut mud; used in drilling reports.

hogging *n*: the distortion of the hull of an offshore drilling rig when the bow and the stern are lower than the middle, caused by wave action or unbalanced or heavy loads. Compare *sagging*.

hoist *n*: 1. an arrangement of pulleys and wire rope or chain used for lifting heavy objects; a winch or similar device. 2. the drawworks. *v*: to raise or lift.

hoisting *n*: the process of lifting.

hoisting cable *n*: the cable that supports drill pipe, swivel, hook, and traveling block on a rotary drilling rig.

hoisting components *n pl*: drawworks, drilling line, and traveling and crown blocks. Auxiliary hoisting components include catheads, catshaft, and air hoist.

hoisting drum *n*: the large flanged spool in the drawworks on which the hoisting cable is wound. See *drawworks*.

hoisting engine *n*: the relatively small pneumatically operated engine on an air hoist, or tugger.

hoisting plug *n*: also called a lifting sub or a lifting nipple. See *lifting nipple*.

hoisting system *n*: the system on the rig that performs all the lifting on the rig, primarily the lifting and lowering of

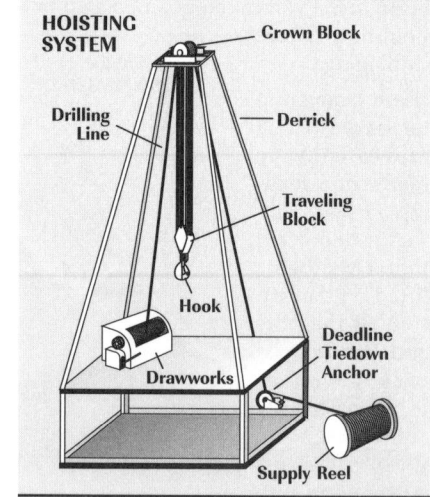

drill pipe out of and into the hole. It is composed of drilling line, traveling block, crown block, and drawworks. See also *hoisting components.*

hold-down *n*: a mechanical arrangement that prevents the upward movement of certain pieces of equipment installed in a well. A sucker rod pump may use a mechanical hold-down for attachment to a seating nipple.

hold-down button *n*: a hydraulic, toothed device in a packer that uses differential pressure across the packer to grip casing and prevent upward packer movement.

hold-down pressure *n*: hydrostatic pressure developed by the weight of the drilling fluid exerted on the bottom of the hole that tends to prevent cuttings from moving up the annulus.

holding current *n*: the minimum current required to maintain a switching device, such as a silicon-controlled rectifier, in a closed, or conducting, state after it is energized or triggered.

hole *n*: 1. in drilling operations, the wellbore or borehole. See *borehole, wellbore.* 2. in electronics, the space left in atoms as free electrons flow from one atom to another. Like free electrons, holes also flow from atom to atom but in the opposite direction from electrons. In electronics, conventional current flow is shown as hole flow and not electron flow. Consequently, circuit diagrams for electronic devices show current flowing from the positive terminal to the negative terminal.

hole angle *n*: the angle at which a hole deviates from vertical.

hole cleaning *n*: the lifting and removing of cuttings made by a bit's cutters by the drilling fluid coming out the bit.

hole collapse *n*: see *caving.*

hole drift *n*: the amount a wellbore is deflected from vertical.

hole-electron pair *n*: in a p-type semiconductor, that which occurs when an impurity atom with only three valence electrons is placed in a semiconducting material with four valence electrons. The fourth electron is left without an electron to pair with, which results in a hole. The hole conducts current.

hole geometry *n*: the shape and size of the wellbore.

hole opener *n*: a device used to enlarge the size of an existing borehole. It has teeth arranged

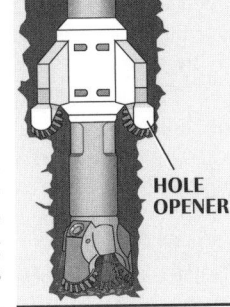
HOLE OPENER

on its outside circumference to cut the formation as it rotates.

holiday *n*: a gap or void in coating on a pipeline or in paint on a metal surface.

holiday detector *n*: an electrical device used to locate a weak place, or holiday, in coatings on pipelines and equipment. Also called jeep.

hollow carrier gun *n*: a perforating gun consisting of a hollow, cylindrical metal tube into which are loaded shaped charges or bullets. On detonation, debris caused by the exploding charges falls into the carrier to be retrieved with the reusable gun.

homestead property *n*: property such as land, house, outbuildings, and tools, that cannot be seized to pay general debts.

homocline *n*: a series of beds dipping in the same direction.

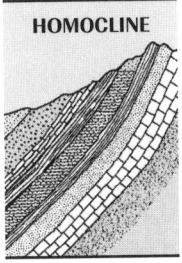
HOMOCLINE

homogeneous *adj*: of uniform or similar nature throughout; a substance or fluid that has at all points the same property or composition.

homogeneous rock *n*: rock whose characteristics (permeability, porosity, density, and so on) are generally similar throughout.

honeycomb formation *n*: a stratum of rock that contains large void spaces, i.e., a cavernous or vuggy formation.

hook *n*: a large, hook-shaped device from which the swivel is suspended. It is designed to carry maximum loads ranging from 100 to 650 tons (90 to 590 tonnes) and turns on bearings in its supporting housing. A strong spring within the assembly cushions the weight of a stand (90 feet, about 27 metres) of drill pipe, thus permitting the pipe to be made up and broken out with less damage to the tool joint threads. Smaller hooks without the spring are used for handling tubing and sucker rods. See also *stand, swivel.*

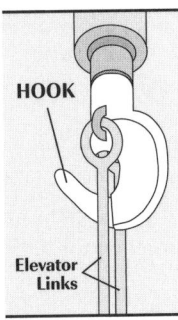
HOOK
Elevator Links

hook block *n*: block with a hook attached used in lifting service; may have a single sheave for double or triple line or multiple sheaves for four or more parts.

hook-block combination *n*: see *combination hook-block.*

hook load *n*: the weight of the drill stem that is suspended from the hook.

hook load capacity *n*: the nominal rated load capacity of a portable hoist and mast

arrangement, usually calculated by an API formula.

hook positioner *n*: a device in which a hook can be rotated to a required position and locked in place. Also called an automatic positioner.

hook roller *n*: on an offshore pedestal crane, one of a series of roller bearings on which the crane's superstructure (including the operator's cab) revolves. The rollers are installed in such a manner that they not only allow the crane superstructure to revolve, but also prevent the superstructure from separating from the pedestal when the crane performs lifts.

hookup wire *n*: wire or cable used to connect components in an electronic device.

hook-wall packer *n*: a packer equipped with friction blocks or drag springs and slips and designed so that rotation of the pipe unlatches the slips. The friction springs prevent the slips from turning with the pipe and assist in advancing the slips up a tapered sleeve to engage the wall of the outside pipe as weight is put on the packer. Also called a wall-hook packer. See *packer.*

hoop stress *n*: a force in the earth surrounding a borehole that stresses the hole somewhat like barrel hoops stress barrel staves. Hoop stress can cause the hole to collapse.

hopper *n*: a large funnel- or cone-shaped device into which dry components (such as powdered clay or cement) can be poured to mix uniformly with water or other liquids. The liquid is injected through a nozzle at the bottom of the hopper. The resulting mixture may be drilling mud to be used as the circulating fluid in a rotary drilling operation, or it may be cement slurry to be used in bonding casing to the borehole.

horizon *n*: distinct layer or group of layers of rock.

horizontal drilling *n*: deviation of the borehole at least 80° from vertical so that the borehole penetrates a productive formation in a manner parallel to the formation.

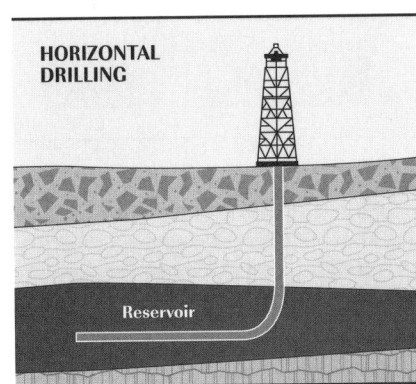
HORIZONTAL DRILLING
Reservoir

A single horizontal hole can effectively drain a reservoir and eliminate the need for several vertical boreholes.

horizontal heater-treater *n*: a heater-treater of cylindrical shape lying parallel to the ground.

horizontal permeability *n*: the permeability of reservoir rock parallel to the bedding plane. Compare *vertical permeability*.

horizontal release *n*: a clause in an oil or gas lease that excludes, or releases, non-producing acreage from the lease at the end of a specified period.

horizontal separator *n*: a separator of cylindrical shape lying parallel to the ground.

horizontal station *n*: a preestablished location in the horizontal plane at ground level along the tank circumference. Compare *vertical station*.

horizontal tension *n*: horizontal component of mooring line tension.

horsehead *n*: the generally horsehead-shaped steel piece at the front of the beam of a pumping unit to which the bridle is attached in sucker rod pumping.

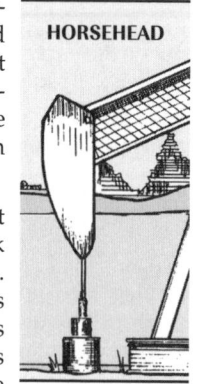

HORSEHEAD

horsepower (hp) *n*: a unit of measure of work done by a machine. One horsepower equals 33,000 foot-pounds per minute. (Kilowatts are used to measure power in the international, or SI, system of measurement.)

horst *n*: a block of the earth's crust that has been raised (relatively) between two faults. Compare *graben*.

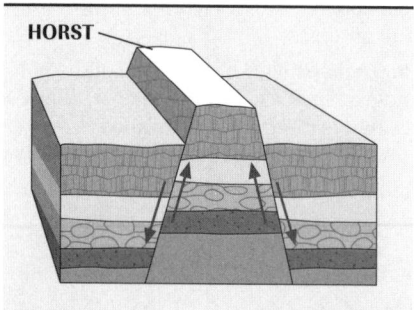

HORST

hose bundle *n*: in drilling from floating offshore drilling rigs that employ a subsea blowout preventer stack, a collection (a bundle) of several hoses enclosed in a heavy-duty protective cover. A hose bundle has several electrical and hydraulic control lines that run to the control pods on the subsea blowout preventers.

When the driller or other person actuates a blowout preventer control on the rig at the surface, the control signal goes through the hose bundle to the control pod on the blowout preventer stack.

hose bundle reel *n*: on offshore floating rigs, a large spool, or reel, onto which the hose bundle is wrapped. See *hose bundle*.

hose storage reel *n*: see *hose bundle reel*.

hot carbonate process *n*: a process for removing the bulk of acid gases from a gas stream by contacting the stream with a water solution of potassium carbonate at a temperature in the range of 220°F to 240°F (104°C to 116°C).

hot-carrier diode *n*: see *Schottky diode*.

hot oil *n*: 1. absorption or other oil used as a heating medium. 2. oil produced in violation of state regulations or transported interstate in violation of federal regulations, or oil that is stolen.

hot-oil treatment *n*: the treatment of a producing well with heated oil to melt accumulated paraffin in the tubing and the annulus.

hot pass *n*: the second pass made on a weld. The hot pass follows the root, or stringer, bead and precedes the filler pass and cap.

hot spot *n*: an abnormally hot place on a casing coupling when a joint is being made up. It usually indicates worn threads on the pipe and in the coupling.

hot tap *v*: to make repairs or modifications on a tank, pipeline, or other installation without shutting down operations.

hot terminal *n*: see *positive terminal*.

hot tie-in *n*: a weld made on a pipeline already in service. The gas from the line is purposely ignited at the point where the welding is to be done, thereby eliminating the chance that a spark could cause an explosion of gas and air.

hot waterflooding *n*: a method of thermal recovery in which water at the boiling point is injected into a formation to lower the viscosity of the oil and allow it to flow more freely toward producing wells. Although generally less effective than steam injection because of lower heat, hot waterflooding may be preferable under certain conditions, such as formation sensitivity to fresh water or high pressures.

hot-wire anemometer *n*: a wind-speed measuring device that has a thin platinum wire that electrical current heats to about 1,000 degrees Celsius (over 1,800 degrees Fahrenheit). A special instrument measures wind speed by examining either the wire's change in resistance or the amount of current required to maintain the temperature of the wire when it is under the influence of wind.

housing *n*: something that covers or protects, as the casing for a moving mechanical part.

hp *abbr*: horsepower.

HP *abbr*: hydrostatic pressure.

HPNS *abbr*: high-pressure nervous syndrome.

HPU *abbr*: hydraulic pressure unit.

HRC valve™ *n*: trade name for a remote-controlled valve on the choke line that controls the flow of fluids through the line from the well to the choke manifold.

HSE *abbr*: health, safety, and environment.

hubbed connection *n*: a fitting for joining two pipes or devices; a metal disk (a hub) on each end of the pipes or devices to be connected are put together and bolted to form a pressure-tight seal.

huff 'n' puff *n*: (slang) cyclic steam injection.

hull *n*: the framework or body of a vessel including all decks, plating, and columns, but excluding machinery.

human-machine interface (HMI) *n*: the apparatus that allows a human operator to work and interact with nonhuman equipment.

humic acid *n*: organic acids of indefinite composition in naturally occurring leonardite lignite. The humic acids are the most valuable constituent. See *lignins*.

humidity *n*: a measure of moisture content in dry air; an expression of the amount of water dispersed as vapor in a quantity of dry air or other gas, usually expressed as grains of water per pound of dry air. See *absolute humidity, relative humidity*.

hunting *n*: a surge of engine speed to a higher number of revolutions per minute (rpm), followed by a drop to normal speed without manual movement of the throttle. It is often caused by a faulty or improperly adjusted governor.

hurricane *n*: a tropical storm with high winds and heavy rains that occurs in the North Atlantic, the Caribbean, the Gulf of Mexico, and in the Eastern Pacific and Central Pacific. Pacific hurricanes, however, are called typhoons.

hurricane warning *n*: an announcement warning that hurricane-force winds are in the area and advising inhabitants in that area to seek shelter.

hurricane watch *n*: an announcement warning that hurricane-force winds may pose a threat to an area and cautioning inhabitants in that area to listen for subsequent advisories.

HWDP *abbr*: heavy-walled drill pipe or heavyweight drill pipe.

hybrid bits *n pl*: combine natural and synthetic diamonds and sometimes tungsten carbide inserts on a fixed-head bit.

Hydrafrac© *n*: the copyrighted name of a method of hydraulic fracturing for increasing productivity.

hydrate *n*: a hydrocarbon and water compound that is formed under reduced temperature and pressure in places where gas that contains water vapor occurs. For example, hydrates can form in gathering, compression, and transmission facilities for gas. They can also form in deepwater drilling where low temperatures occur on or near the seafloor and where gas containing water vapor may be encountered by the borehole. Hydrates can accumulate in troublesome amounts and impede fluid flow. They resemble snow or ice and decompose at atmospheric pressure. *v*: to enlarge by taking water on or in.

hydrated lime *n*: calcium hydroxide, $Ca(OH)_2$, a dry powder obtained by treating quicklime with enough water to satisfy its chemical affinity for water.

hydrate seal *n*: in drilling from floating rigs that use a subsea blowout preventer stack and marine riser system, a groove in the lower body of a hydraulic connector that forms a seal with the outside of the wellhead.

hydration *n*: 1. a chemical reaction in which molecular water is added to the molecule of another compound without breaking it down. 2. reaction of powdered cement with water. The cement gradually sets to a solid as hydration continues.

hydraulic *adj*: 1. of or relating to water or other liquid in motion. 2. operated, moved, or effected by water or liquid.

hydraulic actuator *n*: a cylinder or fluid motor that converts hydraulic power into useful mechanical work; mechanical motion produced may be linear, rotary, or oscillatory.

hydraulic area *n*: the area available for flow at a restriction.

hydraulic balancing *n*: booster stations located along the trunkline assure that the work load for each station is approximately equal.

hydraulic bond *n*: a bonding of cement to the casing or to the formation that blocks the migration of fluids. It is usually determined by applying increasing amounts of liquid pressure at the pipe-cement or formation-cement interface until leakage occurs.

hydraulic bonnet operating system *n*: in a ram preventer, the devices in the ram housing (the bonnet) that open or close the rams when activated with hydraulic pressure from the blowout preventer operating unit (accumulator).

hydraulic brake *n*: also called hydrodynamic brake or Hydromatic® brake. See *hydrodynamic brake*.

hydraulic connector *n*: in drilling from offshore floating rigs, a device that attaches (connects) the blowout preventer stack to the wellhead and to the lower marine riser package. It is controlled from the surface. Hydraulic connectors consist of a lower body, an upper body, a cam ring, locking dogs (or segments), a seal groove, and a hydraulic system. See *lower marine riser package*.

hydraulic control pod *n*: a device used on floating offshore drilling rigs to provide a way to actuate and control subsea blowout preventers from the rig on the surface. Hydraulic and electrical lines from the rig enter the pods, through which fluid and electrical signals are sent to the preventer. Usually two pods, one yellow and one blue, are used, each to safeguard and back up the other. Also called blue pod, yellow pod.

hydraulic coupling *n*: a fluid connection between a prime mover and the machine it drives; it uses the action of liquid moving against blades to drive the machine. Also called fluid coupling.

hydraulic fluid *n*: a liquid of low viscosity (such as light oil) that is used in systems actuated by liquid (such as the brake system in a modern passenger car).

hydraulic force *n*: force resulting from pressure on water or other hydraulic fluid.

hydraulic fracturing *n*: an operation in which a specially blended liquid is pumped down a well and into a formation under pressure high enough to cause the formation to crack open, forming passages through which oil can flow into the wellbore. Sand grains, aluminum pellets, glass beads, or similar materials are carried in suspension into the fractures. When the pressure is released at the surface, the fractures partially close on the proppants, leaving channels for oil to flow through to the well. Compare *explosive fracturing*.

hydraulic governor *n*: a governor on an engine that operates by means of oil inside its housing. Unlike a mechanical

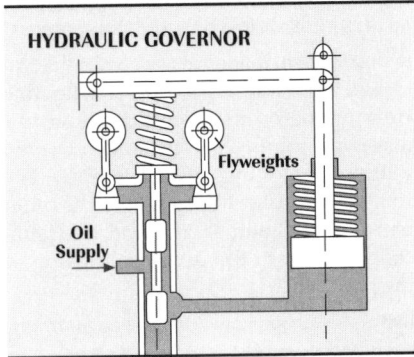

HYDRAULIC GOVERNOR

Flyweights

Oil Supply

governor, which is mechanically linked to the engine's speed control, a hydraulic governor operates the speed control with oil pressure inside the governor. See *governor*. Compare *mechanical governor*.

hydraulic hammer effect *n*: see *fluid knock*, *water hammer*.

hydraulic head *n*: the force exerted by a column of liquid expressed by the height of the liquid above the point at which the pressure is measured. Although "head" refers to distance or height, it is used to express pressure, since the force of the liquid column is directly proportional to its height. Also called head or hydrostatic head. Compare *hydrostatic pressure*.

hydraulic heat-exchanger module *n*: equipment associated with a high-pressure rotating head that circulates hydraulic fluid to cool and lubricate the bearings on which the rotating head's stripper rubber rotates. The hydraulic heat-exchanger system transfers heat generated by the load on the rotating head and cools the bearing pack.

hydraulic holddown *n*: an accessory or integral part of a packer used to limit the packer's upward movement under pressure.

hydraulic horsepower (hhp) *n*: a measure of the power of a fluid under pressure.

hydraulic jar *n*: a type of mechanical jar in which a fluid moving through a small opening slows the piston stroke while the crew stretches the work string. After the hydraulic delay, a release mechanism in the jar trips to allow a mandrel to spring up and deliver a sharp blow. Compare *mechanical jar*, *rotary jar*.

hydraulic jet perforating *n*: the use of sand pumped through horizontally positioned orifices installed in the tubing string to jet holes in the casing and cement sheath into the formation.

hydraulic jet pump *n*: a specialized form of hydraulic pump, used in artificial lift, whose main working parts are nozzle, throat, and diffuser. The nozzle converts the high-pressure, low-velocity energy of the power fluid to high-velocity, low-pressure energy. The power fluid is then mixed with the low-pressure pump intake fluid in the throat to produce a low-pressure stream with a velocity less than that of the nozzle exit, but a high velocity, nevertheless. The velocity energy of this mixed stream is then converted to static pressure in the diffuser to provide the pressure necessary to lift fluid from the well. The power fluid may be either oil or water.

hydraulic junction box *n*: on a blowout preventer control unit (accumulator), a housing (box) with several inlet and outlet ports. The inlet ports receive the accumulator's operating fluid; the outlets send the fluid to the yellow and blue control pods, which, in turn, operate the blowout preventers.

hydraulic module *n*: see *hydraulic heat-exchanger module*

hydraulic pressure unit *n*: see *blowout preventer control unit.*

hydraulic pump *n*: 1. a device that lifts oil from wells without the use of sucker rods. See *hydraulic pumping.* 2. a device that creates pressure on a fluid, usually special hydraulic fluid, to move the fluid.

hydraulic pumping *n*: a method of pumping oil from wells by using a downhole pump without sucker rods. Subsurface hydraulic pumps consist of two reciprocating pumps coupled and placed in the well. One pump functions as an engine and drives the other pump (the production pump). The downhole engine is usually operated by clean crude oil under pressure (power oil) that is drawn from a power-oil settling tank by a triplex plunger pump on the surface. If a single string of tubing is used, power oil is pumped down the tubing string to the pump, which is seated in the string, and a mixture of power oil and produced fluid is returned through the casing-tubing annulus. If two parallel strings are used, one supplies power oil to the pump while the other returns the exhaust and produced oil to the surface. A hydraulic pump may be used to pump several wells from a central source and has been used to lift oil from depths of more than 10,000 feet (3,048 metres).

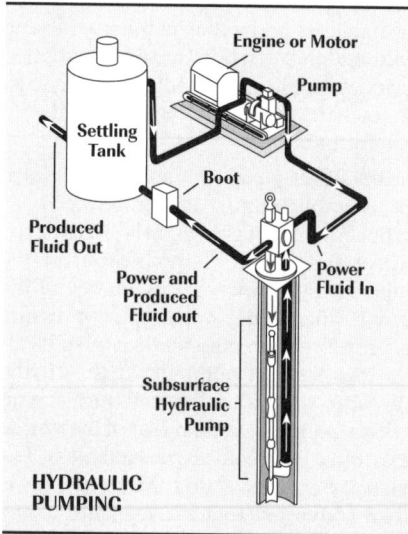

HYDRAULIC PUMPING

hydraulic ram *n*: a cylinder and piston device that uses hydraulic pressure for pushing, lifting, or pulling. It is commonly used to raise portable masts from a horizontal to a vertical position, for leveling a production rig at an uneven location, or

for closing a blowout preventer against pressure.

hydraulics *n*: 1. the branch of science that deals with practical applications of water or other liquid in motion. 2. the planning and operation of a rig hydraulics program, coordinating the power of circulating fluid at the bit with other aspects of the drilling program so that bottomhole cleaning is maximized.

hydraulic-set packer *n*: a packer that is set using pressure applied either by pistons actuated by formation pressure or using externally applied pump pressure.

hydraulic snubber *n*: on a drilling rig's hook, a device inside the hook that is filled with hydraulic fluid. It dampens the tendency of the hook to bounce (move rapidly up and down) when crew members break out a joint of drill pipe. As the last engaged threads of the tool joint pin clears the box threads, the release may be sudden and, if not dampened, the pin could easily strike the box and damage the threads on both.

hydraulic snubbing unit *n*: a self-contained machine powered by hydraulic fluid that puts pipe into the hole or takes it out of the hole when pressure in the hole is great enough to force the pipe out of the hole if the snubbing unit did not hold back the pipe. A hydraulic snubbing unit consists of control heads (special BOPs), stationary and traveling snubbers, operating manifold, rotary table, and the hydraulic system. Hydraulic jacks control the movement of the pipe out of or into the hole.

hydraulic starter *n*: on an engine, a device used to start the engine that uses hydraulic fluid under pressure to operate a motor on the starter. When engaged, the starter motor turns the engine's flywheel to make the engine start.

hydraulic sucker rod pumping unit *n*: a sucker rod pumping unit that uses reciprocal hydraulic pumps. Few are still in operation.

hydraulic supply line *n*: piping that carries hydraulic operating fluid from a storage container to a device that is operated by hydraulic fluid—for example, the piping from an accumulator to a blowout preventer.

hydraulic-tensioner ring *n*: a device attached to the telescopic joint on a floating rig's marine riser system that allows the telescopic joint to be run and retrieved without removing tensioner lines. The ring is hydraulically latched to the outer barrel as the joint is run, and hydraulically latched to the diverter housing as the joint is being retrieved.

hydraulic torque converter *n*: see *hydraulic coupling.*

hydraulic torque wrench *n*: a hydraulically powered device that can break out or make up tool joints and assure accurate torque. It is fitted with a repeater gauge so that the driller can monitor tool joints as they go downhole, doubly assuring that all have the correct torque.

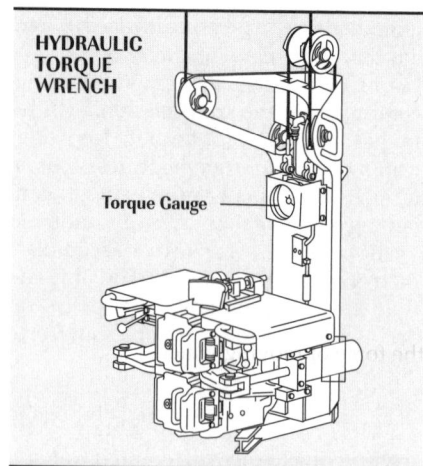

HYDRAULIC TORQUE WRENCH

Torque Gauge

hydraulic workover *n*: a series of hydraulic rams to restrain and pull tubing under well pressure, temporarily attached to the wellhead for workover.

hydraulic workover preventer *n*: a series of hydraulic rams to restrain and pull tubing under well pressure, temporarily attached to the wellhead for workover.

Hydril *n*: the registered trademark of a prominent manufacturer of oilfield equipment, including annular and ram blowout preventers and casing and tubing joints.

hydrocarbon pore volume *n*: the volume of the pore space in a reservoir that is occupied by oil, natural gas, or other hydrocarbons (including nonhydrocarbon impurities). It may be expressed in acre-feet, barrels, or cubic feet.

hydrocarbons *n pl*: organic compounds of hydrogen and carbon whose densities, boiling points, and freezing points increase as their molecular weights increase. Although composed of only two elements, hydrocarbons exist in a variety of compounds because of the strong affinity of the carbon atom for other atoms and for itself. The smallest molecules of hydrocarbons are gaseous; the largest are solids. Petroleum is a mixture of many different hydrocarbons.

hydrochloric acid *n*: an acid compound, HCl, commonly used to acidize carbonate rocks. It is prepared by mixing hydrogen chloride gas in water. Also known as muriatic acid.

hydrocrackate *n*: cracked naphtha for making gasoline.

hydrocracking *n*: cracking in the presence of low-pressure hydrogen, consuming a net amount of hydrogen in the process.

hydrocyanic acid *n*: see *hydrogen cyanide*.

hydrocyclone *n*: a cone-shaped separator for separating various sizes of particles and liquid by centrifugal force. See *desander, desilter*.

hydrodynamic brake *n*: a device mounted on the end of the drawworks shaft of a drilling rig. It serves as an auxiliary to the mechanical brake when pipe is lowered into the well. The braking effect is achieved by means of an impeller turning in a housing filled with water. Sometimes called hydraulic brake or Hydromatic® brake.

hydrodynamic trap *n*: a petroleum trap in which the major trapping mechanism is the force of moving water.

hydrodynamic washpipe *n*: a special washpipe assembly used in swivels and top drives that uses externally mounted hydraulic cylinders and high-pressure hydraulic lines. The lines conduct hydraulic oil to the washpipe and puts external pressure on the washpipe packing. The external pressure forces the packing away from the washpipe so that a thin film of hydraulic oil exists between the packing and the washpipe. The packing therefore does not directly contact the washpipe and, consequently, wear is reduced. When drilling with mud pump pressures higher than 3,000 pounds per square inch (20,685 kilopascals), wear on conventional washpipe packing accelerates rapidly. Hydrodynamic washpipe can be used with pump pressures as high as 7,500 pounds per square inch or 51,713 kilopascals to markedly reduce washpipe packing wear.

hydroelectric *adj*: driven by water power.

hydrofluoric acid *n*: a strong, poisonous liquid acid compound of hydrogen and fluorine (HF). Often mixed with hydrochloric acid, it is used mainly to remove mud from the wellbore and surrounding formation pores. Also called mud acid.

hydrofluoric-hydrochloric acid *n*: a mixture of acids used for removal of mud from the wellbore. See *mud acid.*

hydroforming *n*: a process of petroleum refining in which straight-run, cracked, or mixed naphthas are passed over a solid catalyst at elevated temperatures and moderate pressures in the presence of added hydrogen or hydrogen-containing gases. The main chemical reactions are dehydrogenation and aromatization of the nonaromatic constituents of the naphtha to form either high-octane motor fuel or high-grade aviation gasoline high in

aromatic hydrocarbons such as toluene and xylenes. Ninety percent of the sulfur contained in the naphtha is removed.

hydrogas *n*: another term for liquefied petroleum gas (LPG).

hydrogen (H) *n*: a flammable, colorless, odorless, tasteless gas whose chemical symbol is H and that usually occurs as the diatomic molecule H_2. It is lighter than air; indeed, it is the lightest element in the universe.

hydrogenate *v*: to combine or treat with or expose to hydrogen.

hydrogen cyanide *n*: an extremely poisonous compound of hydrogen, carbon, and nitrogen (HCN), with a boiling point of 79°F (26°C) and having the odor of bitter almonds. It is water-soluble in all proportions. Also called formonitrile, hydrocyanic acid, or prussic acid.

hydrogen embrittlement *n*: low ductility of a metal caused by its absorption of hydrogen gas. Also called acid brittleness.

hydrogen ion concentration *n*: a measure of the acidity or alkalinity of a solution, normally expressed as pH. See *pH*.

hydrogen patch probe *n*: an instrument that, when attached to the exterior of a vessel that has been corroded by hydrogen sulfide, senses the hydrogen content in the steel and records the rate of corrosion.

hydrogen richness *n*: a formation's hydrogen content.

hydrogen sulfide *n*: a flammable, colorless gaseous compound of hydrogen and sulfur (H_2S), which in small amounts has the odor of rotten eggs. Sometimes found in petroleum, it causes the foul smell of petroleum fractions. In dangerous concentrations, it is extremely corrosive and poisonous, causing damage to skin, eyes, breathing passages, and lungs and attacking and paralyzing the nervous system, particularly that part controlling the lungs and heart. In large amounts, it deadens the sense of smell. Also called hepatic gas or sulfureted hydrogen.

hydrogen sulfide corrosion *n*: see *hydrogen sulfide cracking*.

hydrogen sulfide cracking *n*: a type of corrosion that occurs when metals are exposed to hydrogen sulfide gas; it is characterized by minute cracks that form just under the metal's surface.

hydrologic cycle *n*: the complete sequence of events repeated again and again (a cycle) through which water passes from the oceans, through the atmosphere, to the land, and back to the ocean.

hydrolysis *n*: the decomposition of a chemical compound by its reaction with water.

hydrolyze *v*: to undergo hydrolysis; a chemical process of decomposition involving splitting a bond and adding the elements of water.

Hydromatic® brake *n*: trade name for a type of hydrodynamic brake.

hydrometer *n*: an instrument with a graduated stem, used to determine the gravity of liquids. The liquid to be measured is placed in a cylinder, and the hydrometer dropped into it. It floats at a certain level in the liquid (high if the liquid is light, low if it is heavy), and the stem markings indicate the gravity of the liquid.

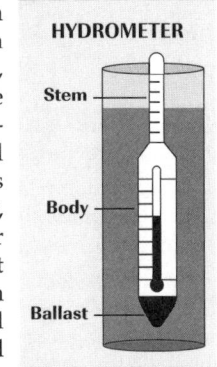

hydrophilic *adj*: tending to adsorb water.

hydrophilic-lipophilic balance (HLB) *n*: an expression of the relative attraction of an emulsifier for water and oil, determined largely by the chemical composition and ionization characteristics of a given emulsifier. The HLB of an emulsifier is not directly related to solubility, but it determines the type of emulsion that tends to be formed. It is an indication of the behavioral characteristics and not an indication of emulsifier efficiency.

hydrophobic *adj*: tending to repel water.

hydrophone *n*: a device trailed in an array behind a boat in offshore seismic exploration that is used to detect sound reflections, convert them to electric current, and send them through a cable to recording equipment on the boat.

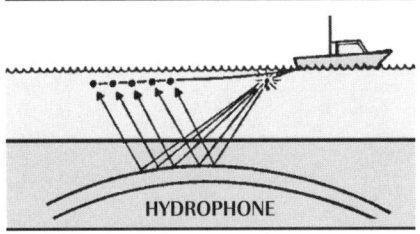

hydropneumatic cylinder *n*: a component of the riser tensioning system on a floating drilling rig that contains hydraulic fluid and high-pressure air and opposes heave placed on the system by ocean movements. See *hydropneumatic riser-tensioning system.*

hydropneumatic riser-tensioning system *n*: the components (the system) that provide upward force (tension) on a marine riser's tensioner lines. The system consists of hydropneumatic cylinders and sheave assemblies, accumulators and system air-pressure vessels, a control panel and piping manifold, high-pressure air compressors, and standby air-pressure vessels.

hydroscopic *adj*: able to absorb water from the air.

hydroset tool *n*: a wireline pressure setting tool for setting permanent downhole tools.

Hydrostarter™ *n*: a brand name of an air-motor starter. See *air-motor starter*.

hydrostatic *adj*: relating to, involving, or typical of fluids that are at rest and the forces and pressures they exert.

hydrostatic bailer *n*: a sand bailer that consists of a sealed cylinder with a rubber disk and a piercing tool at the bottom. A sharp blow on the bottom ruptures the disk, allowing the hydrostatic pressure of the well fluid to force sand into the bailer.

hydrostatic head *n*: see *hydrostatic pressure*.

hydrostatic junk basket *n*: a hydraulically powered fishing tool that uses pressure differential to trap and retrieve junk from the bottom of the hole.

hydrostatic pressure (HP) *n*: the force exerted by a body of fluid at rest. It increases directly with the density and the depth of the fluid and is expressed in many different units, including pounds per square inch or kilopascals. The hydrostatic pressure of fresh water is 0.433 pounds per square inch per foot (9.792 kilopascals per metre) of depth. In drilling, the term refers to the pressure exerted by the

drilling fluid in the wellbore. In a water drive field, the term refers to the pressure that may furnish the primary energy for production.

hydrostatic properties table *n*: on an offshore rig, a list that includes the rig's draft in feet or metres, its displacement, its water plane area in square feet or square metres, and its vertical center of buoyancy. Such tables provide the starting point for calculating the buoyancy, stability, and trim of an offshore floating rig.

hydrostatic testing *n*: the most common final quality-control check of the structural soundness of a pipeline. The line is filled with water and then pressured to a designated point. This pressure is maintained for a specified period of time, and any ruptures or leaks revealed by the test are repaired. The test is repeated until no problems are noted.

hydrotest *v*: to apply hydraulic pressure to check for leaks in tubing or tubing couplings, usually as the tubing is being run into the well. If water leaks from any place in the tubing, the joint of tubing, the coupling, or both are replaced.

hydrotreating *n*: a method of removing sulfur, nitrogen, and metals from crude oil fractions.

hydrotrip pressure sub *n*: a sub with a ball seat run on top of a hydraulically set packer to provide a means to set the packer.

hydroxide *n*: a designation that is given for basic compounds containing the hydroxide (OH) radical. When these substances

are dissolved in water, they increase the pH of the solution. See *base*.

hygrometer *n*: an instrument used to measure water vapor in the air. See *psychrometer, sling psychrometer*.

hygroscopic *adj*: absorbing or attracting moisture from the air.

hyperbaric *adj*: relating to or utilizing greater than normal pressure.

hypercapnia *n*: excessive amount of carbon dioxide in the blood, often resulting from an excessive carbon dioxide partial pressure in a diver's breathing supply. Also called carbon dioxide excess.

hypotenuse *n*: in a right triangle, the side opposite the right angle. See *right triangle*.

hypothermia *n*: reduced body temperature caused by overexposure to chilling temperatures.

hypoxia *n*: see *anoxia*.

hysteresis *n*: 1. the difference between the indications of a measuring instrument when the same value of the quantity measured is reached by increasing and decreasing the quantity. 2. the lagging of changes in the magnetization of a substance behind changes in the magnetic field as the magnetic field is varied.

hysteresis loss *n*: in a transformer, the conversion of electrical energy to heat in the core of the transformer's winding because of hysteresis (the lagging of changes in the core's magnetic field as the transformer's magnetic field is varied). See *hysteresis*.

Hz *sym*: hertz.

I *abbr*: moment of inertia.

I *sym*: current.

IACS *abbr*: International Association of Classification Societies.

IADC *abbr*: International Association of Drilling Contractors.

I-bar *n*: a steel bar that has a cross-sectional shape that resembles the letter I.

IBOP *abbr*: inside blowout preventer.

ICC *abbr*: Interstate Commerce Commission.

iceberg *n*: a large, floating mass of ice detached from a glacier.

ice fog *n*: a fog, composed of ice crystals, that occurs in moist air during cold, calm conditions in winter. It may be produced artificially by the combustion of fuel in cold air. This type of fog is common in the Arctic.

ice ridge *n*: floating sea ice formed by the merging of floes.

ice scour *n*: the abrasion of material in contact with moving ice in a sea, ocean, or other body of water.

ICP *abbr*: initial circulating pressure; used in drilling reports.

ICS *abbr*: incident command system.

ID *abbr*: inside diameter.

ideal gas *n*: 1. a gas whose molecules are infinitely small and exert no force on each other. 2. a gas that obeys Boyle's law and Joule's law. Also called a perfect gas.

ideal gas law *n*: the equation of the state of an ideal gas, showing a close approximation to real gases at sufficiently high temperature and low pressures:

$$PV = RT$$

where—

   P = pressure
   R = gas constant
   T = temperature
   V = volume per mole of gas.

ideal pressure *n*: see *kinetic pressure*.

identification inscriptions *n pl*: all of the words, letters, numbers, and marks carried by a measuring instrument to indicate its origin, destination, operation, characteristics, method of use, and so on.

identification plate *n*: on a flow recorder, the record of the static and differential spring ranges.

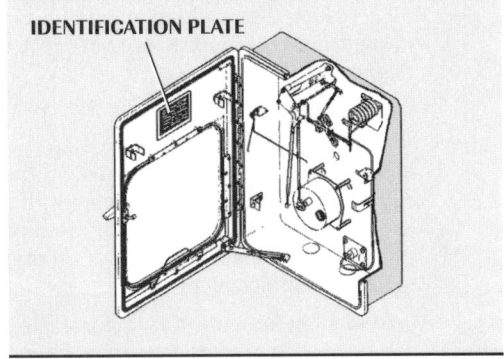

IDENTIFICATION PLATE

idiot stick *n*: (slang) 1. a shovel. 2. a thread profile gauge.

idle *v*: to operate an engine without applying a load to it.

idler *n*: a pulley or sprocket used with belt or chain drives on machinery to maintain desired tension on the belt or chain.

IDLH *abbr*: immediately dangerous to life or health.

I-ES *abbr*: induction-electric survey.

I-EUE *abbr*: internal-external upset end.

IFP *abbr*: Institut Francais du Petrole.

IFR *abbr*: instrument flight rules.

if-then operation *n*: a statement that asserts that if one thing is true then another thing is also true. Mathematically, the operation is, "If p, then q."

ig *abbr*: igneous; used in drilling reports.

IGBT *abbr*: insulated-gate bipolar transistor.

igneous rock *n*: a rock mass formed by the solidification of magma within the earth's crust or on its surface. It is classified by chemical composition and grain size. Granite is an igneous rock.

ignitability *n*: the ability of a substance to catch fire, burn.

ignition coil *n*: a coil in the ignition system of an engine that stores energy in a magnetic field slowly and releases it suddenly to cause a spark that ignites the fuel mixture in a cylinder.

ignition quality *n*: the ability of a fuel to ignite when it is injected into the compressed-air charge in a diesel engine cylinder. It is measured by an index called the cetane number.

ignition temperature *n*: the temperature at which a particular vapor burns.

ignorant end *n*: (slang) the heavier end of any device (such as a length of pipe or a wrench).

IHTS *abbr*: immersion heater temperature switch.

ilmenite *n*: an iron-black mineral of composition $FeTiO_3$ or $Feo \bullet TiO_x$, with a specific gravity of about 4.67, sometimes used for increasing the density of oilwell cement slurries.

imaging radar *n*: radar carried on airplanes or orbital vehicles that forms images of the terrain.

imbibition relative permeability *n*: displacement of hydrocarbons or fluids in a reservoir by increasing the wetting phase saturation. See *wetting phase*.

imbrication *n*: in sedimentary rocks, the arrangement of pebbles in a flat, overlapping pattern like bricks in a wall. Often shown by stream gravel deposits.

IMCO *abbr*: Intergovernmental Maritime Consultative Organization.

IMBRICATION

immediately dangerous to life or health (IDLH) *adj*: (HAZ-WOPER) having an atmospheric concentration of any toxic, corrosive, or asphyxiant substance that poses an immediate threat to life or that would cause irreversible or delayed adverse health effects or that would interfere with an individual's ability to escape from a dangerous atmosphere. *n*: 1. a substance that is immediately dangerous to life or health. 2. a situation in which immediate danger conditions exist.

immersion heater temperature switch (IHTS) *n*: a temperature-sensitive device installed on engines in extremely cold climates, such as in Alaska and Siberia. Before the engine is started, the switch senses very cold coolant and turns on an immersion heater in the coolant tank, thus ensuring that the engine warms up quickly.

immiscible *adj*: not capable of mixing (like oil and water).

immiscible gas injection *n*: immiscible gas injected into injection wells in alternating steps with water to improve recovery.

IMO *abbr*: International Maritime Organization.

IMP *abbr*: Instituto Mexicano del Petróleo (Mexican Petroleum Institute).

impact pressure *n*: pressure exerted by a moving fluid on a plane perpendicular to its direction of flow.

impedance (Z) *n*: in an alternating current (AC) circuit, the circuit's total opposition to current flow. Impedance is similar to resistance in a direct current (DC) circuit. Just as resistance is measured in ohms, so too is impedance. When connecting certain AC components, such as a microphone to an amplifier, it is important for both components to have the same, or very close to the same, impedance; otherwise, the connected components may not operate properly.

impeller *n*: a set of mounted blades used to impart motion to a fluid (e.g., the rotor of a centrifugal pump).

impending blowout *n*: early manifestation or indication of a blowout.

impermeable *adj*: preventing the passage of fluid. A formation may be porous yet impermeable if there is an absence of connecting passages between the voids within it. See *permeability*.

impervious *adj*: see *impermeable*.

implication *n*: see *if-then operation*.

implied covenant *n*: an obligation or benefit not specified in an oil and gas lease but held by the courts to be implicit in such lease; for example, an obligation on the part of the lessee to drill an initial well.

importer *n*: (OSHA) the first business with employees within the customs territory of the United States that receives hazardous chemicals produced in other countries for the purpose of supplying them to distributors or employers within the United States.

impregnated diamond bit *n*: a fixed-head bit that contains small particles (grit) of natural diamonds embedded throughout a tungsten carbide matrix. The diamonds are generally smaller than those used in conventional natural diamond bits. Impregnated diamond bits drill in a similar fashion to natural diamond bits, but, when the diamonds become worn and torn out of the matrix, new ones are continually being exposed. Impregnated diamond bits can drill the hardest, most abrasive formations at high rpm.

impressed-current anode *n*: anode to which an external source of positive electricity is applied (such as from a rectifier or DC generator). The negative electricity is applied to a pipeline, casing, or other structure to be protected by the impressed-current method of cathodic protection.

impressed voltage *n*: in an electrical circuit, voltage applied to the circuit or to a device in the circuit.

impression block *n*: a block with lead or another relatively soft material on its bottom. It is made up on drill pipe or tubing at the surface, run into a well, and set down on the object that has been lost in the well. The block is retrieved and the impression is examined. The impression is a mirror image of the top of the fish; it also indicates the fish's position in the hole, i.e., whether it is centered or off to one side. From this information, the correct fishing tool can be selected.

impression tool *n*: a lead-filled cylindrical device used to ascertain the shape of the top of a fish.

improved plough (plow) steel (IPS) *n*: used in wire rope construction definitions.

improved recovery *n*: the introduction of artificial drive and displacement mechanisms into a reservoir to produce a portion of the oil unrecoverable by primary recovery methods. See *enhanced oil recovery*.

impulse coupling *n*: a special coupling between the camshaft and the magneto of a large engine. At cranking speeds the magneto is made to rotate in quick spurts

to produce large pulses of voltage at the time of ignition for each cylinder.

impulse factor *n*: the increase in the weight in sucker rods in fluid plus the weight of the fluid, caused by the effects of inertia or acceleration. For example, if an increased pumping rate causes an inertial effect of 30% of the weight of the rods in fluid plus the weight of the fluid, the impulse factor will be 1.3.

in. *abbr*: inch.

in.^{2} *abbr*: square inch.

in.^{3} *abbr*: cubic inch.

incandescent lamp *n*: an ordinary light bulb.

inches (millimetres) of mercury *n*: a scale for measuring small increases or decreases in pressure, usually in conjunction with a manometer, a U-tube containing mercury. Pressure on one side of the U-tube causes the mercury to move upward on the other side. The movement is measured in inches or millimetres. See *manometer*.

inches (millimetres) of water *n*: a scale for measuring small increases or decreases in pressure, usually in conjunction with a manometer, a U-tube containing water. Pressure on one side of the U-tube causes the water to move upward on the other side. The movement is measured in inches or millimetres. See *manometer*.

inch-pound (in.-lb) *n*: one-twelfth of a foot-pound and a measure of small amounts of twisting force, or torque. Compare *foot-pound (ft-lb)*.

incident command system (ICS) *n*: a written plan that describes the chain of command in an emergency response and designates senior officials and their authority for a given site.

incipient blowout *n*: see *impending blowout*.

inclination *n*: see *drift angle*.

inclining experiment *n*: a process to determine the position of an offshore rig or vessel's center of gravity. A weight having a known value is placed on board the vessel and is moved a measured distance perpendicular to the vessel's centerline plane. Then, the resulting angle of list is measured to determine the vessel's center of gravity.

inclinometer *n*: an instrument that measures the angle of a hole.

incompetent *n*: a person judged by the court to be incapable of managing his or her own affairs by reason of insanity, imbecility, or feeblemindedness (referred to as non compos mentis).

incompetent formation *n*: a formation composed of materials that are not bound together. It may produce sand along with hydrocarbons if preventive measures are not taken, or it may slough or cave around the bit or drill stem when a hole is drilled into it.

incremental pricing account *n*: account maintained by a pipeline that contains the cost subject to pass-through to certain users.

indefinite pricing provision *n*: any provision of any contract that provides for the establishment or adjustment of the price for natural gas delivered under such contract by reference to other prices for natural gas, crude oil, or refined petroleum products, or that allows for the establishment or adjustment of the price of natural gas delivered under any contract by negotiation between the parties.

independent *n*: a nonintegrated oil company or an individual whose operations are in the field of petroleum production, excluding transportation, refining, and marketing.

independent boom hoist and swing *n*: an offshore crane that is constructed in a manner that allows the crane operator to raise and lower the crane boom and to swing (revolve) the crane independently of other crane functions.

independent inspector *n*: a person or organization acting independently, but on behalf, of one or more parties involved in the transfer, storage, inventory, or analysis of a commodity or the calibration of land or marine vessels for the purpose of determining the quantity, capacity, and/or quality of a commodity.

independent marketer *n*: functions at only one of several levels in the marketing chain.

Independent Petroleum Association of America (IPAA) *n*: an organization of independent oil and gas producers concerned with the relationships between the oil industry and the public and government. Its official publication is *Petroleum Independent*. Address: 1201 15th Street NW, Suite 300; Washington, DC 20005; 202-857-4722.

independent surveyor *n*: often used to mean independent inspector, but usually implying a person or organization capable of total quantity and quality inspection and calibration of shore, truck, rail, and marine vessels, meter proving, and physical properties determinations.

independent tank *n*: a tank the boundaries of which are not part of the hull structure of a barge.

independent wire rope center (IWRC) *n*: see *independent wire rope core.*

independent wire rope core (IWRC) *n*: a core for wire rope consisting of a strand

of steel wires with a spiral winding that is opposite that of the outer strands of the rope. Also called independent wire rope center and king wire.

index *n*: a fixed or movable part of a measuring instrument's indicating device (pointer, liquid surface, recording stylus, etc.) the position of which with reference to the scale, indicates the value of the measured quantity.

indexing valve *n*: operates on the same principle as an annular valve, except it requires pipe rotation for opening and closing operations.

index shaft *n*: in a simple gas meter, a shaft that revolves as fluid flows past it; by counting the number of revolutions of the shaft, gas volume can be inferred.

indicated demand *n*: the sum of crude oil production and imports less changes in crude oil stocks.

indicated throughput *n*: the difference between the opening meter reading and the closing meter reading.

indicated volume *n*: the change in meter reading that occurs during receipt or delivery of a liquid product.

indicated volume with factor *n*: the indicated volume multiplied by the meter factor for the particular liquid and operating conditions under which the meter was proved.

indicating instrument *n*: a measuring instrument in which the value of the measured quantity is visually indicated but not recorded.

indicator *n*: 1. a dial gauge used on the rig to measure the hookload. 2. substances in acid-base titrations that, in solution, change color or become colorless as the hydrogen ion concentration reaches a definite value, these values varying with the indicator. In other titrations, such as chloride, hardness, and other determinations, these substances change color at the end of the reaction. Common indicators are phenolphthalein and potassium chromate.

indirect heater *n*: apparatus or equipment in which heat from a primary source, usually the combustion of fuel, is transferred to a

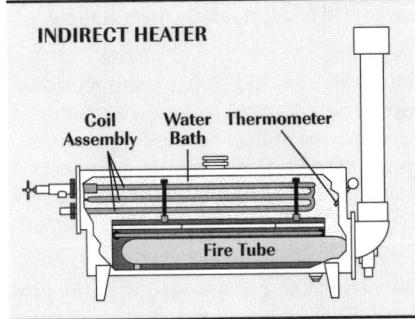

INDIRECT HEATER

Coil Assembly   Water Bath   Thermometer

Fire Tube

fluid or solid, which acts as the heating medium. The substance to be heated does not come into contact with the heat from the primary source. Compare *direct heater.*

indirect measurement *n*: a measurement that produces a final result by calculation using results from one or more direct measurements.

individual riser joints *n pl*: see *riser pipe.*

induced current *n*: the flow of an electric charge, measured in amperes, that is the result of induced voltage. See *induced emf.*

induced emf *n*: electromotive (voltage, or electricity) force that is developed in a conductor by moving the conductor within a magnetic field or moving the magnetic field in relation to the conductor.

induced fracture *n*: a fracture in a formation caused by artificial outside forces. The opposite of a natural fracture.

induced magnetism *n*: developed in a magnetic material by bringing it into a magnetic field.

induced voltage *n*: see *induced emf.*

inductance (L) *n*: the property of an electric circuit or of two neighboring circuits wherein an electromotive force is generated by induction in one circuit by a change of current in itself or in the other. See *induction.*

inductance coil *n*: see *inductor.*

induction *n*: the production of an electric charge, magnetism, or electromotive force in an electric conductor, a magnetizable body, or an electric circuit by the proximity of a similarly energized body or by the variation of the magnetic flux without contact.

induction-electric survey *n*: see *induction survey.*

induction log *n*: see *induction survey.*

induction motor *n*: an alternating-current motor in which the primary winding on a part (usually the stator) is connected to a source of power, and the secondary winding (usually on the rotor) carries current induced by the magnetic field of the primary winding.

induction survey *n*: an electric well log in which the conductivity of the formation rather than the resistivity is measured. Because oil-bearing formations are less conductive of electricity than water-bearing formations, an induction survey, when compared with resistivity readings, can aid in determination of oil and water zones.

inductive coupling *n*: in electronics, the transfer of energy from one circuit to another through the inductance provided by a transformer.

inductive filter *n*: see *LC filter*.

inductive reactance *n*: the retarding effect on the passage of alternating current through a circuit because of inductance. See *inductance*.

inductor *n*: in an electrical circuit, a coil of wire that introduces inductance or magnetic flux into the circuit. It may also react mechanically to changing magnetic flux. Also called electric coil, inductance coil.

indurated *adj*: hardened, as in indurated steel.

inelastic collision *n*: the collision of a neutron and the nucleus of an atom in which the total energy of the neutron is absorbed by the nucleus.

inert gas *n*: 1. the part of a breathing medium, such as helium, that serves as a transport for oxygen and is not used by the body as a life-support agent. Its purpose is to dilute the flow of oxygen to the lungs, thereby preventing oxygen toxicity. 2. in chemistry, gases that have a filled outer electron shell and thus do not easily react with other substances. Examples are helium, argon, neon, and xenon.

inertia *n*: the tendency of an object having mass to resist a change in velocity.

inertia brake *n*: a brake that utilizes the energy of a heavy, turning member to actuate the braking action.

inerting *n*: 1. the process of pressurizing the vapor space of a vessel with an inert gas blanket (usually exhaust gas) to prevent the formation of an explosive mixture. 2. a procedure used to reduce the oxygen content of a vessel's cargo spaces to 8% or less by volume by introducing an inert gas such as nitrogen or carbon dioxide or a mixture of gases such as flue gas.

inferential mass meter *n*: a volume meter with the addition of a densitometer from which mass flow is inferred.

infill drilling *n*: drilling wells between known producing wells to exploit the resources of a field to best advantage.

infilling well *n*: a well drilled between known producing wells to exploit the reservoir better.

inflatable packer *n*: a packer with an element that inflates by means of gas or liquid pumped from the surfae through a line. It is deflated by means of slots that can be opened to allow the gas or liquid to flow out. They are used when a temporary packer is needed in a hole with weakened casing that could be damaged by mechanical slips.

inflatable straddle tool *n*: consists of two packers spaced exactly as the straddle-packer DST. It is usually set by pump pressure instead of mechanically set.

inflow *n*: see *feed in*.

inflow performance relationship (IPR) *n*: the relation between the midpoint pressure of the producing reservoir and the liquid inflow rate of a producing well.

inflow performance relationship curve *n*: a productivity curve plotted when a well has two-phase flow, both oil and gas, based on well tests or a combination of computations and well tests. Also called IPR curve.

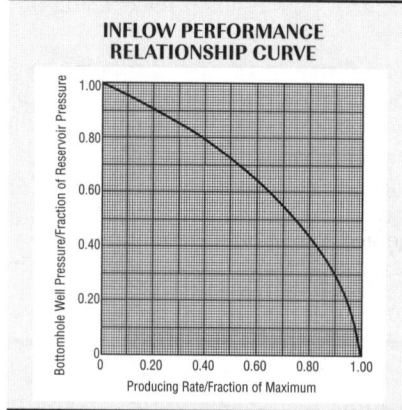

influx *n*: an intrusion of formation fluids into the borehole, i.e., a kick.

Information to Lessees and Operators (ITL) *n*: an MMS document sent to offshore lessees and operators to clarify or supplement operational guidelines.

INGAA *abbr*: Interstate Natural Gas Association of America.

inhalation hazards *n pl*: substances that, on being inhaled, adversely affect health or life.

inherent motor temperature protection *n*: overload or short-circuit protection built into the motor windings.

inhibited acid *n*: an acid that has been chemically treated before the acidizing or acid fracture of a well to lessen its corrosive effect on the tubular goods and yet maintain its effectiveness. See *acid fracture, acidize*.

inhibited mud *n*: a drilling fluid to which chemicals have been added to prevent it from causing clay particles in a formation to swell and thus impair the permeability of a productive zone. Salt is a mud inhibitor.

inhibitor *n*: an additive used to retard undesirable chemical action in a product. It is added in small quantities to gasolines to prevent oxidation and gum formation, to lubricating oils to stop color change, and to corrosive environments to decrease corrosive action.

initial circulating pressure (ICP) *n*: the pressure at which a well that has been closed

in on a kick is circulated when well-killing procedures are begun.

initial gel *n*: see *initial gel strength*.

initial gel strength *n*: the maximum reading (deflection) taken from a direct-reading viscometer after the fluid has been quiescent for 10 seconds. It is reported in pounds/100 square feet. See *gel strength*.

initial potential *n*: the early production of an oilwell, recorded after testing operations and recovery of load oil and used as an indicator of the maximum ability of a well to produce on completion without subsequent reservoir damage.

initial pressure (p_i) *n*: pressure measured at the bottom of a well after it is originally completed and before production starts. Knowing a well's initial pressure assists reservoir engineers in determining the amount of hydrocarbons in the reservoir.

initial set *n*: the point at which a cement slurry begins to harden, or set up, and is no longer pumpable.

initial stability *n*: in a ship or offshore drilling rig, the condition in which the vessel maintains equilibrium (is stable) when its metacentric height is above its center of gravity.

initial stretch *n*: the permanent lengthening that occurs to new wire rope when it is put into service.

injection *n*: the process of forcing fluid into something. In a diesel engine, the introduction of high-pressure fuel oil into the cylinders.

injection gas *n*: 1. a high-pressure gas injected into a formation to maintain or restore reservoir pressure. 2. gas injected in gas-lift operations.

injection line *n*: strong steel tubing that conducts fuel from the fuel tanks to fuel injectors on the engine.

injection log *n*: a survey used to determine the injection profile, that is, to assign specific volumes or percentages to each of the formations taking fluid in an injection well. The injection log is also used to check for casing or packer leaks, bad cement jobs, and fluid migration between zones.

injection pattern *n*: the spacing and pattern of wells in an improved recovery project, determined from the location of existing wells, reservoir size and shape, the cost of drilling new wells, and the oil recovery expected from various patterns. Common injection patterns include line drive, five spot, seven spot, nine spot, and peripheral.

injection profile *n*: specific volumes or percentages assigned to each of the intervals in a well that is taking fluid.

injection pump *n*: a chemical feed pump that injects chemical reagents into a flow-line system to treat emulsions at a rate proportional to that of the flow of the well fluid. Operating power may come from electric motors or from linkage with the walking beam of a pumping well.

injection timing *n*: the injection of fuel at precisely the right moment in an engine's operating cycle. Proper injection timing is vital, because it gives maximum power from the fuel, good fuel economy, and clean combustion.

injection valve *n*: a poppet spring-loaded subsurface valve run in on wireline, landed in a profile, to shut the well in if injection ceases.

injection water *n*: water that is introduced into a reservoir to help drive hydrocarbons to a producing well.

injection well *n*: a well through which fluids are injected into an underground stratum to increase reservoir pressure and to displace oil. Also called input well.

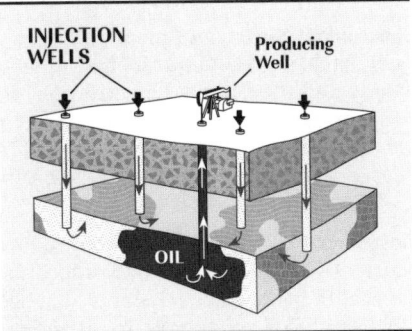

injector *n*: a fuel atomizing device that injects (puts) a fine spray of fuel into the combustion chamber of an engine.

injector head *n*: a control head for injecting coiled tubing into a well that seals off the tubing and makes a pressure-tight connection.

injector pump *n*: 1. on fuel injection systems in engines that do not use a carburetor, a special pump that puts fuel into

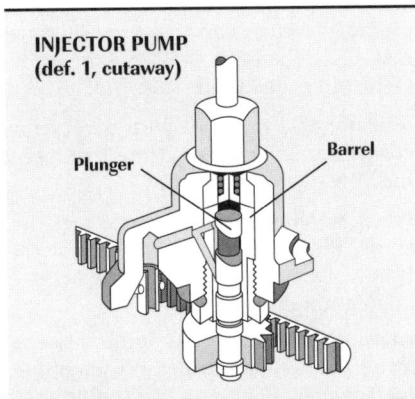

INJECTOR PUMP
(def. 1, cutaway)

Plunger   Barrel

the engine's combustion chamber. 2. a chemical feed pump that injects chemical reagents into a flow-line system to treat emulsions at a rate proportional to that of the flow of the well fluid. Operating power may come from electric motors or from linkage with the walking beam of a pumping well.

in-kind *n*: the taking by an owner of a share of gas or liquids for separate marketing or disposition rather than permitting that share to be disposed of jointly with gas or products belonging to other owners.

inland barge rig *n*: a floating offshore drilling structure consisting of a barge on which the drilling equipment is constructed. When moved from one location to another, the barge floats. When stationed on the drill site, the barge can be anchored in the floating mode or submerged to rest on the bottom. Typically, inland barge rigs are used to drill wells in marshes, shallow inland bays, and areas where the water is not too deep. Also called swamp barge. See *floating offshore drilling rig*.

INLAND BARGE RIG

inlet manifold *n*: the passage that leads from the air filter to the cylinders of an engine. In a diesel engine, air only is introduced on the intake stroke.

in-line proportioner *n*: a venturi tube that draws foam concentrate into a water stream at a predetermined proportion.

in-line system *n*: a type of automatic sampling system where the probe and extractor are located directly on the pipeline and the sample is withdrawn from the liquid stream. Compare *sample loop*.

innage *n*: the height of a liquid in a tank as measured from the bottom (datum plate) of the tank to the liquid surface.

innage bob *n*: a weight attached to the end of a gauge tape. The tape and bob are used to measure liquid height in a tank. The bob keeps the tape vertical so that an accurate reading can be obtained from the tape.

innage gauge *n*: a measure of the liquid in a tank from the bottom of the tank or a fixed datum plate to the surface of the liquid.

innage measurement *n*: a measure of the liquid height in a tank from the bottom of the tank or datum plate to the top of the liquid.

innage tape-and-bob procedure *n*: a method of measuring the height of the liquid in a tank using a pointed gauger's bob and a graduated tape. In this method, the liquid height is measured from the bottom of the liquid to the top. Compare *outage tape-and-bob procedure*.

inner bearings *n pl*: in a bit, the bearings that lie inside of and near the end, or nose, of the cone of a roller cone bit. Typically, the inner bearings include the bearings in the nose of the cone and the bearings near the middle of the cone. These bearings may be a combination of roller, ball, or journal bearings. Compare *outer bearings*.

inner string cementing *n*: a cementing method in which casing is cemented by running drill pipe or tubing to the casing shoe and then pumping the cement down the drill pipe or tubing to the shoe rather than down the casing. The cement exits the shoe and fills the annulus outside the casing as usual.

inorganic compounds *n pl*: chemical compounds that do not contain carbon as the principal element (excepting that in the form of carbonates, cyanides, and cyanates). Such compounds make up matter that is not plant or animal.

in phase *n*: in an AC circuit, voltage and current are in phase when they begin, reach a maximum in the same direction, and return to zero at exactly the same time. Two or more alternating emfs of the same frequency are in phase when their polarities and instantaneous values occur at the same time.

input PLC module *n*: a functional module of a programmable logic controller that accepts discrete or analog signals from a process.

input shaft *n*: the transmission shaft for the drawworks that is driven directly by the compounding transmission on a mechanical-drive rig and is connected to it with the master clutch; or, on an electric-drive rig, the shaft driven directly by the electric motors. The input shaft drives the jackshaft or output shaft.

input signal *n*: a signal applied to a device, element, or system.

input well *n*: an injection well used for injecting fluids into an underground stratum to increase reservoir pressure.

inrush current *n*: the heavy current that develops when an induction motor is started at full voltage. It usually decreases gradually as the speed increases and drops sharply when the motor reaches full speed.

in./sec *abbr*: inches per second.

insert *n*: 1. a cylindrical object, rounded, blunt, or chisel-shaped on one end and usually made of tungsten carbide, that is inserted in the cones of a bit, the cutters of a reamer, or the blades of a stabilizer to form the cutting element of the bit or the reamer or the wear surface of the stabilizer. Also called a compact. 2. a removable part molded to be set into the opening of the master bushing so that various sizes of slips may be accommodated. Also called a bowl. 3. a removable, hard-steel, serrated piece that fits into the jaws of the tongs and firmly grips the body of the drill pipe or drill collars while the tongs are making up or breaking out the pipe. Also called die.

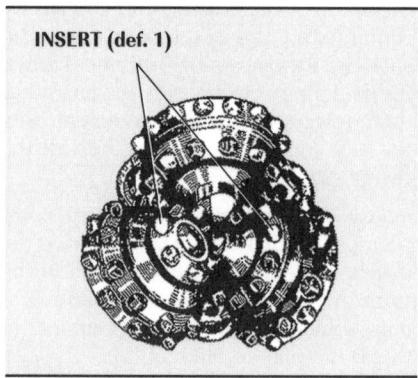

INSERT (def. 1)

insert bit *n*: see *tungsten carbide bit*.

insert bowl *n*: a two-piece steel device with a tapered interior surface that fits into a one-piece or a hinged master bushing. It provides a place in the master bushing for crew members to set the slips. See *master bushing, slips*.

insert moorings *n pl*: moorings where wire and/or chain sections are added during a conventional mooring operation.

insert pump *n*: a sucker rod pump that is run into the well as a complete unit.

inside blowout pre-venter (IBOP) *n*: any one of several types of valve installed in the drill stem or in a

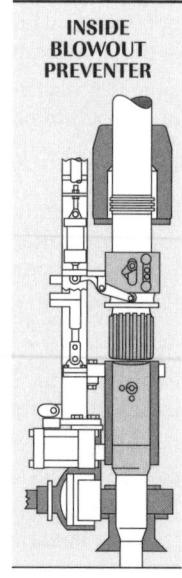

INSIDE BLOWOUT PREVENTER

top drive to prevent high-pressure fluids from flowing up the drill stem and into the atmosphere. Also called an internal blowout preventer.

inside BOP *abbr*: inside blowout preventer.

inside cutter *n*: see *internal cutter*.

inside diameter (ID) *n*: distance across the interior of a circle, especially in the measurement of pipe. See *diameter*.

in situ combustion *n*: a method of improved recovery in which heat is generated within the reservoir by injecting air and burning a portion of the oil in place. The heat of initial combustion cracks the crude hydrocarbons, vaporizes the lighter hydrocarbons, and deposits the heavier hydrocarbons as coke. As the fire moves from the injection well in the direction of producing wells, it burns the deposited coke, releases hot combustion gases, and converts connate water into steam. The vaporized hydrocarbons and the steam move ahead of the combustion zone, condensing into liquids as they cool and moving oil by miscible displacement and hot waterflooding. Combustion gases provide additional gas drive. Heat lowers the viscosity of the oil, causing it to flow more freely. This method is used to recover heavy, viscous oil. Also called fire flooding.

insolation *n*: solar radiation received at the earth's surface.

installation *n*: an offshore production platform.

instantaneous value *n*: the value of current or voltage at any given instant in a cycle of an AC circuit.

Instituto Mexicano del Petróleo (Mexican Petroleum Institute) (IMP) *n*: a decentralized public-interest body created by the Mexican government. Its main objective is to carry out research and technological development as required by the petroleum, petrochemical, and chemical industries; to provide technical services for those industries; and to train personnel involved with the Mexican petroleum industry. Address: Apartado Postal 14.805; México, 14, D.F., México.

instrument *n*: a device for measuring, and sometimes recording and controlling, the value of a quantity under observation.

instrumentation *n*: a device or assembly of devices designed for one or more of the following functions: to measure operating variables (such as pressure, temperature, rate of flow, and speed of rotation); to indicate these phenomena with visible or audible signals; to record them; to control them within a predetermined range; and to stop operations if the control fails. Simple instrumentation might consist of

an indicating pressure gauge only. In a completely automatic system, desired ranges of pressure, temperature, and so on are predetermined and preset.

instrumented pig *n*: an electronic device inserted into a pipeline that senses and records irregularities in the line that could be corroded spots. Also called smart pig.

instrument flight rules (IFR) *n pl*: in aviation, rules promulgated by the United States Federal Aviation Administration (FAA) that allow an aircraft to be flown in weather conditions that do not meet the minimum requirements for visual flight rules (VFR). In such conditions, pilots control the aircraft by watching the aircraft's flight instruments and not by looking out the aircraft's windows. The pilot must have an instrument rating and the aircraft must be equipped and certified for instrument flight. Compare *visual flight rules*.

instrument hanger *n*: a hanger used to lock instruments, such as pressure and temperature bombs, into a seating nipple.

Instrument Society of America (ISA) *n*: a group that sets standards for instruments made and used in the United States. Its official publications are *InTech* and *Programmable Controls*. Address: Box 12277; Research Triangle Park, NC 27709; 919-549-8411; fax 919-549-8288.

instrument transformer *n*: used to reduce current or voltage to a known proportion of that in the circuit. A meter is attached to the reduced voltage or current and the voltage or current in the circuit is computed from the meter reading.

insulated-gate bipolar transistor (IGBT) *n*: in semiconductors, a transistor that combines low forward voltage drop, gate-controlled turnoff, and high switching speed. It is used in high powered converters.

insulating flange *n*: a flange equipped with plastic pieces to separate its metal parts, thus preventing the flow of electric current. Insulating flanges are often used in cathodic protection systems to prevent electrolytic corrosion and are sometimes installed when a flow line is being attached to a wellhead.

insulator *n*: a material with a very low concentration of free electrons that resists the flow of electric current.

intact stability *n*: for offshore rigs and ships, the condition in which a vessel is in a stable state—that is, the vessel is maintaining equilibrium.

intake manifold *n*: on an engine, a special fitting through which air (and sometimes fuel) enters the engine. On diesel engines, only air enters the intake manifold; on gasoline and other spark-ignition engines, both fuel and air usually enter the manifold.

intake port *n*: an opening in a cylinder wall through which gas flows into the cylinder when the intake port is uncovered by the piston.

intake stroke *n*: the downward movement of a piston in a cylinder that creates an area of low pressure inside the cylinder. The low pressure draws in air from the atmosphere (or from a blower).

intake valve *n*: 1. the cam-operated mechanism on an engine through which air and sometimes fuel are admitted to the cylinder. 2. on a mud pump, the valve that opens to allow mud to be drawn into the pump by the pistons moving in the liners.

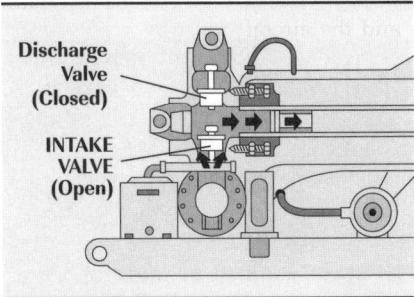

intangible development cost *n*: expense of an item that does not have a salvage value, such as site costs, rig transportation, rig operation, drilling fluid, formation tests, cement, and well supplies.

integer *n*: any positive or negative number including 0.

integral *n*: 1. a solution to a differential equation. 2. the sum of variables in a control. 3. the damping effect place in a control system.

integral control *n*: in an automatic process control system utilizing feedback, a method that improves performance in regulating the process by gaining control at a faster rate. Also termed proportional plus reset.

integral gain *n*: in automatic control, the amount of error multiplied by the amount of time the error has been detected, which determines the severity of the problem. Process system operators use integral gain as a multiplier to give them control over how aggressive he wants this control to correct the problem.

integral hull tank *n*: tank the boundaries of which are the bottom, side, deck, or bulkhead of the barge hull.

integrated circuit *n*: an interconnected arrangement of active and passive electronic elements incorporated (integrated) with a single semiconductor substrate or deposited on the substrate by a continuous series

of compatible processes, and capable of performing at least one complete electronic circuit function.

integrating amplifier *n*: an operational amplifier with a feedback, or shunt, capacitor in which the waveform at the output is the integral (usually over time) of the input.

integrating network *n*: a circuit or network whose output waveform is the time integral of its input waveform. Also known as integrator.

integrating orifice meter *n*: an orifice meter with an automatic integrating device. It is constructed so that the product of the square roots of the differential and static pressures is recorded on the chart. The products are continuously totaled and shown on a counter index. When the product total is multiplied by the orifice flow constant, the rate of flow is determined directly.

integrating wattmeter *n*: see *watthour meter*.

integration department *n*: the department of an oil and gas company that receives orifice meter charts and integrates, or averages, the static and differential pressure readings recorded on the chart. Today, computerized measurement techniques have all but eliminated the need for such a department.

integrator *n*: a computer device that approximates the mathematical process of integration. See *integrating network*.

intensifier *n*: a pressure-multiplier-type well servicing mobile pump.

Inteq Kock-Out™ isolation valve *n*: see *isolation valve*.

interest *n*: pertaining to real estate, a right or a claim to property.

interface *n*: 1. the contact surface between two boundaries of liquids (e.g., the surface between water and the oil floating on it). 2. a means for coupling unlike equipment or functions so that they may communicate and work in unison.

interface unit *n*: a device that buffers, isolates, or changes the character of control signals in an automatic controller.

interfacial tension *n*: the surface tension occurring at the interface between two liquids that do not mix, such as oil and water. Interfacial tension is caused by the difference in fluid pressures of the liquids.

interfit *n*: the distance that the ends of one bit cone extend into the grooves of an adjacent one in a roller cone bit. Also called intermesh.

Intergovernmental Maritime Consultative Organization (IMCO) *n*: an international organization that regulates maritime practices, including possible pollution

of the oceans by tanker-cleaning effluent. Later became International Maritime Organization.

interlocking *n*: automatic process by which an operation cannot be initiated until certain requirements have been met.

INTERMAR *abbr*: Office of International Activities and Marine Minerals (MMS).

intermediate casing string *n*: the string of casing set in a well after the surface casing but before production casing is set to keep the hole from caving and to seal off troublesome formations. In deep wells, one or more intermediate strings may be required. Sometimes called protection casing.

intermediate gears *n pl*: a system of gears that transmits rotary motion.

intermediate string *n*: see *intermediate casing string*.

intermesh *n*: see *interfit*.

intermittent-flow gas lift *n*: see *gas lift*.

intermittent sample *n*: a pipeline sample withdrawn by equal increments at a rate of less than one increment per minute.

intermittent sampler *n*: a sampler that transfers equal increments of oil from a pipeline to the sample container at a uniform rate.

intermitter *n*: a regulation device used in production of a flowing well. The well flows wide open (or through a choke) for short periods several times a day and is then closed in. Also used in some gas-lift installations.

internal blowout preventer *n*: also called inside blowout preventer. See *inside blowout preventer*.

internal-combustion engine *n*: a heat engine in which the pressure necessary to produce motion of the mechanism results from the ignition or burning of a fuel-air mixture within the engine cylinder.

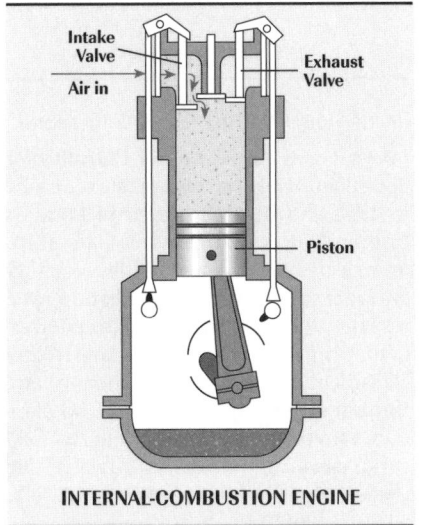

INTERNAL-COMBUSTION ENGINE

internal cutter *n*: a fishing tool containing metal-cutting knives that is lowered into the inside of a length of pipe stuck in the hole to cut the pipe. The severed portion of the pipe can then be returned to the surface. Compare *external cutter*.

internal energy *n*: energy possessed by a body by reason of its molecular forces, independent of its potential or kinetic energy. A match or explosive has internal energy.

internal-external upset end (I-EUE) *n*: on tubing, casing, or drill pipe, the thickening at each end of the joint such that the internal diameter and the external diameter of the joint increase in thickness at each end. Compare *external upset end, internal upset end*.

internal line-up clamp *n*: an alignment clamp used on the inside of pipe. It uses a number of small expandable blocks, or shoes, to grip the inside surfaces of both pipe joints and hold them in place. The clamp can also act as a swab to clean the inside of the pipe. Compare *external line-up clamp*.

internal phase *n*: the fluid droplets or solids that are dispersed throughout another liquid in an emulsion. Compare *continuous phase*.

internal preventer *n*: see *inside blowout preventer*.

internal upset end (IUE) *n*: an extra-thick inside wall on the end of tubing or drill pipe at the point where it is threaded to compensate for the metal removed in threading. Unlike external upset drill pipe, which has the extra thickness on the outside, drill pipe with internal upset has the extra thickness inside and a uniform, straight wall outside. Compare *external upset end, internal-external upset end*.

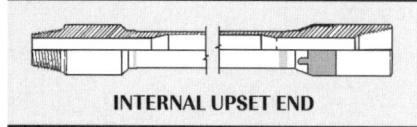

INTERNAL UPSET END

internal-upset pipe *n*: see *internal upset end*.

International Association of Classification Societies (IACS) *n*: an organization formed in 1968 and headquartered in London, England, it is dedicated to safe ships and clean seas. IACS provides technical support, compliance verification, and research and development in the interests of maritime safety and regulation. IACS's classification design, construction, and through-life compliance rules and standards cover more than 90% of the world's cargo carrying tonnage. Address: 5, Old Queen Street, London, SW1H 9JA, UK; www.iacs.org.uk.

International Association of Drilling Contractors (IADC) *n*: an organization of drilling contractors that sponsors or conducts research on education, accident prevention, drilling technology, and other matters of interest to drilling contractors and their employees. Its official publication is *The Drilling Contractor*. Address: 10370 Richmond Ave. Suite 760; Houston, TX 77042; 713-292-1945; fax 713-292-1946.

International Convention for Safety of Life at Sea *n*: see *safety of life at sea rules*.

International Maritime Organization (IMO) *n*: an agency of the United Nations that provides an international forum for nations to discuss international cooperative efforts to improve marine safety and to protect the ocean environment. International treaties on safety (SOLAS) and environmental protection (MARPOL) have been negotiated at IMO. The governing body of the IMO is the Assembly, which meets once every two years and is open to all 136 member states (countries). IMO is headquartered in London. Address: 4 Albert Embankment; London SE 1 75R; 71 7357611.

international standard *n*: a standard recognized by an international agreement to serve as the basis for fixing the value of all other standards of the given quantity.

International System of Units (SI) *n*: a system of units of measurement based on the metric system, adopted and described by the Eleventh General Conference on Weights and Measures. It provides an international standard of measurement to be followed when certain customary units, both U.S. and metric, are eventually phased out of international trade operations. The symbol SI (Le Système International d'Unités) designates the system, which involves seven base units: (1) metre for length, (2) kilogram for mass, (3) second for time, (4) Celsius for temperature, (5) ampere for electric current, (6) candela for luminous intensity, and (7) mole for amount of substance. From these units, others are derived without introducing numerical factors.

interpulse spacing *n*: variations in meter pulse width/space, normally expressed as a percentage.

Interstate Commerce Commission (ICC) *n*: a federal agency that until 1977 had jurisdiction over oil pipelines engaged in interstate commerce. See *Federal Energy Regulatory Commission*.

interstate gas *n*: gas that travels across state lines.

Interstate Natural Gas Association of America (INGAA) *n*: an association of gas transmission companies, producers, and distributors. Address: 10 G Street NE, Suite 700; Washington, DC 20002; 202-216-5900.

Interstate Oil and Gas Compact Commission (IOGCC) *n*: a commission that comprises governors of oil-producing states. It works cooperatively with the EPA to assist states in improving regulatory programs. Address: P. O. Box 53127; Oklahoma City, OK 73152-3127; 405-525-3556.

interstate pipeline *n*: a natural gas pipeline company that is engaged in the transportation, by pipeline, of natural gas across state boundaries. The pipeline is subject to FERC jurisdiction under the Natural Gas Act and under the Natural Gas Policy Act.

interstice *n*: a pore space in a reservoir rock.

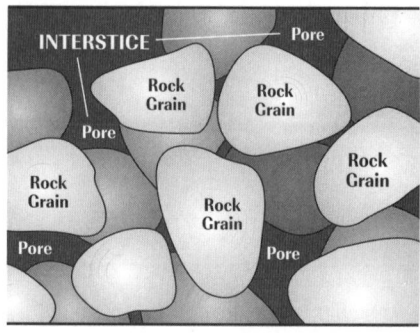

interstitial water *n*: water contained in the interstices, or pores, of reservoir rock. In reservoir engineering, it is synonymous with connate water. Compare *connate water*.

intertropical convergence zone *n*: the low-pressure frontal zone that serves as the dividing line between the northeast and southeast trade winds.

intestate *adj*: without leaving a will. A person may be said to have died intestate.

intrastate gas *n*: gas that does not cross state lines as it is piped from producer to consumer.

intrastate pipeline *n*: a natural gas pipeline company that is engaged in the transportation, by pipeline, of natural gas not subject to FERC jurisdiction under the Natural Gas Act and under the Natural Gas Policy Act.

intrinsic and unique *adj*: a phrase applied to operations that are necessary and uniquely associated with oil and gas exploration and production operations. In order to be intrinsic and unique to E&P operations, they must be associated with efforts to locate oil or gas deposits; to remove oil or natural gas from the ground; or to remove impurities from substances as an integral part of primary field operations.

intrusive rock *n*: an igneous rock that, while molten, penetrated into or between other rocks and solidified.

invaded zone *n*: an area within a permeable rock adjacent to a wellbore into which a filtrate (usually water) from the drilling mud has passed, with consequent partial or total displacement of the fluids originally present in the zone.

invar *n*: a proprietary low-expansion metal alloy consisting of 36 percent nickel, 0.35 percent manganese, and 63.65 percent iron that contains a minimum of carbon and other elements; used for watch parts and the measuring guides of accurate instruments.

invert-emulsion mud *n*: an oil mud in which fresh or salt water is the dispersed phase and diesel, crude, or some other oil is the continuous phase. See *oil mud*.

inverter *n*: an electronic device that converts direct current into alternating current.

inverter circuit *n*: a logic circuit (also called a NOT circuit) with one input and one output that inverts the input signal at the output—that is, the output signal is a logical 1 if the input signal is a logical 0 and vice versa.

inverting input *n*: in an operational amplifier, a type of input wherein if the input signal is changed, the output is also changed. For example, if the inverting input increases in voltage, the output voltage inverts, or decreases. Also called minus input. Compare *noninverting input*.

invert oil-emulsion *n*: a water-in-oil emulsion in which fresh or salt water is the dispersed phase and diesel, crude, or some other oil is the continuous phase. Water increases the viscosity and oil reduces the viscosity.

invert-oil mud *n*: see *invert-emulsion mud*.

iodine number *n*: the number indicating the amount of iodine absorbed by oils, fats, and waxes, giving a measure of the unsaturated linkages present. Generally, the higher the iodine number, the more severe the action of the oil on rubber.

IOGCC *abbr*: Interstate Oil and Gas Compact Commission.

ion *n*: an atom or a group of atoms charged either positively (a cation) or negatively (an anion) as a result of losing or gaining electrons.

ionization *n*: the process by which a neutral atom becomes positively or negatively charged through the loss or gain of electrons.

ionize *v*: to give a neutral atom either a positive or negative charge; after being ionized, the neutral atom becomes an ion. See *ion*.

IPAA *abbr*: Independent Petroleum Association of America.

IPR *abbr*: inflow performance relationship.

IPR curve *n*: see *inflow performance relationship curve*.

IPS *abbr*: improved plough (plow) steel.

IR drop *n*: voltage drop, as determined by the formula $E = IR$.

iron (Fe) *n*: 1. a heavy, magnetic, malleable, and ductile element occurring in the earth's crust in many types of ores. 2. a loose term used in the oilfield for any tool or device generally made of metal, whether it contains iron or not.

iron-cored coil *n*: in a coil of wire that carries electric current, the placement of a length of soft iron in the central interior part of the coil.

iron count *n*: a measure of iron compounds in the product stream, determined by chemical analysis, which reflects the occurrence and the extent of corrosion.

iron monoxide *n*: see *ferrous oxide*.

iron pyrite *n*: see *pyrite*.

Iron Roughneck™ *n*: a manufacturer's name for a floor-mounted combination of a spinning wrench and a torque wrench. The Iron Roughneck moves into position hydraulically and eliminates the manual handling involved with suspended individual tools.

iron sponge process *n*: a method for removing small concentrations of hydrogen sulfide from natural gas by passing the gas over a bed of wood shavings that have been impregnated with a form of iron oxide (iron sponge). The hydrogen sulfide reacts with the iron oxide, forming iron sulfide and water.

iron sulfide *n*: see *ferrous sulfide*.

ISA *abbr*: Instrument Society of America.

isobar *n*: a line drawn on a weather map connecting points of equal atmospheric pressure.

isobutane *n*: 1. a hydrocarbon of the paraffin series with the formula C_4H_{10} and having its carbon atoms branched. 2. in commercial transactions, a product meeting the GPA specification for commercial butane and, in addition, containing a minimum of 95% liquid volume isobutane.

isochore map *n*: a contour map that shows the thickness of a pay section in a reservoir. Used for estimating reservoir content. See *contour map*.

isochronal test *n*: a short-time back-pressure test for low-permeability reservoirs that otherwise require excessively long times for pressure stabilization when wells are shut in.

isochronous governor *n*: a governor that maintains a constant speed of the prime mover regardless of the load applied, within the capacity of the prime mover.

isogonic chart *n*: a map that shows the isogonic lines joining points of magnetic declination, which is the variation between magnetic north and true north. For example, in Los Angeles, California, when the compass needle is pointing toward north, true north actually lies 15° east of magnetic north. See *declination*.

isogonic line *n*: an imaginary line on a map that joins places on the earth's surface at which the variation of a magnetic compass needle from true north is the same. This variation, which may range from 0 to 30 or more degrees either east or west of true north, must be compensated for to obtain an accurate reading of direction.

isokinetic sample *n*: a sample taken from a pipeline in which the linear velocity of the fluid through the opening of the sample probe is equal to the linear velocity in the pipeline and is in the same direction as the bulk of the fluid in the pipeline approaching the probe.

isolate *v*: to pack off above and below a zone of interest.

isolation test plug *n*: a special blocking device (a plug) that is used in pressure testing a blowout preventer. The plug is run and landed in a high-pressure wellhead housing, where it forms a pressure-tight seal. This seal protects (isolates) the casing below the wellhead from the high pressures used to test the blowout preventers, which are located above the wellhead and casing.

isolation valve *n*: a special downhole valve installed in the tubing string that is used to protect the formation from the well's hydrostatic pressure after a gravel pack. The valve has a flapper, which closes and can then be broken out. An isolation valve protects the reservoir from hydrostatic pressure during gravel packing.

isolith map *n*: a map of a formation on which points of similar lithology are connected by a series of contours.

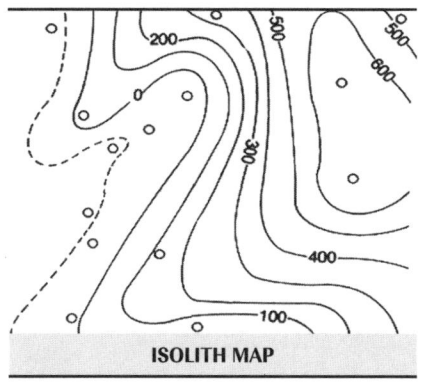

ISOLITH MAP

isomerization *n*: in petroleum refining, the process of altering the fundamental arrangement of the atoms in the molecule without adding or removing anything from the original material. Straight-chain hydrocarbons are converted to branched-chain hydrocarbons with a substantially higher octane rating in the presence of a catalyst at moderate temperatures and pressures. The process is basic to the conversion of normal butene into isobutane.

isometric diagram *n*: a drawing of a three-dimensional object in which lines parallel to the edges are drawn to scale without perspective or foreshortening.

isopach map *n*: a geological map of subsurface strata showing the various thicknesses of a given formation as a series of contours. It is widely used in calculating reserves and in planning improved recovery projects.

isopachous line *n*: a contour line drawn on a map joining points of equal thickness in a stratigraphic unit.

isoparaffin *n*: a material that irritates the skin very little and that is used as an odorless solvent, reaction diluent, and in some proprietary formulations.

isostasy *n*: equilibrium between large segments of the earth's crust, which float on the denser mantle in such a way that thicker segments extend higher and deeper than thinner segments, and lighter blocks rise higher than denser blocks.

ISOPACH MAP

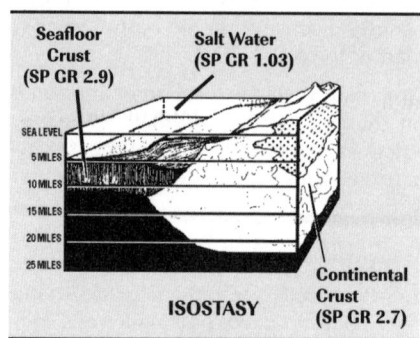

ISOSTASY

isotherm *n*: constant temperature curve.

isothermal compression *n*: the compression of air or gas that exists when the interchange of heat between the air or gas and surrounding bodies (i.e., cylinders or pistons) takes place at a rate exactly sufficient to maintain the air or gas at constant temperature as the pressure increases.

isotope *n*: a form of an element that has the same atomic number as its other forms but has a different atomic mass. Isotopes of an element have the same number of protons but different numbers of neutrons in the nucleus.

I^2R loss *n*: an energy loss in an electric circuit caused by the resistance in ohms of the copper to the flow of current. Equal to the current squared times the resistance in ohms. Also called copper loss.

ITCZ *abbr*: intertropical convergence zone.

ITL *abbr*: Information to Lessees and Operators.

IUE *abbr*: internal upset end.

IWRC *abbr*: independent wire rope core.

J *sym*: 1. joule. 2. productivity index, often used in engineering reports during well testing; also abbreviated PI.

jack *n*: 1. an oilwell pumping unit that is powered by an internal-combustion engine, electric motor, or rod line from a central power source. The walking beam of the pumping jack provides reciprocating motion to the pump rods of the well. See *walking beam*. 2. a device that is manually operated to turn an engine over for starting. *v*: to raise or lift.

jack board *n*: a device used to support the end of a length of pipe while another length is being screwed onto the pipe. Sometimes referred to as a stabbing jack.

jacket *n*: 1. a tubular piece of steel in a tubing liner-type of sucker rod pump, inside of which is placed an accurately bored and honed liner. In this type of sucker rod pump, the pump plunger moves up and down within the liner and the liner is inside the jacket. 2. the support structure of a steel offshore production platform; it is fixed to the seabed by piling, and the superstructure is mounted on it.

jacket water *n*: water that fills, or is circulated through, a housing that partially or wholly surrounds a vessel or machine to remove, add, or distribute heat and thereby to control the temperature within the vessel or machine.

jackhammer *n*: 1. a rock drill that is pneumatically powered and usually held by the operator. 2. an air hammer.

jackknife mast *n*: a structural steel, open-sided tower raised vertically by special lifting tackle attached to the traveling block. See *mast*. Compare *standard derrick*.

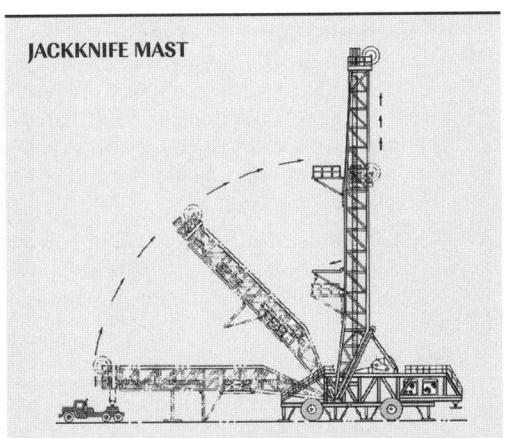

JACKKNIFE MAST

jackknife rig *n*: a drilling rig that has a jackknife mast instead of a standard derrick.

jackshaft *n*: a short shaft that is usually set between two machines to provide increased or decreased flexibility and speed; term applied to an intermediate shaft.

jackup *n*: a jackup drilling rig.

jackup drilling rig *n*: a mobile bottom-supported offshore drilling structure with columnar or open-truss legs that support the deck and hull. When positioned over the drilling site, the bottoms of the legs penetrate the seafloor. A jackup rig is towed or propelled to a location with its legs up. Once the legs are firmly positioned on the bottom, the deck and hull height are adjusted and leveled. Also called self-elevating drilling unit.

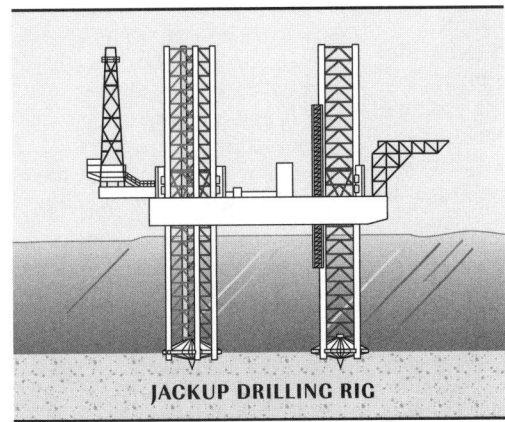

JACKUP DRILLING RIG

jar *n*: a percussion tool operated manually or hydraulically to deliver a heavy downward blow to fish stuck in the borehole. *v*: to apply a heavy blow to the drill stem by use of a jar or bumper sub.

jar accelerator *n*: a hydraulic tool used in conjunction with a jar and made up on the fishing string above the jar to increase the power of the jarring force.

jar intensifier *n*: see *jar accelerator*.

jaw *n*: see *tong jaw*.

jaw clutch *n*: a positive-type clutch in which one or more jaws mesh in the opposing clutch sections.

jaying-up *n*: the act of getting ready to use a J-slot packer or tool.

J-chaser *n*: a chaser used to snag wire or chain mooring sections; usually hung over the stern roller on an anchor-handling vessel.

J curve *n*: the configuration of pipe when it enters the water from an inclined ramp on the stern of a lay barge instead of from a stinger. The J curve eliminates overbend, which can stress the pipe.

JEDEC *abbr*: JEDEC Solid State Technology Association.

JEDEC Solid State Technology Association (JEDEC) *n*: originally known as the Joint Electron Device Engineering Council (JEDEC), this association is the semiconductor engineering standardization body of the Electronic Industries Alliance (EIA), a trade association that represents all areas of the electronics industry. JEDEC was originally created in 1960 as a joint activity between EIA and the National Electrical Manufacturing Association (NEMA), to cover the standardization of discrete semiconductor devices and later expanded in 1970 to include integrated circuits. In spite of the name change, the association is usually abbreviated as JEDEC. Address: 2500 Wilson Blvd., Arlington, VA 22201; 703-907-7534; www.jedec.org.

jeep *n*: see *holiday detector*.

jerk line *n*: a wire rope, one end of which is connected to the end of the breakout tongs and the other end of which is attached to the breakout cathead. When the driller activates the cathead, the cathead pulls on the jerk line with great force to apply torque to break out a tool joint (or to tighten a drill collar connection).

jet *n*: 1. a hydraulic device operated by a centrifugal pump used to clean the mud pits, or tanks, and to mix mud components. 2. in a perforating gun using shaped charges, a highly penetrating, fast-moving stream of exploded particles that forms a hole in the casing, cement, and formation.

jet bit *n*: a drilling bit having replaceable nozzles through which the drilling fluid is directed in a high-velocity stream to the bottom of the hole to improve the efficiency of the bit. See *bit*.

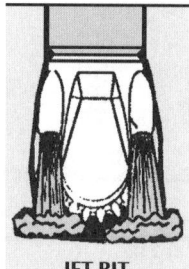

JET BIT

jet bottomhole cutter *n*: a fishing tool that fires a shaped charge downhole to break up junk so that it can be retrieved. It is run into the hole on drill pipe and collars.

jet compressor *n*: a device employing a venturi nozzle through which a high-pressure stream creates a lower pressure or a vacuum into which the gas to be compressed flows. The gas is discharged from the nozzle with the expanded high-pressure medium.

jet cutoff *n*: a procedure for severing pipe stuck in a well by detonating special shaped-charge explosives similar to those used in jet perforating. The explosive is lowered into the pipe to the desired depth and detonated. The force of the explosion makes radiating horizontal cuts around the pipe, and the severed portion of the pipe is retrieved.

jet cutter *n*: a fishing tool that uses shaped charges to sever casing, tubing, or drill pipe stuck in the hole. See *jet cutoff*. Compare *chemical cutter*.

jet deflection bit *n*: a special jet bit that has a very large nozzle used to deflect a hole from the vertical. The large nozzle erodes one side of the hole so that the hole is deflected off vertical. A jet deflection bit is especially effective in soft formations.

jet gun *n*: an assembly, including a carrier and shaped charges, that is used in jet perforating.

jet hopper *n*: a device to hold or feed drilling-mud additives. See *mud-mixing devices*.

jet mixer *n*: a cement mixing system that combines dry cement with a jet of water.

The turbulence from the water thoroughly mixes the cement slurry.

jet out *v*: to use a jet for cleaning out mud tanks, cellar, and other areas.

jet-perforate *v*: to create holes through the casing with a shaped charge of high explosives instead of a gun that fires projectiles. The loaded charges are lowered into the hole to the desired depth. Once detonated, the charges emit short, penetrating jets of high-velocity gases that make holes in the casing and cement for some distance into the formation. Formation fluids then flow into the wellbore through these perforations. See *bullet perforator, gun-perforate*.

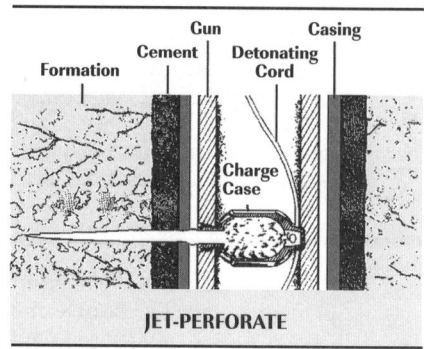

JET-PERFORATE

jet-powered junk basket *n*: see *reverse-circulation junk basket*.

jet pump *n*: a pump that operates by means of a jet of steam, water, or other fluid that imparts motion and subsequent pressure to a fluid medium.

jet siphon *n*: a special pipe installed near the bottom of a mud tank that turns upward from the tank's bottom and is inserted into the end of a larger pipe. When mud is pumped through the small pipe, the pressure lifts mud out of the tank through the larger pipe. This lifting effect is called a siphon effect.

jet sled *n*: in pipeline construction offshore, a pipe-straddling device fitted with nozzles on either side that is lowered by a bury barge. As water is pumped at high pressure through the nozzles, spoil from beneath the pipe is removed and pumped to one side of the trench. The line then sags naturally into position in the trench.

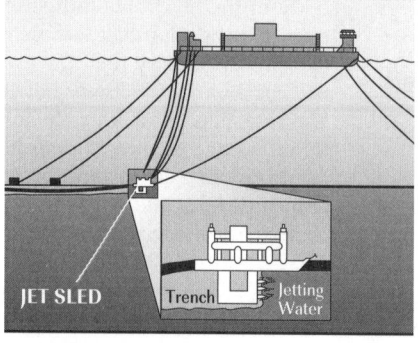

JET SLED   Trench   Jetting Water

jet stream *n*: a relatively narrow band of winds in the upper troposphere, with speeds as high as 300 knots and traveling from west to east.

jet sub *n*: a tool used in the drill string when drilling with air or gas. A short length of pipe—a sub—is made up in the drill string to minimize surging. Surging occurs when air or gas in the liquid drilling fluid comes out of the liquid in large slugs, usually near the surface where pressure is low. A jet bit nozzle is screwed into

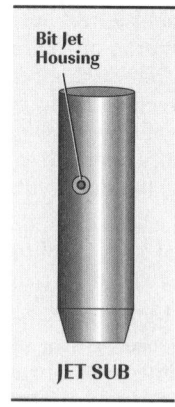

Bit Jet Housing

JET SUB

a housing on the sub. A jet is selected that allows about 20 percent of the drilling fluid and gas to go into the annulus when drilling at equilibrium.

jet the well in *v*: circulating a lower-density fluid to underbalance the well's formation pressure to initiate flow.

jetting *n*: the process of periodically removing a portion or all of the water, mud, and/or solids from the pits, usually by means of pumping through a jet nozzle arrangement.

JFET *abbr*: junction-type field-effect transistor.

jib *n*: the projecting arm of a crane.

J-lay *n*: in laying pipe offshore from a pipe laying vessel, a system in which joints of pipe are welded in a vertical orientation to eliminate the need for a stinger. J-laying allows the pipe to be laid in deeper waters; also, a greater selection of vessels can be used. It is a slower process than the conventional S-lay method because of the pipe's vertical welding position. Compare *S-lay*. See *stinger*.

J-lock chaser *n*: a chaser that can only be pulled in one direction down a section of chain. In the other direction, it will lock.

jobber *n*: a wholesaler who buys gasoline for resale to retailers.

joint *n*: 1. in drilling, a single length (from 16 feet to 45 feet, or 5 metres to 14.5 metres, depending on its range length) of drill pipe, drill collar, casing, or tubing that has threaded connections at both ends. Several joints screwed together constitute a stand of pipe. 2. in pipelining, a single length (usually 40 feet—12 metres) of pipe. 3. in geology, a crack or fissure produced in a rock by internal stresses. 4. in sucker rod pumping, a single length of sucker rod that has threaded connections at both ends.

Joint Electron Device Engineering Council (JEDEC) *n*: a now defunct consortium of manufacturers that devised standards mostly for computer memory modules. Replaced in 1998 by the JEDEC Solid State Technology Association.

joint identifier *n*: a gauge for determining whether the connections of drill collars and tool joints match.

joint movement *n*: the shipment of a tender of oil through the facilities of one or more pipeline companies.

joint of pipe *n*: a length of drill pipe or casing. Both come in various lengths. See *range length*.

joint operating agreement *n*: a contract by which two or more co-owners of the operating rights in a tract of land join to share the costs of exploration and possible development. Compare *farmout*.

joint strength *n*: the amount of hanging weight that can be placed on a connection without failure.

joint tariff *n*: a rate sheet, issued jointly by two or more companies, setting forth charges for moving oil over the facilities of each.

joint tenants *n pl*: two or more persons who are granted lands or tenements to hold in fee simple, fee tail for life, for years, or at will, whose joint title is created by one and the same deed or will. The survivor receives the whole on the death of the other.

joint venture *n*: a business undertaking, usually of more limited scope and length than a partnership, in which control, profits, losses, and liability are all shared.

Jones effect *n*: the net surface tension of salt solutions first decreases with an increase of concentration, passes through a minimum, and then increases as the concentration is raised.

joule (J) *n*: the unit used to measure heat, work, and energy in the metric system. It is the amount of energy required to move an object of 1 kilogram mass to a height of 1 metre. Also called a newton-metre.

Joule's law *n*: a law that states that the number of units of heat that develop in a circuit is proportional to the circuit's resistance, to the square of the strength of the current, and to the time that the current lasts.

Joule-Thomson effect *n*: the change in gas temperature that occurs when the gas is expanded adiabatically from a higher pressure to a lower pressure. The effect for most gases, except hydrogen and helium, is a cooling of the gas.

journal *n*: the part of a rotating shaft that turns in a bearing.

journal angle *n*: the angle formed by lines perpendicular to the axis of the journal and the axis of the bit. Also called pin angle.

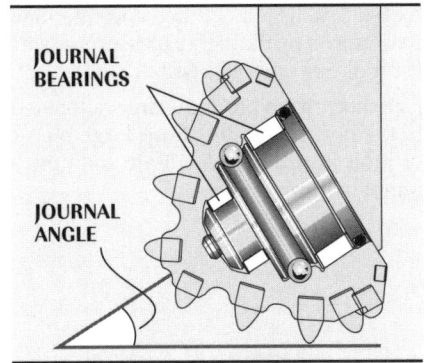

journal bearing *n*: a machine part in which a rotating shaft (a journal) revolves or slides. Also called a plain bearing. Compare *ball bearing*, *roller bearing*.

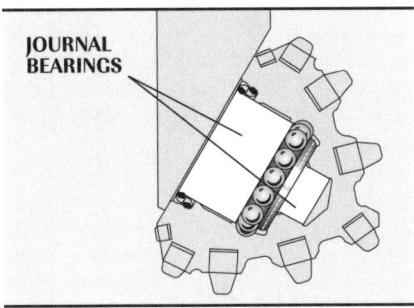

J-slot *n*: a type of mechanism in a packer or a tool in which tubing rotation moves the tool's mandrel through a series of motions, similar to a letter J, to set and release the tool.

J-tool *n*: a sleeve receptacle that has a fitted male element and pins that fit into milled J-shaped slots of the sleeve. The short sides of the J-slots provide a shoulder for supporting weight on the pins of the male element. When the male element is lowered and turned relative to the sleeve, the pins slide in the slot toward the long side of the J, which is open-ended. The pins may thus be raised out, releasing weight that may be supported by the sleeve. The releasing procedure is called "unjaying the tool."

J-tube method *n*: a method for joining a pipeline to a subsea riser on a platform. In the J-tube method, the pipe is lifted off the ocean floor when it reaches the platform and is then fed up to the surface through a guide tube. During this process, the pipe assumes a J configuration. Compare *reverse J-tube method*.

judicial determination *n*: see *judicial transfer*.

judicial transfer *n*: transfer by the court of an interest or of interests in real property. When ownership of land is concerned, a judicial transfer usually involves the appointment of a receiver by the court. The receiver can then act, for example, to execute an oil and gas lease on the property. The citation may arise when the landowner is missing or unknown, after foreclosures, or after tax sales. Also called judicial determination.

jug *n*: see *geophone*.

jug hustler *n*: (slang) the member of a seismograph crew who places the geophones.

jumbo burner *n*: a flare for burning waste gas when the volume of gas is very small or when no market is readily available.

jumbo tank cars *n pl*: tank cars having capacities of 30,000 gallons (114 cubic metres) or more. Standard tank cars have a capacity of 10,000 to 11,000 gallons (38 to 42 cubic metres).

junction *n*: in solid-state electronics, a region of transition between two semiconducting regions in a semiconducting device—for example, the region where the p and n materials come into contact.

junction box *n*: a housing (a box) that provides a place for electrical, hydraulic, or pneumatic connections to be made from one part of the system to another. For example, a junction box on an accumulator provides a point for connecting (interfacing) the remote electrical panels, as well as the accumulator unit's pod select valve, pilot valves, and various system pressures.

junction transistor *n*: a transistor in which emitter and collector barriers are formed between the semiconductor regions of opposite conductivity (the region where the p and n materials come into contact). Also called a junction-type field-effect transistor (JFET).

junction-type field-effect transistor (JFET) *n*: see *junction transistor*.

junior orifice fitting *n*: a one-piece orifice fitting without flanges.

junk *n*: metal debris lost in a hole. Junk may be a lost bit, pieces of a bit, milled pieces of pipe, wrenches, or any relatively small object that impedes drilling or completion and must be fished out of the hole. *v*: to abandon (as a nonproductive well).

junk basket *n*: a device made up on the bottom of the drill stem or on wireline to catch pieces of junk from the bottom of the hole. Circulating the mud or reeling in the wireline forces the junk into a barrel in the tool, where it is caught and held.

When the basket is brought back to the surface, the junk is removed. Also called a junk sub or junk catcher.

junk boot *n*: see *boot sub*.

junk cement *n*: a quantity of cement that is left over after an adequate amount of cement has been pumped into a well during a cement job.

junk mill *n*: a mill used to grind up junk in the hole. See *mill*.

junk pusher *n*: a scraper device run below retainers or packers to clean away debris from casing ID.

junk retriever *n*: a special tool made up on the bottom of the drill stem to pick up junk from the bottom of the hole. Most junk retrievers are designed with ports that allow drilling fluid to exit the tool a short distance off the bottom. This flow of fluid creates an area of low pressure inside the tool so that the junk is lifted and caught in the retriever by the higher pressure outside the tool. See *junk, junk basket*.

junk shot *n*: an explosive charge detonated in the borehole to break up large pieces of junk in order to facilitate the junk's removal from the hole.

junk slot *n*: a groove in the side of a diamond or PDC bit that creates an area of low velocity where relatively small pieces of junk will be lifted and then sent up the hole in the drilling fluid.

junk sub *n*: see *boot sub*.

JU stripper head *n*: an inexpensive and widely used pressure-control head for normal workover operations. It holds well pressure at the surface while allowing tubing to be stripped into or out of the well. It also allows the work string to be rotated if necessary.

k *abbr*: 1. permeability. 2. millidarcys.

K *abbr*: 1. dielectric constant. 2. keel; this abbreviation is mainly used in buoyancy, stability, and trim calculations for offshore rigs. 3. Kelvin.

K *sym*: potassium.

K4 *n*: a type of high-strength chain grade.

Kalrez™ *n*: a trademark for a specially compounded fluoroelastomer for extreme temperature, pressure, and hostile environment service. See *elastomer, fluoroelastomer.*

kaolinite *n*: $Al_2Si_2O_5(OH)_4$, a light-colored clay mineral.

KB *abbr*: kelly bushing; used in drilling reports.

K capture *n*: an interaction in which a nucleus captures an electron from the K shell of atomic electrons (shell nearest the nucleus) and emits a neutrino.

kcmil *abbr*: one thousand circular mils. See *circular mil.*

keel (K) *n*: a centerline strength tube running fore and aft along the bottom of a ship or a floating offshore drilling rig and forming the backbone of the structure.

keeper *n*: 1. an exploration well intended for completion. 2. a device or latch that locks something in place. 3. a soft iron bar placed between the poles of a permanent magnet when it is not being used. A keeper protects the magnet's poles from being demagnetized if the magnet is dropped or struck with a hard object.

keep whole *n*: provision in gas processing agreements that essentially allows the producer to receive at least an amount equal to the proceeds the producer would have been entitled to had he or she sold the gas at the wellhead without processing.

kelly *n*: on drilling rigs that do not use a top drive to rotate the bit, a heavy steel tubular device, four- or six-sided, suspended from the swivel through the rotary table and connected to the top joint of drill pipe to turn the drill stem as the rotary table turns. It has a bored passageway that permits fluid to be circulated into the drill stem and up the annulus, or vice versa. Kellys manufactured to API specifications are available in four- or six-sided versions, are either 40 or 54 feet (12 to 16 metres) long, and have diameters as small as 2.5 inches (6 centimetres) and as large as 6 inches (15 centimetres).

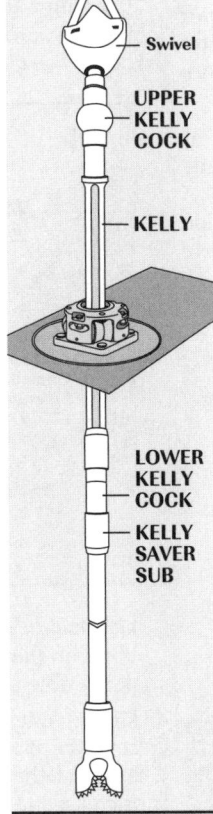

Swivel
UPPER KELLY COCK
KELLY
LOWER KELLY COCK
KELLY SAVER SUB

kelly bushing (KB) *n*: a special device placed around the kelly that mates with the kelly flats and fits into the master bushing of the rotary table. The kelly bushing is designed so that the kelly is free to move up or down through it. The bottom of the bushing may be shaped to fit the opening in the master bushing or it may have pins that fit into the master bushing. In either case, when the kelly bushing is inserted into the master bushing and the master bushing is turned, the kelly bushing also turns. Since the kelly bushing fits onto the kelly, the kelly turns, and since the kelly is made up to the drill stem, the drill stem turns. Also called the drive bushing. See *kelly, master bushing.*

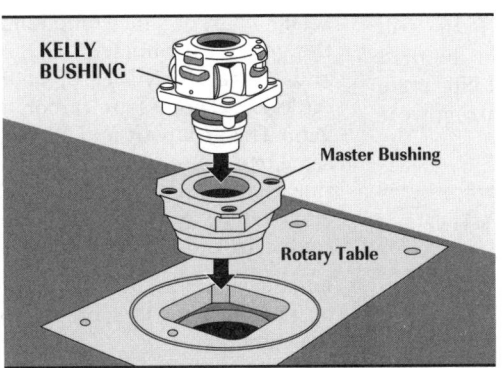

KELLY BUSHING
Master Bushing
Rotary Table

kelly bushing lock assembly *n*: a feature on four-pin kelly bushings installed on floating offshore rigs (which employ a conventional rotary table assembly) that secures the kelly bushing's pins to the master bushing's corresponding drive holes so that the kelly bushing will not separate from (lift off of) the master bushing as the rig heaves up and down with wind and wave motion. See *kelly bushing, master bushing.*

kelly bushing rollers *n pl*: rollers in the kelly bushing that roll against the flat sides of the kelly and allow it to move freely upward or downward. Also called drive rollers.

kelly bypass *n*: a system of valves and piping that allows drilling fluid to be circulated without the use of the kelly.

kelly cock *n*: originally, a term that referred only to the heavy-duty valve made up between the kelly and the swivel, which, when closed, kept back-pressure that was flowing up the drill stem from reaching the swivel and rotary hose. Today, on rigs that use a kelly and rotary table system to rotate the drill stem and bit, two kelly cocks are often employed: one between the kelly and the swivel—the upper kelly cock—and the other between the kelly and the first joint of drill pipe—the lower kelly cock. The lower kelly cock is also called a drill stem safety valve because, when closed and the kelly is removed from the drill stem, it keeps mud from falling out of the kelly. In any case, when a high-pressure backflow occurs inside the drill stem, and the kelly is made up in the drill stem, either valve may be closed to keep pressure off the swivel and rotary hose. See *lower kelly cock, upper kelly cock.*

kelly drive bushing *n*: see *kelly bushing*.

kelly driver *n*: in a rotating head, a device that fits inside the head and inside of which the kelly fits. The kelly driver rotates with the kelly. See *rotating head*.

KELLY DRIVER

kelly flat *n*: one of the flat sides of a kelly. Also called a flat.

kelly hose *n*: see *rotary hose*.

kelly joint *n*: see *kelly*.

kelly saver sub *n*: a heavy and relatively short length of pipe that fits in the drill stem between the kelly and the drill pipe. The threads of the drill pipe mate with those of the sub, minimizing wear on the kelly.

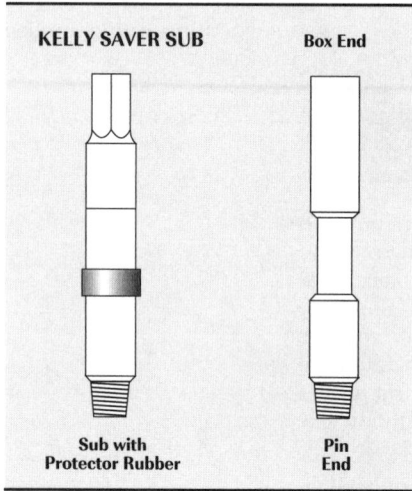

KELLY SAVER SUB   Box End

Sub with   Pin
Protector Rubber   End

kelly spinner *n*: a pneumatically operated device mounted on top of the kelly that, when actuated, causes the kelly to turn or spin. It is useful when the kelly or a joint of pipe attached to it must be spun up, that is, rotated rapidly for being made up.

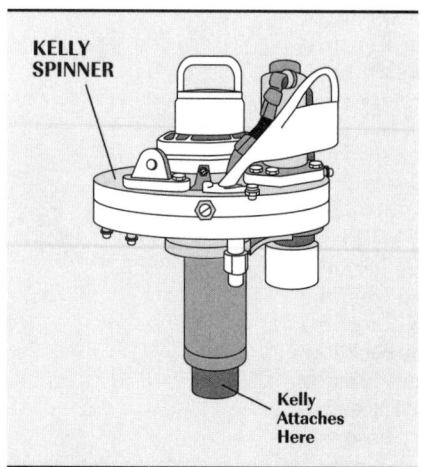

KELLY
SPINNER

Kelly
Attaches
Here

kelly sub *n*: see *kelly saver sub*.

Kelvin (K) *n*: the fundamental unit of thermodynamic temperature in the metric system. See *Kelvin temperature scale*.

Kelvin temperature scale *n*: a temperature scale with the degree interval of the Celsius scale and the zero point at absolute zero. On the Kelvin scale, water freezes at 273.16° and boils at 373.16°. See *absolute temperature scale*.

kenter link *n*: a type of chain connecting link.

kerosene *n*: a light, flammable hydrocarbon fuel or solvent.

ketone *n*: an organic compound having the general formula RR′CO, where R and R′ are alkyl, aryl, or heterocyclic radicals. (C and O are carbon and oxygen.) The groups R and R′ may be the same or different, or incorporated into a ring. Ketones such as acetone and methyl ethyl ketone are used as solvents and to synthesize organic compounds.

Kevlar™ *n*: a type of aramid fiber made by Dupont used in synthetic rope construction.

key *n*: 1. a hook-shaped wrench that fits the square shoulder of a sucker rod and is used when rods are pulled or run into a pumping oilwell. Usually used in pairs; one key backs up and the other breaks out or makes up the rod. Also called a rod wrench. 2. a slender strip of metal that is used to fasten a wheel or a gear onto a shaft. The key fits into slots in the shaft and in the wheel or gear. *v*: to use a cotter key to prevent a nut from coming loose from a bolt or a stud.

keyseat *n*: 1. an undergauge channel or groove cut in the side of the borehole and parallel to the axis of the hole. A keyseat results from the rotation of pipe on a sharp bend in the hole. 2. a groove cut parallel to the axis in a shaft or a pulley bore.

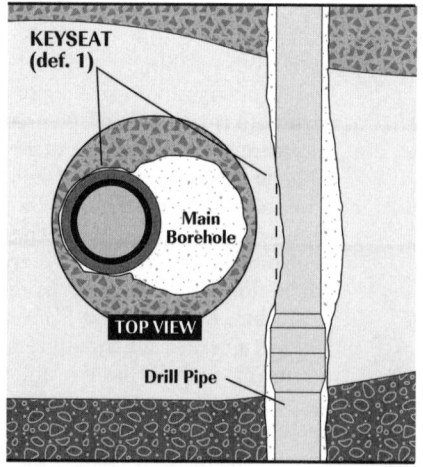

KEYSEAT
(def. 1)

Main
Borehole

TOP VIEW

Drill Pipe

keyseat barge *n*: a barge in which the mast is placed over a channel cut out of the side of the barge and through which drilling or workover operations are performed.

keyseat reamer *n*: see *keyseat wiper*.

keyseat wiper *n*: a device used to ream out a hole where keyseating has occurred. Usually a bumper sub is used first to loosen stuck pipe from a sharp bend, and then a keyseat wiper is used to enlarge the hole at the keyseat caused by the pipe. Also called a keyseat reamer. See *keyseat*.

key valve *n*: see *shutoff valve*.

keyway *n*: a slot in the edge of the barge hull of a jackup drilling unit over which the drilling rig is mounted and through which drilling tools are lowered and removed from the well being drilled. Compare *cantilevered jackup*.

K factor *n*: nominal pulses per unit volume. Used in design of meters.

kg *abbr*: kilogram.

KGS *abbr*: known geologic structures.

kick *n*: an entry of water, gas, oil, or other formation fluid into the wellbore during drilling, workover, or other operations. It occurs because the pressure exerted by the column of drilling or other fluid in the wellbore is not great enough to overcome the pressure exerted by the fluids in a formation exposed to the wellbore. If prompt action is not taken to control the kick, or kill the well, a blowout may occur. See *blowout*.

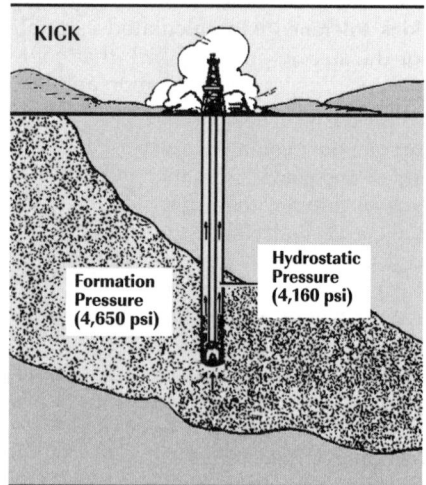

KICK

Formation
Pressure
(4,650 psi)

Hydrostatic
Pressure
(4,160 psi)

kick fluids *n pl*: oil, gas, water, or any combination that enters the borehole from a permeable formation.

kick intensity *n*: a relative measure of the severity of a kick, which is based on the amount that the mud weight must be increased to control the kick.

kick off *v*: 1. to bring a well into production; used most often when gas is injected into a gas lift well to start production. 2. in workover operations, to swab a well to restore it to production. 3. to deviate a wellbore from the vertical, as in directional drilling.

kickoff point (KOP) *n*: the depth in a vertical hole at which a deviated or slant hole is started; used in directional drilling.

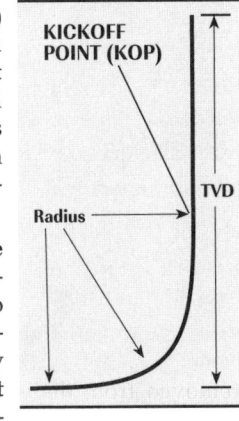

kickoff pressure *n*: the gas pressure required to kick off a gas-lift well, usually greater than that required to maintain the well in production. See *gas lift*.

kickoff tool *n*: see *deflection tool*.

kick out *n*: a phenomenon that occurs in paints or coatings when an incorrect solvent is added to the paint or coating; it is characterized by the separation of the paint's or coating's constituents.

kick-out sub *n*: on a lower marine riser package (LMRP), which is part of the subsea marine riser system, a fitting to which the kill- and choke-line hoses are attached. The sub has an angle of about 45 degrees, which extends (kicks out) the hoses so that they can fit around the LMRP components.

kickstand valve *n*: see *flapper valve*.

kick tolerance *n*: a calculated estimate of the size of potential kick that could fracture an exposed formation and lead to serious well-control problems.

kick volume *n*: the amount of formation fluids that have entered the well after the well has been shut in on the kick; usually measured in barrels, litres, or cubic metres.

kill *v*: 1. in drilling, to control a kick by taking suitable preventive measures (e.g., to shut in the well with the blowout preventers, circulate the kick out, and increase the weight of the drilling mud). 2. in production, to stop a well from producing oil and gas so that reconditioning of the well can proceed. Production is stopped by circulating a kill fluid into the hole.

kill fluid *n*: mud or other fluid in a wellbore whose weight, or density, creates pressure great enough to equal or exceed the pressure exerted by formation fluids.

kill line *n*: a pipe, or conduit, usually attached to openings in the blowout preventer stack below the ram blowout preventers, that allows drilling mud or other fluid to be pumped into the well to control a well when it is not possible to pump down the drill string.

kill mud *n*: see *kill fluid*.

kill rate *n*: the speed, or velocity, of the mud pump used when killing a well. Usually measured in strokes per minute, it is considerably slower than the rate used for normal operations.

kill rate pressure *n*: the pressure exerted by the mud pump (and read on the standpipe or drill pipe pressure gauge) when the pump's speed is reduced to a speed lower than that used during normal drilling. A kill rate pressure or several kill rate pressures are established for use when a kick is being circulated out of the wellbore.

kill sheet *n*: a printed form that contains blank spaces for recording information about killing a well. It is provided to remind personnel of the necessary steps to take to kill a well.

KILL SHEET

kill string *n*: small-diameter tubing that is used inside production tubing for continuous injection of specialized fluids such as corrosion inhibitors or kill fluids. Sometimes used to refer to the drill string through which kill fluids are circulated in drilling.

kilogram *n*: the metric unit of mass equal to 1,000 grams.

kilogram calories *n pl*: the amount of heat required to raise the temperature of 1 kilogram of water 1°C.

kilogram-metre (kg-m) *n*: in the metric system, a unit of measure of twisting force, or torque. The amount of energy required to move 1 kilogram 1 metre vertically.

kilograms per cubic centimetre (kg/cm³) *n*: a measure of the density, or weight, of a fluid (such as drilling mud).

kilohm (kohm, kΩ) *n*: 1,000 ohms.

kiloline *n*: in magnetic circuits, an expression of the flux density of a coil's core in thousands of lines; usually expressed as kilolines per square inch.

kilometre (km) *n*: a metric unit of length equal to 1,000 metres. One kilometre equals .62 miles.

kilopascal (kPa) *n*: 1,000 pascals. The SI metric unit of measurement for pressure and stress and a component in the measurement of viscosity. A pascal is equal to a force of 1 newton acting on an area of 1 square metre.

kilovolt-ampere (kVA) *n*: a rating applied to transformers in which a transformer's power is expressed as 1,000 volt-amperes (VA).

kilowatt (kW) *n*: a metric unit of power equal to approximately 1.34 horsepower; 1,000 watts. See *watt*.

kilowatt-hour (kW-h) *n*: the power of 1 kilowatt applied for 1 hour.

kinematic viscosity *n*: the absolute viscosity of a fluid divided by the density of the fluid at the temperature of the viscosity measurement. Usually expressed in square metres/second.

kinetic energy *n*: energy possessed by a body because of its motion. It is equal to one-half the mass of the body times the square of its speed.

kinetic momentum principle *n*: the principle that recognizes the fact that, when a sample of fluid is vibrated by a weight in contact with the fluid, both the mass of the fluid and the mass of the weight increase.

kinetic pressure *n*: the pressure exerted by a gas under ideal conditions.

kinetic theory *n*: describes the behavior of gas molecules in a confined space.

king wire *n*: in a wire rope, the center wire of a strand; other wires in the strand are laid around this wire. See *independent wire rope core*.

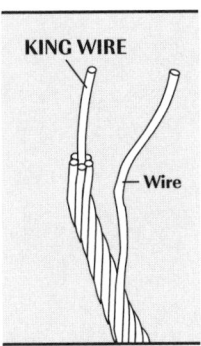

kink *n*: a loop in a wire rope that, having been pulled tight, causes permanent distortion.

kip *n*: a unit of weight or force equal to 1,000 pounds (4,448 newtons).

Kirchoff's second law *n*: the law stating that, at each instant of time, the increase in voltage around a closed loop in a network is equal to the algebraic sum of the voltage drops.

Klinkenberg correction *n*: used to convert air permeability values to equivalent liquid permeability values to obtain an accurate reading of permeability in laboratory analysis of core samples.

Klinkenberg effect *n*: the difference in the flow of a gas and of a liquid through a formation. The difference occurs because gas molecules flow uniformly through small, interconnected pores. Liquid molecules tend to move faster through the center of a pore than along the sides.

kn *abbr*: knot.

knee *n*: on a graph that plots magnetizing force versus flux density for coil cores, the point on the curve of a particular material where further increases in flux density fails to result in significant gains in magnetizing force.

knock *n*: a phenomenon that occurs in an internal combustion engine when the fuel-air mixture in the engine cylinder explodes rather than burning evenly.

knockout *n*: any liquid condensed from a stream by a scrubber following compression and cooling.

knockout drops *n pl*: (slang) chemical used to treat an oilfield emulsion when it is being tested to determine the amount of treating chemical required to break the emulsion.

knockout plug *n*: a plugging device used to effect a dry tubing during run in. It is opened by knocking it out of the tubing; used with retainer and packers.

knockout pot *n*: a vessel that is placed in a pipeline or pipeline sample line and arranged to remove entrained liquids or solids by gravitational means.

knot (kn, kt) *n*: a unit of speed equal to 1 nautical mile (1.852 kilometres or about 1½ statute miles) per hour.

knowledge box *n*: (slang) the cupboard or desk in which the driller keeps the various records pertaining to a drilling operation.

knuckle joint *n*: a deflection tool, placed above the drill bit in the drill stem, with a ball and socket arrangement that allows the tool to be deflected at an angle; used in directional drilling. It is useful in fishing operations because it allows the fishing tool to be deflected to the side of the hole, where a fish may have come to rest.

KO *abbr*: kicked off; used in drilling reports.

kohm *abbr*: kilohm.

Kolite™ *n*: coarsely ground hydrocarbon materials.

KOP *abbr*: kickoff point.

kPa *abbr*: kilopascal.

kΩ *abbr*: kilohm or 1,000 ohms.

K shell *n*: the shell of electrons nearest the nucleus in an atom.

Ksi *abbr*: one thousand pounds per square inch.

kt *abbr*: knot.

kV *abbr*: kilovolt.

kVA *abbr*: kilovolt ampere.

K value *n*: see *vapor-liquid equilibrium ratio*.

kW *abbr*: kilowatt.

kwh, KWH, kw-h *abbr*: kilowatt-hour.

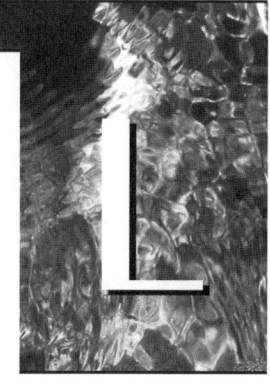

L *abbr*: 1. henry. 2. length. 3. litre.

Labrador Current *n*: cold ocean current that flows south along the Labrador coast through the Davis Strait to the Grand Banks, where it divides and flows into the North Atlantic and the Gulf of St. Lawrence.

LACT *abbr*: lease automatic custody transfer.

LACT unit *n*: an automated system for measuring, sampling, and transferring oil from a lease gathering system into a pipeline. See *lease automatic custody transfer*.

lacustrine delta *n*: a collection of sediment in a lake at the point at which a river or stream enters. When the flowing water enters the lake, the encounter with still water absorbs most or all of the stream's energy, causing its sediment load to be deposited.

ladder logic *n*: a method used to describe the arrangement of electromechanical relays in a control system. Usually, a schematic drawing of such relays consists of vertical lines called rails and horizontal lines called rungs. Consequently, the schematic resembles the rails and rungs of a ladder.

LAER *abbr*: lowest achievable emission rate.

lagging edge *n*: the positive-to-negative transition of an electric signal.

laggings *n pl*: removable and interchangeable drum spool shells for changing hoist drum diameter to provide variation in rope speeds and line pulls.

lag time *n*: the length of time it takes for an adjustment made on a surface choke to appear on the drill pipe (standpipe) pressure gauge when a well is shut in and a choke is being used to maintain back-pressure on the well. A rule of thumb for lag time is 1 second per 1,000 feet (300 metres). For example, in a 5,000-foot well, an adjustment made on the choke does not appear on the drill pipe pressure gauge until 10 seconds after the adjustment was made. The choke adjustment takes 5 seconds to reach the bottom and 5 seconds to reach the surface again. In underbalanced drilling with various fluids besides drilling mud, the rule of thumb does not hold because of the nature of the various gases, liquids, and mixtures of the two in the well. For example, in a 10,000-foot hole that is being drilled underbalanced, lag times can vary from 30 seconds to 60 minutes or more.

laminar flow *n*: a smooth flow of fluid in which no turbulence or cross flow of fluid particles occurs between adjacent stream lines. See *Reynolds number*.

land *n*: 1. the area of a partly machined surface (as with grooves or indentation) that is left smooth. 2. the area on a piston between the grooves into which the rings fit. *v*: to seat tubular goods (such as casing or tubing) into a special device designed to support them from the surface.

land a wellhead *v*: to attach casingheads and other wellhead equipment not already in place at the time of well completion.

land casing *v*: to install casing so that it is supported in the casinghead by slips.

land department *n*: that section or unit of an oil company that seeks out and acquires oil and gas leases.

land disposal *n*: (RCRA) placement in or on the land; placement in a landfill, surface impoundment, waste pile, injection well, land treatment facility, salt dome formation, salt bed formation, underground mine or cave, or in a concrete vault or bunker intended for disposal purposes.

landed cost *n*: the price of imported crude oil at the port of discharge. Includes purchase price at the foreign port plus charges for transporting and insuring the crude oil from the purchase point to the port of discharge; does not include import tariffs or fees, wharfage charges, or demurrage costs.

landform *n*: a recognizable, naturally formed physical land feature having a characteristic shape, such as a plain, alluvial fan, valley, hill, or mountain.

landing depth *n*: the depth to which the lower end of casing extends in the hole when casing is landed.

landing equipment *n*: any downhole tool placed in the tubing string that facilitates the later placement of special tools at that point.

landing nipple *n*: a device machined internally to receive the movable locking devices used to position, lock, and seal subsurface production controls in tubing. A landing nipple provides a seat at a known depth into which various types of retrievable flow control equipment can be set. Also called a seating nipple.

landing shoulder *n*: on conductor casing, a recessed, machined area on the conductor housing or similar fitting on which a wellhead is placed (landed).

landing string *n*: offshore, to land the wellhead on casing beneath the water on the seafloor, it is run in a landing string (drill pipe or tubing) to seafloor or inside casing.

landman *n*: a person in the petroleum industry who negotiates with landowners for oil and gas leases, options, minerals, and royalties and with producers for joint operations relative to production in a field. Also called a leaseman.

land rig *n*: any drilling rig that is located on dry land. Compare *offshore rig*.

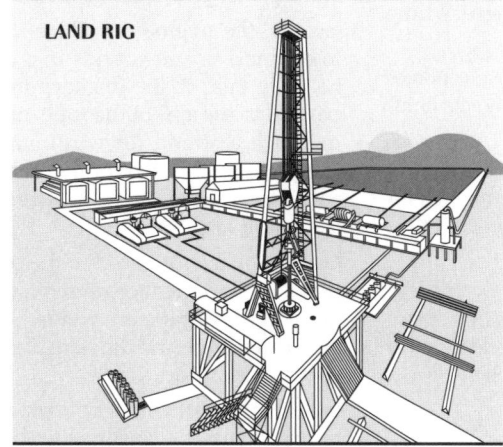

LAND RIG

Landsat *n*: an unstaffed earth-orbiting NASA satellite that transmits multispectral images to earth receiving stations; formerly called ERTS (Earth Resource Technology Satellite).

landspreading *n*: the process of disking, or tilling, low-toxicity wastes into surface soil; land farming.

lang lay *n*: a type of wire rope construction in which the wires that make up the wire rope strands are twisted in the same direction as the strands themselves.

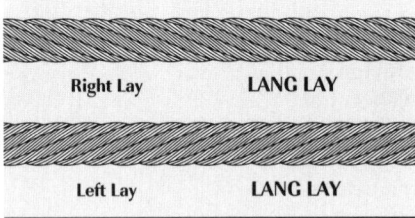

lantern ring *n*: the middle ring in a liner packing in a reciprocating pump. A notch on the lantern ring lines up with the grease fitting so that when lubricant is injected, the ring distributes it between the gland packing rings.

lap *n*: an interval in the cased hole where the top of a liner overlaps the bottom of a string of casing. See *liner lap*.

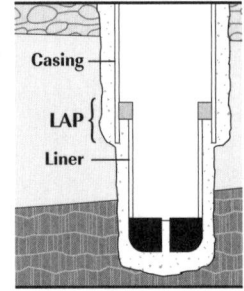

large structure *n*: a relative term that implies the existence of a geologic formation bigger than is usual for a particular area.

last engaged thread *n*: the last pipe thread that is actually screwed into the coupling thread in making up a joint of drill pipe, drill collars, tubing, or casing. If the pipe makes up perfectly, it is also the last thread cut on the pipe.

latch circuit *n*: an electronic circuit that reverses and maintains its state each time power is applied.

latch on *v*: to attach elevators to a section of pipe to pull it out of or run it into the hole.

latch sub *n*: a device, usually with segmented threads, run with seal subs on the bottom of a tubing string and latched into a permanent packer to prevent tubing movement.

lateral curve *n*: in conventional electric logging, the curve drawn on the log by a device on an electric logging tool that is designed to measure formation resistivity about 36 feet (11 metres) from the wellbore.

lateral fault *n*: see *strike slip*.

lateral focus log *n*: a resistivity log taken with a sonde that focuses an electrical current laterally, away from the wellbore, and into the formation being logged. Allows more precise measurement than was possible with earlier sondes. The lateral focus log is a useful means of distinguishing thin rock layers. Also called Laterolog™.

Laterolog™ *n*: trade name for a Schlumberger guard or lateral focus resistivity log, but so commonly used as to be almost a generic term.

latex cement *n*: an oilwell cement composed of latex, cement, a surfactant, and water and characterized by its high-strength bond with other materials and its resistance to contamination by oil or drilling mud.

latitude *n*: an imaginary line joining points on the earth's surface that are of equal distance north or south of the equator.

lattice *n*: the regular geometrical arrangement of points or objects over an area in space—for example, the arrangement of atoms in a crystal. See *crystal*.

latticed boom *n*: boom of open construction with angular or tubular lacing between main corner members in the form of a truss.

Laurasia *n*: the northern part of the supercontinent Pangaea, comprising the future land masses of North America, Greenland, and Eurasia.

lava *n*: magma that reaches the surface of the earth.

law of conservation of energy *n*: the principle that energy cannot be created or destroyed, although it can be changed from one form to another, such as mechanical energy to electrical energy.

law of corresponding states *n*: law that states that when, for two substances, any two ratios of pressure, temperature, or volume to their respective critical properties are equal, the third ratio must equal the other two.

law of partial pressures *n*: see *Dalton's law*.

lay *n*: 1. the manner in which the wires in a strand or the strands in a rope are helically laid. 2. the distance measured parallel to the axis of the rope (or strand) in which a strand (or wire) makes one complete helical convolution about the core (or center). Lay is also referred to as lay length or pitch.

lay barge *n*: a barge used in the construction and placement of underwater pipelines. Joints of pipe are welded together and then lowered off the stern of the barge as it moves ahead.

lay-barge construction *n*: a pipe-laying technique used in swamps and marshes

in which the forward motion of the barge sends the pipe down a ramp and into the water. Also called marine lay.

lay down pipe *v*: to pull drill pipe or tubing from the hole and place it in a horizontal position on a pipe rack. Compare *set back*.

layer *n*: a bed, or stratum, of rock.

laying down *n*: the operation of laying down pipe. See *lay down pipe*.

lb *abbr*: pound.

lb/bbl *abbr*: pounds per barrel.

lb/ft³ *abbr*: pounds per cubic foot.

LCD *abbr*: liquid crystal display.

LC filter *n*: a low-pass filter that smooths the DC output voltage of a rectifier. It consists of one or more sections in series, each section consisting of an inductor on one of the pair of conductors in series with a capacitor between the conductors. Also known as inductive filter, pi (π) filter.

LCM *abbr*: lost circulation material.

LDC *abbr*: local distribution company.

leaching *n*: in geology, the removal of minerals from rock by solution in water or another solvent.

lead (Pb) (pronounced "led") *n*: a chemical element, the symbol of which is Pb, which is a heavy, soft, malleable, and ductile bluish white metal. It is often used in batteries, solder, and radioactivity shields.

lead (pronounced "leed") *n*: in electronics, a wire, usually insulated, which, when placed in an electrical circuit, connects two points in the circuit.

lead acetate test *n*: a method for detecting the presence of hydrogen sulfide in a fluid by discoloration of paper that has been moistened with lead acetate solution.

lead-acid battery *n*: a storage battery in which the electrodes are grids of lead and lead peroxide that change in composition during charging and discharging, and the electrolyte is dilute sulfuric acid.

lead-acid cell *n*: an electrical device, often part of a battery, that consists of lead and lead peroxide plates immersed in a liquid solution (an electrolyte) that is usually sulfuric acid. Such a cell produces voltage and current.

lead impression tool *n*: see *impression tool*.

leading edge *n*: the negative-to-positive transition of an electric signal.

lead line (pronounced "leed") *n*: the pipe through which oil or gas flows from a well to additional equipment on the lease.

lead-tong hand (pronounced "leed") *n*: the crew member who operates the lead tongs when drill pipe and drill collars are being handled. Also called lead-tong man.

lead-tong man *n*: see *lead-tong hand*.

lead tongs (pronounced "leed") *n pl*: the pipe tongs suspended in the derrick or mast and operated by a chain or a wire rope connected to the makeup cathead or the breakout cathead. Personnel call the makeup tongs the lead tongs if pipe is going into the hole; similarly, they call the breakout tongs the lead tongs if pipe is coming out of the hole. Compare *backup tongs*.

leakage current *n*: in an electrical component, such as a capacitor, that employs a dielectric (insulator) to separate conductive elements in the component and prevent the flow of electricity between the conductive elements, the relatively small amount of electricity that flows between the conductors because no dielectric is a perfect insulator. Consequently, some current flows, or leaks, from one conductive element to another.

leakage reactance *n*: in a coil, or winding, in which electrical current is flowing, the current that escapes, or leaks, from it. This leakage, in turn, tends to hold back, or impede, the flow of current through the winding. In this case, the impedance to flow is termed reactance.

leak-off point *n*: in a leak-off test, the pressure at which drilling mud begins to leak off, or enter, the formation from the borehole.

leak-off rate *n*: the rate at which a fracturing fluid leaves the fracture and enters the formation surrounding the fracture. Generally, it is desirable for fracturing fluids to have a low leak-off rate (i.e., very little fluid should enter the formation being fractured), so that the fracture can be better extended into the formation.

leak-off test *n*: a gradual pressurizing of the casing after the blowout preventers have been installed to permit estimation of the formation fracture pressure at the casing seat.

lean amine *n*: amine solution that has been stripped of absorbed acid gases, giving a solution suitable for recirculation to the contactor.

lean gas *n*: 1. residue gas remaining after recovery of natural gas liquids in a gas processing plant. 2. unprocessed gas containing few or no recoverable natural gas liquids. Also called dry gas.

lean glycol *n*: glycol from which water has been removed.

lean oil *n*: a hydrocarbon liquid, usually lighter in weight than kerosene and heavier than paint thinner. In a gas processing plant, lean oil is used in an absorber to remove heavier hydrocarbons from natural gas.

lease *n*: 1. a legal document executed between a landowner, as lessor, and a company or individual, as lessee, that grants the right to exploit the premises for minerals or other products; the instrument that creates a leasehold or working interest in minerals. 2. the area where production wells, stock tanks, separators, LACT units, and other production equipment are located.

lease automatic custody transfer (LACT) *n*: the measurement, sampling, and transfer of oil from the producer's tanks to the connected pipeline on an automatic basis without a representative of either the producer or the gathering company having to be present. See *LACT unit*.

lease bonus *n*: usually, the cash consideration that is paid by the lessee for the execution of an oil and gas lease by a landowner. It is usually based on a per acre basis.

lease broker *n*: an independent landman who may work with several operators or companies.

lease condensate *n*: a natural gas liquid recovered from gas well gas in lease separators or natural gas field facilities. Consists primarily of pentanes and heavier hydrocarbons. Usually blended with crude oil for refining.

leasee *n*: see *lessee*.

lease facility *n*: facility such as a dehydrator, compressor, or separator installed to serve only a single lease.

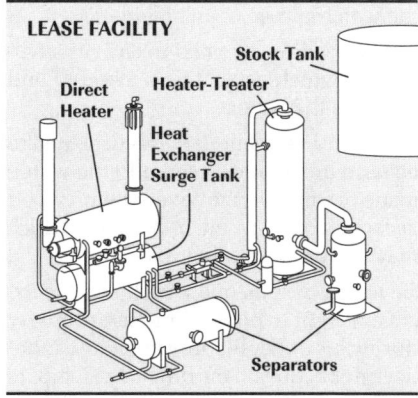

LEASE FACILITY

leasehold *n*: the estate in real property created by a lease. A leasehold is held by a lessee, usually for a fixed period.

leasehold interest *n*: all or a fractional part of the interest of a lessee (grantee) under an oil and gas lease. Such interest includes the lessee's right to search for, drill, and produce oil and gas from a lease tract subject to royalty payments. The term usually refers to the remaining leasehold or working interest exclusive of any nonoperating interests created and assigned therefrom, such as overriding royalty interests and

production payments. Also called operating interest, working interest.

lease hound *n*: (slang) a landman who procures leases on tracts of land for petroleum exploration and production.

leaseman *n*: see *landman*.

lease metering site *n*: the point on a lease where the volume of oil produced from the lease is measured, usually automatically.

lease operator *n*: the oil company employee who attends to producing wells. He or she attends to any number of wells, ensures steady production, prepares reports, tests, gauges, and so forth. Also called a custodian, pumper, or switcher.

lease purchase agreement *n*: an agreement between companies for the purchase by one company of a block of the other's leases. Also used between lease brokers and companies.

lease stipulations *n pl*: special stipulations, or requirements, that are often included in OCS oil and gas leases in response to concerns raised by coastal states, fishing groups, federal agencies, and others. The stipulations may require biological surveys of sensitive seafloor habitats, environmental training for operations personnel, special waste-discharge procedures, archaeological resource reports to determine the potential for the encounter of historic or prehistoric resources, special operating procedures near military bases or their zones of activity, and other restrictions on OCS oil and gas operations. Lease stipulations are legally binding, contractual provisions.

lease superintendent *n*: the oil company employee who supervises one or more lease operators.

lease tank *n*: see *production tank*.

LED *abbr*: light-emitting diode.

ledge *n*: a wellbore irregularity caused by penetration of alternating layers of hard and soft formations, where the soft formation is washed out and causes a change of diametrical size.

lee *n*: the side of a ship, vessel, or floating offshore rig away from the source of the wind. For example, if the wind is blowing from the north, the lee is on the south side of the ship, vessel, or rig. Compare *windward*.

leeward *adj*: the side of a rig that is away from the effects of wind and waves. Also called downwind.

leg *n*: in crane operations, a single length of wire-rope sling, one end of which is attached to the load and the other end of which is attached to the crane's hook. Usually, but not always, slings with at least two legs are used.

legal effect clause *n*: in an oil and gas lease, the clause that binds the parties and declares the lease effective for the lessor when he or she signs the instrument.

length (l) *n*: 1. the distance along something, such as a line, from one end to the other. 2. a measure of the distance along something.

lens *n*: 1. a porous, permeable, irregularly shaped sedimentary deposit surrounded by impervious rock. 2. a lenticular sedimentary bed that pinches out, or comes to an end, in all directions.

lens-type trap *n*: a hydrocarbon reservoir consisting of a porous, permeable, irregularly shaped sedimentary deposit surrounded by impervious rock. See *lens*.

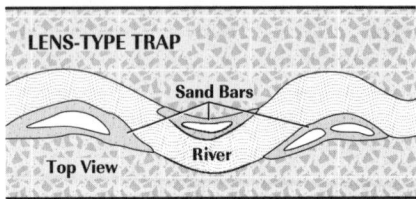

LENS-TYPE TRAP

Sand Bars

Top View   River

lenticular trap *n*: see *lens-type trap*.

Lenz's law *n*: a rule that states that an induced electromotive force (voltage) generates a current that induces a counter magnetic field that opposes the magnetic field generating the current.

leonardite *n*: a naturally occurring oxidized lignite. See *lignins*.

LEPC *abbr*: Local Emergency Planning Committee.

lessee *n*: the recipient of a lease (such as an oil and gas lease). Also called leasee.

lessor *n*: the conveyor of a lease (such as an oil and gas lease).

letter of indemnity *n*: an agreement in which a party receiving gas sales proceeds agrees to refund part of such proceeds in the event FERC orders the party to return part of the gas sales proceeds to the purchaser because a portion of the rate is deemed unjustified.

letter of protest *n*: a letter issued by any participant in a custody transfer citing any condition with which issue is taken. This serves as a written record that a particular action or finding was questioned at the time of occurrence. Also called notice of apparent discrepancy.

Letter to Lessees and Operators (LTL) *n*: an MMS document sent to offshore lessees and operators to clarify or supplement operational guidelines.

levee *n*: 1. an embankment that lies along the sides of a sea channel, a canyon, a river, or a valley. 2. the low ridge sometimes deposited by a stream along its sides.

level *n*: 1. the height or depth at which the top of a column of fluid is located (the level of fluid in a well). 2. horizontally even surface. 3. a device used to determine whether a surface is horizontal.

level control *n*: in a tank or vessel containing a liquid, a method of maintaining the height or depth of the liquid within specified limits. It is accomplished with sensors that monitor the incoming or outgoing liquid and which, in turn, activate a switch to turn the flow either on or off.

level-control assembly *n*: a system of devices whose purpose is to maintain the volume of liquid in a tank or vessel at a predetermined height.

leveling *n*: the act of a coat of paint flowing over the surface to which it is applied in a manner that produces a coat of uniform thickness.

level wind device *n*: a piece of equipment, usually attached to a wire winch, that allows the wraps of wire to be spooled on uniformly.

LGS *abbr*: low gravity solids.

Liberian certification *n*: a license and certification obtained by those who work at sea from the Liberian Maritime Registry System, which adheres to STCW standards. See *Standards of Training, Certification, and Watchkeeping*.

lifeboat *n*: a small boat hoisted on davits or carried on one of the upper decks of a vessel which can be quickly lowered into the water in case of an emergency.

lifeline *n*: a line attached to a diver's helmet by which he or she is lowered and raised in the water.

life raft *n*: a very buoyant raft designed to be used by people forced into the water; made of a metal tube covered with wood or canvas, balsa wood, or of rubber which may be automatically inflated.

life tenant *n*: someone who holds the exclusive right to possess and use property during his or her lifetime but who cannot devise or bequeath the property. Compare *remainderman*.

lifter *n*: a device in an engine against which a cam rotates as the engine runs. As the high point of the cam rotates, it pushes against (lifts) the lifter. The lifter, in turn, actuates a push rod or other device to open a valve or similar device. Sometimes called a cam follower.

lifter-roof tank *n*: a tank whose roof rises and falls with the changes of pressure in the tank but does not float on the product stored in it.

lifting costs *n pl*: the costs of producing oil from a well or a lease.

lifting nipple *n*: a short piece of pipe with a pronounced upset, or shoulder, on the upper end, screwed into drill pipe, drill collars, or casing to provide a positive grip for the elevators. Also called a lifting sub or a hoisting plug.

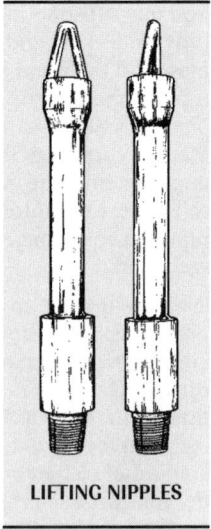

LIFTING NIPPLES

lifting sling *n*: an arrangement of special hooks that fit into receptacles in a master bushing or an insert bowl and chains or wire rope that are connected to a rig's air hoist, all of which enables crew members to insert or remove the master bushing and insert bowls as required.

LIFTING SLING

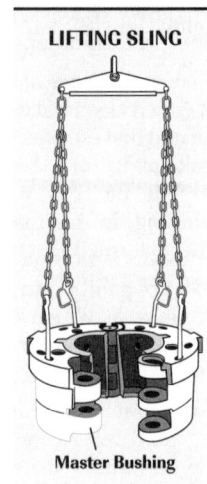

Master Bushing

lifting sub *n*: also called hoisting plug or lifting nipple. See *lifting nipple*.

light crude oil *n*: a crude oil of relatively high API gravity (usually 40° or higher).

light displacement *n*: on mobile offshore drilling rigs, the weight of the rig with all permanently attached equipment but without fuel, supplies, crew, ballast, drill pipe, and so forth.

light-emitting diode (LED) *n*: a semiconductor device that has two electrodes or terminals (a diode) and emits light when voltage is applied to it. LEDs are used in electronic displays such as calculators, wrist watches, and so on.

light ends *n pl*: the lighter hydrocarbon molecules that comprise gasoline, light kerosene, heptane, natural gas, and so forth.

lightening hole *n*: a hole cut into a strengthening member that reduces its weight but does not significantly affect its strength.

lighter *n*: a large, usually flat-bottomed, barge used in unloading or loading ships. *v*: to convey by a lighter.

light fuels *n pl*: fuels, such as gasoline, that have relatively high volatility.

light hydrocarbons *n pl*: the low molecular weight hydrocarbons such as methane, ethane, propane, and butanes.

light hydrogen *n*: see *protium*.

lightning arrester *n*: a device incorporated into an electrical system to prevent damage by heavy surges of high-voltage electricity, such as a stroke of lightning or voltage surges resulting from mishaps in operations.

light products *n pl*: petroleum fractions, such as ethylene, that are relatively light in molecular weight.

lightship displacement *n*: the weight of the steel structure of an offshore rig and its fixed equipment, but without ballast, fuel, consumables, deck loads, bulk material, stores, drilling tubulars, loose equipment, and the like.

lightweight additives *n pl*: reduce the weight of the slurry so the cement can flow past low-pressure zones or soft formations without losing part of the slurry or damaging the wellbore.

lightweight cement *n*: a cement or cement system that handles stable slurries having a density less than that of neat cement. Lightweight cements are used in low-pressure zones where the hydrostatic pressure of long columns of neat cement can fracture the formation and result in lost circulation.

lightweight gear *n*: all diving equipment less complex than the standard dress. This equipment employs face masks or helmets, protective clothing, and swim fins or boots.

lightweight ton *n*: a measure of a ship or offshore floating rig's actual displacement (in long tons) when it is not carrying cargo or expendable supplies and equipment. A rig's lightweight is made up of the hull and upper deck as well as fixed equipment, meaning equipment that is part of the structure and is not normally removed, such as a derrick or crane.

lignins *n pl*: naturally occurring special lignites, e.g., leonardite, that are produced by strip mining from special lignite deposits. Used primarily as thinners and emulsifiers.

lignosulfonate *n*: an organic drilling fluid additive derived from by-products of a papermaking process using sulfite. It minimizes fluid loss and reduces mud viscosity.

limber hole *n*: a hole cut in a structural member of a ship or offshore drilling rig, usually in a tank, to allow water to pass through freely.

lime *n*: a caustic solid that consists primarily of calcium oxide (CaO). Many forms of CaO are called lime, including the various chemical and physical forms of quicklime, hydrated lime, and even calcium carbonate. Limestone is sometimes called lime.

lime hydrate *n*: an alkaline chemical also known as lime, slaked lime, or calcium hydroxide, $Ca(OH)_2$.

lime mud *n*: 1. a calcite-rich sediment that may give rise to shaly limestone. 2. a drilling mud that is treated with lime to provide a source of soluble calcium in the filtrate to obtain desirable mud properties for drilling in shale or clay formations.

limestone *n*: a sedimentary rock rich in calcium carbonate that sometimes serves as a reservoir rock for petroleum.

limited-entry technique *n*: a fracturing method in which fracturing fluid is injected into the formation through a limited number of perforations (i.e., fluid is not injected through all the perforations at once; rather, injection is confined to a few selected perforations). This special technique can be useful when long, thick, or multiple producing zones are to be fractured.

limited exposure *n*: a generic term to describe certain types of packers where the packing element is positioned in such a fashion as to limit wellbore media exposed to the tool's setting or releasing mechanisms.

limiter *n*: see *clipper*.

limiting *n*: see *clipper*.

line *n*: 1. any length of pipe through which liquid or gas flows. 2. rope or wire rope. 3. electrical wire.

lineament *n*: a linear topographic or tonal feature on the terrain and on images and maps of the terrain that is thought to indicate a zone of subsurface structural weakness.

linear caliper *n*: see *caliper*.

linear chart *n*: a direct-reading chart used on the flow recorder of an orifice metering system on which are recorded static and differential pressures during a particular time period. It must have the same static and differential ranges as the flow recorder.

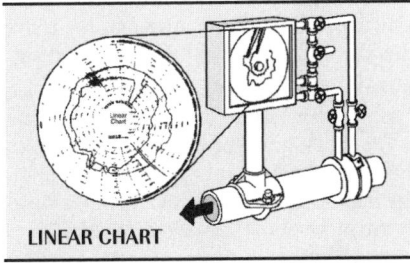

LINEAR CHART

linearity *n*: in electronics, the relationship between two quantities that occurs when a change in one of the quantities produces a directly proportional change in the other quantity.

linearity of a meter *n*: an expression of a meter's deviation from accuracy.

linear meter range *n*: the flow range over which the meter factor does not deviate from specified limits.

linear polarization *n*: a technique used to measure instantaneous corrosion rates by changing the electrical potential of a structure that is corroding in a conductive fluid and measuring the current required for that change.

linear taper *n*: in a potentiometer with multiple windings, the uniform winding of the resistance wire that makes up the potentiometer's control. A linear taper provides the potentiometer with directly proportional control—that is, turning the potentiometer's control directly varies the potentiometer's resistance. For example, turning the control one-fourth turn decreases or increases the resistance by one fourth. Compare *logarithmic taper*.

line circulation *n*: petroleum or other liquid delivered through a pipeline system into a receiving vessel or tank to ensure that the section of pipeline from the source tank to the receiving tank is full in order to minimize the amount of air in the pipeline.

line contractor *n*: control device that makes and breaks the power circuit to the motor.

line displacement *n*: an operation to replace previous material in a pipeline.

line drive *n*: in waterflooding, a straight-line pattern of injection wells designed to advance water to the producing wells in the form of a nearly linear frontal movement. See *waterflooding*.

line drop *n*: opening (venting to atmosphere) in a vessel's piping system so as to allow, to the extent possible, drainage into a tank where the material may be gauged and accounted for.

line loss *n*: 1. the reduction in the quantity of natural gas flowing through a pipeline that results from leaks, venting, and other physical and operational circumstances. 2. the reduction in (or the loss of) electrical energy that occurs when electricity flows through a conductor (a line).

line pack *n*: see *line press*.

line parts *n pl*: see *parts of line*.

line pipe *n*: a steel or plastic pipe used in pipelines, gathering systems, flow lines, and so forth.

line press *n*: the recorded difference in a tank's gauges taken both while the tank's valves are closed (off-line), and while the tank's valves are opened (on-line) into a closed system. All downstream valves of the line section to be pressed are opened while the terminating valves remain closed. Also called line pack.

line pressure *n*: the force per unit area exerted on a pipe wall by a fluid inside the pipe.

liner *n*: 1. a string of pipe used to case open hole below existing casing. A liner extends from the setting depth up into another string of casing, usually overlapping about 100 feet (30.5 metres) above the lower end of the intermediate or the oil string. Liners are nearly always suspended from the upper string by a hanger device. 2. a relatively short

LINER (def. 2)

length of pipe with holes or slots that is placed opposite a producing formation. Usually, such liners are wrapped with specially shaped wire that is designed to prevent the entry of loose sand into the well as it is produced. They are also often used with a gravel pack. 3. in jet perforation guns, a conically shaped metallic piece that is part of a shaped charge. It increases the efficiency of the charge by increasing the penetrating ability of the jet. 4. a replaceable tube that fits inside the cylinder of an engine or a pump. See *cylinder liner.*

line rate *n*: the rate of flow of a fluid through a line.

liner barrel *n*: a pump barrel used for either tubing pumps or rod (insert) pumps. A full-cylinder barrel consists of a steel jacket inside of which is a full-length tube of cast iron or special alloy. The inner surface of the barrel is polished to a mirrorlike finish to permit a fluid-tight seal between it and the plunger. In a sectional liner barrel, the tube placed inside the steel jacket consists of a series of sections placed end to end and held firmly in place by means of threaded collars on the ends of the steel jacket.

liner cementing *n*: the process of cementing a liner string in the hole. See *cementing, liner.*

liner completion *n*: a well completion in which a liner is used to obtain communication between the reservoir and the wellbore.

liner hanger *n*: a slip device that attaches the liner to the casing. See *liner.*

liner lap *n*: the distance that a liner extends into the bottom of a string of casing. See *lap.*

liner patch *n*: a stressed-steel corrugated tube that is lowered into existing casing in a well to repair a hole or leak in the casing. The patch is cemented to the casing with glass fiber and epoxy resin.

liner shoe *n*: a casing shoe attached to liner rather than to casing.

liner top isolation device *n*: a completion tool installed at the top of a production liner that isolates the reservoir from the tubing and can hold pressure both from the tubing or from the reservoir.

line scraper *n*: see *pig.*

lines of force *n pl*: see *magnetic lines of force.*

line speed indicator *n*: an instrument calibrated to show the speed in feet per minute at which wireline is raised or lowered in the hole. The line speed indicator may be driven mechanically by the measuring wheel and placed next to the depth odometer, or it may be driven by a motor.

line spooler *n*: a device fitted on the drawworks and used to cause the fastline to reverse its direction on the drawworks drum or spool when a layer of line is completed on the drum and the next layer is started.

line travel applied coating *n*: in pipeline construction, the coating applied to pipe over the ditch. Coal tar enamels are particularly effective.

line-up clamps *n pl*: used to align the ends of pipe prior to welding.

line-up station *n*: where the end of a joint of pipe is aligned with the joint ahead of it.

link-block bolt *n*: a bolt that secures elevators in the bail. Manufactured with the nut end toward the derrickhand so that workers can monitor the nut pins when latching and unlatching elevators during a trip.

link ear *n*: a steel projection on the drilling hook by means of which the elevator links are attached to the hook.

link locking arm *n*: on a drilling rig's hook, a device that firmly secures and locks the elevator link into the link ear's opening.

link-tilt arms *n pl*: on a top drive, a device that, when actuated by the driller, tilts the unit's built-in elevators into a position to make it easy for crew members to latch the elevators onto a joint of pipe stored in the mousehole. See *top drive.*

lipophile *n*: a substance usually colloidal and easily wet by oil.

lipophilic *adj*: having an affinity for lipids, a class of compounds that includes most hydrocarbons.

liquefaction *n*: the process whereby a substance in its gaseous or solid state is liquefied.

liquefiable hydrocarbons *n pl*: hydrocarbon components of natural gas that can be extracted and saved in liquid form.

liquefied natural gas (LNG) *n*: a liquid composed chiefly of natural gas (i.e., mostly methane). Natural gas is liquefied to make it easy to transport if a pipeline is not feasible (as across a body of water). Not as easily liquefied as LPG, LNG must be put under low temperature and high pressure or under extremely low (cryogenic) temperature and close to atmospheric pressure to liquefy.

liquefied natural gas carrier (LNGC) *n*: a vessel such as a ship used to transport liquefied natural gas.

liquefied petroleum gas (LPG) *n*: a mixture of heavier, gaseous, paraffinic hydrocarbons, principally butane and propane. These gases, easily liquefied at moderate pressure, may be transported as liquids and converted to gases on release of the pressure. Thus, liquefied petroleum gas is a portable source of thermal energy that finds wide application in areas where it is impractical to distribute natural gas. It is also used as a fuel for internal-combustion engines and has many industrial and domestic uses. Principal sources are natural and refinery gas, from which the liquefied petroleum gases are separated by fractionation.

liquefied refinery gas (LRG) *n*: liquid propane or butane produced by a crude oil refinery. It may differ from LP gas in that propylene and butylene may be present.

liquid *n*: a state of matter in which the shape of the given mass depends on the containing vessel, but the volume of the mass is independent of the vessel. A liquid is a fluid that is almost incompressible.

liquid and solid ROB (remaining on board) *n*: the measurable material remaining on board a vessel after a discharge. Includes measurable sludge, sediment, oil, and water or oil residue lying on the bottom of the vessel's cargo compartments and in associated lines and pumps.

liquidate *v*: where participants resell their positions.

liquid calibration procedure *n*: a method of determining a tank's capacity by filling and withdrawing from the tank accurately determined volumes of liquid.

liquid crystal display (LCD) *n*: an electronic digital display that consists of two sheets of glass separated by a sealed-in, normally transparent, liquid crystal material. The outer surface of each glass sheet has a transparent conductive coating etched into character-forming segments that have leads going to the edges of the display. Voltage applied between front and back electrode coatings disrupts the orderly arrangement of molecules, which darkens the liquid enough to form visible characters although no light is generated.

liquid cut *n*: the mark on the gauger's tape indicating the height of the liquid.

liquid desiccant *n*: a hygroscopic liquid, such as glycol, used to remove water from other fluids.

liquid displacement *n*: the crude oil or product that is forced into the hatch or up around the edges of a floating roof by the weight of the roof.

liquid drilling *n*: in underbalanced drilling, the use of a liquid drilling fluid (mud) to drill the well. See *underbalanced drilling*.

liquid head stress *n*: the hydrostatic pressure of the liquid in a stock or holding tank, which causes the shell to expand and contract as liquid is added or removed from the tank.

liquid hydrocarbons *n pl*: liquids that have been extracted from natural gas, such as propanes, butanes, pentanes, and heavier products. Liquid hydrocarbons extracted in gas processing plants are often referred to as "plant products."

liquid-in-glass thermometer *n*: a type of thermometer in which the sensitive part of the instrument consists of a liquid like mercury or alcohol contained in an envelope of glass.

liquid level *n*: the height, or depth, of a liquid in a tank or vessel; liquid level is controlled by level controllers, measured with liquid-level gauges, and viewed with liquid-level indicators.

liquid-level controller *n*: any device used to control the liquid level in a tank by actuating electric or pneumatic switches that open and close the discharge valve or the intake valve, thus maintaining the liquid at the desired level.

liquid-level gauge *n*: any device that indicates the level or quantity of liquid in a container.

liquid-level indicator *n*: a device connected to a vessel and coupled with either a float in the vessel or directly with the fluid therein. It is calibrated to give a visual indication of the liquid level.

liquid nitrogen *n*: nitrogen that occurs as a liquid at atmospheric pressure but at a temperature of –383°F (–195°C). Under normal conditions of pressure and temperature, nitrogen occurs as a gaseous fluid. Liquid nitrogen is sometimes transported to the rig and used as a drilling fluid in underbalanced drilling. See *underbalanced drilling*.

liquid phase *n*: in drilling fluids, that part of the fluid that is liquid. Normally, the liquid phase of a drilling fluid is water, oil, or a combination of water and oil.

lis pendens *n*: notice that a suit has been filed in a court of law and that the property owned by the defendant may be liable to judgment.

list *n*: the position of a ship or offshore drilling rig that heels to one side because of a shift in cargo, machinery, or supplies. Also called heel.

list correction *n*: the correction applied to the volume or gauge observed in a vessel's tank when the vessel is listing, provided that liquid is in contact with all bulkheads in the tank. List correction may be accomplished by referring to the list correction tables for each of the vessel's tanks or by mathematical calculation.

List of Endangered and Threatened Wildlife and Plants *n*: a list published periodically in the *Federal Register* that names endangered and threatened wildlife and plants.

liter (L) *n*: the unit of capacity in the metric system. It is equal to about 0.9464 quarts. Also spelled litre in the SI system.

lithification *n*: the conversion of unconsolidated deposits into solid rock.

lithofacies map *n*: a facies map showing lithologic variations within a formation. It shows the variations of selected lithologic characteristics within a stratigraphic unit.

lithology *n*: 1. the study of rocks, usually macroscopic. 2. the individual character of a rock in terms of mineral composition, structure, and so forth.

lithostatic pressure *n*: see *geostatic pressure*.

litmus *n*: a water soluble blue powder obtained from various lichens, but especially *Variolaria lecanora* and *V. rocella*. Litmus turns red in acid solutions and blue in base (alkaline) solutions and is therefore often used as an acid-base indicator.

litre (L) *n*: an acceptable unit of measure in the SI system of capacity equal to the volume occupied by 1 kilogram of water at 4°C and at the standard atmospheric pressure of 760 millimetres. It is spelled litre in the SI system.

Little Big Inch *n*: a 20-inch (50.8-centimetre) products line constructed during the same period as Big Inch as part of the World War II effort. See *Big Inch*.

liveboating *n*: a diving operation involving the use of a boat or vessel that is under way.

live end *n*: the end of a flexible steel brake band that is attached to a brake lever by means of a linkage. Moving the brake lever pulls the live end down, and the whole band tightens around the flange. This slows or stops the drum by friction.

live oil *n*: crude oil that contains gas and has not been stabilized or weathered. It can cause gas cutting when added to mud and is a potential fire hazard.

live roller circle *n*: an assembly of multiple swing rollers free to roll between the revolving superstructure and the mounting.

Lizzy *n*: an on-site patented fire-suppressor system that uses monoammonium phosphate to put out the fire on a burning well.

Lloyd's Registry of Shipping *n*: an independent risk management organization that provides risk assessment and risk mitigation solutions and management systems certification around the world. Their goal is to improve their clients' quality, safety, environmental, and business performance. Founded in 1760. Address: 71 Fenchurch Street, London EC3M 4BS, UK; www.lr.org. .

lm *abbr*: lime; used in drilling reports.

LMRP *abbr*: lower marine-riser package.

LMRP connector *n*: used to remotely connect the BOP assembly to the wellhead, and the LMRP to the BOP.

LNG *abbr*: liquefied natural gas.

LNGC *abbr*: liquefied natural gas carrier.

load *n*: 1. in mechanics, the weight or pressure placed on an object. The load on a bit refers to the amount of weight of the drill collars allowed to rest on the bit. See *weight on bit*. 2. in reference to engines, the amount of work that an engine is doing; for example, 50 percent load means that the engine is putting out 50 percent of the power that it is able to produce. 3. the amount of gas delivered or required at any specified point or points on a system; load originates primarily at the gas-consuming equipment of the customer. 4. in electronics, a device that consumes electric power. *v*: 1. to engage an engine so that it works. Compare *idle*. 2. to set a governor to maintain a given pressure as the rate of gas flow through the governor varies. Compare *demand*.

load binder *n*: a chain or cable with a latching device, used to secure loads (usually of pipe) on trucks. Also called a boomer.

load block *n*: see *hook block*.

load capacity *n*: the amount of weight a device can safely carry or support.

loaded on top (LOT) procedure *n*: see *LOT (loaded on top) procedure*.

loader *n*: the individual who handles the filling of tank cars, ships, barges, or transport trucks.

load guy *n*: see *guy line*.

load guy line *n*: the wireline attached to a mast or derrick to provide the main support for the structure. Compare *wind guy line*.

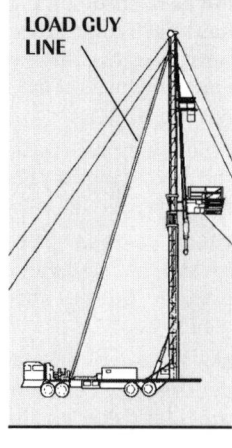

LOAD GUY LINE

loading *n*: occurs when the amount of water or condensate production is too great to be lifted by the velocity of the gas.

loading line *n*: see *suction line*.

loading rack *n*: the equipment used for transferring crude oil or petroleum products into tank cars or trucks.

Load King connector *n*: a brand of bolted-flange connector used in marine riser systems to join (connect) riser pipe. This connector is designed for drilling in water depths of 7,000 feet (2,000 metres) or deeper, and has a tensile load rating of 3.5 million pounds (1.6 million kilograms). See *riser pipe*.

load line *n*: a line, painted or cut on the outside of a floating rig or ship's hull, which marks the maximum waterline when the rig or ship is loaded with the greatest amount of cargo that it can safely carry.

load loss *n*: 1. loss from a cargo of oil. Usually the losses occur because the light ends escape during transport. 2. in a transformer, the reduction (loss) of electrical energy that occurs when the transformer is loaded—that is, when electricity is being drawn from it.

load oil *n*: the crude or refined oil used in fracturing a formation to stimulate a well, as distinguished from the oil normally produced by the well.

load on top *n*: the shipboard procedure of collecting and settling water and oil mixtures. It results from ballasting and tank-cleaning operations (usually in a special slop tank or tanks) and the subsequent loading of cargo on top and the pumping of the mixture ashore at the discharge port. *v*: to commingle on-board quantity with cargo being loaded.

load vessel ratio (LVR) *n*: the total calculated volume (TCV) by vessel measurement on sailing, less onboard quantity (OBQ), divided by the TCV by shore measurement at loading:

$$LVR = TCV \text{ sailing volume} + OBQ/TCV$$
received from shore at loading.

load voltage *n*: see *voltage drop*.

lobed impeller *n*: a rounded blade on a displacement meter that separates the liquid stream into discrete quantities.

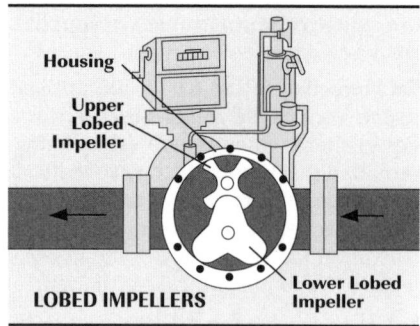

Housing
Upper Lobed Impeller
Lower Lobed Impeller

LOBED IMPELLERS

LOC *abbr*: location; used in drilling reports.

local distribution company (LDC) *n*: a company that sells natural gas to homes, offices, and businesses in towns, cities, or relatively small rural areas.

Local Emergency Planning Committee (LEPC) *n*: a local community committee required under SARA. Committee members are approved by the SERC and include elected state and local officials; police, fire, and public health officials; environmental advocates; hospital and transportation officials; industry representatives; community groups; and the media. The function of the LEPC is to develop an emergency response plan that will be implemented if a hazardous material is released in the community.

lo-cap probe *n*: in electronics, a probe that attenuates (reduces) the strength of a signal being measured by the probe. See *attenuation probe*.

location *n*: the place where a well is drilled. Also called *well site*.

location damages *n pl*: compensation paid to the surface owner for actual and potential damage to the surface and crops in the drilling and operation of a well.

locator sub *n*: a device, larger than the bore of a permanent packer, which is run with seal subs on the bottom of a tubing string and used to locate the top of a permanent packer.

locator tubing seal assembly *n*: a device made up in the tubing string and run inside a packer to prevent fluid and pressure from escaping between the tubing and the packer.

lock assembly *n*: see *kelly bushing lock assembly*.

locking device *n*: on the telescopic joint of a floating offshore rig, a mechanism that secures (locks) the joint's inner barrel to its outer barrel in a fully retracted position. Locking the joint fully retracted allows for easier handling when it is being

shipped, when it is being picked up on the rig, and when the blowout preventer stack is being run or retrieved.

locking dog *n*: on a lower marine riser package's connector, a device with a tapered locking profile on its inside diameter that mates with a tapered locking profile on the outside diameter of the wellhead.

locking mandrels *n pl*: slickline tools with slips and rubber cups to contain pressure and pack-off tubing in wells not equipped with landing nipples.

lock-in thermometer *n*: a dial-indicating temperature instrument with an automatic locking device for the indicator, which ensures that the indication cannot change until the reading has been taken and the instrument reset.

lock nut *n*: a special type of threaded fastener (a nut) that, when screwed down firmly against another nut or a washer, fastens (locks) firmly in place.

lock segment *n*: a device to lock a packer's mandrel to its dragblock housing.

Lockset™ (lokset) *n*: a trademark for a packer with bidirectional slips used in completion.

lodestone *n*: a kind of naturally occurring iron oxide that has magnetic properties and therefore attracts iron or steel. Also called *magnetite*.

loess *n*: unstratified, homogeneous accumulation of silt, often containing small amounts of clay or sand and redeposited by wind from glacial outwash or deserts.

log *n*: a systematic recording of data, such as a driller's log, mud log, electrical well log, or radioactivity log. Many different logs are run in wells to discern various characteristics of downhole formation. *v*: to record data.

log *abbr*: logarithm.

logarithm *n*: the exponent that indicates the power to which a number is raised to produce a given number. For example, the logarithm of 100 to the base 10 is 2.

logarithmic *adj*: of or pertaining to a logarithm, as a logarithmic scale. See *logarithm*.

logarithmic scale *n*: a range of values—a scale—on which actual distances from the origin are proportional to the logarithms of the corresponding scale numbers. Unlike an arithmetic scale, where distances from the origin relate directly to the number, values on a logarithmic scale indicate a proportional relation. For example, the Richter scale of earthquake intensity is logarithmic; thus, an earthquake with an intensity of 4.0 on the Richter scale is much more than twice as intense as an earthquake with an intensity of 2.0.

logarithmic taper *n*: in a potentiometer, the winding of the resistance wire in such a manner that turning the potentiometer's control knob varies the resistance nonproportionately. That is, turning the control knob one-quarter turn increases or decreases the resistance by a factor different from one-fourth—for example, two or four times more. Compare *linear taper*.

log a well *v*: to run any of the various logs used to ascertain downhole information about a well.

logbook *n*: a book used by station engineers, dispatchers, and gaugers for keeping notes on current operating data.

log cross section *n*: a cross section of a reservoir or part of a reservoir constructed with electric or radioactivity logs.

log deflection *n*: the movement of the curve on a log away from a reference, or base, line.

logging devices *n pl*: any of several electrical, acoustical, mechanical, or radioactivity devices that are used to measure and record certain characteristics or events that occur in a well that has been or is being drilled.

logging while drilling (LWD) *n*: logging measurements obtained by measurement-while-drilling techniques as the well is being drilled.

logic *n*: 1. the study of the principles of reasoning, especially of the structure of propositions as distinguished from their content; logic also includes the study of method and validity in deductive reasoning. 2. the nonarithmetic operations performed by a computer, such as sorting, comparing, and matching that involve yes-no decisions.

logical AND operation *n*: a process in logical reasoning whereby two related statements, or premises, must be true before a correct conclusion can be drawn. For example, if one premise is "that dog is panting," and the other premise is that "the temperature is 98°," then the conclusion, "when the temperature is 98°, that dog pants," may be correctly drawn.

logical NOT operation *n*: a process in logical reasoning whereby a true statement is made false by making a false statement.

logical OR operation *n*: a process in logical reasoning whereby if either or both of two statements, or premises, are true, then a correct conclusion can be drawn. For example, the statement that "if 50 tons of iron or 50 tons of coal overloads a truck, then the truck is overloaded," is a correct conclusion. Either 50 tons of iron or 50 tons of coal overload the truck.

logic gate *n*: an electronic circuit that provides the action of comparing, selecting, making references, matching, sorting, and merging where yes-or-no (logic) quantities are involved.

log sheet *n*: for pipelines, a daily report sheet on which operating data are entered by gaugers, dispatchers, and station operators.

longitude *n*: the arc or portion of the earth's equator intersected between the meridian of a given place and the prime meridian (at Greenwich, England) and expressed either in degrees or in time.

long-range forecast *n*: a weather forecast covering more than a week in advance (sometimes for a month or a season).

long rotary slips *n pl*: slips designed to fit a square-drive master bushing, and whose taper length is 8¹³⁄₁₆ inches (224 millimetres). Compare *extra-long rotary slips*. See also *slips, square-drive master bushing*.

longshore current *n*: movement of seawater parallel to the shore.

long string *n*: 1. the last string of casing set in a well. 2. the string of casing that is set at the top of or through the producing zone, often called the oil string or production casing.

long substrate (LS) bond failure *n*: the failure of the adhesion (bond) between the tungsten carbide stud and the tungsten carbide of a polycrystalline diamond compact (PDC) bit cutter. As a result, the PDC layer and the tungsten carbide layer below the PDC layer fall off, leaving only the tungsten carbide stud on the bit.

long thread *n*: any one of two types of thread that are cut on casing as standardized by API. The other is the short thread.

long ton *n*: a unit of measure of mass (weight) that is equal to 2,240 pounds (1,016.06 kilograms). Compare *short ton*.

long-wave radiation *n*: energy that is emitted from a source in the form of waves the lengths of which are longer than 545 metres. Long-wave radiation is not visible. Compare *shortwave radiation*.

long way *n*: displacing fluid from the tubing up the annulus. Compare *short way*.

look box *n*: that portion of the gauge housing that contains a shielded or glassed-in opening through which a gauge is observed.

looping *n*: the technique of laying an additional pipeline alongside an existing one when additional capacity is needed.

loose emulsion *n*: an emulsion that is relatively easy to break. Compare *tight emulsion*.

loss of circulation *n*: see *lost circulation*.

loss of head *n*: friction loss. See *pressure-drop loss*.

lost circulation *n*: the quantities of whole mud lost to a formation, usually in cavernous, fissured, or coarsely permeable beds. Evidenced by the complete or partial failure of the mud to return to the surface as it is being circulated in the hole. Lost circulation can lead to a blowout and, in general, can reduce the efficiency of the drilling operation. Also called lost returns.

lost circulation additives *n pl*: materials added to the mud in varying amounts to control or prevent lost circulation. Classified as fiber, flake, or granular.

lost circulation material (LCM) *n*: a substance added to cement slurries or drilling mud to prevent the loss of cement or mud to the formation. See *bridging materials*.

lost circulation plug *n*: cement set across a formation that is taking excessively large amounts of drilling fluid during drilling operations.

lost hole *n*: a well that cannot be drilled further or produced because of a blowout, unsuccessful fishing job, and so forth.

lost pipe *n*: drill pipe, drill collars, tubing, or casing that has become separated in the hole from the part of the pipe reaching the surface, necessitating its removal before normal operations can proceed; i.e., a fish.

lost returns *n pl*: see *lost circulation*.

lost time accident *n*: an incident in the workplace that results in an injury serious enough that causes the person injured to be unable to work for a day or more.

LOT (loaded on top) procedure *n*: a procedure in which tank-cleaning operations are carried out on board ships and the resulting water/oil mixture is collected in a tank and allowed to separate. The relatively clean water is then pumped out of the vessel, and part of the next cargo is loaded on top of the remaining cargo/water-cleaning residue. This residue is called slops.

Louisiana Independent Oil and Gas Association (LIOGA) *n*: a group of independent oil producers and persons who own royalties from oil and gas wells in Louisiana. Address: 1 American Pl.; 301 Main St.; Ste. 1030; Baton Rouge, LA 70825; 225-388-9525.

low-capacitance probe *n*: see *lo-cap probe*.

low clay-solids mud *n*: heavily weighted muds whose high solids content (a result of the large amounts of barite added) necessitates the reduction of clay solids.

low drum drive *n*: the drawworks drum drive used when hoisting loads are heavy.

lower fairlead *n*: a wire or chain fairlead that is mounted on a lower column of a semisubmersible.

lower hull *n*: the part of a semisubmersible rig that is always submerged or mainly submerged below the waterline. Ballast, drill water, and fuel oil are stored in the lower hull.

lowering-in *n*: the process of laying pipe in a ditch in pipeline construction. Pipe can be lowered into the ditch as part of the coating operation or lowered separately by a lowering-in crew.

lowering-up *n*: in pipeline construction, the process of raising pipe and placing it on vertical support members in parts of the world where frozen earth prevents normal burial of the line. Lowering-up is the counterpart of lowering-in in more temperate climates.

lower kelly cock *n*: on rigs that use a kelly and rotary table system to rotate the bit and drill stem, a heavy-duty valve installed between the kelly and the first joint of drill pipe. Usually, the valve is open so that drilling fluid can flow out of the kelly and down the drill stem. When the mud pump is stopped, the valve can be manually closed with a special wrench to prevent pressurized fluids in the drill string from flowing into the kelly. When closed, the valve also prevents mud in the kelly from spilling onto the rig floor when the kelly is broken out of the drill string. Also called a drill stem safety valve, mud saver valve.

lower kelly valve *n*: see *drill stem safety valve, lower kelly cock.*

lower marine-riser package (LMRP) *n*: the equipment that attaches the bottom part of the marine riser to the subsea blowout preventer (BOP) stack. It includes the BOP connector, a flexible joint to compensate for side-to-side movement, and the marine riser connector. See *riser connector, riser pipe, subsea blowout preventer.*

lower marine-riser package (LMRP) connector *n*: a hydraulically actuated device that is used to remotely connect the blowout preventer stack to the wellhead, and the lower marine riser package to the blowout preventer stack.

lower range value (LRV) *n*: in instrumentation, the measure of a process variable, such as pressure or temperature, over the desired range of measurement at its minimum point. Compare *upper range value.*

lower sample *n*: in tank sampling, a spot sample obtained at a level of ⅚ the depth of liquid below the surface.

lower tier *n*: a category of oil production for purposes of price control. It refers to old oil, i.e., oil contained in reservoirs that were being produced during 1972.

lower troposphere *n*: the part of the earth's atmosphere that starts at sea level and extends to an altitude of about 18,000 to 19,000 feet (about 5.5 to 5.8 kilometres).

lowest achievable emission rate (LAER) *n*: an air emission standard determined by the states on a case-by-case basis. LAER is applied in nonattainment areas, which do not meet NAAQS, and is more stringent than the BACT applied in attainment areas. Consequently, cost and economic impact are not taken into consideration in determining LAER.

low gravity solids *n pl*: in drilling, nonreactive particles in the drilling fluid that are either intentionally added to give the drilling fluid desirable properties or are picked up by the drilling fluid from a formation being drilled. Usually, low gravity solids that get into the drilling fluid as a formation is drilled are removed by mud cleaning equipment at the surface.

low-head drilling *n*: see *near-balance drilling.*

low-pressure area *n*: an area of low atmospheric pressure.

low-pressure distribution system *n*: a system in which the gas pressure in the mains and the service lines is substantially the same as that delivered to the customers' appliances; ordinarily, a pressure regulator is not required on individual service lines.

low-pressure squeeze cementing *n*: a squeeze cementing technique in which the pressure used to place the cement always remains below fracture pressure. See *squeeze cementing.*`

low-solids fluid *n*: see *low-solids mud.*

low-solids mud *n*: a drilling mud that contains a minimum amount of solid material (sand, silt, and so on) and that is used in rotary drilling when possible because it can provide fast drilling rates.

low-temperature fractionation *n*: separation of a hydrocarbon fluid mixture into components by fractionation, wherein the reflux condenser is operated at temperatures requiring refrigeration. See *pod (Podbielniak) analysis.*

low-temperature processing *n*: gas processing conducted below ambient temperatures.

low tide *n*: a tide at its lowest level or the time of day when it occurs. Compare *high tide.* See *tide.*

low vapor pressure liquid *n*: a liquid that, at the measurement or proving temperature of the meter, has a vapor pressure less than atmospheric pressure. Compare *high vapor pressure liquid.*

low-yield clay *n*: commercial clay chiefly of the calcium montmorillonite type and having a yield of approximately 15 barrels per ton (2,385 litres per tonne).

LPG *abbr*: liquefied petroleum gas.

LRG *abbr*: liquefied refinery gas.

LRV abbr: lower range value.

ls *abbr*: limestone; used in drilling reports.

lse *abbr*: lease; used in drilling reports.

L-10 chart *n*: a chart used in measuring gas with orifice meters on which are recorded differential and static pressures. It uses a logarithmic scale from 0 to 10. The readings must be converted, but it can be used on any flow recorder, regardless of range. Also called square-root chart.

LTL *abbr*: Letter to Lessees and Operators.

LTX unit *n*: low-temperature separator. A mechanical separator that uses refrigeration obtained by expansion of gas from high pressure to low pressure to increase recovery of gas-entrained liquids.

lubricant *n*: a substance—usually petroleum-based—that is used to reduce friction between two moving parts.

lubricate *v*: 1. to apply grease or oil to moving parts. 2. to lower or raise tools in or out of a well with pressure inside the well. The term comes from the fact that a lubricant (grease) is often used to provide a seal against well pressure while allowing wireline to move in or out of the well.

lubricator *n*: a specially fabricated length of casing or tubing usually placed temporarily above a valve on top of the casinghead or tubing head. It is used to run swabbing or perforating tools into a producing well and provides a method for sealing off pressure and thus should be rated for highest anticipated pressure.

lubricator stack *n*: see *lubricator.*

lubricator valve *n*: positioned just below the slash zone. Its function is to provide pressure containment of the landing string production bore when wireline or coil tubing activities are being deployed through the surface tree swab valve.

lug *n*: a projection on a casting to which a bolt or other part may be fitted.

lugging power *n*: the torque, or turning power, delivered to the flywheel of a diesel engine.

lunar tide *n*: that portion of a tide that is due to the gravitational attraction of the moon on the earth.

LVR *abbr*: load vessel ratio.

LWD *abbr*: logging while drilling.

lyophilic *adj*: having an affinity for the suspending medium, such as bentonite in water.

lyophilic colloid *n*: a colloid that is not easily precipitated from a solution and is readily dispersible after the precipitation by an addition of the solvent. See *colloid.*

lyophobic colloid *n*: a colloid that is readily precipitated from a solution and cannot be redispersed by an addition of the solution. See *colloid.*

m *abbr*: metre.

m² *abbr*: square metre.

m³ *abbr*: cubic metre.

mA *abbr*: milliampere.

M *abbr*: metacenter; this abbreviation is used mostly in buoyancy, stability, and trim calculations for offshore drilling rigs.

MAC *abbr*: mobile arctic caisson.

macaroni rig *n*: a workover rig, usually lightweight, that is specially built to run a string of ¾-inch or 1-inch (1.9- or 2.54-centimetre) tubing. See *macaroni string*.

macaroni string *n*: a string of tubing or pipe, usually ¾ or 1 inch (1.9 or 2.54 centimetres) in diameter.

mackerel sky *n*: a banded arrangement of cirrocumulus clouds.

MACT *abbr*: maximum achievable control technology.

magma *n*: the hot fluid matter within the earth's crust that is capable of intrusion or extrusion and that produces igneous rock when cooled.

Magnaflux™ *n*: trade name for the equipment and processes used for detecting cracks and other surface discontinuities in iron or steel. A magnetic field is set up in the part to be inspected, and a powder or paste of magnetic particles is applied. The particles arrange themselves around discontinuities in the metal, revealing defects.

magnet *n*: a metal that has the property of attracting ferrous (iron) and certain other metals to it.

magnetic attraction *n*: the principle that the opposite (north and south) poles of a magnet attract each other.

magnetic brake *n*: see *electrodynamic brake*.

magnetic circuit *n*: the path magnetic lines of force (flux lines) take in an electrical device that employs magnets in its operation. A magnetic circuit is analogous to direct current flowing in a conductor such as wire or cable.

magnetic deviation *n*: see *declination*.

magnetic field *n*: the space that magnetic lines of force produced by a magnetic or current-carrying conductor.

magnetic field generator *n*: a source of electricity that converts mechanical energy into electrical energy by the action of a magnetic field on a conductor when one is moving in relation to the other.

magnetic flowmeter *n*: a device that senses and indicates the amount of fluid flowing in a line. It operates on the principle that a magnetic field induces voltage in an electrical conductor moving through the magnetic field. The conductor is the fluid flowing in the line. Field coils mounted on the sides of a special section of line create the magnetic field.

magnetic flux *n*: see *flux*.

magnetic iron ore *n*: see *magnetite*.

magnetic lines *n pl*: see *magnetic lines of force*.

magnetic lines of force *n pl*: the magnetism that emanates from a magnet's north pole and travels in the atmosphere around the magnet to its south pole.

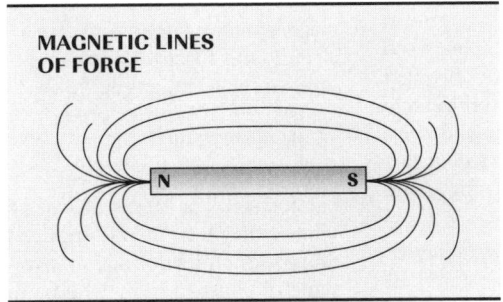

MAGNETIC LINES OF FORCE

magnetic meter *n*: a meter used to measure the electrical conductivity of liquids.

magnetic north *n*: a region located in the north polar region of the earth toward which magnetic compasses point. Its location is different from true, or geographic, north. Compare *true north*.

magnetic particle inspection (MPI) *n*: a method that can be used to find surface and near surface flaws in ferromagnetic materials such as steel and iron. The technique uses the principle that magnetic lines of force (flux) will be distorted by the presence of a flaw in a manner that will reveal its presence. The flow, i.e., a crack, is located from the "flux leakage," following the application of fine iron particles, to the area under examination. There are variations in the way the magnetic field is applied, but they are all dependent on the above principle. The iron particles can be applied dry or wet, suspended in a colored or fluorescent liquid. While magnetic particle inspection is primarily used to find surface breaking flaws, it can also be used to locate subsurface flaws. But its effectiveness quickly diminishes depending on the flaw depth and type.

magnetic pole *n*: 1. either of two regions that are located in the polar areas of the northern and southern hemispheres and toward which a compass needle points from any direction throughout adjacent regions. Compare *geographic pole*. 2. either of the poles of a magnet.

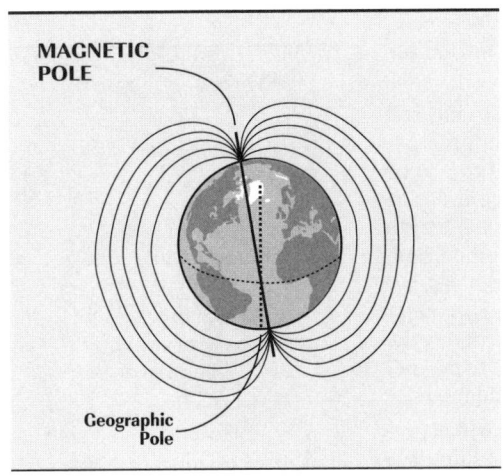

MAGNETIC POLE

Geographic Pole

magnetic repulsion *n*: the phenomenon that like poles (north-north or south-south) of a magnet repel, or repulse, each other.

magnetic survey *n*: an exploration method in which an instrument that measures the intensity of the natural magnetic forces existing in the earth's subsurface is passed over the surface or through the water. The instrumentation detects deviations in magnetic forces, and such deviations may indicate the existence of underground formations that favor the entrapment of hydrocarbons.

magnetic surveying instrument *n*: a device used to determine the direction and drift angle of a deviated wellbore. It uses a magnetic compass to measure magnetic direction, which differs from true direction by the amount of local declination.

magnetic survey tool *n*: a device that measures and records the amount a borehole has deviated from vertical and, sometimes, the direction of the deviation. Such tools use magnetic compasses and therefore must be run inside a nonmagnetic drill collar to eliminate any magnetic interference from the drill string.

magnetic testing *n*: a method of testing for defects in steel parts that is carried out by magnetizing the steel and sprinkling a magnetic powder on the surface to detect flaws or cracks. See *magnetic particle inspection*.

magnetism *n*: physical phenomena that include a magnet's or lodestone's attraction for iron and the forces associated with moving electricity; magnets and electric current exhibit these phenomena, which are characterized by fields of force.

magnetite *n*: the mineral form of black iron oxide, Fe_3O_4, that often occurs with magnesium, zinc, and manganese and is an important ore of iron. Also called lodestone, magnetic iron ore.

magnetizing current *n*: the part of exciting current that is 90 degrees out of phase with the impressed voltage on the circuit of a transformer and that produces magnetic flux in the transformer's winding. See *exciting current*.

magnetizing force (H) *n*: in an electric coil, or solenoid, the ampere-turns per inch of the length of the coil's core to be magnetized. Just as a voltage drop, or loss of voltage, occurs along the length of a conductor that is carrying current, a magnetomotive drop occurs along the length of the magnetic circuit. The longer the magnetic circuit, the greater is the magnetomotive force required to magnetize the core to the desired amount.

magneto *n*: a device (an alternator) that produces alternating current for distribution to the ignition system of some internal-combustion engines. See *alternator*.

magnetometer *n*: an instrument used to measure the intensity and direction of a magnetic field, especially that of the earth.

magnetomotive force (mmf) *n*: in a magnetic circuit, the force that creates magnetic lines of force (magnetic flux); it is analogous to electromotive force, or voltage, in an electric circuit.

magnetostrictive transducer *n*: a tightly banded scroll of special steel that vibrates when a magnetic field is applied to it. The vibration of the scroll sets off a transmitter, which is alternately switched on and off at 15 to 60 times per second. The transmitter causes compressional sound waves to travel through the formations surrounding the borehole. Each formation material will exhibit its own characteristic effect on the elastic wave propagation.

magnetotellurics *n pl*: operates on the theory that rocks of differing composition have different electrical properties.

magnet wire *n*: wire used as windings in electromagnets, transformers, motors, and generators. Its insulation must be thin so that it can be wound several times around a core. Thick insulation would not allow it to be wound an adequate number of times. So, instead of a plastic coating, varying grades of enamel or varnish are baked onto magnet wires to provide good insulating properties.

main bearing *n*: in an engine, a large circular friction-reducing device installed on the engine's crankshaft. Main bearings are mounted in the engine's crankcase and the crankshaft rotates on them. Engine oil lubricates them as the engine runs. Most main bearings are plain bearings, in that their wear surface is flat or plain and do not have balls or rollers. Compare *ball bearing, roller bearing*.

main brake *n*: two bands fitted with brake pads; the bands fit over the two rims of the drawworks drum. When the driller

engages this brake, the pads press down on the rims to stop the drum from hoisting or from letting out drilling line. The main brake also keeps the drum from rotating (and therefore holds the drill stem stationary) when making up or breaking out drill pipe.

main deck *n*: the principal continuous deck of a ship or offshore drilling rig running from fore to aft from which the freeboard is determined.

mainline *n*: a large-diameter pipeline between distant points, i.e., a trunk line.

mainline plant *n*: a plant that processes the gas that is being transported through a cross-country transmission line. Also called on-line, pipeline, or straddle plant.

mainline valve *n*: a device for controlling fluid flow in a mainline.

main tank *n*: in a drilling rig's fuel-supply system, a large tank in which diesel fuel is stored. If the rig has a day tank, it is filled from the main tank; otherwise, the main tank supplies fuel to the engine. Compare *day tank*.

major *n*: a large oil company, such as ExxonMobil, Chevron, or BP Amoco, that not only produces oil, but also transports, refines, and markets it and its products.

majority carrier *n*: the type of carrier, either electron or hole, that makes up more than half the carriers in a semiconductor.

make a connection *v*: to attach a joint of drill pipe onto the drill stem suspended in the wellbore to permit deepening the wellbore by the length of the joint (usually about 30 feet, or 9 metres).

make a hand *v*: (slang) to become a good worker.

make a trip *v*: to hoist the drill stem out of the wellbore to perform one of a number of operations, such as changing bits or taking a core, and so forth, and then to return the drill stem to the wellbore.

make hole *v*: to deepen the hole made by the bit, i.e., to drill ahead; to run casing or pipe.

make mud *v*: to create a fluid with the consistency of mud that occurs when clear water is circulated in a soft formation.

make up *v*: 1. to assemble and join parts to form a complete unit (e.g., to make up a string of drill pipe). 2. to screw together two threaded pieces. Compare *break out*. 3. to mix or prepare (e.g., to make up a tank of mud). 4. to compensate for (e.g., to make up for lost time).

makeup *adj*: added to a system (e.g., makeup water used in mixing mud).

make up a joint *v*: to screw a length of pipe into another length of pipe.

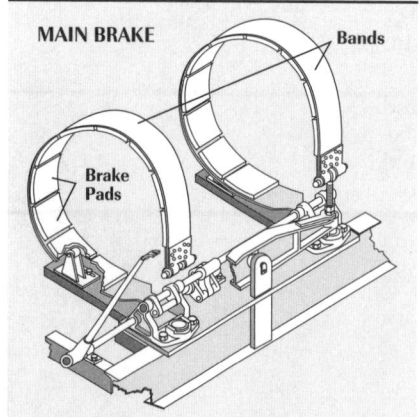

MAIN BRAKE   Bands

Brake Pads

makeup cathead *n*: a device that is attached to the shaft of the drawworks and used as a power source for screwing together joints of pipe. It is usually located on the driller's side of the drawworks. Also called spinning cathead. See *cathead*.

makeup gas *n*: 1. gas that is taken in succeeding years and has been paid for previously under a take-or-pay clause in a gas purchase contract. The contract will normally specify the number of years after payment in which the purchaser can take delivery of makeup gas without paying a second time. 2. gas injected into a reservoir to maintain a constant reservoir pressure and thereby prevent retrograde condensation. 3. in gas processing, the gas that makes up for plant losses. During processing there is a reduction in gas volume because of fuel and shrinkage. Some agreements between gas transmission companies and plant owners require plant losses to be made up or to be paid for.

makeup tongs *n pl*: tongs used for screwing one length of pipe into another for making up a joint. Compare *breakout tongs*. See also *tongs*.

makeup water *n*: the water used as a base in water-based drilling muds. It may be fresh, brackish, or salty.

male connection *n*: a pipe, coupling, or tool that has threads on the outside so that it can be joined to a female connection. Compare *female connection*.

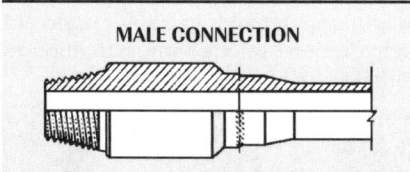

MALE CONNECTION

mammas *n pl*: breast-shaped protuberances appearing on the undersides of clouds, indicating the approach of a storm.

mandrel *n*: a cylindrical bar, spindle, or shaft around which other parts are arranged or attached or that fits inside a cylinder or tube.

manhole *n*: a hole in the top or side of a tank through which a person can enter.

manifold *n*: 1. an accessory system of piping to a main piping system (or another conduit) that serves to divide a flow into several parts, to combine several flows into one, or to reroute a flow to any one of several possible destinations. 2. a pipe fitting with several side outlets to connect it with other pipes. 3. a fitting on an internal-combustion engine made to receive exhaust gases from several cylinders.

manifold capacity *n*: the number of cones needed in a mud cleaner to process the full volume of circulating mud.

manifold pressure gauge *n*: a device installed on an engine's intake manifold that indicates the pressure inside the manifold. Manifold pressure is a measure of the airflow into the engine cylinders at a given speed and intake air temperature. Thus, when combined with the engine's rpm, manifold pressure is an indication of the engine's power output.

manifold valve *n*: a valve placed in a series of piping (a manifold). For example, in a choke manifold on a drilling rig, an adjustable choke, which is placed in the manifold, is one of several types of manifold valve that may be present.

manipulated variable *n*: in instrumentation, an action, such as rate of flow, that changes the control variables. For example, the rate of flow can be sensed and manipulated by electronic devices to control other process variables, such as pressure, temperature, and liquid level. When used in this manner, rate of flow is the manipulated variable.

man-machine interface (MMI) *n*: see *human-machine interface*.

manometer *n*: a U-shaped piece of glass tubing containing a liquid (usually water or mercury) that is used to measure the pressure of gases or liquids. When pressure is applied, the liquid level in one arm rises while the level in the other drops. A set of calibrated markings beside one of the arms permits a pressure reading to be taken, usually in inches or millimetres.

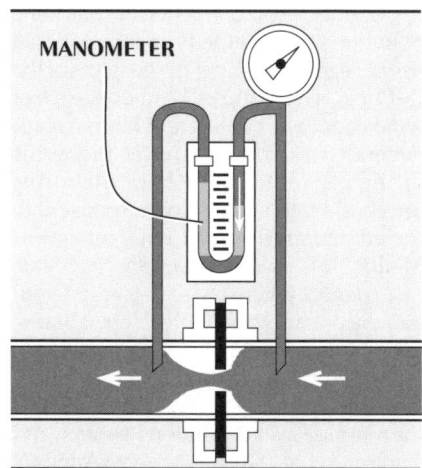

MANOMETER

mantle *n*: the hot, plastic part of the earth that lies between the core and the crust. It begins 5 to 30 miles (10 to 50 kilometres) beneath the surface and extends to 1,800 miles (2,900 kilometres).

manual control *n*: in instrumentation, an operator-controlled process whereby the operator makes adjustments to maintain the desired value without any automatic feedback.

manual welding *n*: a welding process in which an electric arc melts and fuses the pipe ends with the metal of an electrode held by the welder. Also called stick welding.

manufactured gas *n*: see *liquefied petroleum gas*.

manufacturer *n*: 1. under TSCA, anyone who manufactures or imports for commercial purposes over 10,000 pounds per year of a substance listed on the TSCA inventory. 2. under HAZCOM, production and drilling companies are considered manufacturers of crude oil, natural gas, and other products and must maintain an MSDS for these products.

manway *n*: see *manhole*.

MARAD *abbr*: Maritime Administration.

marginal well *n*: a well that is approaching depletion of its natural resource to the extent that any profit from continued production is doubtful.

marigraph *n*: a gauge that makes a continuous graphic record of tide height in relation to time.

marine conductor *n*: a string of casing used in drilling from offshore bottom-supported drilling rigs. It extends from the bottom of the conductor hole to a point into the air directly under the rig. The blowout preventer stack is nippled up on top of the marine conductor. It also serves to guide drilling and subsequent casing strings into the hole.

marine delta *n*: 1. a triangular sea-level extension of land shaped like the Greek letter Δ. 2. a depositional environment in which riverborne sediments accumulate as the flow energy of the river is dissipated in the ocean.

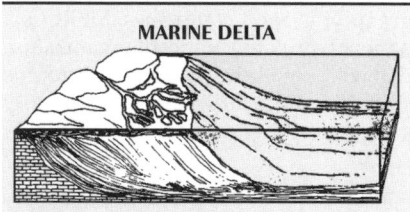

MARINE DELTA

marine lay *n*: see *lay-barge construction*.

Marine Mammal Protection Act *n*: congressional act that provides for protection of all marine mammals.

Marine Plastic Pollution Research and Control Act of 1987 *n*: a congressional act that implements Annex V of the International Convention for the Prevention of Pollution from Ships (MARPOL). Under the provisions of this act, all ships and watercraft (including all commercial and recreational fishing vessels) are prohibited from dumping plastics at sea. The law also severely restricts dumping other vessel-generated garbage and solid waste both at sea and in U.S. navigable waters.

Marine Protection, Research, and Sanctuaries Act of 1973 *n*: congressional act that identifies marine environments of special national significance, provides authority for conservation of these areas, and enhances public awareness and wise use of the marine environment through educational programs and research. Sometimes called the Marine Sanctuaries Act or the Ocean Dumping Act.

marine-riser connector *n*: a fitting on top of the subsea blowout preventers to which the riser pipe is connected.

marine-riser pipe *n*: see *riser pipe*.

marine-riser system *n*: see *riser pipe*.

marine-riser tensioning system *n*: an assembly of devices installed on a floating offshore drilling rig to maintain a constant upward force (tension) on the riser pipe, despite vertical motions (heave) caused by ocean movements.

Mariner's Weather Log *n*: a publication of the United States National Weather Service (NWS) that contains articles, news, and information about marine weather events and phenomena, storms at sea, and weather forecasts. It is published two times a year with a spring-summer issue and a fall-winter issue. It may be obtained from Superintendent of Documents, P.O. Box 371954, Pittsburgh, PA 15250-7954. The Web site URL is www.nws.noaa.gov/om/mwl/mwl.htm.

Marine Sanctuaries Act *n*: see *Marine Protection, Research, and Sanctuaries Act of 1973*.

Marine Spill Response Corporation (MSRC) *n*: an oil spill cooperative formed by a group of U.S. oil companies. MSRC is a national, private, nonprofit organization that will provide response to catastrophic spills in certain U.S. waters. It is primarily geared to respond to large, open water spills.

maritime *adj*: of or relating to the sea, shipping, sailing in ships, or living and working at sea; situated or living close to the sea.

Maritime Administration (MARAD) *n*: an agency of the U.S. Department of Transportation (USDOT) whose mission is to strengthen the U.S. maritime transportation system to meet the economic and security needs of the United States. MARAD programs promote the development and maintenance of an adequate,

well-balanced United States merchant marine, sufficient to carry the nation's domestic waterborne commerce and a substantial portion of its waterborne foreign commerce, and capable of service as a naval and military auxiliary in time of war or national emergency. MARAD also seeks to ensure that the United States maintains adequate shipbuilding and repair services, efficient ports, effective intermodal water and land transportation systems, and reserve shipping capacity for use in time of national emergency. Address: 400 7th Street SW; Washington, DC 20590; (800) 996-2723; www.marad.dot.gov.

marker bed *n*: a distinctive, easily identified rock stratum, especially one used as a guide for drilling or correlation of logs.

marketable title *n*: see *merchantable title*.

marketing costs *n pl*: costs incurred in making gas merchantable, such as compression, dehydration, or treating, and costs of transporting gas to the point of delivery to the purchaser. Sometimes called handling costs.

market-out *adj*: a provision in a natural gas sales contract that allows one or both parties to renegotiate the sales price or terminate the contract if the contractual sales price no longer reasonably reflects the current market. There are many forms of market-out provisions with differing rights and effects.

marl *n*: a semisolid or unconsolidated clay, silt, or sand.

MARPOL 73/78 *n*: the short name for the International Convention for Prevention of Pollution from Ships, 1973 and 1978. (MARPOL is not an abbreviation or acronym; rather it is a contraction of "maritime pollution.") MARPOL is an international treaty negotiated under the auspices of the IMO and is made up of 5 Annexes, each of which addresses a different kind of ship-generated pollution. (Under the terms of the convention, offshore platforms are considered ships.) The responsibility for administration and enforcement of MARPOL is split between the flag state (the country where the ship is registered) and the coastal state (in the United States, the USCG is responsible). MARPOL has led to the promulgation and implementation of several U.S. laws. For example, the Marine Plastic Pollution Research and Control Act of 1987 implements Annex V of MARPOL.

marsh buggy *n*: a tractor-like vehicle the wheels of which are fitted with extra-large rubber tires for use in swamps.

Marsh funnel *n*: a calibrated funnel used in

MARSH FUNNEL

field tests to determine the viscosity of drilling mud.

Marsh funnel viscosity *n*: see *funnel viscosity, kinematic viscosity*.

Martin-Decker™ *n*: a trademarked term for a rig weight indicator. See *weight indicator*.

MASP *abbr*: maximum allowable surface pressure.

mass *n*: the quantity of matter a substance contains, independent of such external conditions as the buoyancy of the atmosphere or the acceleration caused by gravity.

mass flowmeter *n*: an instrument that measures the rate of flow of a fluid in a system, based on the mass of the fluid. The measurement is expressed as mass per unit of time where mass is volume times density. This expression is then converted to volume per unit of time, such as gallons per minute, barrels per hour, or cubic metres per hour.

mass flow rate *n*: a calculation of gas flow through an orifice meter where the quantity of the gas (independent of external conditions like temperature, pressures, or gravity) is determined; usually expressed in pounds mass per unit time.

mass spectrometer *n*: a device that measures the specific gravity or relative density of a gas by means of the separation of gaseous ions according to their differing mass and charge.

mass-transfer zone *n*: the depth of the solid desiccant bed, in a solid desiccant dehydration system, from saturation to initial adsorption.

mast *n*: a portable derrick that is capable of being raised as a unit, as distinguished from a standard derrick, which cannot be raised to a working position as a unit. For transporting by land, the mast can be divided into two or more sections to avoid excessive length extending from truck beds on the highway. Oil workers and manufacturers often use the words "mast" and "derrick" interchangeably. Compare *derrick*.

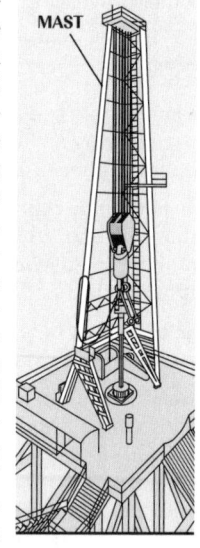
MAST

master bushing *n*: a device that fits into the rotary table to accommodate the slips and drive the kelly bushing so that the

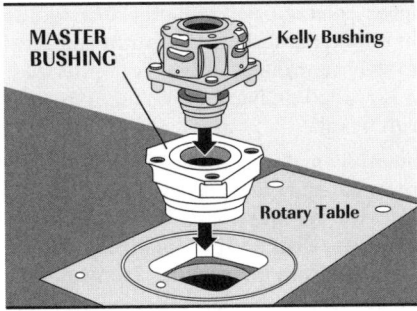

rotating motion of the rotary table can be transmitted to the kelly. Also called rotary bushing. See *kelly bushing, slips.*

master choke line valve *n*: the valve on the choke line that is nearest to the preventer assembly. Its purpose is to control the flow through the choke line. See *HRC valve.*

master clutch *n*: the clutch that connects the compounding transmission on a mechanical-drive rig to the input shaft of the drawworks; disengages the prime mover from all motions of the crane.

master control panel *n*: on a drilling rig, the primary station that controls the operation of the blowout preventers and other well-control equipment. Also called primary control panel. Backup control panels are usually installed should the master panel fail, or should it become inaccessible. See *driller's BOP control panel.*

master control station (MCS) *n*: a computer equipped with a modem to communicate with the surface control module.

master gate *n*: see *master valve.*

master meter *n*: a meter that is proved using a certified prover and then used to calibrate other provers or to prove other meters.

master oscillator *n*: an oscillator that establishes the carrier frequency of the output of an amplifier or transmitter. See *oscillator.*

master switch *n*: an electrical component, usually wired in parallel with other switches, that controls the flow of electricity in the circuit. When closed, the switch allows electricity to flow through the circuit. When open, no electricity can flow. When the master switch is open, other switches in the circuit do not operate.

master tape *n*: a measuring tape that has been certified accurate at a specified temperature and tension by the NIST. It is used to calibrate a working tape. See *strap, working tape.*

master valve *n*: 1. a large valve located on the Christmas tree and used to control the flow of oil and gas from a well. Also called master gate. 2. the blind or blank rams of a blowout preventer (obsolete).

mastic *n*: a heavy-bodied, paste-like coating often applied with a trowel to produce a thick film.

material balance *n*: a calculation used to inventory the fluids produced from a reservoir and the fluids remaining in a reservoir. Using the equations of the calculation, the volume of original oil in gas in place, the amount of drive-water influx, forecasts of production rates for several years, and estimates of ultimate recovery can be made.

material grade *n*: in oilfield equipment and tools, a measure of the equipment or tool's strength—that is, its ability to be bent, stretched, compressed, or loaded without deformation or breaking.

material safety data sheet (MSDS) *n*: a reference document prepared by the chemical manufacturer and required under HAZCOM to be kept with the material during shipment and distributed in the workplace for any employee who may work with that material. The information on the MSDS may take a variety of forms but it must include the identity of the chemical, chemicals, or mixtures as listed on the container label; the physical and chemical characteristics of the material; the physical hazards of the material; the health hazards of the material; the primary routes of entry into the human body; the permissible exposure limits, if available; whether the material has been found to be a carcinogen, a potential carcinogen, or is regulated by OSHA; any generally applicable precautions for safe handling; any generally applicable control measures; emergency first aid procedures; the date the MSDS was prepared; and the name, address, and telephone number of the material source or another responsible party who can provide additional information and emergency procedures for the material if necessary.

materials coordinator *n*: a person who oversees the acquisition of supplies, equipment, and other resources for a company or organization. This person's duties may also include managing and controlling warehouse operations, ordering materials, and managing inventory. Also called materialsman, warehouse supervisor.

matrix *n*: 1. in rock, the fine-grained material between larger grains in which the larger grains are embedded. A rock matrix may be composed of fine sediments, crystals, clay, or other substances. The solid portions of a porous rock not including the pores themselves. 2. the material in which the diamonds on a diamond bit are set. 3. in mathematics, a rectangular array of numbers from vector space.

matrix acidizing *n*: an acidizing treatment using low or no pressure to improve the permeability of a formation without fracturing it. See *wellbore soak.* Compare *fracture acidizing.*

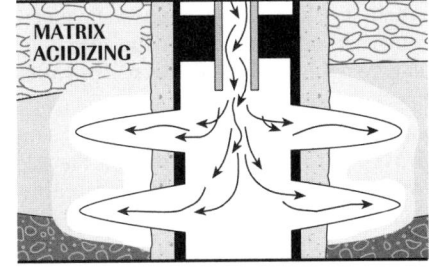

mat-supported jackup rig *n*: a type of bottom support for the legs of a jackup, generally used in areas where the seafloor is very soft. Each leg of the jackup is connected to the mat, which is a steel frame that is relatively wide. The mat, since it is wide, prevents the jackup's legs from penetrating the soft seafloor.

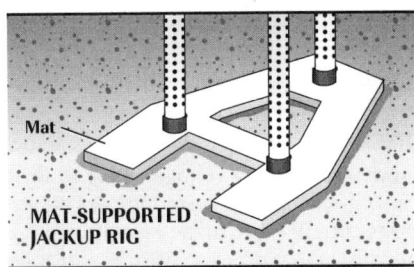

matter *n*: anything that occupies space and can be perceived by one or more senses; a physical body, a physical substance, or the universe as a whole; something that has mass and exists as a solid, liquid, gas, or plasma.

maximum achievable control technology (MACT) *n*: air emission standards under CAA that require the best pollution prevention methods available in specific industry categories for major sources of hazardous airborne pollutants.

maximum allowable pressure *n*: the greatest pressure that may safely be applied to a structure, pipe, or vessel. Pressure in excess of this amount leads to failure or explosion.

maximum allowable surface pressure (MASP) *n*: the maximum amount of pressure that is allowed to appear on the casing pressure gauge during a well-killing operation. This pressure is often determined by a leak-off test and if exceeded can lead to formation fracture at the casing shoe and a subsequent underground blowout or broaching.

maximum capacity *n*: the maximum output of a system or unit (such as a refinery, gasoline plant, pumping unit, or producing well).

maximum efficiency rate (MER) *n*: the producing rate of a well that brings about maximum volumetric recovery from a reservoir with a minimum of residual-oil saturation at the time of depletion. It is often used to mean the field production rate that will achieve maximum financial returns from operation of the reservoir. The two rate figures seldom coincide, however.

maximum lawful price (MLP) clause *n*: the provision in a gas sales contract that allows the price of gas to track the monthly increases allowed under the Natural Gas Policy Act of 1978.

maximum loading gauge *n*: the maximum permissible gauge measurement to which a railroad tank car may be loaded according to Interstate Commerce Commission regulations. Also called stop gauge.

maximum value *n*: the maximum voltage or current in a cycle of an AC circuit.

maximum water *n*: in oilwell cementing, the maximum ratio of water to cement that will not cause the water to separate from the slurry on standing.

maximum wave height *n*: the maximum height of waves observed over a given period.

maximum working voltage rating *n*: in a capacitor, the largest amount of voltage that can be applied to the capacitor under normal operating conditions without damaging or destroying it.

Mbopd *abbr*: thousands of barrels of oil per day.

MBTA *abbr*: Migratory Bird Treaty Act.

MC *abbr*: metal clad.

Mcf *abbr*: 1,000 cubic feet of gas, commonly used to express the volume of gas produced, transmitted, or consumed in a given period.

Mcf/d *abbr*: 1,000 cubic feet of gas per day.

McLeod gauge *n*: an instrument that measures very low pressures by measuring the height of a column of mercury supported by the gas whose pressure is to be measured. It is capable of measuring pressures as low as 1 micrometre of height in the column of mercury. It is often used to check the accuracy of other instruments.

MCS *abbr*: master control station.

md *abbr*: millidarcy (*pl*, millidarcys).

MD *abbr*: measured depth.

mean *n*: the average of two or more observed values.

mean draft *n*: a measurement of draft resulting from the average of several draft readings taken between the bow and stern of a floating vessel. If the vessel has a straight keel, mean draft occurs at the midpoint of the waterline length. See *draft*.

mean environmental load *n*: mean force acting on a rig as a result of the effects of wind, waves, and current conditions.

measurand *n*: a physical quantity, property, or condition that exists or is to be measured. This term is preferred to "input," "parameter to be measured," "physical phenomenon," "stimulus," and "variable."

measured depth (MD) *n*: the total length of the wellbore, measured in feet along its actual course through the earth. Measured depth can differ from true vertical depth, especially in directionally drilled wellbores. Compare *true vertical depth*.

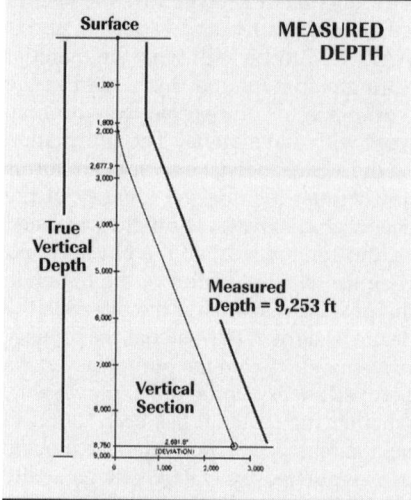

measured gallons *n pl*: the quantity in gallons in a railroad tank car at ambient temperature.

measured signal *n*: the electrical, mechanical, pneumatic, or other variable applied to the input of a device. It is the analog of the measured variable produced by a transducer.

measured variable *n*: the physical quantity, property, or condition that is to be measured. Common measured variables are temperature, pressure, rate of flow, thickness, and velocity.

measure in *v*: to obtain an accurate measurement of the depth reached in a well by measuring the drill pipe or tubing as it is run into the well.

measurement error *n*: the discrepancy between the result of the measurement and the value of the quantity measured.

measurement technician *n*: a person employed by an oil or gas transmission company whose primary responsibility is the maintenance and repair of metering equipment.

measurement ticket *n*: paper or readouts in a meter station that is automated, remotely controlled, and/or computerized. Also called delivery ticket, receipt ticket, run ticket.

measurement while drilling (MWD) *n*: 1. directional and other surveying during routine drilling operations to determine the angle and direction by which the wellbore deviates from the vertical. 2. any system of measuring downhole conditions during routine drilling operations.

measurement-while-drilling system *n*: a system in which downhole conditions are monitored during the drilling of a well.

measure out *v*: to measure drill pipe or tubing as it is pulled out of the hole, usually to determine the depth of the well or the depth to which the pipe or tubing was run.

measuring chamber *n*: the portion of a displacement meter that contains the measuring element.

measuring device *n*: a special reel and power arrangement for single-stranded wireline to make depth measurements in a well. A calibrated wheel and roller assembly is used to measure the footage of wireline as it is lowered into the well.

measuring element *n*: 1. the portion of a displacement meter that moves within the measuring chamber so as to divide the liquid into measured segments as the liquid passes through the meter. 2. the rotating member of a turbine meter, commonly referred to as the rotor.

measuring means *n pl*: in process control, the elements of a controller that sense, measure, and transmit to the controlling means either the value of the controlled variable or its deviation from the desired point.

measuring range *n*: the difference between the maximum and minimum values of a measuring instrument or of a quantity to be measured. This range is obtained under normal conditions and should not be affected by an error exceeding the maximum permissible error.

measuring station *n*: part of the pipeline system in which are found the orifice meter and recorder, volume controller, and temperature recorder.

measuring tank *n*: a calibrated tank that, by means of weirs, float switches, pressure switches, or similar devices, automatically measures the volume of liquid run in and then released. Measuring tanks are sometimes used in LACT systems. Also called metering tanks or dump tanks.

mechanic *n*: an optional crew member who is an all-around repairer for the rig's mechanical components.

mechanical actuator *n*: in process control, a device that exerts a physical force to produce motion or movement to a different position. For example, a mechanical actuator in a process valve opens or closes the valve, or positions the valve's opening by physically moving the valve's stem.

mechanical brake *n*: a brake that is actuated by machinery (such as levers or rods) that is directly linked to it.

mechanical displacement prover *n*: see *pipe prover*.

mechanical-drive rig *n*: see *mechanical rig*.

mechanical governor *n*: a speed-control device on an engine. Mechanical governors consist of flyweights, springs, and mechanical connections to the engine's speed control. Compare *electrically-actuated governor*, *hydraulic governor*. See *governor*.

mechanical integrity test (MIT) *n*: a test required prior to initial injection and once every five years thereafter (at a minimum) for certain types of injection wells under SDWA. These tests evaluate the operational integrity of the well so that underground sources of drinking water will not be endangered. EPA defines mechanical integrity as no significant leak in the casing, tubing, and packer and no significant fluid movement into a USDW through vertical channels adjacent to the injection wellbore.

mechanical jar *n*: a percussion tool operated mechanically to give an upward thrust to a fish by the sudden release of a tripping device inside the tool. If the fish can be freed by an upward blow, the mechanical jar can be very effective. Also called a hydraulic jar.

mechanical log *n*: a log of, for instance, rate of penetration or amount of gas in the mud, obtained at the surface by mechanical means. See *driller's log*, *mud log*.

mechanical nozzle *n*: used in a foam system where the venturi device and the agitation device are in the same nozzle.

mechanical perforating *n*: a procedure used to punch one hole at a time into the tubing string to circulate or to kill a well. Hydraulic jarring action fires a powder charge with enough force to drive a small button through the tubing, but not through the casing.

mechanical rig *n*: a drilling rig in which the source of power is one or more internal-combustion engines and in which the power is distributed to rig components through mechanical devices (such as chains, sprockets, clutches, and shafts). Also called a power rig. Compare *electric rig*.

mechanical riser handling tool *n*: a special device used to lower and pull riser joints on a floating offshore drilling rig using a subsea marine riser system. The tool duplicates a riser-joint connection, and is made up on the riser joint in the same way as another riser joint. The upper section of the handling tool is a standard drill pipe tool joint that is supported by the elevators during raising and lowering operations. See *riser pipe*.

mechanical-set packer *n*: a packer that is actuated by raising, lowering, or rotating the work string on which it is run. Such packers are often deployed in multizone completions, during well tests, and for water shutoff. See *packer*.

mechanical snubbing unit *n*: equipment that uses the hoisting equipment on the rig to snub pipe in or out of the hole. Snubbing is putting pipe in the hole or taking it out of the hole when the pressure in the wellbore is so high that it can force the pipe out of the hole if it is not restrained (snubbed). Basic components of a mechanical snubbing unit are the control heads (special BOPs) and operating manifold, stationary and traveling snubbers, snub line, and balance weight.

mechanical sticking *n*: a condition in which solid material such as sloughing shale, sand, or shale driven uphole by a blowout, or junk in the hole causes the drill stem to become stuck against the wall of the wellbore.

mechanical wireline *n*: see *nonconductive wireline*.

median *n*: a statistical measure of the midmost value, such that half the values in a set are greater and half are less than the median.

medical surveillance *n*: the process of obtaining baseline physicals for employees involved in emergency response or hazardous waste cleanup and monitoring their health at specified intervals. Under HAZWOPER, medical surveillance is applied to members of organized HAZMAT teams and hazardous materials specialists and any emergency response employee who exhibits signs or symptoms of exposure, or wears a respirator for thirty days or more a year, or is exposed to hazardous substances at or above the permissible exposure limits.

Mediterranean front *n*: in meteorology, the front that separates cold air masses over Europe from warm air masses over North Africa. See *front*.

medium-pressure distribution system *n*: see *high-pressure distribution system*.

medium-range forecast *n*: a weather forecast covering a week or less.

megajoule (MJ) *n*: the metric unit of service given by a hoisting line in moving 1,000 newtons of load over a distance of 1,000 metres.

megapascal *n*: one million pascals.

megger *n*: an instrument for measuring large amounts of resistance in a circuit or insulation. Its name comes from Mega Ohmmeter, a proprietary trade name.

megohms (Mohms, MΩ) *n*: 1,000,000 ohms. Sometimes expressed as 10^6 ohms.

melting point (mp) *n*: the temperature at which the solid and liquid states of a substance coexist in equilibrium. The melting point temperature of a substance is usually determined at normal atmospheric pressure.

membrane generated nitrogen *n*: nitrogen that is recovered from the atmosphere by sending air through several bundles of membranes that resemble straws. Oxygen and water vapor quickly penetrate the membrane and escape, which leaves only nitrogen to exit from the end of the membrane. Thousands of membranes are placed inside a stainless steel operating bundle, or canister.

meniscus *n*: the curved upper surface of a liquid column, concave when the containing walls are wet by the liquid (negative meniscus) and convex when not (positive meniscus).

MER *abbr*: maximum efficiency rate.

mercaptan *n*: a compound chemically similar to alcohol, but in which sulfur replaces oxygen in the chemical structure. Many mercaptans have an offensive odor; thus they are used as odorants in natural gas.

merchantable oil *n*: a crude oil in which the S&W content is not in excess of that allowed for purchase; therefore, it is salable.

merchantable title *n*: a title free from material defects or grave doubts and reasonably free from litigation, which can be sold or successfully defended in court; a court of equity will compel the vendor to accept such a title as sufficient. Also called marketable title or good title.

mercury barometer *n*: a barometer that uses mercury to measure atmospheric pressure. Cistern barometers and siphon barometers are the two types of mercury barometer.

mercury bell *n*: a device that directly measures differential pressure between two ports. In a sealed chamber, an inverted bell housing floats in a pool of mercury. Differential pressure moves the bell upward in the mercury to operate a pressure indicator.

mercury pump *n*: a device used to measure bulk volume of core samples by measuring the amount of mercury displaced by a core sample.

mercury vapor lamp *n*: a high-intensity discharge lamp in which the gas is vaporized mercury. Starts with the arcing of argon gas between the main electrodes and gives off a blue-white light.

meridian *n*: a north-south line from which longitudes and azimuths are reckoned.

meridonal *adj*: along, belonging to, relating to, or like a meridian. See *meridian*.

mesh *n*: a measure of the openings of a woven material, screen, or sieve; e.g., a 200-mesh sieve has 200 openings per linear inch. A 200-mesh screen with a wire diameter of 0.0021 inch (0.0533 millimetre) has an opening of 0.074 millimetre (74 microns), or will pass a particle of 74 microns. See *micron*.

Mesozoic era *n*: a span from 230 to 65 million years ago, the era of the dinosaurs and the first mammals.

metacenter (M) *n*: a point located somewhere on a line drawn vertically through the center of buoyancy of the hull of a floating vessel with the hull in one position (e.g., level) and then another (e.g., inclined). When the hull inclines to a new position, the center of buoyancy of the hull also moves to a new position. If a second line is drawn vertically through the new center of buoyancy, it intersects the first line at a point called the metacenter. Location of the metacenter is important because it affects the stability of floating vessels (such as mobile offshore drilling rigs).

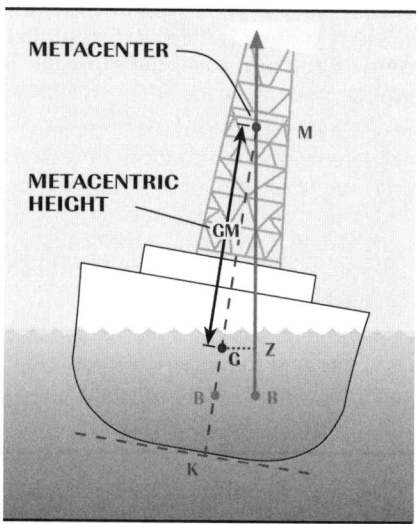

metacentric height (GM) *n*: the vertical distance between an offshore floating vessel's center of gravity and its metacenter. See *metacenter*.

metal *n*: opaque crystalline material, usually of high strength, that has good thermal and electrical conductivity, ductility, and reflectivity.

metal-clad (MC) cable *n*: in electricity, a bundle of several wire conductors, each of which is individually insulated, that is surrounded by a strong metal covering.

metal-edge strainer *n*: in an engine, a fuel filter that consists of several very thin metal discs stacked inside a housing. As fuel flows through the strainer, foreign matter in the fuel is trapped by the very small spaces between the discs.

metallic area *n*: the sum of the cross-sectional areas of all the wires in a wire rope or in a strand.

metallic circuit *n*: the path of electric current through the metallic portions of a corrosion cell.

metallic oxide semiconductor field-effect transistor (MOSFET) *n*: a field-effect transistor whose gate is insulated from the semiconductor's substrate by a thin layer of silicon dioxide.

metallic oxide varistor (MOV) *n*: a semiconductor with two electrodes whose resistance depends on the voltage flowing through it. Its resistance decreases as applied voltage increases. Also called a voltage-dependent resistor.

metal-petal basket *n*: see *cementing basket*.

metamorphic rock *n*: a rock derived from preexisting rocks by mineralogical, chemical, and structural alterations caused by heat and pressure within the earth's crust. Marble is a metamorphic rock.

metamorphism *n*: the process in which rock may be changed by heat and pressure into different forms.

meteorologist *n*: a scientist who studies the Earth's atmosphere and its patterns of climate and weather.

meter *n*: 1. a device used to measure and often record volumes, quantities, or flow rates of gases, liquids, or electric currents. 2. in the metric system of measurement, the unit of length; it equals about 3.28 feet, 39.37 inches, or 100 centimetres. *v*: to measure quantities or properties of a substance.

meter accumulator *n*: see *meter combinator*.

meter accuracy factor *n*: see *meter factor*.

meter bank *n*: fluid meters coupled in parallel. The sum of the fluid measured by the meter bank represents the total fluid measured.

meter calibration *n*: 1. the operation by which meter readings are compared with an accepted standard. 2. adjustment of a meter so that its readings conform to a standard.

meter calibration adjuster *n*: a device to enable the adjusting of a meter register to indicate true volume within acceptable tolerance.

meter capacity *n*: the maximum or minimum rate of flow recommended by the manufacturer to maintain a designated accuracy.

meter case *n*: the outer portion of a meter, which encloses the measuring chamber and other working parts.

meter characteristic *n*: see *meter factor*.

meter chart *n*: in gas measurement, a circular chart of special paper that shows the range of differential pressure and static pressure and is marked by the recording pens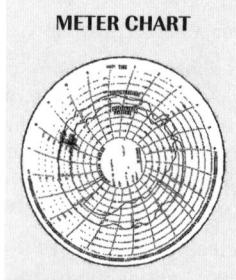

METER CHART

of a flow recorder in an orifice metering system. Charts are rapidly being replaced by electronic flow recorders.

meter combinator *n*: a device or system for accumulating the registration of two or more meters so that their total may be shown on a single readout device. Also called meter accumulator.

meter counter *n*: a counting device, electrical or mechanical, coupled to the measuring element to register the indicated volume that has passed through the meter.

meter difference *n*: difference in a volume determined from separate meters measuring the same gas stream, occasioned by the fact that gas volumes in field operations are not precisely determinable.

meter factor *n*: a number used to correct a meter's inaccuracy. The factor is derived by dividing the actual volume of liquid passed through a meter during proving by the volume registered by the meter. For subsequent metering operations, the actual throughput, or gross measured volume, is determined by multiplying the indicated volume registered on the meter by the meter factor.

meter fitting *n*: see *orifice fitting*.

meter flow rate *n*: the maximum rate of flow recommended by the meter manufacturer or authorized by a regulatory body. The maximum rate is determined by considerations of accuracy, durability, and pressure drop.

metering *n*: in process control, the instruments or recorders that monitor the quantity of variables in a process such as temperature, flow, level, and pressure.

metering manifold *n*: a collection of pipe and fittings, a manifold, in which is mounted a meter for measuring fluid flow through the manifold. Usually a meter is mounted in a manifold so fluid flow is diverted into it for measurement.

metering separator *n*: a complete separator and volume meter integrated into a single vessel. Two-phase units separate oil and gas and meter the oil; three-phase units separate oil, water, and gas and meter the oil and water.

metering tank *n*: see *measuring tank*.

meter installation *n*: in gas measurement, the orifice plate, orifice fitting, manifold, and flow recorder. Also called a meter run, meter station, orifice meter installation.

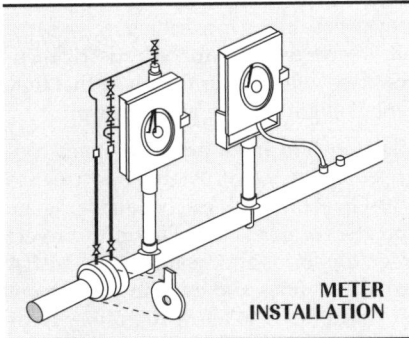

meter K-factor *n*: pulses per barrel. See *pulse generator*.

meter-line thermometer *n*: a thermometer located in a pipeline either immediately upstream or immediately downstream of the meter; used to determine the temperature of the liquid flowing through the meter.

meterman *n*: see *measurement technician*.

meter performance *n*: a general expression for the relationship between the volume registered by a meter and the actual volume that has passed through it. May refer to meter error, meter factor, meter accuracy, and so on.

meter pickup *n*: a device for converting meter rotor movement into an electrical output signal.

meter proof *n*: the multiple passes or round trips of the displacer in a prover for purposes of determining a meter factor. *v*: to establish the meter factor by comparing meter throughput to a prover of known volume.

meter prover *n*: a device used to determine the accuracy of a meter. Two popular types are pipe provers and open tank provers. A pipe prover has a calibrated pipe that contains a known volume. A sphere displaces oil or liquid product through the pipe while the meter being proved registers the amount of oil or liquid displaced. The meter's reading is then compared to the known volume in the pipe. An open tank prover is a calibrated tank that contains a known volume. Oil or liquid product is put into the tank while the meter being proved registers the amount. The meter's reading is then compared to the known volume in the tank.

meter proving *n*: the procedure required to determine the relationship between the true volume of liquid measured by a meter and the volume indicated by the meter.

meter proving counter *n*: a device in a pipe prover that, when actuated by the displacer sphere, registers pulses sent from a pulse generator on the meter being proved. By reading the number of pulses the counter registers, the amount of oil or liquid product the meter registered as having passed through the prover can be determined. The amount the meter registered is then compared to the known volume of the prover to determine the meter's accuracy. See *meter prover, pipe prover*.

meter proving run *n*: any single prover volume measurement in a set of prover volume measurements required to prove a meter.

meter reading *n*: the instantaneous display of the number of units of volume or equivalent thereof read directly from a meter register.

meter receiver instrument *n*: an instrument that receives signals from a transmitter.

meter register *n*: a device that accumulates and displays the indicated volume passed through a meter.

meter run *n*: see *meter installation*.

meter run point *n*: the point in a gas gathering system at which a field measuring meter and accessories are situated.

meter skid *n*: a small platform fabricated from pipe or bar stock and steel plate onto which a meter and a manifold are mounted.

meter slippage *n*: the volume of liquid at a given flow rate that passes through a meter without being measured.

meter station *n*: see *meter installation*.

meter tube *n*: an important part of the primary element of an orifice meter installation that must create a known flow pattern for the fluid as it reaches the plate. It is the straight upstream pipe of the same size between the orifice and nearest pipe fitting and the similar downstream pipe between the orifice and nearest pipe fitting.

metes and bounds *n pl*: a method of describing a piece of land that measures the boundaries by beginning at a well-marked reference point and following the boundaries of the land all the way around to the beginning point again. The description relies heavily on reference to natural or artificial but permanent objects (such as roads and streams).

methane *n*: a light, gaseous, flammable paraffinic hydrocarbon, CH_4, that has a boiling point of -25°F (and is the chief component of natural gas and an important basic hydrocarbon for petrochemical manufacture).

methane series *n*: the paraffin series of hydrocarbons.

methanol (methyl alcohol) *n*: the lightest alcohol, having the chemical formula CH_3OH. Also called wood alcohol.

methylene blue *n*: a dye that colors the reactive clays (bentonite and/or drilled solids) in a drilling mud sample for determining the percentage of clay in the sample.

methyl orange alkalinity *n*: see *M1*.

methyl tertiary butyl ether (MTBE) *n*: lead-free antiknock compound added to gasoline. Compare *ethyl tertiary butyl ether*.

metre (m) *n*: the fundamental unit of length in the international system of measurement (SI). It is equal to about 3.28 feet, 39.37 inches, or 100 centimetres.

metric system *n*: a decimal system of weights and measures based on the meter as the unit of length, the gram as the unit of weight, the cubic meter as the unit of volume, the liter as the unit of capacity, and the square meter as the unit of area. The international system of measurement (SI) is based on the metric system.

metric ton *n*: a measurement equal to 1,000 kilograms or 2,204.6 avoirdupois. In some oil-producing countries, production is reported in metric tons. One metric ton is equivalent to about 7.4 barrels (42 U.S. gallons = 1 barrel) of crude oil with a specific gravity of .0184, or 36° API. In the SI system it is called a tonne.

μF *abbr*: microfarad.

mF *abbr*: microfarad.

MFD *abbr*: manufacturer's abbreviation for microfarad; often stamped on the body of the capacitor.

MFE™ *abbr*: a trademark name for multiple formation evaluation. See *drill stem test*.

mg *abbr*: milligram.

mH *abbr*: millihenry.

mho *n*: unit of conductivity. Compare *ohm*.

mica *n*: a silicate mineral characterized by sheet cleavage; i.e., it separates into thin sheets. Biotite is ferromagnesian black mica, and muscovite is potassic white mica. Sometimes mica is used as a lost circulation material in drilling, and as a dielectric (insulator) in electrical components such as capacitors.

mica capacitor *n*: a capacitor whose dielectric (insulating material) is made from mica. See *capacitor, mica*.

micellar *adj*: capable of forming micelles.

micellar-polymer flooding *n*: a method of improved oil recovery in which chemicals dissolved in water are pumped into a reservoir through injection wells to mobilize oil left behind after primary or secondary recovery and to move it toward production wells. The chemical solution includes surfactants or surfactant-forming chemicals that reduce the interfacial and capillary forces between oil and water, releasing the oil and carrying it out of the pores where it has been trapped. The solution may also contain cosurfactants to match the viscosity of the solution to that of the oil to stabilize the solution and to prevent its adsorption by reservoir rock. An electrolyte is often added to aid in adjusting viscosity. Injection of the chemical solution is followed by a slug of water thickened with a polymer, which pushes the released oil through the reservoir, decreases the effective permeability of established channels so that new channels are opened, and serves as a mobility buffer between the chemical solution and the final injection of water.

micelle *n*: a round cluster of hydrocarbon chains formed when the amount of surfactant in an aqueous solution reaches a critical point. The micelles are able to surround and dissolve droplets of water or oil, forming an emulsion.

microannulus *n*: a space that is left as casing contracts back to its normal size after pulling away from cement in the annulus.

microcaliper log *n*: a special caliper log combined with a Microlog™. See *caliper log, Microlog™*.

microemulsion *n*: a stable, translucent micellar emulsion of oil, one or more surface-active agents, water, and sometimes an electrolyte. A microemulsion is classed as an emulsion because it is a mixture of immiscible substances (oil and water) and because it can have oil-in-water and water-in-oil phases. It can also have a phase in which neither oil nor water is dispersed, however, but are alternated in layers. A microemulsion has certain properties that are like a solution rather

than an emulsion: it is optically clear rather than clouded, and it is stable, i.e., the oil and water do not separate. Micro-emulsions are used for chemical flooding of reservoirs. See *chemical flooding, emulsion, solution*.

microfarad (mF, μF) *n*: one-millionth (10^{-6}) of a farad. See *farad*.

Microlaterolog *n*: a guard-electrode log that measures the resistivity of the flushed zone.

Microlog™ *n*: trade name for a special electric survey method in which three closely-spaced electrodes are pressed against the wall of the borehole to obtain a measurement of formation characteristics next to the wall of the hole.

micrometer *n*: 1. a caliper for making precise measurements; a spindle is moved by a screw thread so that it touches the object to be measured. 2. an instrument used with a telescope or microscope to measure minute distances. 3. in the metric system of measurement, one-millionth of a meter.

micron (μ) *n*: a unit of length equal to one millionth part of a metre, or one thousandth part of a millimetre, or 0.000039 inch. Also called a micrometre.

micropaleontology *n*: paleontology dealing with fossils of microscopic size.

micropore paper *n*: heavy-duty paper perforated with several very small holes (pores). Folded in accordion pleats, micropore paper often serves as a secondary filter element in an engine's fuel system. Fuel passes through the pores, while the unperforated part of the paper stops dirt.

microprocessor *n*: a single silicon chip on which the arithmetic and logic functions of a computer are placed.

microresistivity log *n*: a resistivity logging tool consisting of a spring device and a pad. While the spring device holds the pad firmly against the borehole sidewall, electrodes in the pad measure resistivities in mud cake and nearby formation rock. See *resistivity well logging*.

microsecond *n*: one millionth of a second.

MicroSurvey *n*: a proprietary name for a logging tool that measures resistivity of the flushed zone of the formation adjacent to the wellbore.

MICT *abbr*: moving in cable tools; used in drilling reports.

middle distillate *n*: hydrocarbons in the middle range of refinery distillation, e.g., kerosene, light and heavy diesel oil, heating oil.

middle sample *n*: in tank sampling, a spot sample obtained from the middle of the tank contents.

middle spot sample *n*: taken on tanks that have a capacity larger than 1,000 barrels (159,000 litres) and that contain 10 feet (3 metres) or less of crude oil. One middle spot sample should be taken as near the center of the vertical column of oil as possible.

migration *n*: the movement of oil from the area in which it was formed to a reservoir rock where it can accumulate.

migratory bird *n*: a bird that moves (migrates) from one place to another, usually in connection with changing seasons. During times of migration, such birds, especially those attracted to bodies of water, may choose to rest in a drilling rig's reserve pit. To prevent harm to the birds, rig owners place netting over the pit so that they cannot alight in the liquid in the pit.

Migratory Bird Treaty Act (MBTA) *n*: a congressional act that established treaties with Great Britain, Mexico, Canada, Japan, and the former Soviet Union to protect migratory birds and their habitats. MBTA authorizes fines and even imprisonment for operators who allow migratory birds to become injured in pits or open-topped tanks that contain oil, oil products, caustic materials, or contaminants, such as arsenic or boron, that are poisonous or hazardous to migratory birds.

mil *n*: a distance equal to 0.001 (1/1,000, or 10^{-3}) inches.

MIL *abbr*: military standard.

mil-foot *n*: a measurement of the length of a wire or conductor based on its cross-sectional diameter; a wire that is 1 foot long and has a diameter of 0.001 inches is 1 mil-foot.

military standard (MIL) *n*: specifications for equipment and procedures purchased by U.S. military forces. One source of MILs is Global Engineering Documents, 15 Inverness Way East, Englewood, CO 80112: 303-397-7956 or 800-854-7179; www.global.his.com.

milk emulsion *n*: see *oil-emulsion water*.

mill *n*: a downhole tool with rough, sharp, extremely hard cutting surfaces for removing metal, packers, cement, sand, or scale by grinding or cutting. Mills are run on drill pipe or tubing to grind up debris in the hole, remove stuck portions of drill stem or sections

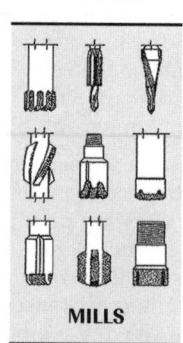

MILLS

of casing for sidetracking, and ream out tight spots in the casing. They are also called junk mills, reaming mills, and so forth, depending on their use. *v*: to use a mill to cut or grind metal objects that must be removed from a well.

mill-coated pipe *n*: pipe coated at the mill as opposed to pipe coated over the ditch in pipeline construction.

milled bit *n*: also called a milled-tooth bit or a steel-tooth bit. See *steel-tooth bit*.

milled-tooth bit *n*: also called milled bit or steel-tooth bit. See *steel-tooth bit*.

mill extension *n*: a special mill made up on the bottom of an overshot to grind away the top of a damaged fish so that the overshot can engage the fish. See *overshot*.

millibar *n*: a unit of atmospheric pressure equal to one thousandth of a bar. See *bar*.

millidarcy (md) (*pl*, **millidarcys**) *n*: one-thousandth of a darcy.

millihenry (mH) *n*: one-thousandth of a henry, or 10^{-3} henrys.

milli-inch *n*: a measure of the diameter of an electrical wire or cable expressed in thousandths of an inch. For example, if a wire's diameter is 0.1 inch, it is also 0.1 × 1,000 or 100 milli-inches.

millilitre *n*: in the SI system of measurement, one-thousandth of a litre. In analyzing drilling mud, this term is used interchangeably with cubic centimetre. A quart equals 964 millilitres.

millimetre (mm) *n*: a measurement unit in the SI system equal to 10^{-3} metre (0.001 metre). It is used to measure pipe and bit diameter, nozzle size, liner length and diameter, and cake thickness.

millimho *n*: one-thousandth of an mho.

milling shoe *n*: see *burn shoe, rotary shoe*.

milling tool *n*: the tool used in the operation of milling. See *mill*.

millisec *abbr*: millisecond.

millivolt *n*: one-thousandth of a volt.

mill out *v*: to use a mill on the end of a workstring to remove a permanent tool or fish.

mill-out extension *n*: a pinned-end pup joint used to provide additional length and inside diameter necessary to accommodate a standard milling tool.

mill scale *n*: thin, dense oxide scale that forms on the surface of newly manufactured steel as it cools. Mill scale can become cathodic to its own steel base, forming galvanic corrosion cells.

mil size *n*: see *circular mil*.

min *abbr*: minute.

mined (humic) acid lignins *n pl*: naturally occurring special lignite, e.g., leonardite, that is produced by strip mining from special lignite deposits. The active ingredient is the humic acids. Mined lignins are used primarily as thinners, which may or may not be chemically modified; however, they are also widely used as emulsifiers.

mineral *n*: 1. a naturally occurring inorganic crystalline element or compound with a definite chemical composition and characteristic physical properties such as crystal shape, melting point, color, and hardness. Most minerals found in rocks are not pure. 2. broadly, a naturally occurring homogeneous substance that is obtained from the ground for human use (e.g., stone, coal, salt, sulfur, sand, petroleum, water, natural gas).

mineral acre *n*: the full mineral interest and rights in one acre of land.

mineral deed *n*: the legal instrument that conveys minerals in place together with the rights to search for and produce them.

mineral estate *n*: rights and interests in the minerals found on or beneath the surface of land, created when the owner severs or separates his or her interests in the property.

mineral owner *n*: owner of the rights and interests in a mineral estate (where interests in a landed estate have been severed). Compare *surface owner*.

mineral rights *n pl*: the rights of ownership, conveyed by deed, of gas, oil, and other minerals beneath the surface of the earth. In the United States, mineral rights are the property of the surface owner unless disposed of separately.

Minerals Management Service (MMS) *n*: an agency of the U.S. Department of the Interior that establishes requirements through the Code of Federal Regulations (CFR) for drilling while operating on the Outer Continental Shelf of the United States. The agency regulates rig design and construction, drilling procedures, equipment, qualification of personnel, and pollution prevention. Address: 1849 C Street NW, Mail Stop 4230; Washington, DC 20240; 202-208-3500.

Mine Safety and Health Administration (MSHA) *n*: a U.S. government agency that evaluates research in the causes of occupational diseases and accidents. It is responsible for administration of the certification of respiratory safety equipment. Address: 4015 Wilson Blvd.; Arlington, VA 22203; 703-235-1385.

miniaturized completion *n*: a well completion in which the production casing is

less than 4.5 inches (11.43 centimetres) in diameter. Compare *conventional completion*.

miniconnector *n*: on floating offshore rigs using a marine riser system and a subsea blowout preventer stack, a device used to connect the choke and kill lines from the blowout preventer to the lower marine riser package's connector. A miniconnector has a pressure rating of 15,000 psi (105,000 kPa) and an inside diameter of 3 inches (76.2 millimetres). See *lower marine riser package*.

minilog *n*: see *pad resistivity device*.

minimum acceptance strength *n*: a measure of the strength of a wire rope. Minimum acceptance strength is 2 percent lower than the catalog or nominal strength. See *nominal strength*.

minimum internal yield pressure *n*: the lowest internal pressure at which permanent distortion of pipe occurs.

minimum pipeline velocity *n*: the velocity that exists at the lowest operating flow rate, excluding those rates that occur infrequently or for periods of less than five minutes.

minimum polished rod load *n*: the lowest load imposed on the polished rod throughout a complete sucker rod pump cycle.

minimum royalty *n*: a royalty payment amount to be made regardless of the rate of production. The excess of such payments over regular royalty is chargeable against future production, if any, accruing to the royalty interest.

minimum yield strength *n*: the stress level of tubular drill pipe above which permanent deformation will occur.

mining partnership *n*: a form of joint venture very similar to an oil and gas joint operating agreement. Profits, losses, operations, and ownership are all shared, and the partners are jointly as well as severally (separately) liable.

minuend *n*: in mathematics, the quantity from which another quantity is to be subtracted. For example, in the expression 38 − 23 = 15, 38 is the minuend.

minus input *n*: see *inverting input*.

minutes per repeat *n*: in process control, the frequency, in minutes, with which a proportional-plus-reset control system changes the final control element to correct for differences in the desired result. The proportional-plus-reset control system constantly monitors the process for deviation from the set point and performs corrections in discrete time intervals to maintain the set point.

MIR *abbr*: moving in rig; used in drilling reports.

miscibility *n*: the capability of mixing together to form a single homogeneous phase. For example, alcohol and oil are miscible, but water and oil are immiscible (not miscible). Certain fluids, such as carbon dioxide and reservoir hydrocarbons, are immiscible on first contact but can become miscible after multiple contacts under suitable reservoir conditions of temperature and pressure. This situation is termed conditional, dynamic, or multiple-contact miscibility.

miscibility pressure *n*: the pressure at which a gas injected into a reservoir vaporizes hydrocarbons from crude oil to form a miscible transition zone between the gas and the crude oil. This liquid transition zone moves reservoir oil toward production wells. Miscibility pressure depends on the characteristics of the gas and the temperature and gravity of the reservoir crude oil. For carbon dioxide, hydrocarbon vaporization begins at pressures between 1,000 pounds per square inch (6,895 kilopascals) and 2,000 pounds per square inch (13,790 kilopascals) at temperatures below 200°F (93.3°C) with oils above 30°API. Several hundred additional points per square inch are required to reach the optimum miscibility pressure, at which sufficient hydrocarbons are vaporized to produce a profitable amount of oil.

miscible *adj*: 1. capable of being mixed. 2. capable of mixing in any ratio without separation of the two phases.

miscible displacement *n*: see *miscible drive*.

miscible drive *n*: a method of enhanced recovery in which various hydrocarbon solvents or gases (such as propane, LPG, natural gas, carbon dioxide, or a mixture thereof) are injected into the reservoir to reduce interfacial forces between oil and water in the pore channels and thus displace oil from the reservoir rock. See *chemical flooding, gas injection, micellar-polymer flooding*.

miscible flood *n*: see *miscible drive*.

mist *n*: an air or gas drilling fluid that consists of 97 to 99 percent air or gas and 1 to 3 percent liquid (usually water). The liquid is not emulsified in the air or gas.

mist drilling *n*: a drilling technique that uses air or gas to which a foaming agent has been added. Also called foam drilling. Underbalanced drilling in which mist is used as the drilling fluid. See *mist, underbalanced drilling*.

mist extractor *n*: a metal device used to remove small droplets of moisture or condensable hydrocarbons from a gas

stream in an oil and gas separator. The small droplets of moisture collect on the metal surface to form larger drops, which are removed from the separator along with other separated liquids.

MIT *abbr*: mechanical integrity test.

mixed-base crude oils *n pl*: the heavier fractions of some crude oils that contain considerable amounts of both paraffins and asphalt.

mixed butane *n*: see *field-grade butane*.

mixed fluid reservoir *n*: 1. a geological formation that contains gas entrained in (mixed with) liquids, such as oil and water, in the formation. 2. on an accumulator (blowout preventer operating unit), one of two containers (reservoirs). One reservoir stores the fluid concentrate and the other stores the mixed fluid. Using two reservoirs doubles the storage capacity of the accumulator.

mixed gas *n*: in diving, oxygen and one or more inert gases.

mixed-gas diving *n*: diving in which a diver uses a breathing medium of oxygen and one or more inert gases synthetically mixed.

mixed naphtha *n*: a combination of straight-run and cracked naphtha.

mixed string *n*: a combination string. See *graded string*.

mixed tide *n*: a tide that has characteristics of both semidiurnal and diurnal tides.

mixing mud *n*: preparation of drilling fluids from a mixture of water and other fluids and one or more of the various dry mud-making materials such as clay and chemicals.

mixing tank *n*: any tank or vessel used to mix components of a substance (as in the mixing of additives with drilling mud).

mixing valve *n*: a valve to which fluids with different characteristics are routed to obtain a fluid that combines the characteristics of the original fluids. For example, a cold fluid and a hot fluid may flow through a mixing valve to achieve a median temperature between the two extremes.

mix mud *v*: to prepare drilling fluids from a mixture of water or other liquids and any one or more of the various dry mud-making materials (such as clay, weighting materials, and chemicals).

mixture *n*: 1. a physical combination of two or more elements or components that maintain their chemical identity; no chemical reaction is involved. 2. under OSHA, any combination of two or more chemicals if the combination is not, in whole or in part, the result of a chemical reaction.

MJ *abbr*: megajoule.

ml *abbr*: millilitre.

mm *abbr*: millimetre.

MM bopd *abbr*: millions of barrels of oil per day.

MMBtu *abbr*: one million Btus.

MMcf *abbr*: million cubic feet; a common unit of measurement for large quantities of gas.

mmf *abbr*: magnetomotive force.

MMI *abbr*: man-machine interface.

MMS *abbr*: Minerals Management Service.

MMscf *abbr*: million standard cubic feet. The standard referred to is usually 60°F and 1 atmosphere (14.7 pounds per square inch) of pressure, but it varies from state to state.

MMscf/d *abbr*: million standard cubic feet per day.

mm² *abbr*: square millimetre.

mm³ *abbr*: cubic millimetre.

MO *abbr*: moving out; used in drilling reports.

mobile arctic caisson (MAC) *n*: a submersible offshore drilling rig designed to drill in the ice-choked waters of the arctic. It consists of a caisson base, which is a large concrete or steel tube, that rests on the seafloor. The drilling equipment is installed on top of the caisson above the water (ice) line. The caisson is built to withstand the enormous force of moving pack ice.

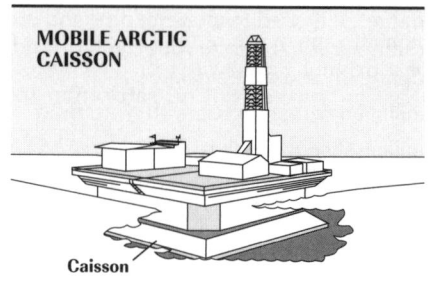

MOBILE ARCTIC CAISSON

Caisson

mobile formation *n*: a formation that squeezes into the wellbore because overburden forces compress it. When a hole is drilled through such a formation, overburden pressure can cause the formation to reduce the size of the wellbore if the hydrostatic pressure of the drilling fluid is not sufficient to prevent it.

mobile offshore drilling unit (MODU) *n*: a drilling rig that drills offshore exploration and development wells. It floats on the surface of the water when being moved from one well site to another, but it may or may not float once drilling begins. Two basic types of mobile offshore drilling units are used to drill most offshore wildcat wells: bottom-supported offshore drilling rigs and floating drilling rigs. Bottom-supported units include jackups, and floating rigs include semisubmersibles and drill ships.

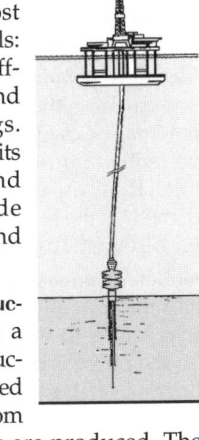

MODU

mobile offshore production unit (MOPU) *n*: a movable, reusable structure, such as a converted jackup drilling rig, from which offshore wells are produced. The main advantage over a conventional fixed or compliant platform is that MOPUs are less expensive.

mobility *n*: a measure of the ease with which a fluid moves through a reservoir. Mobility is measured by dividing a formation's permeability to a fluid by the fluid's viscosity.

mobility buffer *n*: a polymer-water solution used in micellar-polymer flooding as a zone of viscosity transition between the initial injection of surfactant-water solution and the final injection of drive water. The lead edge of the polymer solution has a mobility equal to or less than that of the surfactant solution, and the trailing edge has a mobility close to that of water. This buffer zone prevents the drive water from fingering past the surfactant solution and thus reducing sweep efficiency. See *micellar-polymer flooding.*

mobility ratio *n*: the ratio of the mobility of a driving fluid (water or a chemical solution) at residual oil saturation to the mobility of the driven fluid (oil) at connate water saturation. The mobility ratio affects the sweep efficiency of an improved recovery project.

mode *n*: in process control, a given condition of functioning; a status. For example, a process system may be operating in the dump mode, wherein the system is removing liquid from a vessel to achieve a desired lower level.

model *n*: a description or analogy used to help visualize something that cannot be observed directly, such as the structures, porosities, and permeabilities of a geological formation.

Model 70 connector *n*: a brand of a hydraulic connector used in subsea blowout preventer stacks that uses individual hydraulic cylinders located in the connector's lower body to operate an actuator ring. The connector has nine cylinders, six

of which are primary, and three of which are secondary.

moderate gale *n*: in the Beaufort wind scale, a wind whose speed is from 28 to 33 knots (32 to 38 miles per hour or 52 to 61 kilometres per hour). See *Beaufort scale.*

modified cement *n*: a cement whose properties, chemical or physical, have been altered by additives.

MODU *abbr*: mobile offshore drilling unit.

modular-spaced workover rig *n*: work-over equipment designed in equipment packages or modules that are light enough to be lifted onto an offshore platform by a platform crane. In most cases, the maximum weight of a module is 12,000 pounds (5,443 kilograms). Once lifted from the work boat, the rig can be erected and working within twenty-four to thirty-six hours.

modulate *v*: in electronics, to vary the amplitude, frequency, or phase of a signal's wave. Modulating a wave with electronic information enables the wave to carry the information over a distance. A receiver, when tuned to the correct frequency, can demodulate the wave to make the information available for interpretation and understanding.

modulus of elasticity *n*: the ratio of the increment of some specified form of stress to the increment of some specified form of strain, such as bulk modulus, shear modulus, or Young's modulus. Also called coefficient of elasticity, elasticity modulus, elastic modulus.

Mohm *abbr*: megohm, or 1,000,000 ohms.

MΩ *abbr*: megohm, or 1,000,000 ohms.

MOJO connector *n*: a type of connector used at wire/chain interfaces used in a combination wire/chain system.

mol *sym*: mole.

molar volume *n*: the volume occupied by one mole of a substance.

molasse *n*: a thick sequence of soft, ungraded, cross-bedded, fossiliferous marine and terrestrial conglomerates, sandstones, and shales derived from the erosion of growing mountain ranges.

molded depth *n*: the vertical distance from the baseline of a ship or offshore vessel to the underside of the deck plating at the side, measured at the mid-length of the vessel.

molded dimensions *n pl*: the dimensions of a vessel to the molded lines.

molded draft *n*: the depth of the vessel below the waterline, measured vertically from the baseline to the waterline.

molded lines *n pl*: the lines defining the geometry of a hull as a surface without thickness.

molded volume *n*: the volume of a compartment without deduction for internal structure or fittings.

mole *n*: the fundamental unit of mass of a substance. A mole of any substance is the number of grams or pounds indicated by its molecular weight. For example, water, H_2O, has a molecular weight of approximately 18. Therefore, a gram-mole of water is 18 grams of water; a pound-mole of water is 18 pounds of water. See *molecular weight.*

molecular sieves *n pl*: synthetic zeolites packaged in bead or pellet form for (1) use in recovering contaminants or impurities from liquid and vapor product streams by selective adsorption and (2) use as a catalyst.

molecular weight *n*: the sum of the atomic weights in a molecule. For example, the molecular weight of water, H_2O, is 18, because the atomic weight of each of the hydrogen molecules is 1 and the atomic weight of oxygen is 16. See *mole.*

molecule *n*: the smallest particle of a substance that retains the properties of the substance. It is composed of one or more atoms.

mole percent *n*: the ratio of the number of moles of one substance to the total number of moles in a mixture of substances, multiplied by 100 (to put the number on a percentage basis).

moment *n*: 1. a turning effect created by a force, F, acting at a perpendicular distance, S, from the center of rotation. 2. the product of a force and a distance to a particular axis or point. 3. torque.

moment of force *n*: in relation to offshore floating rigs, a measure of the effectiveness of a force that consists of the product of the force and the perpendicular distance from the line of action of the force to the axis of rotation.

moment of inertia (I) *n*: the sum of the products formed by multiplying the mass (or sometimes, the area) of each element of a figure by the square of its distance from a specified line. Also called rotational inertia.

M1 *n*: the methyl orange alkalinity of the filtrate, reported as the number of millimetres of 0.02 normal (N/50) acid required per millimetre of filtrate to reach the methyl orange end point (pH 4.3). M1 is a measure of the alkalinity of a drilling fluid. Usually, drilling fluids should have a high pH value (alkalinity) to ensure that they perform as they should.

Monel steel *n*: a nickel-base alloy containing copper, iron, manganese, silicon, and carbon. Corrosion-resistant parts are often made of this material.

monitor *n*: an instrument that reports the performance of a control device or signals if unusual conditions appear in a system. For example, an S&W monitor provides a mechanical means of preventing contaminated oil from entering the pipeline by detecting the presence of excessive water and actuating valves to divert the flow back to dehydration facilities.

monkeyboard *n*: the derrickhand's working platform. As pipe or tubing is run into or out of the hole, the derrickhand must handle the top end of the pipe, which may be as high as 90 feet (27 metres) or higher in the derrick or mast. The monkeyboard provides a small platform to raise the derrickhand to the proper height for handling the top of the pipe.

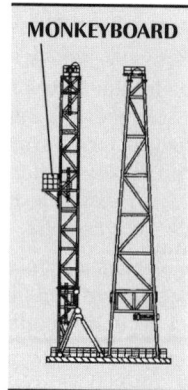

MONKEYBOARD

monocline *n*: rock strata that dip in one direction only. Compare *anticline, syncline.*

MONOCLINE

monoethylene glycol *n*: a chemical (an antifreeze) that is injected into lines and vessels to prevent hydrates from forming in the lines and vessels where hydrocarbon gases and water vapor are present at low temperatures. See *hydrate.*

monomer *n*: a simple molecule capable of linking together as a repeating structural unit to form a polymer.

monsoon *n*: a seasonal wind caused primarily by differences in temperature between a land mass and an adjacent oceanic region. The wind blows from land to sea in the winter months and from sea to land in the summer months.

monsoon fog *n*: fog that may occur in summer monsoons where high humidity and large differences in temperature between land and surrounding oceans are present.

montmorillonite *n*: a clay mineral often used as an additive to drilling mud. It is a hydrous aluminum silicate capable of reacting with such substances as magnesium and calcium. See *bentonite.*

moon pool *n*: a walled hole or well in the hull of a drill ship, ship-shape barge, or semisubmersible drilling rig (usually in the center) through which the drilling assembly and other assemblies pass while a well is being drilled, completed, or abandoned.

MOON POOL

moon pool installed *adj*: of equipment installed through the moon pool of an offshore floating vessel. Using the rig to install subsea equipment eliminates the need for an additional lift vessel.

moor *v*: to fix a boat, ship, floating offshore rig, or other vessel to one place in the water with cables, chains, and anchors.

mooring analysis *n*: calculation performed as part of the design of a mooring system to establish mooring loads, tensions, and safety factors.

mooring line *n*: a cable, chain, or a combination of cable and chain that links a floating offshore rig or a ship to an anchor that grips the seafloor while the vessel is stationary. Offshore rigs usually employ several mooring lines and anchors.

mooring line tension *n*: force imposed on a mooring line in the direction of a mooring line as it enters a vessel's fairlead.

MOP *abbr*: margin of overpull.

MOPU *abbr*: mobile offshore production unit.

Morison's equation *n*: a mathematical formula that is of particular importance in evaluating hydrodynamic forces on structures in fluid media, such as water. Among other factors, the equation considers inertia, drag, the density of acceleration, and speed of water particles.

morning report *n*: see *daily drilling report.*

morning tour (pronounced "tower") *n*: see *graveyard tour.*

mortgage *n*: an estate created by a conveyance absolute in form but intended to secure the performance of some act, such as the payment of money, and to become void if the act is performed in agreement with terms. Compare *deed of trust.*

MOSFET *abbr*: metallic oxide semiconductor field-effect transistor.

mosquito bill *n*: a tube mounted at the bottom of a sucker rod pump and inside a gas

anchor to provide a conduit into the pump for well fluids that contain little or no gas.

mother hubbard *n*: (slang) also called a mud box or mud saver. See *mud box.*

Mother Hubbard clause *n*: a clause in an oil and gas lease that includes the lease lands that may be owned by the lessor and inadvertently omitted from the legal description. These are usually oddly-shaped bits owned by the lessor and adjoining the described tract. Also called coverall clause.

motion compensator *n*: any device (such as a bumper sub or heave compensator) that serves to maintain constant weight on the bit in spite of vertical motion of a floating offshore drilling rig.

motor *n*: any of various power units, such as hydraulic, air, or electric devices, which develop energy or impart motion. Compare *engine.* See *electric motor.*

motor-generator rig *n*: see *electric rig.*

motorhand *n*: the crew member on a rotary drilling rig, usually the most experienced rotary helper, who is responsible for the care and operation of drilling engines. Also called motorman. Compare *fireman.*

motorman *n*: see *motorhand.*

motor slip *n*: see *slip.*

motor valve *n*: a valve operated by power other than manual (i.e., hydraulic, electric, or mechanical).

mousehole *n*: an opening in the rig floor, usually lined with pipe, into which a length of drill pipe is placed temporarily for later connection to the drill string.

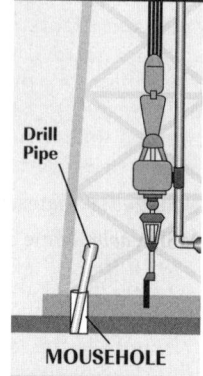

Drill Pipe

MOUSEHOLE

mousehole connection *n*: the procedure of adding a length of drill pipe or tubing to the active string. The length to be added is placed in the mousehole, made up to the kelly, then pulled out of the mousehole

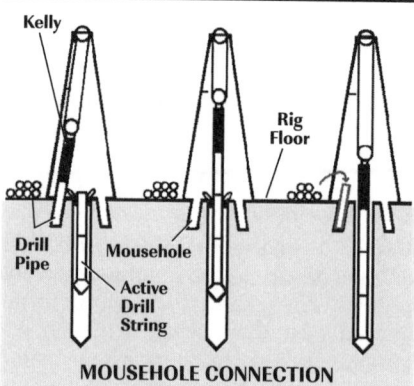

Kelly

Rig Floor

Drill Pipe   Mousehole

Active Drill String

MOUSEHOLE CONNECTION

and subsequently made up into the string. Compare *rathole connection*.

mousetrap *n*: a fishing tool used to recover a parted string of sucker rods from a well.

mousse *n*: the semisolid, gel-like masses of water-in-oil emulsions that form from heavier crude oils after a spill. Sometimes called "chocolate mousse." The masses may contain up to 80% water and are more viscous than the component oil. Mousse weathers slowly, with the oil retaining its initial toxicity, and may become tar balls that float at or near the sea surface and strand on beaches. The transport of mousse and tar balls can increase the area and duration over which a spill is felt.

MOV *abbr*: metallic oxide varistor.

Moyno pump *n*: a positive-displacement pump in which a helical rod (rotor) rotates within a body that has an eccentrically shaped chamber (stator). As the rod rotates, small pockets of fluid in the stator are moved out of the pump. It is often used in downhole motors. See *positive-displacement downhole mud motor*.

MPa *abbr*: megapascal.

mph *abbr*: miles per hour.

MPI *abbr*: magnetic particle inspection.

MPL *abbr*: multiposition lock.

MPPRCA *abbr*: Marine Plastic Pollution Research and Control Act of 1987.

ms *abbr*: microsecond.

Mscf/bbl *abbr*: thousand standard cubic feet per barrel.

Mscf/D *abbr*: thousand standard cubic feet per day.

MSDS *abbr*: material safety data sheet.

MSHA *abbr*: Mine Safety and Health Administration.

MSRC *abbr*: Marine Spill Response Corporation.

MTBE *abbr*: methyl tertiary butyl ether.

mud *n*: the liquid circulated through the wellbore during rotary drilling and workover operations. In addition to its function of bringing cuttings to the surface, drilling mud cools and lubricates the bit and the drill stem, protects against blowouts by holding back subsurface pressures, and deposits a mud cake on the wall of the borehole to prevent loss of fluids to the formation. Although it originally was a suspension of earth solids (especially clays) in water, the mud used in modern drilling operations is a more complex, three-phase mixture of liquids, reactive solids, and inert solids. The liquid phase may be fresh water, diesel oil, or crude oil and may contain one or more conditioners. See *drilling fluid*.

mud acid *n*: a mixture of hydrochloric and hydrofluoric acids and surfactants used to remove wall cake from the wellbore.

mud additive *n*: any material added to drilling fluid to change some of its characteristics or properties.

mud agitator *n*: see *agitator*.

mud analysis *n*: examination and testing of drilling mud to determine its physical and chemical properties.

mud analysis logging *n*: a continuous examination of the drilling fluid circulating in the wellbore for the purpose of discovering evidence of oil or gas regardless of the quantities entrained in the fluid. When this service is utilized, a portable mud logging laboratory is set up at the well. Also called mud logging.

mud anchor *n*: a large-diameter pipe installed outside a gas anchor to reduce or eliminate the entrance of solids into a sucker rod pump.

mud balance *n*: a beam balance consisting of a cup and a graduated arm carrying a sliding weight and resting on a fulcrum. It is used to determine the density or weight of drilling mud.

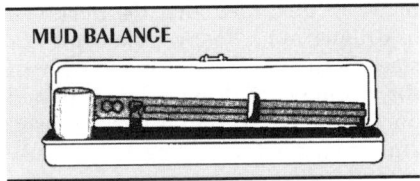

MUD BALANCE

mud booster line *n*: in drilling from floating offshore drilling rigs that use a subsea blowout preventer stack, a line (pipe) sometimes provided on riser joints to increase the return velocity of the mud in the riser. It helps move heavy cuttings up the riser when normal pump circulation is inadequate. The mud booster line is connected to a pump on the rig that enables the driller to circulate drilling fluid from the surface to the bottom of the riser.

mud box *n*: a hinged, cylindrical metal device placed around a joint of pipe as it is being broken out during a trip out of the hole. It keeps mud from splashing beyond the immediate area. Also called mother hubbard, mud saver, splash box, or wet box.

mud cake *n*: the sheath of mud solids that forms on the wall of the hole when liquid

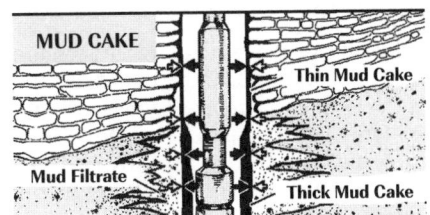

MUD CAKE — Thin Mud Cake — Mud Filtrate — Thick Mud Cake

from mud filters into the formation. Also called filter cake or wall cake.

mud cap *n*: a slug of heavyweight mud injected into the annulus of a well to prevent fluids in the annulus from escaping to the surface. The cap does not fill the annulus; instead, it remains in the annulus as a relatively short length of heavy mud. Because the cap does not fill the annulus, bottomhole pressure is not increased to the extent that it would be if the annulus were full of the mud. Mud caps can minimize or prevent lost circulation and other downhole problems.

mud-cap drilling *n*: drilling with a mud cap in the annulus. See *mud cap*.

mud centrifuge *n*: a device that uses centrifugal force to separate small solid components from liquid drilling fluid.

mud circulation *n*: the process of pumping mud downward to the bit and back up to the surface in a drilling or workover operation. See *normal circulation*, *reverse circulation*.

mud cleaner *n*: a cone-shaped device, a hydrocyclone, designed to remove very fine solid particles from the drilling mud.

mud column *n*: the borehole when it is filled or partially filled with drilling mud.

mud conditioning *n*: the treatment and control of drilling mud to ensure that it has the correct properties. Conditioning may include the use of additives, the removal of sand or other solids, the removal of gas, the addition of water, and other measures to prepare the mud for conditions encountered in a specific well.

mud density *n*: see *mud weight*.

mud density recorder *n*: a device that automatically records the weight or density of drilling fluid as it is being circulated in a well.

mud engineer *n*: an employee of a drilling fluid supply company whose duty it is to test and maintain the drilling mud properties that are specified by the operator.

mud filtrate *n*: the liquid which is expelled from drilling mud during the formation of mud cake.

mud-flow indicator *n*: a device that continually measures and may record the flow rate of mud returning from the annulus and flowing out of the mud return line. If the mud does not flow at a fairly constant rate, a kick or lost circulation may have occurred.

mud-flow sensor *n*: see *mud-flow indicator*.

mud-gas separator *n*: a device that removes gas from the mud coming out of a well when a kick is being circulated out or when the well is being drilled underbalanced and gas must be removed from the liquid part of the drilling fluid.

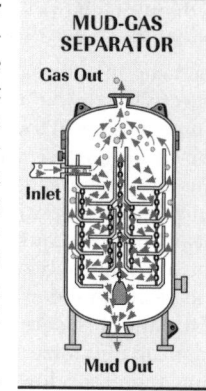

MUD-GAS SEPARATOR

Gas Out

Inlet

Mud Out

mud gradient *n*: pressure exerted with depth by drilling fluid. Often expressed in pounds per square inch per foot. Also called pressure gradient.

mud gun *n*: a device that shoots a jet of drilling mud under high pressure into the mud pit to mix additives with the mud or to agitate the mud.

mud hopper *n*: see *hopper*.

mud hose *n*: also called kelly hose or rotary hose. See *rotary hose*.

mud hound *n*: see *mud engineer*.

mud house *n*: structure at the rig to store and shelter sacked materials used in drilling fluids.

mud inhibitor *n*: a substance, such as salt, potassium chloride, or calcium sulfate, added to drilling mud to minimize the hydration (swelling) of formations with which the mud is in contact.

Mud-kil™ *n*: trade name for a chemical additive for portland cement that reduces the effect of contamination of cementing slurries by the organic chemicals commonly found in drilling muds.

mud-level recorder *n*: a device that measures and records the height (level) of the drilling fluid in the mud pits. The level should remain fairly constant during the drilling of a well. If it rises, the possibility of a kick exists. Conversely, if it falls, loss of circulation may have occurred.

mud line *n*: 1. in offshore operations, the seafloor. 2. a mud return line.

mud log *n*: a record of information derived from examination of drilling fluid and drill bit cuttings. See *mud logging*.

mud logger *n*: an employee of a mud logging company who performs mud logging.

mud logging *n*: the recording of information derived from examination and analysis of formation cuttings made by the bit and of mud circulated out of the hole. A portion of the mud is diverted through a gas-detecting device. Cuttings brought up by the mud are examined

under ultraviolet light to detect the presence of oil or gas. Mud logging is often carried out in a portable laboratory set up at the well.

mud-making *n*: the ability of a formation to mix with clear water being circulated in a wellbore and create a natural drilling mud.

mud man *n*: see *mud engineer*.

mud manifold *n*: see *pump manifold*.

mud misting *n*: a type of mist drilling where a thin mud slurry instead of clear water carries the foamer.

mud-mixing devices *n pl*: any of several devices used to agitate, or mix, the liquids and solids that make up drilling fluid. These devices include jet hoppers, paddles, stirrers, mud guns, and chemical barrels.

mud motor *n*: see *downhole motor*.

mud-off *v*: 1. to seal the hole against formation fluids by allowing the buildup of wall cake. 2. to block off the flow of oil into the wellbore.

mud pit *n*: originally, an open pit dug in the ground to hold drilling fluid or waste materials discarded after the treatment of drilling mud. Today, a mud pit is a steel rectangular tank, usually open at the top, in which drilling fluid is placed on the rig. Several mud pits are used and are named according to their use in the circulating system. For example, the mud pump takes in mud from a suction pit, sediments in the mud fall out in a settling pit, and mud is stored in a storage pit. Although mud pits are steel tanks, they are often referred to as pits. However, "mud tanks" is the preferred terminology.

mud pit volume measuring device *n*: an instrument that transmits either a pneumatic or electrical signal from the mud pits to recorders and alarm devices at the driller's station. It monitors the level of mud in the pits and detects fluid gain or loss.

mud program *n*: a plan or procedure, with respect to depth, for the type and properties of drilling fluid to be used in drilling a well. Some factors that influence the mud program are the casing program and such formation characteristics as type, competence, solubility, temperature, and pressure.

mud pulse *n*: a very small surge in pressure in the drilling mud as it returns to the surface. In measurement-while-drilling systems, mud pulses carry downhole information—much as radio waves carry sound—for interpretation at the surface.

mud pump *n*: a large, high-pressure reciprocating pump used to circulate the mud on a drilling rig. A typical mud pump is a single- or double-acting, two- or three-cylinder piston pump whose pistons travel in replaceable liners and are driven by a crankshaft actuated by an engine or a motor. Also called a slush pump.

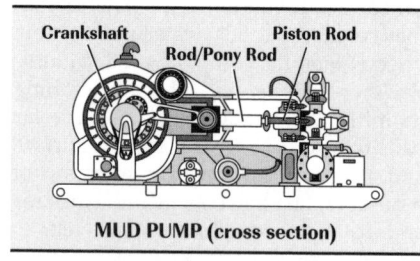

Crankshaft   Piston Rod
Rod/Pony Rod

MUD PUMP (cross section)

mud report *n*: a special form that is filled out by the mud engineer to record the properties of the drilling mud used while a well is being drilled.

mud return line *n*: a trough or pipe that is placed between the surface connections at the wellbore and the shale shaker and through which drilling mud flows on its return to the surface from the hole. Also called flow line.

mud saver *n*: also called mud box or mother hubbard. See *mud box*.

mud saver valve *n*: a lower kelly cock. See *drill stem safety valve, lower kelly cock*.

mud scales *n pl*: see *mud balance*.

mud screen *n*: see *shale shaker*.

mud seal *n*: a synthetic rubber, ring-shaped washer that fits between parts of a device that are exposed to drilling mud and parts that need to be protected from drilling mud.

mud solids *n pl*: the solid components of drilling mud. They may be added intentionally (barite, for example), or they may be introduced into the mud from the formation as the bit drills ahead. The term is usually used to refer to the latter.

mud still *n*: instrument used to distill oil, water, and other volatile materials in a mud to determine oil, water, and total solids contents in volume-percent.

mudstone *n*: 1. a massive, blocky rock composed of approximately equal proportions of clay and silt, but lacking the fine lamination of shale. 2. in general, rock consisting of an indefinite and variable mixture of clay, silt, and sand particles.

mud suction pit *n*: see *suction pit*.

mud system *n*: the composition and characteristics of the drilling mud used on a particular well.

mud tank *n*: one of a series of open tanks, usually made of steel plate, through which the drilling mud is cycled to remove sand and fine sediments. Additives are mixed with the mud in the tanks, and the fluid is temporarily stored there before being pumped back into the well. Modern rotary drilling rigs are generally provided with three or more tanks, fitted with built-in piping, valves, and mud agitators. Also called mud pits.

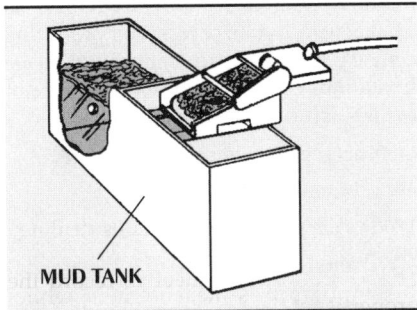

MUD TANK

mud-up *v*: to add solid materials (such as bentonite or other clay) to a drilling fluid composed mainly of clear water to obtain certain desirable properties.

mud volume totalizer *n*: see *pit-level indicator*.

mud weight *n*: a measure of the density of a drilling fluid often expressed as a weight per unit volume, such as pounds per gallon, pounds per cubic foot, or kilograms per cubic metre. Mud weight is directly related to the amount of pressure the column of drilling mud exerts at the bottom of the hole.

mud-weight equivalent *n*: see *equivalent circulating density*.

mud weight indicator *n*: an instrument that automatically determines a mud's weight (density) and displays it on a readout for rig personnel.

mud weight recorder *n*: an instrument installed in the mud pits that has a recorder mounted on the rig floor to provide a continuous reading of the mud weight.

muffler *n*: a device installed on an engine to quiet the barking sound produced by exhaust gases exiting through the exhaust pipe of the engine. One type of muffler is a steel cylinder with baffle plates. The baffle plates, flat steel sheets welded inside the cylindrical body of the muffler, change the direction of exhaust gas flow. Changing the direction of flow allows the gases to expand gradually, rather than all at once. Gradual expansion is quieter than rapid expansion. Sometimes called an exhaust silencer.

mule-head hanger *n*: see *horsehead*.

mule shoe *n*: a sub part of which is formed in the shape of a horseshoe and used to orient the drill stem downhole.

multihull rig *n*: in general, a submersible or semisubmersible drilling rig that has more than one framework or body (hull) on which supporting columns are built. See *hull*.

multimeter *n*: an electronic testing instrument that features several ranges for measuring voltage, current, and resistance. Also called a volt-ohm-milliammeter.

multimeter register *n*: a register that indicates the combined registration of two or more meters.

multipay zone *n*: two or more hydrocarbon producing formations that are penetrated by a single wellbore.

multiple-capacity system *n*: see *capacity* (def. 3).

multiple completion *n*: an arrangement for producing a well in which one wellbore penetrates two or more petroleum-bearing formations. In one type, multiple tubing strings are suspended side by side in the production casing string, each a different length and each packed to prevent the commingling of different reservoir fluids. Each reservoir is then produced through its own tubing string. Alternatively, a small-diameter production casing string may be provided for each reservoir, as in multiple miniaturized or multiple tubingless completions. See *dual completion*.

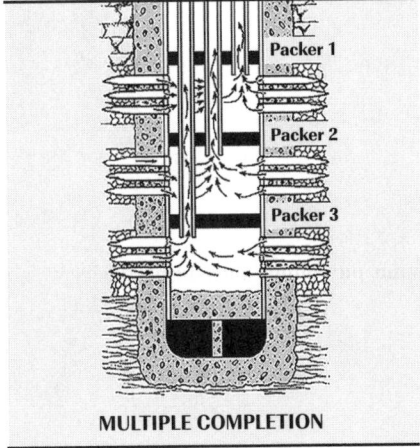

Packer 1

Packer 2

Packer 3

MULTIPLE COMPLETION

multiple miniaturized completion *n*: see *multiple completion*.

multiple-tank composite sample *n*: a mixture of individual samples from several compartments of a ship or barge, each of which contains the same grade of petroleum material. The mixture is blended in proportion to the volume of material in each compartment.

multiple tubingless completion *n*: see *multiple completion*.

multiple well pumping system *n*: a method of lifting oil out of several wells in a field. A pump is placed at every well; however, all of the pumps are powered by a single prime mover (engine or motor) instead of each pump's being powered individually.

multiplex electronic control (MUX) system *n*: an arrangement of equipment employed on floating rigs drilling in water depths greater than 5,000 feet (1,500 metres) to overcome delays in the transmission of signals to close and open the subsea blowout preventers. Surface electronics transmit electronic command signals through cable reels to subsea multiplex electronics packages, which decode and deliver the commands to solenoid valves.

multiplicand *n*: in mathematics, the quantity that is multiplied by another quantity. For example, if 6 is multiplied by 2, 6 is the multiplicand.

multiplier *n*: 1. in mathematics, the quantity that multiplies another quantity. For example, if 6 is multiplied by 2, 2 is the multiplier. 2. in electronics, an indication that one quantity should be multiplied by another quantity, usually in multiples of ten. For example, if an electronic component is stamped 2k (or, sometimes, 2K), the k (or K) stands for kilo and indicates that 2 should be multiplied by 1,000 to arrive at 2,000. (The lower case k is the preferred abbreviation for kilo, but manufacturers sometimes use K.) An M (short for mega) after a number indicates that the preceding number should be multiplied by 1 million; a G (short for giga) after the number indicates the preceding number should be multiplied by 1 billion. (The capital letters M and G are used to abbreviate mega and giga.)

multiposition lock (MPL) *n*: on Hydril ram blowout preventers, a hydraulically operated device that automatically secures (locks) the rams closed after they are closed normally by the blowout preventer control unit (accumulator). It consists of two clutch plates with serrated teeth that allow the rams to be opened only when hydraulic pressure is applied to separate the plates.

multipump injection system *n*: in a diesel engine, a system that uses several fuel pumps to take fuel from a supply tank and send it to the engine's fuel injectors. The pumps may be separate or combined into a single housing.

multishot survey *n*: a directional survey that provides a record of drift angle and direction at various depths in the hole. See *directional survey*.

multistage cementing *n*: the action of pumping cement into the well in stages or separate batches behind a casing string; a procedure used in wells that have critical fracture gradients or that require good cement jobs on long casing strings.

multistage cementing tool *n*: a device that permits cement to be displaced in stages at two or more points above the bottom of the string. Instead of all the cement being pumped out the bottom, a stage tool allows some of the cement to be pumped out at several points above the bottom. Used in cases in which a long column of cement might cause formation breakdown if the cement were displaced from the bottom of the string.

multistage centrifugal pump *n*: a centrifugal pump that develops pressure by means of impellers operating in series. Also called pump stage. See *electric submersible pumping.*

multistage pump *n*: see *multistage centrifugal pump.*

multistrand *n*: a type of synthetic or wire rope construction.

multistrand wire rope *n*: wire rope constructed of many strands.

multitrace recorder *n*: in electronics, an instrument that senses (traces) and records two or more signals generated by a component to which the recorder is attached.

multivariable flow metering *n*: in process control, the regulation and control of a flowing fluid based on two or more process variables such as level, temperature, or pressure.

multivariable transmitter *n*: in process control, a transmitter that monitors several process parameters at one time from different input ports and produces a calculated result. For example, a gas flow multivariable transmitter accepts inputs of temperature, differential pressure, and line pressure to compute the gas flow rate.

multivibrator oscillator *n*: a special type of oscillator that uses two tubes, transistors, or other electronic devices in which the output of each tube, transistor, or device is coupled to the input of the other with resistors and capacitors (or other devices) to obtain in-phase feedback voltage. Multivibrator oscillators are used in oscilloscopes to produce a large range of sweep frequencies.

muriatic acid *n*: see *hydrochloric acid.*

must-take gas *n*: natural gas supplies committed to a purchaser under terms such as drainage protection or reservoir protection clauses, or other provisions that place limitations on the purchaser's ability not to take natural gas from the supplier.

MUX *abbr*: multiplex electronic control.

mV *abbr*: millivolt.

MWD *abbr*: measurement while drilling.

MWD directional survey *n*: a directional survey that uses measurement-while-drilling techniques to determine drift angle and azimuth.

N *sym*: newton.

N *abbr*: normal.

N₂ *form*: nitrogen.

NO₂ *form*: nitrogen dioxide.

NAAQS *abbr*: National Ambient Air Quality Standards.

NACE *abbr*: National Association of Corrosion Engineers.

nail pin *n*: a pin shaped like a carpenter's nail and placed in a pressure relief valve. When the pin shears, it opens the valve to relieve pressure inside a vessel. Even though a nail pin is shaped like a nail, a carpenter's nail should never be substituted for a nail pin. See *shear pin*.

NAME *abbr*: National Association of Maritime Educators.

NAND circuit *n*: a logic circuit whose output signal is a logical 1 if all of its inputs is a logical 0, and whose output signal is a logical 0 if all of its inputs are a logical 1.

nanofarad (nF) *n*: one-billionth (10^{-9}) of a farad. See *farad*.

naphtha *n*: a volatile, flammable liquid hydrocarbon distilled from petroleum and used as a solvent or a fuel.

naphthene-base oil *n*: a crude oil that is characterized by a low API gravity and a low yield of lubricating oils and that has a low pour point and a low viscosity index (compared to paraffin-base oils). It is often called asphalt-base oil, because the residue from its distillation contains asphaltic materials but little or no paraffin wax.

naphthene series *n*: the saturated hydrocarbon compounds of the general formula C_nH_{2n} (e.g., ethene or ethylene, C_2H_4). They are cycloparaffin derivatives of cyclopentane (C_5H_{10}) or cyclohexane (C_6H_{12}) found in crude petroleum.

NAPHTHENE SERIES

Cyclohexane
(C_6H_{12})

Cyclopentane
(C_5H_{10})

nappe *n*: a large body of rock that has been thrust horizontally over neighboring rocks by compressive forces, as during the collision of two continents.

National Ambient Air Quality Standards (NAAQS) *n pl*: standards listed under the CAA for six major, or criteria, pollutants: ozone, carbon monoxide, sulfur dioxide, lead, nitrogen dioxide, and particulate matter. Areas that exceed the recommended levels for these pollutants are nonattainment areas, or areas with poor air quality; areas that meet or fall below the NAAQS levels are attainment areas, or areas with good air quality.

National Association of Corrosion Engineers (NACE) *n*: organization whose function is to establish standards and recommended practices for the field of corrosion control. Its publications are *Corrosion, Corrosion Abstracts,* and *Materials Performance.* Address: P.O. Box 201009; Houston, TX 77216-1009; 281-228-6200; Fax (281) 228-6300.

National Association of Maritime Educators (NAME) *n*: Address: 124 N. Van Ave.; Houma, LA 70363-5895; 985-879-3866.

National Electrical Manufacturers Association (NEMA) *n*: an organization created in 1926 by the merger of the Electric Power Club and the Associated Manufacturers of Electrical Supplies. NEMA provides a forum for the standardization of electrical equipment to ensure the manufacture of safe, effective, and compatible electrical products. NEMA also makes contributions to the electrical industry by shaping public policy development and operating as a central confidential agency for gathering, compiling, and analyzing market statistics and economics data. Address: 1300 North 17th Street, Suite 1847, Rosslyn, VA 22209; 703-841-3200; www.nema.org.

National Electric Code (NEC) *n*: a document sponsored by the National Fire Protection Agency (NFPA), the National Electric Code (NEC) establishes standards for safe electrical installations. Electricians, electrical contractors, engineers, and inspectors use the NEC as a guide when installing electrical wiring. Most cities and other authorities in the U.S. require that houses, buildings, factories, etc., be wired to meet NEC standards.

National Emissions Standards for Hazardous Airborne Pollutants (NESHAP) *n pl*: emissions standards set forth under CAA for airborne pollutants that are immediately hazardous to human health or that cause cancer, gene mutation, or reproductive harm.

National Environmental Policy Act (NEPA) *n*: a congressional act that forms the basic national charter for protection of the environment. Ensures that no agency of the federal government will take action that will significantly affect the quality of the human environment.

National Fire Protection Agency (NFPA) *n*: founded in 1896, an organization that provides fire, electrical, and life safety to the public. The mission of the international nonprofit NFPA is to reduce the worldwide burden of fire and other hazards on the quality of life by providing and advocating scientifically-based consensus codes and standards, research, training, and education. Address: 1 Batterymarch Park, Quincy, MA 02169; www.nfpa.org.

National Fishing Enhancement Act of 1984 *n*: congressional act that provides for the development of a National Artificial Reef Plan to promote and facilitate responsible and effective efforts to establish artificial reefs.

National Historic Preservation Act of 1966 *n*: congressional act that provides for the protection of historic and prehistoric archaeological resources.

National Institute for Occupational Safety and Health (NIOSH) *n*: an organization established by the OSH Act within the Department of Health, Education and Welfare to develop and establish recommended occupational safety and health standards and to conduct research. The OSH Act requires NIOSH to publish an annual listing of all known toxic substances and the concentrations at which such toxicity is known to occur.

National Oceanic and Atmospheric Administration (NOAA) *n*: an agency of the U.S. Department of Commerce that establishes national policies for and manages and conserves our oceanic, coastal, and atmospheric resources and applies its managerial, research, and technical expertise to provide practical services and essential scientific information. Address: Department of Commerce, 14th & Constitution Avenue NW; Washington, DC 20230; 301-713-4000.

National Ocean Service (NOS) *n*: part of the National Oceanic and Atmospheric Administration (NOAA), NOS measures and predicts coastal and ocean phenomena, protects large areas of the oceans, and works to ensure safe navigation. Web site address: www.nos.noaa.gov.

National Park Service (NPS) *n*: a service under the DOI that has the responsibility of forming, overseeing, maintaining, and developing national parks and monuments. The National Park Service was also made protector of the nation's historic and natural heritage with the Water Conservation Act, the Wilderness Act (1964), the Historic Preservation Act (1966), and the Wild and Scenic Rivers Act (1968). Address: 1849 C St. NW; Washington, DC 20240; 202-208-6843.

National Pollutant Discharge Elimination System (NPDES) *n*: a permit system set up under the CWA and implemented by qualified state governments to regulate pollutant discharges. NPDES permits specify the types of control equipment required and the discharges allowed for each facility. The permits specify levels of performance, and failure to achieve these levels must be reported. All permits can be reviewed by the EPA, the Army Corps of Engineers, and the Fish and Wildlife Service.

National Response Center (NRC) *n*: the USCG headquarters for emergency incidents. The NRC is operated 24 hours a day by the USCG in cooperation with 14 other federal agencies. Whenever a hazardous material release occurs in American waters, it must be reported to the NRC under federal law. Address: 2100 2nd Street, Room 2611; Washington, DC 20593; 800-424-8802.

national standard *n*: a standard recognized by an official national decision as the basis for fixing the value, in a country, of all other standards of the given quantity. In general, the national standard in a country is also the primary standard.

National Strike Force Coordination Center (NSFCC) *n*: the USCG national response system headquartered in Elizabeth City, North Carolina. The NSFCC is charged with maintaining a comprehensive list of spill removal resources, personnel, and equipment that is available worldwide and to make that list available to federal and state agencies and the public. The NSFCC provides technical assistance, equipment, and other resources required by the federal on-scene coordinator; coordinates the use of private and public personnel and equipment to remove a worst-case discharge and mitigate or prevent a substantial threat of such discharge; administers the coast guard strike teams; and provides technical assistance in preparing area contingency plans. The NSFCC maintains on file all area contingency plans and is required to review each of those plans that affects its responsibility. Address: 1461 North Road St.; Elizabeth City, NC 27909; 252-331-6006.

National Weather Service (NWS) *n*: an office of the National Oceanic and Atmospheric Administration that provides weather information and warnings to every county, parish, and major metropolitan area in the country. Address: 1325 East-West Highway; Silver Spring, MD 20910; (301) 713-0622. Regional office: 819 Taylor St.; Fort Worth, TX 76102; 817-978-1111, ext. 140.

native gas *n*: see *formation gas*.

natural aspiration *n*: see *naturally aspirated*.

natural clays *n pl*: clays that are encountered when drilling various formations; they may or may not be incorporated purposely into the mud system.

natural drive energy *n*: see *reservoir drive mechanism*.

natural gas *n*: a highly compressible, highly expansible mixture of hydrocarbons with a low specific gravity and occurring naturally in a gaseous form. Besides hydrocarbon gases, natural gas may contain appreciable quantities of nitrogen, helium, carbon dioxide, hydrogen sulfide, and water vapor. Although gaseous at normal temperatures and pressures, the gases making up the mixture that is natural gas are variable in form and may be found either as gases or as liquids under suitable conditions of temperature and pressure.

natural gas liquids (NGL) *n pl*: those hydrocarbons liquefied at the surface in field facilities or in gas processing plants. Natural gas liquids include propane, butane, and natural gasoline.

natural gasoline *n*: the liquid hydrocarbons recovered from wet natural gas, i.e., casinghead gasoline.

natural gas plant *n*: see *natural gas processing plant*.

Natural Gas Policy Act (NGPA) *n*: a congressional act started to deregulate the price charged by producers for "new" gas—gas discovered since 1977—at the wellhead. This act was designed to achieve deregulation gradually, with the total release of price controls effective January 1, 1985.

natural gas processing plant *n*: an installation in which natural gas is processed for recovery of natural gas liquids, the heavier hydrocarbon components of natural gas, including liquefied petroleum gases (such as butane and propane), and natural gasoline.

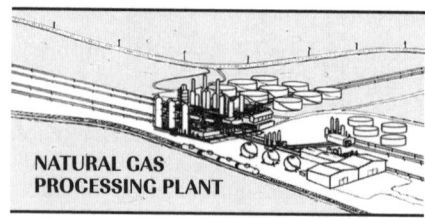

NATURAL GAS PROCESSING PLANT

naturally aspirated *adj*: term used to describe an internal combustion engine in which air flows into the engine by means of atmospheric pressure only. Compare *supercharge*.

naturally aspirated engine *n*: an engine in which the air enters the engine's cylinders by the simple action of the pistons moving downward in the cylinders. Compare *supercharged engine*.

naturally occurring *adj*: found in nature or occurring naturally without the aid of a chemical process.

naturally occurring radioactive materials (NORM) *n pl*: radioactive materials that occur naturally and that are brought to the surface during drilling operations or that contaminate drilling fluids, pipe scale, soil, or equipment. Although the EPA regulates NORM only in fertilizer (as of June 1992), it is likely that the regulation of other NORM material will follow in the near future. Currently, only Louisiana has state NORM regulations, but many other states are in the process of initiating regulations or have expressed an interest in doing so.

natural magnet *n*: a metallic ore, such as magnetite, that has a natural magnetism.

natural mud *n*: a drilling fluid containing essentially clay and water; no special or expensive chemicals or conditioners are added. Also called conventional mud.

natural period of motion *n*: in a ship or offshore floating rig, the duration of one free oscillation of the vessel caused by the effects of wind, waves, and currents on the vessel.

natural period of roll *n*: the time required for a ship or floating offshore drilling rig to roll from one side to the other and back one time.

nautical mile *n*: a unit of length used in sea and air navigation that is based on the length of 1 minute of arc of a great circle, especially an international and U.S. unit equal to 1,852 metres (about 6,076 feet).

naval architect *n*: a person who designs ships, barges, offshore rigs, and other vessels that operate in the water.

naval architecture *n*: the study of the physical characteristics and the design and construction of buoyant structures, such as ships, boats, barges, submarines, offshore rigs and floats that operate in water. The discipline also includes the construction and operation of power plants and other mechanical equipment of the structures.

Naval Oceanographic Office (NAVOCEANO) *n*: part of the United States Navy, this office maintains a fleet of survey vessels that collect data from all parts of the world. Once the data is collected, NAVOCEANO analyzes it and provides information for the military as well as civilian interests. Address: John C. Stennis Space Center, Stennis Space Center, MS 39529; (228) 688-3344; www.ssc.nasa.gov.

navigable waters *n pl*: any waters that are deep enough and wide enough to support a ship. The term has been broadly defined in recent years to include any waters that can float a canoe. Under this definition, drainage ditches, gullies, and small sinks can temporarily become navigable when it rains and thus may be defined as navigable even when dry.

NAVOCEANO (pronounced "nav-oci-ano") *abbr*: Naval Oceanographic Office.

NDT *abbr*: nondestructive testing methods.

neap tide *n*: a tide that has the smallest range between high and low tide at a given location. This type of tide occurs when the sun and moon lie at right angles to each other (during first and third quarters).

near-balance drilling *n*: drilling in which the hydrostatic head of the fluid column in the wellbore is in balance with, or only slightly greater than, formation pressure. Near-balance, or low-head drilling, when properly applied, does not allow hydrocarbons or other formation fluids to enter the wellbore.

near future *n*: in marketing, the next three years.

neat cement *n*: a cement with no additives other than water.

NEC *abbr*: National Electric Code.

neck down *v*: to taper to a reduced diameter. A pipe becomes necked down when it is subjected to excessive longitudinal stress.

necking *n*: the tendency of a metal bar or pipe to taper to a reduced diameter at some point when subjected to excessive longitudinal stress. See *bottleneck*.

needle valve *n*: a device that controls the rate of flow in a line (a valve), which contains a sharp-pointed, needlelike plug that is driven into and out of a cone-shaped seat to control accurately a relatively small rate of fluid flow.

negative charge *n*: a phenomenon that occurs at the atomic level of elements whereby the element has a surplus of electrons—that is, it has a lower electric potential. Compare *positive charge*.

negative equilibrium *n*: see *negative state*.

negative feedback *n*: 1. in process control, an arrangement wherein a portion of the output of a circuit, device, or machine is fed back 180° out of phase with the input signal. Negative feedback decreases amplification to stabilize it with respect to time or frequency. Decreased amplification reduces distortion and noise. Also known as inverse feedback, reverse feedback, and stabilized feedback. See *feedback*. 2. the return of a portion of the output current of a circuit to its input (feedback) in which the portion returned is 180 degrees out of phase with the input signal. This out-of-phase feedback decreases the amount of amplification in the circuit and stabilizes it with respect to time or frequency. Negative feedback reduces distortion and noise in the amplifier.

negative gravity anomaly *n*: on the record made by a gravity survey, salt domes are less dense than the overlying sedimentary rocks, which may indicate a negative gravity anomaly

negative state *n*: in a ship or floating offshore rig. the condition that occurs when the vessel's metacenter is below the vessel's center of gravity. In a negative state, the vessel is not stable and can capsize. See *center of gravity, metacenter*.

negative terminal *n*: the terminal of a battery or other voltage source away from which electrons flow and toward which holes flow. See *holes, terminal*.

NEMA *abbr*: National Electrical Manufacturers Association.

neoprene *n*: a synthetic rubber made by the polymerization of chloroprene; it resists being broken down when exposed to various chemicals and oil. Neoprene can be used for an annular preventer's packing element when oil-based muds are being used at temperatures ranging from –30°F to 170°F.

NEPA *abbr*: National Environmental Policy Act.

NESHAP *abbr*: National Emissions Standards for Hazardous Airborne Pollutants.

net area *n*: a term applied to grated deck flooring on offshore installations that refers to the area of the flooring that does not include areas of grating that are removed to clear obstructions or to allow pipe, ducts, columns, and the like to pass through the grating. Compare *gross area*.

net observed volume *n*: the total volume of all petroleum liquids, excluding S&W and free water, at observed temperature and pressure.

net-oil computer *n*: a system of electronic and mechanical devices that automatically determines the amount of oil in a water and oil emulsion. The water and oil do not have to be separated to measure the volume of the oil.

net-oil volume *n*: the amount, or volume, of oil in the produced fluids of a well or wells, usually as measured by a net-oil computer.

net production *n*: the amount of oil produced by a well or a lease, exclusive of its S&W content. Net production is also called working-interest oil (i.e., the oil produced by all of its wells multiplied by the working interest of a company in the wells).

net profits lease *n*: a lease agreement in which the lessor shares in the net proceeds of production after the lessee has recovered the initial investment. The various expenses allowed as deductions from gross proceeds are points of negotiation. Also applies to working-interest relationships.

net registered tonnage (NRT) *n*: the volume of the interior of a ship measured in tons (100 cubic feet or 2.83 cubic metres), excluding the space occupied by fuel, the engine room, navigation equipment, and the crew quarters.

net revenue interest *n*: the portion of oil and gas production money out of which operating and development costs are paid (i.e., the portion remaining after deduction of royalty interests).

net standard volume (NSV) *n*: the total volume of all petroleum liquids, excluding S&W and free water, corrected by the appropriate temperature correction factor (C_{tl}) for the observed temperature and API gravity, relative density, or density to a standard temperature such as 60°F or 15°C, and also corrected by the applicable pressure correction factor (C_{pl}) and meter factor.

net standard weight (NSW) *n*: the total weight of all petroleum liquids, excluding S&W and free water, determined by applying the appropriate weight conversion factor to the net standard volume.

net throughput *n*: the indicated volume corrected for meter errors, for volume differences due to metering pressure and temperature differing from reference conditions, and for S&W content, where applicable.

net tonnage *n*: the gross tonnage of a ship or a mobile offshore drilling rig less all spaces that are not or cannot be used for carrying cargo, expressed in tons equal to 100 cubic feet.

net volume *n*: the total volume of liquid in a tank after adjustments have been included for S&W content, temperature, and density. Compare *gross volume.*

net working interest *n*: the share of production remaining to the working-interest owners after all royalties, overriding royalties, production payments, and other reservations or assignments have been deducted.

neutral *n*: 1. position of the rig's weight indicator where hook load is zero. 2. position of an engine's transmission when the engine's flywheel or other power transmission device is not being turned by the engine. 3. of or relating to a particle, an object, or a system that has neither a positive nor a negative electric charge; of or relating to a particle, object, or system that has a net electric charge of zero.

neutral equilibrium *n*: the condition when the righting arm is zero and consequently the righting couple is zero. See *neutral state.*

neutralization *n*: a reaction in which the hydrogen ion of an acid and the hydroxyl ion of a base unite to form water, the other ionic product being salt.

neutral occlusion *n*: a weather phenomenon formed when a cold front overtakes a warm or stationary front and occurs when the temperature of the cold and warm air masses are virtually the same. Compare *cold occlusion, warm occlusion.*

neutral packer *n*: a packer that, unlike compression or tension packers, requires neither set-down weight nor upstrain pull to remain set.

neutral state *n*: in a ship or floating offshore drilling rig, the condition that occurs when the vessel's metacenter and center of gravity are in the same position. In the neutral state, the vessel is not as stable as it is when it is in a positive state. See *center of gravity, metacenter, positive state.*

neutrino *n*: a neutral particle ejected from the nucleus of an atom when the neutron-to-proton ratio is too high. Neutrinos are difficult to detect, because of their penetrating power, and are not recorded in logging.

neutron *n*: an electrically neutral subatomic particle in the nucleus of an atom, whose mass is about 1,845 times that of the electron. A neutron is stable when bound in an atomic nucleus. It and the proton form nearly the entire mass of atomic nuclei. A part of the nucleus of all atoms except hydrogen. Under certain conditions, neutrons can be emitted from a substance when its nucleus is penetrated by gamma particles from a highly radioactive source. This phenomenon is used in neutron logging.

Neutron Lifetime Log *n*: a pulsed-neutron survey.

neutron log *n*: a radioactivity well log used to determine formation porosity. The logging tool bombards the formation with neutrons. When the neutrons strike hydrogen atoms in water or oil, gamma rays are released. Since water or oil exists only in pore spaces, a measurement of the gamma rays indicates formation porosity. See *radioactivity well logging.*

neutron radiation *n*: radiation produced by nuclear disintegration, as when a nucleus is penetrated by gamma rays from a highly radioactive source. Neutron radiation can penetrate several feet of lead and is so difficult to observe that it remained undiscovered long after alpha, beta, and gamma rays were well known.

New Source Performance Standards (NSPS) *n*: CAA standards that set forth allowable emissions for new major pollution sources and major modifications to existing sources. NSPS can extend to pollutants not included in NAAQS and NESHAP and may include VOCs and hydrogen sulfide.

newton (N) *n*: an SI unit that expresses force. One newton equals 1 metre-kilogram per second per second ($m \cdot kg/s_2$), which is the force required to move 1 kilogram a distance of 1 metre at a velocity of 1 second squared.

Newtonian flow *n*: see *Newtonian fluid.*

Newtonian fluid *n*: a fluid in which the viscosity remains constant for all rates of shear if constant conditions of temperature and pressure are maintained. Most drilling fluids behave as non-Newtonian fluids, as their viscosity is not constant but varies with the rate of shear.

newton-metre *n*: see *joule.*

nF *abbr:* nanofarad.

NFPA *abbr:* National Fire Protection Agency.

NGL *abbr:* natural gas liquids.

NGPA standards *n pl:* natural gas pricing categories created by the Natural Gas Policy Act. The NGPA divides all natural gas into a large number of sections, each subject to different maximum lawful pricing rules.

Ni *sym:* nickel.

Nichrome *n*: the trademarked name of a nickel-chromium alloy that has high electrical resistance and the ability to withstand high temperatures; it is often used as the elements in electric heaters.

nickel (Ni) *n*: a silver-gray, ductile, malleable, and tough metal used in alloys, plating, coins, ceramics, and electronic circuits.

night toolpusher *n*: an assistant toolpusher whose duty hours are typically during nighttime hours on a mobile offshore drilling unit.

nimbostratus *n*: a dark gray, thick, low-level cloud that is composed of water droplets.

NIOSH *abbr:* National Institute for Occupational Safety and Health.

nipple *n*: a tubular pipe fitting threaded on both ends and less than 12 inches (30 centimetres) long, used for making connections between pipe joints and other tools.

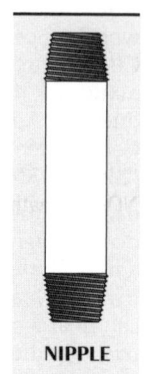

NIPPLE

nipple chaser *n*: (slang) a crew member who procures and delivers tools and equipment for a drilling rig.

nipple up *v*: in drilling, to assemble the blowout preventer stack on the wellhead at the surface.

NIST *abbr:* National Institute of Standards and Technology.

nitridize *v*: to combine with nitrogen.

nitrile rubber *n*: a synthetic rubber formed by polymerization of acrylonitrile with butadiene. Nitrile rubber can be used as a packing element in an annular preventer where oil-based mud is being used at temperatures ranging from 30°F to 180°F.

nitrogen (N_2) *n*: a colorless inert gas used in underbalanced drilling. Although more expensive than air or natural gas, because it is inert, it does not promote corrosion and can therefore save money in corrosion costs.

nitrogen dioxide (NO₂) *n*: a reddish-brown gas sometimes used in workover operations to clean out (jet) a well.

nitrogen narcosis *n*: the intoxicating or narcotic effect of gaseous nitrogen experienced by a diver breathing air at greater than 100 feet (30.5 metres) of depth. The effect increases with depth, impairing a diver's ability to think and act effectively.

nitrogen precharge pressure *n*: in blowout preventer closing systems (accumulators), the pressure, usually at 1,000 pounds per square inch or 6,895 kilopascals, placed inside an accumulator bottle by injecting nitrogen gas into the bottle. Usually, a flexible bladder inside the bottle separates the nitrogen from the operating fluid that is used to open and close the blowout preventers. Separating the two fluids prevents nitrogen from dissolving in the hydraulic operating fluid. See *accumulator*, definition 2.

nitro shooting *n*: a formation-stimulation process first used about a hundred years ago in Pennsylvania. Nitroglycerine is placed in a well and exploded to fracture the rock. Sand and gravel or cement is usually placed above the explosive charge to improve the efficiency of the shot. Nitro shooting has been largely replaced by formation fracturing.

NLPGA *abbr*: National LP-Gas Association.

NOAA (pronounced "noah") *abbr*: National Oceanic and Atmospheric Administration.

NOAA Weather Radio (NWR) *n*: a nationwide network of radio stations broadcasting continuous weather information from a nearby National Weather Service (NWS) office. NWR broadcasts NWS warnings, watches, forecasts, and other information 24 hours a day. NWR includes more than 800 transmitters, covering all 50 states, coastal waters, Puerto Rico, the U.S. Virgin Islands, and the U.S. Pacific Territories. NWR requires a special radio receiver or scanner capable of picking up the signal. Broadcasts occur at seven radio frequencies: 162.400 megahertz (MHz), 162.425 MHz, 162.450 MHz, 162.475 MHz, 162.500 MHz, 162.525 MHz, and 162.550 MHz.

noble metal *n*: any of the metals with low reactive tendencies at the upper end of the electrochemical series.

no-go *n*: a device of a known and precise dimension that is lowered into a well to determine the dimensions of another device or opening already in the well.

no-go nipple *n*: a special nipple made up in the tubing, casing, or drill pipe string the configuration of which is such that a tool contacting it can pass through only if the tool is in the proper position or configuration.

noise (electrical) *n*: 1. an unwanted component of a signal that obscures the information content. 2. any spurious voltage or current arising from external sources and appearing in the circuits of a device.

no-load loss *n*: in a transformer, losses in electrical energy that occur in the primary, secondary, and core of the transformer; no-load losses occur whether or not the transformer is loaded.

nominal size *n*: a designated size that is very close to, but that may be different from, the actual size.

nominal strength *n*: wire rope strength that the manufacturer calculates using a standard procedure established by the wire rope industry. Also called catalog strength. Compare *breaking strength, minimum acceptance strength*.

nominal volume *n*: the quantity assigned to a tank or vessel for the purpose of identification only; the exact volume may be somewhat different from the nominal volume.

nominations *n pl*: the amount of oil or gas a purchaser expects to take from a field as reported to a state regulatory agency.

nomograph *n*: a chart that presents an equation containing a number of variables in the form of several straight lines. The straight lines are scaled with values of the variables. To use it, a straight edge is placed across the scaled lines at the appropriate values. A nomograph can be easier to use than solving the equation.

nonabsolute ownership *n*: the legal view that states that minerals such as oil and gas cannot be owned in place but rather must be brought to the surface before they can be owned. Also called nonownership, nonownership in place.

nonane *n*: a paraffin hydrocarbon, C_9H_{20}, that is liquid at atmospheric conditions. Its boiling point is about 303.5°F (150.8°C) at 14.7 pounds per square inch (332.5 kilopascals per metre).

nonassociated gas *n*: gas in a reservoir that contains no oil.

nonattainment areas *n pl*: areas in the United States that fail to meet NAAQS and that are subject to tighter emission controls and more complicated requirements than are attainment areas. Major new or modified air pollution sources in nonattainment areas must get a permit to build, offset the new emission increase

by obtaining reductions from existing sources, and must meet the lowest achievable emission rate.

nonbleed relay *n*: in process control, a pneumatic device (a relay) whose operating air pressure is retained and sealed in its chamber. Relays transmit, or relay, signals to other components in a process system.

nonconductive mud *n*: any drilling fluid, usually oil-base or invert-emulsion muds, the continuous phase of which does not conduct electricity, e.g., oil.

nonconductive wireline *n*: wireline used for operations that do not require that electric signals be relayed from downhole to the surface or from the surface to a downhole device.

nonconductor *n*: a material through which little or no free electrons can flow.

nonconformity *n*: a buried landscape in which sediments were deposited on an eroded surface of igneous or metamorphic rock. Compare *unconformity*.

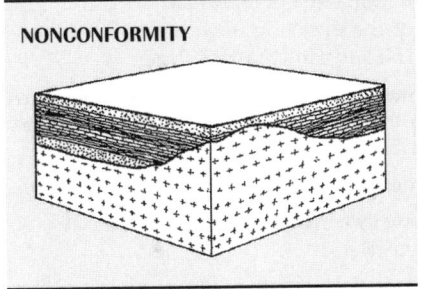

NONCONFORMITY

nonconsent/consent *n*: a provision in a joint operating agreement that allows parties who consent to later operations to penalize parties who do not consent. The penalty can be arranged in various ways but, to be effective, it assumes the productiveness of the proposed operation.

nondestructive testing *n*: in pipeline construction, testing designed to evaluate the quality of both production and field welds without altering their basic properties or affecting their future usefulness. The most common nondestructive testing is radiographic, or X-ray, testing. Compare *destructive testing*.

nondestructive testing methods (NDT) *n pl*: testing methods, such as dye injection and photography, used by divers to find flaws in the structural members of offshore platforms.

nondispersed mud *n*: a drilling mud that includes lightly treated, low-weight muds and spud muds. No thinners or other chemicals are added to prevent components in the mud from clumping. It is usually employed at the top of the hole and in shallow wells.

nondrying oil *n*: an oil that does not take up oxygen from the air and thereby lose its liquid characteristics. A nondrying oil can be added to a paint to slow the paint's drying.

nonferrous alloy *n*: alloy containing less than 50% iron.

nonfoliated metamorphic rock *n*: metamorphic rock that appears massive and homogeneous, i.e., without the layered look of foliated metamorphics.

noninverting input *n*: in an operational amplifier, a type of input wherein if the input signal is changed, the output is also changed. For example, if the noninverting input increases in voltage, the output voltage also increases. Also called plus input. Compare *inverting input*.

nonlocator *n*: terminology to describe the passage entry of seal assemblies into a packer seal bore not locking into place.

nonmagnetic drill collar *n*: a drill collar made of an alloy that does not affect the readings of a magnetic compass placed within it to obtain subsurface indications of the direction of a deviated wellbore. Used in directional drilling.

nonoperator *n*: a working-interest owner other than the one designated as operator of the property.

nonownership *n*: see *nonabsolute ownership*.

nonownership in place *n*: see *nonabsolute ownership*.

nonparticipating royalty owner *n*: a person who owns a severed portion of a royalty interest but who does not execute leases, participate in bonuses or rentals, or have rights of exploration and production.

nonporous *adj*: containing no interstices; having no pores and therefore unable to hold fluids.

nonpositive-displacement compressor *n*: a turbine compressor.

nonpressure tank *n*: a tank of conventional shape intended primarily for the storage of liquids at or near atmospheric pressure.

nonreactive *adj*: as applied to drilling mud components, those components that are unaffected by and do not react with other components in the mud. Compare *reactive*.

nonreactive phase *n*: that part of a liquid drilling mud that consists of solids or other chemicals that do not react with the liquid (or other chemicals in the liquid) part of the mud. Barite, for example, is nonreactive.

nonrecoverable usage *n*: well production that is burned in support facilities such as generators and boilers.

nonrising stem *n*: on a gate valve, the steel threaded rod (the stem) that, when rotated by an operating wheel, moves the valve's opening up or down in the valve body to open or close the valve, but does not move into the valve body. A nonrising stem prevents increases in valve body pressure that would occur if the stem entered the body.

nonrotating meter *n*: any metering device for which the meter pulse output is not derived from mechanical rotation as driven by the flowing stream.

nonstandard gas sales contract *n*: any combination of the fixed-rate and percentage of proceeds type of contract.

nonvolatile vehicle *n*: in paints, a substance in the liquid phase of the paint that does not readily vaporize at relatively low temperatures.

nonwetting phase *n*: the liquid, when two liquids are present in a pore, that has only slight, if any, attraction to the sides of the pore and that is repelled by the surface of the rock. Usually, oil is the nonwetting phase and water is the wetting phase. See *wetting phase*.

NOR circuit *n*: an electrical circuit in which output voltage appears only when a signal is absent from all its input terminals.

NORM *abbr*: naturally occurring radioactive materials.

normal butane *n*: in commercial transactions, a product meeting GPA specification for commercial butane and, in addition, containing a minimum of 95 liquid volume percent normal butane. Chemically, normal butane is an aliphatic compound of the paraffin series with the chemical formula C_4H_{10} and all of its carbon atoms joined in a straight chain.

normal circulation *n*: the smooth, uninterrupted circulation of drilling fluid down the drill stem, out the bit, up the annular space between the pipe and the hole, and back to the surface. Compare *reverse circulation*.

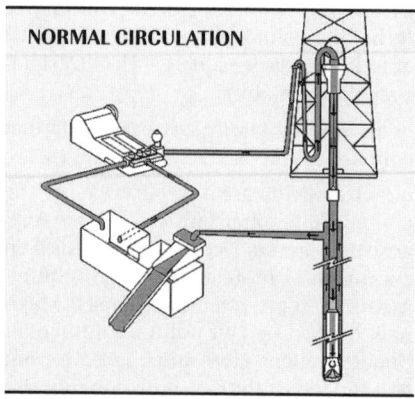

NORMAL CIRCULATION

normal curve *n*: a resistivity curve made by a conventional electric logging device.

normal fault *n*: a dip-slip fault along which the hanging wall has subsided relative to the footwall.

normal formation pressure *n*: formation fluid pressure equivalent to about 0.465 pounds per square inch per foot (10.5 kilopascals per metre) of depth from the surface. If the formation pressure is 4,650 pounds per square inch (32,062 kilopascals) at 10,000 feet (3,048 metres), it is considered normal.

normality *n*: a condition whereby a solution has a concentration of 1 gram equivalent of solute per litre.

normalizing *n*: heat-treating applied to metal tubular goods to ensure uniformity of the grain structure of the metal.

normal log *n*: see *conventional electric log*.

normal pore pressure *n*: pore pressure where the pressure of any water in the rock equals the hydrostatic pressure of a column of salt water of the same depth.

normal pressure gradient *n*: the pressure developed by a column of fluid as the depth of the column increases when the column contains a fluid of normal density. This gradient varies from area to area, but along the Gulf Coast of the United States, it is considered to be 0.465–0.468 psi/foot (10.53–10.59 kPa/metre), which is the pressure developed by the salt water that naturally occurs in the formations of this area. See *normal formation pressure*.

normal solution *n*: a solution that contains 1 gram-equivalent (0.0353 ounce-equivalent) of a substance per litre (3.8 gallons) of solution.

norther *n*: a strong northerly wind that has different effects in various parts of the world.

north magnetic pole *n*: a point on the earth's surface near the north polar region to which a magnetic compass points regardless of the compass's location on earth. The north magnetic pole is located several miles from the North Pole (the north geographic pole); consequently, navigators must correct for this difference when plotting courses or precisely locating a particular area.

North Sea Brent *n*: crude oil produced from the Brent field in the British sector of the North Sea. Its price is often quoted, because it is a benchmark for world oil prices.

nose *n*: the pointed end of the cone of a roller cone bit or the rounded portion of the head of a diamond bit in which the diamonds are embedded.

nose button *n*: a hard-metal projection that is placed on the end of the pilot pin of a roller cone bit to absorb some of the wear created by outward thrusts as the bit rotates.

nosing *n*: in grated deck flooring on offshore installations, an L-shaped length of steel affixed to the leading edge of a stair tread or to the head of the stair.

notch fatigue *n*: metal fatigue concentrated by surface imperfection, either mechanical (such as a notch) or metallurgical (such as a defect in the metal itself).

NOT circuit *n*: see *inverter circuit*.

notice of apparent discrepancy *n*: see *letter of protest*.

Notice to Lessees and Operators (NTL) *n*: an MMS formal document that provides clarification, description, or interpretations of a regulation or OCS standard; guidelines on the implementations of a special lease stipulation or regional requirement; a better understanding of the scope and meaning of a regulation by explaining MMS interpretations of a requirement; or administrative information.

nozzle *n*: 1. a passageway through jet bits that causes the drilling fluid to be ejected from the bit at high velocity. The jets of mud clear the bottom of the hole. Nozzles come in different sizes that can be interchanged on the bit to adjust the velocity with which the mud exits the bit. 2. the part of the fuel system of an engine that has small holes in it to permit fuel to enter the cylinder. Properly known as a fuel-injection nozzle, but also called a spray valve. The needle valve is directly above the nozzle.

NPDES *abbr*: National Pollutant Discharge Elimination System.

npn transistor *n*: a transistor in which n-type material is placed on each side of p-type material. One side of the n-type material is an emitter and the other side is a collector. The p-type material sandwiched in between the two n-type materials is the base. The emitter is negative with respect to the base, while the collector is positive with respect to the base.

NPS *abbr*: National Park Service.

NRC *abbr*: National Response Center; Nuclear Regulatory Commission.

NRT *abbr*: net registered tonnage.

Ns *abbr*: nimbostratus.

NS *abbr*: no show; used in drilling reports.

NSFCC *abbr*: National Strike Force Coordination Center.

NSPS *abbr*: New Source Performance Standards.

NTL *abbr*: Notice to Lessees and Operators.

n-type structure *n*: in a semiconductor material such as silicon, the arrangement of atoms in the material in such a manner that electron flow occurs between the atoms and thus creates a negative charge. Compare *p-type structure*.

nuclear log *n*: see *radioactivity log*.

nuclear tracer *n*: a gas, liquid, or solid material that emits gamma rays.

nucleus *n*: the positively charged central region of an atom, composed of protons and neutrons and containing almost all of the mass of the atom.

number *n*: 1. in mathematics, a member of the set of positive integers; one of a series of symbols of unique meaning in a fixed order that can be derived by counting. A member of any of the further sets of mathematical objects, such as negative integers and real numbers. 2. a symbol or a word used to represent a number.

number 1 oil and **number 2 oil** *n*: two grades of domestic fuel oil.

numerator *n*: in a mathematical fraction, the term or number that is divided by the other term or number (called the denominator), and is written above the line. For example, in the fraction ⅝, 8 is the numerator.

nutating meter *n*: a flowmeter that operates on the principle of the positive displacement of fluid by incorporating the wobbling motion of a piston or a disk. See *positive-displacement meter*.

NVK4 *n*: a type of high strength chain certified by DNA. DNV certification is a division of Det Norske Veritas (DNV), an independent foundation established in 1864 as a ship classification society.

NWR *abbr*: NOAA Weather Radio.

NWS *abbr*: National Weather Service.

Nymex WTI price *n*: the price per barrel of West Texas intermediate-grade crude oil as listed on the New York Mercantile Exchange. This price is a benchmark because it is an indicator of oil prices in the United States.

O *sym*: oxygen.

O&GCM *abbr*: oil- and gas-cut mud; used in drilling reports.

O&SW *abbr*: oil and salt water; used in drilling reports.

objective depth *n*: the depth at which a borehole is drilled to encounter the formation of interest.

oblate spheroid *n*: a three-dimensional object shaped like a sphere; however, instead of the sphere's being round, the sphere's diameter through its equator is greater than its diameter from pole to pole.

oblique slip *n*: slip at an angle between the dip and the strike in a fault plane.

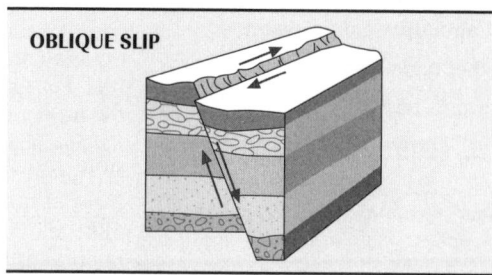

OBLIQUE SLIP

OBQ *abbr*: on-board quantity.

observed reference height *n*: the distance actually measured from the tank bottom or datum plate to the established reference point.

observed values *n pl*: 1. values observed at temperatures other than the specified reference temperature. 2. hydrometer readings observed at temperatures other than the specified reference temperature.

observed volumes *n pl*: those volumes of liquid or gas that are indicated by the register on the meter. Compare *true volume*.

obsidian *n*: an extrusive igneous rock that cooled so rapidly that no crystals formed at all, i.e., volcanic glass.

OC *abbr*: oil-cut; used in drilling reports.

occluded front *n*: formed as a cold front overtakes a warm front or a stationary front and forces the warm air upward.

Occupational Safety and Health Act of 1970 (OSH Act) *n*: a congressional act passed "to assure so far as possible every working man and woman in the nation safe and healthful working conditions and to preserve our human resources." The OSH Act establishes the Occupational Safety and Health Administration as a federal agency and authorizes it to promulgate, modify, or revoke occupational safety and health standards.

Occupational Safety and Health Administration (OSHA) *n*: a U.S. government agency that conducts research into the causes of occupational diseases and accidents. It is responsible for administration of the certification of respiratory safety equipment. Address: Department of Labor; 200 Constitution Avenue, NW; Washington, DC 20210; 202-576-6339.

Ocean Dumping Act *n*: see Marine Protection, Research, and Sanctuaries Act of 1973.

oceanic crust *n*: heavy rock that forms when molten rock cools. Oceanic crust is thin—about 5 to 7 miles (8 to 11 kilometres).

OCM *abbr*: oil-cut mud; used in drilling reports.

OCRE tool *n*: see *full-bore isolation valve*.

OCS *abbr*: Outer Continental Shelf.

OCSIP *abbr*: OCS Oil and Gas Information Program.

OCSLA *abbr*: Outer Continental Shelf Lands Act.

OCS Oil and Gas Information Program (OCSIP) *n*: an MMS program that generates reports and documents to assist state and local officials in planning for potential impacts resulting from offshore oil and gas exploration, development, and production activities.

OCS orders *n pl*: rules and regulations, set by the Minerals Management Service (MMS) of the U.S. Department of the Interior, that govern oil operations in U.S. waters on the Outer Continental Shelf. Now supplanted by rules published in the Code of Federal Regulations (CFR), Part 250.

octal number *n*: a number in a base 8 numbering system that is represented by the digits 0 through 7. Usually, an octal number is written with the subscript 8 to indicate that it is an octal number. For example, 6270_8 is an octal number.

octane *n*: a paraffinic hydrocarbon, C_8H_{18}, that is a liquid at atmospheric conditions and that has a boiling point of 258°F (125.5°C) at 14.7 pounds per square inch (101 kilopascals).

octane level *n*: see *octane rating*.

octane rating *n*: a classification of gasoline according to its antiknock qualities. The higher the octane number, or rating, the greater the antiknock qualities of the gasoline.

OD *abbr*: outside diameter.

odorant *n*: a chemical, usually a mercaptan, that is added to natural gas so that the presence of the gas can be detected by the smell.

odorizer *n*: a device used to impart an odor to natural gas, which is naturally odorless. See *odorant*.

OF *abbr*: open flow; used in drilling reports.

off-center alignment *n*: see *cone alignment*.

off-center wear *n*: a type of bit wear in which the cutters on the cones wear in an uneven pattern because of the whirling action of the bit as it drills. Whirling is the motion a bit makes when it does not rotate around the center; instead it drills with a spiral motion.

off-driller's side *n*: the side of the drawworks opposite the driller. Compare *driller's side*.

off-gas *n*: a by-product of refining used as boiler fuel, but with value as petrochemical feedstock. Composed mainly of ethane, ethylene, propane, propylene, butane, and hydrogen.

Office of International Activities and Marine Minerals (INTERMAR) *n*: an MMS office located in Herndon, Virginia, that functions as a liaison for agency involvement in international activities and provides policy direction for management of mineral resources on the OCS. Address: 381 Elden Street, MS-4030; Herndon, VA 20170-4817; 703-787-1300.

Office of Pipeline Safety (OPS) *n*: part of the Research and Special Programs Administration under DOT that is charged with determining procedures to ensure pipeline safety. Address: 400 7th Street SW, Room 7128; Washington, DC 20590; 202-366-4595.

off-lease gas *n*: gas used on a lease other than the lease from which the gas was produced.

off-production *adj*: shut in or temporarily unable to produce (said of a well).

offset *n*: the amount of distance the rig or vessel is displaced from a centrally located point in a mooring.

offset drilling rule *n*: rule applied by the courts (especially in states that have adopted nonabsolute ownership views of oil and gas) that states that landowners whose property is being drained by wells on neighboring tracts can protect themselves only by drilling wells on their own and producing oil or gas as quickly as they can. Compare *rule of capture*.

offset link *n*: in transmission chain, a combination of roller link and pin link used when a chain has an odd number of pitches.

offset misalignment *n*: a type of chain alignment in which the ends of the shafts, and therefore the sprockets, are not in line with each other. Some sprockets can slide on the shaft, so it is possible for them to be misaligned even if the shafts are not. Offset misalignment alternately stresses the link plates on one side of the chain and then those on the other side. Compare *angular misalignment*.

offset roller cone bit *n*: a roller cone bit each cone of which displays offset from a center point when a line is extended through the middle of each cone. Offset roller cone bits are usually employed to

OFFSET ROLLER CONE BIT

Offset

Direction of Rotation

drill soft formations, because the offset causes the teeth on the cone to gouge and scrape the formation. Gouging and scraping action is required to penetrate soft formations.

offset well *n*: a well drilled in the vicinity of other wells to assess the extent and characteristics of the reservoir and, in some cases, to drain hydrocarbons from an adjoining lease or tract.

offset-well data *n pl*: information obtained from wells that are drilled in an area close to where another well is being drilled or worked over. Such information can be very helpful in determining how a particular well will behave or react to certain treatments or techniques.

offshore *n*: that geographic area that lies seaward of the coastline. In general, the term "coastline" means the line of ordinary low water along that portion of the coast that is in direct contact with the open sea or the line marking the seaward limit of inland waters.

offshore drilling *n*: drilling for oil or gas in an ocean, gulf, or sea, usually on the Outer Continental Shelf. A drilling unit for offshore operations may be a mobile floating vessel with a ship or barge hull, a semisubmersible or submersible base, a self-propelled or towed structure with jacking legs (jackup drilling rig), or a permanent structure used as a production platform when drilling is completed. In general, wildcat wells are drilled from mobile floating vessels or from jackups, while development wells are drilled from platforms or jackups.

offshore installation manager (OIM) *n*: a qualified and certified person with marine and drilling knowledge who is in charge of all operations on a MODU.

offshore pipeline construction *n*: pipeline construction in water depths of 100 feet (30.5 metres) or more.

offshore production platform *n*: an immobile offshore structure from which wells are produced.

offshore rig *n*: any of various types of drilling structures designed for use in drilling wells in oceans, seas, bays, gulfs, and so forth. Offshore rigs include platforms, jackup drilling rigs, semisubmersible drilling rigs, submersible drilling rigs, and drill ships. Compare *land rig*.

OFFSHORE RIGS

O&G *abbr*: oil and gas; used in drilling reports.

OH *abbr*: open hole; used in drilling reports.

ohm (Ω) *n*: a unit for measuring electrical resistance. One ohm is equal to the resistance through which a current of 1 ampere will flow when a potential difference of 1 volt exists across a circuit.

ohmic *adj*: of or relating to the resistance of an object or a material in terms of ohms. See *ohm*.

ohmmeter *n*: an electrical instrument that measures the resistance of a conductor in ohms. It may be calibrated to read in ohms or megohms.

ohm•metre *n*: a unit for measuring electrical resistance. If a container with sides of 1 metre each is filled with a solution and a resistance of 1 ohm is measured when a current is passed through the container from one face to the opposite face, the resistivity of the unit volume is 1 ohm•metre.

ohms-adjust potentiometer *n*: an adjustable resistor (a potentiometer) on an ohmmeter that, when turned, varies the resistance in the meter's circuit. Varying the resistance adjusts the meter's pointer to indicate the appropriate resistance of the circuit or component being tested.

Ohm's law *n*: a rule that explains the behavior of electrical flow through a conductor. It is stated as R = $^E/_I$, where R = resistance in ohms, E = voltage in volts, and I = current in amperes. The law states that resistance, voltage, and current are interrelated and that modifying or changing one value in an electrical circuit modifies or changes the others. It is therefore used to calculate resistance, voltage, and current in electrical circuits.

ohms/mil-ft *abbr*: ohms per mil-foot.

ohms per mil-foot (ohms/mil-ft) *n*: in electronics, a measure of the resistance of a wire based on the wire's gauge and length; in general, the heavier (thicker) and longer the wire is, the more resistance it presents. See *mil-foot*.

oil *n*: a simple or complex liquid mixture of hydrocarbons that can be refined to yield gasoline, kerosene, diesel fuel, and various other products.

oil and gas separator *n*: an item of production equipment used to separate liquid components of the well fluid from gaseous

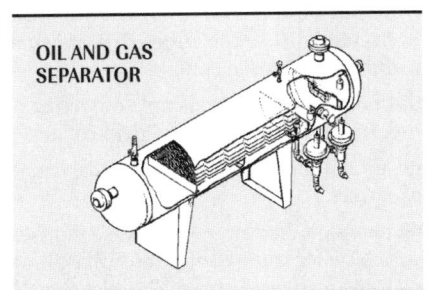

OIL AND GAS SEPARATOR

elements. Separators are either vertical or horizontal and either cylindrical or spherical. Separation is accomplished principally by gravity, the heavier liquids falling to the bottom and the gas rising to the top. A float valve or other liquid-level control regulates the level of oil in the bottom of the separator.

oil-base mud *n*: a drilling or workover fluid in which oil is the continuous phase and which contains from less than 2 percent and up to 5 percent water. This water is spread out, or dispersed, in the oil as small droplets. See *invert-emulsion mud, oil mud.*

oil bath *n*: a type of lubrication in high-speed chain-and-sprocket drives in which a portion of the chain passes through an oil bath (sump), which coats all of the chain on each revolution to lubricate it.

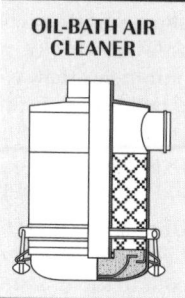

OIL-BATH AIR CLEANER

oil-bath air cleaner *n*: on an engine, a canister that has a relatively small amount of lubricating oil in the bottom and which filters air entering an engine. The running engine draws air through the cleaner, where the air passes through an element and over the oil bath. Dust and dirt particles in the intake air are removed by the element and by the oil bath.

oil-bath cleaner *n*: see *oil-bath air cleaner.*

oil-bath reservoir *n*: in rotary table assemblies, a compartment in the base of the rotary table assembly that contains oil of a specified weight and viscosity and through which parts of the assembly move to be lubricated.

oil breakout *n*: oil that was formerly emulsified in a drilling mud that comes out of the mud and rises to the surface in a mud tank.

oil content *n*: the amount of oil in volume-percent in a drilling fluid.

oil cooler *n*: on an engine, a device through which engine lubricating oil is circulated to reduce its temperature to acceptable levels. Some oil coolers depend on the surrounding air to reduce the temperature, but on most large engines, the oil is circulated through tubes that are surrounded by engine coolant.

oil-country tubular goods *n pl*: oilwell casing, tubing, drill pipe, and drill collars.

oil depletion allowance *n*: see *depletion allowance.*

oil-emulsion mud *n*: a water-base mud in which water is the continuous phase and oil is the dispersed phase. The oil is spread out, or dispersed, in the water in small droplets, which are tightly emulsified so that they do not settle out. Because of its lubricating abilities, an oil-emulsion mud increases the drilling rate and ensures better hole conditions than other muds. Compare *oil mud.*

oil-emulsion water *n*: the water contained in an emulsion of oil and water.

oilfield *n*: the surface area overlying an oil reservoir or reservoirs. The term usually includes not only the surface area, but also the reservoir, the wells, and the production equipment.

oilfield emulsion *n*: a combination of oil and water in which droplets of water are dispersed and suspended in the oil or droplets of oil are dispersed and suspended in the water. Most oilfield emulsions are water-in-oil emulsions; that is, the water is dispersed in the oil. To remove the water from the oil, the emulsion must be treated with heat, chemicals, and/or electricity.

oil-filled capacitor *n*: a capacitor whose dielectric (insulating material) is paper but which is enclosed by an oil-filled metal container. See *capacitor.*

oil filter *n*: on an engine, a device used to remove foreign particles from the engine's lubricating oil. The engine's oil pump forces the oil through one or more oil filters. Special elements in the filter trap the particles and prevent them from circulating through the engine and possibly damaging it. Many types of filter are available.

oil, gas, and mineral lease *n*: the instrument used to grant the rights to develop the resources of the land.

oil in place *n*: crude oil that is estimated to exist in a reservoir but that has not been produced.

oil-in-water emulsion *n*: see *emulsion, reverse emulsion.*

oil-in-water emulsion mud *n*: any conventional or special water-base mud to which oil has been added. The oil becomes the dispersed phase and may be emulsified into the mud either mechanically or chemically. Also called oil-emulsion mud.

oil measurements group *n*: this group has the job of preserving the quality and quantity of oil delivered to the pipeline by the shipper.

oil mud *n*: a drilling mud, e.g., oil-base mud and invert-emulsion mud, in which oil is the continuous phase. It is useful in drilling certain formations that may be difficult or costly to drill with water-base mud. Compare *oil-emulsion mud.*

oil operator *n*: see *operator.*

oil outlet *n*: see *oil sales outlet.*

oil pan *n*: on an engine, a steel trough-like attachment fitted to the bottom of the engine's crankcase. The engine's oil pump picks up oil from the pan and circulates it through the engine.

oil patch *n*: (slang) the oilfield.

oil payment *n*: a nonoperating interest in oil and gas from one or more leases that provides its owner a fractional share of the oil and gas produced, free of the costs of production, and that terminates when a specified dollar amount or volume of production has been realized. Oil payments may be created and reserved when a lease or royalty interest is assigned, or they may be carved out of a leasehold or royalty interest and assigned to another party.

Oil Pollution Act (OPA) of 1990 *n*: a congressional act passed to reduce the risk and damages caused by large oil spills on navigable waters. OPA provides for increased federal response authority, increased civil penalties, emphasis on worst-case contingency planning, and a billion-dollar trust fund to help clean up severe spills. The objective of OPA is to prevent discharges of oil into federal waters from vessels and facilities and to ensure that, if such spills occur, owners and operators have the resources to clean them up.

oil pool *n*: a loose term for an underground reservoir where oil occurs. Oil is actually found in the pores of rocks, not in a pool.

oil pressure gauge *n*: a device through which engine lubricating oil is circulated to reduce its temperature to acceptable levels. Some oil coolers depend on the surrounding air to reduce the temperature, but on most large engines, the oil is circulated through tubes that are surrounded by engine coolant.

oil pump *n*: a special pump, usually of the gear type, that moves oil through an engine. The pump's intermeshing gears rotate to build pressure and circulate the oil.

oil rig quality (ORQ) *n*: a type of chain grade.

oil ring *n*: the ring or rings that are located on the lower portion of a piston. They prevent excessive oil from being drawn into the combustion space during the suction stroke.

oil sales outlet *n*: a drain located about a foot from the bottom of a stock tank. It is used to take oil out of the tank for sale to a pipeline company or other shipper. The space below the outlet allows room for any remaining sediment and water to settle out of the oil. Also called oil outlet, sales outlet.

oil sand *n*: 1. a sandstone that yields oil. 2. (by extension) any reservoir that yields oil, whether or not it is sandstone.

oil saturation (S_o) *n*: the amount of oil in the pores of a rock.

oil saver *n*: a gland arrangement that mechanically or hydraulically seals by pressure. It is used to prevent leakage and waste of gas, oil, or water around a wireline (as when swabbing a well).

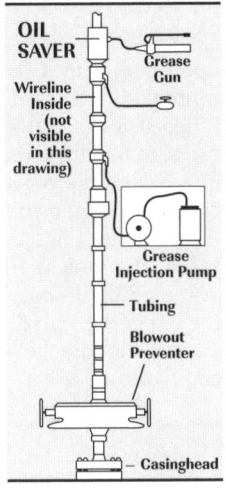

oil scout *n*: a person who gathers data on new oil and gas wells and other industry developments.

oil seep *n*: a surface location where oil appears, the oil having permeated its subsurface boundaries and accumulated in small pools or rivulets. Also called oil spring.

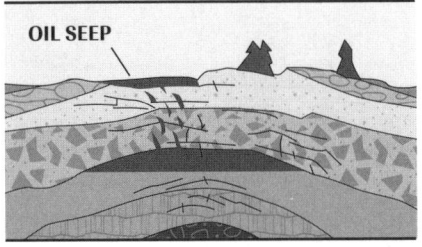

oil shale *n*: a shale containing hydrocarbons that cannot be recovered by an ordinary oilwell but that can be extracted by mining and processing. The cost of mining and treatment of oil shale is generally too great to compete with the cost of oilwell drilling.

oil slick *n*: a film of oil floating on water; considered a pollutant.

oil spill *n*: a quantity of oil that has leaked or fallen onto the ground or onto the surface of a body of water.

oil spill cleanup fund *n*: a pool of money set up under OPA to provide for increased federal response to oil spills. A $1 billion oil spill cleanup fund (created with an existing $.05/barrel fee on oil) was established under OPA. Half of the fund can be used for a single spill, and spill victims can use the fund when a spiller's liability limit has been reached, when the spiller is unknown, or when settlement is delayed.

oil spotting *n*: pumping oil, or a mixture of oil and chemicals, to a specific depth in the well to lubricate stuck drill collars.

oil spring *n*: see *oil seep*.

oil strainer *n*: a wire-mesh screen, usually located at the point where the oil pump picks up oil from an engine's oil pan. The screen traps large foreign particles that may not have settled to the bottom of the pan. Compare *oil filter*.

oil string *n*: the final string of casing set in a well after the productive capacity of the formation has been determined to be sufficient. Also called the long string or production casing.

oil temperature gauge *n*: a device that shows the temperature of the oil in an engine's lubricating system. Oil whose temperature is too high cannot lubricate properly.

oil-water contact *n*: the plane (typically a zone several feet thick) at which the bottom of an oil sand contacts the top of a water sand in a reservoir, i.e., the oil-water interface.

oil-water emulsion *n*: see *emulsion, reverse emulsion*.

oil-water interface *n*: the boundary between the oil on top and the water below when both are contained in the same vessel.

oilwell *n*: a well from which oil is obtained.

oilwell cement *n*: cement or a mixture of cement and other materials for use in oil, gas, or water wells.

oilwell pump *n*: any pump—surface or subsurface—that is used to lift fluids from the reservoir to the surface. See *hydraulic pumping, submersible pump, sucker rod pumping*.

oil-wet reservoir *n*: a hydrocarbon reservoir in which the grains of rock are coated not with water but with oil (occurs only rarely).

oil-wet rock *n*: see *wettability*.

oil window *n*: see *petroleum window*.

oil zone *n*: a formation or horizon of a well from which oil may be produced. The oil zone is usually immediately under the gas zone and on top of the water zone if all three fluids are present and segregated.

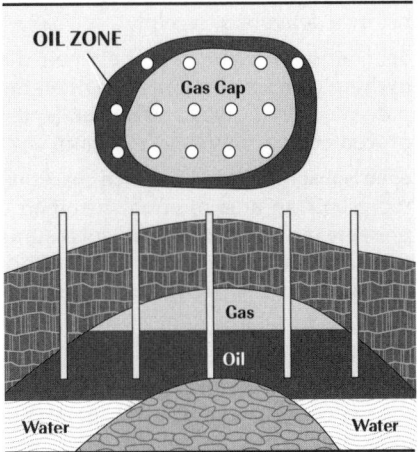

olefins units *n pl*: the units in a refinery that produce ethylene and propylene, both for the polymer units (to make polyethylene and polypropylene) and for sale to other chemical and plastics producers.

on-board quantity (OBQ) *n*: the measurable or estimatable materials—including water, oil, slops, oily residue, oil/water emulsion, sludge, and sediment—remaining on board in vessel cargo tanks and pipelines prior to loading.

on-center alignment *n*: see *cone alignment*.

on-deck *adj*: present on a ship or rig deck and exposed to weather.

one-eyed jet bit *n*: see *jet deflection bit*.

one hundred-year storm *n*: a storm with a 1/100 probability of occurrence in any year.

ones column *n*: in mathematics, the column of numbers written in the far right-hand column in a group of numbers having more than one digit. For example, in the number 3,583, three is in the ones column. It is a way of denoting that the number 3,583 contains 3 ones. In addition, 3,583 contains 8 tens, 5 hundreds, and 3 thousands. Numbers with only one digit have only a ones column.

one-step grooving system *n*: a pattern of drum spooling in which the wire rope is controlled by grooves to move parallel to drum flanges for 70 to 80 percent of the circumference and then to cross over to start the next wrap.

one-trip *adj*: said of a tool or device that is placed in a well and is not retrievable.

one-way valve *n*: see *check valve*.

on-lease gas *n*: gas used in the operation of the lease from which the gas was produced.

on-line plant *n*: see *mainline plant*.

on-off control *n*: in process control, a method that uses discrete devices to turn operating devices on or off. These discrete devices include switches that only have two distinct states or conditions.

on-off tool *n*: a tool used to open or close a downhole valve; a tool used to set or release a downhole tool, such as a retrievable bridge plug.

on-scene incident commander level *n*: a training level achieved by any employee who has been HAZWOPER trained to assume command of an emergency response situation. On-scene incident commanders must know and be able to implement the employer's emergency response plan; know how to implement the employer's incident command system; know and understand the hazards and associated risks for employees working in chemical protective clothing; know how to implement the local emergency response plan; know of the state emergency response plan and of the federal regional response team; and know and understand the importance of decontamination procedures.

on station *n*: when a drill ship or barge is on the drill site.

on-stream *adj*: 1. of a pump or pump station, moving oil by pumping. 2. of a gas processing plant or refinery, in operation or running.

on-suction *adj*: of a tank, open to pump suction.

on-the-beam *adj*: of a well, being pumped by a beam pumping unit.

on-the-horn *adj*: said of a person speaking on a two-way radio or a telephone.

on-the-line *adj*: of a tank, being emptied into a pipeline.

on-the-pump *adj*: of a well, being pumped.

on-vacuum *adj*: said of any pressure-tight vessel or container when the internal pressure is lower than atmospheric pressure.

ool *abbr*: oolitic; used in drilling reports.

oolite *n*: an ovoid, sandlike particle that is formed when calcite accretes on a smaller particle.

oolith *n*: rounded sandlike grains of calcite.

OPA *abbr*: Oil Pollution Act of 1990.

opacity *n*: the quality or state of being opaque. In paint technology, opacity is a measure of the hiding power of the paint.

op-amp *abbr*: operational amplifier.

OPEC *abbr*: Organization of Petroleum Exporting Countries.

open *adj*: 1. of a wellbore, having no casing. 2. of a hole, having no drill pipe or tubing suspended in it.

open circuit *n*: in an electrical circuit, a phenomenon that occurs when a component, such as a wire, breaks or otherwise fails so that the component can no longer carry current. For example, a broken wire in a circuit causes an open circuit

open-circuit diving system *n*: a diving life-support system in which the diver's exhalation is vented completely to the water.

open-circuit regulator *n*: see *demand regulator*.

open-cut crossing *n*: a road crossing in which the pipeline ditch cuts through the road instead of being bored under it. An open-cut crossing is generally used in sparsely populated areas where the

right-of-way crosses little-used dirt or gravel roads. Open-cut crossings are more convenient than bored crossings and also hold down costs.

open drip-proof enclosure *n*: motor enclosure that allows ample outside air to move through the motor to cool it while affording protection against mild weather conditions. The angle of protection is 15° from the vertical.

open flow potential *n*: the theoretical maximum capacity of a gas well as determined by a test conducted under limiting conditions. The method of determining this potential varies from state to state.

open flow test *n*: a test made to determine the volume of gas that will flow from a well during a given time span when all surface control valves are wide open.

open formation *n*: a petroleum-bearing rock with good porosity and permeability.

open hole *n*: 1. any wellbore in which casing has not been set. 2. open or cased hole in which no drill pipe or tubing is suspended. 3. the portion of the wellbore that has no casing.

open-hole completion *n*: a method of preparing a well for production in which no production casing or liner is set opposite the producing formation. Reservoir fluids flow unrestricted into the open wellbore. An open-hole completion has limited use in rather special situations. Also called a barefoot completion.

open-hole fishing *n*: the procedure of recovering lost or stuck equipment in an uncased wellbore.

open-hole log *n*: any log made in uncased, or open, hole.

opening-closing plug *n*: see *bottom plug*, *wiper plug*.

opening gauge *n*: the measurement in a tank before a delivery or receipt. Compare *closing gauge*.

opening page *n*: the measurement in a tank before a delivery or receipt.

opening ratio *n*: the ratio between the hydraulic pressure required to open the preventer and the well pressure under the preventer's packing device or ram.

open isobar *n*: a line drawn on a chart that represents an area of equal barometric pressure (an isobar) that does not assume a circular shape and therefore does not indicate circulation around a center of low pressure. Compare *closed isobar*.

open loop *n*: a control system that does not incorporate automatic feedback; instead, the system uses manual adjustments to maintain the desired function.

open-loop amplifier *n*: an electronic device that increases, or amplifies, a signal without feedback from any part of the circuit. See *feedback*.

open tank prover *n*: a relatively small calibrated tank, open to the atmosphere. Often used to prove meters at loading racks. In use, the meter to be proved is set at zero and the prover tank is filled to a calibrated height while the meter registers the amount of liquid flowing through it. When the tank is filled, the calibrated volume in it is compared to the volume the meter registered. If the meter is inaccurate, it is adjusted or a factor is used to compensate for the inaccuracy.

OPEN TANK PROVER

operand *n*: in computer science, any one of the quantities entering into or arising from an operation. See *operation*.

operating agreement *n*: see *joint operating agreement*.

operating company *n*: see *operator*.

operating condition *n*: the condition of a vessel engaged in drilling operations where drilling and/or associated operations are proceeding with the drilling risers connected.

operating interest *n*: see *leasehold interest*.

operating pressure *n*: in a blowout preventer stack, the amount of hydraulic pressure supplied by the blowout preventer control unit that is required to open and to close a blowout preventer. See *blowout preventer control unit*.

operating tension *n*: tension set in mooring lines at which a rig or vessel will operate after anchors are proof-loaded.

operation *n*: in computer science, a process or procedure that obtains a unique result from any permissible combination of operands. See *operand*.

operational amplifier (op-amp) *n*: an electronic device that increases the magnitude or power level of an electric current (an amplifier) and that has high direct-current stability and high immunity to oscillation, which is achieved by a high amount of negative feedback in the circuit in which the amplifier is installed. See *negative feedback*.

operational draft *n*: on a semisubmersible drilling rig, the draft that yields the best motion characteristics when the rig is in the drilling mode with a maximum deck load. See *draft*.

operator *n*: the person or company, either proprietor or lessee, actually operating an oilwell or lease, generally the oil company that engages the drilling, service, and workover contractors. Compare *unit operator*.

OPS *abbr*: Office of Pipeline Safety.

optical device *n*: an optical plummet or a theodolite equipped with a precision level. See *theodolite*.

optical pyrometer *n*: an instrument that determines the temperature of a very hot surface from its incandescent brightness; the image of the surface is focused in the plane of an electrically heated wire, and current through the wire is adjusted until the wire blends into the image of the surface. Also known as disappearing filament pyrometer.

optical scanner *n*: a device in which a moving spot of light controlled mechanically or electronically scans a meter chart. The reflected light generates signals that are used to process information from the chart.

optimization *n*: the manner of planning and drilling a well so that the most usable hole will be drilled for the least money.

optimum rate of flow *n*: that rate of flow of fluid from a well that will provide maximum ultimate recovery of fluid from the reservoir.

optimum water *n*: the amount of water used to give a cement slurry the best properties for a particular application.

orange peel *n*: a phenomenon that occurs when paint is improperly sprayed onto a surface. It is characterized by a pock-marked appearance that resembles an orange peel; it occurs when the paint fails to achieve a level and uniform surface. See *banana peel*.

Order 500 *n*: federal regulation adopted in August 1987 to provide mechanisms for resolving take-or-pay liability of gas producers.

ordinate *n*: the coordinate obtained by measuring parallel to the y-axis. Compare *abscissa*.

Ordovician *adj*: of or relating to the geologic period from approximately 500 million to 430 million years ago, during the early part of the Paleozoic era, or relating to the rocks formed during this period.

ore *n*: a mineral from which a valuable substance such as a metal can be extracted.

OR function *n*: an instruction that performs the logical operation "or" on a bit-by-bit basis for its two or more quantities entering into or arising from an operation (its operand), usually storing the result in one of the operand locations. See *operand*.

organic compounds *n pl*: chemical compounds that contain carbon atoms, either in straight chains or in rings, and hydrogen atoms. They may also contain oxygen, nitrogen, or other atoms.

organic rock *n*: rock materials produced by plant or animal life (coal, petroleum, limestone, and so on).

organic theory *n*: an explanation of the origin of petroleum that holds that the hydrogen and the carbon that make up petroleum come from land and sea plants and animals. The theory further holds that more of this organic material comes from very tiny swamp and sea creatures than comes from larger land creatures.

Organization of Petroleum Exporting Countries (OPEC) *n*: an organization of the countries of the Middle East, Southeast Asia, Africa, and South America that produce oil and export it. Members as of 1997 are Algeria, Indonesia, Iran, Iraq, Kuwait, Libya, Nigeria, Qatar, Saudi Arabia, the United Arab Emirates, and Venezuela. The organization's purpose is to negotiate and regulate oil prices.

organophilic clay *n*: a clay treated so that it is not hydrophilic for use in oil muds. Also called amine clay, organic clay.

orientation *n*: the process of positioning a deflection tool so that it faces in the direction necessary to achieve the desired direction and drift angle for a directional hole.

oriented core *n*: a core obtained from a precise angle or direction in a formation to obtain information about formation dip and strike, the direction of deposition, the direction of permeability, the direction of fluid migration, and hole deviation. The location of the core has been pinpointed, or oriented, in the reservoir.

oriented drill pipe *n*: drill pipe run in a well in a definite position, often a requisite in directional drilling.

oriented perforating *n*: a perforating technique that uses sensing instruments in a perforating gun to make perforations in a specific direction. It is often used in completions involving multiple production casing strings to perforate one string without damaging another. Also called directional perforating.

orifice *n*: 1. an opening, mouth, or a vent. 2. an open space allowing passage, such as an aperture, hole, mouth, opening, outlet, or vent. 3. an opening of a measured diameter that is used for measuring the flow of fluid through a pipe or for delivering a given amount of fluid through a fuel nozzle. In measuring the flow of fluid through a pipe, the orifice must be of smaller diameter than the pipe diameter. It is drilled into an orifice plate held by an orifice fitting.

orifice fitting *n*: a device specifically designed to hold an orifice plate in a meter installation. Several types are available, including the flange, junior, simplex, and senior.

orifice-flange tap *n*: on a flanged fitting, the threaded holes on either side of the orifice plate. Small pipes are screwed into the taps to connect the fitting to a flow recorder. Taps are used so pressure differential on either side of the plate can be measured and recorded. See *orifice meter*.

orifice flow constant (C′) *n*: a factor used in the calculation of gas volume flow through an orifice meter. Mathematical accounting for variations in pressure, temperature, density, and so on, of a gas as it flows through an orifice of a particular size.

orifice meter *n*: an instrument used to measure the flow of fluid through a pipe. The orifice meter is an inferential device that measures and records the pressure differential created by the passage of a fluid through an orifice of critical diameter placed in the line. The rate of flow is calculated from the differential pressure and the static, or line, pressure and other factors such as the temperature and density of the fluid, the size of the pipe, and the size of the orifice.

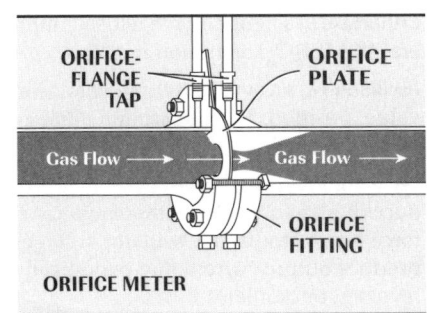

orifice meter chart *n*: a circular chart with a printed scale on which are recorded time, static and differential pressure, and sometimes temperature.

orifice meter installation *n*: see *meter installation*.

orifice pipe tap *n*: see *pressure tap*.

orifice plate *n*: a sheet of metal, usually circular, in which a hole of specific size is made for use in an orifice fitting.

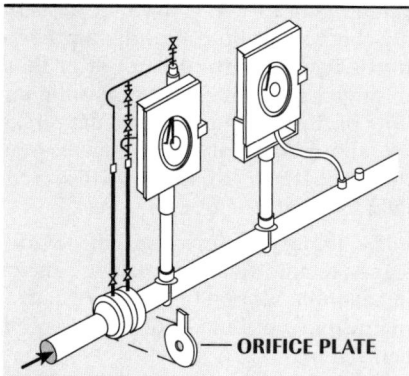

— ORIFICE PLATE

orifice pressure drop *n*: the pressure differential that occurs across an orifice plate.

orifice valve *n*: a device (a valve) that controls the flow of fluid in a line, which contains an opening (an orifice) the size of which is varied by a mechanically, hydraulically, pneumatically, or electrically operated piece that contacts the orifice.

orifice well tester *n*: a device, including orifice plates, a hose, and a manometer, used to measure the gas flow from a well. Static pressure differences measured on either side of a sharp-edged orifice are converted to flow values. It is used primarily for estimating the amount of gas flowing during a drill stem test, when a high degree of accuracy is not required.

O-ring *n*: a circular seal common in the oilfield. O-rings may be made of elastomer, rubber, plastic, or stainless steel. To seal properly, they all require enough pressure to make them deform against a sealing surface.

ORP *abbr*: oxidation-reduction potential.

ORQ *abbr*: oil rig quality.

orthoclase *n*: a light-colored feldspar mineral ($KA1Si_3O_8$), common in granite.

oscillation *n*: an even, rhythmic change in value, position, or state around a mean, or average, value.

oscillator *n*: an electronic circuit that produces high frequencies without rotational force being required. Oscillator voltages produce output current that periodically reverses, or oscillates.

oscilloscope *n*: a test instrument that uses a cathode-ray tube to make visible on a fluorescent screen the instantaneous values and waveforms of electrical quantities that rapidly vary as a function of time or other quantity.

OSHA *abbr*: Occupational Safety and Health Administration.

OSH Act *abbr*: Occupational Safety and Health Act of 1970.

osmosis *n*: passage of a pure liquid, such as water, into a solution, such as salt water, through a membrane that allows the water to pass through but not the salt. See *reverse osmosis*.

osmotic pressure *n*: the force that moves a liquid through a barrier; it is powerful enough to raise sap from the roots to the tops of trees.

Ouija board *n*: before the advent of measurement while drilling (MWD) tools and techniques, a device used by directional drillers to determine which way to turn deflection tools to get the desired drift angle in a hole. This device comprises a protractor and various straight-edged scales that pivot around the protractor to figure the drift angle.

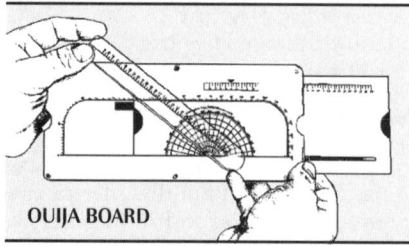

OUIJA BOARD

outage *n*: see *ullage*.

outage bob *n*: a graduated plumb bob that is attached to a metal gauge tape and that is of sufficient weight to keep the tape taut.

outage gauge *n*: a measure of the volume of liquid in a storage tank determined by measuring from a reference point at the top of the tank to the surface of the liquid in the tank and subtracting that measurement from the gauge height of the tank. Used when direct measurement of the liquid level is not possible. Also called ullage gauge.

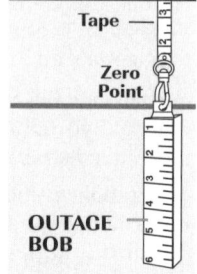

Tape

Zero
Point

OUTAGE
BOB

outage measurement *n*: a measure of the space between the top of the liquid in a tank and the reference point on the hatch. The liquid level in the tank is determined by subtracting the outage measurement from the gauging height of the tank.

outage tape-and-bob procedure *n*: a method of measuring the liquid level in a tank by measuring the space between the top of the liquid in a tank and the reference point on the hatch. The liquid level in the tank is determined by subtracting the outage measurement from the gauging height of the tank. Compare *innage tape-and-bob procedure*.

outboard *adv*: away from the center of the hull or toward the side of an offshore drilling rig.

outcrop *n*: part of a formation exposed at the earth's surface. *v*: to appear on the earth's surface (as a rock).

outcrop map *n*: horizontal representation of the rock types at the surface or just beneath the layer of soil.

outer barrel *n*: in the telescopic joint of a marine riser assembly, a pipe that attaches to the top joint of the marine riser assembly and to which are attached the riser tensioner lines. The inner barrel of the telescopic joint moves up and down within the outer barrel as the floating offshore rig heaves.

outer bearings *n pl*: the main bearings of a roller cone bit that reside inside the largest part of the cone. The outer bearings are either roller bearings or journal bearings. Compare *inner bearings*.

Outer Continental Shelf (OCS) *n*: the land seaward from areas subject to state mineral ownership to a depth of roughly 8,000 feet (2,500 metres), beyond which mineral exploration and development are not, at present, feasible. Boundaries of the OCS are set by law. For example, Louisiana owns 3 miles (4.8 kilometres) seaward from the shoreline. In general, the term is used to describe federally controlled areas.

Outer Continental Shelf Lands Act (OCSLA) *n*: congressional act that authorizes the secretary of the interior to grant mineral leases and to regulate oil and gas activities on OCS lands.

Outer Continental Shelf orders *n pl*: see *OCS orders*.

outflow *n*: the flow system of a well between the bottom of the hole and the storage tanks.

outlet sample *n*: taken at the level of the bottom of the tank outlet (either fixed or swing pipe) but not higher than 1 metre (0.9144 yard) above the bottom of the tank.

outlier *n*: a result that differs considerably from the main body of results in a set.

out-of-gauge bit *n*: a bit that is no longer of the proper diameter.

out-of-gauge hole *n*: a hole that is not in gauge; that is, it is smaller or larger than the diameter of the bit used to drill it.

outpost well *n*: a well located outside the established limits of a reservoir, i.e., a step-out well.

output *n*: a signal transmitted from a device.

output PLC module *n*: equipment in a PLC system that sends commands in the form of current, voltage, or switch closure to operate other equipment. The output module receives its commands from the PLC processor in the form of digital signals. See *programmable logic controller*.

output shaft *n*: the transmission shaft nearest the machinery to be driven. It is driven by the input shaft driven near the power source.

outrigger *n*: a projecting member run out at an angle from the sides of a portable mast or a land crane to the ground to provide stability and to minimize the possibility of having the mast or the crane overturn.

outrun *v*: 1. for fluid, to fall down a well faster than it can be pumped. 2. to try to pull out of the well faster than the wireline tools are being blown upwards by unexpected pressure. 3. to attempt to pump out a gas influx before the expansion of gas reduces pressure enough to allow the well to kick.

outside cutter *n*: see *external cutter*.

outside diameter (OD) *n*: the distance across the exterior circle, especially in the measurement of pipe. See *diameter*.

outside siphon *n*: see *water outlet*.

outward axial thrust *n*: an outward force created along a centerline drawn through a cone (the axis) as a roller cone bit rotates.

outwash *n*: sediment deposited by meltwater streams beyond an active glacier.

over and short station *n*: a pump station where one or more tanks are floating on the line. See *floating tank*.

overbalance *n*: the extent to which the hydrostatic pressure of the mud column exceeds formation pressure.

overbalanced drilling *n*: drilling in which the hydrostatic pressure of the mud column exceeds formation pressure.

overboard vent line *n*: in a diverter system, a pipe (a line) that is permanently attached to side outlets on the diverter's housing. Usually, two vent lines are provided so that well fluids can be safely vented away from the rig, regardless of the wind's direction. See *diverter*.

overburden *n*: the strata of rock that overlie the stratum of interest in drilling.

overburden pressure *n*: the pressure exerted by the rock strata on a formation of interest. It is usually considered to be about 1 pound per square inch per foot (22.621 kilopascals per metre).

overcapacity *n*: capacity beyond what is normal, allowed, or desirable.

overconvey *v*: in regard to land, to convey (intentionally or from ignorance) a larger fraction of interest in property than the owner actually has a right to convey.

overdelivery *n*: the amount by which the actual volume that has passed through the meter exceeds the indicated volume registered by a meter. See *absolute error*.

overflow *n*: the effluent of a cone-shaped centrifuge that passes up the inside of the cone and leaves through the vortex finder.

overflow pipe *n*: a pipe installed at the top of a tank to enable the liquid within it to be discharged to another vessel when the tank is filled to capacity.

overflush *n*: an excess quantity of fluid used to push acid out of the tubing or casing when an acid mixture is put into a well, thus directing the acid to the desired place in the well.

overgauge hole *n*: a hole whose diameter is larger than the diameter of the bit used to drill it. An overgauge hole can occur when a bit is not properly stabilized or does not have enough weight put on it.

overhaul *n*: ability of a weight on the end of a hoist line to unwind cable from the drum when the brake is released.

overhead *n*: the vapor stream leaving the top of a tower, column, or vessel.

overhead product *n*: in distillation, the product that condenses above the fractionation column.

overhead stream *n*: in distillation, the stream of products that condense above the fractionating column.

overpressure *n*: see *abnormal pressure*.

overproduced *adj*: said of a well that has produced more than its allowable.

overpull *v*: to pull on pipe with enough force to exceed the pipe's weight.

overranging *adj*: when applied to metering, indicates that the maximum permitted flow of the meter is being exceeded.

overregistration *n*: the amount by which the indicated volume registered by a meter exceeds the actual volume that has passed through the meter. Determined by means of a suitable standard device. Compare *underregistration*.

overriding royalty *n*: an interest carved out of the lessee's working interest. It entitles its owner to a fraction of production free of any production or operating expense, but not free of production or severance tax levied on production. An overriding royalty may be created by grant or by reservation. Commonly, an override is reserved by the assignor in a farmout agreement or other assignment. An override's duration corresponds to that of the lease from which it was created.

overrunning clutch *n*: 1. a special clutch that permits a rotating member to turn freely under certain conditions but not under others. 2. a clutch that is used in a starter and transmits cranking effort but overruns freely when the engine tries to drive the starter.

overshot *n*: a fishing tool that is attached to tubing or drill pipe and lowered over the outside wall of pipe or sucker rods lost or stuck in the wellbore. A friction device in the overshot, usually either a basket or a spiral grapple, firmly grips the pipe, allowing the fish to be pulled from the hole.

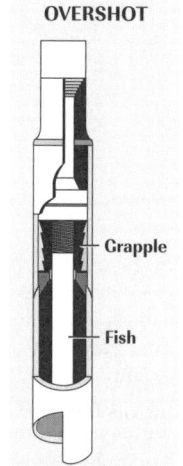

OVERSHOT — Grapple — Fish

overspeed governor *n*: a special type of engine governor that prevents an engine from running too fast (overspeeding) by shutting down the engine if it overspeeds. See *governor*.

overspeeding *n*: an engine's running beyond the maximum speed for which it was designed. Overspeeding an engine can severely damage it.

overspeed trip *n*: see *overspeed governor*.

overspeed trip device *n*: a special type of engine governor that prevents an engine from running too fast and possibly destroying itself if the regular governor fails.

overspray *n*: when spray painting, the paint that misses the surface to be coated.

overthrust fault *n*: a low-dip angle (nearly horizontal) reverse fault along which a large displacement has occurred. Some overthrusts, such as many of those in the Rocky Mountain Overthrust Belt, represent slippages of many miles.

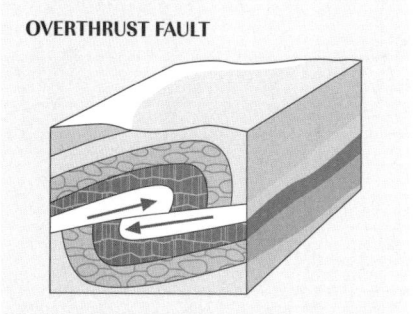

OVERTHRUST FAULT

overtreatment *n*: adding too much chemical in batch treating, resulting in an emulsion that is very difficult to break.

overturned fold *n*: a rock fold that has become slanted to one side so that the layers on one side appear to occur in reverse order (younger layers beneath older).

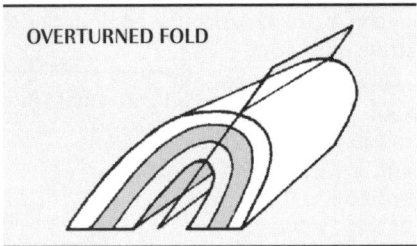

OVERTURNED FOLD

ownership in place *n*: see *absolute ownership*.

oxidation *n*: a chemical process in which oxygen combines with a compound and causes the compound to lose electrons and gain a more positive charge. For example, when exposed to air (which contains about 21 percent oxygen), iron rusts, which means that part of the iron chemically combines with the oxygen in the air and oxidizes, or becomes rusty. The rust is typically a red iron oxide, which gives iron rust its characteristic color.

oxidation-reduction potential (ORP) *n*: the difference in voltage shown when an insert electrode is immersed in a reversible oxidation-reduction system. It is the measurement of the system's state of oxidation. Also called Eh, redox potential.

oxide *n*: a chemical compound in which oxygen is joined with a metal or a nonmetal.

oxide film *n*: a thin layer of material (a film) that is coated with metallic oxides and is often used as a dielectric in capacitors. See *capacitor, dielectric*.

oxidize *v*: 1. to combine with oxygen. 2. to remove one or more electrons from an atom, ion, or molecule.

oxyacetylene welding *n*: see *acetylene welding*.

oxygen (O) *n*: a nonmetallic element constituting 21 percent of the atmosphere by volume that occurs as a diatomic gas, O_2, and in many compounds such as water and iron ore. It combines with most elements, is essential for plant and animal respiration, and is required for nearly all combustion.

oxygenated *adj*: having or containing oxygen.

oxygenation *n*: combining or supplying with oxygen.

oxygen-concentration cell *n*: a corrosion cell formed by differing concentrations of oxygen in an electrolyte.

oxygen corrosion *n*: the eating away of metallic surfaces that occurs when the metal is exposed to oxygen. Oxygen corrosion is accelerated when the metal is exposed to oxygen in the presence of water. In the oilfield, oxygen corrosion is often characterized by red iron oxide (rust) appearing on the surface of the corroded tool.

oxygen toxicity *n*: a harmful reaction experienced by divers breathing extremely high partial pressures of oxygen. Divers may suffer from two forms: one affects the central nervous system; the other affects the pulmonary muscles. Because of these dangers, pure oxygen is not used as a breathing medium below 50 feet (15.2 metres) in commercial operations.

oz *abbr*: ounce.

ozone *n*: at ground level, one of the primary ingredients in smog and a poisonous form of pure oxygen. Smog ozone is created by sunlight acting on nitrogen oxides and volatile organic compounds in the air. Ozone high in the atmosphere forms the earth's main shield against the sun's ultraviolet radiation.

p *abbr*: average reservoir pressure, used in engineering reports.

p$_i$ *abbr*: initial pressure; used in engineering reports.

Pa *sym*: pascal.

PAC *abbr*: polyanionic cellulose.

P&A *abbr*: plug and abandon.

packed column *n*: a fractionation or absorption column filled with small objects that are designed to have a relatively large surface per unit volume (the packing), instead of bubble trays or other devices, to give the required contact between the rising vapors and the descending liquid. In the reboiler still column of a glycol dehydration unit, entrained water flashes into steam when inflowing rich glycol flows over the hot packing.

packed-hole assembly *n*: a bottomhole assembly consisting of stabilizers and large-diameter drill collars arranged in a particular configuration to maintain drift angle and direction of a hole. This assembly is often necessary in crooked-hole country. See *crooked-hole country*.

packed pendulum assembly *n*: a bottomhole assembly in which pendulum-length collars are swung below a regular packed-hole assembly. The pendulum portion of the assembly is used to reduce hole angle. It is then removed, and the packed-hole assembly is run above the bit. See *packed-hole assembly, pendulum assembly*.

packer *n*: a piece of downhole equipment that consists of a sealing device, a holding or setting device, and an inside passage for fluids. It is used to block the flow of fluids through the annular space between pipe and the wall of the wellbore by sealing off the space between them. In production, it is usually made up in the tubing string some distance above the producing zone. A packing element expands to prevent fluid flow except through the packer and tubing. Packers are classified according to configuration, use, and method of setting and whether or not they are retrievable (that is, whether they can be removed when necessary, or whether they must be milled or drilled out and thus destroyed).

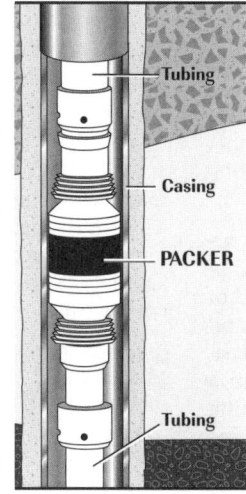

packer-bore receptacle *n*: a retrievable device anchored on the top of a production packer. It receives a tubing seal assembly. See *tubing seal assembly*.

packer flowmeter *n*: a tool for production logging that employs an inflatable packer. It ensures that all the fluid from the well passes through the measuring devices built into the tool.

packer fluid *n*: a liquid, usually salt water or oil, but sometimes mud, used in a well when a packer is between the tubing and the casing. Packer fluid must be heavy enough to shut off the pressure of the formation being produced, must not stiffen or settle out of suspension over long periods of time, and must be noncorrosive.

packer mill *n*: see *mill*.

packer squeeze method *n*: a squeeze cementing method in which a packer is set to form a seal between the working string (the pipe down which cement is pumped) and the casing. Another packer or a cement plug is set below the point to be squeeze-cemented. By setting packers, the squeeze point is isolated from the rest of the well. See *packer, squeeze cementing*.

packer test *n*: a fluid-pressure test of the casing. Also called a cup test.

pack ice *n*: a solid pack of ice that covers more than half of the visible sea surface.

packing *n*: 1. a material used in a cylinder on rotating shafts of an engine in the stuffing box of a valve, or between flange joints to maintain a leakproof seal. 2. the specially fabricated filling in packed fractionation columns and absorbers.

packing assembly *n*: the arrangement of the downhole tools used in running and setting a packer.

packing element *n*: 1. a dense rubber, washer-shaped element encircling a packer, which expands against casing or formation face to seal off the annulus. 2. in an annular blowout preventer, the elastomer sealing piece that, when closed, forms a pressure-tight seal around virtually any tool in the wellbore, as well as on open hole—hole with nothing in it.

packing gland *n*: the metal part that compresses and holds packing in place in a stuffing box. See *gland, stuffing box*.

packing ring *n*: piston ring.

pack-off *n*: a device with an elastomer packing element that depends on pressure below the packing to effect a seal in the annulus. Used primarily to run or pull pipe under low or moderate pressures, this device is not dependable for high differential pressures. Also called a stripper.

pack off *v*: to place a packer in the wellbore and activate it so that it forms a seal between the tubing and the casing.

pack-off nut *n*: in a subsea wellhead assembly, a device that, when rotated, actuates a seal assembly to close the annulus between two casing strings.

pack-off (stripper) preventer *n*: a preventer having a unit of packing material whose closure depends on well pressure coming from below. It is used primarily to strip pipe through the hole or allow pipe to be moved with pressure on the annulus.

pad *n*: 1. a device on a logging tool that contacts the wellbore when the log is being taken. 2. fluid placed in a well to serve a special purpose. 3. a concrete foundation on which a drilling rig rests.

padding *n*: screened or sifted dirt, clean gravel, or foam placed in a ditch to protect pipe from damage caused by rocky or rough soils.

padeye *n*: a steel plate with an opening (the eye) that is usually welded to a heavy object and to which a sling or other lifting line is attached to facilitate the object's being moved, lifted, or lowered.

pad resistivity device *n*: a device (e.g., a minilog) designed to measure the resistivity of small volumes of formation near the borehole. It consists of three electrodes embedded in the center of an insulated fluid-filled rubber pad that is held against the side of the borehole. The electrodes produce two curves—one 1.5 inches (3.8 centimetres) in depth and one 4 inches (10.2 centimetres) in depth. The separation between the two curves shows the difference in resistivity between the mud cake and the formation immediately behind the mud cake.

pad volume *n*: the amount of fluid placed in a well to serve as a pad. See *pad*.

paid-up lease *n*: an oil and gas lease for which all delay rentals are paid along with the cash bonus and on which no further action is required during the primary term.

paint body *n*: the consistency of a paint—for example, its thickness or thinness.

pair production *n*: 1. the process in which a ray or wave reacts with the nucleus of an atom, converting the wave's energy into mass to produce an electron and a positron. The positron immediately reacts with an electron within the radioactive material. In the interaction, two new gamma rays are produced with energy levels less than half the original gamma ray. 2. the conversion of a photon into an electron and a positron when the photon traverses a strong electric field, such as that surrounding a nucleus or an electron.

paleo-comb form*: ancient; early; long ago; primitive.

paleogeography *n*: geography of a specified geologic past.

paleontology *n*: the science that concerns the life of past geologic periods, especially fossil forms and the chronology of the earth.

Paleozoic era *n*: a span of time from 600 million to 230 million years ago during which a great diversity of life forms developed.

panel diagram *n*: a diagram of a block of earth in which a series of cross sections are joined and viewed obliquely from above, to give a three-dimensional view.

It is useful in showing how formation structure and stratigraphic thickness vary both horizontally and vertically. Also called fence diagram.

Pangaea *n*: the supercontinent comprising all of the principal continental masses near the beginning of the Mesozoic era.

paper capacitor *n*: a capacitor whose dielectric (insulating material) is made from paper. See *capacitor*.

parabolic profile *n*: the shape of a PDC bit's head that resembles a parabolic arc, a curve that looks very much like an arch over a doorway. Compare *shallow-cone profile*, *short parabolic profile*.

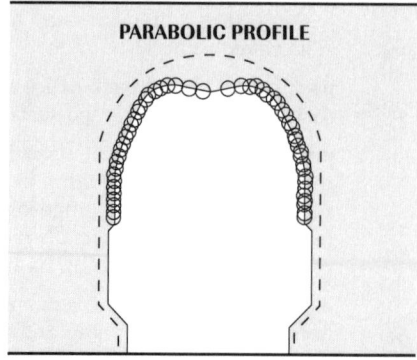

PARABOLIC PROFILE

paraffin *n*: a saturated aliphatic hydrocarbon having the formula C_nH_{2n+2} (e.g., methane, CH_4; ethane, C_2H_6). Heavier paraffin hydrocarbons (i.e., $C_{18}H_{38}$) form a waxlike substance that is called paraffin. These heavier paraffins often accumulate on the walls of tubing and other production equipment, restricting or stopping the flow of the desirable lighter paraffins.

paraffin-base oil *n*: a crude oil characterized by a high API gravity, a high yield of low-octane gasoline, and a high yield of lubricating oil with a high pour point and a high viscosity index. Popularly, and according to an early classification system, a paraffin-base oil is a crude oil containing little or no asphalt and yielding a residue from distillation that contains paraffin wax. Compare *naphthene-base oil*.

paraffin-deposition interval (PDI) *n*: an interval in the production tubing string where heavy paraffin hydrocarbons are deposited on the inside walls of the tubing. Below or above certain temperatures, depending on the characteristics of the paraffins, paraffin will not form.

paraffin hydrocarbon *n*: see *paraffin*.

paraffin inhibitor *n*: a chemical that, when injected into the wellbore, prevents or minimizes paraffin deposition.

paraffin scraper *n*: a tube with guides around it to keep it centered in the hole, and a cylindrical piece with blades attached.

Spaces between the blades allow drilling fluid to pass through and carry away the scrapings.

paraffin wax *n*: a solid substance resembling beeswax but composed entirely of hydrocarbons. It is obtained from the crude wax that results from the solvent dewaxing or cold pressing of light paraffin distillates. The refined product is of relatively large crystalline structure, is white and brittle, and has little taste or odor.

parallel *adj*: of an electrical circuit in which the components are assembled such that current flows through two or more paths. Compare *series*.

parallel circuit *n*: a circuit with two or more paths through which the current flows. The total current equals the sum of the currents through the various paths; the voltage is the same throughout the circuit. Compare *series circuit*.

parallel flow *n*: see *laminar flow*.

parallel strand *n*: a type of synthetic rope construction.

parallel strings *n pl*: in a multiple completion, the arrangement of a separate tubing string for each zone produced, with all zones isolated by packers.

paramagnetic *adj*: only slightly magnetic. Paramagnetic materials do not react to any significant extent to a magnetic field. See *diamagnetic*, *ferromagnetic*.

parameters *n pl*: the values that characterize and summarize the essential features of measurements.

parasite string *n*: in underbalanced drilling, a length of tubing that is strapped to the outside of a casing string and is used to minimize or prevent an air- or gas-drilled well from heading, or surging. A hole in the casing at the point where the tubing enters the casing allows gas to be injected. Air or gas is injected only through the tubing and not the casing. The injected air or gas lightens the fluid column above the point of injection, which helps prevent surging.

PARALLEL STRINGS

parol evidence *n*: evidence given verbally rather than in writing.

parted rods *n pl*: sucker rods that have been broken and separated in a pumping well because of corrosion, improper loading, damaged rods, and so forth.

partial immersion thermometer *n*: a thermometer with a specific length of the bulb and stem immersed in the liquid and the scale above the surface for ease of reading.

partial pressure *n*: the pressure exerted by one specific component of a gaseous mixture.

participating royalty owner *n*: one who owns all or part of the mineral estate and has the executive rights of a mineral estate owner.

particle *n*: 1. one of the extremely small subdivisions of matter, as an atom or molecule. 2. a small quantity or fragment, as a single crystal or a substance that is made up of vast quantities of crystals.

parts of line *n pl*: the number of times the drilling line is passed (reeved) between a rig's crown block and its traveling block. Typically, a line may have from four to 12 parts. In the case of a four-part line, the line is passed two times between the crown and traveling blocks; a 12-part line is passed six times between the two blocks. The greater the number of parts, the more weight the line is able to support. Also called falls.

pascal (Pa) *n*: the SI metric unit of measurement for pressure and stress and a component in the measurement of viscosity. A pascal is equal to a force of 1 newton acting on an area of 1 square metre.

passivation *n*: the process of rendering a metal surface chemically inactive, either by electrochemical polarization or by contact with passivating agents.

passive margin *n*: an area that develops when a growing ocean basin causes continents to drift apart.

patch *n*: a material used to cover, fill up, or mend a hole or weak spot. A metal piece extending half-way around a pipe and welded to it is a half-sole patch. Two half-sole patches make a full-sole patch.

patch tool *n*: see *casing-patch tool.*

patent *n*: in the case of land, an instrument by means of which a government transfers a fee simple estate to another party.

pawl *n*: notches or slots machined into the table part of a rotary table assembly into which a bar on the rotary table assembly's locking device fits to keep the table from turning. See *rotary locking device, rotary table assembly.*

pay *n*: see *pay sand.*

pay formation *n*: see *pay sand.*

paying quantity *n*: see *commercial quantity.*

payload *n*: the load carried by a ship or floating offshore rig exclusive of that which is needed for its operation.

payout *n*: the point at which the operator of a well has recovered the costs of drilling, completing, and operating the well and can begin to show a profit.

pay sand *n*: the producing formation, often one that is not even sandstone. Also called pay, pay zone, and producing zone.

pay string *n*: see *production casing.*

pay thickness *n*: an expression of the vertical height of the formation yielding hydrocarbons in commercial amounts.

pay zone *n*: see *pay sand.*

Pb *sym*: lead.

PB *abbr*: plugged back; used in drilling reports.

PBR *abbr*: polished bore receptacle.

PCB *abbr*: polychlorinated biphenyl.

PCC *abbr*: permanent chain chaser.

pcf *abbr*: pounds per cubic foot.

PDC *abbr*: polycrystalline diamond compact.

PDC bit *n*: a special type of diamond drilling bit that does not use roller cones. Instead, polycrystalline diamond inserts are embedded into a matrix on the bit. PDC bits are often used to drill very hard, abrasive formations, but also find use in drilling medium and soft formations.

PDC log *abbr*: perforating depth control log.

PDC BIT

P-Δ-P *n*: the difference in pressure, commonly referred to as between casing annulus and tubing.

PDI *abbr*: paraffin-deposition interval.

PD meter *n*: positive-displacement meter. It measures the single-phase volume of gas or fluid by filling and emptying chambers of a specific volume.

PDVSA *abbr*: Petróleos de Venezuela S.A.

peak inverse voltage *n*: see *peak reverse voltage.*

peak polished rod load *n*: the highest load imposed on the polished rod throughout a complete sucker rod pump cycle.

peak reverse voltage *n*: the maximum instantaneous anode-to-cathode voltage in the reverse direction that is applied to a diode in an operating circuit.

peak voltage *n*: the instantaneous increase (peak) in voltage that occurs when voltage flows in an alternating current (AC) circuit. In an AC circuit, the voltage not only reverses direction, but also increases for a moment (peaks) at the point when it reverses direction.

pear-shaped link *n*: chain connecting link used to connect different chain sizes together. It has a larger jaw or opening at one end.

Pearson holiday detector *n*: a holiday detector that checks for coating defects as well as for any metal debris near a buried pipeline. See *holiday.*

peat *n*: an organic material that forms by the partial decomposition and disintegration of vegetation in tropical swamps and other wet, humid areas. It is believed to be the precursor of coal.

pebble puppy *n*: (slang) a field geologist's (or rock hound's) assistant.

pedestal crane *n*: relatively large stationary crane mounted on an offshore unit.

peen *n*: a wedge or spherical-shaped end of a hammer head, which is usually opposite a flat face on the head. *v*: to enlarge, straighten, or smooth with a peen.

peening *n*: permanent distortion in the outer wires of a wire rope, often caused by the wire rope's pounding against a sheave or a machine member, or by heavy operating pressure between the wire rope and a sheave, the wire rope and a drum, or the wire rope and an adjacent wrap of rope.

peg model *n*: an analog model of three dimensions used to study the structure and stratigraphy of a subsurface area. It is made by placing pegs of varying heights into a flat platform to represent the structural contours of strata.

PEL *abbr*: permissible exposure limit.

pelican hook *n*: (nautical) a wire rope attached to an anchor and sometimes to the anchor chain and used to pull and lower the anchor. The ends of the pendant not on the anchor are attached to buoys on the surface of the water.

pendant *n*: see *guy rope.*

pendant (pennant) buoy *n*: buoy used to suspend an anchor pendant.

pendulum assembly *n*: a bottomhole assembly composed of a bit and several large-diameter drill collars and stabilizers placed to allow the bottom drill collar to bend toward the vertical. The assembly works on the principle of the pendulum effect and is used to decrease drift angle. See *pendulum effect.*

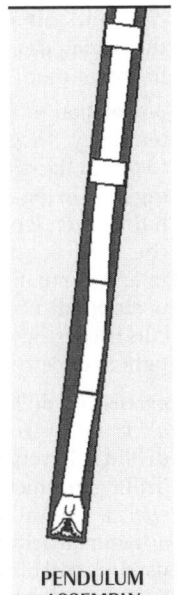

PENDULUM ASSEMBLY

pendulum effect *n*: the tendency of the drill stem—bit, drill collars, drill pipe, and kelly—to hang in a vertical position due to the force of gravity.

penetration rate *n*: see *rate of penetration*.

Pennsylvanian period *n*: a geologic time period in the Paleozoic era, from 320 to 280 million years ago. Also, the latter part of the Carboniferous period. It was named for the outcrops of coal in Pennsylvania.

pentane *n*: a liquid hydrocarbon of the paraffin series, C_5H_{12}.

pentane-plus *n*: a hydrocarbon mixture consisting mostly of normal pentane (C_5H_{12}) and heavier components that are extracted from natural gas.

peptization *n*: increased dispersion of solids in a liquid caused by the addition of electrolytes or other chemical substances. See *deflocculation, dispersion*.

peptized clay *n*: a clay to which an agent has been added to increase its initial yield.

per *abbr*: permeability; used in drilling reports.

percentage chart *n*: a chart for reading differential and static pressures. Its readings reflect a percentage of full scale from 0 to 100 for both static and differential pressure. Readings must be converted, but can be used on any flow recorder, regardless of range.

percentage of proceeds (POP) contract *n*: a gas sales contract under which the buyer processes the gas for recovery of liquid products (ethane and heavier hydrocarbons) and sells the residue under a buyer's contract with the pipelines.

percentage timer *n*: a switch that turns a motor on for a set percentage of the revolution of a cam and off for the remainder of the revolution.

percolation *n*: the tendency for gas to rise in the drilling mud in the annulus, even when the well is shut in and circulation is stopped. It occurs because gas is light in density.

percussion drilling *n*: 1. cable-tool drilling. 2. rotary drilling in which a special tool called a hammer drill is used in combination with a roller cone bit.

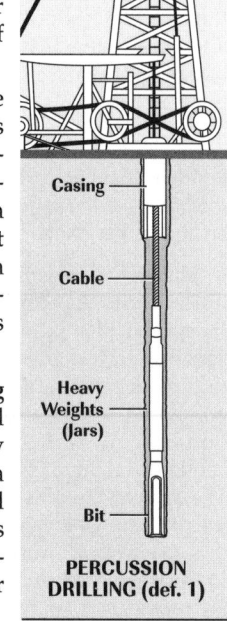

PERCUSSION DRILLING (def. 1)

percussion drilling tool *n*: see *hammer drill*.

perf *abbr*: perforated; used in drilling reports.

perfect gas *n*: see *ideal gas*.

perforate *v*: to pierce the casing wall and cement of a wellbore to provide holes through which formation fluids may enter or to provide holes in the casing so that materials may be introduced into the annulus between the casing and the wall of the borehole. Perforating is accomplished by lowering into the well a perforating gun, or perforator, which fires electrically detonated bullets or shaped charges. See *bullet perforator, jet-perforate, perforating gun*.

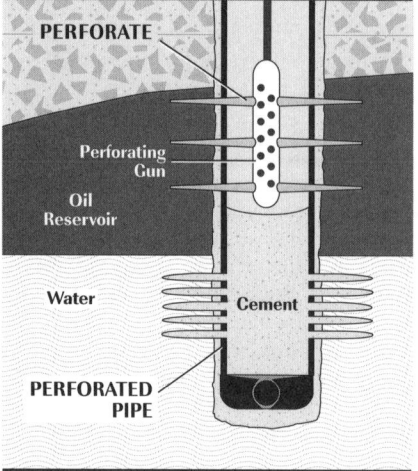

perforated completion *n*: 1. a well completion method in which the producing zone or zones are cased through, cemented, and perforated to allow fluid flow into the wellbore. 2. a well completed by this method.

perforated liner *n*: a liner that has had holes shot in it by a perforating gun. See *liner*.

perforated pipe *n*: sections of pipe (such as casing, liner, and tail pipe) in which holes or slots have been cut before it is set.

perforated spacer tube *n*: a ported, extended production tube used as an alternative path for wireline measuring devices.

perforate underbalanced *v*: to perforate the well with a column of fluid in the wellbore, which exerts less pressure on bottom than the formation does, to cause formation fluids to flow into the wellbore immediately after the casing is perforated. The method is also called reverse-pressure perforating. Its purpose is to force debris in the perforation to flow into the wellbore and not restrict flow within the perforation.

perforating gun *n*: a device fitted with shaped charges or bullets that is lowered to the desired depth in a well and fired to create penetrating holes in casing, cement, and formation.

perforating truck *n*: a special vehicle designed to allow control of a perforating operation within it.

perforation *n*: a hole made in the casing, cement, and formation through which formation fluids enter a wellbore. Usually several perforations are made at a time.

perforation depth control log (PDC log) *n*: a special type of nuclear log that measures the depth of each casing collar. Knowing the depth of the collars makes it easy to determine the exact depth of the formation to be perforated by correlating casing-collar depth with formation depth.

perforation washer *n*: a device utilizing rubber cups run on the tubing string and used to wash, with water, the perforations of wells completed in unconsolidated sands.

perforator *n*: see *perforating gun*.

performance curve *n*: see *accuracy curve of a volume meter*.

perfs *n pl*: perforations in casing for the inflow of hydrocarbons and gas.

period of pitch *n*: the time required for the bow or the stern of a floating offshore drilling rig to start at its lowest position, rise with a wave, and return to its lowest position.

period of roll *n*: the time required for a floating offshore drilling rig to roll from one side to the other and back.

perlite *n*: a volcanic rock that can be extended to many times its original volume by crushing and heating under pressure. Release of the pressure causes expansion when the water in the rock turns to steam. It is used as an extender in cement.

permafrost *n*: perennially frozen subsoil.

permanent chain chaser (PCC) *n*: a device that is located around the mooring wire and/or chain that is used to deploy and recover a mooring leg.

permanent completion *n*: a well completion in which production, workover, and recompletion operations can be performed without removing the wellhead.

permanent guide base *n*: a structure attached to and installed with the foundation pile when a well is drilled from an offshore floating drilling rig. It is seated in the temporary guide base and serves as a wellhead housing. Guidelines are

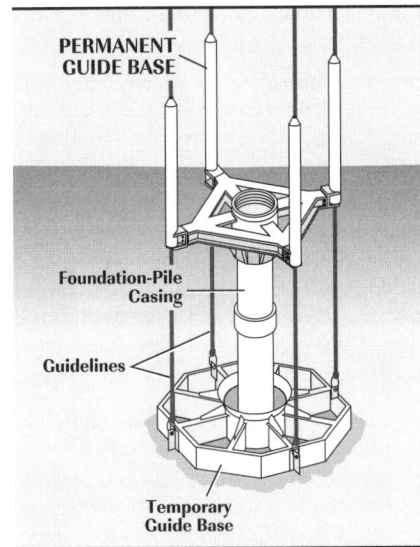

attached to it so that equipment, such as the blowout preventers, may be guided into place on the wellhead.

permanent magnet *n*: a magnet usually made of hardened steel that is strongly magnetized and retains its magnetism after the removal of the magnetizing force.

permanent packer *n*: a nonretrievable type of packer that is very reliable but must be drilled or milled out for removal. Most permanent packers are seal-bore packers or wireline-set packers.

permeability (k) *n*: 1. a measure of the ease with which a fluid flows through the connecting pore spaces of rock or cement. The unit of measurement is the millidarcy. 2. fluid conductivity of a porous medium. 3. ability of a fluid to flow within the interconnected pore network of a porous medium. See *absolute permeability, effective permeability, relative permeability.* 4. on offshore drilling rigs, the percentage of a given space in a vessel that can be occupied by water. 5. the property of a magnetizable substance that determines the degree to which it modifies the magnetic lines of force (the flux) in the region occupied by the substance in a magnetic field; specifically, it is the ratio of the induction to the magnetizing force in the substance.

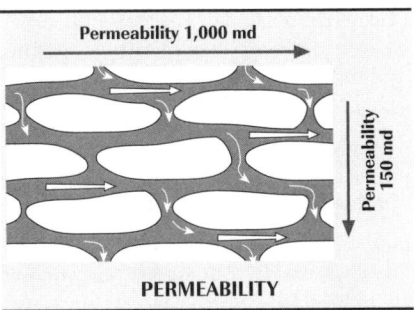

permeability barrier *n*: a hindrance to movement of fluids within the formation rock of a reservoir. Permeability barriers include obvious problems, such as shale lenses and calcite or clay deposits, and less obvious ones, such as porosity changes.

permeable *adj*: allowing the passage of fluid. See *permeability.*

permeable rock *n*: a porous rock formation in which the individual pore spaces are connected, allowing fluids to flow through the formation.

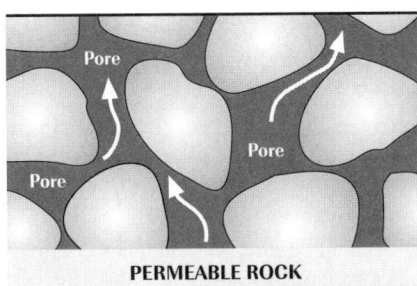

PERMEABLE ROCK

permeameter *n*: a device used to determine the permeability of a core sample by forcing dry air through the sample at the flow rate determined by the pressure differential across the sample. The inlet air pressure and the air flow rate are recorded and, with the dimensions of the sample, are used to calculate permeability.

permeator *n*: the chamber in a reverse osmosis watermaker where reverse osmosis takes place.

Permian Basin *n*: a prolific oil- and gas-producing area in West Texas and eastern New Mexico. So called because many of the rocks from which the oil and gas are withdrawn are of Permian age.

Permian period *n*: the last geologic time period in the Paleozoic era, 280 million to 225 million years ago.

permissible exposure limit (PEL) *n*: the concentration of toxic gas to which it is believed all workers may be repeatedly exposed without adverse effects. Set by the American Conference of Governmental and Industrial Hygienists.

permissible operating temperature *n*: in a solid-state diode, a measure of the diode's ability to handle heat that is generated by voltage and current that occurs where the diode is connected, or joined, to a circuit.

persistence *n*: the durability or longevity of inhibitors used in corrosion control.

personal protective equipment (PPE) *n*: equipment worn by workers to protect themselves from chemical or substance hazards while working.

personnel basket *n*: a net attached to a floatable ring on which personnel ride when being transferred from boat to rig on offshore locations. It is usually rigged to a crane.

personnel net *n*: see *personnel basket.*

petcock *n*: a small valve or faucet for letting out air or for draining liquids.

petrochemical *n*: a chemical manufactured from petroleum and natural gas or from raw materials derived from petroleum and natural gas.

petrol *n*: (British) gasoline.

Petróleos de Venezuela S.A. (PDVSA) *n*: the state-owned oil company of Venezuela.

petroleum *n*: a substance occurring naturally in the earth in solid, liquid, or gaseous state and composed mainly of mixtures of chemical compounds of carbon and hydrogen, with or without other nonmetallic elements such as sulfur, oxygen, and nitrogen. In some cases, especially in the measurement of oil and gas, petroleum refers only to oil—a liquid hydrocarbon—and does not include natural gas or gas liquids such as propane and butane. The API Measurement Coordination Department prefers that petroleum mean crude oil and not natural gas or gas liquids.

petroleum geology *n*: the study of oil- and gas-bearing rock formations. It deals with the origin, occurrence, movement, and accumulation of hydrocarbon fuels.

Petroleum Industry Training Service (P.I.T.S.) (Canada) *n*: an industry-controlled and industry-operated training organization maintained specifically to assist Canadian companies with their training. Address: 1538 25th Avenue NE; Calgary, Alberta T2E 8Y3, Canada; 403-250-9606.

petroleum reservoir *n*: a rock formation that holds oil and gas.

petroleum rock *n*: sandstone, limestone, dolomite, fractured shale, and other porous rock formations where accumulations of oil and gas may be found.

petroleum window *n*: the conditions of temperature and pressure under which petroleum will form. Also called oil window.

petroliferous *adj*: containing petroleum (said of rocks).

petrology *n*: a branch of geology dealing with the origin, occurrence, structure, and history of rocks, principally igneous and metamorphic rocks. Compare *lithology.*

pf *abbr*: power factor.

pF *abbr*: picofarad.

P_f *abbr*: the phenolphthalein alkalinity of the filtrate, reported as the number of millilitres of 0.02 Normal (N/50) acid required per millilitre of filtrate to reach the phenolphthalein end point. P_f is a measure of the alkalinity of the drilling mud. Usually, muds should have a high alkalinity (pH) to ensure adequate performance.

PFD *abbr*: manufacturer's abbreviation for picofarad; sometimes imprinted on the body of a capacitor.

pH *abbr*: an indicator of the acidity or alkalinity of a substance or solution, represented on a scale of 0–14, 0–6.9 being acidic, 7 being neither acidic nor basic (i.e., neutral), and 7.1–14 being basic. These values are based on hydrogen ion content and activity.

phase *n*: 1. a portion of a physical system that is liquid, gas, or solid, that is homogeneous throughout, that has definite boundaries, and that can be separated from other phases. The three phases of H_2O, for example, are ice (solid), water (liquid), and steam (gas). 2. in physics, the stage or point in a cycle to which a rotation, oscillation, or variation has advanced. 3. in electronics, the part of a period through which the time variable of a periodic quantity, such as alternating current, has moved. Phase is measured at any point in time from an arbitrary time origin. In the waveform of alternating current, where the waveform is sinusoidal in shape and where positive polarity is above a zero line and negative polarity is below the zero line, phase is measured at the last point at which the quantity passed through a zero position from a negative to a positive direction.

phase angle *n*: the number of electrical degrees between corresponding points on the sine wave of two or more emfs or currents of the same frequency, or between current and voltage in a circuit.

phase shift *n*: a change in the phase of a periodic quantity, such as the waveform generated by alternating current. See *phase*.

pH control agent *n*: a chemical added to the drilling fluid to control or increase the pH of the mud. Normally, the mud should have a pH higher than seven so that it is alkaline.

pH factor *n*: see *pH value*.

phenol *n*: C_6H_5OH, a white, crystalline, water-soluble, poisonous mass obtained from coal tar, or a hydroxyl derivative of benzene.

phenolics *n pl*: thermosetting plastic materials formed by the condensation of phenols (containing C_6H_5OH) with aldehydes (containing CHO) and used as protective coatings for oilfield structures.

phosphate *n*: 1. generic term for any compound that contains phosphorous and oxygen in the form of a phosphate, which is chemically abbreviated as PO_4. 2. a salt or ester of phosphoric acid.

phosphorite *n*: a rock of biochemical origin, composed largely of calcium phosphate from bird droppings and vertebrate remains.

photoelectric effect (Pe) *n*: the absorption of gamma rays that results in ejection of electrons from an atom. Photoelectric effect occurs when light of sufficient energy falls on an atom and causes it to lose electrons. The energy of the light actually tears electrons away from the atoms of a substance.

photogeology *n*: the practice of examining aerial and satellite photography of the earth to identify and assess the significance of surface features.

photographic survey instrument *n*: a device used to determine the angle and direction by which a wellbore deviates from the vertical. After the instrument is run into the hole, it takes a photograph that is then brought to the surface, developed, and analyzed. Photographic survey instruments can be single-shot or multishot and magnetic or gyroscopic.

photon *n*: a quantum, or unit, of electromagnetic radiation energy.

photooxidation *n*: oxidation introduced by sunlight; a very slow process that helps to break down hydrocarbons exposed to sunlight into other components and, eventually, into carbon dioxide and water.

photosynthesis *n*: a process by which chlorophyll-bearing plants produce simple sugars from carbon dioxide and water using the energy of sunlight.

photovoltaic cell *n*: a device that detects or measures electromagnetic radiation by generating voltage (potential) at the junction between two types of material when radiant energy (as from the sun or other light source) strikes it.

photovoltaic diode (PVD) *n*: a special solid-state diode to which has been added a special impurity that causes the diode to produce voltage when exposed to bright light, such as the sun.

pH value *n*: a unit of measure of the acid or alkaline condition of a substance. A neutral solution (such as pure water) has a pH of 7; acid solutions are less than 7; basic, or alkaline, solutions are more than 7. The pH scale is a logarithmic scale. A substance with a pH of 4 is more than twice as acid as a substance with a pH of 5. Similarly, a substance with a pH of 9 is

more than twice as alkaline as a substance with a pH of 8.

physical absorption *n*: in removing contaminants, acid gases physically dissolve in the liquid absorbent, and the natural gas does not.

physical hazard *n*: (OSHA) a chemical for which there is scientifically valid evidence that it is a combustible liquid, a compressed gas, explosive, flammable, an organic peroxide, an oxidizer, pyrophoric, unstable (reactive), or water reactive.

PI *abbr*: productivity index.

pickle *n*: a cylindrical or spherical device that is affixed to the end of a wireline just above the hook to keep the line straight and to provide weight. *v*: to soak metal pieces in a chemical solution to remove dirt and scale from the metal's surface.

pickup *n*: see *transmitter*.

pick up *v*: 1. to use the drawworks to lift the bit (or other tool) off bottom by raising the drill stem. 2. to use an air hoist to lift a tool, a joint of drill pipe, or other piece of equipment.

pickup-and-laydown system *n*: a boom that lifts pipe from a horizontal position at the V-door and tilts it to vertical and moves it over the mousehole or well center.

pickup meter *n*: a device for converting meter rotor movement into an electrical output signal.

pickup position *n*: the point at which the floor crew of a drilling rig can latch the elevators around the pipe when it is coming out of the hole.

picofarad (pF) *n*: 1 trillionth (10^{-12}) of a farad. See *farad*.

PI curve *n*: see *productivity index curve*.

PID *abbr*: proportional, integral, and derivative.

PID input module *n*: a module that accepts an input reference signal from within a PLC that establishes the desired operating point, compares it to a sensor from the process, and then produces an output signal from the same module to control an output device such as an actuator or proportional valve.

PID loop control *n*: a method of control that utilizes proportional, integral, and derivative techniques to maximize performance and response of the process.

piercement dome *n*: see *diapir*.

piezoelectric crystal *n*: a crystal that generates electrical current as a result of the application of mechanical stress.

piezoelectricity *n*: electricity or electric polarity created by pressure in a crystalline substance, such as quartz.

piezometer ring *n*: a device that converts pressure into an electrical signal through the piezoelectric crystal principle. See *piezoelectricity*.

pi (π) filter *n*: see *LC filter*.

pig *n*: 1. a scraping tool that is forced through a pipeline or flow line to clean out accumulations of wax, scale, and debris from the walls of the pipe. It travels with the flow of product in the line, cleaning the pipe walls by means of blades or brushes affixed to it. Also called a line scraper or a go-devil. 2. a batching cylinder with neoprene or plastic cups on either end and used to separate different products traveling in the same pipeline. 3. a neoprene displacement spheroid, automatically launched and received, used to displace liquid hydrocarbons from natural gas pipelines. 4. in hydrostatic testing of a pipeline, a scraper used inside the line to push air out ahead of the test water and to push water out after the test. *v*: to force a device called a pig through a pipeline or a flow line for the purpose of cleaning the interior walls of the pipe, separating different products, or displacing fluids.

piggyback *v*: (nautical) to install anchors behind each other in tandem on the same mooring line.

pig iron *n*: (slang) a piece of oilfield equipment made of iron or steel.

pigment *n*: in paints, a substance, generally in fine powder form, which is practically insoluble in the paint's base and which is dispersed in the binder to impart specific physical and chemical properties, such as optical, protective, and decorative.

pig run *n*: the trip of a pig through a pipeline.

pigtail *n*: in the brush holder of a generator, a device composed of several braided copper wires that conduct electricity from a generator's brush to the generator's wiring.

pigtail lead *n*: on an electronic component, such as a composition resistor, a single wire (a lead) that exits the component from either end and to which the electrical connection to a circuit is made.

pill *n*: a gelled viscous fluid placed at a specific depth in the well, usually to overcome or minimize a problem at that depth. For example, a special bentonite pill may be spotted (placed) at or near a formation to which drilling fluid is being lost. The pill may seal the formation and reduce or stop the lost circulation problem.

pilot *n*: a rodlike or tubelike extension below a downhole tool, such as a mill, that serves to guide the tool into or over another downhole tool or fish.

pilot balloon *n*: in meteorology, a small balloon whose ascent is followed by a special optical instrument (a theodolite) that measures horizontal and vertical angles in order to obtain data for the computation of the speed and direction of upper level winds. See *theodolite*.

pilot bit *n*: a bit placed on a special device called a hole opener that serves to guide the device into an already existing hole that is to be opened (made larger in diameter). The pilot bit merely guides, or pilots, the cutters on the hole opener into the existing hole so that the hole-opening cutters can enlarge the hole to the desired size.

pilot chart *n*: in marine navigation, a special map of given offshore areas that contains information about safe routes, ocean currents, winds, sea ice, storm tracks, and the like. Ships and mobile offshore drilling units use pilot charts when moving from one offshore location to another.

piloted mill *n*: see *pilot mill*.

pilot hole *n*: in pipeline construction, the hole drilled as the first step of a directionally drilled river crossing. It establishes a pathway for the pipeline.

pilot manifold *n*: a device on a floating offshore drilling rig's accumulator that receives regulated pressure of 2,000 psi (14,000 kPa) and that contains several ¼-in. (6.35-mm) pilot control valves. The control valves send a pilot signal to the subsea control pods to operate the subsea blowout preventers.

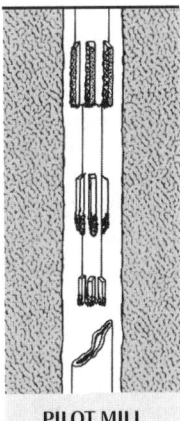

pilot mill *n*: a special mill that has a heavy tubular extension below it called a pilot or stinger. The pilot, smaller in diameter than the mill, is designed to go inside drill pipe or tubing that is lost in the hole. It guides the mill to the top of the pipe and centers it, thus preventing the mill from by-passing the pipe. Also called a piloted mill.

PILOT MILL

pilot pin *n*: the machined extension on the very end of the bearing pin that fits into the nose of the cone of a roller cone bit.

pilot regulator *n*: a special valve (a regulator) on a floating rig's accumulator that sends variable pressure pilot signals, from 0 to 3,000 psi (21,000 kPa), to regulators mounted on the subsea control pod. A pilot regulator is pneumatically operated and has a 30 to 1 ratio, which means

that the hydraulic output pressure is 30 times the air pilot signal.

pilot reports (PIREPS) *n pl*: the recorded observations of aircraft pilots and crewmembers of such phenomena as icing, turbulence, and weather and sky conditions. The National Oceanographic and Atmospheric Administration (NOAA) distributes the reports on http://adds.aviationweather.noaa.gov/pireps. They update the reports every two hours and keep data for six hours.

pilot string *n*: joints of small-diameter pipe attached to the drill assembly and used to bore a pilot hole when laying pipe. After the route has been established, the pilot string is replaced by the work string.

pilot testing *n*: a method of predicting behavior of mud or cement systems by mixing small quantities of mud and mud additives or cement and cement additives and then testing the results.

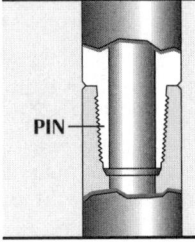

pin *n*: 1. the male threaded section of a tool joint. 2. on a bit, the threaded bit shank. 3. one of the pegs that is fitted on each side into the link plates (side bars) of a chain link

PIN

of roller chain and that serve as the stable members onto which bushings are press-fitted and around which rollers move. See *wrist pin*.

pin angle *n*: see *journal angle*.

pinch bar *n*: a steel lever with a pointed projection at one end; it is used to lift a heavy load.

pinched bit *n*: a roller cone bit on which the legs of the bit have been forced inward by a great force, such as that caused by jamming the bit into undergauge hole.

pinch in *v*: to decrease the size of the opening of an adjustable choke when a kick is being circulated out of a well.

pinch out *v*: to end or terminate by a narrowing and tapering off. When a formation pinches out, it narrows and tapers off.

pinch-out *n*: an oil-bearing stratum that forms a trap for oil and gas by narrowing and tapering off within an impervious formation.

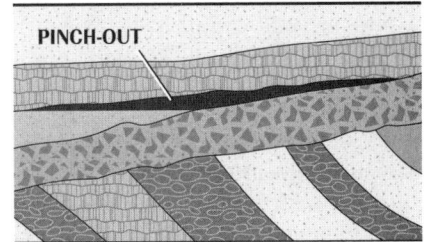

PINCH-OUT

pin connector *n*: in floating offshore drilling operations, a tool that connects the marine riser to the subsea casinghead on the seafloor.

pin diameter *n*: in roller chain, about ⁵⁄₁₆ of the pitch. See *pitch*.

pin-drive master bushing *n*: a master bushing that has four drive holes corresponding to the four pins on the bottom of the pin-drive kelly bushing.

pinger *n*: a sound-emitting device that is used to detect an underwater site or object.

pinholes *n pl*: tiny holes in a dry paint film that form during the application and drying of paint.

pinion *n*: 1. a gear with a small number of teeth designed to mesh with a larger wheel or rack. 2. the smaller of a pair or the smallest of a train of gear wheels.

pin link *n*: a link of roller chain consisting of four parts—two side bars and two pins. The pins are press-fitted into the side bars (pin link plates).

pin link plate *n*: in roller chain, the plate into which the pin link pins ends are immovably fixed. The pins are riveted to the link plate on one side and either riveted or fixed with cotters to the other pin link plate.

pin packer *n*: a packer in which the packing element is held in position by brass or steel pins. When weight is put on the packer, two metal sleeves telescope, shearing the pins and allowing the element to fold and pack off.

pin tap *n*: a short, threaded device made up on the bottom of drill pipe or tubing and used to screw into the box of a stand of drill pipe or drill collars lost in the hole. Once the pin tap is engaged, the lost pipe can be retrieved.

pipe *n*: a long, hollow cylinder, usually steel, through which fluids are conducted. Oilfield tubular goods are casing (including liners), drill pipe, tubing, or line pipe.

pipe bending *n*: the process of bending joints so that a pipeline will conform to the topography of a right-of-way. Pipe bends are made by the cold-work process.

pipe-bending machine *n*: in pipeline construction, a track-mounted hydraulic

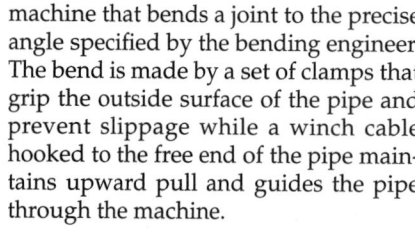

PIPE

machine that bends a joint to the precise angle specified by the bending engineer. The bend is made by a set of clamps that grip the outside surface of the pipe and prevent slippage while a winch cable hooked to the free end of the pipe maintains upward pull and guides the pipe through the machine.

pipe coating *n*: a special material that coats pipe for pipelines and prevents water from coming into contact with the steel of the pipe. The most widely used types of pipe coatings are bituminous enamels, epoxy resins, and tapes.

pipe dolly *n*: any device equipped with rollers and used to move drill pipe or collars. It is usually placed under one end of the pipe while the pipe is being lifted from the other end by an air hoist line.

pipe fitting *n*: an auxiliary part (such as a coupling, elbow, tee, or cross) used for connecting lengths of pipe.

pipe gang *n*: in pipeline construction, the workers responsible for positioning the pipe, aligning it, and making the initial welds. The pipe gang sets the pace that determines the progress of the rest of the spread.

pipe handler *n*: in a top drive, the power and spinning wrenches built into the unit that spins, makes up, breaks out, and backs up the pipe. See *top drive*.

pipe hanger *n*: 1. a circular device with a frictional gripping arrangement used to suspend casing and tubing in a well. 2. a device used to support a pipeline.

pipe heavy *n*: in underbalanced drilling, the condition in which the weight of the drill stem in the wellbore is great enough to overcome well pressure from below. In a pipe heavy situation, the pipe can be stripped into or out of the hole because well pressure is not great enough to force it out of the hole.

pipe jack *n*: a hand tool used to lift and move a stand of pipe that is set back in the derrick. It has a handle on one end and two semicircular pieces on the other end that are designed to fit under the shoulder of a joint of pipe and avoid damage as the pipe is lifted with the tool.

pipe light *n*: in underbalanced drilling, the condition in which the weight of the drill stem in the wellbore is not great enough to overcome well pressure from below. In a pipe light situation, the pipe must be snubbed into or out of the hole against pressure in the wellbore.

pipeline *n*: a system of connected lengths of pipe, usually buried in the earth or laid on the seafloor, that is used for transporting

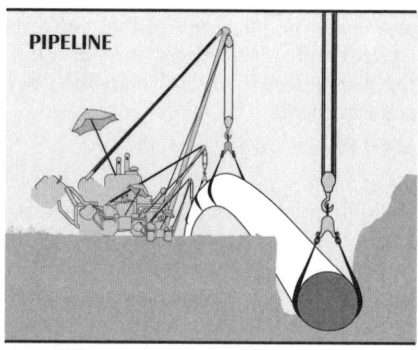

PIPELINE

petroleum, petroleum products, chemicals, and natural gas. A pipeline serves as both a conveyor and a temporary container.

pipeline bundle *n*: a collection of flow lines, umbilical cables, and other devices accompanying a pipeline that are packaged inside an external carrier, which can also be filled with an insulating material.

pipeline connection *n*: the outlet from a well or a tank by which oil or gas is transferred to a pipeline for transportation away from the field.

Pipe Line Contractors Association (PLCA) *n*: a national trade organization of American pipeline contractors. Address: 1700 Pacific Ave., Suite 4100; Dallas, TX 75201-4675; 214-969-2700.

pipeline dispatcher *n*: the person who communicates with other points on the pipeline system. Using a computer console, the dispatcher controls pumping stations hundreds of miles (kilometres) away. Sometimes called a line operator or oil movement controller.

pipeline gas *n*: gas that meets the minimum specifications of a transmission company.

pipeline gauger *n*: an employee of a pipeline company who measures the quantity and quality of crude oil in a tank before the oil is pumped into a pipeline.

pipeline oil *n*: a crude oil whose S&W content is low enough to make the oil acceptable for pipeline shipment.

pipeline pack *n*: the volume of gas in a pipeline.

pipeline patrol *n*: a watch, usually maintained from an airplane, to check the route of a pipeline for leaks or other abnormal conditions.

pipeline plant *n*: see *mainline plant*.

pipeline scheduler *n*: the person who slates a shipment to leave a certain point of origin on a set date and to be delivered to a destination point on a set date.

pipeline testing *n*: the process of proving the structural soundness of an installed pipeline and its capability to fulfill safely the function for which it was designed. The most common testing method is hydrostatic testing.

pipe locator *n*: a device used in leakage surveys to locate affected pipe. It sends out radio signals underground and traces the signal the length of the pipe. Works for both steel and plastic pipe, as plastic pipe has a metal wire running its entire length.

pipe protector *n*: prevents drill pipe from rubbing against the hole or against the casing.

pipe prover *n*: 1. an accurately calibrated, usually U-shaped, pipe used to check, or prove, the accuracy of a meter on a LACT unit or other oil measuring installation. 2. a type of continuous-flow volumetric prover comprising a length of pipe from which a known volume is displaced by a displacer to or from a meter being proved at normal operating conditions.

pipe rack *n*: a horizontal support for tubular goods.

pipe racker *n*: 1. (obsolete) a worker who places pipe to one side in the derrick. 2. a pneumatic or hydraulic device often used on drill ships that, on command from an operator, either picks up pipe from a rack and lifts it into the derrick or takes pipe from out of the derrick and places it on the rack. It eliminates the need to stand pipe in the derrick or mast while it is out of the hole, which is desirable for maintaining the vessel's center of gravity as low as possible and for minimizing the possibility of capsizing.

pipe-racking fingers *n pl*: extensions within a pipe rack for keeping individual pipes separated.

pipe ram *n*: a sealing component for a blowout preventer that closes the annular space between the pipe and the blowout preventer or wellhead.

pipe ram preventer *n*: a blowout preventer that uses pipe rams as the closing elements. See *pipe ram*.

pipe repair clamp *n*: a clamp used to make a temporary repair of a leak in a pipeline.

pipe saddle *n*: a fitting made in parts to clamp onto a pipe to stop a leak or to provide an outlet.

pipe spinner *n*: see *spinning wrench*.

pipe tap *n*: in an orifice meter installation, the threaded hole into which is screwed a small pipe to connect the orifice fitting to the flow recorder. Two taps are usually employed: the upstream tap center is located two and one-half times the published inside pipe diameter upstream of the nearest plate face; the downstream tap center is located eight times the published inside pipe diameter downstream

of the nearest plate face. See *orifice-flange tap*, *orifice meter*.

pipe tensioner *n*: a braking device used on a lay barge to control the descent rate of the pipe. Tensioners also support the entire submerged weight of the pipe as it approaches the bottom.

pipe tongs *n pl*: see *tongs*.

pipe-trenching barge *n*: see *bury barge*.

pipe upset *n*: that part of the pipe that has an abrupt increase of dimension.

pipe wiper *n*: a flexible disk-shaped device, usually made of rubber, with a hole in the center through which drill pipe or tubing passes. It is used to wipe off mud, oil, or other liquid from the pipe as it is pulled from the hole.

pipe wrapping *n*: material applied on top of pipeline coating to protect the coating from damage. Materials used for wrapping include felt, fiberglass, fiberglass-reinforced felt, and kraft paper.

Pirani gauge *n*: a gauge that measures inferential pressure and is concerned with certain electrical effects that are observed in an environment of rarefied air or other gas. The gauge consists of a battery, ammeter, and a pressure-controlled glass bulb containing a quantity of resistance wire whose resistance is a function of temperature. Current flowing in the resistance element causes it to heat up, thus changing its resistance.

PIREPS *abbr*: pilot reports.

piston *n*: a cylindrical sliding piece that is moved by or that moves against fluid pressure within a confining cylindrical vessel.

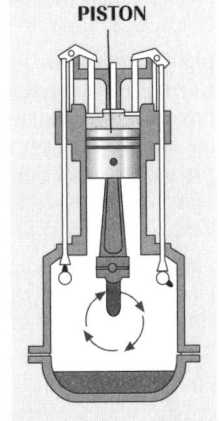
PISTON

piston body *n*: the metal body of a piston over which the piston rubber fits.

piston crown *n*: see *crown*.

piston displacement *n*: the actual volume displaced by a piston in a cylinder as it moves from the beginning of the compression stroke to the end of it.

piston engine *n*: see *reciprocating engine*.

piston pin *n*: a pin that forms a flexible link between the piston and the connecting rod. This bearing area has the highest load per square inch (square millimetre) of any in an engine, perhaps as high as 50,000 pounds per square inch (345 megapascals). Also called a wrist pin.

piston ring *n*: a yielding ring, usually metal, that surrounds a piston and maintains a tight fit inside a cylinder.

piston rod *n*: 1. a metal shaft that joins the piston to the crankshaft in an engine. 2. a metal shaft in a mud pump, one end of which is connected to the piston and the other to the pony rod.

piston rubber *n*: the flexible rubber or synthetic sealing element of a piston.

piston stroke *n*: the length of movement, in inches (millimetres), of the piston of an engine from top dead center (TDC) to bottom dead center (BDC).

pit *n*: 1. a temporary containment, usually excavated earth, for wellbore fluids. 2. a mud tank. 3. a reserve pit.

pitch *n*: 1. in wireline spooling, the degree of slope that the wireline travels in going from one wrap to the next. 2. in roller chain, the distance (in inches or millimetres) between the centers of two members next to one another, i.e., the distance between the centers of the bushings or rollers. 3. on offshore floating rigs, the up-and-down movement of the hull from the bow to the stern.

pitch diameter *n*: root diameter of drum, lagging, or sheave, plus the diameter of the rope.

pit gain *n*: an increase in the average level of mud maintained in each of the mud pits, or tanks. If no mud or other substances have been added to the mud circulating in the well, then a pit gain is an indication that formation fluids have entered the well and that a kick has occurred.

pit level *n*: height of drilling mud in the mud tanks, or pits.

pit-level indicator *n*: one of a series of devices that continuously monitor the level of the drilling mud in the mud tanks. The indicator usually consists of float devices in the mud tanks that sense the mud level and transmit data to a recording

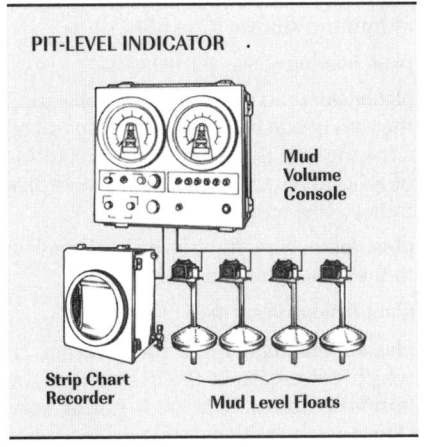
PIT-LEVEL INDICATOR
Mud Volume Console
Strip Chart Recorder   Mud Level Floats

and alarm device (a pit-volume recorder) mounted near the driller's position on the rig floor. If the mud level drops too low or rises too high, the alarm sounds to warn the driller of lost circulation or a kick.

pit-level recorder *n*: see *pit-level indicator*.

pitman *n*: the arm that connects the crank to the walking beam on a pumping unit by means of which rotary motion is converted to reciprocating motion.

Pitot tube *n*: an open-ended tube arranged to face the current of a stream of fluid. It is used in measuring the velocity of a flowing medium.

Pitot-tube meter *n*: a meter that uses a Pitot tube and a manometer or other differential-pressure mechanism to measure flowing fluids. The difference between the pressure on the Pitot tube and the static pressure is the velocity of the flow, which is directly related to the rate of flow.

P.I.T.S. *abbr*: Petroleum Industry Training Service (Canada).

pit-volume recorder *n*: the gauge at the driller's position that records data from the pit-level indicator.

Pit Volume Totalizer™ (PVT) *n*: trade name for a type of pit-level indicator. See *pit-level indicator*.

pit watcher *n*: on offshore drilling rigs, the derrickhand or an assistant who monitors the mud tanks and drilling mud properties, mixes mud, and transfers mud from one tank to another.

pixel *n*: the basic and smallest part of the composition of an image on a television screen, computer monitor, or similar cathode-ray tube.

pk *abbr*: pink; used in drilling reports.

pkr *abbr*: packer; used in drilling reports.

pl *abbr*: pipeline; used in drilling reports.

plagioclase *n*: a common rock-forming mineral varying in composition from sodium aluminum silicate ($NaAlSi_3O_8$) to calcium aluminum silicate ($CaAl_2Si_2O_8$).

plain bearing *n*: see *journal bearing*.

planimeter *n*: an instrument for measuring the area of a plane figure. As the point on a tracing arm is passed along the outline of a figure, a graduated wheel and disk indicate the area encompassed.

plant losses *n pl*: light hydrocarbons lost to the atmosphere in a gas plant.

plant residue sales meter *n*: see *tail gate*.

plastic deformation *n*: a phenomenon in which a decrease in the diameter of the borehole occurs because a plastic substance, such as salt, is being drilled. Since

the temperature of the salt is higher than normal for a given depth (high temperature causes the salt to expand) and since mud weight is usually decreased when drilling such formations to maintain high penetration rates (thus decreasing the hydrostatic pressure exerted on the salt), the section of borehole in the salt zone can expand and possibly stick the drill stem. Also called salt squeeze.

plastic flow *n*: see *plastic fluid*.

plastic fluid *n*: a complex, nonNewtonian fluid in which the shear force is not proportional to the shear rate. Most drilling muds are plastic fluids.

plasticity *n*: the ability of a substance to retain a shape that it has attained after being deformed. Compare *elasticity*.

plasticizer *n*: chemical added, especially to rubbers and resins, to impart flexibility, workability, and stretchability.

plastic squeezing *n*: the procedure by which a quantity of resinous material is squeezed into a sandy formation to consolidate the sand and to prevent its flowing into the well. The resinous material is hardened by the addition of special chemicals, which creates a porous mass that permits oil to flow into the well but holds back the sand at the same time. See *sand consolidation*.

plastic viscosity (PV) *n*: an absolute flow property indicating the flow resistance of certain types of fluids. It is a measure of shearing stress.

plat *n*: a map of a particular tract, group of tracts, or area of land.

plate clutch *n*: a type of friction clutch that has metal plates that may or may not have a friction material on them and an air-operated diaphragm that presses the plates together. Springs push the plates apart when the clutch is disengaged. Compare *drum clutch*.

plate seal *n*: a device that creates a non-leaking union between orifice plate and fitting; keeps gas from leaking around the plate.

plate tectonics *n pl*: movement of great crustal plates of the earth on slow currents in the plastic mantle, similar to the movement of boxes on a conveyor belt. Today geologists believe that the earth's crust is divided into six major plates and several smaller ones atop some of which

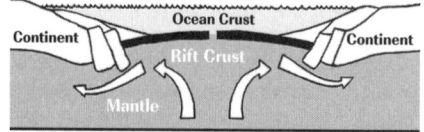

the continents are carried away from a system of midocean ridges and toward another system of deep-sea trenches. The theory of plate tectonics explains most of the mysteries that confounded earlier geologists. Compare *continental drift*.

platform *n*: see *platform rig*.

platform jacket *n*: a support that is firmly secured to the ocean floor and to which the legs of a platform are anchored.

platform rig *n*: an immobile offshore structure from which development wells are drilled and produced. Platform rigs may be built of steel or concrete and may be rigid or compliant. Rigid platform rigs, which rest on the seafloor, are the caisson-type platform, the concrete gravity platform, and the steel-jacket platform. Compliant platform rigs, which are used in deeper waters and yield to water and wind movements, are the guyed-tower platform, the tension-leg platform, and the compliant piled tower platform.

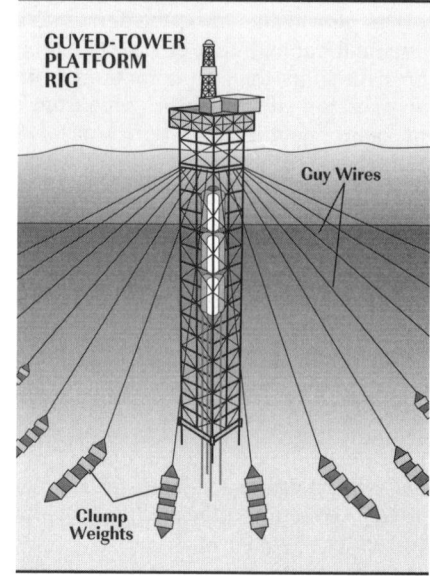

play *n*: 1. the extent of a petroleum-bearing formation. 2. the activities associated with petroleum development in an area.

playa *n*: the flat bottom of an undrained desert basin that at times becomes a shallow lake. Also called sebkha or sabkha.

PLC *abbr*: programmable logic controller.

PLCA *abbr*: Pipe Line Contractors Association.

Pleistocene *adj*: 1. of or relating to the geologic epoch from about 2.5 million to 10 thousand years ago, the first part of the Quaternary period of the Cenozoic era, sometimes called the Ice Age, which extended from the end of the Tertiary period until the last retreat of the northern continental ice sheets. 2. of or relating to the rocks or sediments formed during this epoch.

plenum *n*: an enclosure in which air or other gas is at a pressure higher than that outside the enclosure.

Plimsoll mark *n*: a mark placed on the side of a floating offshore drilling rig or ship to denote the maximum depth to which it may be loaded or ballasted. The line is set in accordance with local and international rules for safety of life at sea (SOLAS).

plow (plough) steel *n*: high-quality, high-strength steel used to make the wire for wire rope.

plug *n*: any object or device that blocks a hole or passageway (such as a cement plug in a borehole).

plug and abandon (P&A) *v*: to place cement plugs into a dry hole and abandon it.

plug back *v*: to place cement in or near the bottom of a well to exclude bottom water, to sidetrack, or to produce from a formation higher in the well. Plugging back can also be accomplished with a mechanical plug set by wireline, tubing, or drill pipe.

plug-back cementing *n*: a secondary-cementing operation in which a plug of cement is positioned at a specific point in the well and allowed to set. Compare *squeeze cementing*.

plug choke *n*: 1. a device used to plug the tubing at some point along the string. A plug choke may be set to test for packer leakage, to stop production when wellhead equipment must be removed, to test for tubing corrosion or other damage, or to separate zones for production or stimulation treatments. 2. a type of well-control choke in which hydraulic pressure moves a cylinder, the plug, into an upset in a 2-in. pipe. The choking plug is on the end of a rod controlled by a pressure balance diaphragm. On various models, the plug may not be pressure tight when closed but generally retracts to a full-open position. Failure of the operating mechanism generally leaves the choke fully open.

plug container *n*: see *cementing head*.

plug flow *n*: a fluid moving as a unit in which all shear stress occurs at the pipe wall and hole wall. The stream thus assumes the shape of several telescopic layers of fluid with lowest velocities near the pipe and hole walls and the fastest in the middle.

plugging material *n*: a substance used to block off zones temporarily or permanently while treating or working on other portions of the well.

plug plucker *n*: a tool used with a mill to retrieve the milled debris.

plug trap *n*: a geologic configuration in which an intrusive body of rock has

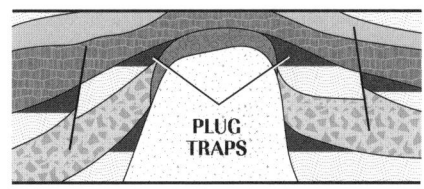

penetrated and deformed one or more surrounding rock layers in such a way that conditions exist for hydrocarbons to accumulate.

plug valve *n*: see *valve*.

plumb bob *n*: see *plumb line*.

plumb line *n*: a string from which a weight (a plumb bob) is suspended and which is used to indicate vertical direction.

plunger *n*: 1. a basic component of the sucker rod pump that serves to draw well fluids into the pump. 2. the rod that serves as a piston in a reciprocating pump. 3. the device in a fuel-injection unit that regulates the amount of fuel pumped on each stroke.

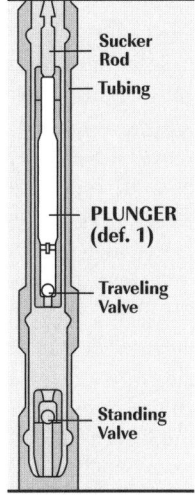
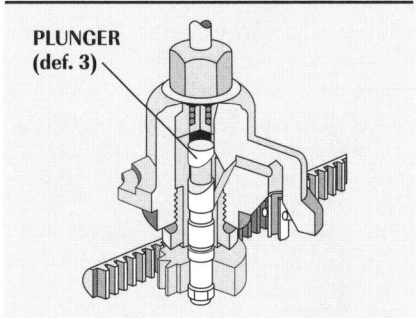

plunger lift *n*: a method of artificial lift that utilizes a plunger that travels up and down inside the tubing. Internally, the plunger contains a bypass valve that is opened when it hits the top of the tubing and closed when it hits the bottom. The plunger's fit with the tubing reduces liquid slippage

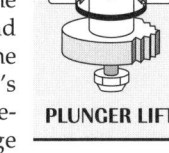

back through the well fluid and the gas that is propelling it. Plunger lift is used in less than 1 percent of all artificially produced wells. It is particularly applicable, however, in high gas-liquid ratio oilwells

or in gas wells with low bottomhole pressure and low productivity.

plunger overtravel *n*: an increase in the effective stroke length of the plunger of a sucker rod pump caused by the elongation of the rod string due to the dynamic loads imposed by the pumping cycle.

plunging fold *n*: a fold of rock whose long axis is not horizontal.

plus input *n*: see *noninverting input*.

pluton *n*: a large subterranean body of igneous rock.

P$_m$ *abbr*: the phenolphthalein alkalinity of the mud reported as the number of millilitres of 0.02 Normal (N/50) acid required per millilitre of mud. P$_m$, like P$_f$, is a measure of the alkalinity of drilling muds. Usually, muds require high pH values (alkalinity) to perform well.

pneumatic *adj*: operated by air pressure.

pneumatic control *n*: a control valve that is actuated by air. Several pneumatic controls are used on drilling rigs to actuate and control rig components (such as clutches, hoists, engines, and pumps).

pneumatic line *n*: any hose or line, usually reinforced with steel, that conducts air from an air source (such as a compressor) to a component that is actuated by air (such as a clutch).

pneumatic sucker rod pumping unit *n*: a reciprocal pumping unit in which one or more power cylinders are placed over the wellhead and a counterbalance system using compressed air or gas is provided. On the upstroke, high-pressure gas is applied to the underside of the pistons of the inside power cylinders, and low-pressure gas above the pistons is exhausted to the sales line. On the down-stroke, high-pressure gas is applied to the top of the pistons, and the low-pressure gas below the pistons is exhausted to the sales line.

pneumatic tube fire detector *n*: a system that can detect fires in open spaces, consisting of a length of flexible plastic or metal tubing that forms a loop around the outside of the structure it protects.

pneumofathometer *n*: a depth-indicating instrument used by a diver.

pnp transistor *n*: a transistor in which p-type material is placed on each side of n-type material. One side of the p-type material is an emitter and the other side is a collector. The n-type material is sandwiched in between the two p-type materials in the base. The emitter is positive with respect to the base, while the collector is negative with respect to the base.

pod *n*: see *hydraulic control pod*.

pod (Podbielniak) analysis *n*: an analytical procedure for hydrocarbon gases and liquids whereby the various components are quantitatively separated by low-temperature distillation for identification and measurement.

pod latch mechanism *n*: in a marine riser system on floating offshore rigs, a device on a retrievable control pod that allows the pod to be remotely connected and disconnected from the lower marine riser package's baseplate.

pod regulator *n*: a special valve (a regulator) that reduces and maintains (regulates) accumulator supply pressure to a subsea blowout preventer's normal working pressure. Two pod regulators exist, one designated as the BOP manifold regulator, and one designated as the annular regulator.

pod-select valve *n*: on a floating offshore drilling rig using a subsea blowout preventer stack and marine riser system, a control device (a valve) that distributes accumulator fluid from the surface to a selected subsea control pod located on the lower marine riser package.

point bar *n*: 1. an accumulation of sediment on the inside of a river bend, usually consisting of sand, silt, and clay. 2. a stratigraphic trap for petroleum formed by the burial and lithification of such a sedimentary deposit.

point-reaction force *n*: a force that counteracts another force at a single point.

points *n pl*: 1. value that indicates hook load or force, read from a weight indicator; 1 point = 1,000 pounds, or 500 decanewtons. 2. in mud density measurement, the portion of the mud's weight expressed in decimal points. For example, a mud that weighs 12.2 pounds per gallon (1,461.8 kilograms per cubic metre) weighs two points more than 12 pounds (1,438 kilograms). Also, if a 12.2 pound-per-gallon mud's density is increased to 12.5 pounds per gallon (1,497.8 kilograms per cubic metre), the density has been raised by 3 points.

poise *n*: the viscosity of a liquid in which a force of 1 dyne (a unit of measurement of small amounts of force) exerted tangentially on a surface of 1 square centimetre of either of two parallel planes 1 centimetre apart will move one plane at the rate of 1 centimetre per second in reference to the other plane, with the space between the two planes filled with the liquid.

Poisson ratio (μ) *n*: when a rod is stretched by applying force at each end of the rod and the force is parallel to the rod's axis, the ratio of the transverse contracting strain to the elongation strain.

Poisson's effect *n*: movement that results from structural compression caused by pressure exerted on the ends of the male (pin) portion of a box-and-pin connection.

polar circumference *n*: circumference measured through the north and south poles of a sphere.

polar compound *n*: a compound (such as water) with a molecule that behaves as a small magnet with a positive charge on one end and a negative charge on the other.

polar front n: see *Antarctic front, Arctic front.*

polar front jet stream *n*: a 300-mile-wide current of wind (a jet stream) that lies at an altitude of 40,000 feet (12,000 metres), and which blows at speeds of 217 knots and faster. Generally, it stays near 40° latitude in the summer months and moves south in the winter months.

polarity *n*: the quality of being either negative or positive; in magnetism, the north or south pole of a magnet.

polar molecule *n*: a molecule whose atoms are arranged so that all positive and negative share a common orientation within the molecule.

polar reaction n: see *electrovalent reaction.*

pole *n*: 1. either end of an axis through a sphere. 2. either of the regions bordering the extremities of the earth's rotational axis, which are the North Pole and the South Pole. 3. a magnetic pole. 4. in electricity, either of two oppositely charged terminals, as in an electric cell or battery. 5. a long, slender, and usually cylindrical object, such as fishing pole.

pole mast *n*: a portable mast constructed of tubular members. A pole mast may be a single pole, usually of two different sizes of pipe telescoped together to be moved or extended and locked to obtain maximum height above a well. Double-pole masts give added strength and stability. See *mast.*

polished bore receptacle *n*: a device made up in the casing string into which the bottom of the tubing string is placed, or landed.

polished rod *n*: the topmost portion of a string of sucker rods. It is used for lifting fluid by the rod-pumping method. It has a uniform diameter and is smoothly polished to seal

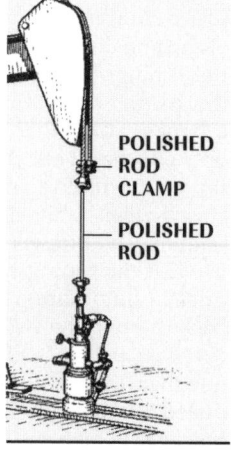

POLISHED ROD CLAMP

POLISHED ROD

pressure effectively in the stuffing box attached to the top of the well.

polished rod clamp *n*: fastening device for connecting the polished rod to the bridle of a beam pumping unit.

polished rod dynamometer *n*: a device that indicates the variation in load on the polished rod as the rod string reciprocates. A continuous record of the result of forces acting along the axis of the polished rod is provided on a dynamometer card, from which the performance of the well-pumping equipment is analyzed.

pollution credits *n pl*: a state practice of trading the air emissions from new facilities for the reduction or elimination of air emissions from old facilities. Offset emissions are bought from other sources to balance the emissions from the new facility. The object is to improve air quality while allowing for new business. This encourages the use of the best technology at new sites and the retrofitting of old sites to improve air quality.

polyacrylamide *n*: a polymer whose basic repeating unit, or monomer, is a combination of carbon, hydrogen, oxygen, and nitrogen. Polyacrylamides are used to adjust the viscosity of water slugs during chemical flooding operations. They can also adhere to the walls of rock pores, decreasing the effective permeability of established channels, forcing the injection fluid into new channels, and thus improving sweep efficiency.

polyanionic cellulose (PAC) *n*: a chemical compound used to reduce water loss in muds that are affected by salt contamination.

polychlorinated biphenyl (PCB) *n*: a family of highly toxic chemical compounds consisting of two benzene rings in which chlorine takes the place of two or more hydrogen atoms; known to cause skin diseases and suspected of causing birth defects and cancer.

polycrystalline diamond compact (PDC) *n*: a disk (a compact) of very small synthetic diamonds, metal powder, and tungsten carbide powder that are used as cutters on PDC bits. Compare *thermally stable polycrystalline diamond bit.*

polycrystalline diamonds *n pl*: very small synthetic diamonds formed by applying great heat and pressure to carbon.

polyester *n*: a thermosetting or thermoplastic material formed by esterification of polybasic organic acids with polyhydric acids.

polymer *n*: a substance that consists of large molecules formed from smaller molecules in repeating structural units (monomers). In oilfield operations, various types of polymers are used to thicken drilling mud, fracturing fluid, acid, water, and other liquids. See *micellar-polymer flooding, polymer mud*. In petroleum refining, heat and pressure are used to polymerize light hydrocarbons into larger molecules, such as those that make up high-octane gasoline. In petrochemical production, polymer hydrocarbons are used as a feedstock for plastics.

polymer flooding *n*: a type of miscible drive in which a polymer is injected into an injection well to allow oil and water to mix and flow to a producing well. See *miscible drive*. Compare *alkaline (caustic) flooding, waterflooding*.

polymerization *n*: the bonding of two or more simple molecules to form larger molecular units.

polymer mud *n*: a drilling mud to which a polymer has been added to increase the viscosity of the mud.

polymer units *n pl*: units in a refinery that polymerize propylene and ethylene in the presence of a catalyst and in a liquid solvent. Polypropylene and polyethylene are produced. Following polymerization, the product is removed from the reactor and processed through a purification facility for removal of catalyst residue and separation of the solvent. The powdered product is then conveyed in a pneumatic conveyor to the finishing area, where additives are blended with powder and the mixture is processed through an extruder to form pellets.

polyphase *n*: in electronics, a phenomenon in which a circuit has several alternating electromotive forces (emfs) of the same frequency and sine wave form. A polyphase generator has two or more circuits in the field windings. *adj*: having or utilizing two or more phases of an alternating-current power line. See *phase*.

polysaccharide *n*: a carbohydrate composed of many monosaccharides. Polysaccharides are used to adjust the viscosity of water slugs in chemical flooding operations.

pontoon *n*: 1. an attachment, added to a stinger, that is flooded to lower pipeline toward the seafloor at an angle that will not overstress it. 2. on many semisubmersible drilling rigs, one of two or more structures, or hulls, on which the rigs floats. Columns rise from the top of the pontoons and support the main deck and equipment. The pontoons may be ballasted or deballasted (water may be added or removed) from the pontoons to adjust the depth at which the rig floats.

pony rod *n*: 1. a sucker rod, shorter than usual, used to make up a sucker rod

string of desired length. Pony rods are usually placed just below the polished rod. 2. the rod joined to the connecting rod and piston rod in a mud pump.

POOH *abbr*: pull-out-of-hole.

pool *n*: a reservoir or group of reservoirs. The term is a misnomer in that hydrocarbons seldom exist in pools, but, rather, in the pores of rock. *v*: to combine small or irregular tracts into a unit large enough to meet state spacing regulations for drilling.

pooling *n*: the combining of small or irregular tracts into a unit large enough to meet state spacing regulations for drilling. Compare *unitization*.

pooling and unitization clause *n*: in an oil and gas lease, the clause that permits the lessee to pool or unitize the leased tract.

poor boy *v*: to make do; to do something on a shoe-string. *adj*: homemade.

poor boy degasser *n*: usually, a mud-gas separator that is fabricated by the personnel on a drilling rig's location, or by welders employed by the drilling contractor in the contractor's storage yard. It is a steel, air-tight cylinder into which drilling mud is piped to provide a space for gas in the mud to escape.

poor boy gravel pack *n*: a bradenhead pack; a method of gravel packing in which no packer is used.

poor boy junk basket *n*: see *finger-type junk basket*.

POP *abbr*: putting on the pump; used in drilling reports.

POP contract *n*: see *percentage of proceeds contract*.

popcorn *adj*: (slang) having the quality of being substandard, unsafe, or cheap.

poppet valve *n*: a device that controls the rate of flow of fluid in a line or opens or shuts off the flow of fluid completely. When open, the sealing surface of the valve is moved away from a seat; when closed, the sealing surface contacts the seat to shut off flow. The direction of movement of the valve is usually perpendicular to the seat. Poppet valves are used extensively as pneumatic (air) controls on drilling rigs and as intake and exhaust valves in most internal-combustion engines.

pop valve *n*: a spring-loaded safety valve that opens automatically when pressure exceeds the limits for which the valve is set. It is used as a safety device on pressurized vessels and other equipment to prevent damage from excessive pressure. Also called a relief valve, safety relief valve, or safety valve.

por *abbr*: porosity or pores; used in drilling reports.

porcupine bit *n*: name for a polycrystalline diamond compact (PDC) bit that has a particularly large number of PDC cutters installed on the head of the bit.

pore *n*: an opening or space within a rock or mass of rocks, usually small and often filled with some fluid (water, oil, gas, or all three). Compare *vug*.

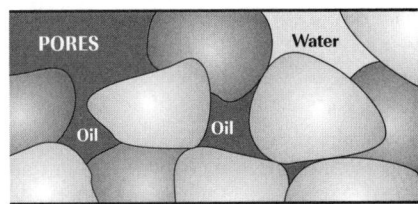

pore pressure *n*: see *formation pressure*.

pore throats *n pl*: connections between pores.

pore volume (PV) *n*: the total volume of pore space in a reservoir formation.

porosimeter *n*: a device used to determine porosity of a formation by measuring a known volume of gas at a known pressure compressed into a core sample. The gas is measured at atmospheric pressure and at an elevated pressure. These measurements are used to calculate gain volume and pore volume.

porosity *n*: 1. the condition of being porous (such as a rock formation). 2. the ratio of the volume of empty space to the volume of solid rock in a formation, indicating how much fluid a rock can hold. See *absolute porosity, effective porosity, pore*.

porous *adj*: having pores, or tiny openings, as in rock.

porous rock *n*: a rock or rock formation containing small openings or spaces within the rock. The spaces are often filled with fluid (such as water, oil, gas, or all three).

port *n*: 1. (nautical) left side of vessel (determined by looking toward the bow). 2. the opening in the side of a liner in a two-stroke cycle engine.

portable mast *n*: a mast mounted on a truck and capable of being erected as a single unit. See *telescoping mast*.

ported sub *n*: a device made up in a tubing or workover string that has openings (ports) through which fluid can be circulated.

portland cement *n*: the cement most widely used in oilwells. It is made from raw materials such as limestone, clay or shale, and iron ore.

Ports and Waterways Safety Act *n*: congressional act that protects navigational safety and the resources of navigable waters.

Under the act, vessel operators must notify port authorities of any "hazardous condition," which includes fire, leaking cargoes, or anything that could adversely affect the environmental quality of any U.S. port, harbor, or navigable water.

positioner *n*: a device used where friction between a valve stem and its packing causes unsatisfactory response. Where friction becomes troublesome, a positioner and an actuator ensure that the valve is positively positioned at the setting required by the process under control.

positioning wire *n*: solid or stranded wire or cable that connects a float to a tape or that drives a dial indicator or transmitter.

position-reference system *n*: any system or method by which surveillance is maintained on the position of a floating offshore drilling rig in relation to the subsea wellhead when the rig is dynamically positioned. The rig should always be directly over the wellhead to minimize wear on subsea equipment and to facilitate operations involved with the equipment. See *acoustic position reference, tautline position-reference system*.

positive charge *n*: a phenomenon that occurs at the atomic level of elements whereby the element has a deficiency of electrons—that is, it has a higher electric potential. Compare *negative charge*.

positive choke *n*: a choke in which the orifice size must be changed to change the rate of flow through the choke.

positive clutch *n*: a clutch in which jaws or claws interlock when pushed together, e.g., the jaw clutch and the spline clutch.

positive crankcase ventilation (PCV) valve *n*: a valve installed on an engine that, when open, directs gases from inside the engine through piping to the intake manifold. At the intake manifold, the crankcase gases enter the engine's combustion chamber and are burned. Burning crankcase gases in this manner not only cuts down on air pollution, but also scavenges corrosive fumes from the engine to prevent sludge from forming.

positive-displacement compressor *n*: see *reciprocating compressor*.

positive-displacement downhole mud motor *n*: a device used to rotate the bit without rotating the drill stem. Basically, the motor comprises a spiral rod that is housed inside a helical-shaped chamber—the rod and chamber are a pump. Circulating mud down the drill stem and to the motor causes the rod to rotate. Since the bit is mechanically connected to the motor, as the motor rotates, so does the bit. This method of bit rotation eliminates the need to rotate the entire drill stem and thus is especially useful in directional drilling.

positive-displacement meter *n*: a mechanical fluid-measuring device that measures by filling and emptying chambers of a specific volume. The displacement of a fixed volume of fluid may be accomplished by the action of reciprocating or oscillating pistons, rotating vanes or buckets, nutating disks, or tanks or other vessels that automatically fill and empty. Also called a volume meter or volumeter.

positive-displacement motor *n*: see *Dyna-Drill*.

positive-displacement pump *n*: a reciprocating or a rotary pump that moves a measured quantity of liquid with each stroke of a piston or each revolution of vanes or gears.

positive gravity anomaly *n*: on the record made by a gravity survey, basement rock (usually denser than overlying sedimentary rock) shows up as a positive gravity anomaly.

positive-pressure self-contained breathing apparatus (SCBA) *n*: a face mask and portable air tank used for the purpose of breathing in contaminated atmospheres.

positive state *n*: the position a ship or offshore floating rig assumes when the vessel's metacenter is above its center of gravity. When the vessel is in a positive state, it is stable and is able to right itself. See *center of gravity, metacenter*.

positive terminal *n*: the terminal of a battery or other voltage source toward which electrons flow and from which holes flow. See *holes, terminal*.

positive-volume prover *n*: a relatively small tank with an accurately calibrated volume used to prove a meter. Liquid is flowed from the meter and into the tank until the calibrated level is achieved. The volume of liquid in the tank is then compared with the volume the meter recorded. If required, the meter is adjusted or a special factor is used to correct the meter. Also called a closed stationary tank prover.

positron *n*: a particle similar to an electron but carrying a positive charge of the same mass and magnitude as the charge of an electron. Positrons are emitted when there is an excess of protons in the nucleus of an atom.

possessory estate *n*: the mineral estate regarded as fee ownership of the minerals in place.

possum belly *n*: 1. a receiving tank situated at the end of the mud return line. The flow of mud comes into the bottom of the device and travels over baffles to control mud flow over the shale shaker. 2. a metal box under a truck bed that holds pipeline repair tools.

posted barge submersible rig *n*: a mobile submersible drilling structure consisting of a barge hull that rests on bottom, steel posts that rise from the top of the barge hull, and a deck that is built on top of the posts, well above the waterline. It is used to drill wells in water no deeper than about 30 to 35 feet (9 to 10.7 metres). Most posted barge submersibles work in inland gulfs and bays. See *submersible drilling rig*.

posted field price *n*: announced price the purchaser will pay for crude oil with a specified gravity in a particular field or area.

posted gauge height *n*: the vertical distance from the bottom of the tank or datum plate to the reference point on the hatch; normally engraved on or near the reference point.

postemergency response operation *n*: the cleanup, removal of contaminated materials, and removal of hazardous substances that remain after an emergency response operation is terminated; that portion of the emergency response performed after the immediate threat of a release has been stabilized or eliminated and cleanup of the site has begun.

posthole digger *n*: (slang) a small or makeshift drilling rig.

posthole well *n*: (slang) a relatively shallow well. Sometimes, a dry hole.

potash *n*: potassium carbonate (K_2CO_3).

potassium *n*: one of the alkali metal elements with a valence of 1 and an atomic weight of about 39. Potassium compounds, most commonly potassium hydroxide (KOH), are sometimes added to drilling fluids to impart special properties, usually inhibition.

potential *n*: 1. the maximum volume of oil or gas that a well is capable of producing, calculated from well test data. 2. See *electric potential*.

potential difference *n*: see *electromotive force*.

potential energy *n*: energy possessed by a body because of its position or configuration. A wound spring or a raised weight has potential energy.

potential hill *n*: in solid-state semiconductors, an energy barrier created by the semiconductor's space-charge region, which must be overcome by the application of external voltage to the semiconductor to make current flow across the junction where an n-type and p-type semiconductor are joined. See *n-type semiconductor, p-type semiconductor, space-charge region*.

potential test *n*: a test of the rate at which a well can produce oil or gas by measuring formation pressures. See *potential*.

potentiometer *n*: 1. a resistor having a continuously adjustable sliding contact that is usually mounted on a rotating shaft. By rotating the shaft, the voltage flowing through the potentiometer is varied as desired. A potentiometer is mainly a voltage divider. 2. a device that measures an electromotive force by comparing it with a known potential difference. See *resistor*.

potentiometric surface *n*: 1. a surface representing the hydrodynamic pressure gradient of groundwater flowing through an aquifer. 2. the level to which unconfined flowing water would rise.

pound equivalent *n*: a laboratory unit used in pilot testing. One gram or pound equivalent, when added to 350 millilitres of fluid, is equivalent to 1 pound/barrel.

pounds per cubic foot (pcf) *n*: a measure of the density of a substance (such as drilling fluid).

pounds per gallon (ppg) *n*: a measure of the density of a fluid (such as drilling mud).

pounds per 100 square feet *n*: a measure of force, for example, of the gel strength of drilling mud.

pounds per square inch (psi) *n*: an English measure of the amount of pressure on an area that is 1 inch square.

pounds per square inch gauge (psig) *n*: the pressure in a vessel or container as registered on a gauge attached to the container. This reading does not include the pressure of the atmosphere outside the container.

pounds per square inch per foot (psi/ft) *n*: a measure of the amount of pressure in pounds per square inch that a column of fluid (such as drilling mud) exerts on the bottom of the column for every foot of its length. For example, 10 pounds per gallon mud exerts 0.52 pounds per square inch per foot (11.8 kilopascals per metre), so a column of 10 pounds per gallon (1,198.2 kilograms per cubic metre) mud that is 1,000 feet (304.8 metres) long exerts 520 pounds per square inch (3.6 megapascals) at the bottom of the column. See *pressure gradient*.

pour point *n*: the lowest temperature at which a fuel will flow. For oil, the pour point is a temperature 5°F (–15°C) above the temperature at which the oil is solid.

power *n*: 1. the source or the means of providing energy. 2. the time rate at which work is done. 3. an exponent, as the 3 in x^3.

power cable *n*: see *cable*.

power distribution system *n*: an arrangement of many pieces of equipment, such as transformers, conductors, switches, heat sinks, fans, poles, and towers, which serve to deliver (distribute) electrical power from its source to its point of use. Power sources include hydroelectric generation by dams, natural gas (or other fuel) fired steam turbines to drive electrical generators, wind generators, and solar generators.

power-driven mud pump *n*: a reciprocating pump for circulating drilling fluids. It operates through cranks and connecting rods by power supplied to its crank-shaft from an electric motor or internal-combustion engine. It may be a duplex (with two cylinders) or a triplex (with three cylinders). Some mud pumps have double-acting pistons, and some have single-acting pistons that function as plungers.

powered mixer *n*: device that depends on an external source of power for the energy required to mix a fluid.

power end *n*: the portion or end of a mud pump that contains the parts involved in producing the mechanical force that moves the pistons in liners to move liquid. The power end is opposite the fluid end, which has the parts that move the liquid mud.

power factor (pf) *n*: the ratio of true power to apparent power in an alternating current circuit. True power is always less than apparent power. See *apparent power*.

power fluid *n*: in subsurface hydraulic pumping, the crude oil that is produced by a well, cleaned, and pumped back into the well to power the subsurface pump.

power law *n*: determines the power a circuit is capable of producing. It consists of three variables: voltage, E or V, measured in volts; current, I, measured in amperes; and power, P, measured in watts. In equation form, the power law is $P = EI$ or $P = VI$.

power mixers *n pl*: in a sample container, a device that is moved by an electric motor to thoroughly mix the liquid in the container.

power of attorney *n*: a legal instrument that authorizes one person to act for another, usually specifically. It ends on the death of either of the parties unless specific language allows it to survive.

power oil *n*: in subsurface hydraulic pumping, the crude oil that is produced by a well, cleaned, and pumped back into the well to energize the subsurface pump.

power rating *n*: 1. rating given by a manufacturer of an engine operating at its most efficient output. 2. in electricity, the power available at the output terminals of a component or a piece of equipment that is operated according to the manufacturer's specifications. 3. in electronics, the amount of power a component, such as a resistor, is capable of handling without malfunctioning or failing. Rating are usually given in watts so that a component with a power rating of 2 watts means that it is capable of handling 2 watts of power.

power rig *n*: see *mechanical rig*.

power rod tongs *n pl*: tongs that are actuated by air or hydraulic fluid and are used for making up or breaking out sucker rods.

power-shift transmission *n*: clutches and gears used with an engine for automatically and smoothly changing power ratios. The device is usually used in conjunction with a hydraulic torque converter.

power side *n*: the back of the drawworks, that is, the side nearest the engines that supply power to the drawworks.

power slips *n pl*: see *slips*.

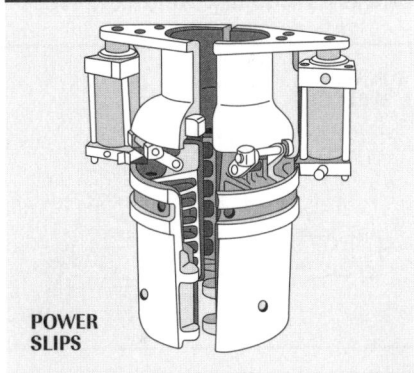

POWER SLIPS

power stroke *n*: downward movement of an engine piston caused by the rapid expansion of the burning fuel-air mixture on top of the piston.

power sub *n*: a hydraulically powered device used in lieu of a rotary to turn the drill pipe, tubing, or casing in a well.

power supply *n*: a source of electrical energy, such as a battery, an electrical line, or a set of components with a transformer in an electronic device, that furnishes electrical power to the semiconductor devices of an electronic circuit with the proper voltages and currents for their operation.

power swivel *n*: see *top drive*.

power takeoff (PTO) *n*: a gearbox or other device that serves to relay the power of a prime mover to auxiliary equipment.

power-tight coupling *n*: a coupling screwed on casing tightly enough to be leakproof at the time of makeup.

power tongs *n pl*: see *power wrench*.

power tools *n pl*: equipment operated hydraulically or by compressed air for making up and breaking out drill pipe, casing, tubing, rods, nuts, and so on.

power transformer *n*: 1. a transformer that takes power from a generating station and steps up the voltage for transmission, or that takes power from the transmission line and steps down the voltage to the primary distribution level of 7,200 or 12,400 volts. 2. a relatively small transformer used in an electronic circuit to obtain the proper voltage required by the circuit to operate. See *distribution transformer*.

power triangle *n*: a right triangle derived from a vector diagram of inductive and capacitive VAR (VAR$_H$ and VAR$_C$) and watts. A power triangle allows relationships between reactive power and real power to be easily seen. See *reactive power, real power, var*.

power wrench *n*: a wrench that is used to make up or break out drill pipe, tubing, or casing on which the torque is provided by air or fluid pressure. Conventional tongs are operated by mechanical pull provided by a jerk line connected to a cathead.

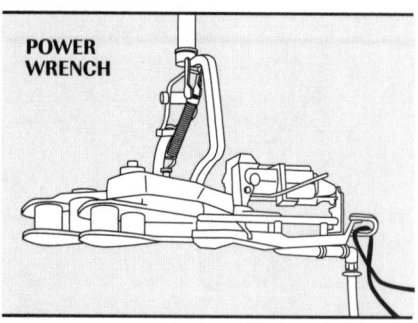

POWER WRENCH

pozzolan *n*: a natural or artificial siliceous material commonly added to portland cement mixtures to impart certain desirable properties. Added to oilwell cements, pozzolans reduce slurry weight and viscosity, increase resistance to sulfate attack, and influence factors such as pumping time, ultimate strength, and watertightness.

pozzolan-cement mixture *n*: a mixture of pozzolan and cement.

pozzolan-lime reaction *n*: the reaction between pozzolan and lime in the presence of water, wherein a cementitious material primarily composed of hydrated calcium silicates is formed.

PPE *abbr*: personal protective equipment.

ppg *abbr*: pounds per gallon.

ppm *abbr*: parts per million.

prairie dog plant *n*: a small, comparatively simple refinery located in a remote area.

Precambrian era *n*: a span of 4 billion years from the earth's beginning until 600 million years ago, during which the earth was devoid of all but the most primitive life forms.

precharge pressure *n*: see *nitrogen precharge pressure*.

precipitate *n*: a substance, usually a solid, that separates from a fluid because of a chemical or physical change in the fluid. *v*: to separate in this manner.

precipitation *n*: the production of a separate liquid phase from a mixture of gases (e.g., rain), or of a separate solid phase from a liquid solution, as in the precipitation of calcite cement from water in the interstices of rock.

precision of error measurement *n*: the degree of conformity to each other of measurements repeated under specified conditions, irrespective of whether they are close or far from the true value.

prefab *n*: a windbreak used around the rig floor, engines, substructure, and other areas to protect the crew from cold winds during winter operations. Windbreaks are constructed of canvas (tarp), wood, or metal.

preferential right of purchase *n*: a prior right of purchase reserved to buy an oil and gas interest by meeting the terms of a proposed sale of the interest to any other party.

preferred resistance values *n*: in electronics, a set of standard values for resistors that manufacturers use as a guide for making resistors of various tolerances. Preferred values range from a low of 1 percent for E96 resistors, to a high of 20 percent for E6 resistors, with values of 5 and 10 percent in between the extremes.

preflush *n*: 1. an injection of water prior to chemical flooding that is used to induce reservoir conditions favorable to the surfactant solution by adjusting reservoir salinity and reducing ion concentrations. A preflush may also be used to obtain advance information on reservoir flow patterns. 2. fluid injected prior to the acid solution pumped into a well in an acid-stimulation treatment; sometimes called a spearhead. Compare *overflush.*.

preforming *n*: a process in the manufacture of wire rope that crimps the strands, giving the rope a permanent set and controlling its flexibility.

pregranted abandonment *n*: a provision of a certificate of public convenience and necessity that authorizes abandonment on a future subsequent or on a date certain.

preheater *n*: a heater used to warm the sample before and during centrifuging to determine suspended S&W content.

preheating *n*: in pipeline construction, the process of heating pipe ends before welding. Preheating is usually necessary in areas where ambient temperatures are below 40°F (4°C) or where there is

overnight condensation of moisture on the pipe. Wagon-wheel heaters are used for preheating.

preignition *n*: a condition in an internal-combustion engine characterized by a knocking sound and caused by the fuel-air mixture's having been ignited too soon because of an abnormal condition.

preloading *n*: the tightening of blowout preventer (BOP) components to final torque under pressure high enough to simulate well pressures. Preloading prevents separation of the components from well pressures and, in the case of subsea BOP stacks, from bending forces caused by currents and rig movements.

premium connections *n pl*: proprietary connections generally distinguished from standard API connections by any of the following features: metal seals, torque shoulders, specialized thread forms, O-ring seals, and tight machining tolerances.

Preservation of Historical and Archaeological Data Act of 1974 *n*: congressional act that protects historical or archaeological data.

preservative *n*: any material, but often paraformaldehyde, used to prevent starch or other substances in a drilling fluid from fermenting through bacterial action.

preset instrument *n*: a measuring instrument fitted with a device that automatically terminates the measurement when it reaches a value fixed in advance.

preset (prelaid) mooring *n*: mooring where the majority of components are installed without the rig or vessel being on site.

press-fit tolerance *n*: the amount of tolerance allowed when a device, such as a bearing, is pressed into a machined receptacle.

pressure *n*: the force that a fluid (liquid or gas) exerts uniformly in all directions within a vessel, pipe, hole in the ground, and so forth, such as that exerted against the inner wall of a tank or that exerted on the bottom of the wellbore by a fluid. Pressure is expressed in terms of force exerted per unit of area, as pounds per square inch, or in kilopascals.

pressure-actuated thermometer *n*: a thermometer in which the pressure developed by thermal changes in the thermometric filling medium actuates an indicating or recording device calibrated in terms of degrees of temperature.

pressure base *n*: see *base pressure*.

pressure base factor *n*: a factor used in the formula recommended prior to August 1992 for the calculation of gas volume flow. The factor adjusted the volume to a standard atmospheric pressure regardless of the location.

pressure buildup plot *n*: a logarithmic plot of bottomhole buildup pressure versus time. Pressure buildup plots are useful in analyzing formation test data for values used in calculating reservoir permeability, formation damage, radius or investigation during the formation test, reservoir depletion, and permeability barriers and other flow irregularities near the wellbore.

pressure compensator *n*: a device with sealed and lubricated bearings that is installed on the leg of a roller cone bit. Its function is to maintain equal pressure inside and outside the bit's bearings in spite of the fact that the drilling mud in the hole can exert very high pressure outside the bit.

pressure control *n*: 1. the act of preventing the entry of formation fluids into a wellbore. 2. the act of controlling high pressures encountered in a well.

pressure controller *n*: an electronic or pneumatic device, such as a pressure-operated valve, that maintains a constant pressure at a specific point in a process.

pressure coring *n*: a coring process in which nitrogen is used to maintain a pressure of 5,000 to 7,000 pounds per square inch (34.5 to 48.3 megapascals) on the core as it is retrieved from the borehole. Pressure coring is used to minimize the escape of fluid from the core during recovery and to prevent mud invasion and flushing.

pressure coupling *n*: 1. a joining device, such as a Squnch joint, in which the force of weight is used to make the connection. 2. a pipe or line coupling that is capable of maintaining a tight, leak-free connection under high pressures.

pressure depletion *n*: the method of producing a gas reservoir that is not associated with a water drive. Gas is removed and reservoir pressure declines until all the recoverable gas has been expelled.

pressure differential *n*: see *differential pressure*.

pressure drawdown *n*: the reduction in a well's bottomhole pressure. See *drawdown*.

pressure drop *n*: a loss of pressure that results from friction sustained by a fluid passing through a line, valve, fitting, or other device.

pressure-drop loss *n*: see *friction loss*.

pressure extension *n*: in gas measurement with orifice meters, a mathematical expression derived from the flow-rate equation. The pressure extension is the square root of the differential pressure in inches of water (h_w) times the static pressure in pounds per square inch absolute.

pressure gauge *n*: an instrument that measures fluid pressure and usually registers the difference between atmospheric pressure and the pressure of the fluid by indicating the effect of such pressures on a measuring element (e.g., a column of liquid, pressure in a Bourdon tube, a weighted piston, or a diaphragm).

pressure gradient *n*: 1. a scale of pressure differences in which there is a uniform variation of pressure from point to point. For example, the pressure gradient of a column of water is about 0.433 pounds per square inch per foot (9.794 kilopascals per metre) of vertical elevation. The normal pressure gradient in a formation is equivalent to the pressure exerted at any given depth by a column of 10 percent salt water extending from that depth to the surface (0.465 pounds per square inch per foot, or 10.518 kilopascals per metre). 2. the change (along a horizontal distance) in atmospheric pressure. Isobars drawn on weather maps display the pressure gradient.

pressure hazard *n*: explosive and compressed gas (as defined by the Code of Federal Regulations).

pressure head *n*: see *head*.

pressure-integrity test *n*: a method of determining the amount of pressure that is allowed to appear on the casing pressure gauge as a kick is circulated out of a well. In general, it is determined by slowly pumping mud into the well while it is shut in and observing the pressure at which the formation begins to take mud.

pressure lock *n*: a manually operated, semiautomatic, self-enclosed gauging device that is used for the prevention of vapor losses in the gauging of atmospheric pressure, variable vapor space, and high-pressure tanks.

pressure loss *n*: 1. a reduction in the amount of force a fluid exerts against a surface, such as the walls of a pipe. It usually occurs because the fluid is moving against the surface and is caused by the friction between the fluid and the surface. 2. the amount of pressure indicated by a drill pipe pressure gauge when drilling fluid is being circulated by the mud pump. Pressure losses occur as the fluid is circulated.

pressure loss drop *n*: the differential pressure in the flowing liquid stream (which will vary with flow rate) between the inlet and outlet of a meter, flow straightener, valve, strainer, length of pipe, and so on.

pressure lubrication *n*: in a chain-and-sprocket drive, using a pump to force oil continuously through pipes with holes in them, through nozzles, or through holes in the sprockets. Oil sprays out directly onto the chain with pressure. On some rigs, the oil sprays onto the outside of the chain, and on others onto the inside of the chain. Compare *disk lubrication*, *drip lubrication*.

pressure maintenance *n*: a method of increasing ultimate oil recovery by injecting gas, water, or other fluids into the reservoir to reduce or eliminate a decline in pressure before reservoir pressure has dropped appreciably, usually early in the life of the field.

pressure management *n*: the efforts made to maintain reservoir drive at efficient levels for maximum recovery of oil and gas. Pressure management involves controlled production rates based on the daily rate of withdrawal and monthly production allowed for each well and takes into account any pressure maintenance methods in use.

pressure parting *n*: a phenomenon in which a rock formation is broken apart along bedding planes or in which natural cracks are widened by the application of hydraulic pressure. It is sometimes called breaking or cracking the formation, earth lifting, or formation fracturing.

pressure probe *n*: a diagnostic tool used to ascertain whether there is a gas leak in the tubing of a gas-lift well. If there is a tubing leak, the pressure on the annulus will equal the pressure on the tubing.

pressure rating *n*: the operating (allowable) internal pressure of a vessel, tank, or piping used to hold or transport liquids or gases.

pressure-reducing valve *n*: a valve that lowers the pressure of a fluid flowing through it.

pressure regulator *n*: a device for maintaining pressure in a line, downstream from the device.

pressure-relief fitting *n*: a device that is set to open at a preset pressure to provide a point for excess pressure to exit to the atmosphere or to a suitable receptacle for later disposal. On the drive shaft of a rotary table assembly, for example, relief fittings are provided to unload excess grease in the shaft.

pressure-relief valve *n*: a valve that opens at a preset pressure to relieve excessive pressures within a vessel or line. Also called a pop valve, relief valve, safety valve, or safety relief valve.

pressure sensor *n*: a device that senses the process variable of pressure and produces a proportional signal, usually in electrical form. Typical pressure sensors are capacitors and resistive strain gauges.

pressure setting assembly *n*: the devices used to set permanent tools in a producing well; the assembly is lowered into the well on wireline. When the desired setting depth is reached, an explosive is detonated to set the tool.

pressure sink *n*: a condition in which the pressure at the wellbore is less than the reservoir pressure. Flow of oil from the reservoir to the wellbore occurs because of this pressure differential.

pressure storage tank *n*: a storage tank constructed to withstand pressure generated by the vapors inside. Such a tank is often spherical, has a wall thickness greater than that of the usual storage tank, and has a concave or a convex top.

pressure surge *n*: a sudden and usually of short-duration increase in pressure. When pipe or casing is run into a hole too rapidly, an increase in the hydrostatic pressure results, which may be great enough to create lost circulation.

pressure switch *n*: on an accumulator, a control housed in an explosion proof junction box that is energized by hydraulic pressure. Hydraulic pressure turns the switch on or off.

pressure tap *n*: in an orifice fitting, the threaded holes on each side of the orifice plate. Small pipes are screwed into the holes to connect the fitting with a flow recorder. The taps are used so that pressure differential on either side of an orifice plate can be recorded. See *orifice meter*.

pressure transducer *n*: an electronic device that senses fluid pressure and converts the pressure to electrical voltage. This voltage, in turn, can actuate other devices, such as meters on remote panels, which provide pressure readouts of all system pressures. See *transducer*.

pressure transient analysis *n*: a method of determining reservoir characteristics such as permeability, skin damage, and average pressure.

pressure-type tank *n*: a tank specially constructed for the storage of volatile liquids under pressure. Tanks may be spheroidal, spherical, hemispherically ended, or other special shapes.

pressure vessel *n*: any container designed to contain fluids at a pressure substantially greater than atmospheric.

pressure, volume, and temperature (PVT) analysis *n*: an examination of reservoir fluid in a laboratory under various pressures, volumes, and temperatures to determine the characteristics and behavior of the fluid.

pressurize *v*: to increase the internal pressure of a closed vessel.

pressurized cooling system *n*: an engine cooling system in which a pressure-tight seal is maintained, usually by a special cap placed on the radiator's opening. The pressure-tight seal keeps the pressure on the cooling system slightly above atmospheric pressure. Maintaining pressure slightly above that of the atmosphere raises the boiling point of water (often the main ingredient in the engine's coolant), eliminates evaporation of coolant from the system, and permits a higher operating coolant temperature, which results in more effective heat transfer from the coolant to the air.

prestart lubrication system *n*: an assembly of devices, including a special oil pump that works separately from the engine, that allows an engine operator to circulate lubricating oil through an engine prior to starting it. Pressuring up the oil in an engine before it starts ensures that a good lubricating oil film forms on the engine parts, thus reducing wear on them.

pretesting *n*: in marketing, analysis of a product under simulated market conditions.

prevailing environment *n*: the resultant force of wind, waves, or current that is prevailing on the rig.

prevailing wind *n*: a wind pattern of the lower troposphere that persists throughout the year with some seasonal modification.

preventer *n*: shortened form of blowout preventer. See *blowout preventer*.

preventer packer *n*: in annular and ram blowout preventers, the rubber or rubberlike material that contacts itself or drill pipe to form a seal against well pressure.

preventer stack *n*: see *stack*.

prevention of significant deterioration (PSD) *n*: a provision under the CAA that prevents deterioration of air quality in areas where the air is already better than NAAQS. Under PSD, major new emission sources must demonstrate that they will not degrade air quality any more than the NAAQS allow in that particular area. Major air emitters in attainment areas must obtain PSD permits and install the best-available control technology.

preventive maintenance *n*: a system of conducting regular checks and testing of equipment to permit replacement or repair of weakened or faulty parts before equipment failure results.

Primacord *n*: a textile-covered fuse with a core of very high explosive.

primary *n*: see *primary winding*.

primary cell *n*: a unit that produces electricity from the chemical reaction of conductors

of dissimilar metals through an electrolyte. In a primary cell, the chemical energy has its origin within the cell, and the chemical reaction is not reversible. Also, the electrolyte is not liquid; rather, it is a moist solid, often called "dry." See *dry cell, wet cell*.

primary cementing *n*: the cementing operation that takes place immediately after the casing has been run into the hole. It provides a protective sheath around the casing, segregates the producing formation, and prevents the undesirable migration of fluids.

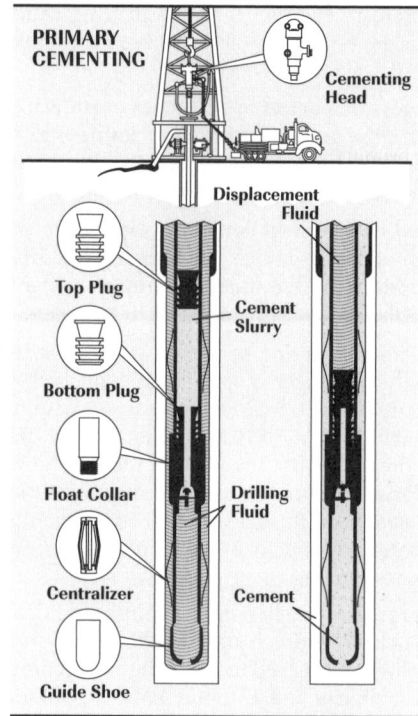

primary control panel *n*: see *master control panel*.

primary detector *n*: see *sensing element*.

primary element *n*: that part of an orifice meter installation that creates the pressure drop in the pipeline necessary for gas measurement; includes the meter tube, orifice plate, fitting, and the pressure taps.

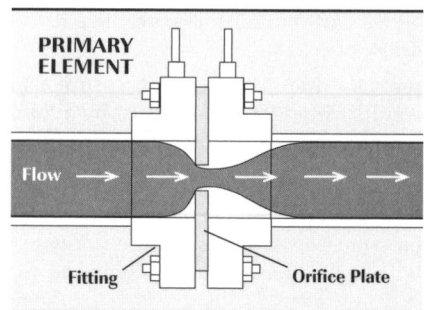

primary line *n*: a power line from the electric utility supplier to the lease distribution point.

primary loss *n*: in a transformer, the reduction (loss) in electrical energy that occurs in the primary winding of the transformer.

primary maximum capacity *n*: in a transformer, the product of the maximum volts and amperes of the primary.

primary migration *n*: movement of hydrocarbons out of source rock into reservoir rock.

primary porosity *n*: natural porosity in petroleum reservoir sand or rocks, i.e., the porosity developed during the original sedimentation process. Formations having this property (such as sand) are usually granular.

primary production *n*: see *primary recovery*.

primary pump *n*: see *booster pump*.

primary recovery *n*: the first stage of oil production in which natural reservoir drives are used to recover oil, although some form of artificial lift may be required to exploit declining reservoir drives.

primary sediment *n*: sediment that comes with the water in an untreated emulsion sample when it is centrifuged with a slugging compound. Compare *secondary sediment*.

primary standard *n*: a particular measure with the highest metrological qualities in a given field. Never used directly for measurement other than for comparison with duplicate standards or reference standards. Compare *secondary standard*, *working standard*.

primary term *n*: the specified duration of an oil and gas lease (e.g., three years), within which time a well must be drilled to keep the lease in effect.

primary well control *n*: prevention of formation fluid flow by maintaining a hydrostatic pressure equal to or greater than formation pressure. Compare *secondary well control*.

primary winding *n*: in a transformer, the first of two windings (coils) to which electrical energy flows to induce electrical energy in the second coil, which is called the secondary winding. Compare *secondary winding*.

prime *v*: to fill a pump with drilling fluid before starting it up in order to prevent air from entering the system.

prime meridian *n*: the line of longitude that is 0 degrees. All other longitudes are measured either east or west of the prime meridian. See *longitude*.

prime mover *n*: an internal-combustion engine or a turbine that is the source of power for driving a machine or machines.

primer *n*: the first coat of a painting system that helps bind subsequent coats to the substrate and which may inhibit its deterioration.

probe *n*: any small device that, when brought into contact with or inserted into a system, can make measurements on that system. In corrosion, probes can measure electrical potential or the corrosivity of various substances to determine a system's corrosive tendencies.

processed gas *n*: gas handled through a plant for the extraction of liquefiable hydrocarbons.

processing agreement *n*: agreement between a producer and a plant owner that provides for the processing of the producer's gas in a plant for a fee, either in cash or products in-kind.

processing rights *n pl*: provision in gas purchase contracts in which the producer reserves the right to separate and extract liquefiable hydrocarbons, except methane and nonhydrocarbon substances, from the natural gas.

processor *n*: in PLC systems, the primary device that contains the microprocessor, memory, logic functions, and timing and scanning elements that create software programs, read input conditions, and issue output commands.

process reaction rate *n*: the speed with which a chemical or industrial process produces a change in the original composition of components.

process stream *n*: a charge or stream of liquids or gases moving through different processes in a refinery or fractionating plant.

process variables *n pl*: quantities such as pressure, temperature, flow, and level in a process system.

producer *n*: 1. a well that produces oil or gas in commercial quantities. 2. an operating company or individual in the business of producing oil; commonly called the operator.

Producers 88 *n*: any of a wide variety of lease forms used in the midcontinent and Gulf regions.

producing horizon *n*: see *pay sand*.

producing interval *n*: see *pay sand*.

producing platform *n*: an offshore structure accommodating a number of producing wells.

producing zone *n*: the zone or formation from which oil or gas is produced. See *pay sand*.

product *n*: 1. in mathematics, the quantity that results when two or more quantities are multiplied together. For example, in the equation $2 \times 8 = 16$, 16 is the product. 2. hydrocarbon or other fluid produced by processing or refining. For example, gasoline is a product shipped in pipelines.

production *n*: 1. the phase of the petroleum industry that deals with bringing the well fluids to the surface and separating them and with storing, gauging, and otherwise preparing the product for the pipeline. 2. the amount of oil or gas produced in a given period.

production casing *n*: the last string of casing set in a well, inside of which is usually suspended a tubing string.

production liner *n*: a liner that functions as production casing. See *liner*, *production casing*.

production liner lap *n*: the amount of lap between surface or intermediate casing or liner and the production liner. See *lap*.

production log *n*: see *spinner survey*.

production maintenance *n*: the efforts made to minimize the decline in a well's production. It includes, for example, acid-washing of casing perforations to dissolve mineral deposits, scraping or chemical injection to prevent paraffin buildup, and various measures taken to control corrosion and erosion damage.

production master valve *n*: see *master valve* (def. 1).

production packer *n*: any packer designed to make a seal between the tubing and the casing during production.

production payment *n*: a cost-free percentage of the working interest that ends when a specified amount of money or number of barrels has been reached.

production platform *n*: see *platform rig*.

production ramp *n*: a special conveyor belt onto which joints of concrete-coated pipe are hoisted by a heavy-duty crane.

production-related costs *n pl*: most notably related to FERC Order 94, production-related costs refer to all costs associated with bringing natural gas or LPG to a marketable point of title transfer. The more common production-related costs include, but are not limited to, compression, dehydration, gathering, processing, treating, liquefaction, conditioning, or transporting.

production rig *n*: a portable servicing or workover outfit, usually mounted on wheels and self-propelled. A well-servicing unit consists of a hoist and engine mounted on a wheeled chassis with a self-erecting mast. A workover rig is basically the same, with the addition of a substructure with rotary, pump, pits, and auxiliaries to permit handling and working a drill string.

production riser *n*: pipe and special fittings used to connect a subsea wellhead to a floating vessel, such as a tanker.

production seal unit *n*: see *seal nipple assembly.*

production swab valve *n*: a special valve installed on Christmas trees. When closed while the well is producing, it directs well flow to the tree. When open during workover operations, it shuts off production and allows tools to be placed into the well. Normally, this valve is opened only during workover operations. It can either be hydraulically operated via the workover control system or manually operated by a remotely operated vehicle.

production tank *n*: a tank used in the field to receive crude oil as it comes from the well. Also called a flow tank or lease tank.

production tax *n*: a state or municipal tax on oil and gas products levied at the wellhead for the removal of the hydrocarbons. Also called severance tax.

production test *n*: a test of the well's producing potential usually done during the initial completion phase.

production tubing *n*: a string of tubing used to produce the well, providing well control and energy conservation.

production well *n*: in fields in which improved recovery techniques are being applied, the well through which oil is produced. See *injection well.*

production wing valve *n*: see *wing.*

productivity index (PI or J) *n*: a well-test measurement indicative of the amount of oil or gas a well is capable of producing. It may be expressed as

$$PI = q \div (P_s - P_f)$$

where—

PI = productivity index (barrels/day or thousand cubic feet/day per pounds per square inch of pressure differential)

q = rate of production (barrels/day or thousand cubic feet/day)

P_s = static bottomhole pressure (pounds per square inch)

P_f = flowing bottomhole pressure (pounds per square inch).

productivity index (PI or J) curve *n*: the curve on a graph that results when the volume of a well's production is plotted against

time. The curve is used to evaluate a well's performance when it has single-phase flow.

productivity test *n*: a combination of a potential test and a bottomhole pressure test the purpose of which is to determine the effects of different flow rates on the pressure within the producing zone of the well to establish physical characteristics of the reservoir and to determine the maximum potential rate of flow. See *bottomhole pressure test, potential test.*

products cycle *n*: the sequence or order in which a number of different products are batched through a pipeline.

products line *n*: a pipeline used to ship refined products.

profile *n*: see *parabolic profile, shallow-cone profile, short parabolic profile.*

profile testing *n*: a technique for simultaneously sampling gas or liquid at several points across the diameter of a pipe to identify the extent of stratification at a proposed location.

prognostic wave chart *n*: a chart showing predicted wave heights and directions. Wind estimates for an area are used to develop this type of chart.

progradation *n*: the seaward buildup of a beach, delta, or fan by nearshore deposition of sediments by a river, by waves, or by longshore currents.

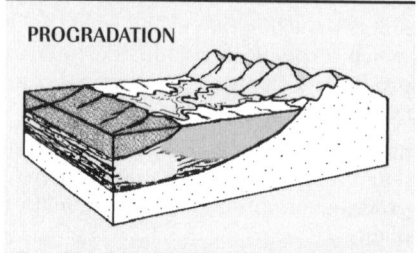

PROGRADATION

programmable logic controller (PLC) *n*: a device used to manage, or control, another device or devices that govern the operation of a system or process. An operator, using an attached computer, can program the controller to maintain a given set of desirable circumstances and to respond to changes or upsets in the system or process using ladder logic, which is a logic system that operates much like the rungs on a ladder—that is, before the next rung on the ladder can be scaled, the controller must determine that certain conditions are met on the current rung.

programming *n*: in instrumentation, software instructions delivered from a personal computer into a PLC processor in a particular format. Programming is the method used to convert ladder logic diagrams of control systems into input

data for PLCs. Programming instructions to the PLC processor can also be in the form of Boolean logic.

program time switch *n*: a switch that turns a motor on and off for preset time intervals, such as 12 hours on, 12 hours off.

progressing cavity pump *n*: a type of subsurface pump in which the rods are rotated instead of reciprocated. The pump consists of an auger-like assembly that rotates inside a cylinder. The reverse screw action forces fluid up through the tubing to the subsurface. Especially useful in heavy crude-high solids content lift applications.

progressive cavity pumping system *n*: a form of sucker rod pumping unit in which the rotor is a chrome-plated steel external helix and the stator is a synthetic elastomer with a double internal helix permanently bonded into a steel housing. Rotation of the rods by a vertical spindle electric motor at the surface causes the rods to stretch a predetermined amount to maintain the rod string in tension. It also causes a cavity containing well fluid to progress upward. The system is a rotary positive-displacement unit.

prong grab *n*: a fishing tool having two or more prongs with inward-facing barbs used to snag and retrieve broken wireline from the wellbore.

proof test *n*: in evaluating a product or piece of equipment, the application of stresses to the equipment to determine whether it has defects in the material of which it is made or other defects in the manufacturing process.

prop *n*: short for proppant. See *propping agent.*

propane *n*: a paraffinic hydrocarbon (C_3H_8) that is a gas at ordinary atmospheric conditions but is easily liquefied under pressure. It is a constituent of liquefied petroleum gas.

propane, commercial *n*: see *commercial propane.*

propene *n*: see *propylene.*

proportional *adj*: in process control systems, of or relating to the closeness of agreement between the desired set point and the actual process level. For example, in a proportional control system, the proportional band determines how much change in the controlled variable is required to operate the final control element (a valve) from fully open to fully closed.

proportional band *n*: in process control systems, the range of values of the controlled variable that causes a controller to operate over its full range.

proportional control *n*: in process control systems, a type of control in which the amount of corrective action is proportional to the amount of error; often controls pressure, flow rate, or temperature in a process system.

proportional controller *n*: in process control systems, a controller whose output is proportional to the error signal.

proportional gain *n*: in process control systems, the ratio of the full range of the final control element to the range of the controlled variable.

proportional, integral, and derivative (PID) *n*: in process control systems, mathematical terms applied to the theory of automated control systems, but which also relate to the more practical terms of proportional, proportional-plus-reset, and proportional-plus-reset-plus-rate. Certain closed-loop controls use proportional, integral, and derivative functions as feedback.

proportional-plus-reset controller *n*: the combined proportional and integral (PI) elements in a control system where the reset, or integral, function returns the controlled variable to the desired set point at regular intervals to eliminate any error in the system.

proportional-plus-reset-plus-rate *n*: a combination of proportional, integral, and derivative control (PID) in a control system where rate (derivative) is added to the PI system to improve speed of correction and response and improve stability.

proportional-speed floating control *n*: see *floating control.*

proportionate reduction clause *n*: in an oil and gas lease, the clause that allows for proportionate reductions in rentals and royalties should the lessor's interests be less than the entire fee simple estate.

proppant *n*: see *propping agent.*

propping agent *n*: a granular substance (sand grains, aluminum pellets, or other material) that is carried in suspension by the fracturing fluid and that serves to keep the cracks open when fracturing fluid is withdrawn after a fracture treatment.

propylene *n*: the chemical compound of the olefin series having the formula C_3H_6. Its official name is propene.

proration *n*: a system, enforced by a state or federal agency or by agreement between operators, that limits the amount of petroleum that can be produced from a well or a field within a given period.

proration unit *n*: see *drilling and spacing unit.*

prospect *n*: 1. an area of land under exploration that has good possibilities of producing profitable minerals. 2. the set of circumstances, both geologic and economic, that justify drilling a wildcat well. *v*: to examine the surface and subsurface of an area of land for signs of mineral deposits.

protection casing *n*: a string of casing set deeper than the surface casing to protect a section of the hole and to permit drilling to a greater depth. Sometimes called intermediate casing string.

protium *n*: an isotope of hydrogen with no neutrons in the nucleus, designated as $_1H^1$. It is the lightest isotope of hydrogen, known as light hydrogen.

proton *n*: the positively charged elementary particle that occurs with the neutron in an atomic nucleus.

prove *v*: to determine the accuracy of a petroleum measurement meter.

proved reserves of crude oil *n pl*: according to API standard definitions, proved reserves of crude oil as of December 31 of any given year are the estimated quantities of all liquids statistically defined as crude oil that geological and engineering data demonstrate with reasonable certainty to be recoverable in future years from known reservoirs under existing economic and operating conditions.

proved reserves of natural gas *n pl*: according to API standard definitions, proved reserves of natural gas as of December 31 of any given year are the estimated quantities of natural gas that geological and engineering data demonstrate with reasonable certainty to be recoverable in future years from known natural gas reservoirs under existing economic and operating conditions.

prover *n*: a device used to determine the accuracy of a petroleum measurement meter.

prover connector valve *n*: a device used to connect the prover to the pipeline either upstream or downstream of the meter to be proved.

prover counter *n*: an electronic device that counts each pulse generated by a meter transmitter.

prover displacer *n*: a spherical or cylindrical object that is a component of a pipe prover. It has an elastic seal that contacts the inner pipe wall of a prover to prevent leakage. Flowing fluid causes it to move through the prover pipe, and as it moves it displaces a known measured volume of fluid between two fixed detecting devices.

prover loop *n*: see *pipe prover.*

prover pass *n*: 1. one movement of the displacer between the detectors in a prover. 2. the volume determined by a displacer traveling between detector switches in a single direction.

prover round trip *n*: 1. the forward and reverse passes in a bidirectional prover. 2. the volume determined by a bidirectional displacer traveling between detector switches in one direction and in the return direction.

prover tank *n*: a tank used to calibrate liquid flow meters. See *open tank prover.*

Proximity Log™ *n*: trade name for a log similar to a Minilog.

prussic acid *n*: see *hydrogen cyanide.*

PSA *abbr*: pressure setting assembly.

PSD *abbr*: prevention of significant deterioration.

pseudocritical properties *n pl*: empirical values for the critical properties (such as temperature, pressure, and volume) of a chemical system made up of multiple components.

pseudo-oil-base mud *n*: see *synthetic-based mud.*

pseudoplastic *adj*: having the capability of changing apparent viscosity with a change in shear rate. Pseudoplastic fluids gain viscosity when subjected to a decrease in shear rate, and lose viscosity when the shear rate is increased. See *shear.*

psi *abbr*: pounds per square inch.

psia *abbr*: pounds per square inch absolute. See *absolute pressure.*

psi/ft *abbr*: pounds per square inch per foot.

psig *abbr*: pounds per square inch gauge.

psychrometer *n*: a device used to measure the amount of water vapor, or relative humidity, of the air. See *sling psychrometer.*

P-tank *n*: see *bulk tank.*

PTO *abbr*: power takeoff.

p-type structure *n*: in a semiconductor material such as silicon, the arrangement of atoms in the material in such a manner that few electrons are able to flow between the atoms; a positive charge is thus created. Compare *n-type structure.*

public domain land *n*: all land and water originally (and still) owned by the United States.

puddling *n*: 1. in cement evaluation work, the agitation of cement slurry with a rod to remove trapped air bubbles. 2. in field practice, the rotation of the casing during or after a primary cementing operation.

Pugh clause *n*: a clause in an oil and gas lease that releases nonproducing acreage (horizontal release) or zones (vertical release) at the end of the primary term or some other specified period. Under the clause, unproductive or untested zones and acreage that are outside a producing pooled unit must be released if drilling or exploration does not occur by the end of the specified time. Also called a Freestone rider.

pull a well *v*: to remove rods or tubing from a well.

pull a well in *v*: to collapse a derrick or mast.

pull back *v*: to raise the drill stem or tubing string in the wellbore.

pull casing *v*: to remove casing from a well.

pull-down *n*: a snubbing unit.

pull dry *v*: to remove the drill stem from the hole without keeping the stem full of mud.

pulley *n*: a wheel with a grooved rim, used for pulling or hoisting. See *sheave*.

pulling tool *n*: a hydraulically operated tool that is run in above the fishing tool and anchored to the casing by slips. It exerts a strong upward pull on the fish by hydraulic power derived from fluid that is pumped down the fishing string.

pulling unit *n*: a well-servicing outfit used in pulling rods and tubing from the well. See *production rig*.

pull it green *v*: to pull a bit from the hole for replacement before it is greatly worn.

pull out *v*: see *come out of the hole*.

pull out of hole *v*: to remove the drill stem or other tools from the wellbore. See *trip out*.

pullout torque *n*: the greatest turning force (torque) under which an electric motor can operate without sharply losing speed.

pull rod *n*: one of several steel rods used to connect pump jacks to a central power source.

pull-rod line *n*: a wire rope used to connect a pump jack to a pull rod.

pull singles *v*: to remove the drill stem from the hole by disconnecting each individual joint.

pull the trigger *v*: to fire a wireline-operated downhole tool from inside the service truck.

pull wet *v*: to remove the drill stem from the hole while keeping the stem full or nearly full of drilling mud.

pulsating flow *n*: flow that is variable; unstable flow.

pulsation dampener *n*: 1. any gas- or liquid-charged, chambered device that minimizes periodic increases and decreases

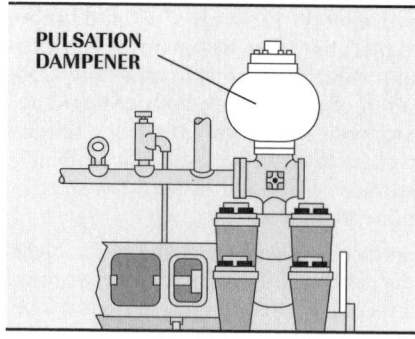

PULSATION DAMPENER

in pressure (as from a mud pump). 2. a device used to reduce pressure pulsations in a flowing stream.

pulsed neutron logging device *n*: a measuring instrument run inside casing to obtain an indication of the presence or absence of hydrocarbons outside the pipe, to determine water saturation in a reservoir behind casing, to detect water movement in the reservoir, to estimate porosity, and to estimate water salinity.

pulsed-neutron survey *n*: a special cased-hole logging method that uses radioactivity reaction time to obtain measurements of water saturation, residual oil saturation, and fluid contacts in the formation outside the casing of an oilwell.

pulse-echo techniques *n pl*: corrosion-detecting processes that, by recording the action of ultrasonic waves artificially introduced into production structures, can determine metal thicknesses and detect flaws.

pulse generator *n*: 1. in electronics, a generator that produces repetitive pulses or signal-initiated pulses. 2. an electrical device that produces very short surges of high-voltage or high-current power by discharging capacitors in parallel or in series.

pulse interpolation *n*: any of the various techniques by which the whole number of meter pulses are counted between two events (such as detector switch closures). Any remaining fraction of a pulse between the two events is calculated.

pulse module *n*: in process control systems, an input PLC module that accepts electrical pulses at a particular rate to measure flow rates of fluids from a pulsing transmitter.

pulse monitor *n*: a device in a pipe prover system that senses and counts the number of pulses put out by a pulse generator on the meter being proved. Pulse count relates to the amount of fluid that flows through the meter.

pulser *n*: see *pulse generator*.

pumice *n*: vesicular obsidian formed from gas-filled lava that cooled rapidly. It is often light enough to float on water.

pump *n*: a device that increases the pressure on a fluid or raises it to a higher level. Various types of pumps include the bottomhole pump, centrifugal pump, hydraulic pump, jet pump, mud pump, reciprocating pump, rotary pump, sucker rod pump, and submersible pump.

pumpability *n*: the physical characteristic of a cement slurry that determines its ability to be pumped.

pump barrel *n*: the cylinder or liner in which the plunger of a sucker rod pump reciprocates. See *sucker rod pump, working barrel*.

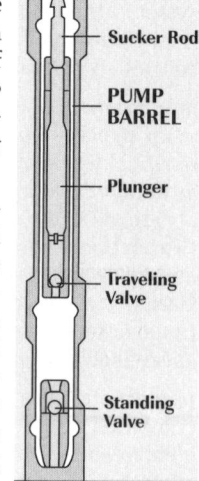

Sucker Rod

PUMP BARREL

Plunger

Traveling Valve

Standing Valve

pump-down *adj*: descriptive of any tool or device that can be pumped down a wellbore. Pump-down tools are not lowered into the well on wireline; instead, they are pumped down the well with the drilling fluid.

pump-down tools *n pl*: for servicing subsea completions, special tools that can be pumped down the flow line and into the tubing string to do a specific job.

pumper *n*: see *lease operator*.

pump house *n*: a building that houses the pumps, engines, and control panels at a pipeline gathering station or trunk station.

pumping cycle *n*: see *cycle*.

pumping tee *n*: a heavy-duty steel, T-shaped pipe fitting that is screwed or flanged to the top of a pumping well. The polished rod works through a stuffing box on top of the tee and in the run of the tee to operate a sucker rod pump in the well. Pumped fluid is discharged through the side opening of the tee.

pumping time *n*: the time required to mix and pump cement slurry down the hole and up the annulus behind the pipe.

pumping unit *n*: the machine that imparts reciprocating motion to a string of sucker rods extending to the positive-displacement

PUMPING UNIT

pump at the bottom of a well. It is usually a beam arrangement driven by a crank attached to a speed reducer.

pump jack *n*: a surface unit similar to a pumping unit but having no individual power plant. Usually, several pump jacks are operated by pull rods or cables from one central power source. Commonly, but erroneously, beam pumping units are called pump jacks. Compare *beam pumping unit*.

pump liner *n*: a cylindrical, accurately machined, metallic section that forms the working barrel of some reciprocating pumps. Liners are an inexpensive means of replacing worn cylinder surfaces, and in some pumps they provide a method of conveniently changing the displacement and capacity of the pumps.

pump manifold *n*: an arrangement of valves and piping that permits a wide choice in the routing of suction and discharge fluids among two or more pumps.

pump off *v*: to pump (a well) so that the fluid level drops below the standing valve of the pump and the pump stops working.

pump-out plug *n*: a device that prevents the entry of fluids into the tubing string while it is being lowered into the well. At the desired packer setting depth, a surface pump is started to elevate the tubing pressure and open the pump-out plug. With the plug open, formation fluids can enter the tubing.

pump plunger *n*: see *sucker rod pump*.

pump pressure *n*: fluid pressure arising from the action of a pump.

pump rate *n*: the speed, or velocity, at which a pump is run. In drilling, the pump rate is usually measured in strokes per minute.

pump room *n*: 1. on an offshore drilling rig, an enclosed area, usually below the main deck, in which the mud pumps are located. 2. an enclosed area, especially on an offshore drilling rig, in which special pumps, such as ballast pumps, are located.

pump speed *n*: the speed, or velocity, at which a pump is run. In drilling, the pump speed is usually measured in strokes per minute.

pump stage *n*: see *multistage centrifugal pump*.

pump station *n*: one of the installations built at intervals along an oil pipeline to contain storage tanks, pumps, and other equipment to route and to maintain the flow of oil.

pump stroke counter *n*: see *pump stroke indicator*.

pump stroke indicator *n*: an instrument that measures pump speed by counting the number of strokes per minute. Also called pump stroke counter.

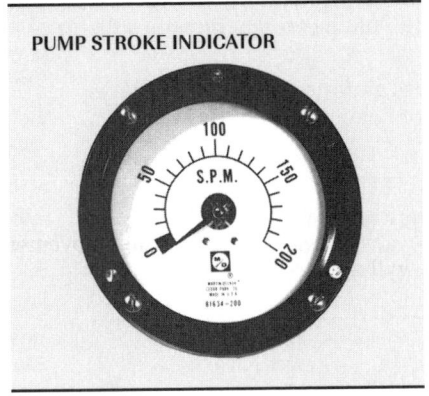

pump valve *n*: any of the valves on a reciprocating pump (such as suction and discharge valves) or on a sucker rod pump (such as a ball-and-seat valve).

pup joint *n*: a length of drill or line pipe, tubing, or casing shorter than range 1 (18 feet or 6.26 metres for drill pipe) in length.

pure fatigue *n*: metal fatigue for which no cause can be determined, such as stress, cracking, or cyclic stress.

pusher *n*: shortened form of toolpusher.

push-in construction *n*: a pipe-laying technique used in swamps and marshes in which the pipe brought into the job site is welded together and then pushed into a water-filled ditch. Compare *lay-barge construction*.

pushrod *n*: a device used to link the valve to the cam in an engine.

pushrod guide *n*: in an engine's valve train, a hollow cylindrical opening or tube through which valve pushrods move up and down. In many engines, the cam on the engine camshaft moves a pushrod that contacts a rocker arm. The rocker arm, in turn, contacts the valve stem, which opens the valve.

put a well on *v*: to make a well start flowing or pumping.

PV *abbr*: 1. plastic viscosity. 2. pore volume.

PVD *abbr*: photovoltaic diode.

PVT *abbr*: 1. Pit Volume Totalizer. 2. pressure, volume, and temperature. See *pressure, volume, and temperature analysis*.

PVT analysis *abbr*: pressure, volume, and temperature analysis.

pycnometer *n*: 1. a vessel of known size, often equipped with a thermometer, used to determine or to compare densities of solids or liquids. 2. in core analysis, the cylindrical steel chamber filled with mercury used to measure the bulk volume of a core sample. When a core sample is placed in the pycnometer, the mercury that it displaces spills into a dish and is weighed. By dividing the weight of the displaced mercury by the density of the mercury, the bulk volume of the sample can be determined.

pyramid *n*: a hump caused when wire rope is spooled onto a drum in a one-step grooving pattern.

pyrite *n*: a hard, yellow, metallic mineral (FeS_2). If encountered during drilling, pyrite can damage equipment and in some cases must be fished from the hole. Also called iron pyrite and fool's gold.

pyroclastic particles *n pl*: particles produced directly by volcanic action when gases within molten lava expand rapidly and the water suddenly flashes into steam, blasting the molten mass into tiny splinters of solidifying glass. The hot particles eventually come to rest in thick blankets of cooling cinders, called ash.

pyroclastic rock *n*: rock formed from pyroclastic particles.

pyrometer *n*: an instrument for measuring temperatures, especially those above the range of mercury thermometers.

pyrophoric *n*: something that ignites spontaneously or that emits sparks when scratched or struck.

Pythagorean theorem *n*: in mathematics, the principle stating that in a right triangle, the square of the length of the hypotenuse is equal to the sum of the squares of the lengths of the other two sides. See *hypotenuse, right triangle*.

q *abbr*: flow rate.

qt *abbr*: quart.

qtz *abbr*: quartz; used in drilling reports.

qtze *abbr*: quartzite; used in drilling reports.

quadruple *n*: see *fourble*.

quantity indicator *n*: an instrument that measures the quantity of a given variable in an electric circuit.

quantity measured *n*: the gross volume metered corrected to agreed reference conditions of pressure and temperature.

quantity meter *n*: a fluid meter that measures the volume of flow.

quantity recorder *n*: an instrument that measures and makes a record of given variables in an electric circuit.

quantum *n*: a unit of energy.

quartz (qtz) *n*: a hard mineral composed of silicon dioxide (silica), a common component in igneous, metamorphic, and sedimentary rocks.

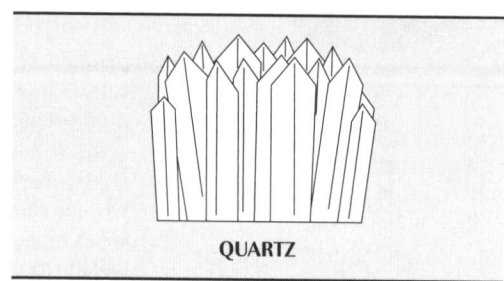

QUARTZ

quartzite (qtze) *n*: a compact granular rock composed of quartz and derived from sandstone by metamorphism.

quasi-static analysis *n*: a mooring analysis that considers static rig offset combined with the effects of both first and second order wave motions.

quebracho *n*: a South American tree that is a source of tannin extract, which was extensively used as a thinning agent for drilling mud, but is seldom used today.

QUEBRACHO

quench *v*: to cool heat-treated metal rapidly by immersion in an oil or water bath.

quench oil *n*: oil injected into the liquid product from a cracking furnace with a view to cooling it rapidly and thus terminating the cracking reaction.

quicklime *n*: calcium oxide, CaO, used in certain oil-base muds to neutralize the organic acid.

quick-look log *n*: a well-site computer log.

quick-opening valve *n*: a specially designed valve, usually hydraulically operated from a location remote from the valve, which, when actuated immediately opens the valve.

quick-setting cement *n*: a lightweight slurry designed to control lost circulation by setting very quickly.

quiescence *n*: the state of being quiet or at rest (being still). Static.

quitclaim deed *n*: a deed that relinquishes to someone else any rights or interests that a person may have in property. The grantor of the quitclaim deed warrants nothing, merely conveys whatever rights, if any, he or she may have.

quotient *n*: in mathematical division, the quantity that results when one quantity is divided by another. For example, where $6 \div 2 = 3$, 3 is the quotient.

r *abbr*: radius.

R *abbr*: 1. Rankine. See *Rankine temperature scale*. 2. rel. 3. resistance.

rabbit *n*: 1. a small plug that is run through a flow line to clean the line or to test for obstructions. 2. any plug left unintentionally in a pipeline during construction (as in, a rabbit that ran into the pipe). 3. a metal device that is placed in the inner core barrel before coring. When all of the core has been removed from the core barrel, the rabbit, or core marker, falls out to indicate that the barrel is empty. *v*: to pull a plug (a rabbit) through a pipe or line to ensure that it is open and free of dents.

race *n*: a groove for the balls in a ball bearing or for the rollers in a roller bearing.

raceway *n*: in electricity, a tube or channel that holds, guides, and protects electric wires.

rack *n*: 1. framework for supporting or containing a number of loose objects, such as pipe. See *pipe rack*. 2. a bar with teeth on one face for gearing with a pinion or worm gear. 3. a notched bar used as a ratchet. *v*: 1. to place on a rack. 2. to use as a rack.

rack-and-pinion *n*: see *rack-and-pinion gear*.

rack-and-pinion gear *n*: a gear comprising a bar with teeth on one face that engages with a pinion (a small gear).

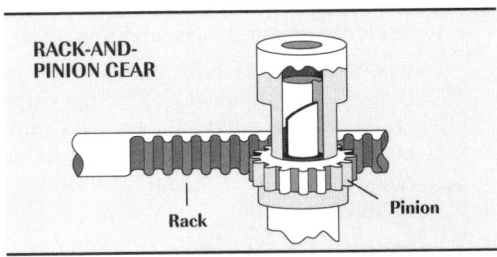

RACK-AND-PINION GEAR

Rack — Pinion

racking platform *n*: a small platform with fingerlike steel projections attached to the side of the mast on a well-servicing unit. When a string of sucker rods or tubing is pulled from a well, the top end of the rods or tubing is placed (racked) between the steel projections and held in a vertical position in the mast.

rack pipe *v*: 1. to place pipe withdrawn from the hole on a pipe rack. 2. to stand pipe on the derrick floor when pulling it out of the hole.

RAD *abbr*: radioactive densitometer.

radar *n*: electronic equipment that transmits and receives high-frequency radio waves to detect, locate, and track distant objects. The word is coined from the phrase RA(DIO) D(ETECTING) A(ND) R(ANGING).

radial drilling *n*: the drilling of several holes in a single plane, all of which radiate from a common point.

radial flow *n*: the flow pattern of fluids flowing into a wellbore from the surrounding drainage area.

radial lead (pronounced "leed") *n*: in an electronic component, a wire (a lead) that exits each end of the component and is arranged so that the leads are attached to the component in a radial manner to the end of the component. Compare *axial lead*.

radially cut grating *n*: on offshore installations with grated deck flooring, a section or panel of grating that is cut into annular segments to fit into circular or annular shaped areas. See *grating*.

radial packer seal *n*: on the telescopic joint of a marine riser system, a flexible, strong rubber ring between the joint's inner and outer barrel that keeps mud in the inner barrel from entering the outer barrel.

radiant heat *n*: the heat from a fire that travels in all directions.

radiate *v*: to send out, or emit, energy—for example, heat or light—in the form of rays or waves.

radiation *n*: 1. energy that is emitted from a source in the form of rays or waves—for example, heat, light, or sound. 2. energy that is emitted in the form of particles by substances—for example, uranium and plutonium, whose atoms are not stable and spontaneously decaying.

radiation logging *n*: see *radioactivity well logging*.

radiator *n*: an arrangement of pipes that contains a circulating fluid and is used for heating an external object or cooling an internal substance by radiation.

radiator core *n*: on an engine, the tube or tubes through which engine coolant is circulated. These tubes bend back and forth several times so that the coolant will have plenty of surface to contact as it flows through them. Very thin metal plates (fins) are attached to the tube, which radiate heat in the coolant to the surrounding air.

radiator fin *n*: on an engine, very thin metal plates that are attached to the radiator's tubes. Because there are hundreds of these plates in contact with the surrounding air, they efficiently radiate heat from the coolant circulating through the tubes.

radiator hose *n*: on an engine, one of usually two flexible, reinforced tubes that conduct coolant in or out of the engine to or from the radiator. One hose is at the top of the engine and conducts hot coolant to the radiator; the other is at the bottom of the engine and conducts cooled coolant from the radiator back into the engine.

radiator tube *n*: on an engine radiator, the piping through which coolant travels, usually from top to bottom, through the radiator. Radiator tubes usually have fins attached to them, which radiate (give up) heat to the surrounding air.

radical *n*: two or more atoms behaving as a single chemical unit, i.e., as an atom, e.g., sulfate (SO_4), phosphate (PO_5), nitrate (NO_4), in which an atom of sulfur (S) and four atoms of oxygen (O), or an atom of phosphorus (P) and five atoms of oxygen, or an atom of nitrogen (N) and four atoms of oxygen behave as though they were a single atom when they combine with another atom or atoms. Ammonium sulfate $(NHSO_4)$, for example, is a compound in which the sulfur and oxygen atoms behave as a single chemical unit when they combine with nitrogen and hydrogen (H).

radioactive *adj*: of, caused by, or exhibiting radioactivity.

radioactive decay *n*: the spontaneous transformation of a radioactive atom into one or more different atoms or particles, resulting in a long-term transformation of the radioactive element into lighter, nonradioactive elements.

radioactive densitometer *n*: a densimeter that measures fluid density by sensing the decay of naturally occurring radioactive elements in the fluid.

radioactive iodine *n*: an isotope of the element iodine, which is radioactive and which is sometimes used as a radioactive tracer.

radioactive tracer *n*: a radioactive material (often carnotite) put into a well to allow observation of fluid or gas movements by means of a tracer survey.

radioactivity *n*: the property possessed by some substances (such as radium, uranium, and thorium) of releasing alpha particles, beta particles, or gamma particles as the substance spontaneously disintegrates.

radioactivity curve *n*: usually, a gamma ray curve, but could also refer to any logging curve obtained by radioactivity logging.

radioactivity log *n*: a record of the natural or induced radioactive characteristics of subsurface formations. Also called nuclear log. See *radioactivity well logging*.

radioactivity well logging *n*: the recording of the natural or induced radioactive characteristics of subsurface formations. A radioactivity log, also known as a radiation log or a nuclear log, normally consists of two recorded curves: a gamma ray curve and a neutron curve. Both help to determine the types of rocks in the formation and the types of fluids contained in the rocks.

radiofacsimile *n*: see *radiofax*.

radiofax *n*: a high-frequency (HF) radio broadcast that transmits weather information to special receivers (radios) on board ships and offshore installations and rigs. The receivers are equipped with facsimile (fax) machines that print out text and graphics of the information. Also called HF FAX, radio facsimile, and weatherfax.

radiographic examination or testing *n*: photographic record of corrosion damage obtained by transmitting X-rays or radioactive isotopes into production structures. It may also be used to produce a shadowgraph of a pipeline weld and reveal any flaws. Also called X-ray testing.

radioisotope *n*: an element, or one of its variants, that exhibits radioactivity.

Radio Manufacturers Association (RMA) *n*: a now defunct association that published manufacturing standards for electronic components. RMA evolved into the Electronic Industry Alliance (EIA). See *Electronic Industry Alliance*.

radiometric *adj*: relating to the measurement of geologic time by means of the rate at which radioactive elements disintegrate.

radiometric dating *n*: a technique for measuring the age of an object or a sample of material by determining the ratio of the concentration of a radioisotope to that of a stable isotope in it.

radiosonde *n*: a balloon-borne instrument used to measure pressure, temperature, and humidity above the earth's surface.

radio triangulation *n*: a surveying method using radio waves to measure a large area of land by establishing a base line from which a network of triangles is built up; in a series, each triangle has at least one side common with each adjacent triangle.

radio tube *n*: see *electron tube*.

radius (r) *n*: 1. in a circle, a straight line that extends from the center of the circle to its edge. 2. in a sphere, a straight line that extends from the center of the sphere to its surface.

raffinate *n*: the remaining mixture that is separated from the extract solution.

rail *n*: a high-pressure manifold for fuel injection systems of some engines.

rake angle *n*: on polycrystalline diamond compact (PDC) bits, the angle at which the studs, on which are mounted the PDC cutters, are placed in the bit's head. If a stud has a rake angle of 90°, the stud is perpendicular to the plane of the bit's head. In most cases, the studs are mounted with back rake angle. See *back rake angle*.

ram *n*: the closing and sealing component on a blowout preventer. One of three types—blind, pipe, or shear—may be installed in several preventers mounted in a stack on top of the wellbore. Blind rams, when closed, form a seal on a hole that has no drill pipe in it; pipe rams, when closed, seal around the pipe; shear rams cut through drill pipe and then form a seal.

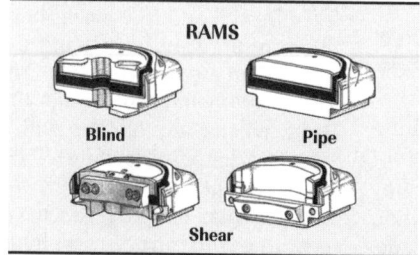

RAMS

Blind   Pipe

Shear

ram block *n*: see *ram*.

ram blowout preventer *n*: a blowout preventer that uses rams to seal off pressure on a hole that is with or without pipe. Also called a ram preventer. Compare *annular blowout preventer*.

ram bonnet *n*: the housing on a ram blowout preventer inside of which the rams and ram operating parts move when the preventer is operated.

ram cavity *n*: see *ram bonnet*.

Ramfor™-link *n*: a type of chain connecting link (Ramnas Mfg.).

ram packer *n*: the sealing device in a ram blowout preventer that is usually made of synthetic rubber. The packer is molded to the ram's upper and lower steel anti-extrusion plates.

ram preventer *n*: see *ram blowout preventer*.

ramp-sweep generator *n*: a device that varies a frequency at a constant rate and incorporates an oscillator that can be programmed to provide an output over a specified frequency range.

random-access memory (RAM) *n*: electronic data stored in such a manner as to allow a user to access and program it, and then read from it, erase it, and reprogram it.

random error *n*: an error that varies in an unpredictable manner in absolute value and in algebraic sign when a large number of measurements of the same value of a quantity are made under effectively identical conditions.

range *n*: 1. the name given to the east-west lines of the rectangular survey system. Compare *township*. 2. in oceanography, the difference in height between consecutive high and low tides. 3. the region between the limits within which a quantity is measured, received, or transmitted, expressed by stating the lower and upper range values.

rangeability *n*: the capability of a meter or flow-measuring device to operate between the minimum and maximum flow range within an acceptable tolerance. Generally expressed as the ratio of maximum flow to minimum flow.

range length *n*: a grouping of pipe lengths. API designation of range lengths is as follows:

	Range 1	Range 2	Range 3
Casing	16–25 ft (4.88–7.62 m)	25–34 ft (7.62–10.36 m)	34–48 ft (10.36–14.63 m)
Drill pipe	18–22 ft (5.48–6.71 m)	27–30 ft (8.23–9.14 m)	38–45 ft (11.58–13.72 m)
Tubing	20–24 ft (6.10–7.32 m)	28–32 ft (8.53–9.75 m)	

range line *n*: an east-west line of the rectangular survey system.

range of load *n*: in sucker rod pumping, the difference between the polished rod peak load on the upstroke and the minimum load on the downstroke.

range of stability *n*: the maximum angle to which a ship or mobile offshore drilling rig may be inclined and still be returned to its original upright position.

range of uncertainty *n*: the interval within which the true value is expected to lie with a stated degree of confidence.

ranging *n*: in process control, the calibration of process measuring instruments between its lower range value (zero) and its upper range value (span).

Rankine temperature scale *n*: a temperature scale with the zero point at absolute zero. On the Rankine scale, water freezes at 491.60° and boils at 671.69°. See *absolute temperature scale, absolute zero temperature*.

rasp *n*: a mill used in fishing operations, before running the fishing tool, to reduce the size of the box or collar on the lost tool.

ratable take *n*: usually relating to contract provisions stating that the pipeline company will attempt to take the seller's fair share with other producers in the same reservoir.

ratchet *v*: to rotate a tool in a series of short movements (like that of a ratchet wrench) to cause the tool to set. Some packers, for example, are set by ratcheting the tubing string in which the packer is made up.

rate meter *n*: see *counting rate meter*.

rate of penetration (ROP) *n*: a measure of the speed at which the bit drills into formations, usually expressed in feet (metres) per hour or minutes per foot (metre).

rate of shear *n*: rate (commonly given in rpm) at which an action resulting from applied forces causes or tends to cause two adjacent parts of a body to slide relative to each other in a direction parallel to their plane of contact.

rate response *n*: in process control systems, the speed with which a controlled variable changes in value.

rathole *n*: 1. a hole in the rig floor, some 30 to 40 feet (9 to 12 metres) deep, which is lined with casing that projects above the floor, into which the kelly and the swivel are placed when hoisting operations are in progress. 2. a hole of a diameter smaller than the main hole and drilled in the bottom of the main hole. *v*: to reduce the size of the wellbore and drill ahead.

rathole connection *n*: the addition of a length of drill pipe or tubing to the active string using the rathole instead of the mousehole, which is the more common connection. The length to be added is placed in the rathole, made up to the kelly, pulled out of the rathole, and made up into the string. Compare *mousehole connection*.

rathole rig *n*: a small, usually truck-mounted rig, the purpose of which is to drill ratholes for regular drilling rigs that will be moved in later. A rathole rig may also drill the top part of the hole, the conductor hole, before the main rig arrives on location.

ratification *n*: approval and confirmation of a contract or other legal instrument, usually by means of a second written instrument.

ratio test *n*: in emulsion treating, a test to determine the best ratio of chemical to emulsion.

raw crude *n*: unrefined crude oil.

raw gas *n*: unprocessed gas or the inlet gas to a plant.

raw gasoline *n*: gasoline extracted from wet natural gas.

raw make *n*: see *raw mix liquids*.

raw mix liquids *n pl*: a mixture of natural gas liquids prior to fractionation. Also called raw make.

raw water *n*: in an engine's heat exchanger, water that circulates outside and around the tubes through which the engine's coolant circulates. Raw water is untreated water that removes heat from the engine's coolant. Offshore, raw water is often seawater. Since raw water does not contact engine parts, it usually does not require treatment.

ray *n*: energy in wave form rather than in particle form.

RBOP™ *abbr*: a trademarked abbreviation for rotating blowout preventer.

RC circuit *n*: see *resistance-capacitance circuit*.

RCRA *abbr*: Resource Conservation and Recovery Act.

RDX *n*: see *cyclonite*.

reactance *n*: in an alternating current circuit, the impedance of current flow caused by capacitive or inductive components.

reaction products *n pl*: the compounds formed as a result of a chemical reaction, such as the reaction of an acid with rock. They may be solids (in which case they are called precipitates), liquids, or gases.

reactive *adj*: as applied to drilling muds, those components in the mud that are affected by and react with other mud components.

reactive hazard *n*: unstable reactive, organic peroxide; also water reactive (as defined by the Code of Federal Regulations).

reactive phase *n*: the part of a drilling mud that consists of microscopic particles that can react with the liquid. The main reactive solids in most drilling muds are clays. Also called the colloidal phase.

reactive power *n*: the value of the power in an electric circuit obtained by multiplying the effective value of the current in amperes, the effective value of the voltage in volts, and the sine of the angular phase difference between current and voltage.

reactive solids content *n*: the amount of water-absorbent material in the drilling fluid.

reactive torque *n*: the tendency of the drill string to turn in a direction opposite that of the bit, a factor for which a driller must compensate when using a downhole motor.

reactivity *n*: a measure of one substance's ability to chemically react with other substances.

read-only memory (ROM) *n*: a device for storing electronic data in permanent, or nonerasable, form—that is, the data cannot be accessed, rewritten, or altered by a user. It is usually a static electronic or magnetic device that allows rapid access to the data.

readout *n*: a device that displays numbers or symbols and incorporates electric or electronic features.

readout device *n*: a device that indicates or registers the value measured by an instrument in practical units.

reagent *n*: a substance used in preparing a product or measuring a component because of its chemical or biological activity.

real gas specific gravity *n*: the ratio of the density of a gas, under the observed conditions of temperature and pressure, to the density of dry air at the same temperature and pressure. The ideal gas specific gravity is the ratio of the molecular weight of the gaseous mixture to the molecular weight of air.

real power *n*: the component of apparent power that represents true work. Real power is expressed in watts and equals volt-amperes multiplied by the power factor.

ream *v*: to enlarge the wellbore by drilling it again with a special bit. Often a rathole is reamed or opened to the same size as the main wellbore. See *rathole*.

reamer *n*: a tool used in drilling to smooth the wall of a well, enlarge the hole to the specified size, help stabilize the bit, straighten the wellbore if kinks or doglegs are encountered, and drill directionally. See *ream*.

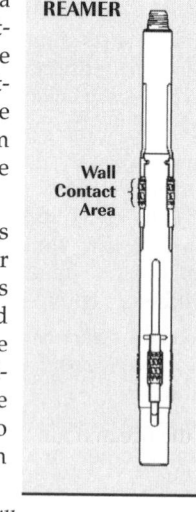

REAMER

Wall Contact Area

reamer pad *n*: on a diamond bit, a flattened place above the bottomhole cutting surfaces whose purpose is to ream the hole above the bottom of the bit.

reaming *n*: a process in which the driller rotates and moves the drill stem up and down in the wellbore while circulating drilling fluid to clear the hole of debris or to prevent the drill stem from getting stuck.

reaming mill *n*: see *mill*.

rear end radius *n*: on an offshore pedestal crane, the distance from the crane's center of rotation to the end of the crane's maximum rear extension; also called tail swing.

reassignment obligation *n*: a part of a farmout agreement or other assignment that stipulates an obligation to reassign earned acreage back to the farmor or assignor before the farmee or assignee allows the lease to expire on the acreage.

reboiler *n*: the auxiliary equipment to a fractionator or other column that supplies heat to the column. The vessel in which glycol that has absorbed water from the stream of gas in the contactor is regenerated by heating to boil off the water.

rec *abbr*: recovered; used in drilling reports.

recap *n*: periodic record of drilling mud performance and characteristics over the course of the drilling project.

receipt ticket *n*: see *measurement ticket*.

receiver *n*: electronic equipment used for receiving modulated radio waves and converting them into the original intelligence, such as into sounds or pictures, or converting to desired useful information as in a radar receiver.

receiver coil *n*: in an induction logging tool, a device that puts out electrical voltage, which is induced in the coil by eddy currents in a formation. The eddy currents are created by a transmitter coil on the logging tool. The amount of voltage induced in the receiver coil depends on the conductivity of the formation.

reciprocal equation *n*: in parallel electronic circuits, the equation

$$R_T = \frac{1}{\frac{1}{R_1} + \frac{1}{R_2} + \frac{1}{R_3} + \frac{1}{R_4} + etc.}$$

which states that total resistance (R_T) in the circuit is equal to 1 divided by the sum of 1 divided by the resistance of each resistor (R) in the circuit, which are indicated as R_1, R_2, R_3, R_4, and so on.

reciprocating compressor *n*: a type of compressor that has pistons moving back and forth in cylinders, suction valves, and discharge valves, i.e., a positive-displacement compressor. Reciprocating compressors are used extensively in the transmission of natural gas through pipelines.

reciprocating engine *n*: a type of engine that has up-and-down (reciprocating) motion of pistons in cylinders. Also called piston engine or displacement engine.

reciprocating motion *n*: back-and-forth or up-and-down movement, such as that of a piston in a cylinder.

reciprocating pump *n*: a pump consisting of a piston that moves back and forth or up and down in a cylinder. The cylinder is equipped with inlet (suction) and outlet (discharge) valves. On the intake stroke, the suction valves are opened, and fluid is drawn into the cylinder. On the discharge stroke, the suction valves close, and the discharge valves open, and fluid is forced out of the cylinder.

reciprocation *n*: a back-and-forth or up-and-down movement (as the movement of a piston in an engine or pump).

recirculating average mixer (RAM) *n*: a recirculating cement mixer, which provides a more homogeneous slurry than a recirculating cement mixer (RCM). See *recirculating cement mixer*.

recirculating cement mixer (RCM) *n*: a cement mixing system in which previously mixed slurry is recirculated and mixed with a partial slurry formed by forcing dry cement into a ring of water. The partial slurry and the recirculated slurry combine to form a more consistent and more homogeneous slurry.

reclaimer *n*: a system in which undesirable high-boiling contaminants of a stream are separated from the desired lighter materials; a purifying still.

recommended standard (RS)-232 *n*: an Electronic Industries Alliance (EIA) standard that describes the requirements for the interface between data processing and data communications equipment. It is widely used to connect computers to peripheral devices.

recompletion *n*: after the initial completion of a well, the action and techniques of reentering the well and redoing or repairing the original completion to restore the well's productivity.

recompression *n*: increasing the ambient pressure on a diver for the primary purpose of treating decompression sickness.

recompression chamber *n*: see *deck decompression chamber*.

record chart *n*: a strip, disk, or sheet on which the indications of the measuring instrument are marked by a pen in the form of a graph.

recorder carrier *n*: a device made up in a drill stem testing assembly that holds pressure and temperature recorders. See *drill stem test*.

recording *n*: 1. the act by which a legal instrument is entered in a book of public record, usually in the county clerk's office. Such recording amounts to legal notice to all persons of the rights of claims specified in the instrument. 2. any means of preserving signals, data, sounds, or other information for future reference or reproduction.

recording gauge *n*: a device, driven by a clockwork mechanism, that provides a chronological record of gauge indications (e.g., by tracing values of pressure, vacuum, voltage) on a paper form.

recording instrument *n*: a measuring instrument that records the value of the measured variable by marking or printing on a removable paper chart, tape, or other suitable recording material.

recording meter *n*: a quantity recorder that keeps a permanent record of given variables in an electrical circuit on a paper strip or circular chart.

record section *n*: a cross section of the earth generated by computer from tapes that have recorded the sound vibrations reflected during seismic exploration. Expert interpretation can reveal what may be a trap for petroleum. Also called seismic section.

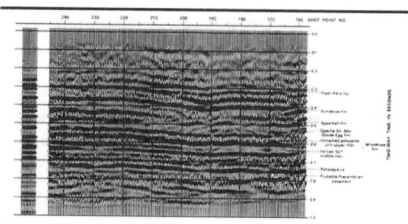

RECORD SECTION

recoverable gas-lift gas *n*: gas-lift gas that has returned to the surface and is not reinjected into the gas-lift system; rather, it is transferred to a pipeline. Sometimes called spent gas-lift gas.

recoverable gas reserves *n pl*: the quantity of natural gas determined to be economically recoverable and available for delivery from a well or wells. Amount may be limited to a specific period of time.

recoverable usage *n*: the injection of oil into a formation to stimulate production. This oil may be recovered in later production.

recovery *n*: the total volume of hydrocarbons that has been or is anticipated to be produced from a well or field.

recovery efficiency *n*: the recoverable amount of original or residual hydrocarbons in place in a reservoir, expressed as a percentage of total hydrocarbons in place. Also called recovery factor.

recovery factor *n*: see *recovery efficiency*.

rectangular survey system *n*: the method of measuring land adopted by the United States in 1785. Under this system, land is measured in squares called congressional townships, which are approximately 6 miles (9.6 kilometres) wide and approximately 6 miles long. The squares are marked off by means of parallel north-south lines called township lines and parallel east-west lines called range lines.

rectifier *n*: an electrical device that converts alternating current into direct current.

rectifier meter *n*: a permanent magnet meter with a rectifier built in to convert AC to DC.

rectify *v*: to change an alternating current to a direct current.

recumbent fold *n*: a fold of rock in which the axial plane of an overturned fold has become horizontal or nearly so.

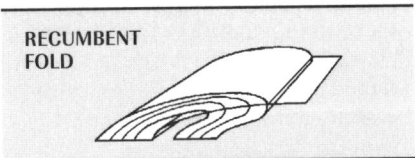

RECUMBENT FOLD

redbed *n*: a layer of sedimentary rock that is predominantly red, especially one of Permian or Triassic age.

redelivery gas *n*: natural gas delivered to a pipeline company either from a gathering system or from another pipeline company rather than to a distribution point. Compare *delivery gas*.

red-lime mud *n*: a water-base clay mud containing caustic soda and tannates to which lime has been added. Also called red mud.

red mud *n*: see *red-lime mud*.

redox potential *n*: see *oxidation-reduction potential*.

reduced circulating pressure (RCP) *n*: the amount of pressure generated on the drill stem when the mud pumps are run at a speed (or speeds) slower than the speed used when drilling ahead. An RCP or several RCPs are established for use when a kick is being circulated out of the hole.

reducing elbow *n*: a fitting that makes an angle between two joints of pipe and that decreases in diameter from one end to the other.

reducing flange *n*: a flange fitting used to join pipes of different diameters.

reducing nipple *n*: a pipe fitting that is threaded on both ends and decreases in diameter from one end to the other.

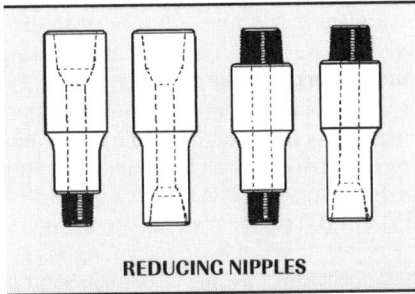

REDUCING NIPPLES

reducing tee *n*: a T-shaped pipe fitting with openings of two different sizes.

reduction *n*: adding one or more electrons to an atom or ion or molecule.

redundancy *n*: any deliberate duplication or partial duplication of circuitry or information to decrease the probability of a system or communication failure. In the transmission of information, the fraction of the gross information content of a message which can be eliminated without loss of essential information.

Redwood viscosity *n*: a unit of viscosity measurement, expressed in seconds, obtained when using a Redwood viscometer. It is the standard of viscosity measurement in Great Britain.

reef *n*: 1. a type of reservoir trap composed of rock (usually limestone) formed from the shells or skeletons of marine animals. 2. a buried coral or other reef from which hydrocarbons may be withdrawn.

reel *n*: a revolving device (such as a flanged cylinder) for winding or unwinding something flexible (such as rope or wire).

reel barge *n*: a lay barge specially outfitted to lay pipe from an immense reel on deck. The pipe is connected and reeled onto the barge on shore. This is an efficient way to lay offshore pipelines of relatively small diameter, although improved techniques

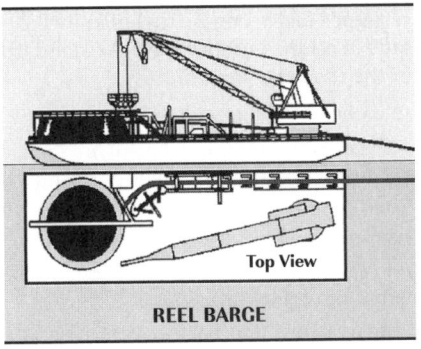

REEL BARGE

Top View

may make it possible to use pipe of up to 30 inches (76 centimetres) in diameter.

reeled tubing *n*: see *coiled tubing*.

reel method *n*: an offshore pipeline construction technique in which the welded, coated, and tested pipe is coiled onto a reel and transferred to a reel barge, where it is paid out at a steady rate onto the ocean floor.

reel ship *n*: see *reel vessel*.

reel ship subsea pipeline offshore installation *n*: necessitates pipe to be prefabricated in a linear onshore facility and then wound onto a reel. Any pipe deformation while being spooled onto the reel is removed during the laying process by passing the pipe through a straightening machine.

reel vessel *n*: a ship or barge specially designed to handle pipeline that is wound onto a large reel. To lay the pipeline, the vessel pays out the pipe off the reel at a steady rate onto the ocean floor. The pipeline has been constructed at an onshore facility where it has been welded, coated, inspected, and wound onto the reel.

reeve *v*: to pass (as a rope) through a hole or an opening; to pass a rope over or through sheaves (pulleys) in a block.

reeve the line *v*: to string a wire rope drilling line through the sheaves of the traveling and crown blocks to the hoisting drum.

reeving *n*: a rope system where the rope travels around drums and sheaves.

reface *v*: to renew a faced surface by recutting or regrinding.

reference circumference *n*: the circumference of the bottom ring of a tank measured by the manual tank strapping method.

reference conditions *n pl*: the conditions of temperature and pressure to which measured volumes are to be corrected.

reference depth *n*: the distance from the reference point to the datum plate or the bottom of the tank. It should be stamped on the fixed benchmark plate or stenciled on the tank roof near the gauging hatch. Also called reference height.

reference fuel *n*: engine fuel of known octane or cetane number used as a standard in the engine testing of fuels.

reference height *n*: see *reference depth*.

reference measuring instrument *n*: a device calibrated to hold or deliver a known volume of liquid.

reference point *n*: see *gauge point*.

refine *v*: to manufacture petroleum products from crude oil.

refiner acquisition cost *n*: cost of crude oil to the refiner, including transportation and fees. The composite cost is the weighted average of domestic and imported crude oil costs.

refinery *n*: the physical plant and attendant equipment used in the process of refining.

refinery gas *n*: the gas produced from certain petroleum refinery operations (such as cracking or reforming). The composition of refinery gas varies in accordance with the process by which it is produced, but it consists essentially of the same paraffin hydrocarbons as natural gas plus olefins (propylene, butylene, and ethylene) not found in natural gas.

refining *n*: fractional distillation of petroleum products, usually followed by other processing such as cracking.

refining catalyst *n*: any chemical used in the refining process that speeds or alters the process without the catalyst's being affected.

reflect *v*: to redirect something that strikes a surface, especially light, sound, or heat usually back toward its point of origin.

reflectance *n*: in cases where a wave of energy encounters a surface, the ratio of the energy reflected from a surface to the energy that falls on the surface. For example, if light strikes a surface that has high reflectance, most of the light reflects from it; conversely, if the surface has low reflectance, very little of the light reflects from it.

reflection *n*: the process of shortwave radiation's being sent back from the earth into space in the form of long-wave radiation. The process occurs when shortwave radiation meets light-colored areas that lack absorptive qualities.

reflux *n*: in a distillation process, that part of the condensed overhead stream that is returned to the fractionating column as a source of cooling. *v*: in distillation extraction of fluids from a core, to use a solvent to flow over a core sample a second time to clean it.

reflux coil *n*: a length of tubing or pipe bent to assume a coiled shape and through which a fluid cooler than the fluid outside the coil is circulated. The cooler fluid is used to cool the warmer fluid. Cooling causes vapors to condense to liquid.

reflux condenser *n*: a device on which reflux condenses in the distillation process. See *reflux*.

reflux ratio *n*: a relative measurement of the volume of reflux in the distillation process. The ratio is commonly expressed as the quantity of reflux divided by the quantity of net overhead product.

reform *v*: to rewrite a contract, guided by principles of equity. Parties who believe, for example, that the written form of an oil and gas lease does not express what was in fact intended or agreed on may sue in hope that the court will agree with them and reform the lease to express the intended facts or circumstances.

reformate *n*: a liquid mixture of hydrocarbons.

reforming *n*: a cracking process in which low-octane naphthas or gasolines are converted into high-octane products. Thermal reforming is carried out at high temperatures and pressures (932°–1,040°F, 500°–560°C, 250 to 1,000 pounds per square inch—1.7 megapascals). Catalytic reforming is carried out at lower temperatures (850°–950°F, 454°–510°C) and much lower pressures. Reforming is usually a once-through process.

reformulated gasoline (RFG) *n*: a special gasoline blend that burns with less hydrocarbon and other pollutant emissions.

refract *v*: to alter the course of a wave of energy that passes into something from another medium, as water does to light entering it from the air.

refraction *n*: deflection from a straight path undergone by a light ray or energy wave in passing from one medium to another in which the wave velocity is different, such as the bending of light rays when passing from air into water.

refractory *n*: any of several heat-resisting materials, usually a ceramic.

refracturing *n*: fracturing a formation again. See *acid fracture, formation fracturing, hydraulic fracturing*.

Refuse Act *n*: see *Rivers and Harbors Appropriations Act of 1899*.

regeneration gas *n*: wet gas that has been heated in a regeneration gas heater to temperatures of 400°–460°F (204°–238°C) in a solid desiccant dehydration system. The gas is passed through a saturated adsorber tower to dry the solid desiccant in the tower and remove the previously adsorbed water.

regional metamorphism *n*: a type of metamorphism that occurs in bodies of rock that have been deeply buried or greatly deformed by tectonic changes.

register *n*: a mechanical device that displays numbers.

registered breadth *n*: the width of the hull of a mobile offshore drilling rig or a ship, measured at its points of greatest width and used to determine its registered tonnage.

registered volume *n*: the amount of a substance as indicated by the register on the meter.

register ton (RT) *n*: a unit of measure of the internal capacity of ships, which equals 100 cubic feet or about 2.8317 cubic metres.

regular cement *n*: see *common cement*.

regular lay *n*: a type of wire rope construction in which the wires in the wire rope strands are twisted in a direction opposite the strands themselves.

regulated mixing valve *n*: a valve used in process control systems that blends two fluids of different temperatures to achieve a desired output temperature.

regulations *n*: see *guidelines*.

regulator *n*: a device that reduces the pressure or volume of a fluid flowing in a line and maintains the pressure or volume at a specified level.

regulator pressure *n*: in process control systems, the pressure that results when a fluid is sent to a pressure regulator to obtain a desired output pressure downstream from the regulator. The regulator is set to obtain the desired pressure and is designed to hold the desired pressure.

regulator station *n*: reduces the high-pressure pipeline gas to a lower usable pressure for the compressor and records the amount of gas consumed as fuel.

Reid vapor pressure *n*: the vapor pressure of a liquid at 100°F (37.78°C, 311°K) as determined by ASTM D 323-58, Standard Method of Test for Vapor Pressure of Petroleum Products (Reid Method).

rel (R) *n*: a proposed unit to express reluctance in a magnetic circuit. The rel has not, however, been universally accepted.

relative density *n*: 1. the ratio of the weight of a given volume of a substance at a given temperature to the weight of an equal volume of a standard substance at the same temperature. For example, if 1 cubic inch of water at 39°F (3.9°C) weighs 1 unit and 1 cubic inch of another solid or liquid at 39°F weighs 0.95 unit, then the relative density of the substance is 0.95. In determining the relative density of gases, the comparison is made with the standard of air or hydrogen. 2. the ratio of the mass of a given volume of a substance to the mass of a like volume of a standard substance, such as water or air.

relative error *n*: the quotient of the absolute error divided by the true value of the measured quantity. This multiplied by 100 gives the relative error as a percentage.

relative humidity *n*: the ratio of the amount of water vapor in the air to the amount it would contain if completely saturated at a given temperature and pressure. See *absolute humidity*.

relative motion *n*: in a marine riser system, movement caused by the structural compression that occurs when pressure is exerted on the ends of the pins in the riser connectors (Poisson's effect), temperature differences between the fluid in the main riser and the fluids in the auxiliary lines, and bending loads imposed by deflections of the riser. Such motion of the riser connectors can cause fatigue cracking of the support flange if an adequate gap is not provided between the support flange and the coupling. See *Poisson's effect*.

relative permeability *n*: the permeability of a reservoir to oil, gas, or water relative to absolute permeability. Example: $K_{ro} = K_o / K$, where K_{ro} is defined as oil relative permeability; K_o is defined as oil permeability; and K is defined as absolute permeability. Similar relationships are true for gas and water. Consequently, in a reservoir where oil, gas, and water usually coexist, the absolute permeability reflects the productivity of the total rock pore environment. Relative permeability defines the ability of the rock to transmit a particular fluid. Relative permeability varies with various rock properties including saturation and wettability. Thus, productivity of oil, gas, and water vary with time as saturations change. Compare *absolute permeability*, *effective permeability*.

relative viscosity *n*: the ratio of absolute viscosity of the fluid being measured to the viscosity of water at 68°F. See *absolute viscosity*.

relaxed invert-emulsion mud *n*: an invert-emulsion mud with a higher oil-to-water ratio than is usual.

relay *n*: a device in an electric circuit that senses electrical variations in the circuit and then makes or breaks one or more connections in the same or another electric circuit.

release *n*: 1. a statement filed by the lessee of an oil and gas lease indicating that the lease has been relinquished. 2. any spilling, leaking, pumping, pouring, emitting, emptying, discharging, injecting, escaping, leaching, dumping, discarding, or disposing into the environment.

reliability *n*: the ability of an item to operate as specified for an indicated time period. Often expressed as mean time between failures (MTBF) or mean time to failure (MTTF).

relief *n*: the elevations or inequalities of a land surface.

relief valve *n*: see *pressure-relief valve*.

relief well *n*: a well drilled near and deflected into a well that is out of control, making it possible to bring the wild well under control. See *wild well*.

reluctance *n*: in electromagnetics, a measure of the opposition presented to magnetic lines of force (flux) in a magnetic circuit; it is analogous to resistance in an electric circuit.

remainder *n*: in mathematical division, a quantity that results when one quantity does not divide evenly into another quantity. For example, 7 ÷ 2 = 3 with a remainder of 1. That is, 2 goes into 7 three times with 1 left over.

remainderman *n*: someone who holds a future interest in property and who will come into possession when the present possessory interest ends (as on the death of a life tenant). Compare *life tenant*.

remaining on board (ROB) *adj* or *n*: usually referred to by its abbreviation. Sometimes used as an adjective (cargo remaining on board), but more often used as a noun (estimating the ROB). ROB includes water, oil, slops, oil residue, oil-water emulsions, sludge, and sediment.

remedial cementing *n*: cement placed in a wellbore after the primary cementing has occurred. Remedial cementing is used to repair holes in the casing, fill voids left behind in casing after primary cementing, stop lost circulation, and so on. Compare *primary cementing*.

remedial stimulation *n*: the restoration of a well's performance using chemicals such as acid and paraffin solvents to dissolve restricting materials in the perforations and near the wellbore.

remediate *v*: to remedy or correct a problem with a situation or area.

remediation *n*: the process of remedying or restoring an area or situation to its natural (or as close to natural as possible) state.

remote-actuated tool *n*: in well service and workover, a piece of equipment installed in the well that does not require mechanical activation by a tool that is run on wireline or coiled tubing. Remote-actuated tools are often used in highly deviated or horizontal wells where mechanical intervention is costly and difficult or impossible to use.

remote BOP control panel *n*: a device placed on the rig floor that can be operated by the driller to direct air pressure to actuating cylinders that turn the control valves on the main BOP control unit, located a safe distance from the rig.

remote choke panel *n*: a set of controls, usually placed on the rig floor, that is manipulated to control the amount of drilling fluid being circulated through the choke manifold. This procedure is necessary when a kick is being circulated out of a well. See *choke manifold*.

remote connection *n*: an offshore pipeline-joining technique in which the connection process is directed from somewhere other than the immediate site, such as from a control panel on the platform deck. Two types of remote connection are the deflect-to-connect and direct pull-in techniques.

remote (secondary) control panel *n*: a system of blowout preventer controls, convenient to the driller, which can be used selectively to actuate valves at the master control panel. Also called secondary control panel. See *driller's BOP control panel*.

remote control station *n*: a centrally located station containing equipment to control and regulate operations in one or more fields.

remotely operated vehicle (ROV) *n*: in offshore operations, an underwater device controlled from a vessel on the water's surface that is used to inspect and operate certain devices on subsea equipment, such as a blowout preventer stack, and that can be used in place of or in conjunction with diving personnel.

remote operating choke *n*: see *automatic choke*.

remote reading gauge *n*: an instrument that provides indications of pressure, vacuum, voltage, and so forth at a point distant from where the indications are actually taken.

remote sensing *n*: refers to using infrared or other means to map an area; has largely replaced aerial photography.

remote station *n*: an auxiliary set of controls for operating the blowout preventers.

remote terminal unit (RTU) *n*: that part of an automated lease that relays signals from the central computer to the end devices and relays status and alarm conditions from the end devices to the computer.

remote transmission and telemetering *n*: a separate or integral instrument system, used in conjunction with some other basic measuring means (such as an automatic tank gauge), that transmits the basic reading to some place other than the point of measurement.

renewable energy *n*: energy obtained from sources that are essentially inexhaustible (unlike, for example, the fossil fuels). Renewable sources of energy include wood, waste, geothermal, wind, photovoltaic, and solar thermal.

repeatability *n*: 1. the closeness of the results of successive measurement of the same quantity carried out by the same method, by the same person with the same measuring instrument at the same location over a short period of time. 2. the ability of a meter and prover system to repeat its registered volume during a series of consecutive proving runs under constant operating conditions.

repeats per minute *n*: in process control systems, the time required for a controller to adjust itself back to its set point when a change in its input or output occurs.

replacement *n*: the act of putting enough drilling fluid in the wellbore to replace the volume of any pipe or other tools that were removed.

reportable quantity (RQ) *n*: 1. the quantity of released hazardous materials that, when exceeded, must be reported under CERCLA. The EPA is charged with setting the allowable limits for releases of each hazardous material. If a release exceeds the EPA reportable quantity, CERCLA requires that the owners or operators of the facility notify the National Response Center immediately. 2. SARA (section 304) also requires facilities that release a reportable quantity of an EHS or CERCLA hazardous substance to notify the SERC and the LEPC immediately. This requirement is similar to the CERCLA requirement to notify the NRC in case of reportable releases. Like CERCLA, SARA exempts federally permitted releases, continuous releases, any release contained on site, as well as other CERCLA exemptions from RQ reporting requirements. In addition, SARA requires a written follow-up report to be submitted to the SERC and the LEPC as soon as practicable after the release.

representative sample *n*: a small portion extracted from the total volume of material that contains the same proportions of the various flowing constituents as the total volume of liquid being transferred. The precision of extraction must be equal to or better than the method used to analyze the sample.

repressure *v*: to increase or maintain reservoir pressure by injecting a pressurized fluid (such as air, gas, or water) to effect greater ultimate recovery.

reproducibility *n*: 1. the closeness of the results of measurements of the same quantity where the individual measurements are made by different observers using different methods with different measuring instruments at different locations after a long period of time, or where only some of the factors listed are different. 2. the ability of a meter and prover system to reproduce results over a long

period of time in service where the range of variation of pressure, temperature, flow rate, and physical properties of the metered liquid is negligibly small.

resaturation effect *n*: in waterflooding, the entrapment of oil in rocks of lower permeability, where gas space has developed during primary production. The resaturated oil is bypassed by the waterflood as the water seeks the path of highest permeability.

rescue boat *n*: in offshore operations, any boat that may be used in saving the lives of personnel at sea. For example, the United States Coast Guard (USCG) possesses a fleet of vessels called cutters that USCG uses for many purposes, ranging from preventing smuggling to breaking ice. However, in times when lives are in peril at sea, any of these vessels could be called to perform rescues. Similarly, the standby boat that is usually tied up next to an offshore facility can also perform rescues if needed.

Research and Special Projects Administration (RSPA) *n*: a DOT agency that oversees the Office of Pipeline Safety, intermodal containers, highway portable tanks, railroad cars, and anything used in interstate or international commerce not regulated by the coast guard. Under OPA, RSPA has authority over onshore oil and hazardous materials pipelines that are used in interstate commerce. Address: 400 7th Street SW; Washington, DC 20590; (202) 366-4433.

reserve buoyancy *n*: the buoyancy above the waterline that keeps a floating vessel upright or seaworthy when the vessel is subjected to wind, waves, currents, and other forces of nature or when the vessel is subjected to accidental flooding.

reserve capacity *n*: capacity in excess of that required to carry peak load.

reserve pit *n*: 1. (obsolete) a mud pit in which a supply of drilling fluid is stored. 2. a waste pit, usually an excavated earthen-walled pit. It may be lined with plastic or other material to prevent soil contamination.

reserves *n pl*: the unproduced but recoverable oil or gas in a formation that has been proved by production.

reserve tank *n*: a special mud tank that holds mud that is not being actively circulated. A reserve tank usually contains a different type of mud from that which the pump is currently circulating. For example, it may store heavy mud for emergency well-control operations.

reservoir *n*: 1. a subsurface, porous, permeable rock body in which oil and/or gas has accumulated. Most reservoir rocks

are limestones, dolomites, sandstones, or a combination. The three basic types of hydrocarbon reservoirs are oil, gas, and condensate. An oil reservoir generally contains three fluids—gas, oil, and water—with oil the dominant product. In the typical oil reservoir, these fluids become vertically segregated because of their different densities. Gas, the lightest, occupies the upper part of the reservoir rocks; water, the lower part; and oil, the intermediate section. In addition to its occurrence as a cap or in solution, gas may accumulate independently of the oil; if so, the reservoir is called a gas reservoir. Associated with the gas, in most instances, are salt water and some oil. In a condensate reservoir, the hydrocarbons may exist as a gas, but, when brought to the surface, some of the heavier ones condense to a liquid. 2. a container or vessel that stores fluid, as a reservoir on an accumulator that holds hydraulic operating fluid.

reservoir damage *n*: see *formation damage*.

reservoir drive *n*: see *reservoir drive mechanism*.

reservoir drive mechanism *n*: the process in which reservoir fluids are caused to flow out of the reservoir rock and into a wellbore by natural energy. Gas drive depends on the fact that, as the reservoir is produced, pressure is reduced, allowing the gas to expand and provide the principal driving energy. Water drive reservoirs depend on water and rock expansion to force the hydrocarbons out of the reservoir and into the wellbore. Also called natural drive energy.

reservoir heterogeneities *n pl*: nonuniformities in the structure of properties of petroleum reservoirs, such as lenticular formation, pinch-outs, faults that cut across the reservoir, shale barriers, and variations in permeability.

reservoir oil *n*: oil in place in the reservoir; unproduced oil. Compare *stock tank oil*.

reservoir pressure *n*: the average pressure within the reservoir at any given time. Determination of this value is best made by bottomhole pressure measurements with adequate shut-in time. If a shut-in period long enough for the reservoir pressure to stabilize is impractical, then various techniques of analysis by pressure buildup or drawdown tests are available to determine static reservoir pressure.

reservoir rock *n*: a permeable rock that may contain oil or gas in appreciable quantity and through which petroleum may migrate.

reservoir simulation *n*: computer model of a reservoir to predict reservoir behavior and to show production over time, allowing decisions about managing the reservoir to be made. The technique consists of digitizing data from appropriate geologic maps and engineering studies and entering them into a computer program, called the model. The model uses past and present values for each parameter of reservoir behavior.

reservoir temperature *n*: the average temperature within the reservoir, measured during logging, drill stem testing, or bottomhole pressure testing using a bottomhole temperature recorder.

reservoir volume factor *n*: see *formation volume factor*.

reset *n*: in process control, the action of maintaining a prescribed output through reset action. *v*: to restore a setting or input to its original value.

resid *n*: shortened form of residual.

residual fuel *n*: see *residuals*.

residual fuel oil *n*: the heavier oils that remain after the distillate fuel oils and lighter hydrocarbons are distilled away in refinery operations that conform to ASTM Specifications D396 and 975.

residual gas saturation *n*: the portion of hydrocarbons that cannot be removed by ordinary producing mechanisms when a porous reservoir has been saturated with hydrocarbons. This value is usually a specific percentage of the pore volume. In the case of gas, the volume, measured at standard conditions, that is retained in a reservoir as residual gas saturation is an inverse function of the pressure, due to the effect of the gas laws.

residual magnetism *n*: the relatively small amount of magnetism that virtually all metallic objects that contain iron retain as a result of the magnetic influences. It affects magnetic compasses by attracting the compass needle away from actual magnetic north to give a false reading.

residual oil *n*: 1. in improved recovery, oil remaining in a reservoir after an improved recovery method has been applied and displacement has occurred. 2. in petroleum refining, the combustible, viscous, or semiliquid bottoms product from crude oil distillation, used as adhesives, roofing compounds, asphalt, low-grade fuels, and sealants.

residuals *n pl*: the heavy refined hydrocarbons that are used as fuels. Bunker C oil, which is sometimes used to fuel ships, is an example of a residual.

residue gas *n*: 1. a natural gas mixture essentially of methane and ethane. 2. natural gas after it has been processed and almost all hydrocarbons other than methane and some ethane removed.

residue gas returned *n*: plant residue gas delivered back to the producer for use in lease operations.

residue pooling *n*: provision in many casinghead gas purchase contracts that provides that all of the seller's properties covered by the contract will be treated as a single property in determining whether the producer has taken more residue gas for use in operations than the quantity to which he or she is entitled.

resin *n*: a semisolid or solid complex, amorphous mixture of organic compounds having no definite melting point or tendency to crystallize. Resins may be a component of compounded materials that can be added to drilling fluids to impart special properties to the fluid.

resin cement *n*: an oilwell cement that is composed of resins, water, and portland cement and that provides an improved cement bond. It is mainly used in remedial operations, because its high cost prohibits its use for routine cementing of casing.

resistance (R) *n*: opposition to the flow of direct current caused by a particular material or device. Resistance is equal to the voltage drop across the circuit divided by the current through the circuit.

resistance-capacitance (RC) circuit *n*: an electrical circuit that has a resistance and a capacitance in series, and in which inductance is negligible.

resistance temperature detector (RTD) *n*: in process control, an electronic device that senses temperature by means of changes in electrical resistance caused by changes in temperature.

resistance thermometer *n*: a thermometer that uses an electrical resistor to detect temperature and electrical means to measure and indicate temperature.

resistance tolerance *n*: a measure of a resistor's accuracy, usually stated in percent. For example, if a 2,400-ohm resistor has a tolerance of 5 percent, its resistance should vary no more than 5 percent of 2,400 ohms, either up or down, or its resistance should be no less than 2,280 or more than 2,520 ohms.

resistivity *n*: the electrical resistance offered to the passage of current; the opposite of conductivity.

resistivity log *n*: a record of the resistivity of a formation. Usually obtained when an electric log is run. See *resistivity well logging*.

resistivity meter *n*: an instrument for measuring the resistivity of drilling fluids and their cakes.

resistivity well logging *n*: the recording of the resistance of formation water to natural or induced electrical current. The mineral content of subsurface water allows it to conduct electricity. Rock, oil, and gas are poor conductors. Resistivity measurements can be correlated to formation lithology, porosity, permeability, and saturation and are very useful in formation evaluation. See *electric well log*.

resistor *n*: a component that tends to impede the flow of electric current, usually without any inductive or capacitive effects.

resolution *n*: the smallest change in the quantity measured to which the instrument will react with an observable change in an analog or digital indication.

resonant frequency *n*: of sucker rods, the frequency at which the periods of resonance are at a maximum.

Resource Conservation and Recovery Act of 1976 (RCRA) *n*: a federal regulatory program designed to ensure responsible management of hazardous waste at all levels of contact, i.e., generation, transportation, treatment, storage, and disposal; regulates the management of on-land disposal of all solid wastes. This comprehensive program is designed to protect human health and the environment from the unintentional exposure to solid waste that is identified as "hazardous waste." Also institutes a "cradle-to-grave" monitoring program that closely tracks and regulates the handling of hazardous waste from the time the waste is first generated until it is disposed.

resources *n pl*: concentrations of naturally occurring liquid or gaseous hydrocarbons in the earth's crust, some part of which are currently or potentially economically extractable. Some categories of resources are (1) from the economic standpoint, economic (recoverable, commercial), marginally economic, and subeconomic; and (2) from the geological standpoint, identified (subdivided into measured, indicated, and inferred), and undiscovered.

response lag *n*: in process control, the difference in time between a command from the set point to the actual change in the controlled variable.

response time *n*: in a thermometer, the time required to indicate 63.2% of the magnitude of a change in the measured temperature.

restoring force *n*: 1. the force exerted by a centralizer against the borehole to keep the pipe away from the borehole wall. 2. horizontal force required by a mooring vessel to counteract the forces of the environment and maintain the vessel on station.

restricted ABS classification *n*: a classification that states that the rig can operate only in a specific geographic area named in its classification documents. Restricted rigs meet modified requirements relating to stability, including stability in severe weather, for a specific region, such as the Gulf of Mexico.

result *n*: the observed value of a variable determined by a single measurement.

resultant force *n*: the force that occurs from the combined effects of two or more forces.

retainer *n*: a cast-iron or magnesium drillable tool consisting of a packing assembly and a back-pressure valve. It is used to close off the annular space between tubing or drill pipe and casing to allow the placement of cement or fluid through the tubing or drill pipe at any predetermined point behind the casing or liner, around the shoe, or into the open hole around the shoe.

retainer head *n*: see *cementing head*.

retainer valve *n*: in subsea completions, a valve located at the lower end of the completion-workover riser. When closed, it keeps fluid within the riser when it is necessary for the surface vessel to disconnect from the riser in an emergency.

retaining ring *n*: a steel ring that holds the cone on the bearing assembly in some journal bearing bits.

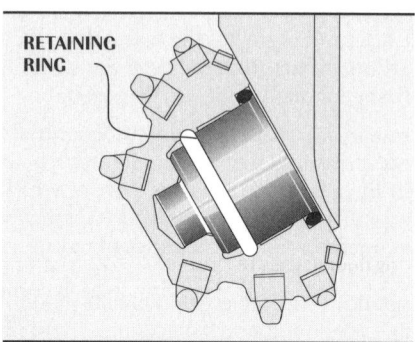

RETAINING RING

retarded cement *n*: a cement in which the thickening time is extended by adding a chemical retarder.

retarder *n*: 1. a substance added to cement to prolong the setting time so that the cement can be pumped into place. Retarders are used for cementing in high-temperature formations. 2. a slow-to-evaporate solvent added to paint, which slows down the speed of paint's drying. A paint retarder improves a paint's application properties or produces a better film.

retentiveness *n*: see *retentivity*.

retentivity *n*: in magnetism, the tendency of an unmagnetized iron or steel bar to retain, or hold, its magnetism after it is magnetized. Iron or steel with low retentivity

loses its magnetism relatively quickly, while an iron or steel with high retentivity keeps its magnetism relatively longer.

retort *n*: an instrument used to distill oil, water, and other volatile materials in a mud to determine oil, water, and total solids contents in volume-percent. Also called a still.

retort oven *n*: a device that measures the saturation, or the amount, of each fluid in a core sample by distilling the fluid from the core with heat. Also called a retort still.

retort still *n*: see *retort oven*.

retrace *n*: in an oscilloscope, the brief instant at which the voltage driving the oscilloscope's beam suddenly decreases and drives the beam to the left of the screen.

retractable bit *n*: a bit that can be changed by wireline operations without withdrawing the drill string. Field tests have indicated its economic feasibility, but its practicability is undetermined.

retrievable float valve *n*: a piston-type valve that can be used to reposition a string float up the hole to avoid the use of multiple string floats. It consists of a retainer sub and a wireline retrievable piston that fit inside the sub. It allows flow down the string but not up. The float can be retrieved with wireline and a second (or third) retainer sub can be set higher into the drill pipe during a connection.

retrievable packer *n*: a packer that can be pulled out of the well when it fails, to be repaired or replaced.

retrievable wireline choke *n*: a bottomhole choke run on wireline and landed in the tubing string.

retrograde condensation *n*: in reservoir mechanics, the formation of liquid droplets in a gas as the well is produced and the pressure drops. Some hydrocarbons exist naturally above their critical temperature in the reservoir; as a result, when pressure is decreased, instead of expanding to form a gas, they condense to form a liquid.

retrograde phenomenon *n*: see *retrograde reservoir*.

retrograde reservoir *n*: a reservoir in which the pressure is high and the hydrocarbon content is completely in a gaseous or supercritical phase at initial conditions. As the pressure drops because of production, the heavier hydrocarbon components condense, forming liquids within the reservoir. Such action is the retrograde phenomenon. If the reservoir pressure is completely depleted, only a small portion of these liquids will revaporize and be recovered.

return bend *n*: a U-shaped section of piping that connects two other pipes parallel to each other.

return permeability *n*: the permeability of a reservoir that is available for production after completion, relative to the theoretical undamaged permeability. It is a measurement similar to skin damage. See *formation damage, skin*.

returns *n pl*: the mud, cuttings, and so forth, that circulate up the hole to the surface.

reverse-acting electric actuator *n*: a device that increases an engine's speed by decreasing the positive voltage going to the electric actuator and governor. As the engine needs more fuel to go faster, the reverse-acting actuator decreases the positive voltage to make the governor increase the fuel. Compare *direct-acting electric actuator*. See *governor*.

reverse-balloon *v*: in reference to tubing under the effects of temperature changes, sucker rod pumping, or high external pressure, to decrease in diameter while increasing in length. Compare *balloon*.

reverse bias *n*: where a p-type and an n-type semiconductor are joined, a condition in which no current flows in the semiconductors because the negative terminal of a battery is connected to the lead of the p-type material and the positive terminal of the battery is connected to the lead of the n-type material. Compare *forward bias*. See *n-type semiconductor, p-type semiconductor*.

reverse breakdown *n*: see *Zener breakdown*.

reverse circulation *n*: the course of drilling fluid downward through the annulus and upward through the drill stem, in contrast to normal circulation in which the course is downward through the drill stem and upward through the annulus. Seldom used in open hole, but

REVERSE CIRCULATION

frequently used in workover operations. Also referred to as "circulating the short way," since returns from bottom can be obtained more quickly than in normal circulation. Compare *normal circulation*.

reverse-circulation junk basket *n*: a fishing tool that is lowered into the hole during normal circulation and then produces reverse circulation to create a vacuum so that junk is sucked inside the tool. Also called a jet-powered junk basket.

reverse combustion *n*: a type of in situ combustion in which the combustion front moves counter to the direction of the injected air. Air is injected into a production well and burning is started near the well. When the combustion zone has advanced a short distance from the well, air injection is stopped from the production well and started from an adjacent injection well. The fire advances toward the injection well, but the oil moves toward the production well. This method may be used with very viscous oils. Compare *forward combustion*.

reverse current relay *n*: a device installed in a diesel-electric system that disconnects a generator from the system when an engine is underloaded and the generator attached to it behaves like a motor. A generator acting like a motor draws current from the system, instead of supplying current. The generator acting like a motor then drives the underloaded engine, which can severely damage the engine.

reverse drilling break *n*: a sudden decrease in the rate of penetration. When drilling with an oil mud and diamond bit, and an abnormally high-pressure formation is penetrated, the penetration rate may decrease rather than increase. An increase is a drilling break.

reverse emulsion *n*: a relatively rare oilfield emulsion composed of globules of oil dispersed in water. Most oilfield emulsions consist of water dispersed in oil.

reverse fault *n*: a dip-slip fault along which the hanging wall has moved upward relative to the footwall. Also called a thrust fault.

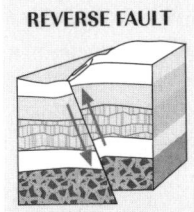
REVERSE FAULT

reverse J-tube method *n*: a method of joining a pipeline to a subsea riser on an offshore platform. In this method the pipe is welded together on the platform itself and then fed down through a guide tube to the seafloor. Compare *J-tube method*.

reverse osmosis *n*: a method of desalting brackish or salt water by passing it through a membrane that is not permeable to salt. See *osmosis*.

reverse out *v*: to displace the wellbore fluid back to the surface; to displace tubing volume back to the pit.

reverse-pressure perforating *n*: see *perforate underbalanced*.

reversible grating *n*: on offshore installations with grated decks, grating so constructed that it may be installed either side up, with no difference in appearance or carrying capacity. See *grating*.

reversible thermometer *n*: a device containing a mercury-in-glass thermometer, which may be inverted after the thermometer has

reached thermal equilibrium with the oil in which it is immersed. The inversion breaks the mercury thread, disconnecting it from the sensing element and allowing it to run down to the other end of the thermometer stem. The instrument is then drawn to the surface and the recorded temperature, which remains unchanged until the instrument is reset, can then be read.

reversing hand *n*: a well-servicing hand who cleans out wellbores.

reversionary interest *n*: a future interest created by law when an estate is, for example, leased. The reversion is not conveyed but is retained to take effect later in favor of the grantor or his or her heirs. See *term minerals*.

rework *v*: to restore production from an existing formation when it has fallen off substantially or ceased altogether. See *work over*.

Reynolds number *n*: a dimensionless number defined as

$$R_e = DuP \div m$$

where—

D = inside diameter of the pipe
u = mean flow velocity
P = density of the fluid
m = dynamic viscosity,

all in consistent units.

RF flanged connector *n*: a brand of hydraulic connector used in subsea marine riser systems. It is similar to Cameron's Load King connector but has a tensile load rating of 2 million pounds (900 tonnes). See *Load King connector*.

rheology *n*: the study of the flow of gases and liquids of special importance to mud engineers and reservoir engineers.

rheostat *n*: a resistor that is used to vary the electrical current flow in a system.

rhyolite *n*: a light-colored, fine-grained volcanic rock; the extrusive equivalent of granite.

ribbon line *n*: a kind of conductor in which two wires are separated a short distance from each other by an insulating material. The conductor thus has the appearance of a flat ribbon. Ribbon line is often used to attach a television antenna to the television set and has an impedance of 300 ohms. See *impedance*.

rich amine *n*: the amine leaving the bottom of the contactor. It is the lean amine plus the acid gases removed from the gas by the lean amine.

rich gas *n*: a gas that is suitable as feed to a gas processing plant and from which products can be extracted.

rich glycol *n*: water-laden glycol.

rich oil *n*: a lean oil that has absorbed heavier hydrocarbons from natural gas.

rich-oil demethanizer *n*: a vessel used in gas processing plants to remove methane from rich oil.

rider *n*: a separately listed provision in a lease. Also called an exhibit or allonge.

ridge set *n*: in diamond bits, the setting of small diamonds on raised ridges on the cutting surface of the bit. Compare *surface set*.

rifle boring *n*: a hole bored through a machined metal piece (as in a steel pin in a traveling block sheave) through which a lubricant travels.

rift zone *n*: the zone along which crustal plates separate because of slowly diverging convection currents in the semisolid, deformable mantle. As the rift widens and the land masses on both sides move apart, new oceanic crust is formed. Since the Mid-Atlantic rift zone opened about 200 million years ago, North and South America have been moving away from Europe and Africa at a rate of 1½-inches a year.

rig *n*: the derrick or mast, drawworks, and attendant surface equipment of a drilling or workover unit.

RIG

Rig Floor

rig crew member *n*: see *rotary helper*.

rig down *v*: to dismantle a drilling rig and auxiliary equipment following the completion of drilling operations. Also called tear down.

rig floor *n*: the area immediately around the rotary table and extending to each corner of the derrick or mast—that is, the area immediately above the substructure on which the rotary table and other equipment rest. Also called the derrick floor, drill floor.

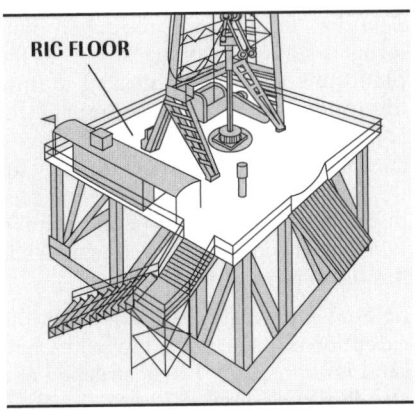
RIG FLOOR

right angle *n*: an angle of 90 degrees.

right-hand rule for coils *n*: the statement of the fact that if the coil is grasped in the right hand with the fingers wrapped around it in the direction of the current flow, the thumb points toward the north pole.

right-hand rule for conductors *n*: the statement of the fact that if the fingers of the right hand are placed around a wire so that the thumb points in the direction of the current flow, the fingers point in the direction of the magnetic field produced by the wire.

righting arm *n*: the horizontal distance between the center of gravity and a vertical line through the center of buoyancy of a floating offshore vessel that is displaced from the upright position. See *center of buoyancy, center of gravity*.

righting energy *n*: the force that causes a floating vessel that is heeled over to attain an upright position.

righting moment *n*: the turning force (torque) that tends to restore a floating vessel that is heeled over to its upright position.

right-of-way *n*: the legal right of passage over public land and privately owned property; also the way or area over which the right exists. The width of a right-of-way varies according to contract specifications and individual easements, but it is generally between 50 and 100 feet (15 and 30 metres).

right-of-way restoration *n*: in pipeline construction, the process of returning a right-of-way to its original condition or better after the pipeline has been completed. Right-of-way restoration depends on legal stipulation in the contract with the pipeline owner and agreements made with individual landowners.

right triangle *n*: in mathematics, a triangle in which one of the angles is a right angle. See *right angle*.

rigid platform rig *n*: an offshore platform rig that does not move with the motion of the wind and sea.

rigid riser *n*: on fixed production platforms, a stationary pipe clamped to the platform's legs and connected to flow lines on the seabed. Because the platform is fixed—it does not move with waves, currents, or wind—it is not necessary for the riser to move, or flex. Wells are produced through a rigid riser and treating fluids are injected into producing wells through a rigid riser.

rig irons *n pl*: the metal parts (with the exception of nails, bolts, guy wires, and sand lines) used in the construction of a standard cable-tool rig.

rig manager *n*: an employee of a drilling contractor who is in charge of the entire drilling crew and the drilling rig, providing logistics support to the rig crew and liaison with the operating company.

rig operator *n*: see *unit operator*.

rig safety and training coordinator (RSTC) *n*: a person in the employ of a drilling contractor whose duty station is on the rig. This person is responsible for coordinating safety and training activities on the rig and supporting the contractor's, the operating company's, and the government authority's safety and training requirements. This person's duties may include monitoring the safety performance of rig personnel, conducting rig-site training courses, maintaining training records, orienting visitors on the rig, and maintaining liaison between the rig and the contractor's administrative offices.

Rigs-to-Reef Program *n*: a program developed under the National Fishing Enhancement Act that led to the state sponsorship of the conversion of rigs to artificial reefs in the Gulf of Mexico.

rig superintendent *n*: see *toolpusher*.

rig supervisor *n*: see *toolpusher*.

rig up *v*: to prepare the drilling rig for making hole, i.e., to install tools and machinery before drilling is started.

RIH *abbr*: run-in-hole.

ring *n*: see *piston ring, shell*.

ring gear *n*: in a rotary table assembly, a circular, ring-shaped device with projections (teeth) that engage a beveled gear (a pinion) on the end of a drive shaft. The drive shaft is usually driven by a chain-and-sprocket arrangement from the drawworks. When engaged, the drive shaft turns the pinion, which meshes with the ring gear, to turn the rotary table.

ring grooves *n pl*: the grooves that hold the piston rings in pistons.

ring-joint flange *n*: a special type of flanged connection in which a metal ring (resting in a groove in the flange) serves as a pressure seal between the two flanges.

ringworm corrosion *n*: a form of corrosion sometimes found in the tubing of condensate wells. It occurs in a ring a few inches from the upset. Cause of ringworm corrosion has been traced to the upsetting process, in which heat required in upsetting causes the heated end to have a different grain structure from the rest of the pipe. Normalizing prevents this condition.

ripper *n*: a claw-shaped, plowlike attachment used on a bulldozer to loosen rock and locate solid formations that may require explosives in clearing a right-of-way for pipeline construction.

ripple *n*: the alternating current component in the output of a direct current power supply, which arises in the power supply because of incomplete filtering or from a direct current generator.

riprap *v*: to space logs and timbers evenly along the length of a pipeline right-of-way to stabilize soil in swamps or wet areas.

riser *n*: a pipe through which liquid travels upward. See *riser pipe*.

riser adapter *n*: a fitting installed on top of the lower marine riser package's (LM-RP's) flex joint. It is a fitting to which the first riser joint is attached to the LMRP.

riser angle indicator *n*: an acoustic or electronic device used to monitor the angle of the flex joint on a floating offshore drilling rig. A small angle should usually be maintained on the flex joint to minimize drill pipe fatigue and wear and damage to the blowout preventers and to maximize the ease with which tools may be run. Also called azimuth angle indicator.

riser assembly *n*: see *riser pipe*.

riser collapse *n*: the caving in of a riser joint or several riser joints that can occur if the pressure inside the riser drops below external seawater pressure.

riser collapse resistance *n*: the ability of the riser pipe to withstand its tendency to cave in if pressure inside the riser drops below external seawater pressure.

riser connector *n*: a fitting attached to the end of each riser joint that allows the joints to be attached to one another.

riser disconnect *n*: the act of removing the riser pipe (marine riser) from its attachment to the subsea blowout preventer stack.

riser flange *n*: on a riser connecter, the flat, ring-shaped fitting that provides a place for riser joints to be attached to each other.

riser handling tool *n*: a special tool that enables riser joints, as well as the telescopic joint, to be raised and lowered into and out of the water. See *hydraulic riser handling tool, mechanical riser handling tool*.

riser joint *n*: see *riser pipe*.

riser joint connector *n*: see *riser connector*.

riserless drilling *n*: an unconventional deepwater offshore drilling technique used on floating drilling rigs. The technique utilizes relatively small diameter pipe as a mud return line from the seafloor instead of a large diameter marine riser. Riserless drilling eliminates the need for a large rig that must be able to handle the enormous weights of riser pipe as well as be able to store it. Moreover, drilling without a riser does not require that the rig be able to store and handle the large amounts of mud needed to circulate through the riser. Finally, riserless drilling eliminates the need to set numerous strings of casing to solve the problem of protecting formations whose fracture pressure is close to the pressure required to prevent kicks.

riser lifting tool *n*: a device used on offshore floating drilling rigs that employ marine riser pipe to pick up or lay down riser pipe joints when the pipe is being run or removed from the seafloor. See *riser pipe*.

riser margin *n*: the slight increase in mud weight used to offset friction losses that occur as the mud is circulated through the riser in a subsea blowout preventer system.

riser pipe *n*: the pipe and special fittings used on floating offshore drilling rigs to establish a seal between the top of the wellbore, which is on the ocean floor, and the drilling equipment, located above the surface of the water. A riser pipe serves as a guide for the 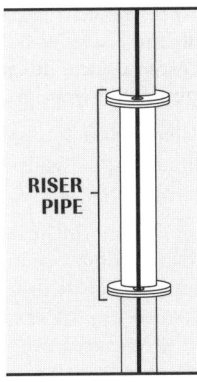 drill stem from the drilling vessel to the wellhead and as a conductor of drilling fluid from the well to the vessel. The riser consists of several sections of pipe and includes special devices to compensate for any movement of the drilling rig caused by waves. Also called marine riser pipe, riser joint.

riser pup joint *n*: a short length—usually, from 5 to 40 feet (1 to 12.2 metres) long—of riser pipe that is used to space out the riser assembly as it is run to various water depths. See *space out*.

riser spider *n*: a device attached to a rig's rotary table that supports riser pipe on a floating offshore drilling vessel.

riser-spider assembly *n*: see *riser-support spider*.

riser string *n*: see *riser pipe*.

riser-support gimbal *n*: in deepwater applications on floating rigs, a device placed between the rotary table and riser-support spider to compensate for offset caused by ocean movements. Hydraulic cylinders, or flex elements, support the weight of the spider, the entire riser string, and BOP assembly. It also cushions shock loads imparted on the system. See *riser-support spider*.

riser-support spider *n*: a fitting placed in the rotary table of an offshore floating rig which supports a joint of riser pipe

as the pipe is being lowered into or pulled from the water. The spider consists of a base plate and top plate with a series of support dogs, or arms, between them. The dogs support the riser joint under the riser joint's support flange. Hydraulic cylinders force the support dogs inward and under the riser joint support flange. They are also retracted hydraulically to allow the riser joints to pass through. On some riser spiders, the support arms are engaged and disengaged manually. The riser and BOP assembly rest on the riser-support spider as the crew picks up and adds each new joint to the string. The riser spider's ID is the same as the rotary table's ID, and the base plate is designed for the specific rotary table with which it is used.

riser system *n*: see *riser pipe*.

riser tensioner *n*: an assembly of strong cables (lines) connected to the outer barrel of the telescopic joint and a hydropneumatic piston-and-cylinder sheave assembly. The tensioning system supports the weight of the riser joints by applying force (tension) to the outer barrel of the telescopic joint. Tensioner lines are connected to a tensioner ring or to fixed padeyes on the outer barrel. The applied tension must remain constant while compensating for vessel heave.

riser tensioner line *n*: a cable that supports the marine riser while compensating for vessel movement. See *riser tensioner*.

risk analysis *n*: the activity of assigning probabilities to the expected outcomes of a drilling venture.

river crossing *n*: a type of special pipeline construction used when a pipeline must cross a river or stream. Types of river crossings include aerial crossings, conventional crossings, and directionally drilled crossings.

Rivers and Harbors Appropriations Act of 1899 (sometimes called the Refuse Act) *n*: congressional act that authorizes the secretary of the army through the U.S. Army Corps of Engineers to issue permits for the placement of structures or the discharge of refuse in navigable waters of the United States.

RMA *abbr*: Radio Manufacturers Association.

rmg *abbr*: reaming; used in drilling reports.

RMS *abbr*: robot maintenance system.

RMS value *abbr*: root-mean-square value.

road crossing *n*: laying of a pipeline under a roadbed or through a road.

ROB *abbr*: remaining on board.

robot maintenance system (RMS) *n*: a system which services underwater equipment.

rock *n*: a hardened aggregate of different minerals. Rocks are divided into three groups on the basis of their mode of origin: igneous, metamorphic, and sedimentary.

rock a well *v*: to initiate flow by alternately bleeding pressure from, and closing off the casing and tubing of, a well that contains liquid.

rock bit *n*: name for the first roller cone bits; now almost obsolete. See *roller cone bit*.

rock cycle *n*: the possible sequences of events, all interrelated, by which rocks may be formed, changed, destroyed, or transformed into other types of rock. The events include formation from magma, erosion, sedimentation, and metamorphism.

rock ditching *n*: excavating a trench in rock or rocky soil.

rock drill *n*: a drill generally powered by compressed air and used to drill holes for explosives. Rock drills may be required in pipeline right-of-way grading and in ditching.

rocker arm *n*: a bell-crank device that transmits the movement of the pushrod to the valves in an engine.

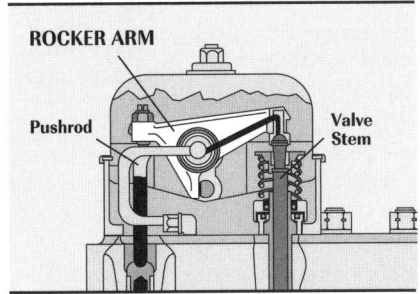

rock hound *n*: (slang) a geologist.

rock oil *n*: see *petroleum*.

rock pressure *n*: see *geostatic pressure*.

rock stratigraphic unit *n*: a distinctive body of rock that can be identified by its lithologic or structural features regardless of its fossils or time boundaries; commonly, a formation. Compare *time stratigraphic unit*.

rock texture *n*: all of the properties relating to the grain-to-grain relationships of a rock. Textural properties include chemical composition, grain shape and roundness, grain size and sorting, grain orientation, porosity, and permeability. See *clastic texture*, *crystalline texture*.

Rockwell hardness test *n*: an arbitrarily defined measure of resistance of a material to indentation under static or dynamic load.

ROD *abbr*: rich-oil demethanizer.

rod *n*: see *piston rod, sucker rod*.

rod back-off wheel *n*: a device used to unscrew rods when the pump is stuck or sanded up and the rods and tubing must be pulled together.

rod bearing *n*: see *connecting rod bearing*.

rod blowout preventer *n*: a ram device used to close the annular space around the polished rod or sucker rod in a pumping well.

rod cap *n*: on an engine's piston rod, a part on the bottom of the rod that is bolted onto the top part of the rod. To install a piston rod onto the crankshaft, one half of the rod bearing is inserted into the top part of the rod and the other half into the rod cap. The top part of the rod and the rod cap are then placed on the crankshaft and the cap bolted to the top part.

rod elevators *n pl*: a device used to pull or to run sucker rods. They have a bail attached to the rod hook.

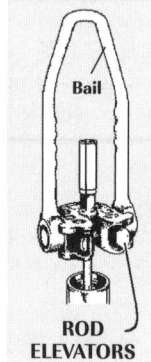

ROD ELEVATORS

rod gap *n*: lightning arrester assembly using two rods end-to-end with the gap separating rods determining arcing voltage.

rod hanger *n*: a device used to hang sucker rods on the mast or in the derrick.

rod hook *n*: a small swivel hook having a fast-operating automatic latch to close the hook opening when weight is suspended from the hook.

rod packing *n*: a special material installed around a pump's piston rod that allows the rod to move back and forth yet maintains a seal to prevent mud from getting onto the rod and scoring it.

rod pump *n*: see *sucker rod pump*.

rod reversal *n*: the action of the sucker rod string as it comes to rest momentarily at the bottom or top of the pump stroke and then begins to move downward (if at the top of the stroke) or upward (if at the bottom of the stroke).

rod rotor *n*: a ratchet mechanism that is actuated by a fixed rod or chain connected to the walking beam of a pumping unit and that provides a slow rate of rotation to the rod string, distributing the wear on both rods and tubing.

rod score *n*: a scratch on the surface of a sucker rod or a piston rod.

rod string *n*: a sucker rod string, that is, the entire length of sucker rods, which usually consists of several single rods

screwed together. The rod string serves as a mechanical link from the beam pumping unit on the surface to the sucker rod pump near the bottom of the well.

rod stripper *n*: a device closed around the rods when the well may flow through the tubing while the rods are being pulled. It is a form of blowout preventer.

rod sub *n*: a short length of sucker rod that is attached to the top of the sucker rod pump.

rod-transfer elevator *n*: a special type of elevator designed to accommodate the end of a sucker rod. It allows the derrickman to transfer the rod to the racking platform from the regular elevator being used to lift the rod out of the well.

rod-transfer equipment *n*: all the devices used to accomplish the moving of sucker rods from the elevators to the racking platform.

rod wax *n*: a paraffin wax that forms on the sucker rod string.

rod whip *n*: the rapid, whiplike motion of the rods in a sucker rod pumping system, caused by vibration of the rod string.

rod wrench *n*: a special wrench designed for spinning up and hammering tight the joints between sucker rods. Also called a key.

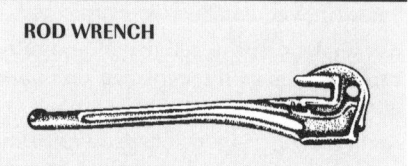

ROD WRENCH

roll *n*: the angular motion of a ship or floating offshore drilling rig as its sides move up and down. *v*: to move from side to side.

roll-dampening tanks *n pl*: the compartments on a floating offshore drilling rig that are filled with water to offset the rig's tendency to roll.

roller *n*: 1. on a kelly bushing, a cylindrical device that fits inside the bushing, whose exterior shape is matched to the kelly's shape so that they mate with the kelly when it is inside the kelly bushing. See *kelly bushing, roller assembly*. 2. part of a roller bearing. See *roller bearing*.

roller assembly *n*: on a kelly bushing, an arrangement of rollers, roller pins, and roller bearings that mate with the kelly as it moves up or down inside the kelly bushing. The roller assembly transfers the turning motion of the kelly bushing to the kelly, and, at the same time, allows the kelly to move up and down freely. See *roller, roller bearing, roller pin*.

roller bearing *n*: a bearing in which a finely machined shaft (the journal) rotates in contact with a number of cylinders (rollers). Compare *ball bearing, journal bearing*.

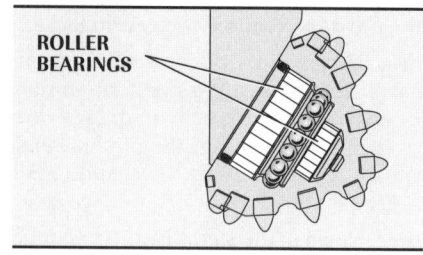

ROLLER BEARINGS

roller bit *n*: see *roller cone bit*.

roller chain *n*: a type of chain that is used to transmit power by fitting over sprockets attached to shafts, causing rotation of one shaft by the rotation of another. Transmission roller chain consists of offset links, pin links, and roller links.

roller chaser *n*: a permanent chain chaser that is specifically used with mooring wire.

roller cone bit *n*: a drilling bit made of two, three, or four cones, or cutters, that are mounted on extremely rugged bearings. The surface of each cone is made of rows of steel teeth or rows of tungsten carbide inserts. Also called rock bit.

ROLLER CONE BIT

roller diameter *n*: in roller chain, the outside diameter of the roller, about ⅝ of the pitch.

roller link *n*: one of the links in a roller chain. It consists of two bushings press-fitted into the link plates (side bars) and two rollers that fit over the bushings. The bushings are locked into the link plates to prevent rotation.

roller path *n*: the surface on which the rollers that support the revolving superstructure travel; may accommodate either cone rollers, cylindrical rollers, or live rollers.

roller pin *n*: on a kelly bushing, a shaft that fits inside each roller on the bushing to affix the rollers to it. See *kelly bushing, roller assembly*.

roller race *n*: a track, channel, or groove in which roller bearings roll.

roll-on coil connector *n*: in coiled tubing operations, a device that attaches a tool string

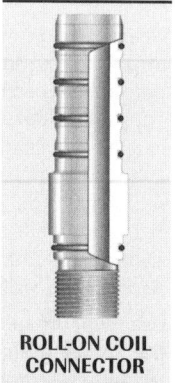

ROLL-ON COIL CONNECTOR

to the coiled tubing or joins two strings of coil tubing. The device features several sets of O-rings and grooves that ensure a high-pressure, high-strength connection.

rollover anticline *n*: an anticline formed when the dip of a growth fault approaches the horizontal at depth and deposition is faster on the down-thrown side, which tends to "roll over" or curl downward.

rollover fault *n*: see *growth fault*.

ROM *abbr*: read-only memory.

root *n*: in mathematics, a number which, when raised to some exponent (or power), equals that number. For example, the root of 4 is 2 because $2^2 = 4$.

root bead *n*: the initial welding pass made in uniting two joints of pipeline. Also called stringer bead.

root-mean-square (RMS) *n*: the square root of the time average of the square of a quantity. For a periodic quantity, the average is taken over one complete cycle.

root-mean-square (RMS) value *n*: the value of an AC current or voltage that produces the same heating effect as a direct current of the same amount; equal to 0.707 times the maximum value. Electrical measuring devices read effective values unless otherwise indicated. Also called effective value.

root pass *n*: the welding pass made after the root bead is made when uniting two joints of pipeline.

Roots blower *n*: a special compressor used on two-stroke engines to supercharge the engine's intake air and scavenge (remove) exhaust gases from the engine cylinders. The blower has two corkscrew-shaped (helical) rotors with blades (lobes) that rotate inside a housing. The engine drives the rotors with gears at about twice the engine's speed. The rotors rotate in opposite directions at the same speed. As they rotate, the lobes compress air drawn through an air cleaner on top of the housing. The compressed air exits the blower from the bottom or the side of the housing and goes into the engine's air intake manifold.

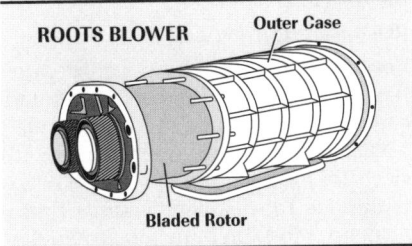

ROP *abbr*: rate of penetration.

rope socket *n*: a device to connect the wireline to the tool string.

rosin *n*: a translucent amber-colored to almost black brittle friable resin obtained by chemical means from oleoresin, the dead wood of pine trees, or from tall oil and used in making varnish, paper, and soap. It is also used as a soldering flux.

rotameter *n*: an instrument used to indicate flow rate. It consists of a float in a tapered tube, or a tapered float that moves in a fixed orifice. Flow passing around an annular space between the float and its container causes the float to rise until its weight counterbalances the pressure drop across it.

rotary *n*: the machine used to impart rotational power to the drill stem while permitting vertical movement of the pipe for rotary drilling. Modern rotary machines have a special component, the rotary or master bushing, to turn the kelly bushing, which permits vertical movement of the kelly while the stem is turning.

rotary bushing *n*: see *master bushing*.

rotary chain *n*: a chain drive powered by the drawworks that drives the rotary table assembly; it runs from the drawworks sprocket to a drive-shaft sprocket. The drive-shaft sprocket turns the drive-shaft assembly that drives the rotary table.

rotary drilling *n*: a drilling method in which a hole is drilled by a rotating bit to which a downward force is applied. The bit is fastened to and rotated by the drill stem, which also provides a passageway through which the drilling fluid is circulated. Additional joints of drill pipe are added as drilling progresses.

rotary drive countershaft *n*: a rotating shaft on the opposite side of the drawworks from the driller's console that gets power from one of the transmissions and sends it to the rotary table. It sits inside its own housing, which may be a part of the main drawworks frame or may be detachable for transportation.

rotary helper *n*: a worker on a drilling or workover rig, subordinate to the driller, whose primary work station is on the rig floor. On rotary drilling rigs, there are at least two and usually three or more rotary helpers on each crew. Sometimes called floorhand, floorman, rig crew member, or roughneck.

rotary hose *n*: a steel-reinforced, flexible hose that is installed between the standpipe and the swivel or top drive. It conducts drilling mud from the standpipe to the swivel or top drive. Also called the kelly hose or the mud hose.

rotary jar *n*: a type of mechanical jar whose jarring force is actuated and determined by rotating the work string; the more torque, the harder the jar. See *mechanical jar*.

rotary line *n*: see *drilling line*.

rotary locking device *n*: on a rotary table, a steel pin, often spring-loaded, that, when engaged, fits into one of several notches machined onto the perimeter of the rotary assembly's turntable. When engaged, the lock prevents the turntable from turning.

rotary pump *n*: a pump that moves fluid by positive displacement, using a system of rotating vanes, gears, or lobes. The vaned pump has vanes extending radially from a rotating element mounted in the casing. The geared rotary pump uses oppositely rotating meshing gears or lobes.

rotary shoe *n*: a length of pipe whose bottom edge is serrated or dressed with a hard cutting material and that is run into the wellbore around the outside of stuck casing, pipe, or tubing to mill away the obstruction. Also called a burn shoe. See *washover pipe*.

rotary-shouldered connection *n*: the threaded and shouldered joint used in rotary drilling to join the various components of the drill stem.

rotary side *n*: see *off-driller's side*.

rotary slips *n pl*: see *slips*.

rotary solenoid *n*: a solenoid that imparts rotary motion to a shaft.

rotary speed *n*: the speed, measured in revolutions per minute, at which the rotary table is operated.

rotary support table *n*: a strong but relatively lightweight device used on some rigs that employ a top drive to rotate the bit. Although a conventional rotary table is not required to rotate the bit on such rigs, crew members must still have a place to set the slips to suspend the drill string in the hole when tripping or making a connection. A rotary support table provides such a place but does not include all the rotary machinery required in a regular rotary table.

rotary system *n*: see *rotating components*.

rotary table *n*: the principal piece of equipment in the rotary table assembly; a turning device used to impart rotational power to the drill stem while permitting vertical movement of the pipe for rotary drilling.

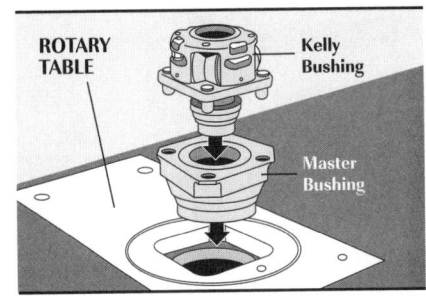

The master bushing fits inside the opening of the rotary table; it turns the kelly bushing, which permits vertical movement of the kelly while the stem is turning. See *kelly bushing, master bushing*.

rotary table assembly *n*: a rotating machine housed primarily inside a rectangular steel box with an opening in the middle for the kelly and the drill pipe that creates and transfers the turning motion for rotary drilling; parts of the assembly include the base, rotary table, master bushing, drive-shaft assembly, drawworks sprockets, drive-shaft sprockets, and locking devices.

rotary table base *n*: a cast steel or reinforced fabricated steel shell that encloses the pinion end of the drive shaft and the rotary table.

rotary table locking device *n*: a small mechanical brake for the rotary table made of an iron rod and a notched wheel; its function is to stop the turning movement of the rotary table and hold it securely while the crew makes or breaks a connection or performs other jobs.

rotary table system *n*: a series of devices that provide a way to rotate the drill stem and bit. Basic components consist of a turntable, master bushing, kelly drive bushing, kelly, and a swivel.

rotary tongs *n pl*: see *tongs*.

rotary torque *n*: the rotational force applied to turn the drill stem.

rotate on bottom *v*: see *make hole*.

rotating annular preventer *n*: an annular preventer used in underbalanced drilling that is designed to rotate as the drill stem rotates and, at the same time, provide a pressure-tight seal against well pressure from below.

rotating blowout preventer *n*: see *rotating head*.

rotating components *n pl*: those parts of the drilling or workover rig that are designed to turn or rotate the drill stem and bit—swivel, kelly, kelly bushing, master bushing, and rotary table.

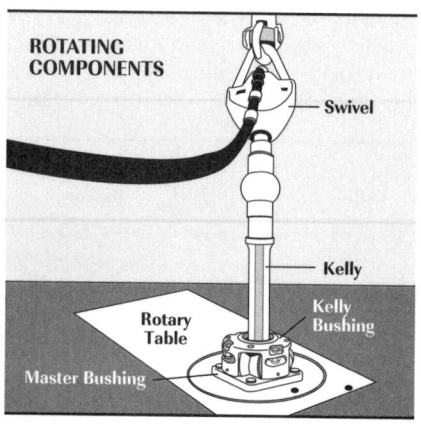

ROTATING COMPONENTS
— Swivel
— Kelly
Rotary Table
Kelly Bushing
Master Bushing

rotating head *n*: a sealing device used to close off the annular space around the kelly in drilling with pressure at the surface, usually installed above the main blowout preventers. A rotating head makes it possible to drill ahead even when there is pressure in the annulus that the weight of the drilling fluid is not overcoming; the head prevents the well from blowing

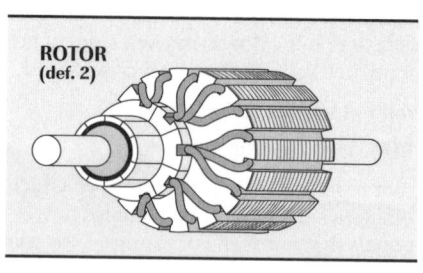

ROTATING HEAD

out. It is used mainly in the drilling of formations that have low permeability. The rate of penetration through such formations is usually rapid.

rotating impeller *n*: in a centrifugal pump, a vanelike device that turns (rotates) to create an area of low pressure in the center of the impeller.

rotating meter *n*: any metering device for which the meter pulse output is derived from mechanical rotation as driven by the flowing stream.

rotating plate choke *n*: a well-control choke that operates by rotating a downstream plate that has a half-moon opening against a fixed upstream plate with a similar opening. The plates can be rotated from completely open to completely closed. A rack and pinion rotates the plates. Either an air-operated hydraulic pump or a manual hydraulic pump operates the rack and pinion. The plates can also be turned manually. Failure of the operating mechanism leaves the choke in the position where it was last set.

rotational flow *n*: see *swirl*.

rotational inertia *n*: see *moment of inertia*.

rotational viscometer *n*: instrument used for assessing mud properties that records values for both plastic viscosity and yield point.

rotation gas lift *n*: a gas-lift system in which the gas that is injected and subsequently produced is recompressed and reinjected into the well, effecting a continuous, closed system that does not require the introduction of additional gas from an extraneous source for operation, except that needed to make up losses in the system.

rotation lock *n*: on a drilling rig's hook, a device that, when unlocked, allows crew members to rotate the hook to make the elevators face in any desired direction. When in the desired position, crew members can lock it there.

rotor *n*: 1. a device with vanelike blades attached to a shaft. The device turns or rotates when the vanes are struck by a fluid directed there by a stator. 2. the rotating part of an induction-type alternating current electric motor. Compare *stator*.

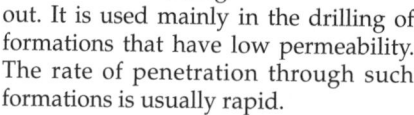

ROTOR (def. 2)

roughneck *n*: see *rotary helper*.

round trip *n*: the procedure of pulling out and subsequently running back into the hole a string of drill pipe or tubing. Also called tripping.

roustabout *n*: 1. a worker on an offshore rig who handles the equipment and supplies that are sent to the rig from the shore base. The head roustabout is very often the crane operator. 2. a worker who assists the foreman in the general work around a producing oilwell, usually on the property of the oil company. 3. a helper on a well servicing unit.

roustabout foreman *n*: head roustabout; supervisor of the roustabout crew.

routine well service *n*: routine repairs including but not limited to sucker rod failures, tubing leaks, subsurface pump repairs, and attendant equipment and material repairs and replacement.

ROV *abbr*: remotely operated vehicle.

ROV intervention system *n*: in a subsea blowout prevention system with a marine riser, the equipment that allows a remotely operated vehicle (ROV) to tie into and operate certain functions, such as closing the shear rams and unlocking the lower marine riser package (LMRP) when the normal surface operating system fails. A ported hydraulic stabbing device in the ROV mates with a receptacle in the LMRP's hydraulic circuit and pumps hydraulic fluid into the circuit to operate the selected component.

ROW *abbr*: right-of-way.

Rowland's laws *n pl*: several rules originated by Henry A. Rowland, an American physicist working in the mid-to-late 1800s, about the behavior of magnetic circuits. Perhaps his best-known law is the one for magnetic circuits that is analogous to Ohm's law for current flow through wire circuits. Rowland's law for magnetic circuits is $\phi = {}^F\!/_R$, where ϕ = magnetic flux, F = magnetomotive force, and R = reluctance.

royalty *n*: the portion of oil, gas, and minerals retained by the lessor on execution of a lease or their cash value paid by the lessee to the lessor or to one who has acquired possession of the royalty rights, based on a percentage of the gross production from the property free and clear of all costs except taxes.

royalty clause *n*: the clause in an oil and gas lease that establishes the percentage of production paid to the lessor.

royalty deed *n*: the legal instrument that conveys a share of oil or gas production. Unlike a mineral deed, a royalty deed does not create a severance of the estate.

royalty owner *n*: a person who owns a royalty interest in production.

RP *abbr*: rock pressure; used in drilling reports.

rpm *abbr*: revolutions per minute.

RQ *abbr*: reportable quantity.

RQ4 *n*: a type of high-strength chain certified by the American Bureau of Shipping (ABS).

RSPA *abbr*: Research and Special Projects Administration (DOT).

RT *abbr*: rotary table; used in drilling reports.

RT *abbr*: register ton.

RTD *abbr*: resistance temperature detector.

RTTS™ *n*: a trademark for a retrievable squeeze tool.

RTU *abbr*: remote terminal unit.

RUCT *abbr*: rigging up cable tools; used in drilling reports.

rugosity *n*: roughness or irregularity of a solid surface.

rule of capture *n*: rule applied by the courts (especially in states that have adopted nonabsolute ownership views of oil and gas) that gives title to oil and gas produced from a tract of land to the party reducing it to possession. The rule has been modified a great deal by state regulatory agencies. Compare *offset drilling rule*.

run *n*: the amount of crude oil sold and transferred to the pipeline by the producer.

runaround *n*: a platform encircling the top of the derrick.

run a tank *v*: to transfer oil from a stock tank into a pipeline.

run casing *v*: to lower a string of casing into the hole. Also called to run pipe.

rungs *n pl*: in a relay ladder logic diagram, the horizontal lines containing input and output devices.

run in *v*: to go into the hole with tubing, drill pipe, and so forth.

runner *n*: the driven element in a hydraulic coupling that is connected to the driven equipment.

running sample *n*: in tank sampling, a sample obtained by lowering an unstoppered beaker or bottle from the top of the oil to the level of the bottom of the outlet connection or swing line and returning it to the top of the oil at a uniform rate of speed so that the beaker or bottle is about three-quarters full when withdrawn from the oil.

running squeeze cementing *n*: a squeeze cementing technique used to repair damaged casing in which cement is pumped slowly and continuously until final pressure is obtained. See *squeeze cementing*.

running start-and-stop method *n*: in meter proving, the method wherein the opening and closing meter readings of the test run are determined at flowing conditions. Compare *standing start-and-stop method*.

running tool *n*: any specialized tool used to run equipment in a well, such as a marine riser joint or a wireline running tool for installing retrievable gas-lift valves. Various tubing-type running tools are also used.

run pipe *v*: to lower a string of casing into the hole. Also called to run casing.

run sheet *n*: a landman's list and brief description of all the documents in the history of ownership of a given tract of land. Compare *takeoff*. See *chain of title*.

run statement *n*: a monthly summary of run tickets given by the purchaser or operator of crude oil, detailing volume gravity, price, and value of each run ticket.

run ticket *n*: a record of the oil transferred from the producer's storage tank to the pipeline. It is the basic legal instrument by which the lease operator is paid for oil produced and sold. Also called delivery ticket, measurement ticket, receipt ticket.

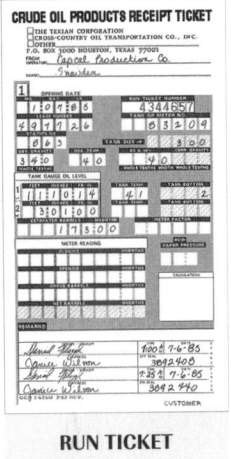

RUN TICKET

RUR *abbr*: rigging up rotary rig; used in drilling reports.

s *sym*: second.

S *sym*: sulfur.

sabkha *n*: see *playa*.

sack *n*: a container for cement, bentonite, ilmenite, barite, caustic, and so forth. Sacks (bags) contain the following amounts:

Cement	94 pounds (42.6 kilograms) (1 cubic foot)
Bentonite	100 pounds (45.5 kilograms)
Ilmenite	100 pounds
Barite	100 pounds

sacrificial anode *n*: in cathodic protection, anodes made from metals whose galvanic potentials render them anodic to steel in an electrolyte. They are used up, or sacrificed.

saddle *n*: see *pipe saddle*.

saddle bearing *n*: the center bearing on a conventional walking beam pumping unit. It is mounted on top of the samson post.

saddle clamp *n*: see *clamp*.

SAFE *abbr*: Safety Award for Excellence program.

Safe Drinking Water Act (SDWA) *n*: a congressional act that provides for the protection of underground sources of drinking water by regulating drinking water systems and injection wells.

Safety Award for Excellence (SAFE) program *n*: an MMS safety award program that recognizes private companies that protect the human and natural environment by avoiding accidents and pollution through adherence to safety and environmental regulations and guidelines.

safety clamp *n*: a clamp placed very tightly around a drill collar that is suspended in the rotary table by drill collar slips. Should the slips fail, the clamp is too large to go through the opening in the rotary table and therefore prevents the drill collar string from falling into the hole. Also called drill collar clamp.

safety factor *n*: a means of gauging the adequacy of the strength of a mooring component. It is expressed as a factor or ratio of the ultimate breaking strength or design load (e.g., a safety factor means that the maximum load seen on a component in a particular condition is half the breaking strength of that component).

safety factor of wire rope *n*: a measurement of load safety for wire rope obtained by using the following formula:

$$\text{Factor of safety} = B/W$$

where—

B = nominal catalog breaking strength of the wire rope, and

W = calculated total static load.

Also called design factor of wire rope.

safety joint *n*: an accessory to a fishing tool, placed above it. If the tool cannot be disengaged from the fish, the safety joint permits easy disengagement of the string of pipe above the safety joint. Thus, part of the safety joint and the tool attached to the fish remain in the hole and become part of the fish.

safety latch *n*: a latch provided on a hook to prevent an object suspended from the hook from accidentally slipping or falling out of it.

SAFETY LATCH

safety margin *n*: see *trip margin*.

safety of life at sea (SOLAS) rules *n pl*: international maritime rules established to set minimum safety standards for ships and mobile offshore drilling units under way or under tow.

safety platform *n*: the monkeyboard, or platform on a derrick or mast on which the derrickhand works while wearing a safety harness (attached to the mast or derrick) to prevent falling.

safety release *n*: a device made up in a tubing string above a packer or other downhole tool that allows a crew on the surface to remove the tubing string should the packer or other tool get stuck.

safety relief valve *n*: see *pop valve*.

safety shoes *n pl*: metal-toed shoes or boots with nonskid, corrosion-resistant soles worn by oilfield workers to minimize falls and injury to their feet.

safety slide *n*: a wireline device normally mounted near the monkeyboard to afford the derrickhand a means of quick exit to the surface in case of emergency. It is usually affixed to a wireline, one end of which is attached to the derrick or mast and the other end to the surface. To exit by the safety slide, the derrickhand grasps a handle on it and rides it down to the ground. Also called a Geronimo.

safety valve *n*: 1. an automatic valve that opens or closes when an abnormal condition occurs (e.g., a pressure relief valve on a separator that opens if the pressure exceeds the set point, or the shutdown valve at the wellhead that closes if the line pressure becomes too high or too low). 2. a control device (a valve) installed at the top of the drill stem, which, when closed, prevents flow out of the drill pipe if a kick occurs during tripping operations.

safety wire *n*: steel cable attached to the monkeyboard and anchored to the ground at some distance from the rig. It is used by the derrickhand to slide clear of danger in an emergency.

safe working load *n*: in crane operations, that portion of a wire rope's nominal strength that can be applied either to move or sustain a load without damaging or breaking the rope. The safe working load of a rope is accurate only when the rope is new and the equipment is in good condition. Because most ropes on an installation quickly become used, the safe working load of a rope is also quickly reduced. For this reason, safe working load is seldom used to denote wire-rope strength. See *nominal strength*.

Saffir-Simpson Hurricane Scale *n*: a five-step scale meteorologists use to rate the intensity of hurricanes. The scale has five categories, with category 1 being the least intense and category 5 being the most. Meteorologists use the scale to estimate potential property damage and flooding expected along the coast from a hurricane's landfall. The hurricane's wind speed determines its category. A category 1 hurricane has sustained wind speeds of from 74 to 95 miles per hour (mph), a category 2 from 96 to 110 mph, a category 3 from 111 to 130 mph, a category 4 from 131 to 155 mph, and a category 5 more than 155 mph. Herbert Saffir, a consulting engineer, and Bob Simpson, director of the National Hurricane Center, formulated the scale in 1969.

sag bend *n*: a temporarily unsupported span of pipe between the stinger and the seabed in marine pipe laying.

sagging *n*: the distortion of the hull of a vessel when the middle is lower than either end because of excessively heavy or unbalanced loads. Compare *hogging*.

sales line *n*: any line through which oil, gas, or products flow to a sales point.

sales outlet *n*: see *oil sales outlet*.

saline drilling fluid *n*: see *salt mud*.

salinity *n*: a measure of the amount of salt dissolved in a liquid.

salinity log *n*: a special nuclear well log that produces an estimate of the relative amounts of oil, gas, or salt water in a formation. This log is electronically adjusted to reflect gamma ray emissions resulting from the collision of neutrons with chlorine atoms in the formations.

SALM *abbr*: single anchor leg mooring.

salt *n*: a compound that is formed (along with water) by the reaction of an acid with a base. A common salt (table salt) is sodium chloride, NaCl, derived by combining hydrochloric acid, HCl, with sodium hydroxide, NaOH. The result is sodium chloride and water, H_2O. This process is written chemically as HCl + NaOH → NaCl + H_2O. Another salt is calcium sulfate, $CaSO_4$, obtained when sulfuric acid, H_2SO_4, is combined with calcium hydroxide, Ca $(OH)_2$.

salt barrel *n*: a 55-gallon (208-litre) drum modified to salt-saturate the water going into circulation to prevent the dissolution of formation salt when building mud volume.

salt-brine cement *n*: a cementing slurry whose liquid phase contains sodium chloride.

salt creep *n*: a phenomenon that sometimes occurs when a salt formation is drilled.

Because some salt formations tend to behave somewhat as a liquid, they flow, or creep, into the borehole after it is drilled. Salt creep reduces the size of the borehole and must be taken into account.

salt dome *n*: a dome that is caused by an intrusion of rock salt into overlying sediments. A piercement salt dome is one that has been pushed up so that it penetrates the overlying sediments, leaving them truncated. The formations above the salt plug are usually arched so that they dip in all directions away from the center of the dome, thus frequently forming traps for petroleum accumulations.

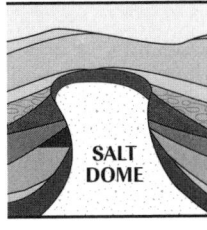

salt mud *n*: 1. a drilling mud in which the water has an appreciable amount of salt (usually sodium or calcium chloride) dissolved in it. Also called saltwater mud or saline drilling fluid. 2. a mud with a resistivity less than or equal to the formation water resistivity.

salt squeeze *n*: see *plastic deformation*.

salt water *n*: a water that contains a large quantity of salt, i.e., brine.

saltwater clay *n*: see *attapulgite*.

saltwater disposal *n*: the method and system for the disposal of salt water produced with crude oil. A typical system is composed of collection centers (in which salt water from several wells is gathered), a central treating plant (in which salt water is conditioned to remove scale- or corrosion-forming substances), and disposal wells (in which treated salt waste is injected into a suitable formation).

saltwater flow *n*: an influx of formation salt water into the wellbore.

saltwater mud *n*: see *salt mud*.

sample bailer *n*: a tool with a sharp pointed end designed to slice into sand and a hollow cylinder to trap sand samples. The sample bailer is slammed into the bottom of the wellbore with enough impact to force sand into it.

sample catcher *n*: in underbalanced drilling, a device installed in the blooey line to retrieve samples of the cuttings (dust) returning to the surface from the bottom of the hole.

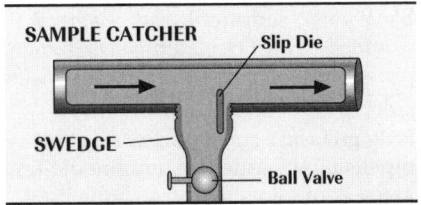

sample conditioning *n*: the mixing required to prepare liquid samples taken from a pipeline or sample loop prior to sample transfer for analysis.

sample controller *n*: on a pipeline or loop, a device for governing the operation of the sample extracting mechanism in proportion to either time or flow.

sample extractor *n*: device for extracting the sample grabs from the pipeline or from the sample loop.

sample grab *n*: the volume of liquid extracted from the pipe by a single actuation of the sample extractor. The sum of all grabs results in a sample.

sample handling *n*: the extraction, conditions, transferring, and transporting of a representative sample from a container to analytical glassware or centrifuge tubes.

sample log *n*: a graphic representative model of the rock formations penetrated by drilling, prepared by the geologist from samples and cores.

sample loop *n*: a relatively small pipe that is a by-pass from the main pipeline. Liquid in the pipeline also flows through the loop, from which the liquid can be sampled.

sample mud *n*: drilling fluid formulated so that it will not alter the properties of the cuttings the fluid carries up the well.

sample proving *n*: the technique used to validate an automatic sampling system.

sampler *n*: a device attached to a pipeline to permit continuous sampling of the oil, gas, or product flowing in the line.

sample receiver *n*: receptacle normally connected to a sampling draw-off connection or pipeline probe and used to receive the sample. When disconnected it may be used as a sample container.

sample rotameter float *n*: a float moving vertically within a linearly tapered tube and exposing a variable area to the flow. Also called variable area flowmeter.

samples *n pl*: 1. the well cuttings obtained at designated footage intervals during drilling. From an examination of these cuttings, the geologist determines the type of rock and formations being drilled and estimates oil and gas content. 2. small quantities of well fluids obtained for analysis.

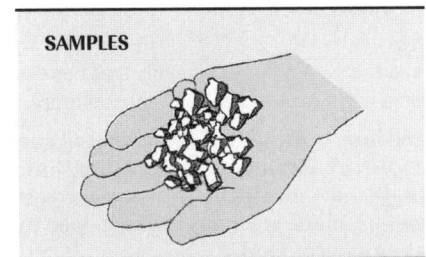

sampling *n*: 1. the taking of a representative sample of fluid from a tank or pipeline to measure its temperature, specific gravity, and S&W content. 2. the process of cutting a core or pieces of core for analysis.

samson post *n*: 1. the part of the surface equipment of a standard cable-tool drilling rig that supports the walking beam. 2. the member of a rod pumping unit that supports the walking beam.

sand *n*: 1. an abrasive material composed of small quartz grains formed from the disintegration of preexisting rocks. Sand consists of particles less than 2 millimetres (0.078 inches) and greater than ⅟₁₆ millimetre (0.062 inches) in diameter. 2. sandstone.

sand bailer *n*: a device used to remove sand from the wellbore. It sucks up sand by creating a partial vacuum and traps the sand by closing at the bottom.

sand consolidation *n*: any one of several methods by which the loose, unconsolidated grains of a producing formation are made to adhere to prevent a well from producing sand but permit it to produce oil and gas.

sand content *n*: the insoluble abrasive solids contents of a drilling fluid rejected by a 200-mesh screen. Usually expressed as the percentage bulk volume of sand in a drilling fluid. This test is an elementary type in that the retained solids are not necessarily silica and may not be altogether abrasive. For additional information concerning the kinds of solids retained on the 200-mesh screen, more specific tests would be required. See *mesh*.

sand control *n*: any method by which large amounts of sand in a sandy formation are prevented from entering the wellbore. Sand in the wellbore can cause plugging and premature wear of well equipment. See *gravel pack, sand consolidation, screen liner*.

sand cutter *n*: a device that ejects sand at a very high velocity to cut casing being salvaged from a plugged and abandoned well.

sand drum *n*: on the drawworks, a spool for the sandline.

sanded-up *adj*: 1. of a well, under restricted production because of sand accumulation in the wellbore. 2. impeded or hindered, especially because of sand accumulation.

sand fill *n*: a column of sand that has entered and accumulated in the wellbore.

sandfrac *n*: method of fracturing subsurface rock formations by injecting fluid and sand under high pressure to increase permeability. Fractures are kept open by the grains of sand.

sanding off *n*: a phenomenon that occurs when sand and sediment settle out of oil being produced from a well and impedes or blocks the flow of produced fluids from the well.

sand lens *n*: see *lens*.

sandline *n*: a wireline used on drilling rigs and well-servicing rigs to operate a swab or bailer, to retrieve cores or to run logging devices. It is usually ⁹⁄₁₆ of an inch (14 millimetres) in diameter and several thousand feet or metres long.

sandline drill *n*: a special bit or mill run on a cable-tool drilling line, a well servicing rig, or the sandline of a rotary drilling rig to mill out tools or drill out downhole debris.

sand out *v*: to plug a well inadvertently with proppants during formation fracturing. Sanding out is usually the result of a slowed fracture-fluid velocity, or screening effect, which allows the proppants to become separated from the fluid instead of being carried away from the wellbore. Also called screening out.

sand reel *n*: a metal drum on a drilling rig or a workover unit around which the sandline is wound. On a drilling rig, it may be attached to the drawworks catshaft and may be used for coring or other wireline operations. When used on a drilling rig, it is sometimes called a coring reel.

sand screen *n*: see *wire-wrapped screen*.

sandstone *n*: a sedimentary rock composed of individual mineral grains of rock fragments between 0.06 and 2 millimetres (0.002 and 0.079 inches) in diameter and cemented together by silica, calcite, iron oxide, and so forth. Sandstone is commonly porous and permeable and therefore a likely type of rock in which to find a petroleum reservoir.

sand-thickness map *n*: a map that shows the thickness of subsurface sands. See *isopach map*.

sand trap *n*: a steel tank placed under the shale shaker into which mud falls after passing through the shale shaker. The shaker removes mainly cuttings from the mud so solids such as sand and other fine particles fall with the mud into the sand trap. Many of the solids fall out of the mud in the sand trap; those that do not settle out are removed with other specialized solids control mud treatment equipment such as desanders and desilters.

S&W *abbr*: sediment and water. API Committee on Petroleum Measurement prefers this abbreviation to the older "BS&W."

S&W probe *n*: a small device inserted in a pipeline to measure the amount of S&W in the liquid flowing through the line.

SAPP *abbr*: sodium acid pyrophosphate.

SARA Title III *n*: Superfund Amendments and Reauthorization Act of 1986. A series of regulations promulgated by the EPA for general industry after the Bhopal disaster in India, where thousands died as a result of the release of a toxic cloud from an insecticide factory. SARA requires oil and gas operators to (1) appoint an emergency coordinator, (2) submit a list or individual Material Safety Data Sheets for all hazardous chemicals present at a facility at or above 10,000 pounds or extremely hazardous substances at or above 500 pounds or the threshold planning quantity, (3) submit an annual inventory of hazardous chemicals or extremely hazardous substances reported on the Material Safety Data Sheets, (4) notify the State Emergency Response Commission (SERC) of the presence of an extremely hazardous substance at or above the threshold planning quantity, (5) report releases of any hazardous or extremely hazardous chemical that leaves the boundaries of the facility at or above the reporting quantity to the SERC and the Local Planning Committee.

sat *abbr*: saturated or saturation; used in drilling reports.

satellite system *n*: a system that is located some distance from a gas plant for which it performs a function. Examples are absorbers or compressors, in which gas is treated or compressed prior to reaching the plant.

satellite well *n*: usually a single well drilled offshore by a mobile offshore drilling unit to produce hydrocarbons from the outer fringes of a reservoir that cannot be produced by primary development wells drilled from a permanent drilling structure (such as a platform rig). Sometimes, several satellite wells will be drilled to exploit marginal reservoirs and avoid the enormous expense of erecting a platform.

saturated Btu *n*: a measure of heating value for natural gas that is fully saturated with water vapor under standard temperature, pressure, and gravity conditions. This is typically a laboratory condition to standardize the amount of water vapor in the gas at a convenient, albeit arbitrary, level. This standard of measure usually has little or nothing to do with the state in which the gas is actually delivered for first sales.

saturated Btu at delivery conditions *n pl*: the number of Btus contained in a cubic foot of natural gas fully saturated with water under actual delivery pressure, temperature, and gravity conditions.

saturated Btu at test conditions *n pl*: the number of Btus contained in a cubic foot of natural gas fully saturated with water at a specified pressure base and 60°F (15.5°C).

saturated core *n*: in magnetic circuits, the instant in the core of a coil whereby the flux density reaches a level beyond which little or no additional density can be achieved by further increases in magnetomotive force.

saturated hydrocarbons *n pl*: hydrocarbon compounds, e.g., in natural gas and natural gas liquids, in which all carbon valence bonds are filled with hydrogen atoms.

saturated liquid *n*: liquid that is at its boiling point or is in equilibrium with a vapor phase in its containing vessel.

saturated salt mud *n*: mud that contains the maximum amount of salt that can be dissolved.

saturated salt water *n*: water that has all the salt dissolved in it that is possible at a given temperature.

saturated solution *n*: a solution that contains at a given temperature as much of a solute as it can retain. At 68°F (20°C) it takes 126.5 pounds (57.38 kilograms) of salt to saturate 1 barrel (0.1592 cubic metres) of fresh water. See *supersaturation*.

saturated steam *n*: steam that exists at a temperature corresponding to its absolute pressure. It may contain, or be free of, water particles.

saturated vapor *n*: vapor at its dew point.

saturation *n*: 1. a state of being filled or permeated to capacity. Sometimes used to mean the degree or percentage of saturation (e.g., the saturation of the pore space in a formation or the saturation of gas in a liquid, both in reality meaning the extent of saturation). 2. the condition of air when it contains the largest amount of water vapor it can possibly hold at a given temperature and pressure.

saturation diving *n*: diving in which a diver's tissues are saturated with an inert gas to a point where no more of the gas can be absorbed by the body. Once a diver is saturated, decompression time remains the same whether he or she stays at the saturated depth for 24 hours or for several days.

saturation point *n*: the point at which, at a certain temperature and pressure, no more solid material will dissolve in a liquid.

saver sub *n*: an expendable substitute device made up in the drill stem to absorb much of the wear between the frequently broken joints (such as between the kelly or top drive and the drill pipe). See *kelly saver sub*.

SAWRS *abbr*: Supplemental Aviation Weather Reporting Station.

sawtooth oscillator *n*: an electronic circuit that converts energy from a direct-current source to a periodically varying electric output; its waveform has the shape of a sawtooth.

sawtooth sweep generator *n*: an electronic device that creates, or generates, an output voltage that has a waveform in the shape of a sawtooth. Such a generator produces sweep voltage for cathode-ray tubes. See *sweep voltage*.

Saybolt Second Universal (SSU) *n*: a unit for measuring the viscosity of lighter petroleum products and lubricating oils. See *Saybolt viscometer*.

Saybolt viscometer *n*: an instrument used to measure the viscosity of fluids, consisting basically of a container with a hole or jet of a standard size in the bottom. The time required for the flow of a specific volume of fluid is recorded in seconds at three temperatures (100°F–37.8°C, 130°F–54.4°C, and 210°F–98.9°C). The time measurement unit is referred to as the Saybolt Second Universal (SSU).

SBHT *abbr*: static bottomhole temperature.

SBM *abbr*: synthetic-based mud.

Sc *abbr*: stratocumulus.

scale *n*: 1. a mineral deposit (e.g., calcium carbonate) that precipitates out of water and adheres to the inside of pipes, heaters, and other equipment. 2. an ordered set of gauge marks together with their defining figures, words, or symbols with relation to which position of the index is observed when reading an instrument.

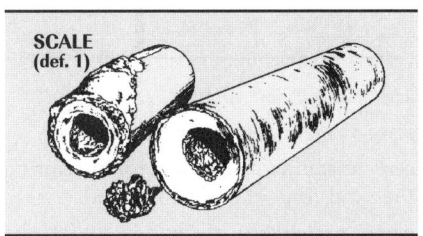

scale base *n*: the line, actual or implied, that passes through the midpoints of the shortest marks on the scale.

scale division *n*: the interval between any two successive scale marks of the scale.

scale length *n*: the linear or curvilinear length measured along the scale base between the centers of the terminal scale marks.

scale mark *n*: a line or other mark on the scale of an indicating device corresponding to one or more defined values of the quantity measured.

scale numbering *n*: the set of numbers marked on a scale that correspond either to the values of the quantity measured, defined by the scale marks, or that indicate only the numerical order of the scale marks.

scale range *n*: 1. the zone included between the scale marks that corresponds to the maximum and minimum values of the scale. 2. the difference between the maximum and minimum values of the scale.

scaling tool *n*: a circular-shaped wire brush that is attached to a pneumatically, hydraulically, or electrically operated tool (such as a grinder) and that is used to remove rust or scale from pipe or other oilfield equipment.

scanner chart *n*: an orifice meter chart with faint lines, intended to be read, or integrated, by a scanner machine.

scantlings *n pl*: (nautical) the dimensions of the structural members in the hull.

scarp *n*: an extended cliff or steep slope, produced by erosion or faulting, that separates two level or gently sloping areas, i.e., an escarpment.

scavenge *v*: to remove exhaust gases from a cylinder by forcing compressed air in and exhaust gases out. Such removal takes place in all two-cycle diesel engines.

scavenging efficiency *n*: in two-strokes-per-cycle engines, a measure of the engine blower's ability to remove exhaust gases from the engine cylinder after combustion.

SCBA *abbr*: positive-pressure self-contained breathing apparatus.

scf *abbr*: standard cubic feet.

scf/d *abbr*: standard cubic feet per day.

scheduler *n*: the employee responsible for establishing pipeline oil movements based on the coordination of all shipper requirements.

schematic *n*: a drawing of an electric circuit that uses several standard symbols to represent electrical components and that shows the arrangement and connections of the conductors and the components in the circuit.

schist *n*: a coarse-grained, foliated metamorphic rock that splits easily into layers. It is formed when shale under deep burial becomes slate and then, with more intense metamorphism, becomes schist.

Schlumberger (pronounced "slumberjay") *n*: one of the pioneer companies in electric well logging, named for the French scientist who first developed the method. Today, many companies provide logging services of all kinds.

Schmitt circuit *n*: see *Schmitt trigger*.

Schmitt trigger *n*: pulse generator capable of assuming either of two stable states (a bistable pulse generator) in which an output pulse of constant amplitude exists only as long as the input voltage exceeds a certain value. Also called a Schmitt circuit. See *bistable*.

Schottky barrier diode *n*: see *Schottky diode*.

Schottky diode *n*: a diode formed by contact between a semiconductor layer and a metal coating. It is used as a special rectifier. Electrons or holes are emitted from the area where the semiconductor and metal coating contact each other (the p-n junction) and move to the metal coating. It provides very fast switching speeds. Also called a hot-carrier diode, Schottky barrier diode.

scintillation counter *n*: detects the light emitted when a gamma ray strikes a scintillator. Measures gamma rays emitted by a formation.

scintillation detector *n*: one of four types of detectors used since the inception of radiation logging. It converts tiny flashes of light produced by gamma rays as they expend themselves in certain crystals into electrical pulses. Pulse size depends on the amount of energy absorbed.

scintillator *n*: special material, either plastic or crystalline, that fluoresces when struck by a gamma ray.

scope *n*: the ratio of the total length of a mooring line (as on a mobile offshore drilling rig) to the depth of the water.

scoping *n*: the NEPA process of identifying the scope and significance of important issues associated with a proposed federal action and identifying alternatives to that action through coordination and consultation with government (federal, state, and local), the public, and any interested individuals or organizations.

SCR *abbr*: silicon-controlled rectifier.

scraper *n*: any device that is used to remove deposits (such as scale or paraffin) from tubing, casing, rods, flow lines, or pipelines.

scraper trap *n*: a specially designed piece of equipment that is installed in a pipeline to launch or receive a pipeline scraper.

scratcher *n*: a device that is fastened to the outside of casing to remove mud cake from the wall of a hole to condition the hole for cementing. By rotating or moving the casing string up and down as it is being run into the hole, the scratcher, formed of stiff wire, removes the cake so

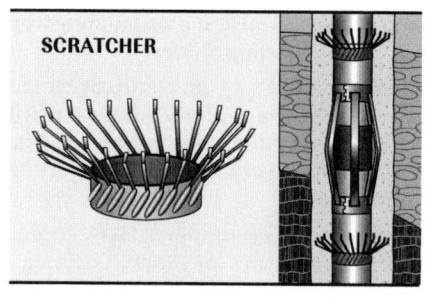

that the cement can bond solidly to the formation.

SCR bridge *n*: an arrangement of silicon-controlled rectifiers into a common electrical bridge circuit.

screen analysis *n*: determination of the relative percentages of substances, e.g., the suspended solids in a drilling fluid that pass through or are retained on a sequence of screens of decreasing mesh size. Also called sieve analysis.

screening effect *n*: the tendency of proppants to separate from fracture fluid when the speed, or velocity, of the fluid is low.

screening out *n*: see *sand out*.

screen liner *n*: a pipe that is perforated and often arranged with a wire wrapping to act as a sieve to prevent or minimize the entry of sand particles into the wellbore. Also called a screen pipe. See *wire-wrapped screen*.

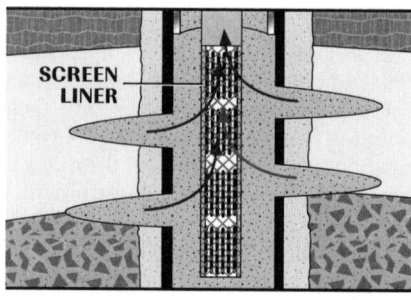

screen out *v*: see *sand out*.

screen pipe *n*: see *wire-wrapped screen*.

screen set *n*: an instrument for measuring the sand content of drilling mud.

screw packer *n*: a packer in which the packing element is expanded by rotating the pipe. It is used when it is not desirable to put tubing weight on the packer.

scrub *v*: to remove certain constituents of a gas by passing it through a scrubber in which the gas is intimately mixed with a suitable liquid that absorbs or washes out the constituent to be removed.

scrubber *n*: 1. a vessel through which fluids are passed to remove dirt, other foreign matter, or an undesired component of the fluid. 2. a vessel with or without internal devices used to separate entrained liquids or solids from gas. It may be used

to protect downstream rotating equipment or to recover valuable liquids from a gas or vapor originating upstream. 3. a unit that removes carbon dioxide from the diver's breathing medium by chemical absorption.

scrubber oil *n*: oil that is recovered from a knockout (scrubber) vessel. Usually associated with a plant that requires compression. Gas is scrubbed prior to compression.

scrubbing *n*: 1. the purification of gas by treatment in a water or chemical wash. Scrubbing also removes the entrained water in the gas. 2. friction wear.

SCSSV *abbr*: surface-controlled subsurface safety valve.

scuba *n*: self-contained underwater breathing apparatus.

scud *n*: see *fractus*.

S curve *n*: the configuration of pipe when it enters the water from a stinger of a lay barge in pipeline construction. The overbend is closest to the barge, and the sag bend is on the seafloor. Compare *J curve*. See *sag bend*.

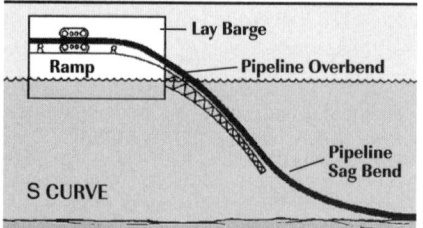

sd *abbr*: sand or sandstone; used in drilling reports.

SDO *abbr*: shut down for orders; used in drilling reports.

SDWA *abbr*: Safe Drinking Water Act.

sdy *abbr*: sandy; used in drilling reports.

SEA *abbr*: suction embedment anchor.

sea *n*: a wave that is still in its fetch, i.e., under the influence of the fetch's wind.

sea anchor *n*: an open-ended parachute-shaped device that creates drag but, because water flows through it, slows the speed of a body attached to it.

seabed *n*: see *seafloor*.

seabed gas diverter (SBGD) *n*: special diverter designed to handle shallow gas kicks that occur offshore. Unlike a normal diverter that the gas from the kick releases at the surface, an SBGD allows the gas to be released from the wellhead on the seafloor. Thus, the gas bypasses the rig. See *diverter*.

seafloor *n*: the bottom of the ocean. Also called seabed.

sea fog *n*: see *advection fog*.

sea ice *n*: any of several types of ice that form on the surface of the sea when the temperature drops below the seawater's freezing point. Types of sea ice include fast ice, or coast ice, which is sea ice attached to the shore; pack ice, which is a solid concentration of ice that covers more than half of the visible sea surface; sheet ice, which is a smooth, thin layer of ice on the water's surface; and brash, which are ice fragments that float on the sea's surface. Sea ice also forms ice fields, which are large, flat expanses of ice on the sea's surface. Similar to ice fields are floes, which are simply small ice fields.

seal assembly *n*: apparatus that seals the annulus between each casing string, and isolates the previous casing from the higher wellbore pressures encountered at greater depths.

seal bore *n*: a smooth, polished bore in pipe or in a packer that is designed to provide a surface against which to make an effective seal.

seal-bore extension *n*: a short tube attached to the packer seal bore that extends the length of the bore; used where excessive tubing expansion or contraction is anticipated.

seal-bore packer *n*: a packer containing a seal bore to receive a tubing seal assembly.

sealed bearing *n*: on a roller cone bit, a type of bearing that is not exposed to the drilling fluid in the wellbore. Instead, a synthetic rubber ring is placed between the outside of each cone and the shirttail of the bit where the cones reside. This seal prevents drilling fluid from entering the space between the cone and shirttail and thus the bearings, which lie inside the cones. Sealed bearing bits must also contain a built-in reservoir of grease since they are not lubricated by the drilling fluid.

Seale strand *n*: a wire-rope strand design commonly used as drilling line. It consists of a single, central wire around which are laid two additional layers of wire. The inner layer of wires are all the same diameter but are smaller in diameter than the central wire. The outer layer of wires are also all the same diameter but they are thicker in diameter than the inner wires. A typical Seale strand consists of 19 wires: 1 central wire, 9 inner wires, and 9 outer wires. Several strands are laid together to make the wire rope.

sealing agent *n*: any of various materials, such as mica flakes or walnut hulls, that cure lost circulation. See *lost circulation, lost circulation materials*.

sealing fault *n*: a fault that contains material of low permeability, such as gouge.

seal nipple assembly *n*: a sealing member placed in the production tubing that is landed (placed) inside the packer's seal bore.

seal-off *n*: the penetration of a drilling fluid into a potentially productive formation, thus restricting or preventing the formation from producing.

seal pot *n*: a small chamber in a gas meter installation that traps water and prevents it from causing inaccurate measurement.

seal sub *n*: a smoothly finished steel tube with rubber or synthetic seal rings run on the bottom of the tubing string and seated in a permanent packer in order to make a pressure seal.

seal unit *n*: an extension with seals that is placed in the tubing string and that moves within a packer bore or a packer-bore extension.

seamless drill pipe *n*: drill pipe that is manufactured in one continuous piece. Most drill pipe is of seamless construction.

sea smoke *n*: see *steam fog*.

sea state *n*: a relative term used to describe the condition of a large body of water, usually an ocean or a sea. Generally, sea states describe the height and intensity of the waves occurring at the time of the observation.

sea suctions *n pl*: valve-controlled pipelines, fitted with pumps, which permit a vessel to take on seawater for ballasting into any of the vessel's tanks.

seat *n*: the point in the wellbore at which the bottom of the casing is set.

seating nipple *n*: a special tube installed in a string of tubing, having machined contours to fit a matching wireline tool with locking pawls. It is used to hold a regulator, choke, or safety valve; to anchor a pump; or to permit installation of gas-lift valves. Also called a landing nipple. A smooth bore tubing nipple, having an inside diameter less than tubing, used to 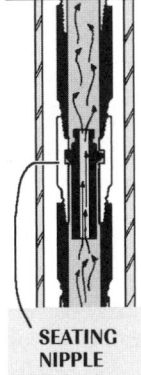 **SEATING NIPPLE** accept the seal assembly of a rod pump, as a stop for a plunger, as a safety stop for tubing swabs.

sea trials *n pl*: the final testing of a ship or offshore drilling vessel before it is put to work.

seawater mud *n*: a special class of saltwater muds in which sea water is used as the fluid phase.

sebkha *n*: see *playa*.

sec *abbr*: 1. secant. 2. second. 3. section; used in drilling reports.

secant (sec) *n*: in a right triangle, the trigonometric function that is the ratio of the hypotenuse to the side adjacent to a given angle.

second *n*: 1. the fundamental unit of time in the metric system. 2. in some countries, the crew member who relieves the toolpusher, or rig manager, and is second in command.

secondary *n*: see *secondary winding*.

secondary cell *n*: a device that produces electricity from the chemical reaction of conductors of dissimilar metals through an electrolyte. In a secondary cell, the electrolyte is a liquid solution and the chemical reaction is reversible—that is, the secondary cell can be recharged. It is often called a wet cell. See *dry cell, wet cell*.

secondary cementing *n*: any cementing operation after the primary cementing operation. Secondary cementing includes a plug-back job, in which a plug of cement is positioned at a specific point in the well and allowed to set. Wells are plugged to shut off bottom water or to reduce the depth of the well for other reasons.

secondary circuit *n*: in a riser connector, a way of routing hydraulic fluid to provide additional unlocking force that may be required to overcome high friction forces generated between the tapers on the cam ring and the locking dogs, or segments.

secondary control panel *n*: see *remote control panel*.

secondary element *n*: that part of an orifice meter installation that contains the recording elements, including the gauge lines, orifice meter chart, flow recorder, clock, static spring, etc.

secondary line *n*: a power line from the lease distribution point to electrical apparatus on the lease.

secondary loss *n*: in a transformer, the reduction (loss) of electrical energy that occurs in the secondary winding of the transformer.

secondary migration *n*: movement of hydrocarbons, subsequent to primary migration, through porous, permeable reservoir rock, by which oil and gas become concentrated in one locality.

secondary porosity *n*: porosity created in a formation after it has formed, because of dissolution or stress distortion taking place naturally, or because of treatment by acid or injection of coarse sand.

secondary recovery *n*: 1. the use of waterflooding or gas injection to maintain formation pressure during primary production and to reduce the rate of decline of the original reservoir drive. 2. waterflooding of a depleted reservoir. 3. the first improved recovery method of any type applied to a reservoir to produce oil not recoverable by primary recovery methods. See *pressure maintenance, primary recovery*.

secondary sediment *n*: sediment that does not show up in an untreated sample of emulsion but that does appear as additional water in a sample treated with a slugging compound. Compare *primary sediment*.

secondary standard *n*: a standard the value of which is fixed by direct or indirect comparison with a primary standard or by means of a reference-value standard. Compare *primary standard, working standard*.

secondary term *n*: the term, usually in a phrase like "so long thereafter as oil or gas is produced in paying quantities," that extends a lease beyond its primary term. Compare *primary term*.

secondary well control *n*: the use of blowout prevention equipment to control a kick. Compare *primary well control*.

secondary winding *n*: in a transformer, the second of two windings (coils) in which electrical energy is induced by the magnetic energy in the primary winding. Compare *primary winding*.

second order motions *n pl*: vessel motions caused by both wave action and impacted by the stiffness of the mooring system.

seconds API *n*: the time in seconds that it takes 1 quart of drilling mud to flow out of a Marsh funnel. It is a measure of the mud's viscosity. See also *Marsh funnel*.

section *n*: 1. a unit of land measurement in the rectangular survey system. Each 6-mile (9.7-kilometre) square, or township, is divided into 36 sections. A section usually is 1 square mile (2,590 square kilometres), or 640 acres (256 hectares). It may be larger or smaller, depending on its position in the township. 2. vertically arrayed data.

section milling *n*: the process by which a portion of pipe, usually casing, is actually removed by cutting with a mill.

sed *abbr*: sediment; used in drilling reports.

sediment *n*: 1. the matter that settles to the bottom of a liquid; also called tank bottoms. 2. in geology, buried layers of sedimentary rocks.

sediment and water (S&W) *n*: a material coexisting with, yet foreign to, petroleum liquid and requiring a separate measurement for reasons that include sales accounting. This foreign material includes free water and sediment (dynamic measurement) and/or emulsified or suspended water and sediment (static measurement). The quantity of suspended material present is determined by a centrifuge or laboratory testing of a sample of petroleum liquid.

sediment and water sample *n*: a sample obtained from the bottom of the tank to determine the amount of nonmerchantable material present.

sedimentary rock *n*: a rock composed of materials that were transported to their present position by wind or water. Sandstone, shale, and limestone are sedimentary rocks.

sedimentation *n*: the process of deposition of layers of clastic particles or minerals that settle out of water, ice, or other transporting media.

sediment clingage *n*: the sediment in oil or other liquid that tends to adhere, or cling, to the side of a tank or vessel.

sedimentology *n*: the science dealing with the description, classification, and interpretation of sediments and sedimentary rock.

seed *n*: an undesirable particle or granule found in a paint or varnish.

seep *n*: the surface appearance of oil or gas that results naturally when a reservoir rock becomes exposed to the surface, thus allowing oil or gas to flow out of fissures in the rock.

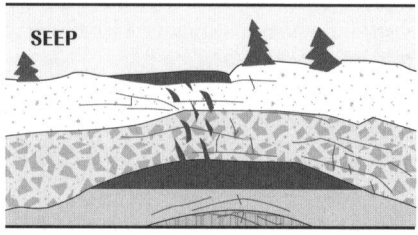

SEG *abbr*: Society of Exploration Geophysicists.

segregation drive *n*: see *gravity drainage*.

seis *abbr*: seismograph; used in drilling reports.

seismic *adj*: of or relating to an earthquake or earth vibration, including those artificially induced.

seismic check-shot survey *n*: see *well velocity survey*.

seismic data *n pl*: detailed information obtained from earth vibration produced naturally or artificially (as in geophysical prospecting).

seismic method *n*: a method of geophysical prospecting using the generation, reflection, refraction detection, and analysis of sound waves in the earth.

seismic option agreement *n*: an agreement that permits seismic exploration of land for a specified price per acre. The company gathering seismic information can, by the terms of the agreement, eventually lease selected acreage, again for an agreed-upon price.

seismic profile *n*: digital recording of the return to geophones of a shock wave emitted from the surface.

seismic sea wave *n*: see *tsunami*.

seismic section *n*: see *record section*.

seismic survey *n*: an exploration method in which strong low-frequency sound waves are generated on the surface or in the water to find subsurface rock structures that may contain hydrocarbons. The sound waves travel through the layers of the earth's crust; however, at formation boundaries some of the waves are reflected back to the surface where sensitive detectors pick them up. Reflections from shallow formations arrive at the surface sooner than reflections from deep formations, and since the reflections are recorded, a record of the depth and configuration of the various formations can be generated. Interpretation of the record can reveal possible hydrocarbon-bearing formations.

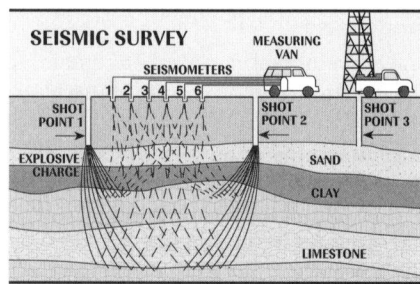

seismic wave *n*: shock wave generated by an earthquake or explosion.

seismogram *n*: the record of an earth tremor by a seismograph.

seismograph *n*: a device that detects vibrations in the earth. It is used in studying the earth's interior and in prospecting for probable oil-bearing structures. Vibrations are created by discharging explosives in shallow boreholes, by striking the surface with a heavy blow, or by vibrating a heavy plate in contact with the ground. The type and velocity of the vibrations as recorded by the seismograph indicate the general characteristics of the section of earth through which the vibrations pass.

seismology *n*: the study of earth vibrations.

seize *v*: to bind the end of a wire rope with fine wire or a metal band to prevent it from unraveling.

seizing *n*: a wrapping of wire around the end of a wire rope, the purpose of which is to prevent the strands of the wire rope from unwrapping.

seizing strand *n*: a small-diameter wire-rope strand, usually made up of seven wires. It is used to seize large-diameter wire rope. See *seizing*.

seizing wire *n*: a wire used to seize the end of wire rope. See *seizing*.

selective-set shear *n*: a feature on some downhole production tools that allows the operator to select the setting depth of the tools.

selective shear *v*: to set a downhole production tool by shearing shear screws or pins that the operator installed before the tool was run.

selective-tank remote gauge *n*: a single receiver used with a remote transmission system to permit the use of one or more tank transmitters so that selective liquid level readings can be obtained by switching from one tank to another. Compare *single-tank remote gauge*.

selective transmission *n*: once the drawworks receives power from either the compound or electric motors, the type of transmission that allows the driller to select how the power is distributed (torque/speed combinations) to various components of the drawworks. The drawworks on both mechanical-drive and electric-drive rigs have a selective transmission.

self-actuated sampler *n*: a sampling device that is operated by streamflow or stream pressure.

self-elevating drilling unit *n*: see *jackup drilling rig*.

self-elevating substructure *n*: a base on which the floor and mast of a drilling rig rests and which, after it is placed in the desired location, is raised into position as a single unit.

self-potential (SP) *n*: see *spontaneous potential*.

self-potential curve *n*: see *spontaneous potential curve*.

self-propelled *adj*: 1. able to move or travel using a built-in power source, rather than requiring an external power source. 2. of or relating to a rig's ability to move or travel with its own power source.

self-propelled unit *n*: see *carrier rig*.

selsyn *n*: a type of electric AC motor used to transmit motion and position.

SEM *abbr*: subsea electronic module.

semiautomatic welding *n*: a welding technique in which the arc is maintained in a continuous stream of gas between an electrode and the pipe being welded. The semiautomatic welding apparatus consists of a spool of coiled wire that provides filler metal for the weld, a pair of driving rollers that guide the wire to the weld, a welding gun, and a supply of shielding gas. Each type of semiautomatic welding is usually identified by the type of shielding gas used; in CO_2 wire welding, for example, carbon dioxide is the shielding gas.

semiclosed circuit *n*: a diving life-support system in which the gas is partially vented and the remainder is recycled, purified, and reoxygenated.

semiconductor *n*: a solid (such as germanium or silicon) whose electrical conductivity lies between that of a conductor and that of an insulator. Semiconductor materials are used in the manufacture of diode rectifiers, transistors, and integrated circuits.

semidiurnal tide *n*: a tide having two high water levels and two low water levels per day.

semiexpendable gun *n*: a perforating gun that consists of a metallic strip on which encapsulated shaped charges are mounted. After the gun is fired, the strip is retrieved. See *gun-perforate*.

semiliquid *n*: a substance that is neither liquid nor solid but that flows slowly as a liquid.

semipermeable *n*: the ability to let small molecules pass or flow through.

semisubmerged *n*: a state in which a specially designed floating drilling rig (a semisubmersible) floats just below the water's surface.

semisubmersible *n*: see *semisubmersible drilling rig*.

semisubmersible drilling rig *n*: a floating offshore drilling unit that has pontoons and columns that, when flooded, cause the unit to submerge to a predetermined depth. Living quarters, storage space, and so forth are assembled on the deck. Semisubmersible rigs are self-propelled or towed to a drilling site and anchored or dynamically positioned over the site, or both. In shallow water, some semisubmersibles can be ballasted to rest on the seabed. Semisubmersibles are more stable than drill ships and ship-shaped barges and are used extensively to drill wildcat

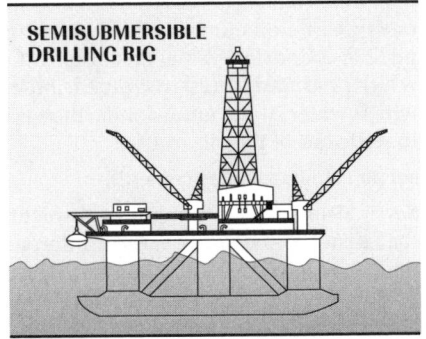

SEMISUBMERSIBLE DRILLING RIG

wells in rough waters such as the North Sea. Two types of semisubmersible rigs are the bottle-type and the column-stabilized. See *floating offshore drilling rig*.

semisubmersible lay vessel *n*: a type of pipe-laying vessel with a submerged pontoon hull and an elevated work area. Semisubmersible lay vessels remain relatively steady in high seas and are used in areas where conditions are expected to be continually rough.

senior emergency response official *n*: the person under HAZWOPER who responds to an emergency and is the individual in charge of the ICS. As such, he or she is responsible for coordinating and controlling all emergency responders and their communications.

senior orifice fitting *n*: a one-piece orifice fitting that allows the orifice plate in it to be changed without depressuring the meter run or shutting off the flow of gas.

sensing element *n*: the part of a measuring device that is responsive to the absolute value or change in a physical quantity, such as temperature, pressure flow rate, pH, light, or sound. It converts that change into an input signal for an information-gathering system.

sensitivity *n*: a measure of an engine governor's ability to sense a change in an engine's speed and make an adjustment to correct the speed change.

separate property *n*: the estate that is owned separately by one spouse before marriage.

separation *n*: the process of removing gas from liquid and liquid from gas.

separation sleeve *n*: a metal sleeve that shuts off tubing-to-annulus flow if a packer's sliding sleeve fails. See *sliding sleeve*.

separator *n*: a cylindrical or spherical vessel used to isolate the components in mixed streams of fluids. See *oil and gas separator*.

separator dump *n*: the discharge from a separator, such as oil or water.

separator gas *n*: gas remaining after being separated from condensate.

SEPLA *abbr*: suction embedment plate anchor.

sequence control *n*: automatic process by which a series of operations are initiated sequentially when certain requirements have been met.

sequence restart timer *n*: a time switch that starts motors in sequence after a power outage or shutdown. It prevents excessive voltage drop, which would occur if several motors were started simultaneously.

sequence stratigraphy *n*: a the sequence, or order, of formations. Geologists take information from seismic surveys and analyze it to deduce the environment that existed when a rock layer first formed.

sequential logic device *n*: a component or element that has at least one input channel, at least one output channel, and at least one internal state variable, so designed and constructed that the output signals depend on the past and present states of the inputs.

sequestering agent *n*: a chemical used with an acid in well treatment to inhibit the precipitation of insoluble iron hydroxides, which form when the acid contacts scales or iron salts and oxides, such as are found in corrosion products on casing.

sequestration *n*: the formation of stable calcium, magnesium, iron compounds by treating water or mud with phosphates.

SERC *abbr*: State Emergency Response Commission.

series *adj*: of or pertaining to an end-to-end arrangement of electrical components in a circuit that forms a single path for current to travel.

series circuit *n*: a circuit that has two or more electrical devices connected to form a single path for the electric current. The current is the same throughout the circuit; the total voltage is the sum of the voltages across each element of the circuit.

series winding *n*: a winding that is connected in series with another winding or device in the same circuit, as in a series motor. See *series circuit*.

serpentine *n*: an igneous rock composed in part of hydrated magnesium silicate. Hydrocarbons may be associated with serpentine, but not usually.

service company *n*: a company that provides a specialized service, such as a well-logging service or a directional drilling service.

service factor *n*: a percentage multiplier of rated horsepower at which a motor may be operated without exceeding the temperature rise limit.

service rig *n*: see *production rig*.

service well *n*: 1. a nonproducing well used for injecting liquid or gas into the reservoir for enhanced recovery. 2. a saltwater disposal well or a water supply well.

servitude estate *n*: the mineral estate regarded as subject to a specified use by one party, even though the surface is owned by another.

servo-mechanism *n*: a device consisting of a sensing element, a power device, and an amplifier that is used to automatically control a mechanical device. The power device is usually called a servomotor.

servomotor *n*: an electric motor or hydraulic piston that powers a servomechanism.

set *n*: an ocean current's direction of motion. *v*: to harden; for example, when cement sets, it hardens.

set back *v*: to place stands of drill pipe and drill collars in a vertical position to one side of the rotary table in the derrick or mast of a drilling or workover rig. Compare *lay down pipe*.

set casing *v*: to run and cement casing at a certain depth in the wellbore. Sometimes called set pipe.

set-down tool *n*: a packer that is set by applying weight from the tubing string.

set pipe *v*: see *set casing*.

set point *n*: 1. in drilling, the depth at which the bottom of the casing extends into the wellbore when it is ready to be cemented. 2. in process control, the desired value or output in a control system. It establishes the reference in a closed-loop control system.

setting depth *n*: the depth at which the bottom of the casing extends in the wellbore when it is ready to be cemented.

setting tool *n*: a tool used to set drillable or permanent tools, such as packers, retainers, or plugs; it can be mechanical, electric, or hydraulic.

settled production *n*: oil and gas production from longtime fields that produce at approximately the same rate every day.

settling *n*: the separation of substances because of different sizes and specific gravities of components in the substances.

settling depth *n*: the depth that sediment achieves when it settles out of a fluid.

settling pit *n*: a pit that is dug in the earth for the purpose of receiving mud returned from the well and allowing the solids in the mud to settle out. Steel mud tanks are more often used today, along with various auxiliary equipment for controlling solids quickly and efficiently.

settling tank *n*: 1. the steel mud tank in which solid material in mud is allowed to settle out by gravity. It is used only in special situations today, for solids control equipment has superseded such a tank in most cases. Sometimes called a settling pit. 2. a cylindrical vessel on a lease into which produced emulsion is piped and in which water in the emulsion is allowed to settle out of the oil.

set up *v*: to harden (as cement).

Seven Sisters *n pl*: a term, now obsolete, that denoted the seven largest international oil companies. The companies were Exxon, Royal Dutch Shell, British Petroleum, Mobil, Gulf, Texaco, and Chevron.

Mergers and name changes have reduced the number to Exxon Mobil, BP Amoco, Chevron-Texaco, and Shell. The future, no doubt, will herald additional changes.

severance *n*: the separation of a mineral or royalty interest from other interests in that land given by grant or reservation.

severance tax *n*: see *production tax*.

severed royalty interest *n*: nonexpense-bearing interest in minerals produced and saved from a tract owned by someone other than the surface owner. Owner of severed royalty interest gets a share of production from wells, but does not have to share the costs of production. The interest may be set up prior or subsequent to the leasing of the land, granted or reserved for years, for life, in fee simple defeasible, or in perpetuity.

sewage treatment plant *n*: a system on offshore locations used to render human and other wastes biologically inert before the wastes are discharged overboard.

S$_g$ *abbr*: gas saturation; used in reservoir engineering.

SG *abbr*: show of gas; used in drilling reports.

SGA *abbr*: Southern Gas Association.

sh *abbr*: shale; used in drilling reports.

shackle *n*: a U- or anchor-shaped fitting with pin. It is attached to one end of a wire rope and is used to attach the wire rope to other ropes or devices.

shake out *v*: to spin a sample of oil at high speed, usually in a centrifuge, to determine its S&W content.

shake-out machine *n*: see *centrifuge*.

shaker *n*: shortened form of shale shaker. See *shale shaker*.

shaker pit *n*: see *shaker tank*.

shaker tank *n*: the mud tank adjacent to the shale shaker, usually the first tank into which mud flows after returning from the hole. Also called a shaker pit.

shale *n*: a fine-grained sedimentary rock composed mostly of consolidated clay or mud. Shale is the most frequently occurring sedimentary rock.

shale oil *n*: see *oil shale*.

shale-out *n*: the termination of a shale bed.

shale shaker *n*: a vibrating screen used to remove cuttings from the circulating fluid in rotary drilling operations. The size

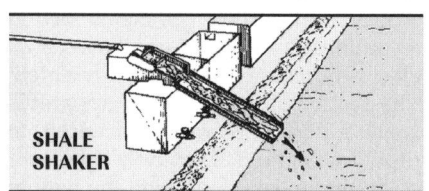

SHALE SHAKER

of the openings in the screen should be selected carefully to be the smallest size possible to allow 100 per cent flow of the fluid. Also called a shaker.

shale shaker screen *n*: a special wire mesh installed in a shale shaker that allows liquid mud to pass through but traps the cuttings and larger solid particles the mud carries from the hole. Often, more than one screen is installed in the shale shaker; also, the screens are usually vibrated to help remove the cuttings and solids on top of the screen.

shale slide *n*: a slanted trough that collects cuttings that fall onto the shaker and vibrate off.

shallow-cone profile *n*: the shape of a PDC bit's head that resembles two short and rounded cones placed side by side. The compacts are placed on the surface of the head. Compare *parabolic profile, short parabolic profile*.

shallow gas *n*: natural gas deposit located near enough to the surface that a conductor or surface hole will penetrate the gas-bearing formations. Shallow gas is potentially dangerous because, if encountered while drilling, the well usually cannot be shut in to control it. Instead, the flow of gas must be diverted. See *diverter*.

shaped charge *n*: a relatively small container of high explosive that is loaded into a perforating gun. On detonation, the charge releases a small, high-velocity stream of particles (a jet) that penetrates the casing, cement, and formation. See *perforating gun*.

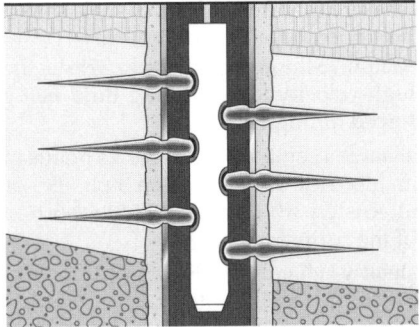

SHAPED CHARGES

shark's jaw *n*: a stoppering device for wire or chain used on the back deck of an AHTS vessel. Usually can be retracted into the deck when not in use.

shear *n*: action or stress that results from applied forces and that causes or tends to cause two adjoining portions of a substance or body to slide relative to each other in a direction parallel to their plane of contact.

shear bit *n*: see *fixed-head bit*.

shear-blind ram *n*: a blind-shear ram. See *shear ram*.

shear bond *n*: mechanical support of casing by cement in the borehole. Shear bond is determined by measuring the force required to initiate pipe movement and dividing that figure by the cement-casing contact surface area.

shear degradation *n*: the breakdown of molecules of a substance such as a polymer when subjected to shear stress. See *shear*.

shear modulus *n*: a coefficient of elasticity that expresses the ratio of the force per unit area that deforms the substance laterally and the shear produced by this force.

shearometer *n*: an instrument used to measure the shear strength, or gel strength, of a drilling fluid. See *gel strength*.

shear pin *n*: a pin that is inserted at a critical point in a mechanism and designed to shear when subjected to excess stress. The shear pin is usually easy to replace.

shear ram *n*: the component in a blowout preventer that cuts, or shears, through drill pipe and forms a seal against well pressure. Shear rams are used in floating offshore drilling operations to provide a quick method of moving the rig away from the hole when there is no time to trip the drill stem out of the hole.

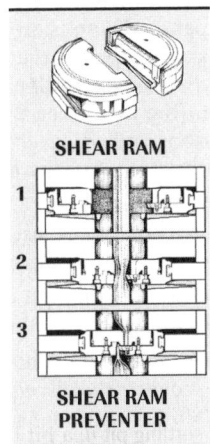

SHEAR RAM

SHEAR RAM PREVENTER

shear ram preventer *n*: a blowout preventer that uses shear rams as closing elements. See *shear ram*.

shear rate *n*: the rate of shear due to stress.

shear relief valve *n*: a type of pressure relief valve in which excess pressure causes a shearing action on a pin to relieve pressure. See *pressure-relief valve*.

shear strength *n*: 1. a unit that is used to measure soil strength (e.g., lbs/ft²). 2. see *gel strength*.

shear stress *n*: force applied to a liquid to cause it to flow.

shear thinning *n*: viscosity reduction of non-Newtonian fluids (e.g., polymers, most slurries and suspensions, and lube oils with viscosity-index improvers) under conditions of shear stress.

shear wave *n*: distortional, equivolumnar, secondary, or transverse wave.

sheave (pronounced "shiv") *n*: 1. a grooved pulley. 2. support wheel over which tape, wire, or cable rides.

sheave gauge *n*: a measuring instrument used to determine the size of sheaves on a block. It has sized metal fingers which are placed inside sheave grooves to calibrate the grooves.

sheave groove *n*: an individual groove in a sheave.

sheer *n*: the longitudinal curve of a vessel's deck in a vertical plane. As a result of sheer, a vessel's deck height above the baseline is higher (or lower) at the ends than amidships.

sheet ice *n*: a smooth, thin layer of ice on the water's surface.

shell *n*: 1. the body of a tank. 2. the horizontal tank on a tank car that contains the liquid being transported. 3. the steel backing of a precision insert bearing on a bit. 4. a hypothetical spherical surface centered on the nucleus of an atom that contains orbiting electrons.

shell full *adj*: used to describe a tank that is filled to its shell capacity.

shell height *n*: the distance between the bottom of the bottom angle of the tank and the top of the top angle.

Shepard's cane *n*: an earth-resistivity meter used to measure the resistance of soil to the passage of electrical current.

shepherd's hook *n*: see *chicken hook*.

shepherd's stick *n*: see *chicken hook*.

shielded-arc welding *n*: a welding technique in which the rod coating involves an inert gas shield that protects the weld from the rapid oxidation caused by contact with oxygen in the atmosphere. Shielded-arc welds are fine grained, free of oxides and nitrides, and have great ductility and toughness.

shielded conductor *n*: single or multiple conductors surrounded by a flexible metal shield for the purpose of preventing spurious signals from being carried.

shielding gas *n*: an inert gas that is used as a shielding medium in pipe welding. Its primary purpose is to prevent oxidation of the weld at the point of contact with the pipe metal by excluding oxygen in the air from the area around the molten metal.

shim *n*: a thin, often tapered piece of material used to fill in space between things (such as for support, leveling, or adjustment of fit).

ship drift *n*: a method that uses a ship itself as an instrument to measure currents. Navigation equipment on board measures the current's effect on the vessel.

shipper *n*: any company that transports oil or gas.

ship-shaped barge *n*: a floating offshore drilling structure that is towed to and from the drilling site. The unit has a streamlined bow and squared-off stern, a drilling derrick usually located near the middle of the barge, and a moon pool below the derrick through which drilling tools pass to the seafloor. It is identical in appearance to a drill ship, but is not self-propelled. It must therefore be towed to the drill site. Ship-shaped barges are most often used for drilling wells in deep, remote waters. See *floating offshore drilling rig.*

ship-shaped drilling rig *n*: drill ship.

shirttail *n*: the part of a drilling bit on which the cone is anchored. Shirttails extend below the threaded pin of the bit and are usually rounded on bottom, thus acquiring the name.

SHIRTTAIL

shock hose *n*: see *vibrator hose.*

shock loading *n*: the sudden application of force by a moving body on a resisting body so that a great deal of stress is produced in the resisting body.

Shock Sub *n*: a vibration dampener.

shoe *n*: a device placed at the end of or beneath an object for various purposes (e.g., casing shoe, guide shoe).

shoestring sand *n*: a narrow, often sinuous, sand deposit, usually a buried sandbar or filled channel.

shoofly *n*: a special access road constructed to link a right-of-way with existing roads. Shooflies are necessary only in remote areas.

shoot *v*: 1. to explode nitroglycerine or other high explosives in a hole to shatter the rock and increase the flow of oil; now largely replaced by formation fracturing. 2. in seismographic work, to discharge explosives to create vibrations in the earth's crust. See *seismograph.*

shooting rock *n*: the process of using explosives to clear rock from a pipeline right-of-way or from the ditch line. Also called blasting.

shoreface *n*: that part of the seashore seaward of the low-tide mark that is affected by wave action.

Shore Protection Act *n*: a congressional act that requires ships transporting garbage and refuse to assure that the garbage and refuse are properly contained on board so that it will not be lost in the water as a result of inclement wind or water conditions.

short *n*: see *short circuit.*

short circuit *n*: a phenomenon that occurs in electrical circuits wherein two components carrying current inadvertently come into contact and interrupt the current path. For example, if two bare wires in a circuit accidentally touch, current flows directly back to the source of electricity without traveling its normal path. Often, because so little resistance exists, current flow is high enough to melt wires and other components.

short-coupled installation *n*: a full-sized ell or tee upstream and downstream of a meter.

short full-pitch crossover *n*: angled grooves on a drum.

short normal curve *n*: in conventional electric logging, measures resistivity in the formation only a short distance from the borehole. Used for shallow investigation of the formation.

short parabolic profile *n*: the shape of a PDC bit's head in which the head resembles a short parabolic arc. A parabolic arc is an arch-shaped curve that looks similar to the arch over a doorway or other structure. Compare *parabolic profile, shallow-cone profile.*

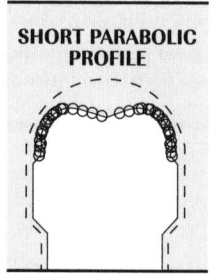

SHORT PARABOLIC PROFILE

short-range forecast *n*: a weather forecast covering 48 hours or less.

short string *n*: in a dual well, the tubing string for the shallower zone.

short ton *n*: a unit of measure of mass (weight) that is equal to 2,000 pounds (907.2 kilograms). Compare *long ton.*

shortwave energy *n*: see *shortwave radiation.*

shortwave radiation *n*: radiation that is visible as light.

short way *n*: the displacing of wellbore fluids from the annulus up the tubing. Compare *long way.*

shot *n*: 1. a charge of high explosive, usually nitroglycerine, detonated in a well to shatter the formation and expedite the recovery of oil. Shooting has been almost completely replaced by formation fracturing and acid treatments. 2. a point at which a photograph is made in a single-shot survey. See *directional survey.*

shot blasting *n*: the use of steel shot (ball bearings) in the drilling mud to hit the bottom of the hole as the shot leaves the bit jet with the mud. Experiments have shown that the rate of penetration is improved in some circumstances, but difficulties with handling the shot in the mud system have precluded wide use.

shot-hole drilling *n*: the drilling of relatively small holes into the earth, the purpose of which is to provide a means for lowering an explosive shot in order to create shock waves for seismic analysis.

shoulder *n*: 1. the flat portion machined on the base of the bit shank that meets the shoulder of the drill collar and serves to form a pressure-tight seal between the bit and the drill collar. 2. the flat portion of the box end or the pin end of a tool joint; the two shoulders meet when the tool joint is connected and form a pressure-tight seal.

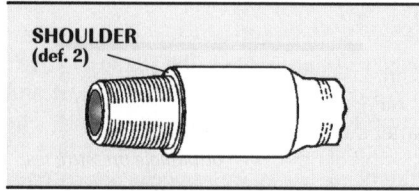

SHOULDER (def. 2)

show *n*: the appearance of oil or gas in drilling fluids, cuttings, samples, or cores from a drilling well.

shrinkage *n*: 1. a decrease in oil volume caused by evaporation of solution gas or by lowered temperature. 2. the reduction in volume or heating value of a gas stream due to removal of some of its constituents. 3. the unaccounted loss of products from storage tanks.

shrink-on tool joint *n*: a tool joint made to fit the pipe by the process of shrinking on, that is, by heating the outer member to expand the bore for easy assembly and then cooling it so that it contracts around the inner member.

shrouded jet nozzle *n*: a special type of jet nozzle that is manufactured with a projection (the shroud), which serves to minimize the erosion of the nozzle by the high-velocity jet of drilling fluid being forced through it.

shunt *n*: a conductor joining two points in an electrical circuit to form a parallel or alternate path through which a portion of the current may pass.

shunt winding *n*: a winding that is connected in parallel with another winding or device in the same circuit, as in a shunt-wound DC motor. See *parallel circuit.*

shut down *v*: to stop work temporarily or to stop a machine or operation.

shutdown *n*: the act of stopping a machine or device from running. For example, engine operators perform a shutdown when they stop an engine.

shutdown rate *n*: a rate provision that is usually contained in a drilling contract and that specifies the compensation to the independent drilling contractor when drilling is suspended at the request of the operator.

shutdown system *n*: on an engine, a series of devices that sense an abnormality in the engine's operation and automatically shut off the fuel and air supply to the engine to make it stop running.

shut in *v*: 1. to close the valves on a well so that it stops producing. 2. to close in a well in which a kick has occurred.

shut-in *adj*: shut off to prevent flow. Said of a well, plant, pump, and so forth, when valves are closed at both inlet and outlet.

shut-in bottomhole pressure (SIBHP) *n*: the pressure at the bottom of a well when the surface valves on the well are completely closed. It is caused by formation fluids at the bottom of the well.

shut-in bottomhole pressure test *n*: a bottomhole pressure test that measures pressure after the well has been shut in for a specified period of time. See *bottomhole pressure test.*

shut-in casing pressure (SICP) *n*: pressure of the annular fluid on the casing at the surface when a well is shut in.

shut-in drill pipe pressure (SIDPP) *n*: pressure of the drilling fluid on the inside of the drill stem. It is used to measure the difference between hydrostatic pressure and formation pressure when a well is shut in after a kick and the mud pump is off and to calculate the required mud-weight increase to kill the well.

shut-in gas *n*: potential natural gas production curtailed because of state conservation orders (prorationing), unfavorable economics, lack of buyers at existing prices, failure of committed buyers to take gas, or other reasons.

shut-in period *n*: an interval during which a well is shut in to allow pressure buildup while pressure behavior is being measured during a formation test.

shut-in pressure (SIP) *n*: the pressure when the well is completely shut in, as noted on a gauge installed on the surface control valves. When drilling is in progress, shut-in pressure should be zero, because the pressure exerted by the drilling fluid should be equal to or greater than the pressure exerted by the formations through which the wellbore passes. On a flowing, producing well, however, shut-in pressure should be above zero.

shut-in royalty *n*: payment to royalty owners in lieu of production, rentals, or other consideration on a shut-in gas well that is capable of producing but does not have a market.

shut-in royalty clause *n*: clause in a lease specifying the payments that must be made on a gas well capable of producing but shut in for lack of a market or pending connection with a pipeline.

shut-in well *n*: usually, a gas well shut in for lack of a market or pending connection with a pipeline.

shut off *v*: to stop or decrease the production of water in an oilwell by cementing or mudding off the water-producing interval.

shutoff valve *n*: a device installed in a line, which, when closed, stops flow in the line and, when open, allows flow.

shuttle valve *n*: a flow control device (a valve) with two inlet ports and one outlet port in an accumulator, which is employed in a subsea blowout preventer system. Hoses from the lower marine riser package's yellow and blue control pods are connected to the shuttle valve's inlet port. The outlet port is connected directly to the blowout preventer's inlet port.

shuttle vessel *n*: a tank ship capable of navigating shallow ports, usually 30,000–70,000 dead weight tons (27,216–63,504 dead weight tonnes). It shuttles between anchored large vessels and port.

Si *sym*: silicon.

SI *abbr*: 1. shut in; used in drilling reports. 2. International System of Units, or metric system.

SIBHP *abbr*: shut-in bottomhole pressure; used in drilling reports.

SIC *abbr*: standard industrial classification.

SICP *abbr*: shut-in casing pressure.

side-door elevators *n pl*: elevators arranged so that they are latched and unlatched from the side instead of in the middle. Compare *center-latch elevators.*

side-door mandrel *n*: see *gas-lift mandrel.*

side entry *n*: an opening in the wall of the riser adapter to which a mud booster line is attached. See *mud booster line, riser adapter.*

side-looking airborne radar (SLAR) *n*: imaging radar used in airplanes.

side outlet *n*: an opening usually located in the blowout preventer stack below a ram preventer, to which a choke line or a kill line is attached. Typically, outlets are provided in pairs, one on each side of the bore. Usually, outlets are provided for a choke line and a kill line.

side pocket *n*: an offset heavy-wall sub in the production string for placing gas-lift valves, and so on.

side-pocket mandrel *n*: see *gas-lift mandrel.*

side rake angle *n*: in a PDC bit, the left-to-right orientation of the cutter with respect to the bit's face; usually, the cutters are angled

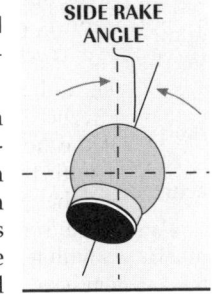
SIDE RAKE ANGLE

toward the outside of the bit to help direct cuttings to the annulus.

side tap *n*: the joining of another pipe to the main body of a pipeline.

sidetrack *v*: to use a whipstock, turbo-drill, or other mud motor to drill around broken drill pipe or casing that has become lodged permanently in the hole.

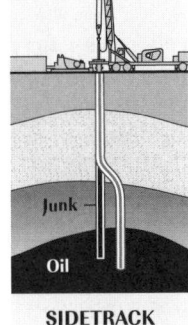
SIDETRACK

side tracking bit *n*: a bit especially designed for use in drilling a hole around an obstruction in the main borehole. Usually, a cement plug is set on top of the obstruction, the side tracking bit is run into the hole on a whipstock or other directional drilling device, and the new borehole begun. Because the new borehole goes off to the side of the plugged borehole, the operation is called side tracking.

sidewall coring *n*: a coring technique in which core samples are obtained from the hole wall in a zone that has already been drilled. A hollow bullet is fired into the formation wall to capture the core and then retrieved on a flexible steel cable. Core samples of this type usually range from ¾ to 1³⁄₁₆ inches (20 to 30 millimetres) in diameter and from ¾ to 4 inches (20 to 100 millimetres) in length. This method is especially useful in soft-rock areas.

sidewall epithermal neutron log *n*: a device in which a neutron source and a detector are in a pad using the same hardware as the density log. The pad is pushed against the side of the borehole wall. The log measures porosity.

sidewall neutron porosity logging tool *n*: a logging device that is held firmly against the wall of the hole and that measures the porosity of a formation.

sidewall sampler *n*: used to obtain a core sample. The sampler is activated to fire several small cylinders into the wall of the hole. The cylinders penetrate a short distance into the rock and cut a small core.

sidewinder unit *n*: a type of wireline unit in which power is transferred by means of a pulley and a belt from the left rear wheel of the wireline truck to the spool of wire that is pulled from the left side of the truck. The spool is used to run the wireline downhole and bring it back up again.

SIDPP *abbr*: shut-in drill pipe pressure; used in drilling reports.

sieve analysis *n*: the determination of the percentage of particles that pass through several screens of graduated fineness.

sieve tray *n*: a tray, installed in an absorber tower or fractionating column, that is similar to a bubble-cap tray, except that it has only holes but no bubble caps through it. This type of tray is more efficient than the bubble-cap tray or the valve tray and less expensive than either; however, it does not operate properly over a wide range of flow rates.

sight draft *n*: a draft or order for payment that must be picked up on the day that it arrives at the drawer's bank. For example, if a lessee pays a lessor a cash bonus by means of a sight draft, the lessee must pick up that draft on the same day that it arrives at his or her bank from the lessor's bank.

sight glass *n*: a glass tube or a glass-faced section on a process line or vessel; used for visual reading of liquid levels or of manometer pressures.

signal *n*: information about a variable that can be transmitted.

signal generator *n*: an electronic test instrument that delivers a sinusoidal output at an accurately calibrated frequency ranging anywhere from audible to microwave frequencies. The frequency and amplitude of the device's output are adjustable over a wide range. The output may be frequency- or amplitude-modulated.

signal integrator *n*: an operational amplifier that senses electronic signals in a circuit and electronically arranges the signals so that they function together in an efficient and logical way. See *operational amplifier*.

signal-to-noise ratio *n*: the ratio of the magnitude of the electrical signal to that of the electrical noise.

significant harm *n*: a classification of larger spills released into navigable waters that can cause substantial environmental damage.

significant wave *n*: a statistical wave that has a height equal to the average of the highest one-third of the waves in a particular area.

silica *n*: a mineral that has the chemical formula SiO_2 (silicon dioxide). It is relatively hard and insoluble. Quartz is a form of silica, but usually contains impurities that give it color.

silica flour *n*: a silica (SiO_2) ground to a fineness equal to that of portland cement.

silica gel *n*: highly adsorbent, gelatinous form of silica used chiefly as a dehumidifying and dehydrating agent.

silicate *n*: crystalline compound composed largely of silicon in chemical combination with aluminum, magnesium, oxygen, and other common elements.

siliceous *adj*: containing abundant silica, or silicon dioxide (SiO_2).

silicon (Si) *n*: a nonmetallic element occurring extensively in the earth's crust in the form of silicon dioxide, or sand. Silicon usually occurs as dark-brown crystals. It is used in glass, steel, semiconducting devices, concrete, brick, and other materials.

silicon-controlled rectifier (SCR) *n*: a semiconductor device (a rectifier) that changes alternating current to direct current by means of a silicon control gate. Normally, an SCR acts as an open circuit, but rapidly switches to a conducting state when an appropriate gate signal is applied to the gate terminal. An SCR conducts current in only one direction. Commonly called an SCR or, erroneously, a thyristor. While one type of thyristor is an SCR, another type exists, which is called a triac. See *triac*.

silicon steel *n*: an iron alloy that contains carbon (a steel) and 0.5 to 4.5 percent silicon. Silicon steel is often used in electric transformer coils. The silicon allows the steel to be drawn into thin, flat plates that are ideal for transformer construction. See *silicon*.

silicon tetrafluoride *n*: a gas that can be readily absorbed by water and that is used to seal off water-bearing formations in air drilling.

silt *n*: a material that exhibits little or no swelling and whose particle size ranges from 2 to 74 microns. Dispersed clays and barite fall into this particle-size range.

siltstone *n*: a fine-grained, shalelike sedimentary rock composed mostly of particles 1/16 to 1/256 millimetre (1.6 to .10 inch) in diameter.

simple circuit *n*: an electrical circuit that consists of a single path from one terminal of the source of electromotive force, through one electrical device, and back to the terminal of the source of electromotive force.

simplex fitting *n*: modified version of flanged fitting. Flange bolts are replaced with a single chamber that holds the plate.

simultaneous filing *n*: a noncompetitive procedure—basically a drawing—used to grant oil and gas leases on federal lands that have not been leased before.

sin *abbr*: sine.

sine (sin) *n*: a trigonometric function equal to the ratio of the side opposite a given acute angle to the hypotenuse.

sine wave *n*: an undulating form (a wave) whose amplitude varies as the sine of a linear function of time. See *sine*.

single *n*: a joint of drill pipe. Compare *double, fourble, thribble*.

single-acting *adj*: the action of a mud pump piston cylinder moving mud only on its forward stroke. Compare *double-acting*.

single anchor leg mooring (SALM) *n*: a type of offshore mooring in which the facility is anchored by a single gravity anchor or by piled anchors; usually does not include a storage facility.

single-capacity system *n*: see *capacity* (def. 3).

single-chamber fitting *n*: in an orifice meter installation, a mechanism for holding the orifice plate in the pipeline. In a single-chamber fitting, the gas flow must be stopped and the meter tube depressured before the orifice plate can be removed. Compare *dual-chamber fitting*.

single-chamber fitting with gearing *n*: in an orifice meter installation, a type of single-chamber fitting that has a lever and gears that can be used to roll the orifice plate up and out of the fitting.

single-grip *adj*: said of packers with only one set of slips for supporting weight and pressure from above.

single-layer strand *n*: a wire rope strand that is made up of one layer of wires laid around one center wire.

single-part sling *n*: a sling that consists of one wire rope in the sling body.

single-phase flow *n*: the flow of only one material—a gas or a liquid—in a pipeline.

single-phase injection well *n*: an injection well in which only one fluid is being injected.

single-phase motor *n*: a motor energized by a single source of alternating voltage.

single-phase voltage *n*: voltage produced by an alternating current circuit that has only two points of entry, or one which, having more than two points of entry, is intended to be so energized that the potential differences between all pairs of points of entry are either in phase or differ by 180 degrees.

single-plane roller assembly *n*: in a kelly bushing, a single set of roller assemblies that are installed in the bushing at the same level, or plane. Compare *double-plane roller assembly*. See also *kelly bushing, roller assembly*.

single-point buoy mooring system (SPBM) *n*: an offshore system to which the production from several wells located on the seafloor is routed and to which a tanker ship ties up to load the produced oil.

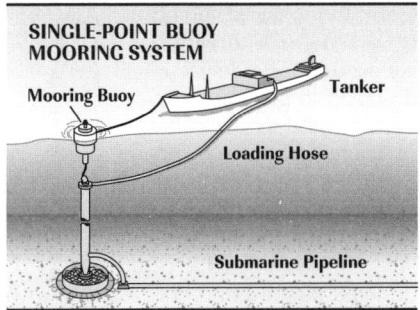

SINGLE-POINT BUOY MOORING SYSTEM

The tanker is moored to a single point on the buoy and is thus free to rotate around it, depending on wind and current directions.

single-pole, double-throw (SPDT) switch *n*: an electrical switch with three terminals that connects one terminal to one of the other two terminals.

single-pole rig *n*: a well-servicing unit whose mast consists of but one steel tube, usually about 65 feet (19.8 metres) long.

single-pole, single-throw (SPST) switch *n*: an electrical switch with two terminals that is arranged so that the switch opens or closes one circuit.

single-shot survey *n*: a directional survey that provides a single record of the drift direction and off-vertical orientation of the hole. See *directional survey*.

single sideband shortwave receiver *n*: a special radio (a receiver) that receives broadcasts from a radio transmitter that sends its signals on a frequency band located either above or below the carrier frequency, within which fall the frequency components of the wave produced by the process of modulation.

single-stage cementing *n*: a common cementing procedure; consists of pumping a calculated volume of slurry into casing after pipe has been landed at the desired depth, and displacing the slurry around the shoe and into the annulus in a circulating mode with another fluid, such as water, mud, or completion fluid.

single-string dual completion *n*: 1. a method of well completion in which two pay zones are produced by directing fluids from one zone through a tubing string and from the other zone through the annulus. 2. the tool assembly used to complete a well by this method.

single-tank composite sample *n*: a blend of liquid samples taken from the upper, middle, and lower sections of a tank. For a tank of uniform cross section, such as an upright cylindrical tank, the blend consists of equal parts of the three samples. For a horizontal cylindrical tank, the blend consists of the three samples in the proportions shown in table 2, API MPMS, ch. 8.1.

single-tank remote gauge *n*: a remote transmission system for determining liquid level in tanks that require a separate receiver for each tank transmitter. Compare *selective-tank remote gauge*.

sinker bar *n*: a heavy weight or bar placed on or near a lightweight wireline tool. The bar provides weight so that the tool will lower properly into the well.

sinking fund *n*: a fund to retire a debt, usually a bond issue.

sinter *v*: to bond metallic powder into a mass by heating it. Tungsten carbide inserts are often bonded to the cones of button bits by sintering when the bits are being manufactured.

sinusoidal *adj*: having a magnitude (size) that varies as the sine of an independent current. See *sine*.

SIP *abbr*: shut-in pressure; used in drilling reports.

siphon *n*: an inverted U-shaped tube or pipe through which a liquid flows from a high level to a lower level at atmospheric pressure. For a siphon to work, it must be filled with liquid, thus reducing pressure inside the tube and allowing atmospheric pressure to force liquid to the lower level.

siphon barometer *n*: an atmospheric pressure measuring device (a barometer) that is made up of a U-shaped glass tube containing mercury. One end of the tube is open, and the other end is closed. High atmospheric pressure pushes against mercury at the open end of the tube, and the mercury rises toward the sealed end of the tube. Low pressure causes the mercury to drop.

SI system *n*: see *International System of Units*.

SIT camera *n*: silicon-intensified light camera used by divers for underwater photography.

six- (6-) strand wire rope *n*: wire rope constructed with six (6) strands.

skid *n*: a low platform mounted on the bottom of equipment for ease of moving, hauling, or storing.

skid off *n*: a jackup drilling rig that can slide its drill floor structure off the jackup barge and onto another structure.

skid shoe *n*: a hard metal pad mounted on a borehole contact logging tool that contacts the wellbore and prevents wear to the rest of the tool.

skid the rig *v*: to move a rig with a standard derrick from the location of a lost or completed hole preparatory to starting a new hole. Skidding the rig allows the move to be accomplished with little or no dismantling of equipment.

skimmer *n*: a special tank into which water with small quantities of oil is piped. When the oil rises to the top, blades on a skimmer direct the oil into a discharge line.

skimmer tank *n*: a piece of equipment in a glycol dehydration unit that performs the same function as a separator; however, it cannot handle as much gas and condensate as a separate unit. See *separator*.

skim pit *n*: an earthen pit, often lined with concrete or other material, into which water with small amounts of oil is pumped. The minute quantities of oil are skimmed off the top of the water in the pit, and the water is disposed of.

skin *n*: 1. the area of the formation that is damaged because of the invasion of foreign substances into the exposed section of the formation adjacent to the wellbore during drilling and completion. 2. the pressure drop from the outer limits of drainage to the wellbore caused by the relatively thin veneer (or skin) of the affected formation. Skin is expressed in dimensionless units: a positive value denotes formation damage; a negative value indicates improvement. Also called skin effect.

skin damage *n*: see *formation damage*.

skin effect *n*: see *skin*.

sky-top mast *n*: a mast on a well servicing unit that utilizes a split traveling block and crown block, which makes it possible to pull 60-foot (18.3-metre) stands with a 50-foot (15.2-metre) mast.

slack looping *n*: in pipeline construction, the process of putting slack in a pipeline to counter the effects of contraction and expansion caused by extreme variations in daily temperature. Instead of laying the pipe in a straight line down the center of a ditch, it is laid so that it alternately touches one side of the ditch or the other, thereby providing a curve or slack in the line.

slack off *v*: to lower a load or ease up on a line. A driller will slack off on the brake to put additional weight on the bit.

slag *n*: cinder.

slaked lime *n*: see *hydrated lime*.

slant-hole rig *n*: a drilling rig used to drill directional wells. See *directional drilling*.

slate *n*: a metamorphic rock formed when shale becomes buried deeply. The heat and pressure fuse individual mineral grains into slate.

S-lay *n*: a method of installing an offshore pipeline in which lengths of pipe are welded together horizontally and launched over the stern of the pipe-lay vessel over a long stinger while the vessel moves slowly forward. S-laying is typically employed in shallow to medium-depth waters. It avoids subjecting the pipeline to excessive stresses. See *stinger*. Compare *J-lay*.

sleeve *n*: a tubular part designed to fit over another part.

sleeve valve *n*: a valve in the bottom of a retainer.

slick *n*: oil that floats on the water's surface.

slick boring *n*: in pipeline construction, a boring technique sometimes used for road crossings in which a large amount of liquid is pumped into the hole outside of the pipe to reduce friction.

slick drill collar *n*: a drill collar whose outer wall is smooth—that is, it does not have a spiral groove machined into it.

slick line *n*: see *wireline*.

slick riser joint *n*: in drilling from deep water floating offshore drilling rigs, a riser joint that does not have buoyant riser modules added to it. See *buoyant riser joint, riser pipe*.

slide-and-guide *n*: a special saddle, or cradle, that holds pipe on a vertical support member in the lowering-up type of pipe laying. The slide-and-guide allows longitudinal movement of the pipe caused by thermal expansion and contraction. See *lowering-up*.

sliding scale royalty *n*: a royalty paid when the percentage varies with the volume of production.

sliding sleeve *n*: 1. a special device placed in a string of tubing that can be operated by a wireline tool to open or close orifices to permit circulation between the tubing and the annulus. It may also be used to open or shut off production from various intervals in a well. Also called circulation sleeve, sliding-sleeve nipple. 2. a device in a packer that prevents fluid from flowing from the tubing into the annulus.

sliding-sleeve nipple *n*: see *sliding sleeve*.

slim-hole drilling *n*: drilling in which the size of the hole is smaller than the conventional hole diameter for a given depth. This decrease in hole size enables the operator to run smaller casing, thereby lessening the cost of completion. See *miniaturized completion*.

slim-line integral design *n*: uses a hydraulic press to cold-form both box and pin, to expand the box and reduce the pin diameters.

sling *n*: 1. in pipeline operations, a wide rubber and fabric apron-like device on the end of the boom cat's hoisting lines that is used for lowering in or handling coated and wrapped pipe. 2. in crane operations, a braided wire-rope device used to attach a load to the crane's hook. Often, several slings are used. 3. a looped line used to carry objects.

slinger disk lubrication *n*: in a chain-and-sprocket drive, a type of oil bath lubrication in which the chain itself does not pass through the sump, but one or two disks rotating with the sprocket pick up oil from the sump and sling it against a plate, which then feeds it to the top of the lower span. Compare *drip lubrication, pressure lubrication*.

sling psychrometer *n*: a hygrometer, or psychrometer, attached to a handle and chain cord, which is whirled rapidly through the air. It also has two thermometers. The bulb of one is kept wet so that the cooling that results from evaporation as the thermometers are whirled makes it register a lower temperature than the dry one. The difference between the dry and wet readings is a measure of the relative humidity of the atmosphere. See *psychrometer*.

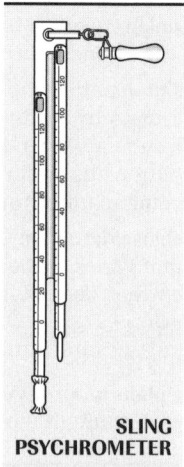

SLING PSYCHROMETER

slingshot substructure *n*: see *self-elevating substructure*.

slip *n*: in an electric motor, the difference between the optimal and the actual output of the motor. *v*: to change the position of the drilling line periodically so that it wears evenly as it is used.

slip-and-cutoff program *n*: a procedure to ensure that the drilling line wears evenly throughout its life. After a specified number of ton-miles (megajoules) of use, the line is slipped—i.e., the traveling block is suspended in the derrick or propped on the rig floor so that it cannot move, the deadline anchor bolts are loosened, and the drilling line is spooled onto the drawworks drum. Enough line is slipped to change the major points of wear on the line, such as where it passes through the sheaves. To prevent excess line from accumulating on the drawworks drum, the worn line is cut off and discarded.

slip bowl *n*: a device in a rotary table or other tool into which tubing or drill pipe slips can be inserted. See also *slips*.

slip elevator *n*: a casing elevator containing segmented slips with gripping teeth inside. Slip elevators are recommended for long strings of casing, because the teeth grip the casing and help prevent casing damage from the weight of long, heavy strings hanging from the elevators. Slip elevators may also be used as slips.

slip-grub screw connector *n*: in coiled tubing operations, a device that attaches the tool

string to the coiled tubing by means of a threaded connection. The device is designed to accommodate torsional and tensile loads. It is available in sizes to match the outside diameter of the coiled tubing string; thus, it is often used in cases where it is not possible or desirable for the connecting device to be larger in diameter than the tubing.

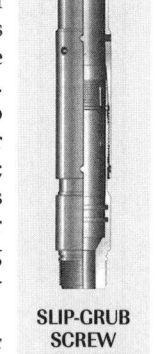

SLIP-GRUB SCREW CONNECTOR

slip joint *n*: see *telescopic joint*.

slippage *n*: liquid that slips between the blade or lobed impeller of a displacement meter and the meter housing instead of being captured between the two blades or impellers and being measured.

slipping and cutoff program *n*: procedure for a given rig in which drilling line is slipped through the system at such a rate that it is evenly worn, then cut off at the drum end just as it reaches the end of its useful life.

slip plane *n*: closely spaced surfaces along which differential movement takes place in rock. Analogous to surfaces between playing cards.

slip ring *n*: a conducting ring that gives current to or receives current from the brushes in a generator or a motor. See *brush*.

slips *n pl*: wedge-shaped pieces of metal with serrated inserts (dies) or other gripping elements, such as serrated buttons, that suspend the drill pipe or drill collars in the master bushing of the rotary table when it is necessary to disconnect the drill stem from the kelly or from the top-drive unit's drive shaft. Rotary slips fit around the drill pipe and wedge against the master bushing to support the pipe. Drill collar slips fit around a drill collar and wedge against the master bushing to support the drill collar. Power slips are pneumatically or hydraulically actuated devices that allow the crew to dispense with the manual handling of slips when making a connection.

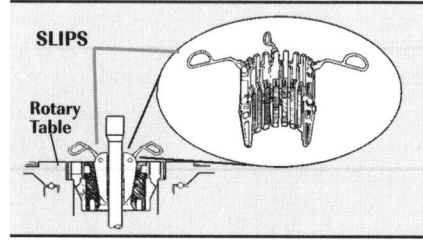

SLIPS

Rotary Table

slip segment *n*: a single part of all the parts, or segments, that make up the slips. See *slips*.

slip speed *n*: see *slip*.

slip stick *n*: a phenomenon that occurs as the drill string is rotated and contacts the wall of the hole. It creates torque and rotational speed oscillations along the entire length of the drill string.

slip test *n*: a procedure to determine whether the slips' inserts (dies) are uniformly contacting the wall of the drill pipe. Paper is wrapped around the pipe and the slips are set. The paper is then unwrapped from the pipe and the pattern the slips made on the paper is examined. If uniform contact is not shown, the slips' inserts are replaced or the insert bowl is repaired or replaced.

slip velocity *n*: 1. the rate at which drilled solids tend to settle in the borehole as a well is being drilled. 2. difference between the annular velocity of the fluid and the rate at which a cutting is removed from the hole.

slop *n*: a term rather loosely used to denote mixtures of oil produced at various places in a plant and requiring rerun or other processing to make it suitable for use. Also called slop oil.

slop oil *n*: see *slop*.

slops *n pl*: oil that has been washed from the tanks of a vessel and is pumped to a special tank where most of the water will be permitted to separate for decanting.

slotted liner *n*: a relatively short length of pipe with holes or slots that is placed opposite a producing formation. Usually, it is wrapped with specially shaped wire that is designed to prevent the entry of loose sand into the well as it is produced. It is also often used with a gravel pack.

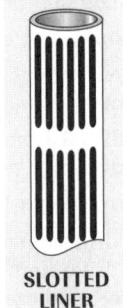

SLOTTED LINER

slough *v*: see *caving*.

sloughing (pronounced "sluffing") *n*: see *caving*.

sloughing hole *n*: a condition wherein shale that has absorbed water from the drilling fluid expands, sloughs off, and falls downhole. A sloughing hole can jam the drill string and block circulation.

slow-release inhibitor *n*: corrosion-preventive substance that is released into production fluids at a slow rate.

slow-set cement *n*: a manufactured cement in which the thickening time is extended by the use of a coarser grind, the elimination of the rapid hydrating components in its composition, and the addition of a chemical retarder. API classes N, D, E, and F are slow-set cements.

sludge *n*: 1. a tarlike substance that is formed when oil oxidizes. 2. a highly viscous mixture of oil, water, sediment, and residue.

slug *n*: a quantity of fluid injected into a reservoir to accomplish a specific purpose, such as chemical displacement of oil.

slugging *n*: the accumulation of a liquid (water, oil, or condensate) in a gas line.

slugging compound *n*: a special chemical demulsifier that is often added to emulsion samples to determine the total amount of sediment and water in the samples. Also called knockout drops.

slug tank *n*: a relatively small separate tank or a small part of a larger tank that holds a small amount of mud for a special purpose. For example, it may hold a small quantity of heavy mud that will be used to slug the drill string. That is, the slug will be pumped into the string to keep mud from falling onto the rig floor when a drill pipe joint is broken out during a trip.

slug the pipe *v*: to pump a quantity of heavy mud into the drill pipe. Before hoisting drill pipe, it is desirable (if possible) to pump into its top section a quantity of heavy mud (a slug) that causes the level of the fluid to remain below the rig floor so that the crew members and the rig floor are not contaminated with the fluid when stands are broken out.

slurry *n*: 1. in drilling, a plastic mixture of cement and water that is pumped into a well to harden. There it supports the casing and provides a seal in the wellbore to prevent migration of underground fluids. 2. a mixture in which solids are suspended in a liquid.

slurry viscosity *n*: the consistency of a slurry, measured in poise.

slurry volume *n*: the sum of the absolute volumes of solids and liquids that constitute a slurry.

slurry weight *n*: the density of a cement slurry, expressed in pounds per gallon, pounds per cubic foot, kilograms per litre, and so on.

slurry yield *n*: the volume of slurry obtained when one sack of cement is mixed with the desired amount of water and additives (such as with accelerators and fluid-loss control agents).

slush pit *n*: the old term for a mud pit. See *mud pit*.

slush pump *n*: see *mud pump*.

slush tank *n*: see *mud tank*.

small angle *n*: the tilt of the ship or rig that is at an angle of 10 degrees or less from the vertical.

small tank *n*: a crude oil storage tank with a 1,000-barrel (159,000-litre) or less capacity.

small-volume prover *n*: a prover with a volume between detectors that does

not permit a minimum accumulation of 1,000 direct (unaltered) pulses from the meter. Small-volume provers require meter pulse discrimination by pulse interpolation counter or other techniques to increase the resolution.

smart mass flow transmitter *n*: an electronic fluid measurement device that contains a microprocessor and a memory, and which produces an analog signal that is proportional to the mass rate of flow of a fluid.

smart pig *n*: see *instrumented pig*.

smart transmitter *n*: a process-control instrument that allows its calibration, ranging, or other data to be changed by using microprocessors and memory devices. When used in conjunction with a programming terminal or interface device, a smart transmitter can have its characteristics modified through software communication.

SMGC data *abbr*: surface marine gridded climatology data.

snake *n*: see *swivel-connector grip*.

snake grip *n*: see *swivel-type stringing grip*.

snap-on meter *n*: see *clamp-on meter*.

snatch block *n*: 1. a block that can be opened to receive wire rope or wireline. 2. a block that is suited for a single sheave and is used for pulling horizontally on an A-frame mast.

SNG *abbr*: synthetic or substitute natural gas.

sniffer *n*: see *explosimeter*.

snipe *n*: see *cheater*.

snub *v*: 1. to force pipe or tools into a high-pressure well that has not been killed (i.e., to run pipe or tools into the well against pressure when the weight of pipe is not great enough to force the pipe through the BOPs). Snubbing usually requires an array of wireline blocks and wire rope that forces the pipe or tools into the well through a stripper head or blowout preventer until the weight of the string is sufficient to overcome the lifting effect of the well pressure on the pipe in the preventer. In workover operations, snubbing is usually accomplished by using hydraulic power to force the pipe through the stripping head or blowout preventer. 2. to tie up short with a line.

snubber *n*: 1. a device that mechanically or hydraulically forces pipe or tools into the well against pressure. 2. a device within some hooks that acts as a shock absorber in eliminating the bouncing action of pipe as it is picked up.

snubbing *n*: the forcing of pipe or tools into a high-pressure well that has not been killed (i.e., running pipe or tools into the well against pressure when the weight of pipe is not great enough to force the pipe through the BOPs). Snubbing usually requires an array of wireline blocks and wire rope that forces the pipe or tools into the well through a stripper head or blowout preventer until the weight of the string is sufficient to overcome the lifting effect of the well pressure on the pipe in the preventer. In workover operations, snubbing is usually accomplished by using hydraulic power to force the pipe through the stripping head or blowout preventer.

snubbing line *n*: 1. a line used to check or restrain an object. 2. a wire rope used to put pipe or tools into a well while the well is closed in. See *snub*.

snubbing unit *n*: either a stand-alone device or a rig-assist device that is used to force pipe into the well when the well is shut in on a kick. When the pipe's weight is not sufficient to overcome the upward force of well pressure, a snubbing unit must be used. Compare *stripping in*.

snub line *n*: a strong wire rope attached to the end of the tongs and to one leg of the derrick to keep the tongs from turning too far when they are being used to make up, break out, or back up drill pipe or drill collars.

snuffer *n*: a tank safety device that seals the vapor vent manually and prevents vapor from escaping into a fire, thus snuffing out the flame.

snuggle exploration and production *n*: exploring for and developing oil and gas reservoirs that occur close to existing transportation, processing, and refining facilities. Snuggling makes such reservoirs commercially attractive and capable of rapid development.

S$_o$ *abbr*: oil saturation; used in reservoir engineering.

SO *abbr*: show of oil; used in drilling reports.

SO$_2$ *form*: sulfur dioxide.

soak phase *n*: in cyclic steam injection, the period between the steam injection phase and the production phase.

soap *n*: the sodium or potassium salt of a high-molecular-weight fatty acid. Commonly used in drilling fluids to improve lubrication, emulsification, sample size, and defoaming.

Society of Exploration Geophysicists (SEG) *n*: official publications are *Geophysics* and *The Leading Edge of Exploration*. Address: Box 702740; Tulsa, OK 74170; 918-493-3516.

Society of Petroleum Engineers (SPE) *n*: organization of registered petroleum engineers. Its official publications are

Journal of Petroleum Technology, SPE Drilling Engineering, SPE Production Engineering, SPE Formation Evaluation, and *SPE Reservoir Engineering*. Address: P. O. Box 833836; Richardson, TX 75083-3836; 972-952-9393; fax 972-952-9435.

socket *n*: 1. a hollow object or open device that fits or holds an object. 2. any of several fishing tools used to grip the outside of a lost tool or a joint of pipe. 3. in crane operations, a device on one end of a wire rope into which the wire rope is inserted and firmly attached.

socket basket *n*: the conical portion of a socket into which a wire rope end is inserted and secured. See *socket*.

sock filter *n*: a cylindrical-shaped filter made of cloth that fits inside a special holder in the line conducting the liquid to be filtered. Often used in glycol dehydration systems to filter glycol.

soda ash *n*: see *sodium carbonate*.

sodium (Na) *n*: one of the alkali metal elements with a valence of 1, an atomic number of about 23. Numerous sodium compounds are used as additives to drilling fluids.

sodium acid pyrophosphate (SAPP) *n*: a thinner used in combination with barite, caustic soda, and fresh water to form a plug and seal off a zone of lost circulation.

sodium bicarbonate *n*: the half-neutralized sodium salt of carbonic acid, $NaHCO_2$, used extensively for treating cement contamination and occasionally other calcium contamination in drilling fluids.

sodium bichromate *n*: $Na_2Cr_2O_7$. Also called sodium dichromate. See *chromate*.

sodium carbonate *n*: Na_2CO_3, used extensively for treating various types of calcium contamination. Also called soda ash.

sodium carboxymethyl cellulose *n*: see *carboxymethyl cellulose*.

sodium chloride *n*: common table salt, NaCl. It is sometimes used in cement slurries as an accelerator or a retarder, depending on the concentration.

sodium chromate *n*: Na_2CrO_4. See *chromate*.

sodium dichromate *n*: see *sodium bichromate*.

sodium hydroxide *n*: see *caustic soda*.

sodium polyacrylate *n*: a synthetic high-molecular-weight polymer of acrylonitrile used primarily as a fluid loss-control agent.

sodium silicate muds *n pl*: special class of inhibited chemical muds using as their bases sodium silicate, salt, water, and clay.

sodium vapor lamp *n*: a high-intensity discharge lamp in which the gas is a mixture of sodium and mercury. It starts

with the arcing of xenon gas between the main electrodes and gives off a yellow-orange color.

soft crossover system *n*: a pattern of drum spooling in which the wire rope travels in a two-step grooving pattern but has flat or level areas for crossing over to act as shock absorbers for the rope and to reduce the rise or hump produced in all multiwrapping of wire rope.

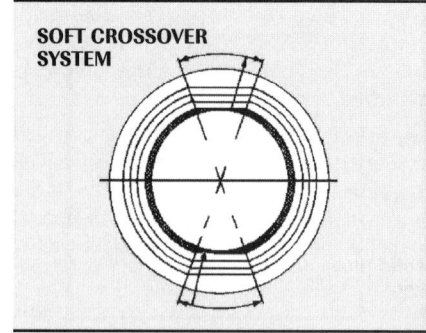

SOFT CROSSOVER SYSTEM

soft eye splice *n*: either hand- or mechanically-made splice in either wire or synthetic rope with no use of thimbles.

soft line *n*: a fiber rope.

soft shut-in *n*: in well-control operations, closing the BOPs with the choke and HCR, or fail-safe, valves open. Compare *hard shut-in*.

software *n*: computer programs.

soft water *n*: water that is free of calcium or magnesium salts. Compare *hard water*.

SO&G *abbr*: show of oil and gas; used in drilling reports.

soil stress *n*: the uneven penetration of pipeline coatings due to changes in soil volume and moisture along the pipeline bed.

soil venting *n*: see *volatilization*.

sol *n*: a colloidal solution. See *colloid*.

solar cell *n*: a device made up of silicon and silicon crystals to convert sunlight directly and efficiently into useful amounts of electricity. Also called a sun battery.

solar radiation *n*: the electromagnetic radiation emitted by the sun.

solar tide *n*: that portion of a tide that is due to the gravitational attraction of the sun on the earth.

SOLAS *abbr*: safety of life at sea.

SOLAS rules *n pl*: see *safety of life at sea rules*.

solenoid *n*: a cylindrical coil of wire that resembles a bar magnet when it carries a current so that it draws a movable core into the coil when the current flows.

solenoid valve *n*: in an accumulator used on a floating offshore drilling rig, a flow control device (a valve) that, when energized by operating a blowout preventer, opens to allow rig air pressure to flow to a pneumatic cylinder, which, in turn, operates the blowout preventer.

solid *n*: one of the three physical states of matter. A solid is somewhat rigid and bounds itself internally in all dimensions; therefore, it does not require a container to retain its shape.

solid-body kelly bushing *n*: a kelly bushing that is cast in a single piece. Compare *split-body kelly bushing*. See also *kelly bushing*.

solid desiccant dehydration system *n*: see *dry-bed dehydrator*.

solid master bushing *n*: a master bushing made in one piece. Usually, solid master bushings have split insert bowls. Compare *hinged master bushing, split master bushing*. See also *insert bowl, master bushing*.

solid phase *n*: also called reactive, or colloidal phase. See *mud solids*.

solids *n pl*: see *mud solids*.

solids concentration *n*: total amount of solids in a drilling fluid as determined by distillation. Includes both the dissolved and the suspended or undissolved solids.

solids content *n*: see *solids concentration*.

solid solution *n*: a homogeneous crystalline structure in which one or more types of atoms or molecules may be partly substituted for the original atoms or molecules without changing the structure.

solids removal equipment *n*: the devices installed in the mud circulating system that removes such solids as sand, silt, other particles that may be in the drilling mud.

solid-state *adj*: relating to the properties, structure, or reactivity of solid material, especially relating to the arrangement and behavior of ions, molecules, electrons, and holes in the crystals of a substance, such as a semiconductor. The term distinguishes semiconducting electronic components that are composed of and rely on solid materials for operation instead of materials such as evacuated glass tubes.

solid waste *n*: a U.S. government term for waste that is in solid, liquid, semisolid, or contained gaseous form. A solid waste can be anything that is discarded or may be discarded.

solid wire *n*: in electronics, a conductor of electricity that is drawn out of a conducting material (such as copper) to form a single cylindrical length of wire. Compare *stranded wire*.

solid wireline *n*: a special wireline made of brittle but very strong steel, usually 0.066 to 0.092 inches (0.17 to 0.23 centimetres) in diameter (as opposed to stranded wirelines, which may be ³⁄₁₆ inch—0.47

centimetre—or larger). Solid, or slick, wirelines are used in depth measurements and in running special devices into a well under pressure.

solubility *n*: the degree to which a substance will dissolve in a particular solvent.

solute *n*: a substance that is dissolved in another (the solvent).

solution *n*: a single, homogeneous liquid, solid, or gas phase that is a mixture in which the components (liquid, gas, solid, or combinations thereof) are uniformly distributed throughout the mixture. In a solution, the dissolved substance is called the solute; the substance in which the solute is dissolved is called the solvent.

solution gas *n*: lighter hydrocarbons that exist as a liquid under reservoir conditions but that effervesce as gas when pressure is released during production.

solution-gas drive *n*: a source of natural reservoir energy in which the solution gas coming out of the oil expands to force the oil into the wellbore.

solution gas-oil ratio *n*: see *gas-oil ratio*.

solvent *n*: 1. a substance, usually liquid, in which another substance (the solute) dissolves. 2. in paint technology, a single or blended liquid that is volatile under normal drying conditions and in which the binder dissolves completely.

solvent extraction *n*: the use of a solvent to selectively dissolve a particular compound and remove it from a mixture of hydrocarbons.

solvent treating *n*: the removal of aromatics, naphthenes, and olefins that decrease the stability of a lubricating oil's viscosity under temperature changes and cause gummy deposits to form.

sonar *n*: (derived from <u>so</u>und <u>na</u>vigation <u>r</u>anging) an apparatus that detects the presence of an underwater object by sending out sonic or supersonic waves that are reflected back to it by the object.

sonde *n*: a logging tool assembly, especially the device in the logging assembly that senses and transmits formation data.

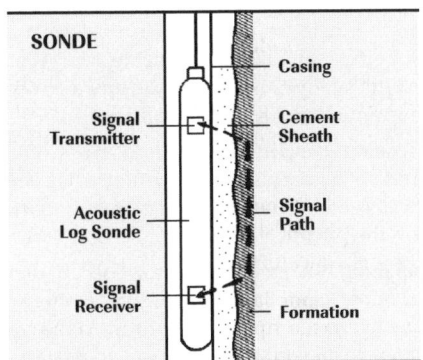

sonic flow nozzle *n*: a specially designed nozzle that is installed in a line through which fluids are flowing. It is used to measure the volume of fluids. It works on the same principle as an orifice plate in that the nozzle causes a pressure drop from which volume of flow can be inferred. It is used where the velocity of flow is particularly high because the pressure loss through the nozzle is lower than through an orifice. See *orifice plate*.

sonic log *n*: a type of acoustic log that records the travel time of sounds through objects, cement, or formation rocks. Often used to determine whether voids exist in the cement behind the casing in a wellbore.

sonic logging *n*: see *acoustic well logging*.

sonic meter *n*: see *ultrasonic meter*.

sorb *v*: to take up and hold by adsorption or absorption.

sorbent *n*: a material that absorbs oil or to which oil adheres.

sorber *n*: a vessel for absorption, adsorption, or desorption.

sorting *n*: a dynamic process by which different-sized sediments are separated from one another and deposited in different locations or layers as relatively uniform deposits. For instance, a high-energy mountain stream may leave only large cobbles and gravel in one location and carry finer sediments downstream to accumulate in a lower-energy environment, such as a lake.

sound generator *n*: a device that makes a low frequency noise; used in seismic exploration.

sour *adj*: containing or caused by hydrogen sulfide or another acid gas (e.g., sour crude, sour gas, sour corrosion).

source *n*: the terminal in a field-effect transistor from which majority carriers flow into the conducting channel in the semiconductor's material. See *field-effect transistor*.

source-detector spacing *n*: the spacing on a neutron logging device between the neutron source and the detector. Total count rate, porosity resolution, and borehole effects influence source-detector spacing. The spacing is selected specifically for the source-detector characteristics exhibited by each device; optimum spacings must be determined for each tool.

source rock *n*: rock within which oil or gas is generated from organic materials.

source-specific natural gas sales contract *n*: a natural gas sales contract that commits the seller to deliver natural gas, usually within a stated maximum and minimum, from specific described and committed natural gas reserves or sources. Such contracts are usually drafted to commit the seller to deliver natural gas only to the extent it can be produced or produced economically from the committed reserves or sources.

source station *n*: a pump station at a pipeline junction from which oil is pumped from a main line into a branch or lateral line.

sour corrosion *n*: embrittlement and subsequent wearing away of metal caused by contact of the metal with hydrogen sulfide.

sour crude *n*: see *sour crude oil*.

sour crude oil *n*: oil containing hydrogen sulfide or another acid gas.

sour gas *n*: gas containing an appreciable quantity of hydrogen sulfide and/or mercaptans.

sour hole *n*: a wellbore or formation known to contain hydrogen sulfide gas.

Southern Gas Association (SGA) *n*: an organization founded to promote the development of the gas distribution and transmission industry, to encourage scientific research affecting the industry, to exchange ideas and information among member companies, and to cooperate with other organizations having mutual objectives. SGA is the largest of four regional gas organizations started in 1908. Address: 3030 LBJ Freeway, Suite 1300, LB60; Dallas, TX 75234; (972) 620-8505; fax (972) 620-8518.

south magnetic pole *n*: a point on the earth's surface near the south polar region with which the south pole of a magnetic compass aligns regardless of the compass's location on earth. The south magnetic pole is located several miles from the South Pole (the south geographic pole).

southwest monsoon *n*: the rainy season in southern Asia.

sovereign *n*: in the case of land, the government that holds and is capable of transferring title.

Soxhlet extractor *n*: a device used to extract oil and water from a core sample. A solvent vaporizes from a half-filled flask at the bottom of the extractor. Vapor rises to the top chamber and condenses there. The condensed vapor moves to the middle chamber until it is filled. Then the solvent is siphoned back through the sample to clean it.

SP *abbr*: spontaneous potential or self-potential.

space-charge region *n*: when p-type and n-type semiconductors are joined, the area (the region) on the semiconductors in which free electrons flow from the n-type semiconductor to the holes of the p-type semiconductor. In this region, the n-type material becomes slightly positive, and the p-type material becomes slightly negative. When this space charge has been established, the junction is stabilized. No more electrons can flow across the junction because they are now repelled by the slight negative charge in the p-type material. See *hole, n-type structure, p-type structure*.

space-out *n*: 1. the installation of one or more riser pup joints to obtain the overall required length of the riser from the blowout preventer stack to the telescopic joint. 2. the act of ensuring that a pipe ram preventer will not close on a drill pipe tool joint when the drill stem is stationary. A pup joint may be made up in the drill string to lengthen it sufficiently.

space out *v*: to position the correct number of feet (or metres) or joints of pipe from the packer to the surface tree, or from the rig floor to the blowout preventer stack.

space-out joint *n*: the joint of drill pipe that is used in hang-off operations so that no tool joint is opposite a set of preventer rams.

space-out measurement tool *n*: in completion technology, a device that helps determine the height of the production casing hanger relative to the lock-down profile in the bore of the wellhead. Such information is vital to the successful installation of downhole production tools.

spacer *n*: 1. a thickened fluid that displaces drilling mud because of the difference in viscosity and weight between the spacer and the mud. Besides carrying lost circulation materials, spacers may be weighted with various inert materials such as fly ash and barite. 2. member of a pipeline construction gang who is responsible for assuring that the exact distance between the beveled pipe ends for a welding process to be used on the joint is maintained. Spacers strike a wedge into the interface between the pipe bevels and then maneuver them to an exact, uniform distance around the entire circumference.

spacing *n*: 1. in electric well logging, refers to the distance between electrodes on a logging tool. Electrode spacing affects the depth of investigation of the tool into the formation. 2. see *well spacing*.

spacing clamp *n*: a clamp used to hold the rod string in pumping position when the well is in the final stages of being put back on the pump.

spacing-out *n*: positioning the correct number of feet or joints of pipe from the packer to the surface tree, or from the rig floor to the stack

spaghetti *n*: tubing or pipe with a very small diameter.

spall *n*: a condition in which the surface materials of a bit bearing separates from the bearing's core material. Bearings spall when the metal of which they are made fatigues because of overuse. *v*: to break off in chips or scales, as on a plain, or journal, bearing.

span adjustment *n*: in process control systems, the calibration of a process transmitter to acquire or set its upper range value.

spark arrestor *n*: a water-spray device installed in the tail pipe of an engine's exhaust system to suppress or prevent small, very hot objects, such as carbon particles, from causing a fire near the engine.

spark ignition (SI) *n*: ignition of a fuel-air mixture by means of a spark discharged by a spark plug.

spark-ignition engine *n*: an internal combustion engine that uses an electrical spark to ignite the fuel-air mixture inside its cylinders. Usually, the engine employs spark plugs to provide the electrical spark.

spark plug *n*: a device that fits into the cylinder of an internal-combustion engine and that provides the spark for ignition of the fuel-air mixture during the combustion stroke of the piston. It carries two electrodes separated by an air gap; current from the ignition system discharges across the gap to form the spark.

SPBM *abbr*: single-point buoy mooring.

SPCC *abbr*: Spill Prevention, Countermeasures and Control plan.

spd *abbr*: spudded; used in drilling reports.

SPDT *abbr*: single-pole, double-throw.

SPE *abbr*: Society of Petroleum Engineers.

spear *n*: a fishing tool used to retrieve pipe lost in a well. The spear is lowered down the hole and into the lost pipe. When weight, torque, or both are applied to the string to which the spear is attached, the slips in the spear expand and tightly grip the inside of the wall of the lost pipe. Then the string, spear, and lost pipe are pulled to the surface.

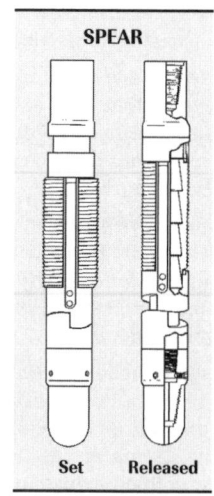

SPEAR

Set   Released

spearhead *n*: see *preflush.*

spearhead overshot *n*: the small wireline-operated overshot designed to latch onto a spear point rope socket so that the wireline below the socket can be retrieved.

spear point rope socket *n*: a rope socket with a point on it that mates with a spearhead overshot.

specific gravity *n*: see *relative density.*

specific heat *n*: the amount of heat required to cause a unit increase in temperature in a unit mass of a substance, expressed as numerically equal to the number of calories needed to raise the temperature of 1 gram of a substance by 1°C.

specific permeability *n*: see *absolute permeability.*

specific viscosity *n*: the ratio of the absolute viscosity of a substance to that of a standard fluid, such as water, with the viscosity of both fluids being measured at the same temperature. See *absolute viscosity.*

specific weight *n*: density times the attraction of gravity.

Spectra® *n*: a type of high modulus polyethylene (HMPE) fiber made by Allied Chemical. Used in rope construction.

spectral wave model *n*: a representation of a pattern of wave development for a specific area. For example, a model can assess the wave conditions, such as heights, periods, and direction, in a coastal area over a given period. From this data, scientists can estimate wave forces at the shoreline.

speed droop *n*: the number of revolutions per minute that an engine slows down from running at maximum no-load speed to running at maximum full-load speed. Speed droop should not exceed 7%.

speed droop governor *n*: a governor that provides a decrease in prime mover speed for any increase in load or an increase in prime mover speed for any decrease in load.

speeder spring *n*: a small spring inside an engine governor that counteracts the force of flyweights in the governor. The speeder spring moves down on a sleeve in the governor to increase the fuel supply to the engine, at the same time, the flyweights move the sleeve up to decrease the fuel supply. Since the flyweights spin at a speed determined by the engine's speed, if engine speed drops, the speeder spring moves down to speed the engine up. See *flyweight, governor.*

speed kit *n*: a dual-speed traveling block, which permits one elevator to pick up stands as they are broken out while the traveling block continues to move.

speed limiter *n*: a type of engine governor that prevents an engine from exceeding a set speed. Usually, limiters simply prevent the engine from running too fast to prevent damage to the engine; they do not shut down the engine.

speed reducer *n*: a set of gears installed between a prime mover and the equipment it drives to reduce the running speed. For example, on a beam pumping unit, the engine may run at a speed of 600 revolutions per minute, but the pumping unit it drives may need to operate at 20 strokes per minute. The speed reducer makes it possible to obtain the correct pump speed.

speedup capacitor *n*: in a flip-flop circuit, a capacitor that speeds up the switching action in the circuit. Usually, two speed-up capacitors are used.

spelter socket *n*: the end termination of wire rope. It can be cast, forged, or prefabricated. Wire is inserted into the socket and secured with zinc alloy or resin. The socket can be "open" or "closed."

spent *adj*: descriptive of a substance whose strength or merit has been exhausted in a process. For example, after a well has been acidized, any acid that remains in the well is said to be a spent acid because its strength has been used up in the acidizing process.

spent gas-lift gas *n*: see *recoverable gas-lift gas.*

sp gr *abbr*: specific gravity.

sphere *n*: 1. a neoprene ball that is run in a pipeline to clean it, to displace liquid hydrocarbons from natural gas pipelines, or to separate batches in liquid shipment. 2. a ball made of neoprene or other material that is put into a pipe prover to activate switches on the prover so that a known volume of petroleum or product is measured.

spherical blowout preventer *n*: see *annular blowout preventer.*

spherical separator *n*: a separator that is a round, ball-shaped vessel.

spheroid *n*: a three-dimensional object that resembles a sphere. For example, rather than a sphere, the earth is an oblate spheroid, which means that, instead of being a perfect sphere, it is flattened at the poles.

spider *n*: 1. a circular steel device that holds slips supporting a suspended string of drill pipe, casing, or tubing. A spider may be split or solid. 2. a riser support spider. See *riser support spider.*

spill point *n*: the level at which trapped oil or gas can begin "spilling" upward and out of the trap.

Spill Prevention, Countermeasures and Control Plan (SPCC) *n*: requirement based on the Water Pollution Control Act Amendments of 1972 and subsequent EPA regulations that applies to owners or operators of nontransportation-related onshore and offshore facilities engaged in drilling, producing, gathering, or consuming oil and that, because of their location, could reasonably be expected to discharge oil in harmful quantities into or on navigable waters. The objective of these regulations is to prevent the discharge of oil in harmful quantities into the navigable waters of the United States or adjoining shorelines. Where such discharge potential exists, an SPCC should be drawn up and should address practices that will prevent oil spills, contain oil spills, train employees in spill prevention/containment, prevent pollutant runoff from a site, and provide secondary containment where applicable. The plan must use "good engineering practices" and must be reviewed and certified by a professional registered engineer.

Spindletop *n*: a small knoll near Beaumont, Texas, on which, in 1901, a well blew in (began producing) at the then phenomenal rate of 80,000 barrels (12.7 million litres) per day. The Spindletop well proved that rotary drilling could be effectively used to drill wells and that subterranean formations could contain large amounts of hydrocarbons. The Spindletop well is often said to have ushered in the modern petroleum era.

SPINDLETOP

spinner survey *n*: a production-logging method that uses a small propeller turned by fluid movement. By use of a recording arrangement, the number of turns of the propeller can be related to the fluid quantity flowing past the instrument to obtain a log of the amount of fluid flowing from a formation.

spinning cathead *n*: see *makeup cathead, spinning chain.*

spinning chain *n*: a relatively short length of chain attached to the tong pull chain on the manual tongs used to make up drill pipe. The spinning chain is attached to the pull chain so that a crew member can

SPINNING CHAIN

wrap the spinning chain several times around the tool joint box of a joint of drill pipe suspended in the rotary table. After crew members stab the pin of another tool joint into the box end, one of them then grasps the end of the spinning chain and with a rapid upward motion of the wrist "throws the spinning chain"—that is, causes it to unwrap from the box and coil upward onto the body of the joint stabbed into the box. The driller then actuates the makeup cathead to pull the chain off of the pipe body, which causes the pipe to spin and thus the pin threads to spin into the box. Spinning wrenches have all but replaced spinning chains.

spinning wrench *n*: air-powered or hydraulically powered wrench used to spin drill pipe in making or breaking connections.

spin time test *n*: a turbine meter field check to determine the relative level of mechanical friction in a meter.

spin-up *n*: the rapid turning of the drill stem when one length of pipe is being joined to another.

spiral grapple *n*: a helically shaped gripping mechanism that is fitted into an overshot to retrieve fish from the borehole. See *grapple*.

spiral heavy-walled drill pipe *n*: heavy pipe that is spiraled—much like a spiral drill collar—with extra-length tool joints but without a center wear pad. Compare *heavy-walled drill pipe*. See *spirally-grooved drill collar*.

spirally-grooved drill collar *n*: a drill collar with a round cross section that has a long continuous groove or flute machined helically into its outer surface. The spiraled groove provides space between the wall of the hole and the body of the collar, minimizing the area of contact between the hole wall and the collar; thus the possibility of differential pressure sticking is reduced.

spiral strand *n*: wire rope constructed from many spirally laid wires. Also called bridge strand.

spirit level *n*: a device consisting of a scribed tube partially filled with a viscous liquid (spirit) that is designed to determine a horizontal line or plane. Since the tube is only partially filled, a bubble forms. By checking where the bubble rests between the scribed lines on the tube, one can determine whether an object is level.

splash box *n*: see *mother hubbard*.

splash-proof enclosure *n*: a motor enclosure that allows some outside air to circulate through the motor for cooling. The angle of protection extends to 100° from the vertical.

splash zone *n*: the area on an offshore structure that is regularly wetted by seawater

but is not continuously submerged. Metal in the splash zone must be well protected from the corrosive action of seawater and air.

splice *v*: to join two parts of a rope or wireline by interweaving individual strands of the line together. Unlike a knot, a splice does not significantly increase the diameter of the line at the point where the parts are joined.

spline clutch *n*: a positive-type clutch that works by means of metal strips that fit into keyways.

split-body kelly bushing *n*: a two-piece kelly bushing that is secured by hold-down nuts and bolts passing from the top of the bushing to the bottom of the bushing. Compare *solid-body kelly bushing*. See also *kelly bushing*.

split master bushing *n*: a master bushing that is made in two pieces. Each half has a tapered surface to accept the slips. Compare *hinged master bushing, solid master bushing*. See also *master bushing, slips*.

splitter *n*: another term for a fractionator, particularly one that separates isomers. For example, a deisobutanizer may be called a butane splitter.

spm *abbr*: strokes per minute.

SPM *abbr*: subsea pilot manipulated.

spoil *n*: excavated dirt.

sponge absorbent *n*: an absorbent for recovering vapors of a lighter absorbent that is used in the main absorption process of a gas processing plant.

sponge absorption unit *n*: a unit wherein the vapors of lighter absorption oils are recovered.

sponge barrel coring *n*: coring performed with the inner core barrel lined with polyurethane sponge. The sponge absorbs oil that bleeds from the core and holds it opposite the formation from which it bled. Both the core and the sponge are analyzed to obtain information on saturation. Each foot of sponge is matched to the core it surrounded.

sponson *n*: on the earliest semisubmersible drilling rigs, a structure projecting from the side of the rig's column that added stability to the rig by effectively increasing its width. One end of the sponson was attached to the column and the other contacted the water's surface. The end contacting the water's surface featured a water-tight pod, which was buoyant. Usually, several sponsons were used. Modern semisubmersible designs eliminate the need for sponsons.

spontaneous potential (SP) *n*: one of the natural electrical characteristics exhibited

by a formation as measured by a logging tool lowered into the wellbore. Also called self-potential or SP.

spontaneous potential (SP) curve *n*: a measurement of the electrical currents that occur in the wellbore when fluids of different salinities are in contact. The SP curve is usually recorded in holes drilled with freshwater-base drilling fluids. It is one of the curves on an electric well log. Also called self-potential curve.

spontaneous potential (SP) deflection *n*: the spikes or peaks in the curve on an SP log, measured with respect to a shale baseline.

spontaneous potential (SP) log *n*: a record of a spontaneous potential curve.

spool *n*: the drawworks drum. Also a casinghead or drilling spool. *v*: to wind around a drum.

spool valve *n*: a slide-type hydraulic valve that has a movable part called a spool.

spot *v*: to pump a designated quantity of a substance (such as acid or cement) into a specific interval in the well. For example, 10 barrels (1,590 litres) of diesel oil may be spotted around an area in the hole in which drill collars are stuck against the wall of the hole in an effort to free the collars.

spot a pill *v*: to place a special mixture of clay and oil, or other materials, at a specific point in the wellbore. The size of the pill is usually several barrels. For example, a rig crew may spot a 20-barrel (3,180-litre) pill of bentonite clay and diesel oil at or near a formation to which drilling mud is escaping. By lowering the drill stem to the desired depth, the pill is slowly pumped down the stem and into the annulus. If all goes well, the pill will migrate into the formation taking the drilling mud and plug it closed.

spot market *n*: the buying and selling of crude oil in a commodities exchange market on a daily basis, just as other commodities, such as gold, wheat, and heating oil, are bought and sold.

spot oil *v*: to circulate oil or some other lubricant down the drill stem and into the annulus to a stuck portion.

spot sale *n*: generally a short-term sale on an interruptible or best efforts basis, usually for one month.

spot sample *n*: in tank sampling, a sample obtained at some specific location in the tank by means of a thief, a bottle, or a beaker.

spotting *n*: the pumping of a substance such as oil into an interval in the well. See *spot*.

SPR *abbr*: Strategic Petroleum Reserve.

spray nozzle valve *n*: see *nozzle*.

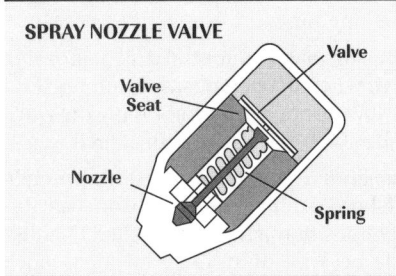

SPRAY NOZZLE VALVE

Valve

Valve Seat

Nozzle

Spring

spread *n*: the equipment and crew needed to build a pipeline. Modern spreads, which are like moving assembly lines, can consist of one hundred pieces of equipment and over five hundred workers.

spreader *n*: see *spreader bar*.

spreader bar *n*: a rod positioned between two lifting lines so that the weight of a joint being lifted off a trailer or a boat is evenly distributed and the pipe does not buckle in the center.

spreading rate *n*: in paint technology, the area in square feet that 1 gallon of paint can cover at a given wet film thickness.

spread mooring *n*: a multileg mooring system emanating from the corners of a vessel.

spread mooring system *n*: a system of rope, chain, or a combination attached to anchors on the ocean floor and to winches on the structure to keep a floating vessel near a fixed location on the sea surface.

spread superintendent *n*: in pipeline construction, the person with responsibility for running the spread. The spread superintendent represents the contractor's interests in the field.

spring collet *n*: a spring-actuated metal band or ring (ferrule) used to expand a liner patch when making casing repairs. See *liner patch*.

spring-loaded centrifugal governor *n*: a mechanical governor that includes a special spring, called a speeder spring, that offsets the centrifugal force of spinning flyweights in the governor. In many governors, the flyweights tend to make the governor slow the engine down, while the spring tends to make the governor speed the engine up. Centrifugal force and spring pressure balance each other to maintain the engine's speed.

spring-loaded lock *n*: in a dog-type riser connector, a device that prevents the actuator screws in the connector from backing out because of vibration. See *dog-type riser connector*.

spring-loaded pin *n*: in an ABB Vetco riser connector, one of several pins that protrude from the connector's body and fit under a support shoulder on the gasket. Hydraulic

force releases the pins and spring force retains the gasket. See *riser connector*.

spring shunt *n*: on a generator's brush holder, a braided copper wire conductor that conducts (shunts) any current that may build up in the brush holder's spring away from the spring.

spring slips *n pl*: slips that set by means of springs when a floorhand stands on them. See *slips*.

SPRING SLIPS

spring tide *n*: a tide that has the highest range at a given location. This type of tide occurs when the sun and the moon are aligned (during new and full moons).

sprocket *n*: a wheel with projections on the periphery to engage with the links of a chain.

sprocket ratio *n*: in a chain-and-sprocket drive, the amount of difference in size between the two sprockets.

SPST *abbr*: single-pole, single throw.

spud *v*: 1. to move the drill stem up and down in the hole over a short distance without rotation. Careless execution of this operation creates pressure surges that can cause a formation to break down, resulting in lost circulation. 2. to force a wireline tool or tubing down the hole by using a reciprocating motion. 3. to begin drilling a well; i.e., to spud in.

spud bit *n*: a special kind of drilling bit with sharp blades rather than teeth. It is sometimes used for drilling soft, sticky formations.

spud can *n*: a cylindrical device, usually with a pointed end, that is attached to the

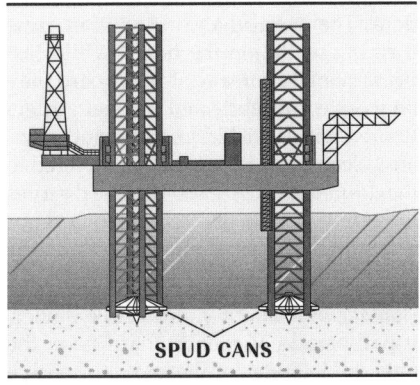

SPUD CANS

bottom of each leg of a jackup drilling unit. The pointed end of the spud penetrates the seafloor and helps stabilize the unit while it is drilling.

spud date *n*: the first boring of a hole in the drilling of a well.

spudder *n*: a portable cable-tool drilling rig, sometimes mounted on a truck or trailer.

spud in *v*: to begin drilling; to start the hole.

spud mud *n*: the fluid used when drilling starts at the surface, often a thick bentonite-lime slurry.

spur line *n*: an oil pipeline that picks up oil from the gathering lines of several oilfields and delivers it to a main line or trunk line.

spurt loss *n*: the initial loss of drilling mud solids by filtration, making formations easier to drill. See *filtration loss, surge loss*.

sq *abbr*: square.

sq ft (ft²) *abbr*: square feet.

sq in. (in.²) *abbr*: square inches.

squ *abbr*: squeeze; used in drilling reports.

squall *n*: in the United States, a wind of 16 knots or higher sustained for at least 2 minutes.

squall line *n*: in meteorology, a series of small storms that occur along a cold front.

square drill collar *n*: a special drill collar, square but with rounded edges, used to control the straightness or direction of the hole; often part of a packed-hole assembly.

square-drive kelly bushing *n*: a kelly bushing that has a square-shaped base that fits into a corresponding square-shaped recess in the master bushing. When the base is engaged in the recess, and the master bushing is rotated, the kelly bushing, the kelly, and the attached drill stem also rotate. See *kelly bushing, master bushing*.

square-drive master bushing *n*: a master bushing that has a square recess to accept and drive the square that is on the bottom of the square-drive kelly bushing. See *kelly bushing, master bushing, square-drive kelly bushing*.

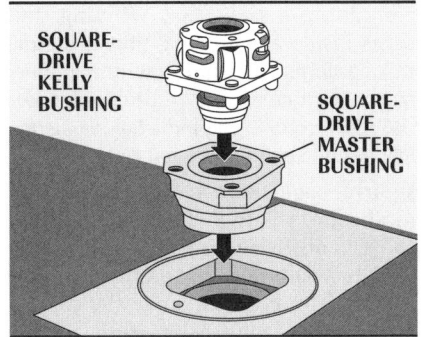

SQUARE-DRIVE KELLY BUSHING

SQUARE-DRIVE MASTER BUSHING

square kelly *n*: a kelly with a four-sided (square) cross section. Compare *hexagonal kelly*.

square metre *n*: an SI unit of measure of an area equal to a square that measures 1 metre on each side.

square root chart *n*: see *L-10 chart*.

squeeze *n*: 1. a cementing operation in which cement is pumped behind the casing under high pressure to recement channeled areas or to block off an uncemented zone. 2. the increasing of external pressure on a diver's body caused by improper diving technique.

squeeze cementing *n*: the forcing of cement slurry by pressure to specified points in a well to cause seals at the points of squeeze. It is a secondary cementing method that is used to isolate a producing formation, seal off water, repair casing leaks, and so forth. Compare *plug-back cementing*.

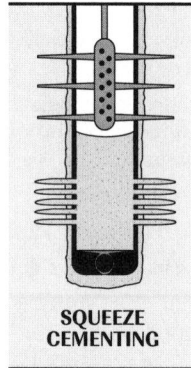

SQUEEZE CEMENTING

squeeze job *n*: a remedial well-servicing activity whereby a cement slurry is pumped into open perforations, split casing, or a fractured formation, to effect a blockage.

squeeze manifold *n*: a type of manifold used in squeeze jobs.

squeeze packer *n*: a downhole permanent, or drillable, packer that is set by lowering some of the weight of the tubing string onto the packer. The weight expands the packer's sealing element to prevent flow between the tubing string and the casing below the packer.

squeeze point *n*: the depth in a wellbore at which cement is to be squeezed.

squeeze tool *n*: a special retrievable packer set at a particular depth in the wellbore during a squeeze cementing job. See also *squeeze cementing*.

squirrel-cage induction motor *n*: an electric motor that uses a rotating device (a rotor) containing a secondary winding that turns inside a stationary unit (a stator) that contains the primary winding. Electric current created by induction makes the motor work. See *induction, induction motor*, and *squirrel-cage rotor*.

squirrel-cage rotor *n*: in an induction motor, the rotating device (rotor) that consists of several copper bars running longitudinally along the outside of the rotor's iron core. Because the copper bars surround the core longitudinally, the

rotor resembles a cage used to exercise squirrels in captivity.

Squnch Joint™ *n*: a special threadless tool joint for large-diameter pipe, especially conductor pipe, sometimes used on offshore drilling rigs. When the box is brought down over the pin and weight is applied, a locking device is actuated to seat the joints. Because no rotation is required to make up these joints, their use can save time when the conductor pipe is being run. Squnch Joint is a registered trademark of the manufacturer.

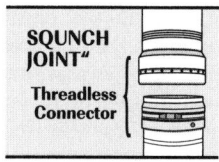

SQUNCH JOINT"

Threadless Connector

ss *abbr*: sand or sandstone; used in drilling reports.

SSO *abbr*: slight show of oil; used in drilling reports.

SSTT *abbr*: subsea test tree.

SSU *abbr*: Saybolt Second Universal.

SSV *abbr*: surface safety valve.

St *abbr*: stratus.

S/T *abbr*: sample tops; used in drilling reports.

stab *n*: a type of connector on tools in which the joining, or mating, of the tools is achieved by inserting a pin on one tool into a receptacle on the other. *v*: to guide and insert the pin of one tool into the receptacle of another, as the pin on a drill pipe's tool joint into the box of another tool joint when making a connection.

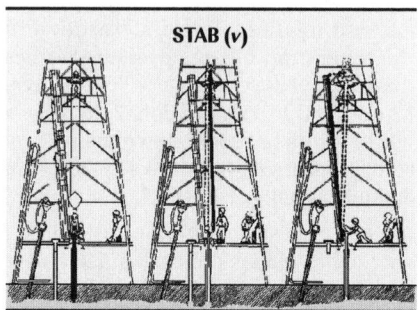

STAB (v)

stabbing board *n*: a temporary platform erected in the derrick or mast some 20 to 40 feet (6–12 metres) above the derrick floor. The derrickhand or another crew member works on the board while casing is being run in a well. The board may be wooden or fabricated of steel girders floored with antiskid material and powered electrically to be raised or lowered to the desired level. A stabbing board serves the same purpose as a monkeyboard but is temporary instead of permanent.

stabbing jack *n*: see *jack board*.

stabbing protector *n*: a protective device, usually made of rubber, that fits on the outside diameter of the box of the pipe

that is in the hole. It has a funnel-shaped top and serves as a cushion and guide for stabbing pipe.

stabbing valve *n*: a special drill stem valve that, when in open position, allows fluid to flow through it, thus allowing the valve to be stabbed into the drill stem.

stability *n*: 1. the ability of a ship or mobile offshore drilling rig to return to an upright position when it has rolled to either side because of an external force (such as waves). 2. the ability of a measuring instrument to maintain its accuracy over a long period.

stability meter *n*: an instrument to measure the amount of voltage needed to break down invert emulsions.

stabilized *adj*: said of a flowing well when its rate of production through a given size of choke remains constant, or, in the case of a pumping well, when the fluid column within the well remains constant in height.

stabilized bottomhole flowing pressure *n*: the bottomhole flowing pressure that has been allowed time to reach a steady state.

stabilized bottomhole shut-in pressure *n*: the bottomhole shut-in pressure that has been allowed time to reach a steady state.

stabilized condensate *n*: condensate that has been stabilized to a definite vapor pressure in a fractionation system.

stabilizer *n*: 1. a tool placed on a drill collar near the bit that is used, depending on where it is placed, either to maintain a particular hole angle or to change the angle by controlling the location of the contact point between the hole and the collars. See *packed-hole assembly*. 2. a vessel in which hydrocarbon vapors are separated from liquids. 3. a fractionation system that reduces the vapor pressure so that the resulting liquid is less volatile.

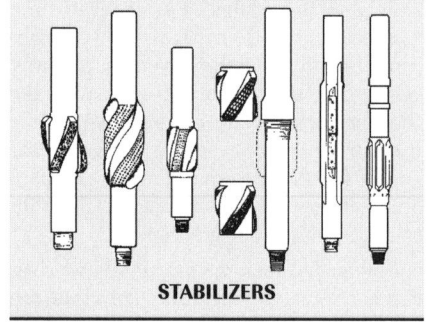

STABILIZERS

stable emulsion *n*: see *emulsion*.

stable equilibrium *n*: the tendency of an offshore floating rig to return to its initial upright position after an external force that heeled the rig to one side is removed.

stack *n*: 1. a vertical arrangement of blowout prevention equipment. Also called preventer stack. See *blowout preventer*. 2. the vertical chimney-like installation that is the waste disposal system for unwanted vapor such as flue gases or tail-gas streams. See *exhaust stack.*

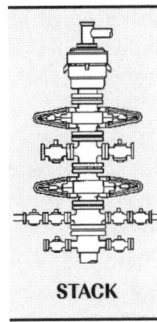

STACK

stack a rig *v*: to store a drilling rig on completion of a job when the rig is to be withdrawn from operation for a time.

stack gas *n*: see *flue gas.*

stage cementing *n*: a primary cementing operation in which cement is pumped into the well in a series of operations or stages.

stage separation *n*: an operation in which well fluids under pressure are separated into liquid and gaseous components by being passed consecutively through two or more separators. The operating pressure of each succeeding separator is lower than the one preceding it. Stage separation is an efficient process in that a high percentage of the light ends of the fluid are conserved.

stage tool *n*: a special tool used in stage cementing.

staging *n*: the placement of compressors, pumps, cooling systems, treating systems, and so forth, in a series with another unit or units of like design to improve operating efficiency and results.

stagnation pressure *n*: see *total pressure.*

stake a well *v*: to locate precisely on the surface of the ground the point at which a well is to be drilled. After exploration techniques have revealed the possibility of a subsurface hydrocarbon-bearing formation, a certified and registered land surveyor drives a stake into the ground to mark the spot where the well is to be drilled.

stand *n*: the connected joints of pipe racked in the derrick or mast when making a trip. On a rig, the usual stand is about 90 feet (about 27 metres) long (three lengths of drill pipe screwed together), or a thribble.

standard *n*: 1. a measuring instrument intended to define, to represent physically, or to reproduce the unit of measurement of a quantity (or a multiple or submultiple of that unit) to transmit it to other measuring instruments by comparison. See also *primary standard, secondary standard, working standard.* 2. a prescribed set of voluntary rules, conditions, or requirements concerned with the definition of terms; classification of components; delineation of procedures;

specification of dimension; construction criteria; materials; performance; design; or operations; measurement of quality and quantity in describing materials, products, systems, services, or practices; or description of fit and measurement of size.

standard air *n*: the accepted density of standard air varies between the U.S., British, and metric systems of measurement. The correct densities have been incorporated in the ASTM-IP Measurement Tables and IP 250/69.

standard brass *n*: brass of a specified density used in fabricating precision balance weights.

standard conditions *n pl*: the standard pressure and temperature to which measurements should be referred. These are (1 bar [101.325kPa]/cm^2), 15°C for the SI metric system, and 14.73 lb/in.^{2}, 60°F for the U.S. and British systems.

standard cubic foot *n*: a gas volume unit of measurement at a specified temperature and pressure. The temperature and pressure may be defined in the gas sales contract or by reference to other standards.

standard derrick *n*: a derrick that is built piece by piece at the drilling location, as opposed to a jackknife mast, which is preassembled. Standard derricks have been replaced almost totally by jackknife masts. Compare *mast.*

standard deviation *n*: the root mean square (rms) deviation of the observed values from the average.

standard dress *n*: diving equipment consisting of brass diving helmet, breastplate, heavy dry suit, weighted boots, weighted belt, hose, compressor, and communications.

standard gas measurement law *n*: a law, specific to each of several states, that defines the pressure and temperature bases under which a standard cubic foot of gas should be measured in the state.

standard industrial classification (SIC) codes *n pl*: numerical codes assigned by the government to groups of industries with similar activities or operations.

standard joint *n*: a 45-foot long tubular that provides pressure integrity and tensile capacity at a given water depth.

standard lateral curve *n*: in conventional electric logging, measures resistivity in the formation a long distance from the borehole. Used for deep investigation of the borehole.

standard pressure *n*: the pressure exerted by a column of mercury 760 millimetres (29.9 inches) high; equivalent to 14.7 pounds per square inch absolute. Compare *base pressure.*

standard riser joint *n*: a joint of riser pipe that is of normal length—that is, it is not a riser pup joint. See *riser pipe, riser pup joint.*

Standards of Training, Certification, and Watchkeeping (STCW) *n*: a standard that resulted from an International Maritime Organization (IMO) Conference in 1978 to establish minimum professional standards for those who work at sea.

standard tape *n*: the measuring tape used to measure the circumference of the ring at the bottom of a tank. It is calibrated by the National Institute of Standards and Technology and is usually 100 feet (30.5 metres) long.

standard temperature *n*: a predetermined temperature used as a basic measurement. The petroleum industry uses 60°F (15.5°C) as its standard temperature during measurement of oil. The volume of a quantity of oil at its actual temperature (assuming it is not 60°F) is converted to the volume the oil would occupy at 60°F. Conversion can be aided by the use of API conversion tables.

standby condition *n*: the condition of a vessel engaged in drilling operations where drilling has been suspended but the risers are not connected.

standing start-and-stop method *n*: in meter proving, when the opening and closing meter readings of the test run are determined at no-flow conditions. Compare *running start-and-stop method.*

standing valve *n*: a fixed ball-and-seat valve at the lower end of the working barrel of a sucker rod pump. The standing valve and its cage do not move, as does the traveling valve. Compare *traveling valve.*

standoff *n*: in perforating, the distance a jet or bullet must travel in the wellbore before encountering the wall of the hole.

standoff problem *n*: difficulty in obtaining proper penetration with a tubing perforating gun in casing due to the casing gun's lying against one side of the casing because of hole deviation.

standpipe *n*: a vertical pipe rising along the side of the derrick or mast, which joins the discharge line leading from the mud pump to the rotary hose and through which mud is pumped into the hole.

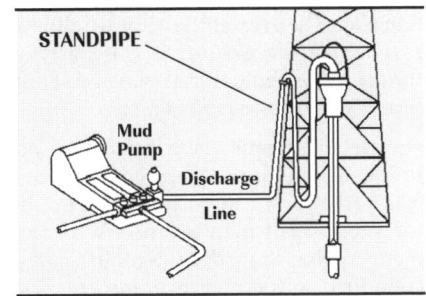

STANDPIPE

Mud Pump

Discharge Line

standpipe manifold *n*: a part of the mud circulation system that contains several valves, piping, tees, and elbows and is part of the standpipe. Manifold valves direct the mud from the mud pumping system through the piping and fittings and allow the rig crew to direct the mud to its desired destination, such as to the rotary hose,

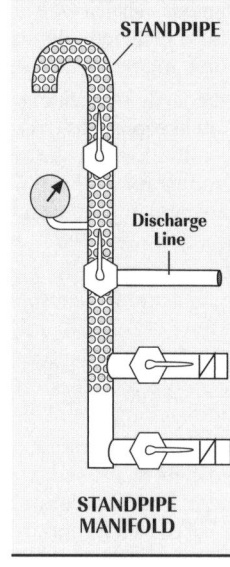

a backup standpipe, or other piece of circulating equipment. See *standpipe*.

standpipe pressure *n*: see *shut-in drill pipe pressure*.

stand tubing *v*: to support tubing in the derrick or mast when it is out of the well rather than to lay it on a rack. Portable workover rigs are usually fitted with a mast that holds stands about 60 feet long (about 18 metres) (doubles).

stand-up title opinion *n*: prepared in the absence of an abstract of title, stand-up opinions are written by title examiners who work from run sheets that they check at the county courthouse before deciding whether they need additional facts. Compare *abstract-based title opinion*.

starboard *n*: (nautical) the right side of a vessel (determined by looking toward the bow).

starch *n*: a complex carbohydrate sometimes added to drilling fluids to reduce filtration loss.

starter *n*: on an engine, an electrical, hydraulic, air, or other device used to rotate the engine's flywheel or pistons so that the fuel, air, and (in some cases) spark can enter the engine and make it begin running on its own.

starting torque *n*: in an electric induction motor, the turning force (torque) that is affected by the resistance of the rotor's winding. The greater the rotor winding's resistance, the greater is the starting torque up to the point at which pullout torque occurs. See *pullout torque*.

start-up differential check *n*: a test used to identify increased friction levels in a rotary meter; it can be used as a guide for subsequent maintenance. With the meter in a bypass and a differential gauge connected to the upstream meter tap, a valve may be opened slowly enough to observe the rise and subsequent fall in the differential as the meter just begins to rotate (the downstream differential tap being vented to atmosphere).

State Emergency Response Commission (SERC) *n*: entity set up by each state health department typically and at the request of the governor under the Emergency Planning and Community Right-to-Know Act of 1986 to designate emergency planning districts at the local level.

state of equilibrium *n*: in process control, a condition in which a process variable is maintained under fixed operating conditions without change occurring.

state standard pressure base *n*: the pressure base that state regulations require for the reporting of gas volumes. The pressure base for reporting to the states of Texas, Oklahoma, and Kansas is 14.65 pounds. Louisiana and California pressure bases are 15.025 pounds and 14.73 pounds, respectively.

state standard volume *n*: a volume of gas determined in accordance with measurement standards prescribed by state regulations.

static *adj*: at rest; not moving. See also *quiescence*. Compare *dynamic*.

static bottomhole pressure *n*: pressure at the bottom of the wellbore when there is no flow of oil.

static electricity *n*: electricity produced by friction.

static fluid level *n*: the level to which fluid rises in a well when the well is shut in.

static load *n*: a nonvarying weight; the force exerted by the weight of a mass at rest.

static lockup pressure *n*: highest pressure to which a meter and its attending system will be subjected. It determines the ANSI design rating of all components in the system.

static mixer *n*: a mixer with no moving parts. Kinetic energy of the moving fluid provides the power for mixing. An example is a jet hopper.

static pen *n*: in gas measurement with orifice meters, the pen on a flow recorder that records the static pressure, usually in blue or black ink, on the chart.

static pressure *n*: 1. the stationary or line pressure existing in a vessel or pipe. 2. the pressure exerted by a fluid on a surface that is at rest in relation to the fluid. 3. the pressure exhibited at the surface or point downhole during the time the well is shut in. 4. surface or bottomhole pressure after sufficient time has elapsed for the pressure to become stable.

static range *n*: the extent of variation between the lowest and highest static pressures that a particular orifice meter can record.

static spring *n*: in gas measurement by orifice meters, the spring in a flow recorder whose size and strength determines the range over which static pressure will be recorded.

static stability *n*: a mobile offshore drilling unit's ability to resist the forces that cause motion when the unit is floating at rest (is in a static, or unmoving, condition).

stationary front *n*: a discontinuity or zone between two adjacent air masses that is not moving or is moving only a little.

station bill *n*: on offshore drilling rigs and production platforms, a poster that gives duties and places for each individual on the rig or platform for various types of emergencies. Every person on the rig or platform should be familiar with the station bill.

stationkeeping *n*: the means used to maintain a rig or vessel at a particular location. This can be with moorings, dynamic positioning or, possibly tugs, or with a combination of the above.

stator *n*: 1. a device with vanelike blades that serves to direct a flow of fluid (such as drilling mud) onto another set of blades (called the rotor). The stator does not move; rather, it serves merely to guide the flow of fluid at a suitable angle to the rotor blades. 2. the stationary (unmoving) part of an induction-type alternating-current electric motor. The stator is the stationary part of a rotating motor, and it contains the stationary parts of the magnetic circuit and associated windings. Compare *rotor*.

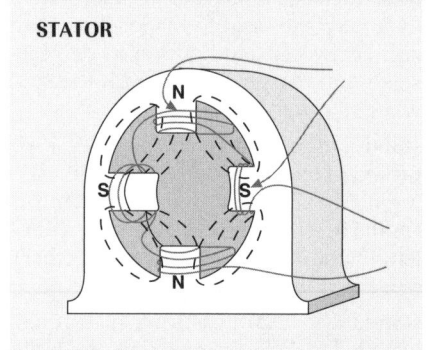

statute law *n*: 1. the descendant of Roman law, which, through the French Napoleonic Code, came to be the basis of law in Louisiana. 2. law enacted by a legislative body. Also called civil law.

statute mile *n*: a unit of linear measurement equal to 5,289 feet, 1,760 yards, or 1.6 kilometres.

statutory unitization *n*: unitization that proceeds without the willing cooperation of all the affected parties. It is authorized by order of a state regulatory agency in accordance with state statute. Also called forced unitization.

STB *abbr*: stock tank barrel.

STB/d *abbr*: stock tank barrels per day.

STCW *abbr*: Standards of Training, Certification, and Watchkeeping.

std *abbr*: standard.

stds *abbr*: stands; used in drilling reports.

steam *n*: water in its gaseous state.

steam boiler *n*: a closed steel vessel or container in which water is heated to produce steam.

steam coil *n*: a pipe, or set of pipes, within an emulsion settling tank through which steam is passed to warm the emulsion and make the oil less viscous. See *fire tube*.

steam cracking *n*: petrochemicals that are produced from hydrocarbons by thermal cracking in the presence of steam.

steam drive *n*: a method of improved recovery in which steam is injected into a reservoir through injection wells and driven toward production wells. The steam reduces the viscosity of crude oil, causing it to flow more freely. The heat vaporizes lighter hydrocarbons; as they move ahead of the steam, they cool and condense into liquids that dissolve and displace crude oil. The steam provides additional gas drive. This method is used to recover viscous oils. Also called continuous steam injection or steam flooding. Compare *cyclic steam injection, thermal recovery*.

steam flooding *n*: see *steam drive*.

steam fog *n*: a fog created when cold, dry air moves over a much warmer body of water. Also called sea smoke.

steam injection *n*: see *steam drive*.

steam methane reforming *n*: a series of chemical reactions that extract hydrogen from steam and methane.

steam rig *n*: a rotary drilling rig on which steam engines operate as prime movers. High-pressure steam is furnished by a boiler plant located near the rig. Steam rigs have been replaced almost totally by mechanical or electric rigs.

steam soak *n*: see *cyclic steam injection*.

stearate *n*: salt of stearic acid that is a saturated, 18-carbon fatty acid. Certain compounds, such as aluminum stearate, calcium stearate, and zinc stearate, are used in drilling fluids as defoamers or lubricants. Stearates can also be used as surfactants in air drilling to prevent small

quantities of water from creating mud that could clog the hole.

steel *n*: a hard, strong, durable, and malleable alloy of iron and carbon, which usually contains between 0.2 and 1.5 percent carbon. Steel often has other constituents such as manganese, chromium, nickel, molybdenum, copper, tungsten, cobalt, or silicon, depending on the desired alloy properties, and widely used as a structural material.

steel cone *n*: see *roller cone bit*.

steel-jacket platform rig *n*: a rigid offshore drilling platform used to drill development wells. The foundation of the platform is the jacket, a tall vertical section made of tubular steel members. The jacket, which is usually supported by piles driven into the seabed, extends upward so that the top rises above the waterline. Additional sections that provide space for crew quarters, the drilling rig, and all equipment needed to drill are placed on top of the jacket. See *platform rig*.

steel-tooth bit *n*: a roller cone bit in which the surface of each cone is made up of rows of steel teeth. Also called a milled-tooth bit or milled bit, although some steel teeth are forged.

steering diode *n*: a diode used in logic circuits to guide (steer) voltages in the circuits from one point to another. See *diode*.

steering tool *n*: a directional survey instrument used in combination with a deflected downhole motor. It shows, on a rig floor monitor, the inclination and direction of a downhole sensing unit.

stem *n*: see *sinker bar, swivel stem*.

stemming *n*: material used to hold back the force of an explosion, such as sand, gravel, or a cement plug placed in a well above a nitroglycerine charge.

step change *n*: in process control, the change of a variable from one value to another in a single process, which takes a negligible amount of time. In automatic controls, the step change can occur as an input change from the operator or process, or the step change can occur from a feedback sensor.

step-down transformer *n*: a transformer in which applied voltage of one value is decreased to a lower value by means of primary and secondary windings, which are coils of conductive wire wound around the transformer's laminated steel core. In a step-down transformer, more windings are made in the primary than in the secondary, which decreases the voltage in the secondary. Compare *step-up transformer*. See *transformer*.

step-out well *n*: a well drilled adjacent to or near a proven well to ascertain the limits of the reservoir; an outpost well.

step-over *n*: a device used in tank strapping for measuring the distance apart along the arc of two points on a tank shell where it is not possible to use a strapping tape directly because of an intervening obstruction, e.g., a protruding fitting.

stepper drive *n*: a means for driving remotely located meter accessories.

step-up transformer *n*: a transformer in which applied voltage of one value is increased to a higher value by means of primary and secondary windings, which are coils of conductive wire wound around the transformer's laminated steel core. In a step-up transformer, fewer windings are made in the primary than in the secondary, which increases the voltage in the secondary. Compare *step-down transformer*. See *transformer*.

stern *n*: the nautical term for the rear part of a ship or vessel.

stern roller *n*: a large rotating roller installed on the stern of AHTS vessels and used to deploy anchors, wire, and chain in a MODU mooring operation.

Stevmanta *n*: a type of VLA (vertically loaded anchor) manufactured by Vryhof.

stick *n*: a solid corrosion inhibitor that can be dropped down a well where it dissolves and mixes with the well fluids. See *corrosion inhibitor*.

stickiness *n*: the characteristic of a soft formation to adhere (stick) to the bit as the formation is drilled.

stick welding *n*: see *manual welding*.

stiff *n*: in naval architecture, a floating vessel's tendency not to heel over much in spite of great wind or other forces that tend to make the vessel heel. Compare *tender*.

stiff drilling assembly *n*: see *packed-hole assembly*.

still *n*: any vessel in which hydrocarbon distillation is effected. See *mud still*.

still column *n*: a vertical column atop the reboiler through which water and glycol vapors rise and are separated by condensation; i.e., the glycol condenses and falls back into the reboiler.

stilling well *n*: see *still pipe*.

still pipe *n*: a vertical cylindrical slotted pipe built into a tank to contain the liquid level-detecting element and arranged to reduce errors arising from turbulence or agitation of the liquid. Also called stilling well.

stillwater level *n*: in oceanography, the mean, or average, between the height of a wave and the depth of the trough between waves. For example, if the wave height is 10 feet (3 metres), the stillwater level is 5 feet (1.5 metres).

stimulation *n*: the action of attempting to improve and enhance a well's performance by the application of horsepower using pumping equipment, placing sand in artificially created fractures in rock, or using chemicals such as acid to dissolve the soluble portion of the rock.

stimulation valve *n*: see *surge valve*.

stinger *n*: 1. a cylindrical or tubular projection, relatively small in diameter, that extends below a downhole tool and helps to guide the tool to a designated spot (such as into the center of a portion of stuck pipe). 2. a device for guiding pipe and lowering it to the water bottom as it is being laid down by a lay barge. It is hinged to permit adjustments in the angle of pipe launch.

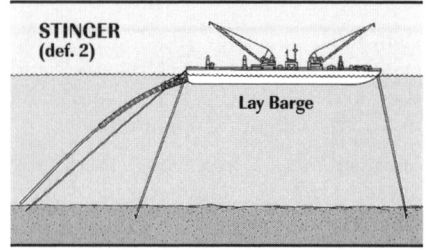

sting in *v*: to lower pipe or tubing into the bore of a downhole tool.

stipulation *n*: a condition, demand, or promise in an agreement or contract.

stn *abbr*: stain; used in drilling reports.

stock build *n*: oil produced but not consumed and therefore in storage.

stock tank *n*: a storage tank for treated crude oil. Compare *production tank*.

stock tank barrel (STB) *n*: a measure of the volume of crude oil in a stock tank on a lease or tank farm.

stock tank oil *n*: oil as it exists at atmospheric conditions in a stock tank. Stock tank oil lacks much of the dissolved gas present at reservoir pressure and temperatures.

stopcock *n*: a valve that shuts off or regulates fluid flow. *v*: to shut in intermittently to allow a buildup of gas pressure in a producing oilwell, which thus effects more efficient recovery.

stop gauge *n*: see *maximum loading gauge*.

storage battery *n*: a series of storage cells that produce electricity by chemical action of acid or alkaline solution on metallic plates. Charging the battery with DC electricity in the opposite direction restores the chemical condition necessary for further output of electricity.

storage capacity *n*: in computer technology, the quantity of data that can be retained simultaneously in a storage device; usually measured in bits, digits, characters, bytes, or words. Also known as capacity and memory capacity.

storage gas *n*: gas that is stored in an underground reservoir.

storage reel *n*: used for storing drilling line. See *supply reel*.

storage tank *n*: a tank in which oil is stored pending transfer by pipeline, truck, or other vehicle for selling.

storm *n*: a disturbance in the air above the earth that is characterized by strong winds that may be accompanied by rain, snow, sleet, or hail, and sometimes lightning and thunder.

Storm Choke *n*: a tubing safety valve.

Stormer viscometer *n*: a rotational shear viscometer used for measuring the viscosity and gel strength of drilling fluids. This instrument has been largely superseded by the direct-indicating viscometer.

storm packer *n*: a heavy-duty squeeze tool that can be closed to allow operations to cease during a severe storm offshore. Later, the tool can be opened to allow operations to continue.

storm plug *n*: a retrievable tool used to suspend drilling temporarily during a storm offshore.

storm surge *n*: a high tide near the shore that is significantly higher than normal because of high winds or storms. Also called storm tide.

storm tide *n*: see *storm surge*.

storm track *n*: a route traversed by low-pressure systems.

storm water discharge *n*: pollutants that are washed into water resources in storm situations. Such discharges are regulated under CWA.

stove pipe *n*: see *conductor casing*.

stovepipe assembly *n*: in laying pipe, an assembly on lay barges in which pipe joints are assembled in a continuous string. Each joint passes through individual work stations spaced along a gently sloping production ramp.

straddle packer *n*: two packers separated by a spacer of variable length. A straddle packer

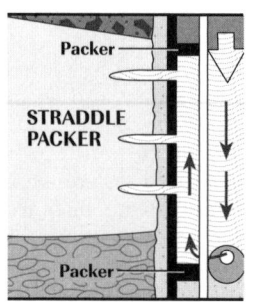

may be used to isolate sections of open hole to be treated or tested or to isolate certain areas of perforated casing from the rest of the perforated section.

straddle plant *n*: see *mainline plant*.

straddle test *n*: selective testing of an interval or formation by the use of two packers, one above and one below the zone being tested.

straight-chain compound *n*: a compound in which the atoms are aligned in a straight chain.

straightening vanes *n pl*: bundles of small-diameter tubing tack-welded together in a concentric pattern and placed in the upstream section of an orifice-meter run for the purpose of reducing considerably the amount of straight pipe required upstream of the orifice. They eliminate swirls and crosscurrents set up by the pipe fittings and valves preceding the meter tube.

straight hole *n*: a hole that is drilled vertically. The total hole angle is restricted, and the hole does not change direction rapidly—no more than 3° per 100 feet (30.48 metres) of hole.

straight-run *adj*: refers to a petroleum product produced by the primary distillation of crude oil.

strain *v*: to effect a change of form or size as a result of the application of a stress.

strainer *n*: 1. a part of a LACT unit that removes foreign particles from the crude, which might disrupt the operation of close-tolerance moving parts. 2. a device placed upstream of a meter to remove foreign material from the stream that might damage the meter or interfere with its operation. The strainer element is generally coarser than a filter designed to remove solid contaminants.

strain gauge *n*: an instrument used to measure minute distortions caused by stress forces in mechanical components.

strand *n*: a group of several wires laid together to form part of a wire rope. Several strands are laid together to make a wire rope.

strand core *n*: the center of a wire-rope strand; it may be made of wires or fibers.

stranded wire *n*: a conductor of electricity that is made by interweaving several wires of relatively small gauge to obtain the effect of a single wire of large gauge. In large gauge sizes, stranded wire is easier to bend than solid wire, so stranded wire is often used where large gauge wire is required and where it must be bent or curved rather sharply to fit into a particular installation.

stranding machine *n*: see *closing machine.*

strand pattern *n*: the manner in which a wire-rope manufacturer braids the individual wires that make up a wire-rope strand. Many patterns exist, including Seale, Warrington, and filler wire. See *filler wire, Seale strand, Warrington strand.*

strap *v*: to measure and record the dimensions of oil tanks to prepare tank tables (gauging tables) for determining accurately the volume of oil in a tank at any measured depth.

strap in *v*: to measure a length of pipe as it is run into the hole.

strapping *n*: the measurement of the external diameter of a vertical or horizontal cylindrical tank by stretching a steel tape around each course of the tank's plates and recording the measurement.

strapping pole *n*: an aluminum pole of adjustable length used on several rings of a tank while marking the working tape's path around the tank.

strapping tables *n pl*: see *gauging tables.*

strapping tape *n*: a measuring tape graduated in units of length and used for taking the measurements for producing a tank calibration table. Also called tank strapping.

strata *n pl*: distinct, usually parallel, and originally horizontal beds of rock. An individual bed is a stratum.

Strategic Petroleum Reserve (SPR) *n*: a large quantity of crude oil stored by the U.S. Federal Government in four subterranean sites in Texas and Louisiana. Established in 1975, the sites can currently store up to 700 million barrels of oil, although inventories vary. The SPR's purpose is to store crude oil for use in emergencies, such as when sources from outside the U.S. are interrupted. For the most part, the gravity of the crude oils are from 30 to 40° API and contain less than 2% total sulfur. The crude is commingled in large caverns associated with salt domes.

stratification *n*: the natural layering or lamination characteristic of sediments and sedimentary rocks.

stratigraphic correlation *n*: the process of comparing geologic formations.

stratigraphic test *n*: a borehole drilled primarily to gather information on rock types and sequence.

stratigraphic trap *n*: a petroleum trap that occurs when the top of the reservoir bed is terminated by other beds or by a change of porosity or permeability within the reservoir itself. Compare *structural trap.*

stratigraphic unit *n*: unit consisting of stratified, mainly sedimentary, rocks

grouped for description, mapping, correlation, or the like.

stratigraphy *n*: a branch of geology concerned with the study of the origin, composition, distribution, and succession of rock strata.

stratocumulus *n*: gray or white clouds that appear in rolls or as a continuous sheet broken into irregular parallel bands. This type of cloud is composed of water droplets.

stratospheric ozone layer *n*: a layer of ozone in the atmosphere that forms the earth's main shield against the sun's ultraviolet radiation, which in large doses increases the risk of cancer (especially skin cancer and melanoma), health problems, and crop damage.

strat test *n*: common name for stratigraphic test.

stratum *n*: singular of strata. A distinct, generally parallel bed of rock.

stratus *n*: a low-level cloud composed of water droplets, which covers the sky in a gray sheet.

stray current *n*: a portion of an electric current that flows over a path other than the intended path, causing corrosion of structures immersed in the same electrolyte.

streaking *n*: in paint technology, the formation of irregular lines—streaks—of various colors in a paint film. It is often caused by contamination or by insufficient or improper incorporation of a colorant.

stream *n*: the liquid or gas contained in any pipeline or flowing line.

stream conditioning *n*: the mixing of pipeline contents, upstream of the sampling location, that is necessary for delivery of a representative sample.

stream day *n*: a day of full plant operation. A basis for calculating plant production that differs from a calendar day, which would be used to give average production for a full year.

streamer *n*: marine recording cable.

streaming potential *n*: the electrokinetic portion of the spontaneous potential electric-log curve that can be influenced significantly by the characteristics of the filtrate and mud cake or the drilling fluid that was used to drill the well.

streamline flow *n*: flow of a fluid in which no turbulence occurs. The fluid follows a well-defined, continuous path.

stress *n*: a force that, when applied to an object, distorts or deforms it.

stress a tank *v*: to fill a tank completely with a liquid that has the same or greater density as the liquid that will normally be in the tank. Stressing the tank prior

to strapping it ensures that the measurements will more accurately reflect the tank's true capacity.

stress concentrator *n*: a notch or pit on a pipe or joint that raises the stress level and concentrates the breakdown of the metal structure. Also called a stress riser.

stress joint *n*: a fitting used in subsea offshore production that provides an interface between the emergency disconnect package and the standard riser. Normally it is tapered and is thickest at its lower end. Tapering provides extra bending capacity at the base of the riser.

stress riser *n*: see *stress concentrator.*

stretch *n*: in crane operations, the permanent deformation of a wire rope or sling caused by the pull of a load too heavy for the rated load capacity of the rope or sling. A stretched rope or sling should be immediately removed from service. Also called elongation.

stretched pin *n*: a tool joint pin which has been subjected to loading which has caused permanent lengthening of the threaded length of the pin. This condition generally results from excessive torque rather than tensile loads.

strike *n*: the horizontal direction of a formation bed or fault plane as measured at a right angle to the dip.

strike plate *n*: an extra piece of metal placed on the bottom of an oil storage tank to protect it from the repeated striking of the plumb bob at the end of the gauger's tape.

strike slip *n*: horizontal displacement along a fault plane. The San Andreas fault in California is a strike-slip fault.

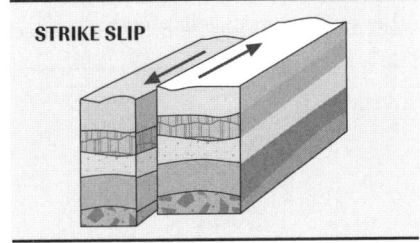

STRIKE SLIP

striking point *n*: a spot on the bottom of a storage tank or on the datum plate that is directly below the reference point on the hatch. This location is where the innage bob comes to rest when the tank is gauged and serves as the zero point for all innage measurements.

string *n*: an assembly of several individual joints, or lengths, of tubulars, such as drill pipe, drill collars, casing, and tubing or other lengths of equipment, such as sucker rods. The tubulars or rods are joined to form a continuous length of pipe or rods.

For example, connecting several single joints of drill pipe form a string of drill pipe that can be thousands of feet or metres long; or connecting several lengths of sucker rods form a string of sucker rods that can also be thousands of feet or metres long.

stringer *n*: 1. an extra support placed under the middle of racked pipe to keep the pipe from sagging. 2. a relatively narrow splinter of a rock formation that is stratigraphically disjointed, interrupts the consistency of another formation, and makes drilling that formation less predictable. A shale formation, for example, may be broken by a stringer of sandstone.

stringer bead *n*: see *root bead*.

string float *n*: see *drill pipe float*.

stringing *n*: in pipeline construction, the process of delivering and distributing line pipe where and when it is needed on the right-of-way. Stringing also includes the delivery of joints of special wall thickness and pipe grade to specific locations, such as road crossings, where heavy wall thickness may be specified by the contract or by regulations. Pipe is strung so that the movement of livestock and vehicles is not impeded.

string shot *n*: an explosive method utilizing Primacord, which is an instantaneous textile-covered fuse with a core of very high explosive. It is used to create an explosive jar inside stuck drill pipe or tubing so that the pipe may be backed off at the joint immediately above where it is stuck.

string-shot back-off *n*: see *string shot*.

string up *v*: to thread the drilling line through the sheaves of the crown block and traveling block. One end of the line is secured to the hoisting drum and the other to the derrick substructure.

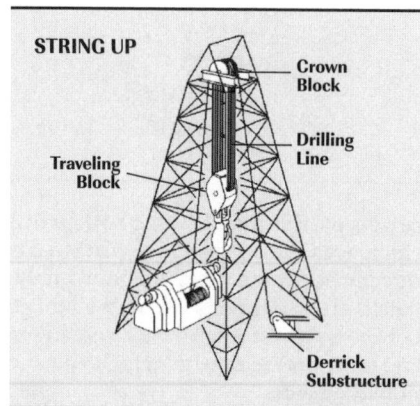

STRING UP

Crown Block

Drilling Line

Traveling Block

Derrick Substructure

strip a well *v*: 1. to pull rods and tubing from a well at the same time—for example, when the pump is stuck. Tubing must be stripped over the rods a joint at a time, and the exposed sucker rod is then

backed off and removed. 2. to move the drill stem, tubing, and other tools into or out of the hole with the well closed in. If the weight of the pipe is sufficient to overcome the upward force of well pressure, then the pipe can be stripped in. Compare *snub*.

strip chart *n*: strip charts are sometimes used in lieu of the circular chart for recording gas flow through an orifice meter.

stripped gas *n*: a processed gas from which liquefied hydrocarbons have been removed.

stripper *n*: 1. a well nearing depletion that produces a very small amount of oil or gas, usually ten barrels per day or less. 2. a stripper head. 3. a column wherein absorbed constituents are stripped from absorption oil. The term is applicable to columns using a stripping medium, such as steam or gas.

stripper head *n*: a blowout prevention device consisting of a gland and packing arrangement bolted to the wellhead. It is often used to seal the annular space between tubing and casing.

stripper oil *n*: see *upper tier*.

stripper rubber *n*: 1. a rubber disk surrounding drill pipe or tubing that removes mud as the pipe is brought out of the hole. 2. the pressure-sealing element of a stripper blowout preventer. See *stripper head*.

stripping *n*: see *strip a well*.

stripping gas *n*: in a glycol dehydration unit, hot gas that is bubbled through rich glycol to strip water from the glycol and carry it away as vapor.

stripping in *n*: 1. the process of lowering the drill stem into the wellbore when the well is shut in on a kick and when the weight of the drill stem is sufficient to overcome the force of well pressure. 2. the process of putting tubing into a well under pressure.

stripping job *n*: the simultaneous pulling of rods and tubing when the sucker rod pump or rods are frozen in the tubing string.

stripping out *n*: 1. the process of raising the drill stem out of the wellbore when the well is shut in on a kick. 2. the process of removing tubing from the well under pressure.

strip pipe *v*: 1. to remove the drill stem from the hole while the blowout preventers are closed. 2. to pull the drill stem and the washover pipe out of the hole at the same time.

strks *abbr*: streaks; used in drilling reports.

stroke *n*: the up and down (reciprocating) movement of a piston in a cylinder.

stroke jar *n*: a mechanical percussion tool that can deliver either upward or downward impact. See *jar*.

stroke length *n*: in a telescopic joint, the distance the inner barrel moves within the outer barrel. It can be from 45 to 65 feet (14 to 20 metres) depending on the rig requirements.

strokes-per-cycle *n pl*: the number of strokes an engine piston makes inside a cylinder to complete one firing cycle. Most engines are either two-strokes-per-cycle or four-strokes-per-cycle. In a two-strokes-per-cycle engine, the engine crankshaft makes two revolutions to complete one cycle. As the crankshaft moves the piston down on the first stroke, fuel is injected and combustion and power occur. As the crankshaft moves the piston up on the second stroke, burned gases go to exhaust, air is forced into the cylinder, and the piston compresses the air as it moves up the cylinder. In a four-strokes-per-cycle engine, the crankshaft makes four revolutions to complete one cycle. As the crankshaft moves the piston down on the first stroke, the piston draws in air (or it is forced in with a blower). As the crankshaft moves the piston up, the piston compresses the air in the cylinder; just before the piston reaches the top of its travel, fuel is injected. Combustion of the fuel-air mixture creates power and moves the piston down. Finally, as the piston moves up on the fourth stroke, the piston pushes the burned gases into the exhaust system.

strokes per minute (spm) *n*: the number of times all the mud pump's pistons move forward and back to complete one stroke.

strong gale *n*: in the Beaufort wind scale, a wind whose speed is from 41 to 47 knots (47 to 54 miles per hour or 76 to 87 kilometres per hour).

structural casing *n*: see *conductor casing*.

structural mast *n*: a portable mast constructed of angular as opposed to tubular steel members. See *jackknife mast*.

structural trap *n*: a petroleum trap that is formed because of deformation (such as folding or faulting) of the reservoir formation. Compare *stratigraphic trap*.

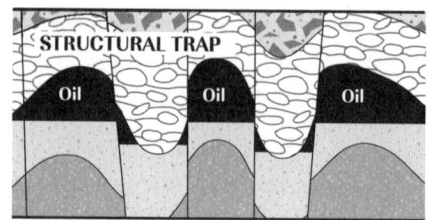

STRUCTURAL TRAP

Oil   Oil   Oil

structure *n*: a geological formation of interest to drillers. For example, if a particular well is on the edge of a structure, the wellbore has penetrated the reservoir (structure) near its periphery.

structure contour map *n*: horizontal representation of elevations of a subsurface rock layer or structure.

structure map *n*: a map of critical horizons—subsurface oil-producing zones—in a given area.

S-tube *n*: in a top drive, the S-shaped pipe that incorporates a gooseneck to which the rotary hose is attached. It conducts drilling mud into the top drive from the rotary hose. See *top drive*.

stublines *n pl*: lines that branch off from the mainlines and carry products to nearby areas.

stuck pipe *n*: drill pipe, drill collars, casing, or tubing that has inadvertently become immovable in the hole. Sticking may occur when drilling is in progress, when casing is being run in the hole, or when the drill pipe is being hoisted.

stuck point *n*: the depth in the hole at which the drill stem, tubing, or casing is stuck. Also called freeze point.

studless chain *n*: (nautical) an anchor chain that does not have studs (a short bar across the center of the chain link). Studless chain is lighter in weight than stud-link chain and thus sometimes used where a rig must be anchored in deep water. Compare *stud-link chain*.

stud-link chain *n*: (nautical) an anchor chain on which each link has a bar, or stud, across the shorter dimension of the link to prevent kinking and deformation under load.

stuffing box *n*: a device that prevents leakage along a piston, rod, propeller shaft, or other moving part that passes through a hole in a cylinder or vessel. It consists of a box or chamber made by enlarging the hole and a gland containing compressed packing. On a well being artificially lifted by means of a sucker rod pump, the polished rod operates through a stuffing box, preventing escape of oil and diverting it into a side outlet to which is connected the flow line leading to the oil and gas separator or to the field storage tank. For a bottomhole pressure test, the wireline goes through a

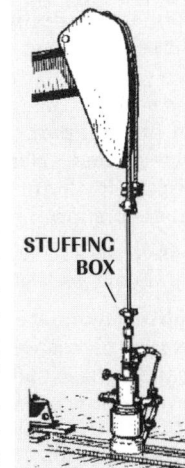

STUFFING BOX

stuffing box and lubricator, allowing the gauge to be raised and lowered against well pressure. The lubricator provides a pressure-tight grease seal in the stuffing box.

stump pressure test *n*: a pressure test of a subsea blowout preventer stack performed on the rig floor on a test stump—a device that allows pressure to be exerted on the stack—to ensure that all the pressure-sealing elements of the stack are working properly.

styolite *n*: an irregular surface, generally parallel to a depositional layer, in which small, toothlike projections on one side of the surface fit into cavities of complementary shape on the other surface.

sub *n*: a short, threaded piece of pipe used to adapt parts of the drilling string that cannot otherwise be screwed together because of differences in thread size or design. A sub (i.e., a substitute) may also perform a special function. Lifting subs are used with drill collars to provide a shoulder to fit the drill pipe elevators; a kelly saver sub is placed between the drill pipe and the kelly to prevent excessive thread wear of the kelly and drill pipe threads; a bent sub is used when drilling a directional hole.

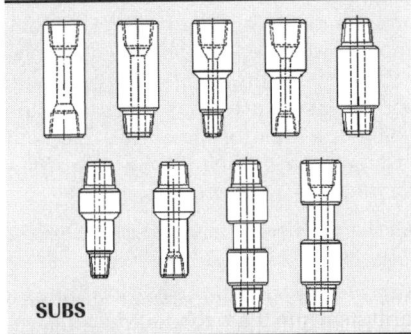

SUBS

subduction zone *n*: a deep trench formed in the ocean floor along the line of convergence of oceanic crust with other oceanic or continental crust when one plate (always oceanic) dives beneath the other. The plate that descends into the hot mantle is partially melted. Magma rises through fissures in the heavier, unmelted crust above, creating a line of plutons and volcanoes that eventually form an island arc parallel to the trench.

sub elevator *n*: a small attachment on the rod-transfer equipment that picks up the rods after they are unscrewed from the string and then transfers them to the rod hanger, or reverses the procedure when going into the hole. See *rod-transfer equipment*.

subgeologic map *n*: a map of the formations directly above an unconformity. Also called a worm's-eye map.

submerged *n*: a state in which a rig that floats on the surface while being moved is in contact with the seafloor when it is in the drilling mode.

submerged-arc welding *n*: an automatic welding process utilizing a continuous wire feed and a shielding medium of fusible granular flux. Submerged-arc welding offers high deposition rates and weld passes of substantial thickness.

Submerged Lands Act *n*: congressional act that grants coastal states jurisdiction over a belt of submerged lands that extend seaward off the coastlines to a distance of 3 geographical miles (9 miles for Texas and Florida). Various state agencies are given authority under this act to enforce state environmental provisions for these lands.

submerged production system (SPS) *n*: a grouping of satellite wells that supply natural gas or crude oil to a centralized production system.

submersible *n*: 1. a two-person submarine used for inspection and testing of offshore pipelines. 2. a submersible drilling rig.

submersible buoy *n*: a buoy used in a mooring leg which is submerged when the mooring leg is connected to a rig.

submersible drilling rig *n*: a mobile bottom-supported offshore drilling structure with several compartments that are flooded to cause the structure to submerge and rest on the seafloor. Submersible rigs are designed for use in shallow waters to a maximum of 175 feet (53.4 metres). Submersible drilling rigs include the posted barge submersible, the bottle-type submersible, and the arctic submersible. See *bottom-supported offshore drilling rig*.

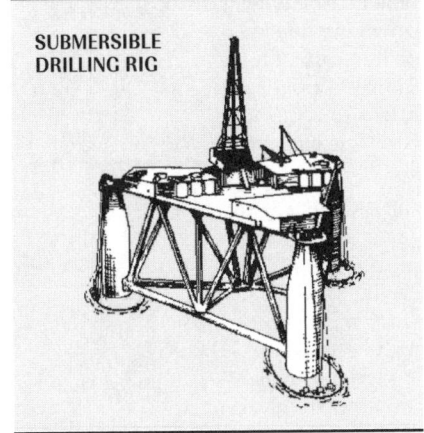

SUBMERSIBLE DRILLING RIG

submersible pump *n*: a pump that is placed below the level of fluid in a well. It is usually driven by an electric motor and consists of a series of rotating blades that impart centrifugal motion to lift the fluid to the surface.

subnormal formation pressure *n*: formation pressure that is below the pressure that is expected at a given depth in a well. In general, subnormal formation pressure exerts less than 0.433 pounds per square inch per foot (9.70 kilopascals per metre).

subordination *n*: in the case of an oil and gas lease, a supplementary agreement that resolves the priority of rights to the leased property and subordinates an earlier instrument (for example, a mortgage) to the oil and gas lease.

subscript *n*: in mathematics, a number or symbol written below and usually to the right of another number or symbol for any of a number of purposes, such as to identify a particular element or elements of a set, to denote a constant value or a variable, or, in a chemical formula, to indicate the number of atoms of a particular kind of molecule.

subsea accumulator *n*: a device mounted on or near a subsea blowout preventer stack, which stores hydraulic operating fluid and, because of its close proximity to the stack, shortens the length of time required to close a preventer. It may also store fluid under high pressure to close a shear ram.

subsea accumulator module *n*: a special tool used in subsea production systems in water depths greater than 1,500 feet (500 metres). It is installed on the completion riser. In an emergency, a crew member on the surface facility activates the subsea accumulator module from a surface control module. This action disconnects the riser from either the subsea Christmas tree or the locked down tubing hanger in less than 60 seconds.

subsea blowout preventer *n*: a blowout preventer placed on the seafloor for use by a floating offshore drilling rig. See *blowout preventer*.

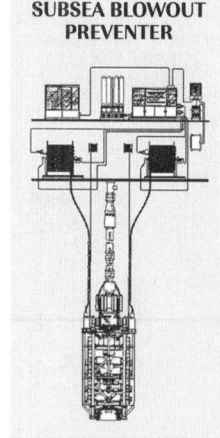

SUBSEA BLOWOUT PREVENTER

subsea choke-line valve *n*: a valve mounted in the choke line of a subsea blowout preventer stack. It serves to regulate the flow of well fluids being circulated through the choke line when the well is closed in.

subsea Christmas tree control system *n*: equipment installed on a subsea Christmas tree that allows Christmas tree valves, downhole safety valves,

chokes, chemical control valves, and metering valves to be controlled from the surface. The control system can also monitor process variables such as pressure, temperature, flow rate, and valve and choke position.

subsea completion *n*: a well completion in which the flow of hydrocarbons from the well is controlled by equipment placed on or below the seafloor. See *well completion*.

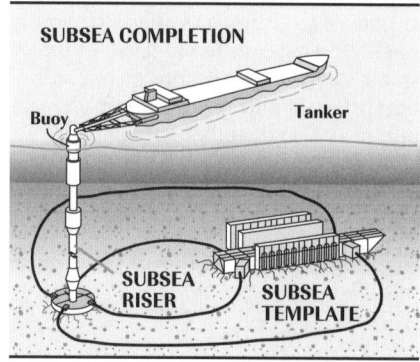

SUBSEA COMPLETION

Buoy   Tanker

SUBSEA RISER   **SUBSEA TEMPLATE**

subsea control pod *n*: see *hydraulic control pod*.

subsea electronic module (SEM) *n*: in drilling from floating drilling rigs in deep water, a device in a multiplex electronic (MUX) control system that contains a programmable logic controller, a modem, and a power supply. A SEM converts signals received from the blowout preventer operating system on the surface to ensure quick operation of the subsea blowout preventer components. See *multiplex electronic (MUX) control system*.

subsea engineer *n*: see *subsea equipment supervisor*.

subsea equipment supervisor *n*: an employee on a floating offshore drilling rig whose main responsibility is running, monitoring, and maintaining such subsea equipment as the blowout preventer stack, the marine riser system, and similar subsea equipment.

subsea mudlift drilling *n*: deepwater offshore drilling from a floating rig using a special system of pumps, valves, and additional equipment. The system, which is installed on the seafloor at the top of the borehole and below the riser pipe, allows the rig to maintain seawater (or other relatively light-density fluid) in the riser and simultaneously circulate drilling mud of a higher density in the borehole. In regular offshore drilling from a floating rig, the rig circulates mud down the drill string inside the riser and borehole and back to the surface through the riser. A subsea mudlift system allows the rig to circulate drilling mud down the drill

string as usual. However, the mud returning from the bottom of the hole does not circulate up the riser pipe. Instead, the subsea mudlift system diverts the mud into riser's choke and kill lines, which transport the mud back to the rig on the water's surface. Using seawater in the riser and drilling mud in the borehole is referred to as *dual gradient drilling*. In deep water, such as 8,000 feet or more, the hydrostatic pressure developed by the fluid in the riser adds a significant amount of pressure to the bottom of the hole. When drilling, this pressure can fracture, or break down, the formations lying just under the seafloor. The conventional solution to the problem is for rig crews to drill through these easily fractured formations with a light-density fluid and then run casing to protect them from the heavy muds they will use to drill the well to total depth. A subsea mudlift drilling system reduces the hydrostatic pressure created in the riser pipe, which, in turn, reduces hydrostatic pressure in the wellbore. This reduced hydrostatic pressure allows the operator to reduce the number of casing strings run in the well.

subsea pilot manipulated (SPM) *adj*: of a device that is operated and controlled (manipulated) by a subsea pilot valve. See *subsea pilot manipulated (SPM) valve*.

subsea pilot manipulated (SPM) valve *n*: on the control pod of a subsea blowout preventer system, a control device (a valve) employed in the operation of the blowout preventers. An SPM valve is a hydraulically operated, three-position, four-way valve that directs regulated pressure to the blowout preventers.

subsea riser *n*: a vertical section of pipe that connects pipeline on the sea bottom to a production platform on the surface. The riser is an integral part of the pipeline and is clamped directly to a leg or brace on the platform.

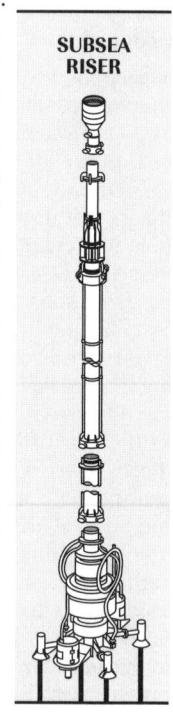

SUBSEA RISER

subsea stack *n*: see *subsea blowout preventer*.

subsea template *n*: a device placed on the seafloor to facilitate the production of wells. When a template is used, the wells are drilled through the template and are completed and produced on it. Since the erection of a platform to produce the wells is not necessary,

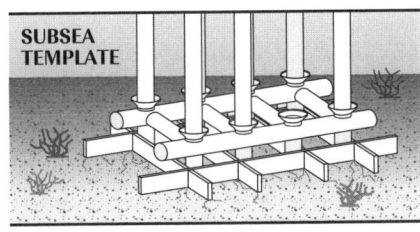

marginal offshore fields can sometimes be produced because the expense of erecting the platform is avoided.

subsea test tree *n*: a device designed to be landed in a subsea wellhead or blowout preventer stack to provide a means of closing in the well on the ocean floor so that a drill stem test of an offshore well can be obtained.

subsea wellhead *n*: the equipment installed at the top of the wellbore, which is below the water's surface and on the seabed, and to which the blowout preventer stack is attached.

subsea wellhead and casing system *n*: in offshore drilling from floating rigs, the equipment and pipe (casing) that forms the foundation of a well and links the wellbore to the subsea blowout preventer stack.

substitute natural gas (SNG) *n*: see *synthetic natural gas.*

substrate *n*: in paint technology, the surface to which a coat of paint or varnish is applied.

substructure *n*: the foundation on which the derrick or mast and usually the drawworks sit. It contains space for storage and well-control equipment.

subsurface *adj*: below the surface of the earth (e.g., subsurface rocks).

subsurface environment *n*: the environment below the surface of the earth; groundwater, deep water, underground formations, etc.

subsurface geology *n*: the study of rocks that lie beneath the surface of the earth.

subsurface safety valve *n*: see *tubing safety valve.*

subsurface sampling *n*: a procedure in which a bottomhole sampler is lowered into the well and filled with a sample that is representative of the reservoir conditions and that contains all the constituents of the fluid in their true proportions. Tests run on this sample help to obtain an accurate knowledge of the physical properties of the reservoir fluid under actual conditions.

subtrahend *n*: in mathematics, the quantity that is subtracted from another quantity. For example, in the equation 18 − 8 = 10, 8 is the subtrahend.

sucker rod *n*: a special steel pumping rod. Several rods screwed together make up the mechanical link from the beam pumping unit on the surface to the sucker rod pump at the bottom of a well. Sucker rods are threaded on each end and manufactured to dimension standards and metal specifications set by the petroleum industry. Lengths are 25 or 30 feet (7.6 or 9.1 metres); diameter varies from ½ to 1⅛ inches (12 to 30 millimetres). There is also a continuous sucker rod (trade name: Corod™).

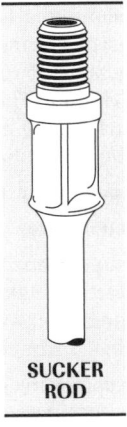

SUCKER ROD

sucker rod coupling *n*: an internally threaded fitting used to join sucker rods.

sucker rod pump *n*: the downhole assembly used to lift fluid to the surface by the reciprocating action of the sucker rod string. Basic components are barrel, plunger, valves, and hold-down. Two types of sucker rod pumps are the tubing pump, in which the barrel is attached to the tubing, and the rod, or insert, pump, which is run into the well as a complete unit.

sucker rod pumping *n*: a method of artificial lift in which a subsurface pump located at or near the bottom of the well and connected to a string of sucker rods is used to lift the well fluid to the surface. The weight of the rod string and fluid is counterbalanced by weights attached to a reciprocating beam or to the crank member of a beam pumping unit or by air pressure in a cylinder attached to the beam.

sucker rod whip *n*: an undesirable whipping motion in the sucker rod string that occurs when the string is not properly attached to the sucker rod pump or when the pump is operated at a resonant speed.

suction *n*: in mud circulation, the drawing in of liquid into a pump.

suction dampener *n*: a steel chamber with an air-charged rubber bladder (or diaphragm) inside that is mounted on a mud pump's intake (suction) line. It absorbs the impact that occurs when the smooth flow of fluid in the suction line coming out of the suction tank meets the intermittent flow caused by fast moving pump pistons.

suction dredge *n*: in pipe laying, a type of trenching machine used on river crossings when the channel cannot be diverted or when the volume of material to be removed is large. A suction pump rapidly forces large amounts of soil into a discharge pipe for deposit on the adjacent bank. A cutterhead can also be used on a suction dredge.

suction embedment anchor (SEA) *n*: an anchor type (usually tubular or cylindrical) embedded into the seafloor using suction techniques, mainly through skin friction with the surrounding soil. SEAs are able to withstand high vertical loads.

suction embedment plate anchor (SEPLA) *n*: a type of plate anchor or VLA embedded into the seafloor using suction techniques, similar to a SEA, but when the suction embedment tool (or follower) is removed, holding capacity is achieved by the soil acting on the plate anchor.

suction head *n*: see *suction pressure.*

suction line *n*: the line that carries a product out of a tank to the suction side of the pumps. Also called the loading line.

suction pit *n*: also called a mud suction pit, suction tank, or sump pit. See *suction tank.*

suction pressure *n*: the pressure of a gas or fluid entering the suction valve of a compressor.

suction screen *n*: see *oil strainer.*

suction-side *adj*: inlet side of a compressor or the side on which the gas enters the compressor.

suction strainer *n*: a sieve-like device installed on the suction pipe of a mud pump that prevents large pieces of foreign material from being ingested by the pump.

suction tank *n*: the mud tank from which mud is picked up by the suction of the mud pumps. Also called a mud suction pit, sump pit, or a suction pit.

suction temperature *n*: the temperature of a gas or fluid entering the suction valve of a compressor.

suction valve *n*: on a mud pump, the intake valve through which mud is drawn into the pump.

suicide squeeze *n*: a type of squeeze cement job in which the cement is pumped (squeezed) through perforations that are made in the tubing above the packer.

suitcase sand *n*: a formation found to be nonproductive. When such a formation is encountered, operations are suspended, and the crews pack their suitcases and move to another job.

sulfamic acid *n*: a crystalline acid (NH_2SO_3H) derived from sulfuric acid that is sometimes used in acidizing.

sulfate *n*: a compound containing the SO_4 group, as in calcium sulfate $(CaSO_4)$.

sulfate-reducing bacteria *n pl*: bacteria that digest sulfates present in water, causing the release of hydrogen sulfide, which combines with iron to form iron sulfide, a troublesome scale.

sulfate resistance *n*: the ability of a cement to resist deterioration by sulfate ions.

sulfide stress cracking *n*: the brittle failure of metals by cracking under the combined action of tensile stress and corrosion in the presence of water and hydrogen sulfide.

sulfur (S) *n*: a pale yellow, nonmetallic chemical element. In its elemental state—free sulfur—it has a crystalline or amorphous form. In many gas streams, sulfur may be found as volatile sulfur compounds—hydrogen sulfide, sulfur oxides, mercaptans, carbonyl sulfide. Reduction of their concentration levels is necessary for corrosion control and, in many cases, necessary for health and safety reasons.

sulfur dioxide *n*: a colorless gaseous compound of sulfur and oxygen (SO_2) with the odor of rotten eggs. A product of the combustion of hydrogen sulfide, it is poisonous and irritating.

sulfureted hydrogen *n*: see *hydrogen sulfide.*

sulfuric acid(H_2SO_4) *n*: a colorless, oily liquid compound of hydrogen, sulfur, and oxygen (H_2SO_4), strongly poisonous and corrosive. It is formed when hydrogen sulfide (H_2S) or sulfur dioxide (SO_2) is mixed with water (H_2O). Also called vitriolic acid. It is often used as an electrolyte in cells and batteries.

sulfur light crude *n*: light crude oil that contains sulfur compounds, often in the form of hydrogen sulfide.

sulfurous acid *n*: a colorless liquid compound of hydrogen, sulfur, and oxygen (H_2SO_3), weakly corrosive, with the odor of sulfur. It is formed when hydrogen sulfide (H_2S) or sulfur dioxide (SO_2) is mixed with water (H_2O).

sulfur plant *n*: a plant that makes sulfur from the hydrogen sulfide extracted from natural gas. One-third of the hydrogen sulfide is burned to sulfur dioxide, which reacts with the remaining hydrogen sulfide in the presence of a catalyst to make sulfur and water.

sul wtr *abbr*: sulfur water; used in drilling reports.

sum *n*: in mathematics, the quantity that results when two or more quantities are added together. For example, the sum of 2 + 2 is 4.

summer valley *n*: the depression that occurs in the summer months in the daily load of a gas-distribution system or pipeline.

sump *n*: a low place in a steel guard or casing that surrounds a moving chain or gear that requires constant lubrication. The sump holds a quantity of oil through which the moving parts travel and thus become lubricated.

sump pit *n*: see *suction pit.*

sun battery *n*: see *solar cell.*

supercharge *v*: to supply a charge of air to the intake of an internal-combustion engine at a pressure higher than that of the surrounding atmosphere.

supercharged engine *n*: an engine in which a compressor raises the pressure of the air and forces it into the engine's cylinders.

supercharger *n*: a device that compresses atmospheric air and forces it into an engine. Raising the pressure of an engine's intake air makes it denser and thus delivers more oxygen into the cylinder. More oxygen produces more power in the combustion process.

supercharging pump *n*: a centrifugal pump that ensures flooded suction to a mud pump by moving a large volume of mud from the suction pit into the mud pump suction.

supercompressibility *n*: a deviation of natural gas from Boyle's and Charles's laws for ideal gas. Natural gas is not an ideal gas in that it is a mixture of several gases. As the pressure increases, the volume of space that a given weight of natural gas would occupy becomes increasingly smaller than the volume calculated by application of Boyle's and Charles's laws.

supercompressibility factor *n*: a term sometimes used to mean compressibility factor, but more often to mean the compressibility factor that is appropriate for high-pressure calculations. See *compressibility factor.*

Superfund Amendments and Reauthorization Act of 1986 *n*: see *SARA Title III.*

Superfund sites *n pl*: sites where hazardous waste has been disposed of improperly and that are targeted for government cleanup under CERCLA.

superposition *n*: 1. the order in which sedimentary layers are deposited, with the oldest layer on the bottom, the youngest layer on top. 2. the process of sedimentary layering.

super pressure *n*: formation pressure that is the result of the density of the fluids in the pores of the formation plus the weight of the rocks (overburden pressure) above the formation. Super pressure exerts about 1 pound per square inch per foot (22.6 kilopascals per metre), which is the equivalent of a mud density of 20 pounds per gallon (2,396.4 kilograms per cubic metre). See *overburden pressure.*

supersaturation *n*: the condition of containing more solute in solution than would normally be present at the existing temperature.

supertanker *n*: a tanker with a capacity over 100,000 deadweight tons (90,720 dead weight tonnes). Supertankers with a capacity larger than 100,000 deadweight tons but less than 500,000 deadweight tons (45,360 deadweight tonnes) are called very large crude carriers. Those with a capacity over 500,000 deadweight tons are called ultralarge crude carriers.

supervisory control *n*: in pipelining, the coordination of all facets of operations to ensure proper handling of oil movements. It begins before the physical movement of a product and continues through recordkeeping of quality and quantity of movement.

Supplemental Aviation Weather Reporting Station (SAWRS) *n*: a facility where weather observations are taken, prepared, and transmitted by a local National Weather Service (NWS) certified operator under federal government supervision.

supplemental gaseous fuel *n*: primarily synthetic natural gas, propane-air, and refinery (still) gas. May also include coke oven gas, biomass gas, manufactured gas, and air injected for Btu stabilization.

supply reel *n*: a spool that holds drilling line.

supply voltage *n*: in an electric circuit, the source of the electromotive force (voltage) in the circuit. For example, a battery provides supply voltage in some circuits.

support agreement *n*: an agreement between petroleum companies in which one contributes money or acreage to another's drilling operation in return for information gained from the drilling.

support flange *n*: on a riser joint of a floating offshore drilling rig, the flat area at each end of the joint to which other joints are connected. The flange supports static and dynamic loads on the riser system, as well as the auxiliary lines and buoyancy modules, if modules are fitted. See *riser pipe.*

support gimbal *n*: in deepwater offshore drilling operations from a floating rig, a device between the rotary table and the riser support spider. The support gimbal has hydraulic cylinders, or flex elements, that support the spider, the riser, and the blowout preventer stack. It compensates for riser offset caused by water motions, yet maintains the riser flange square to the drill floor to facilitate the making of connections. It also cushions shock loads imparted on the system. See *riser support spider.*

sur *abbr*: survey; used in drilling reports.

surface accumulator unit *n*: a storage and operating device on a floating offshore drilling rig that contains hydraulic fluid reservoirs, a fluid-mixing system, pumps, accumulator storage bottles, flowmeter, pod select valve, pilot manifold, hydraulic junction boxes, pressure gauges, and electrical junction boxes required to operate the subsea blowout preventer components. See *accumulator*.

surface-active agent *n*: see *surfactant*.

surface casing *n*: see *surface pipe*.

surface chart *n*: see *weather map*.

surface completion *n*: in an offshore completion, wellhead valves and fittings placed above the surface of the water.

surface decompression *n*: a process used by divers to eliminate inert gases from the tissues, whereby they breathe high partial pressures of oxygen while resting after a dive to reduce the risk of getting decompression sickness.

surface drilling unit *n*: an offshore drilling rig that is either a drill ship or a drilling barge; so called because the rig floats on the surface of the water.

surface estate *n*: rights and interests in the surface of land, created when the owner severs or separates his or her interests in the property.

surface hole *n*: that part of the wellbore that is drilled below the conductor hole but above the intermediate hole.

surface joint *n*: in offshore production systems from floating surface facilities, the uppermost joint of riser to which a Christmas tree and wellhead are attached.

surface marine gridded climatology (SMGC) data *n pl*: information made available by the United States Navy's Fleet Numerical Meteorology and Oceanography (METOC) Detachment on the state of the sea, weather, wind, cloud density, precipitation, visibility, and air and sea temperature in many places of the world.

surface-motion compensator *n*: see *heave compensator*.

surface owner *n*: owner of the rights and interests in a surface estate (where interests in a landed estate have been severed). Compare *mineral owner*.

surface pipe *n*: the first string of casing (after the conductor pipe) that is set in a well. It varies in length from a few hundred to several thousand feet (metres). Some states require a minimum length to protect freshwater sands. Also called surface casing.

surface plug *n*: a concrete plug placed in the surface casing usually when the well is abandoned.

surface pressure *n*: pressure measured at the wellhead.

surface-readout device *n*: an electronic device in which a probe is inserted into the drill stem near a directional drilling deflection tool. The probe sends continuous signals to the surface that show the direction and angle at which the bit is drilling. Readout devices greatly simplify accurate orientation of the drilling assembly so that a number of directional surveys can be eliminated. See *directional drilling*.

surface-readout instrument *n*: a telemetry instrument used in the surveying of a directional well. A downhole sensing unit records directional data, which are converted into signal pulses and then transmitted uphole, via conducting wireline or through the drilling fluid, to a monitor. See *downhole telemetry*.

surface safety valve *n*: a valve, mounted in the Christmas tree assembly, that stops the flow of fluids from the well if damage occurs to the assembly.

surface sample *n*: a sample taken at the surface of a liquid in a tank.

surface-sensing element *n*: the detecting element of a surface-sensing automatic tank gauge.

surface set *n*: in diamond bits, the setting of relatively large diamonds into the cutting surface of the bit; the manufacturer buries about two-thirds of each diamond into the bit's matrix, leaving about one-third of each diamond exposed. Compare *ridge set*.

surface stack *n*: a blowout preventer stack mounted on top of the casing string at or near the surface of the ground or the water. Surface stacks are employed on land rigs and on bottom-supported MODUs.

surface tension *n*: the tendency of liquids to maintain as small a surface as possible. It is caused by the cohesive attraction between the molecules of liquid.

surface test tree *n*: in surface well testing, a temporary set of valves installed for flow control on a wellhead with no Christmas tree.

surface-type drilling unit *n*: see *barge, drill ship, ship-shaped barge*.

surface waste *n*: waste incurred by line leaks, seepage, inexpedient storage, and so forth. Such waste is usually regulated by federal or state agencies.

surface waters of the United States *n pl*: the oceans, rivers, streams, lakes, ponds, water holes, mud flats, prairie potholes, bogs, and swamps of the United States.

surface wave *n*: wave that propagates along the earth's surface, e.g., the Love, Rayleigh, hydrodynamic, or coupled waves.

surface wind *n*: a wind that can be measured at a surface observing station.

surfactant *n*: a soluble compound that concentrates on the surface boundary between two substances such as oil and water and reduces the surface tension between the substances. The use of surfactants permits the thorough surface contact or mixing of substances that ordinarily remain separate. Surfactants are used in the petroleum industry as additives to drilling mud and to water during chemical flooding. See *micellar-polymer flooding, surfactant mud*.

surfactant mud *n*: a drilling mud prepared by adding a surfactant to a water-base mud to change the colloidal state of the clay from that of complete dispersion to one of controlled flocculation. Such muds were originally designed for use in deep, high-temperature wells, but their many advantages (high chemical and thermal stability, minimum swelling effect on clay-bearing zones, lower plastic viscosity, and so on) extend their applicability.

surge *n*: 1. an accumulation of liquid above a normal or average level, or a sudden increase in its flow rate above a normal flow rate. 2. the motion of a mobile offshore drilling rig in a direction in line with the centerline of the rig, especially the front-to-back motion of the rig when it is moored in a seaway.

surge column *n*: see *boot*.

surge dampener *n*: see *pulsation dampener*.

surge disk *n*: see *surge valve*.

surge effect *n*: a rapid increase in pressure downhole that occurs when the drill stem is lowered rapidly or when the mud pump is quickly brought up to speed after starting.

surge loss *n*: the flux of fluids and solids that occurs in the initial stages of any filtration before pore openings are bridged and a filter cake is formed. Also called spurt loss.

surge tank *n*: a tank or vessel through which liquids or gases are passed to ensure steady flow and to eliminate pressure surges.

surge valve *n*: a device employed with a packer or to clean open perforations by allowing surges of pressure to enter the perforations. Also called a surge disk.

surging *n*: 1. a rapid increase in pressure downhole that occurs when the drill stem is lowered too fast or when the mud pump is brought up to speed after starting. 2. in underbalanced drilling, the alternate slugs of liquid and gas that exit the well at the surface when the gas breaks out of the liquid as pressure goes down near the surface. Also called heading.

surrender clause *n*: the clause in an oil and gas lease that specifies the procedure to be followed should the lessee wish to surrender all or part of his or her leased interests.

surveying service company *n*: a business organization that is hired by an operator to measure a well's trajectory. The company takes complete surveys of the hole, usually either before or after setting a string of casing or upon completion of the hole. This type of surveying program is separate from, and independent of, the directional supervisor's surveys.

survival condition *n*: the condition of a vessel when drilling has been suspended. Usually the drilling riser has been disconnected and pulled (or hung off) and possibly the vessel has been abandoned (hurricane condition).

survival draft *n*: the depth to which the pontoons of a semisubmersible rig are submerged below the water's surface during a storm. Survival draft provides an adequate and safe distance (air gap) between the water's surface and the underside of the main deck.

suspended load *n*: 1. in a flowing stream of water, the finer sand, silt, and clay that are carried well off the bottom by the turbulence of the water. Compare *bed load, dissolved load*. 2. the weight of the drill stem when suspended from the hook.

suspended S&W *n*: sediment and water that are suspended in oil and that can be separated only by (1) centrifuge with appropriate solvents or (2) extraction by distillation.

suspending agent *n*: an additive used to hold the fine clay and silt particles that sometimes remain after an acidizing treatment in suspension; i.e., it keeps them from settling out of the spent acid until it is circulated out.

suspension *n*: a mixture of small nonsettling particles of solid material within a gaseous or liquid medium.

suspensoid *n*: a mixture consisting of finely divided colloidal particles floating in a liquid. The particles are so small that they do not settle but are kept in motion by the moving molecules of the liquid (Brownian movement).

SW *abbr*: salt water; used in drilling reports.

swab *n*: a hollow, rubber-faced cylinder mounted on a hollow mandrel with a pin joint on the upper end to connect to the swab line. A check valve that opens upward on the lower end provides a way to remove the fluid from the well when pressure is insufficient to support flow.

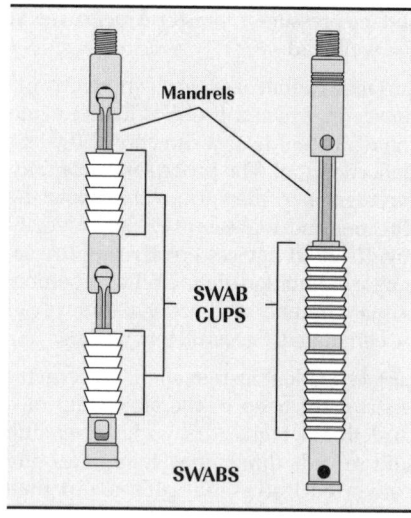

SWAB CUPS

SWABS

v: 1. to operate a swab on a wireline to bring well fluids to the surface when the well does not flow naturally. Swabbing is a temporary operation to determine whether the well can be made to flow. If the well does not flow after being swabbed, a pump is installed as a permanent lifting device to bring the oil to the surface. 2. to pull formation fluids into a wellbore by raising the drill stem at a rate that reduces the hydrostatic pressure of the drilling mud below the bit.

swabbed show *n*: formation fluid that is pulled into the wellbore because of an underbalance of formation pressure caused by pulling the drill string too fast.

swabbing *n*: see *swabbing effect*.

swabbing effect *n*: a phenomenon characterized by formation fluids being pulled, or swabbed, into the wellbore when the drill stem and bit are pulled up the wellbore fast enough to reduce the hydrostatic pressure of the mud below the bit. If enough formation fluid is swabbed into the hole, a kick can result.

swabbing line *n*: see *sandline*.

swab cup *n*: a rubber or rubberlike device on a special rod (a swab), which forms a seal between the swab and the wall of the tubing or casing.

swab off *v*: to pull off during a trip into or out of the hole because of pressure differential. For example, if a packer is run in too quickly, the pressure differential across the packer swabs off the packing elements, making it necessary to trip back out to replace them.

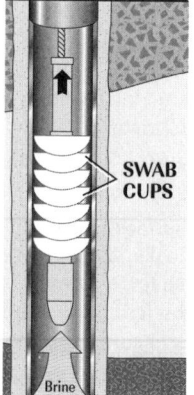

SWAB CUPS

Brine

Swaco underbalanced choke *n*: see *underbalanced choke*.

swag *n*: downward bend in a pipeline to conform to a dip in the surface of the right-of-way or to the contours of a ravine or creek.

swage *n*: a solid cylindrical tool pointed at the bottom and equipped with a tool joint at the top for connection with a jar. It is used to straighten damaged or collapsed casing or tubing and drive it back to its original shape. *v*: to reduce the diameter of a rod, a tube, or a fitting by forging, hammering, or other method.

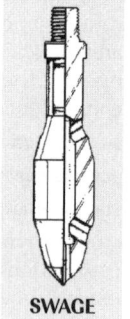

SWAGE

swage nipple *n*: a pipe fitting with external threads of different sizes on each end.

swaging *n*: the tendency of a body, such as a length of drill pipe, to be bent by the action of a tool, such as the slips, applied to it with a great deal of force. In the case of drill pipe, if the slips are allowed to stop the drill stem as the driller lowers it into the hole, the slips can swage, or deform, the body of the pipe by the force they apply against the pipe when they settle into the master bushing.

swallow float *n*: an instrument used to measure an ocean current's velocity. It can be adjusted to be neutrally buoyant and to travel at a specific density level. It is dropped into the water to travel with a current, and its journey is tracked by a pinger.

swamp barge *n*: see *inland barge rig*.

swamper *n*: (slang) a helper on a truck, tractor, or other machine.

sway *n*: the motion of a mobile offshore drilling rig in a linear direction from side to side or perpendicular to a line through the centerline of the rig; especially, the side-to-side motion when the rig is moored in a seaway.

sway braces *n pl*: stiffening ropes or rods with or without turnbuckles, sometimes used on each side of a boom.

swbd *abbr*: swabbed; used in drilling reports.

swbg *abbr*: swabbing; used in drilling reports.

swedge *n*: see *swage*.

sweep efficiency *n*: the efficiency with which water displaces oil or gas in a water drive oil or gas field. Water flowing in from the aquifer does not displace the oil or gas uniformly but channels past certain areas due to variations in porosity and permeability.

sweep-out pattern *n*: the areal pattern of the injection fluid advancing from injection wells to production wells during an improved recovery operation.

sweep voltage *n*: periodically varying voltage applied to the deflection plates of a cathode-ray tube to give a beam displacement that is a function of time, frequency, or other data base.

sweet *adj*: having an absence or near-absence of sulfur compounds, as defined by a given specification standard.

sweet corrosion *n*: the deterioration of metal caused by contact with carbon dioxide in water.

sweet crude *n*: see *sweet crude oil*.

sweet crude oil *n*: oil containing little or no sulfur, especially little or no hydrogen sulfide.

sweeten *v*: to remove sulfur or sulfur compounds from gas or oil.

sweet gas *n*: gas that has no more than the maximum sulfur content defined by (1) the specifications for the sales gas from a plant or (2) the definition by a legal body such as the Railroad Commission of Texas.

swell *n*: a wave that has moved out of its fetch and into weaker winds, where it decreases in height and has regular movement.

swelled box *n*: a box connection on a tool joint that has been belled by too much torque.

swelling *n*: see *hydration*.

swing *n*: rotation of the superstructure for the movement of loads in a horizontal direction about the axis of rotation.

swing bearing *n*: a combination of rings with balls or rollers capable of sustaining radial, axial, and moment loads of the revolving superstructure with boom and load.

swing check *n*: a type of check valve.

swing gear *n*: external or internal gear with which swing pinion on the revolving superstructure meshes to provide swing motion.

swing lease *n*: a lease from which gas production can be used to eliminate over/under delivery balances under a gas exchange arrangement.

swingline *n*: an extension of the suction line that pivots vertically inside an oil tank. It allows an operator to withdraw product from varying heights in the tank.

swing mechanism *n*: the machinery involved in providing dual directional rotation of the revolving superstructure.

swing production *n*: the production of a country, zone, or area that acts as a balancing region. The swing producer needs adequate reserves to meet demand surges and can tolerate low production when demand slows. Since the late 1960s, Saudi Arabia has been considered the world's swing producer.

swing suction *n*: a discharge pipe in a tank whose intake end can be raised or lowered to prevent the discharge of the sediment in the bottom of the tank. Compare *weir*.

swirl *n*: a qualitative term describing tangential motions of liquid flow in a pipe or tube.

swirl pipe *n*: a baffle installed at or near the outlet connection of a liquid storage container to prevent the formation of a vortex and air/vapor entrainment in the liquid. Also called vortex eliminator.

switch *n*: a manually, mechanically, or electronically operated device for making, breaking, or changing the connections in an electric circuit.

switch and control gear *n*: on a diesel-electric drilling rig, the equipment utilized to distribute and transmit electric power to the electric motors from the generators.

switcher *n*: (obsolete) lease operator or pumper. See *pumper*.

switchgear *n*: the aggregate of electronic devices in a circuit that bring other devices into the circuit into an operating or nonoperating state.

swivel *n*: 1. a rotary tool that is hung from the hook and the traveling block to suspend and permit free rotation of the drill stem. It also provides a connection for the rotary hose and a passageway for the flow of drilling fluid into the drill stem. 2. a component used to remove wire and/or chain rotations. It can be connected directly into chain or wire spelter sockets or used in conjunction with shackles.

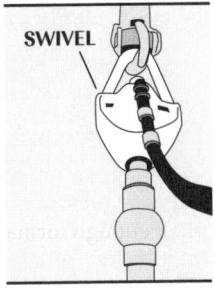

SWIVEL

swivel-connector grip *n*: a braided-wire device used to join the end of one wire rope to the end of another wire rope temporarily. When tension is put on this device, it stretches and grips the wire ropes firmly, allowing the wire rope to be threaded through the blocks. When tension is released, this device relaxes, allowing the rope to be released. Also called a snake or a swivel-type stringing grip.

swivel packing *n*: special rubberized compounds placed in a swivel to prevent drilling fluid under high pressure from leaking out.

swivel stem *n*: a length of pipe inside the swivel that is installed to the swivel's washpipe and to which the kelly (or a kelly accessory, such as the upper kelly cock) is attached. It conducts drilling mud from the washpipe and to the drill stem. See *washpipe*.

swivel sub *n*: a sub containing a swivel joint. It is capable of permitting rotation between its two ends.

swivel tensioner ring *n*: in drilling from floating offshore drilling rigs, a device installed on top of a riser joint to which tensioner lines are manually installed to pad eyes on the ring after the joint is lowered through the rotary table. Tensioner lines must be manually removed from the ring when a riser joint is being raised through the rotary table. See *riser pipe, riser tensioner*.

swivel-type stringing grip *n*: see *swivel-connector grip*.

sx *abbr*: sacks; used in drilling and mud reports.

symmetrical anchor pattern *n*: in mooring an offshore floating rig that is anchored on location, the deployment of the mooring lines and anchors in a manner such that the number of lines deployed from one part of the rig are matched by the same number of lines on another part of the rig that is directly opposite the first part. For example, if four lines are deployed from the stern of the vessel, four lines are also deployed from the bow in a manner identical to the stern lines. Thus, the lines are symmetrically arranged.

sync amplifier *abbr*: synchronization amplifier.

sync control *abbr*: synchronization control.

synchronization amplifier *n*: in an oscilloscope, a device that receives a small portion of the vertical input signal and feeds it to the horizontal-sweep oscillator as a pulse that triggers the oscillator in step with the vertical input signal.

synchronization control *n*: in an oscilloscope, a device that varies the amount of pulse strength that reaches the oscillator and that provides enough pulse to stabilize the pattern displayed on the oscilloscope's screen.

synchronous *adj*: occurring at the same moment. For two or more generators, it means being phased together.

synchronous motor *n*: an AC motor designed to "keep time" with the AC power supply; it rotates at the same speed as the alternator that supplies the current or at a fixed multiple of that speed.

synchronous speed *n*: in a two-pole induction motor, the maximum speed at which such a motor is capable of operating. In a 60-Hz system, the synchronous speed is 3,600 revolutions per minute (rpm). Synchronous speed can be stated as the equation—

$$\text{synchronous speed (rpm)} = \frac{120 \times \text{frequency (hertz)}}{\text{number of poles}}$$

synchroscope *n*: a device used by an engine operator to ensure that all the engines on a rig are running at synchronous speed. Ensuring synchronous speed is necessary to avoid damaging a generator when taking it off line or putting it on line.

syncline *n*: a trough-shaped configuration of folded rock layers. Compare *anticline*.

synergistic effect *n*: the added effect produced by two processes working in combination. It is greater than the sum of the individual effects of each process.

synoptic scale guidance material *n*: information provided by the National Weather Service's Climate Prediction Center on its Web site at www.cpc.ncep.noaa.gov. This material monitors climate and weather in the Western and Eastern Hemispheres.

synoptic wave chart *n*: a chart comprising oceanic wave data collected simultaneously from vessels and computed wave heights for areas lacking reports.

syntactic foam *n*: a material used in the manufacture of buoyant riser modules that contains numerous spheres with hard shells that trap air. The spheres are embedded in a durable foam-like material that is molded into a shape that will fit around a slick riser joint (a joint of riser pipe that does not have buoyant modules added to it). See *buoyant riser joint, riser pipe*.

syntactic foam module *n*: see *buoyant riser joint*.

synthetic-based mud *n*: a drilling fluid containing man-made chemicals that emulate natural oil. Natural oil-base muds require that the cuttings made by the bit be specially handled to prevent damage to the environment; synthetic muds were therefore developed to replace oil-base muds in environmentally sensitive areas. For this reason, synthetic-based muds are sometimes termed pseudo-oil-base mud. See *oil-base mud*.

synthetic mooring rope *n*: a mooring line constructed of synthetic material. Polyester, high modulus polyethylene, and Aramid are the best known materials.

synthetic natural gas (SNG) *n*: a gas that is obtained either by heating coal or by refining heavier hydrocarbons. Hydrogen must be added to the product to make up for deficiencies in the original hydrocarbon source.

systematic error *n*: an error that, in the course of a number of measurements made under the same conditions and of the same value of a given quantity, either remains constant in absolute value and sign, or varies according to a definite law when the conditions change. Thus, it causes a bias.

Système International d'Unités (SI) *n*: the most widely used system of units. It is used for everyday commerce in virtually every country of the world except the United States. SI was selected from the existing Metre-Kilogram-Second system of units (MKS), with the addition of extra units, rather than the older Centimetre-Gram-Second system of units (CGS). SI is sometimes referred to as the metric system (especially in the United States, which has not widely adopted it, and the UK, where conversion is incomplete). However, not all metric units of measurement are accepted as SI units. There are seven base units and several derived units, together with a set of prefixes. Non-SI units can be converted to SI units (or vice versa) according to the conversion of units.

system of units of measurement *n*: a set of base and derived units corresponding to a particular group of quantities.

system resistance *n*: in process control systems, opposition encountered when transferring energy from one source to another, such as that encountered in a hot water system. Any component, or part, of the system that opposes the free transfer of energy between two capacities is a resistance. See *capacity*.

T *abbr*: 1. ton. 2. top of, used in drilling reports.

T *sym*: tesla.

t *sym*: tonne.

TA *abbr*: temporarily abandoned.

tachometer (tach) *n*: an instrument that measures the speed of rotation.

tackle (hoist) *n*: assembly of ropes and sheaves arranged for pulling.

tag *v*: to touch an object downhole with the drill stem (as to tag the bottom of the hole or to tag the top of the fish).

tagging *n*: running pipe or tubing and landing it on a downhole tool.

tag line *n*: in crane operations, a small wire rope attached to the bottom of a load suspended by the crane, which, when grasped by a crew member, allows the crew member to prevent rotation of the load.

tail *n*: the ungraduated lower portion of a mercury-in-glass thermometer.

tail chain *n*: a short length of chain that is attached to the end of a winch line. It is usually provided with a special hook that fastens to objects.

tail gas *n*: the exit gas from a plant.

tail gate *n*: the point in a gas processing plant at which the residue gas is last metered, usually the allocation meter or the plant residue sales meter.

tailing in *n*: guiding a downhole tool into the wellbore or onto the rig floor.

tail line *n*: see *guideline*.

tail out rods *v*: to pull the bottom end of a sucker rod away from a well when laying rods down.

tail pipe *n*: 1. a pipe run in a well below a packer. 2. a pipe used to exhaust gases from the muffler of an engine to the outside atmosphere.

tail roller *n*: a large roller located across the stern of an anchor-handling boat. Pendant lines travel over it when an anchor is being brought in or dropped.

tail shaft *n*: in a ram blowout preventer, the steel rod, or shaft, that protrudes from each side of the preventer's housing and to which the rams are attached. As the rams are operated, the tail shaft rotates to open or close the rams.

taino *n*: a tropical cyclone in Haiti.

take a strain on *v*: to begin to pull on a load.

take-in-kind *n*: see *in-kind*.

takeoff *n*: usually prepared by an abstract company, a takeoff lists and briefly describes the documents relevant to the title of a given piece of property. It costs much less to prepare than an abstract of title and is similar to a landman's run sheet.

take-or-pay *n*: the quantity of gas that a gas purchaser agrees to take, or to pay for if not taken.

take-or-pay clause *n*: a contract clause that guarantees payment to a seller for gas, even though the particular gas volume is not taken during a specified time period. Some contracts stipulate a time period for the buyer to take later delivery without penalty.

take-or-pay credits *n pl*: under Federal Energy Regulatory Commission Order No. 500, credits provided to a transporting pipeline by the producer of the natural gas that the pipeline can apply against any take-or-pay or take-and-pay obligations the pipeline might have to that producer under certain qualifying gas purchase contracts existing on June 23, 1987. Take-or-pay credits may only be received by a pipeline from a producer whose gas it transports and from whom the pipeline has received an "offer of credits" pursuant to Order No. 500. The many and complicated rules about take-or-pay credits are set out in detail in Order No. 500.

take-or-pay quantity *n*: under a take-or-pay clause, the threshold quantity of product that the buyer is obligated to pay for, whether actually taken during the stated period of time. A take-or-pay quantity is usually stated in terms of an absolute quantity or a certain percentage of a total contract quantity over a specific period of time, usually a year.

take out *v*: to remove a joint or stand of pipe from the drill stem.

taking *n*: under ESA, killing, capturing, hunting, harassing, or modifying or destroying the habitat so that breeding, feeding, or sheltering of a species is significantly impaired. Consequently, clearing an area that is the actual or potential habitat of an endangered or threatened species could constitute a violation of the act.

tally *v*: to measure and record the total length of pipe, casing, or tubing that is to be run in a well.

talus *n*: angular pieces of rock produced by weathering that come to rest in a steep slope at the bottom of a mountainside. The rock is broken loose from the larger mass by repeated freezing and thawing.

tan *abbr*: tangent.

tang *n*: a piece that provides an extension of an instrument (such as a file) and serves to form the handle or make a connection for the attachment of a handle.

tangent *n*: in a right triangle, the ratio of the side opposite a given angle to the side adjacent to the angle.

tank *n*: a metal, plastic, or wooden container used to store a liquid. Three types include mud tanks for drilling, production tanks, and storage tanks.

TANK

tankage *n*: the total capacity of the number of tanks in a field.

tank barge *n*: a large, flat-bottomed vessel divided into compartments and used to carry crude or fuel oil.

tank battery *n*: a group of production tanks located in a field to store crude oil.

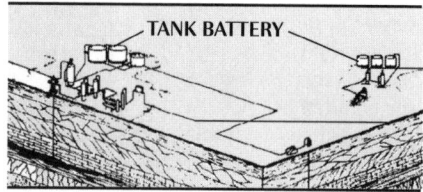

TANK BATTERY

tank bottoms *n pl*: the settlings in the bottom of a storage tank. See *basic sediment and water, bottoms*.

tank calibration *n*: see *strap*.

tank capacity table *n*: see *calibration table*.

tank car *n*: a railroad car used to transport petroleum or petroleum products.

tank course *n*: one circumferential ring of plates in a tank. See *storage tank*.

tank dike *n*: a structure erected to contain petroleum or a petroleum-fed fire in case a storage vessel ruptures or collapses. Usually a dike is built around the petroleum storage tank and a steel or stone wall is put up between the prime movers and the oil pumps in a pipeline pumping station. Also called fire wall.

tanker *n*: a ship designed to transport oil, LPG, LNG, or SNG. Tankers whose capacity is 100,000 deadweight tons (90,720 deadweight tonnes) or more are supertankers, either very large crude carriers or ultralarge crude carriers. Also called a tank ship.

tank farm *n*: a group of large tanks maintained by a pipeline and used to store oil after it has been transferred from the production tanks and before it is transported to the refinery.

tank prover *n*: an open or closed vessel of known capacity designed for the accurate determination of the volume of liquid delivered into or out of it during a meter-proving operation.

tank ship *n*: in offshore operations, a ship designed to transport oil when oil cannot be pipelined to shore.

tankside sample *n*: a spot sample taken from a suitable sample connection to the side of a tank.

tank sludge *n*: settlings that collect at the bottom of storage tanks containing crude oils, residues, and other petroleum products. Such sludge usually contains water.

tank strapper *n*: the person who measures a tank at various levels to see how much it will hold.

tank strapping *n*: the obtaining of several measurements, such as ring circumference, gauge height, and tank height, to determine how much a tank will hold.

tank table *n*: a table giving the number of barrels of fluid contained in a storage tank that correspond to the linear measurement on a gauge line. Tank tables are prepared from tank strapping measurements. See *strap*. Also called calibration table, tank capacity table.

tank truck *n*: a truck designed to transport petroleum or petroleum products.

tannic acid *n*: the active ingredient of quebracho and quebracho substitutes, such as mangrove bark, chestnut extract, and hemlock.

tap *n*: 1. a tool for forming an internal screw thread. It consists of a hardened tool-steel male screw grooved longitudinally so as to have cutting edges. 2. a hole or opening in a line or vessel into which a gauge or valve may be inserted and screwed tight. *v*: 1. to form a female thread by means of a tap. 2. to extract or cause to flow by means of a borehole, e.g., to tap a reservoir.

tape clamp *n*: a quick-release clamp that may be fitted around a strapping tape at any convenient position throughout its length.

tape coating *n*: in pipeline construction, a protective coating of polyethylene, polyvinyl, coal-tar-base, or butyl-mastic tape that is wrapped around pipe to prevent corrosion. Tape coatings are applied on the line, and any defects can be repaired relatively quickly and simply.

tape positioner *n*: a guide that slides freely along strapping tape and that is used to pull and hold the tape in the correct position for taking measurements.

tapered bowl *n*: a fitting, usually divided into two halves, that crew members place inside the master bushing to hold the slips.

tapered hole *n*: 1. a condition wherein the hole diameter narrows with depth owing to wear on the bit gauge caused by drilling abrasive formations. 2. a wellbore whose diameter is larger near the surface than near the bottom.

tapered string *n*: drill pipe, tubing, sucker rods, and so forth with the diameter near the top of the well larger than the diameter below.

taper tap *n*: a tap with a gradually decreasing diameter from the top. It is used to retrieve a hollow fish such as a drill collar and is the male counterpart of a die collar. The taper tap is run into a hollow

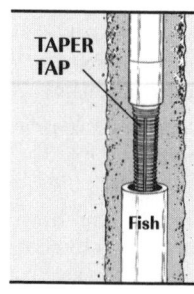

TAPER TAP

Fish

fish and rotated to cut enough threads to provide a firm grip and permit the fish to be pulled and recovered. See *tap*. Compare *die collar*.

tape wrapping *n*: rolls of plastic sheeting with a preapplied adhesive that are used to coat pipelines to prevent corrosion after they are buried.

tap hole *n*: a hole radially drilled in the wall of the meter tube or fitting, the inside edge of which is flush and without any burrs.

taping machine *n*: in pipeline construction, a machine that moves along the pipe, wrapping joints with tape in overlapping segments.

TAPS *abbr*: Trans-Alaska Pipeline System.

tap sampling *n*: in tank sampling, withdrawing liquid samples through sample taps, at least three of which should be placed equidistant throughout the tank height and extending at least 3 feet (1 metre) inside the tank shell. A standard ¼-inch (0.64-centimetre) pipe with suitable valve is satisfactory.

tare *n*: the weight of an empty container subtracted from the gross weight of the container and that which is in the container to ascertain net weight.

target *n*: a bull plug or a blind flange at the end of a tee to prevent erosion at a point where a change in flow direction occurs.

targeted *adj*: said of a fluid piping system in which flow impinges on a lead-filled end (target) or a piping tee when fluid flow changes direction.

targeted substances *n pl*: certain substances under TSCA that present "an unreasonable risk of injury to health or to the environment," for example, asbestos and PCB.

target meter *n*: a meter that balances target force with a feedback force applied above a fulcrum.

tariff *n*: the rate set by pipeline companies for moving oil.

tar sand *n*: a sandstone that contains chiefly heavy, tarlike hydrocarbons. Tar sands are difficult to produce by ordinary methods; thus it is costly to obtain usable hydrocarbons from them.

tattletale *n*: a device on an instrument control panel that indicates the cause of a system shutdown or alarm signal.

tattletale hole *n*: see *weephole*.

taut leg mooring (TLM) *n*: a multileg spread mooring using components in a taut leg configuration.

tautline position-reference system *n*: a system for monitoring the position of a floating offshore drilling rig in relation to the subsea wellhead. A taut steel line is stretched from the rig to the ocean floor. An inclinometer measures the slope of the line at the rig, and, because the line is assumed to have been straight when stretched from the rig to the ocean floor, any angle in the line indicates that the rig has moved. The system's weakness is that the tautline can be distorted by currents and thus give inaccurate readings. Compare *acoustic position reference, position-reference system.*

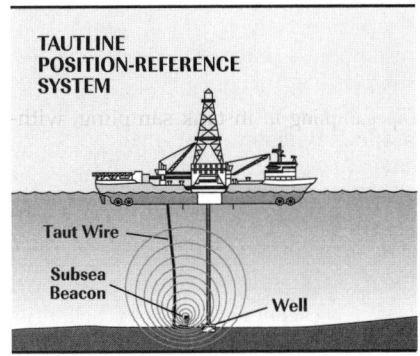

tax-free interest *n*: an interest in production that does not bear its portion of production taxes levied on production from the property. The tax applicable to this interest is borne by other interest owners in the property. Compare *exempt interest.*

Tcf *abbr*: trillion cubic feet.

Tcf/d *abbr*: trillion cubic feet per day.

T connection *n*: connection for transforming three-phase power using two transformers; a variation of the open-delta connection.

TCP *abbr*: tubing conveyed perforator.

TD *abbr*: total depth.

TDC *abbr*: top dead center.

tear down *v*: see *rig down.*

teardrops *n pl*: in paint technology, the drops of paint that collect on the bottom edges of items that are painted by dipping the item into a container of paint.

tectonic *adj*: of or relating to the deformation of the earth's crust, the forces involved in or producing such deformation, and the resulting rock forms.

tectonically stressed formation *n*: a formation in which movements of the earth's crust compress or stretch the formation. When a well is drilled into such a formation, the rock can collapse into the wellbore and cause the drill stem to become stuck. Maintaining the mud weight to a value high enough to offset the tectonic

stresses on the formation can prevent formation collapse.

tee *n*: a pipe fitting that is shaped like the letter T. A pipe or other fitting can be attached to each end of the tee, since the tee ends are threaded.

TEFC *abbr*: totally enclosed fan-cooled motor.

TEG *abbr*: triethylene glycol.

telemetry *n*: the process of gathering data by electronic or other kinds of sensing devices and transmitting those data to remote points. See *downhole telemetry.*

telescopic joint *n*: a device used in the marine riser system of a floating drilling rig to compensate for the vertical motion of the rig caused by wind, waves, or weather. It consists of an inner barrel attached beneath the rig floor and an outer barrel attached to the riser pipe and is an integrated part of the riser system. Also called slip joint.

telescoping derrick *n*: see *telescoping mast.*

telescoping mast *n*: a portable mast that can be erected as a unit, usually by a tackle that hoists the wireline or by a hydraulic ram. The upper section of a telescoping mast is generally nested (telescoped) inside the lower section of the structure and raised to full height either by the wireline or by a hydraulic system. Erroneously but commonly called a telescoping derrick.

telescoping swivel sub *n*: a sub with a telescoping joint used in dual or triple completions for running additional tail pipe.

tell-tale *adj*: terminology used to describe a screen that, when packed off by gels, will give a pressure rise at the surface, thereby "telling" the tool operator that the gel has reached a certain location. Also called tattletale.

telltale hole *n*: a hole drilled into the space between rings of packing material used with a liner in a mud pump. When the liner packing fails, fluid spurts out of the telltale hole with each stroke of the piston, indicating that the packing should be renewed. See *weephole.*

temperature *n*: a measure of heat or the absence of heat, expressed in degrees Fahrenheit or Celsius. The latter is the standard used in countries on the metric system.

temperature bomb *n*: an instrument lowered into a well to record downhole temperature.

temperature coefficient *n*: an expression of the amount resistance increases in a material as the temperature increases. It is stated as per degree Celsius per ohm.

temperature compensator *n*: a device on a meter that automatically converts the flowing temperature of the fluid through the meter to a standard temperature.

temperature correction factor *n*: a factor for correcting volumes of oil or gas to the volume occupied at a specific reference temperature. The reference temperature most commonly used in the petroleum industry is 60°F (15.56°C).

temperature device *n*: a sensor, transmission medium, and readout equipment in an operating configuration used to determine the temperature of a liquid for measurement purposes.

temperature error *n*: the measuring error caused by the temperature of a measurement differing from the pertinent reference value.

temperature gradient *n*: 1. the rate of change of temperature with displacement in a given direction. 2. the increase in temperature of a well as its depth increases.

temperature log *n*: a survey run in cased holes to locate the top of the cement in the annulus. Since cement generates a considerable amount of heat when setting, a temperature increase will be found at the level where cement is found behind the casing.

temperature range *n*: the range of ambient temperatures, given by their extremes, within which a transducer is intended to operate.

temperature regulator *n*: see *thermostat.*

temperature rise *n*: the maximum temperature increase that any part of the motor winding or magnetic structure accessible for measurement with a thermometer will have when the motor is developing its full load.

temperature sensor *n*: a sensing element and its housing, if any, and defined as the part of a temperature device that is positioned in a liquid the temperature of which is being measured.

temperature survey *n*: an operation used to determine temperatures at various depths in the wellbore. It is also used to determine the height of cement behind the casing and to locate the source of water influx into the wellbore.

temperature transmitter *n*: in process control systems, a device that senses the process variable of temperature and converts it to a proportional electrical signal of, usually, 4 to 20 milliamperes. Transmitters can be reranged, or calibrated, over a specific temperature range for better resolution around the temperature of interest.

temper screw *n*: used to regulate the force of the blow delivered to the drill bit on a cable-tool rig. Attached to the walking beam, it controls the feed rate of the drilling tools. Also called temple screw.

template *n*: see *temporary guide base*.

temple screw *n*: see *temper screw*.

temporarily abandoned *adj*: temporarily shut in but not plugged.

temporary guide base *n*: the initial piece of equipment lowered to the ocean floor once a floating offshore drilling rig has been positioned on location. It serves as an anchor for the guidelines and as a foundation for the permanent guide base and has an opening in the center through which the bit passes. It is also called a drilling template.

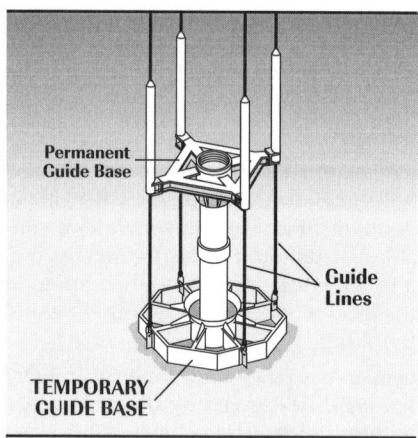

Permanent Guide Base

Guide Lines

TEMPORARY GUIDE BASE

temporary magnet *n*: a magnet usually made of soft iron that loses its magnetism shortly after the magnetizing force is removed from it.

tenants in common *n pl*: see *cotenants*.

tender *n*: 1. the barge anchored alongside a relatively small offshore drilling platform. It usually contains living quarters, storage space, and the mud system. 2. a shipment of oil presented by a shipper to a pipeline for movement. 3. a form required by certain regulatory bodies in some states for their approval of products shipped from plants or other sources. 4. the person responsible for tending to a diver's needs. 5. in naval architecture, the tendency of a floating vessel, such as an offshore rig, to heel easily from wind, current, and wave forces. Compare *stiff*.

tender number *n*: a number assigned to a particular shipment, or tender, of petroleum or petroleum product.

tendons *n pl*: steel tubes that attach a compliant platform to the seafloor. Called a tension-leg platform rig.

ten-minute gel strength *n*: the measured 10-minute gel strength of a fluid is the maximum reading (deflection) taken from a direct-reading viscometer after the fluid has been quiescent for 10 minutes. The reading is reported in pounds/100 square feet. See *gel strength*.

ten round *n*: same as an eight round, except ten threads per inch. See *eight round*.

tens column *n*: in a column of numbers having more than one digit, the column immediately to the left of the ones column. For example, in the number 3,284, 8 is in the tens column. It denotes that the number 3,284 contains 8 tens. Besides 8 tens, the number also contains 3 thousands, 2 hundreds, and 4 ones.

tensile *adj*: of or relating to tension.

tensile load *n*: the amount of longitudinal stress borne by an object or substance.

tensile strength *n*: the greatest longitudinal stress that a metal can bear without tearing apart. A metal's tensile strength is greater than its yield strength.

tensile stress *n*: stress developed by a material bearing a tensile load. See *stress*.

tension *n*: the condition of a string, wire, pipe, or rod that is stretched between two points.

tensioner system *n*: a set of devices installed on a floating offshore drilling rig to maintain constant tension on the riser pipe, despite any vertical motion made by the rig.

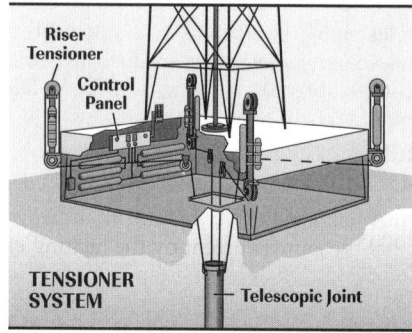

Riser Tensioner

Control Panel

TENSIONER SYSTEM

Telescopic Joint

tensioner unit *n*: part of the riser tensioner system on a floating offshore drilling rig that uses a hydropneumatic cylinder assembly to which is attached the tensioning lines from the telescopic joint. High-pressure air and hydraulic fluid in the cylinder assembly provide tension on the lines as the rig moves up and down (heaves) with ocean movements.

tension gauge *n*: a device that measures and indicates the amount of pulling force (tension) being put on pipe, a tape measure, etc. In tank strapping, the strapping crew attaches a tension gauge to the measuring tape they use to determine a tank's circumference. Knowing how much tension to apply to the tape is essential to accurate measurement because the same amount of tension must be applied to the tape each time they make a measurement.

tension joint *n*: on offshore production systems from floating surface facilities, a riser joint that is positioned directly below the surface joint. It applies additional tension to the riser from within the surface facility's moon pool using the marine riser tensioning system. See *surface joint*.

tension-leg platform rig *n*: a compliant offshore drilling platform used to drill development wells. The platform, which resembles a semisubmersible drilling rig, is attached to the seafloor with tensioned steel tubes. The buoyancy of the platform applies tension to the tubes. See *platform rig*.

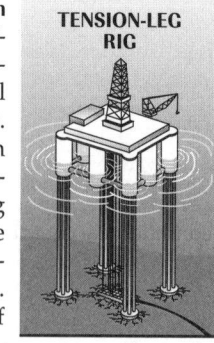

TENSION-LEG RIG

tension packer *n*: a packer that is held in its set position by upstrain, or upward pull.

tension tool *n*: a device on a retrievable or drillable packer that allows the packer to be set with an upward pull (tension); it is often used when not enough pipe weight is available to set the packer by applying downward force (compression).

TENV motor *abbr*: totally enclosed nonventilated motor.

TEOR *abbr*: thermal enhanced oil recovery.

term clause *n*: see *habendum clause*.

terminal *n*: 1. a point to which oil is transported through pipelines. It usually includes a tank farm and may include tanker-loading facilities. 2. a screw, soldering lug, or other point to which electric connections can be made. 3. one of the electrical input or output points of a circuit or component.

termination joint *n*: see *termination spool*.

termination spool *n*: a standard riser joint with a side entry to allow connection of a mud booster line. It is used to overcome a height restriction, and is the first riser joint connected to the lower marine riser package (LMRP). Some spools feature an automatic check valve that allows mud to flow only down the circulation line and up the riser string.

term minerals *n pl*: severed minerals acquired for a certain time and, generally, as long thereafter as production continues.

tertiary recovery *n*: 1. the use of improved recovery methods that not only restore formation pressure but also improve oil displacement or fluid flow in the reservoir. 2. the use of any improved recovery method to remove additional oil after secondary recovery. Compare *primary recovery, secondary recovery.*

tesla *n*: the unit of magnetic-flux density in the metric system.

test cap *n*: on choke and kill lines that are attached to a riser joint, a pressure-tight stopper (a cap) fitted on the top of the choke and kill lines. When the lines are capped, crew members can apply pressure to them and ensure that they do not leak as they run the riser pipe.

tester *n*: a person who tests pipe and casing for leaks.

test gallons *n pl*: see *theoretical gallons.*

test marketing *n*: product analysis conducted under real conditions of the market.

test measures *n pl*: vessels used as secondary standards for volume measurements.

test plug *n*: when pressure testing a blowout preventer stack, a blocking device that is run and landed in the high-pressure wellhead housing. The plug isolates the casing string, which has a lower pressure rating than the blowout preventer stack, from blowout preventer test pressure.

test pressure *n*: the working pressure of a container or vessel times a safety factor. See *working pressure.*

test separator *n*: an oil and gas separator that is used to separate relatively small quantities of oil and gas, which are diverted through the testing devices on a lease.

test voltage *n*: in capacitors, voltage applied to a capacitor that is higher than the capacitor's maximum working voltage rating. Test voltage is applied to ensure that a capacity can perform at its working voltage without failing. See *maximum working voltage rating.*

test well *n*: a wildcat well.

tethered diving *n*: diving in which an umbilical hose is used to connect a diver to the gas supply.

Tethys Sea *n*: an ancient great ocean of 200 million years ago, between what will become Eurasia and Africa. More than half the world's known petroleum accumulated along the margins of this ocean during a brief (30 million-year) interval.

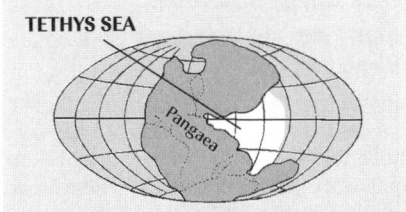
TETHYS SEA
Pangaea

tetraethyl lead *n*: an antiknock compound for motor gasoline. It is being phased out because lead is a pollutant.

Texas deck *n*: the main load-bearing deck of an offshore drilling structure and the highest above the water, excluding auxiliary decks such as the helicopter landing pad. So called because it is the largest deck on the structure.

TFL *abbr*: through-the-flow-line. See *through-the-flow-line equipment.*

thalweg *n*: 1. in geology, the line connecting the lowest points along a stream bed or valley. Also known as a valley line. 2. a line crossing all contour lines on a land surface perpendicularly.

thaw subsidence *n*: the tendency of permafrost to sink when a heavy weight or structure is placed on it. When the structure is placed on frozen ground, any heat that thaws the soil causes it to become soft, so that the structure sinks or subsides into it.

theodolite *n*: an optical instrument used to track rising pilot balloons and thus obtain indications of wind directions and speeds above the earth's surface.

theoretical gallons *n pl*: the content of liquefiable hydrocarbons in a volume of gas, determined from analyses or tests of the gas.

theoretical residue gas remaining *n*: the volume of residue that theoretically remains after volume reductions attributable to the processing of a volume of gas in a plant. It is determined by applying a factor from a table or calculation to the volume of gas delivered to the plant.

therm *n*: a unit of gross heating value equal to 100,000 Btu (1.055056108 joules).

thermal ammeter *n*: a quantity indicator that measures current by the heating effect of the current on a thermocouple.

thermal cracking *n*: the process of making oils of low boiling range (100°–550°F, 37.8°–287.8°C), to be used for motor fuels and burning oils, from oils of high boiling range (550°–800°F, 287.8°–426.7°C), such as gas oil and fuel oil. All modern commercial methods get this breaking-down action by subjecting the high boiling oils to high temperatures. Pressure up to 1,000 pounds per square inch (6,895 kilopascals) is used for producing a dense system and for making sure of good contact within the desired temperature and in an apparatus of reasonable size.

Thermal Decay Time Log *n*: proprietary name for a type of pulsed-neutron survey.

thermal decomposition *n*: the breakdown of a compound or substance by temperature into simple substances or into constituent elements.

thermal enhanced oil recovery (TEOR) *n*: a method of increasing well productivity and recoverable oil by using heat, e.g., steam injection (the huff 'n' puff method), steam flood, or in situ combustion.

thermal equilibrium *n*: the achievement of equal temperatures in two or more substances.

thermalization *n*: the process of reducing the energy of neutrons to thermal energy levels.

thermally stable polycrystalline (TSP) diamond bit *n*: a special type of fixed-head bit that has synthetic diamond cutters that do not disintegrate at high temperatures. Compare *polycrystalline diamond compact.*

thermal neutron *n*: neutron having an average energy level at room temperature of 0.025 electron volts.

thermal neutron population *n*: the number of thermal neutrons around the neutron logging tool. The number of capture gamma rays present at any time is directly proportional to the thermal neutron population.

thermal processes *n pl*: see *thermal recovery.*

thermal recovery *n*: a type of improved recovery in which heat is introduced into a reservoir to lower the viscosity of heavy oils and to facilitate their flow into producing wells. The pay zone may be heated by injecting steam (steam drive) or by injecting air and burning a portion of the oil in place (in situ combustion). See *cyclic steam injection, in situ combustion, steam drive.*

thermal stability *n*: a relative measure of the ability of a substance to withstand heat without disintegrating.

thermal time constant *n*: the time required for a thermometer to indicate 63.2 percent of the magnitude of a change in the measured temperature.

thermistor *n*: a temperature-sensitive element with variable resistivity for overload protection in motors.

thermocouple *n*: a device consisting of two dissimilar metals bonded together, with electrical connections to each. When the device is exposed to heat, an electrical current is generated, the magnitude of which varies with the temperature. It is used to measure temperatures higher than those that can be measured by an ordinary thermometer, such as those in engine exhaust.

thermodynamic equilibrium *n*: property of a system that is in mechanical, chemical, and thermal equilibrium.

thermoelectric device *n*: an electrical component that directly converts heat into electrical energy or directly converts electrical energy into heat.

thermohydrometer *n*: a combination hydrometer-thermometer designed to measure gravity and temperature of a liquid.

thermometer *n*: an instrument that measures temperature. Thermometers provide a way to estimate temperature from its effect on a substance with known characteristics (such as a gas that expands when heated). Various types of thermometers measure temperature by measuring the change in pressure of a gas kept at a constant volume, the change in electrical resistance of metals, or the galvanic effect of dissimilar metals in contact. The most common thermometer is the mercury-filled glass tube, which indicates temperature by the expansion of the liquid mercury.

thermometer thread *n*: the length of mercury in the capillary of a mercury-in-glass thermometer, which indicates the temperature.

thermoplastics *n pl*: any of a variety of materials often used in pipe coatings, the molecular structure of which allows them to soften repeatedly when heated and harden when cooled.

thermoresistive element *n*: a device whose electrical resistance changes with temperature. One thermoresistive element in process control is a resistance temperature detector (RTD) that is constructed of platinum, nickel, or copper.

thermosetting plastics *n pl*: plastics that solidify when first heated under pressure, but whose original characteristics are destroyed when remelted or remolded.

thermosiphon baffle *n*: a device for assisting in controlling the direction of thermal currents in an indirect heater.

thermostat *n*: a control device used to regulate temperature.

thermowell *n*: a well in a process vessel or line used as a thermometer or thermocouple holder.

thickened water *n*: see *polymer*.

thickening time *n*: the amount of time required for cement to reach an API-established degree of consistency, or thickness. Thickening time begins when the slurry is mixed.

thief *n*: a device that is lowered into a tank to take an oil sample at any depth. The sample will subsequently be used to determine the quality and S&W content of the oil in the tank. *v*: to obtain a sample of oil from a tank using a thief.

thief formation *n*: 1. a formation that absorbs drilling fluid as it is circulated in the well. Lost circulation is caused by a thief formation. Also called a thief sand or a thief zone. 2. a low pressure reservoir or zone that "steals" fluids from a higher pressure reservoir or zone and prevents their production to the surface.

thief hatch *n*: a lidded opening on top of a stock tank. It allows access for measuring the oil level and for taking samples with a device that is lowered into a tank to take an oil sample at any depth. The sample will subsequently be used to determine the quality and sediment and water content of the oil in the tank.

thief sample *n*: see *tube sample*.

thief sand *n*: see *thief formation*.

thief zone *n*: see *thief formation*.

thimble *n*: a grooved metal or synthetic fitting to protect the eye or the fastening loop of a wire or synthetic rope. The thimble conforms to the inside shape of the rope's eye and covers the wire rope's strands.

thin *v*: to add a substance such as water or a chemical to drilling mud to reduce its viscosity.

thinning agent *n*: a special chemical or combination of chemicals that, when added to a drilling mud, reduces its viscosity.

30 CFR 250 *abbr*: 30 Code of Federal Regulations, Part 250.

30 Code of Federal Regulations Part 250 *n*: the U.S. Department of the Interior, Minerals Management Service rules and regulations that must be followed by those who drill and produce oil and gas wells located in the Outer Continental Shelf.

thixotropic cement *n*: a blend of portland cement and calcium sulfate hemihydrate designed primarily for cementing lost circulation zones and porous or fractured formations.

thixotropic paint *n*: a paint in which brushing or stirring lowers its viscosity.

thixotropy *n*: the property exhibited by a fluid that is in a liquid state when flowing and in a semisolid, gelled state when at rest. Most drilling fluids must be thixotropic so that cuttings will remain in suspension when circulation is stopped.

thorium *n*: a radioactive metallic element found combined in minerals.

thousand circular mil (kcmil) *n*: a measure of wire diameter stated in thousands of circular mils. See *circular mil*.

thread *n*: a continuous helical rib, as on a screw or pipe.

thread compound *n*: see *dope*.

thread dope *n*: see *dope*.

threaded coupling *n*: a type of connector that has threads on each end, making it possible to screw two pieces of pipe together. Compare *pressure coupling*. See also *coupling*.

thread profile gauge *n*: a device to measure the amount of wear or stretch on pipe threads.

thread protector *n*: a metal or plastic device that is screwed onto or into pipe threads to protect them from damage when the pipe is not in use.

threatened species *n*: a species that is likely to become endangered.

three-mode controller *n*: an instrument that continuously measures the value of signals generated by three variable quantities or conditions and acts to control any deviation from a desired preset value.

three-phase separator *n*: a separator that separates well fluids into oil or emulsion, gas, and water, or, in a glycol dehydration unit, into gas, condensate, and glycol.

three-phase voltage *n*: voltage generated in a circuit that is energized by alternating-current voltages that differ in phase by one-third of a cycle, or 120 degrees.

threshold limit value (TLV) *n*: the average concentration of toxic gas in a breathing atmosphere in which a normal person can be exposed safely for 8 hours per day, 5 days per week; usually established by a regulatory agency such as OSHA.

threshold planning quantities (TPQ) *n pl*: EPA minimum quantities of extremely hazardous substances under SARA that may not be exceeded without meeting reporting requirements. Under SARA (section 302), companies must notify and coordinate emergency plans with the SERC and the LEPC when a facility reaches or exceeds the EPA-stipulated TPQ for any EHS. For example, hydrofluoric acid is an EHS with a TPQ of 100 pounds. If a facility has 100 pounds or more of hydrofluoric acid, it must notify the SERC and the LEPC.

thribble *n*: a stand of pipe made up of three joints and handled as a unit. Compare *double, fourble, single*.

thribble board *n*: the name used for the derrickhand's working platform, the monkeyboard, when it is located at a height in the derrick equal to three lengths of pipe joined together. Compare *double board, fourble board*.

THROT *abbr*: tubing hanger running and orientation tool.

throttling *n*: the choking or failing that occurs when a mud pump fails to deliver a full amount of fluid through one or more of its valves. Throttling is usually caused by improper valve lift.

throttling range *n*: in process control, the full range of control over which the final control element in a process can be operated. See *proportional band*.

throttling valve *n*: a valve, usually only partially open or closed, used to regulate the rate of flow.

throughput volume *n*: the volume of fluid that has flowed through a meter during a specific amount of time.

through-the-flow-line (TFL) equipment *n*: any equipment designed to be pumped down a completed well to effect a repair, modify the well's flow, or for other reasons.

throw *n*: the distance from the centerline of the main bearing of a crankshaft to the centerline of the connecting-rod journal. Two times the throw equals the stroke.

throw the chain *v*: to flip the spinning chain up from a tool joint box so that the chain wraps around the tool joint pin after it is stabbed into the box. The stand or joint of drill pipe is turned or spun by a pull on the spinning chain from the cathead on the drawworks.

THROW THE CHAIN

thrust *n*: 1. a force applied to an object to move it in the desired direction. 2. the force exerted in any direction by a powered screw (propeller) or a fluid jet.

thruster *n*: a special propulsion device used on offshore floating drilling rigs such as drill ships and semisubmersibles that is located on the hull of the vessel and is actuated by a sensing system. A computer to which the system feeds signals directs the thrusters to maintain the rig on location.

thrust fault *n*: see *reverse fault*.

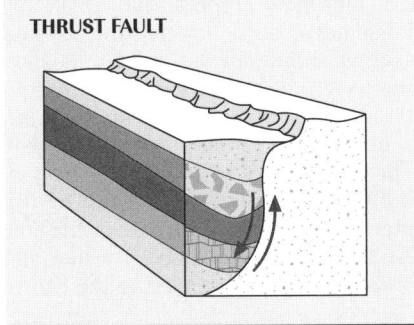

THRUST FAULT

thrust load *n*: a load or pressure parallel to or in the direction of the shaft.

thumper *n*: a hydraulically operated hammer used in obtaining a seismograph in oil exploration. It is mounted on a vehicle and, when dropped, creates shock waves in the subsurface formations, which are recorded and interpreted to reveal geological information.

thunderhead *n*: see *cumulonimbus*.

thyratron *n*: a virtually obsolete electrical device, similar to a vacuum tube, that is a hot-cathode gas tube in which one or more control electrodes initiate but do not limit anode current except under certain operating conditions. Solid-state devices that perform the same function have replaced thyratrons.

thyrector *n*: a solid-state silicon diode that behaves as an insulator up to its rated voltage, and as a conductor above its rated voltage. It protects components in AC circuits from voltage surges.

thyristor *n*: a transistor that initiates but does not limit anode current except under certain operating conditions. As collector current is increased to a critical value, a thyristor gives a high-speed triggering action.

tidal current *n*: the horizontal movement of the tide as it rises and falls and moves toward and away from the shore.

tidal range *n*: the difference in height between a low tide and a high tide.

tidal wave *n*: any unusually high and generally destructive sea wave along a shore. Compare *tsunami*.

tide *n*: the periodic rising and falling of the surface of the oceans and water bodies connected with the oceans (gulfs and bays).

tide gauge *n*: an instrument that measures the height of tides.

tide wave *n*: in oceanography, a long-period wave associated with the tide-producing forces of the sun and the moon, and identified with the rising and falling of tide. Also called a tidal wave. See *tide*.

tie-back string *n*: casing that is run from the top of a liner to the surface.

tie-down *n*: a device to which a guy wire or brace may be attached, such as the anchoring device for the deadline of a hoisting-block arrangement.

tie-in *n*: a collective term for the construction tasks bypassed by regular crews on pipeline construction. Tie-in includes welding road and river crossings, valves, portions of the pipeline left disconnected for hydrostatic testing, and other fabrication assemblies, as well as taping and coating the welds.

tie-in gang *n*: workers responsible for tie-in tasks in the construction of a pipeline.

tier two form *n*: a federal form used by companies or individuals for submitting annual inventory data on hazardous chemicals or extremely hazardous substances. See *SARA Title III*.

tight emulsion *n*: an emulsion that is relatively difficult to break. Compare *loose emulsion*.

tighten up *v*: to add oil to a system, which causes the oil to break out and rise to the surface.

tight formation *n*: a petroleum- or water-bearing formation of relatively low porosity and permeability.

tight hole *n*: 1. a well about which information is restricted for security or competitive reasons. 2. a section of the hole that, for some reason, is undergauge. For example, a bit that is worn undergauge will drill a tight hole.

tight sand *n*: sand or sandstone formation with low permeability.

tight spot *n*: a section of a borehole in which excessive wall cake has built up, reducing the hole diameter and making it difficult to run the tools in and out. Compare *keyseat*.

tight well *n*: a well that produces from a reservoir of low permeability and thus stabilizes slowly. Such wells show slow buildup of bottomhole pressure on a buildup test, and the flowing pressure does not stabilize quickly at a constant value on a flow test.

time *n*: 1. the dimension of the physical universe which, at a given place, orders the sequence of events. 2. a designated instant in this sequence, as the time of day.

time proportional sample *n*: an automatic sample usually obtained from a pipeline and taken at regular timed intervals.

time release *n*: feature built into oilfield inhibitors that allows them to be introduced into production systems and their active ingredients to be released at timed intervals.

time stratigraphic unit *n*: a layer of rock, with or without facies variations, deposited during a distinct geologic time interval. Compare *rock stratigraphic unit*.

timing *n*: the relationship of all moving parts in an engine. Each part depends on another, so all parts must operate in the right relation with each other as the engine turns.

timing circuit *n*: a source of accurately timed pulses, used in synchronizing a digital computer or as a time base in a transmission system. Also called a clock.

timing gear *n*: see *timing gear train*.

timing gear train *n*: a set of gears in an engine, driven by the crankshaft, set to drive the equipment necessary for the engine's operation. Engine equipment driven by the timing gears includes the oil pump, intake and exhaust valve mechanisms, fuel injectors, and magnetos.

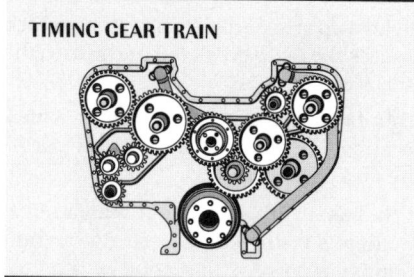

TIMING GEAR TRAIN

Tinkerbell line *n*: see *Geronimo*.

titanium drill pipe *n*: drill pipe made from a metal alloy containing a large amount of titanium. Even though it is more expensive, titanium drill pipe is much stronger (has a higher tensile strength) than conventional steel pipe. Further, in horizontal drilling applications, titanium is almost twice as flexible as steel, which means that titanium pipe can be bent more than conventional pipe without being permanently deformed.

title *n*: a term standing for those facts that, if proved, will enable a person to regain or retain possession of property.

title opinion *n*: the written opinion of a title examiner on the status of the title to a given piece of property. Compare *division order opinion, drill site opinion.*

Title III of the Superfund Amendments and Reauthorization Act *n*: see *SARA Title III.*

titration *n*: a process of chemical analysis by which drops of a standard solution are added to another solution or substance to obtain a desired response—color change, precipitation, or conductivity change—for measurement and evaluation.

TIW valve *n*: inside blowout preventer (so named for one company that manufactured it, Texas Iron Works).

TLM *abbr*: taut leg mooring.

TLV *abbr*: threshold limit value.

toeplate *n*: on offshore installations with grated decks or floors, a flat bar attached flat against the outer edge of a grating or the edge of a stair tread that projects above the top surface of the grating or tread to form a lip or curb.

tolerance *n*: the range of variation permitted in maintaining a specific dimension in machining a part. For example, a shaft measurement could be 2.000 inches ±0.001 (50 millimetres ±0.025). This means that any measurement from 1.999 inches to 2.001 inches (49.975 millimetres to 50.025 millimetres) would be acceptable.

toluene *n*: a liquid aromatic hydrocarbon, C_7H_8, that resembles benzene but is less volatile, flammable, and toxic.

ton (T) *n*: 1. (nautical) a volume measure equal to 100 square feet applied to mobile offshore drilling rigs. 2. a measure of weight equal to 2,000 pounds. 3. (metric) a measure of weight equal to 1,000 kilograms. Usually spelled "tonne."

tong arm *n*: the part of the tongs that extends behind the tong jaws, and to which the jaws are attached. A snub line is attached to the end of the tong arm to prevent the tong from turning too far when it is being used to make up, break out, or back up pipe.

tong counterbalance *n*: a weight placed in the derrick or under the rig floor that is attached by means of wire rope to the tong hanger. Each set of tongs has a counterbalance to ensure that they hang at a convenient height above the rig floor.

tong dies *n pl*: very hard and brittle pieces of serrated steel that are installed in the tongs and that grip or bite into the tool joint of drill pipe when the tongs are latched onto the pipe.

tong hand *n*: the member of the drilling crew who handles the tongs.

tong hanger *n*: a relatively long narrow steel projection bolted to the tong arm to which wire rope is attached; suspends the tongs in the derrick.

tong jaw *n*: on a set of tongs, one of two hinged devices that crew members latch around elements of the drill stem to make up or break out such elements. Tong jaws normally have replaceable serrated inserts (dies) to grip the pipe.

tongman *n*: the member of the drilling crew who handles the tongs.

tong pull line *n*: a length of wire rope one end of which is connected to the end of the tongs and the other end of which is connected to the automatic cathead on the drawworks. When the driller actuates the cathead, it takes in the tong line and exerts force on the tong to either make up or break out drill pipe.

tongs *n pl*: the large wrenches used for turning when making up or breaking out drill pipe, casing, tubing, or other pipe; variously called casing tongs, pipe tongs, and so forth,

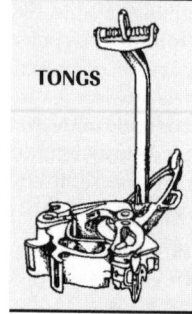

TONGS

according to the specific use. Power tongs or power wrenches are pneumatically or hydraulically operated tools that serve to spin the pipe up tight and, in some instances, to apply the final makeup torque.

ton-mile *n*: the unit of service given by a hoisting line in moving 1 ton of load over a distance of 1 mile. The SI measurement is the megajoule.

tonnage *n*: the size of a ship or vessel or the spaces within a ship or vessel as measured in tons.

tonne (t) *n*: a mass unit in the metric system equal to 1,000 kilograms.

tonnes-per-centimetre (TPC) immersion *n*: in the SI system of measurement, the weight in tonnes that must be loaded or discharged from a floating offshore drilling rig to change the average draft of the rig in seawater by one centimetre. See *draft* (def. 1).

tons-per-inch (TPI) immersion *n*: in the English, or conventional, system of measurement, the weight in short tons that must be loaded or discharged from a floating offshore drilling rig to change the average draft of the rig in seawater by 1 inch. See *draft* (def. 1).

tool dresser *n*: a driller's helper on a cable-tool rig, once responsible for sharpening or dressing the drill bit. Sometimes called a toolie.

tool face *n*: the part of a deflection tool—usually marked with a scribe line—that is oriented in a particular direction to make a predetermined deflection of the wellbore.

tool hand *n*: a worker on a well service or workover rig who helps run packers and other tools into the well.

toolhouse *n*: a building for storing tools.

toolie *n*: (slang) tool dresser.

tool joint *n*: a heavy coupling element for drill pipe made of special alloy steel. Tool joints have coarse, tapered threads and seating shoulders designed to sustain the weight of the drill stem, withstand the strain of frequent coupling and uncoupling, and provide a leakproof seal. The male section of the joint, or the pin, is attached to one end of a length of drill pipe, and the female section, or the box, is attached to the other end. The tool joint is usually friction welded to the end of the pipe.

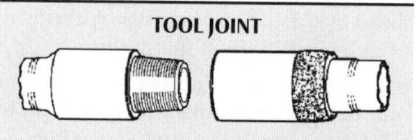

TOOL JOINT

toolpush *n*: Canadian term for toolpusher. See *toolpusher*.

toolpusher *n*: an employee of a drilling contractor who is in charge of the entire drilling crew and the drilling rig. Also called a drilling foreman, rig superintendent, or rig supervisor.

top angle *n*: the angle on a storage tank where the side wall is joined to the roof of the tank; it is usually 90°.

top dead center (TDC) *n*: the position of a piston when it is at the highest point possible in the cylinder of an engine; often marked on the flywheel.

top drive *n*: a device similar to a power swivel that is used in place of the rotary table to turn the drill stem. It also includes power tongs. Modern top drives combine the elevator, the tongs, the swivel, and the hook. Even though the rotary table assembly is not used to rotate the drill stem and bit, the top-drive system retains it to provide a place to set the slips to suspend the drill stem when drilling stops.

TOP DRIVE

top-drive system *n*: see *top drive*.

top hold-down *n*: a mechanism, located at the top of the working barrel, for anchoring a sucker rod pump to the tubing. Compare *bottom hold-down*.

top lease *n*: a lease acquired while a mineral lease to the same property is still in effect. The top lease (held by a different company) replaces the existing lease when it expires or is terminated.

top off *v*: to fill a wellbore with fluid up to the surface.

topographic map *n*: horizontal representation of land elevations as a series of contour lines, each line connecting points of equal elevation.

topography *n*: the configuration of a land surface, including its natural and artificial features with their relative positions and elevations.

top out *v*: to finish filling a tank.

top plug *n*: a cement wiper plug that follows cement slurry down the casing. It goes before the drilling mud used to displace the cement from the casing and separates the mud from the slurry. See *cementing, wiper plug*.

top receiver plate *n*: a flat steel piece (a plate) that houses the choke and kill line connections to the lower marine riser package (LMRP), the blowout preventer's control system receptacles, and the LMRP's guidance devices.

top sample *n*: in tank sampling, a spot sample obtained 6 inches (150 millimetres) below the top surface of the oil.

top seal *n*: in a ram blowout preventer, a piece of synthetic rubber that is molded to the upper and lower steel antiextrusion plates of the ram block.

topset bed *n*: a part of a marine delta that is nearest the shore and that is composed of the heavier, coarser particles carried by the river.

topsoiling *n*: the technique of placing topsoil in a spoil bank separate from the rest of the materials excavated during pipeline construction so that it can be replaced in its original stratum during back-filling operations.

top sub *n*: a component of a packer to which the tubing is connected.

top wiper plug *n*: a device placed in the cementing head and run down the casing behind cement to clean the cement off the walls of the casing and to prevent contamination between the cement and the displacement fluid.

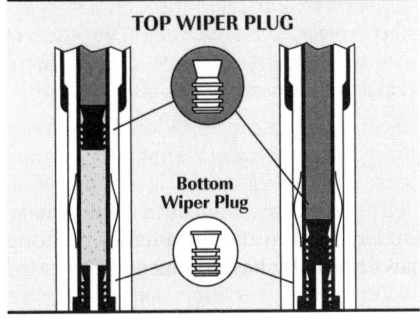

TOP WIPER PLUG

Bottom Wiper Plug

tornado *n*: a usually destructive funnel-shaped rotating column of air that passes over a narrow path of land. Compare *cyclone*.

torque *n*: the turning force that is applied to a shaft or other rotary mechanism to cause it to rotate or tend to do so. Torque is measured in foot-pounds, joules, newton-metres, and so forth.

torque converter *n*: a hydraulic device connected between an engine and a mechanical load such as a compound. Torque converters are characterized by an ability to increase output torque as the load causes a reduction in speed. Torque converters are used on mechanical rigs that have compounds.

torque indicator *n*: an instrument that measures the amount of torque (turning

or twisting action) applied to the drill or casing string. The amount of torque applied to the string is important when joints are being made up.

torque recorder *n*: an instrument that measures and makes a record of the amount of torque (turning or twisting action) applied to the drill or casing string.

torque track *n*: on a portable top-drive unit, the steel rail that is mounted in the derrick and to which the top drive is attached. The top drive moves up or down on the track and, because it is firmly joined to the track, cannot turn as its drive shaft rotates the drill stem. Compare *guide rails, torque tube*. See also *top drive*.

torque tube *n*: a large cylindrical tube mounted in the derrick to which a top drive is attached and on which it can move up and down. Like guide rails, a torque tube prevents the top drive from turning when it is connected to the drill stem and is rotating it. Compare *guide rails*. See also *top drive*.

torsion *n*: twisting deformation of a solid body about an axis in which lines that were initially parallel to the axis become helices. Torsion is produced when part of the pipe turns or twists in one direction while the other part remains stationary or twists in the other direction.

torsional yield strength *n*: the amount of twisting force that a pipe can withstand before twisting off.

torsion balance *n*: one of the earliest gravitational instruments invented; first marketed commercially in 1922.

tortuosity *n*: the relative degree of crookedness of a hydrocarbon flow path through a reservoir.

total calculated volume (TCV) *n*: the total volume of all petroleum liquids and sediment and water, corrected by the appropriate temperature correction (C_{tl}) for the observed temperature and API gravity, relative density, or density to a standard temperature such as 60°F or 15°C, and also corrected by the applicable pressure factor (C_{pl}) and meter factor and all free water measured at observed temperature and pressure. Total calculated volume is equal to the gross standard volume plus free water volume.

total depth (TD) *n*: the maximum depth reached in a well.

total displacement *n*: on a floating offshore vessel, the lightship displacement plus the maximum allowed variable load. See *lightship displacement*.

total disposition *n*: sum of crude oil input to refineries, crude oil exports, crude oil burned as fuel, and crude oil losses.

total immersion thermometer *n*: a thermometer that is designed for the whole length of mercury column to be immersed in liquid. An emergent stem correction has to be applied to the reading to correct for that part of the column that is above the liquid surface during reading.

totalizer *n*: a meter register that receives signals from several other meters and displays or records the sum of the readings from the other meters.

totally enclosed motor *n*: a motor with no provision for the entrance of outside air; heat must be conducted away through the frame. It may be fan-cooled.

total observed volume (TOV) *n*: the total measured volume of all petroleum liquids, sediment and water, and free water at observed temperature and pressure.

total porosity *n*: see *absolute porosity*.

total power *n*: see *apparent power*.

total pressure *n*: the pressure on the surface at which a flowing fluid is brought to rest in excess of the pressure on it when the fluid is not flowing. Also called stagnation pressure.

total S&W *n*: all sediment and water in a vessel's cargo tanks, whether settled or suspended.

total volume *n*: in diesel engine fuel tanks, the full capacity of the tank; usually, not all the fuel in the tank is usable. Compare *useful volume*.

total water *n*: the sum of the dissolved, entrained, and free water in the cargo or the parcel.

toughness *n*: measure of steel's ability to withstand a crack or flaw without fracturing. It can be altered by changing chemical composition, microstructure, and heat treatment. Also called ductility. Compare *brittleness*.

tour (pronounced "tower") *n*: a working shift for drilling crew or other oilfield workers. Some tours are 8 hours; the three daily tours are called daylight, evening (or afternoon), and graveyard (or morning). Often, 12-hour tours are used, especially on offshore rigs; they are called simply day tour and night tour.

towboat *n*: a relatively flat-bottomed boat with a square bow that can push a string of barges carrying tons of petroleum products over great distances.

tow cat *n*: a tractor used to tow equipment up steep grades.

tower *n*: 1. a vertical vessel such as an absorber, fractionator, or still. 2. a cooling tower.

town border station *n*: see *city gate station*.

township *n*: 1. the north-south lines of the rectangular survey system. 2. the square, 6 miles on each side, that is the major unit of land in the rectangular survey scheme of measurement. Compare *range*.

township line *n*: a north-south line of the rectangular survey system.

Toxic Chemical Release Form *n*: a report required under SARA (section 313) that quantifies all releases of toxic chemicals that a facility has had throughout the year. Currently, E&P facilities are not subject to the toxic chemical release reporting requirement, since it is applied only to manufacturers with standard industrial classification codes of 20 through 39.

toxicity *n*: the ability of a substance to be poisonous if inhaled, swallowed, absorbed, or introduced into the body through cuts or breaks in the skin.

Toxic Substance Control Act (TSCA) *n*: a congressional act that regulates the manufacture, processing, distribution in commerce, use, and disposal of chemical substances.

TP *abbr*: tubing pressure; used in drilling reports.

TPC *abbr*: tonnes-per-centimetre.

TPI *abbr*: tons-per-inch.

TPQ *abbr*: threshold planning quantities.

traceability *n*: the relation of a prover or a transducer calibration, through a step-by-step process, to an instrument or group of instruments calibrated and certified by a national or international primary standard.

tracer *n*: a substance added to reservoir fluids to permit the movements of the fluid to be followed or traced. Dyes and radioactive substances are used as tracers in underground water flows and sometimes helium is used in gas. When samples of the water or gas taken some distance from the point of injection reveal signs of the tracer, the route of the fluids can be mapped.

tracer log *n*: a survey that uses a radioactive tracer such as a gas, liquid, or solid having a high gamma ray emission. When the material is injected into any portion of the wellbore, the point of placement or movement can be recorded by a gamma ray instrument. The tracer log is used to determine channeling or the travel of squeezed cement behind a section of perforated casing. Also called tracer survey.

tracer survey *n*: see *tracer log*.

tracking *n*: a rare type of bit-tooth wear occurring when the pattern made on the bottom of the hole by all three cones of a rock bit matches the bit-tooth pattern to such an extent that the bit follows the groove or channel of the pattern and drills ahead very little.

traction motor *n*: a direct current electric motor normally found in railway use, but frequently adapted to power certain drilling rigs.

tractive force *n*: in an electromagnet, the force of attraction that exists between an iron (or other ferrous material) bar placed close to the poles of the electromagnet.

trade wind *n*: a prevailing tropical wind blowing toward the equator from the northeast in the northern hemisphere or from the southeast in the southern hemisphere.

trailer rig *n*: a rig mounted on a wheeled and towed trailer. It has a mast, a rotary, and one or two engines. The rig systems are powered by a prime mover.

trajectory *n*: the path of the wellbore.

trammel (tram) *n*: a metal rod of precise length used to measure distance between two points where accessibility is limited. It is often used to mark crankshaft positions on engines.

Trans-Alaska Pipeline System (TAPS) *n*: the largest long-distance, big-inch pipeline in the non-Communist world. Stretching approximately 800 miles (1,288 kilometres) from Deadhorse (near Prudhoe Bay) to Valdez, Alaska, the 48-inch pipeline was completed in 1977 at a cost of $8 billion. Today, it transports from 1 million to 2 million barrels (159 million to 318 million litres) of oil a day from fields on the North Slope of Alaska to tank farms in Valdez.

transboundary pollution *n*: pollution that crosses political or governmental or geographical boundaries.

transducer *n*: a substance or a device, such as a piezoelectric crystal, a microphone, or a photoelectric cell, that converts input energy of one form into output energy of another. For example, a telephone receiver receives electric power and supplies acoustic power.

transducer gas voucher *n*: a computer-generated readout of a quantity of gas measured by the computer system.

transfer *v*: to lower pipe or tubing onto a downhole tool so that all or part of the pipe's or tubing's weight is transferred from the traveling block's hook to the tool.

transfer lag *n*: in an electronic circuit, the delay that occurs between the time the circuit receives an electrical signal and the time the circuit produces a modified form of this signal at its output.

transfer order *n*: an agreement regarding change of production ownership that indemnifies the pipeline company or the purchaser.

transfer proving *n*: the use of a master meter to calibrate positive-displacement and turbine meters measuring gas flow.

transfer pump *n*: a pump that pressures the crude oil from the stock tank through a LACT unit.

transformer *n*: a device that uses the principle of induction to convert variations of current in a primary circuit into variations of current and voltage in a secondary circuit. It typically consists of two coils with different numbers of turns around a single iron core.

transform fault *n*: a strike-slip fault caused by relative movement between crustal plates.

transistor *n*: a small electronic device containing a small block of semiconducting material to which at least three electrical contacts are made, usually two rectifying contacts and one ohmic, or nonrectifying contact. A transistor is used in circuits as an amplifier, detector, or switch. Physicists at Bell Laboratories, who invented the device in the late 1940s, coined the term by combining the words transfer and resistor—that is, a transistor transfers a signal across a resistor. See *semiconductor*.

transit draft *n*: on a semisubmersible rig, the draft of the pontoons in a deballasted condition. For a ship-shaped drilling rig, the transit draft varies. Compare *operational draft*, *survival draft*.

transition foam *n*: see *foamed mist*.

transition zone *n*: 1. the area in which underground pressures begin to change from normal to abnormally high as a well is being deepened. 2. the areas in the drill stem near the point where drill pipe is made up on drill collars.

translation *n*: the movement of molecules.

transmission *n*: the gear or chain arrangement by which power is transmitted from the prime mover to the drawworks, the mud pump, or the rotary table of a drilling rig.

transmission company *n*: the seller of natural gas.

transmission line *n*: 1. a high-voltage line used to transmit electric power from one place to another. 2. a pipeline used to transmit natural gas or other fluids.

transmitter *n*: in instrumentation, an instrument or a transducer that converts a process variable such as temperature, pressure, level, or flow into a proportional electrical current or voltage.

transportation contract *n*: an agreement setting forth the terms and conditions applicable to transportation service. For interstate pipelines, this contract is usually a Form of Service Agreement stated in the pipeline's FERC gas tariff.

transportation costs *n pl*: the sum of the fixed costs and variable costs of moving gas or oil.

transportation lag *n*: see *distance-velocity lag*.

transporter *n*: the pipeline company that transports natural gas for a shipper.

transverse *n*: something, such as a bar or a beam, that lies crossways (is transverse). *adj*: situated or lying across; crosswise.

transverse bar *n*: on offshore installations with grated decks or floors, a length of steel strip fixed at right angles to the load-bearing bars to provide lateral restraint.

trap *n*: a body of permeable oil-bearing rock surrounded or overlain by an impermeable barrier that prevents oil from escaping. The types of traps are structural, stratigraphic, or a combination of these.

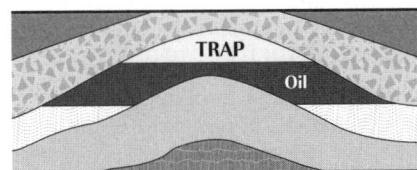

trapped pressure *n*: in well control and underbalanced drilling, pressure that occurs in the BOP stack when the well is closed in at the flow line or choke line without first relieving the pressure in the BOP stack. Closing a ram or an annular can trap high pressure between the closed blowout preventers because the annular and rams displace a certain amount of volume as they are closed. If the fluid in the stack is all or mostly liquid, high pressures can occur that may cause valve failure or incomplete ram closure.

traveling barrel pump *n*: a sucker rod insert pump in which the working barrel travels and the plunger remains stationary. The working barrel is connected to the sucker rod string through a connector and the traveling valve; the standing valve is connected to the top of the plunger, which in turn is connected to the bottom hold-down.

traveling block *n*: an arrangement of pulleys, or sheaves, through which drilling line is reeved and which moves up and down in the derrick or mast. See *block*.

traveling slips *n pl*: see *snubber* defn. 1.

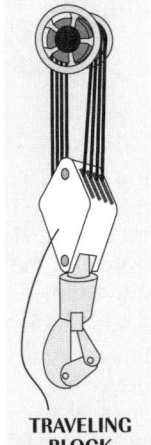

TRAVELING BLOCK

traveling valve *n*: one of the two valves in a sucker rod pumping system. It moves with the movement of the sucker rod string. On the upstroke, the ball member of the valve is seated, supporting the fluid load. On the downstroke, the ball is unseated, allowing fluid to enter into the production column. Compare *standing valve*.

travel time *n*: the time it takes for sound to travel from one receiver on an acoustic sonde to another receiver on the sonde.

tray *n*: a horizontal device in a tower that holds liquid and provides a vapor-liquid contact step. Several common types are bubble-cap, perforated, or valve trays.

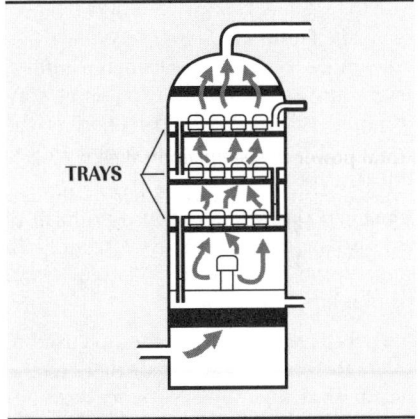

trayed column *n*: a vessel wherein gas and liquid, or two miscible liquids, contact on trays, usually countercurrently.

tread *n*: on offshore installations with grated decks or floors, the grated panel on which personnel step to ascend or descend the stairway.

tread diameter *n*: the diameter of the sheave of a drilling block, measured from groove bottom to groove bottom. It is the diameter of the circle around which the wire rope is bent.

treat *v*: to subject a substance to a process or to a chemical to improve its quality or to remove a contaminant.

treater *n*: a vessel in which oil is treated for the removal of S&W or other objectionable substances by the addition of chemicals, heat, electricity, or all three.

treating plant *n*: a plant that removes contaminants from natural gas.

treatment fluids *n pl*: acids and other fluids used to stimulate reservoirs. See *acidize*, *fracturing*.

treatment, storage, and disposal (TSD) facility *n*: a facility that treats, stores, or disposes of hazardous waste above certain volume limits and beyond certain time periods. If hazardous waste is stored or collected for shipment, generators need to be aware that exceeding volume and time limits can cause a generator site to be classed as a TSD facility. RCRA places rigorous controls on TSD facilities and they must have permits.

tree *n*: short for Christmas tree, which is the valves and fittings placed on top of a flowing well to control production.

tree saver tool *n*: a tubular device employed as an isolation tool inside the Christmas tree that protects the tree from the very high pressures that occur during well stimulation.

trend *n*: the directional line of a rock bed or petroleum deposit.

triac *n*: a type of thyristor that switches current to high levels from a low-level control signal in AC and DC circuits. While triacs are similar to silicon-controlled rectifiers, triacs, unlike SCRs, are capable of conducting current in both directions in the circuit; SCRs can conduct current only in one direction.

triangular coring *n*: a rarely used form of sidewall coring that uses a wall corer with saw-type blades set at 45° angles to cut triangular-shaped cores.

triaxial stress analysis *n*: a method used to calculate the combined loads that a casing or tubing string may experience throughout the service life of the well.

tribal land *n*: land within an Indian reservation or owned by an Indian tribe, group, or band.

trickle charge *n*: the continuous flow of direct current electricity into a cell or battery at a low rate to either recharge the cell or battery or to maintain a charge as the cell or battery is used.

tricone bit *n*: a type of bit in which three cone-shaped cutting devices are mounted in such a way that they intermesh and rotate together as the bit drills. The bit body may be fitted with nozzles, or jets, through which the drilling fluid is discharged.

triethylene glycol *n*: a liquid chemical used in gas processing to remove water from the gas. An organic compound with the formula $HO(C_2H_4O)_3H$, used to dehydrate natural gas to inhibit the formation of hydrates. See *glycol, glycol dehydrator.*

trigonometric function *n*: in mathematics, the relationship, or function, such as the sine, cosine, secant, cosecant, tangent, and cotangent of an arc or angle expressed as the ratios of pairs of the sides of a right triangle. See *function.*

trilobite *n*: one of a class of extinct Paleozoic arthropods.

trim *n*: 1. the difference between fore and aft draft readings on a marine vessel or an offshore drilling rig. 2. the use

of special corrosion-resistant metals in blowout preventers, valves, and other oilfield devices to minimize the effects of hydrogen sulfide. *v*: 1. to minimize the difference between fore and aft readings. 2. to install H_2S corrosion-resistant materials in oilfield equipment used in H_2S environments.

trim correction *n*: a correction applied to cargo ullages when the vessel is not on an even keel. Valid only when liquid completely covers the tank bottom and is not in contact with the underside of the deck.

trimming moment *n*: the movement of a floating vessel's bow or stern either higher or lower in the water, which is created by shifting weight on the vessel toward the bow or stern.

triode *n*: a virtually obsolete electrical component, which is a three-electrode vacuum tube that contains an anode, a cathode, and a control electrode. Triode transistors have almost completely replaced triode tubes.

trip *n*: the operation of hoisting the drill stem from and returning it to the wellbore. *v*: to insert or remove the drill stem into or out of the hole. Shortened form of "make a trip." See *make a trip, overspeed governor.*

trip gas *n*: gas that enters the wellbore when the mud pump is shut down and pipe is being pulled from the wellbore. The gas may enter because of the reduction in bottomhole pressure when the pump is shut down, because of swabbing, or because of both.

trip in *v*: to go in the hole.

triple *n*: see *thribble.*

triplex pump *n*: a reciprocating pump with three pistons or plungers.

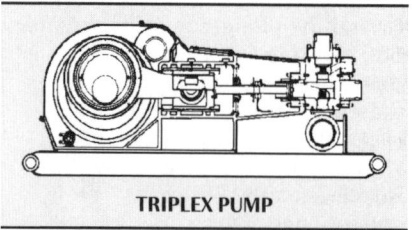

TRIPLEX PUMP

trip margin *n*: the small amount of additional mud weight carried over that needed to balance formation pressure to overcome the pressure-reduction effects caused by swabbing when a trip out of the hole is made.

tripod *n*: a device with three extensions (legs) that is suspended over a manway or other vertical access into a space. It supports hoisting gear and cables used to pull a person from a confined space.

trip out *v*: to come out of the hole.

tripping *n*: the operation of hoisting the drill stem out of and returning it to the wellbore. See *make a trip.*

trip rod *n*: the stem on the end of a core thief that can be bumped or tapped to close and thus catch a sample in the thief.

trip sheet *n*: a record of the measured drilling fluid that displaces the drill string as one pulls the pipe out of the hole or runs the pipe into the hole. See *trip tank.*

trip tank *n*: a small mud tank with a capacity of 10 to 15 barrels (1 to 3 cubic metres), often with 1-barrel or ½-barrel (decalitre or litre) divisions, used to ascertain the amount of mud necessary to keep the wellbore full with the exact amount of mud that is displaced by drill pipe. When the bit comes out of the hole, a volume of mud equal to that which the drill pipe occupied while in the hole must be pumped into the hole to replace the pipe. When the bit goes back in the hole, the drill pipe displaces a certain amount of mud, and a trip tank can be used again to keep track of this volume.

trip tank indicator *n*: a device installed on a trip tank that shows the amount of mud being removed from or added to the trip tank.

tritium *n*: an isotope of hydrogen with three atoms of three times the mass of ordinary hydrogen atoms, designated $_1H^3$.

tropical cyclone *n*: a tropical cyclonic storm with wind speeds of 65 knots or greater. See *baguio, cordonazo, cyclone, hurricane, taino, typhoon, willy-willy.*

tropical depression *n*: a tropical cyclonic system with wind speeds from 28 to 34 knots.

tropical disturbance *n*: a tropical cyclonic system with wind speeds of up to 27 knots.

tropical storm *n*: a tropical cyclonic system with wind speeds from 35 to 64 knots.

tropical waves *n pl*: disturbances on the intertropical convergence zone that travel from east to west. Low-pressure areas travel in tropical waves.

Tropic of Cancer *n*: a line of latitude that is 23.5 degrees north of the equator.

Tropic of Capricorn *n*: a line of latitude that is 23.5 degrees south of the equator.

troposphere *n*: the lowest and densest layer of the atmosphere, which extends from the surface to about 6 to 12 miles (10 to 20 kilometres) and in which most of the earth's weather occurs.

trough *n*: 1. in meteorology, a zone or belt of relatively low pressure. 2. in oceanography, the lowest point of the water surface level between crests.

trough length *n*: in oceanography, the distance between the stillwater level and the end of one wave and the beginning of another. See *stillwater level*.

truck-mounted rig *n*: a well-servicing and workover rig that is mounted on a truck chassis.

true mean draft *n*: the average of the observed drafts. On a floating offshore rig or ship, draft is the vertical distance between the bottom of a vessel floating in water and the waterline. Actual draft varies from the bow to the stern of the vessel because of up-and-down variations in the bottom of the vessel's hull. Thus, several draft measurements are made along the length of the hull, totaled, and averaged. This average is the true mean draft of the vessel.

true north *n*: the direction of the north geographic pole—as opposed to the north magnetic pole, which is several hundred miles away, toward Hudson Bay.

true power *n*: effective power. See *apparent power*.

true-to-gauge hole *n*: a hole that is the same size as the bit that was used to drill it. It is frequently referred to as a full-gauge hole.

true value *n*: the theoretically correct amount. In practice, it is represented by the standard being used for comparison, such as a prover.

true vapor pressure *n*: pressure measured relative to zero pressure.

true vertical depth (TVD) *n*: the depth of a well measured from the surface straight down to the bottom of the well. The true vertical depth of a well may be quite different from its actual measured depth, because wells are very seldom drilled exactly vertical.

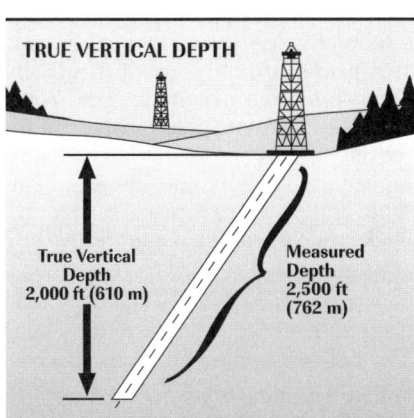

true volume *n*: the volume of fluid flowing through a meter, including adjustments for meter factor, temperature, density, pressure, etc. Compare *observed volumes*.

truncate *v*: 1. to cut the top or end from. 2. to terminate abruptly.

trunkline *n*: a main line.

trunkline station *n*: the station that relays oil from the gathering station to refineries or shipping terminals.

trunnion *n*: metal pins protruding from the sides of a device that serve to mount the device onto another part.

trunnion cup *n*: in a centrifuge, the container into which a sample is placed for centrifuging.

trust *n*: a right in property held by one party for the benefit of another. The trustee holds the legal interest or title, and the beneficiary holds the equitable interest or title and receives the benefits.

trustee *n*: person who holds the legal title to property in trust for the benefit of another. With the title go specified powers and duties relating to the property.

truth table *n*: a table listing the results obtained from a logical proposition (a truth value) that result from all the possible combinations of the truth values of its components. See *truth value*.

truth value *n*: the result of logical propositions, which is either true or false.

TSCA *abbr*: Toxic Substance Control Act.

TSCA inventory *n*: a chemical inventory set up under the EPA that is a listing of chemicals that are, or once were, used in commerce. New chemicals that are not on the inventory are reviewed as produced. Operators and employers may not use any chemical that has not been listed on the inventory; however, it is important to note that some substances are excluded from the inventory because they are regulated by other laws or agencies.

TSD *abbr*: treatment, storage, and disposal.

TSP *abbr*: thermally stable polycrystalline. See *thermally stable polycrystalline diamond bit*.

T-square *n*: a ruler with a crosspiece or head at one end used in making parallel lines.

tstg *abbr*: testing; used in drilling reports.

tsunami *n*: a long-period wave produced by a submarine earthquake or explosion. If it strikes land, it can be very destructive. Also called seismic sea wave.

tube *v*: to run tubing in a well. *n*: see *electron tube*.

tube-and-fin spacing *n*: in an engine radiator, the distance between the tubes through which coolant flows and the fins on the tubes that radiate heat into the air. The manufacturer must put enough tubes and fins on the radiator to adequately cool the engine on which the radiator will be mounted, yet not so many that some tubes block the flow of air across other tubes.

tube bundle *n*: the inner piping of a condenser, heat exchanger, or indirect heater typically consisting of a pipe that is bent several times and placed inside the shell of a tank.

tube sample *n*: a liquid sample obtained with a sampling tube or special thief, as either a core sample or a spot sample from a specified point in the tank or container.

tube sheet *n*: a metal plate through which the tubes in the tube bundles are placed for support, effecting a pressure-tight connection between the tubes and the heads of a condenser or heat exchanger.

tubing *n*: relatively small-diameter pipe that is run into a well to serve as a conduit for the passage of oil and gas to the surface.

tubing anchor *n*: a device that holds the lower end of a tubing string in place by means of slips, used to prevent tubing movement when no packer is present. Especially for sucker rod pumping applications.

tubing broach *n*: a tool with three consistently larger spools that are used to enlarge tubing diameter gradually. The tubing broach may be used either to enlarge the diameter of obstructed tubing or to cut paraffin or other substances that may be inside the tubing.

tubing-conveyed perforating *n*: a system in which the tubing is used to convey the perforating guns to the target reservoir and charges are detonated either by dropping a bar or by the application of external pressure.

tubing-conveyed perforator *n*: a perforating gun that is lowered into the well on the tubing string. See also *perforate*.

tubing coupling *n*: a special connector used to connect lengths of tubing.

tubing elevators *n pl*: a clamping apparatus used to pull tubing. The elevators latch onto the pipe just below the top collar. The elevators are attached to the hook by steel links or bails.

tubing end locator *n*: a tool containing a spring-loaded finger, or dog, that pops out when the tool reaches the bottom end of the tubing string. The extended dog prevents the tool from being raised back into the tubing, thus indicating to an operator on the surface that the bottom of the tubing has been reached. The tool is removed from the tubing by jarring upward with enough force to shear a brass pin and allow the dog to rotate downward and into a slot in the body of the tubing end locator.

tubing gauge *n*: a tool designed to cut away scale, calcium, small deposits of paraffin, and other debris in an obstructed length of tubing. The tool is also used to measure, or gauge, the internal diameter of tubing.

tubing hanger *n*: an arrangement of slips and packing rings used to suspend tubing from a tubing head.

tubing hanger handling and test tool *n*: enables the tubing hanger to be handled safely and pressure tested; should be able to interface with the tubing hanger and seal at the required pressure.

tubing hanger running and orientation tool (THROT) *n*: used to run, orientate, and lock the tubing hanger and completion into the wellhead through the drilling riser and BOP.

tubing head *n*: a flanged or threaded fitting that supports the tubing string, seals off pressure between the casing and the outside of the tubing, and provides a connection that supports the Christmas tree.

tubing head swivel *n*: a special pipe sub made up at the upper end of the tubing test string to allow tubing to rotate during a well test, or remedial operations.

tubing jewelry *n*: downhole equipment attached to the tubing string that is used to control or modify flow through the tubing (for example, gas-lift mandrels and subsea safety valves).

tubing job *n*: the act of pulling tubing out of and running it back into a well.

tubingless completion *n*: a method of producing a well in which only production casing is set through the pay zone, with no tubing or inner production string used to bring formation fluids to the surface. This type of completion has its best application in low-pressure, dry-gas reservoirs and to low reserve medium-pressure reservoirs.

tubing movement *n*: changes in tubing length due to changes in pressure, temperature, or drag from sucker rod strokes. Also called breathing.

tubing pressure (TP) *n*: pressure on the tubing in a well at the wellhead.

tubing pump *n*: a sucker rod pump in which the barrel is attached to the tubing. See *sucker rod pump*.

tubing safety valve *n*: a device installed in the tubing string of a producing well to shut in the flow of production if the flow exceeds a preset rate. Tubing safety valves are widely used in offshore wells to prevent pollution if the wellhead fails for any reason. Also called subsurface safety valve.

tubing seal assembly *n*: a device encircled with rubber rings and made up on tubing to fit into a seal-bore packer.

tubing slips *n pl*: slips designed specifically to be used with tubing.

tubing spider *n*: a device used with slips to prevent tubing from falling into the hole when a joint of pipe is being unscrewed and racked.

tubing sub *n*: any type of sub used in the tubing string. See *sub*.

tubing tester *n*: a valve operated by tubing rotation that is used to shut off formation pressure above a packer, thus testing all connections from the packer to the tree.

tubing test string *n*: a string of tubing run in a well for a production test in which a well test surface package is used. A tubing test string is similar to a normal tubing string except that it may contain special valves and other tools needed for the well test.

tubing tongs *n pl*: large wrenches used to break out and make up tubing. They may be operated manually, hydraulically, or pneumatically.

tubing work string *n*: see *work string*.

tubular goods *n pl*: any kind of pipe. Oilfield tubular goods include tubing, casing, drill pipe, and line pipe. Also called tubulars.

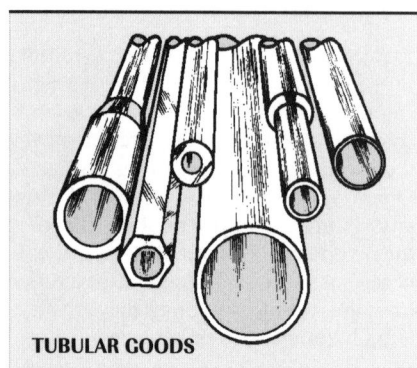

TUBULAR GOODS

tubular heater *n*: a pipeline heating device consisting of a resistance wire surrounded by an insulating material such as magnesium oxide and contained in a metal tube.

tubulars *n pl*: shortened form of tubular goods.

tugboat *n*: a comparatively small strongly built boat designed to push or pull other vessels in harbors, on inland waterways, and in coastal waters.

tugger *n*: see *hoisting engine*.

tun *n*: 1. a large cask for beer or wine. 2. a measure of liquid volume, especially one for wine equal to 252 gallons or 954 litres.

tundra *n*: the treeless plain characteristic of much of the Arctic. Although the tundra remains frozen for much of the year,

it is highly sensitive to environmental disturbance, and special pipeline construction techniques are required there.

tungsten carbide *n*: a fine, very hard, gray crystalline powder, a compound of tungsten and carbon. This compound is bonded with cobalt or nickel in cemented carbide compositions and used for cutting tools, abrasives, and dies.

tungsten carbide bit *n*: a type of roller cone bit with inserts made of tungsten carbide. Also called tungsten carbide insert bit.

tungsten carbide insert *n*: a very hard, and relatively small, cylinder made of tungsten carbide bonded with nickel or cobalt. One end of the insert is rounded or tapered; the other end is truncated (cut off) and inserted into a bit cone. The round or tapered end contacts the formation and drills it. Several inserts make up the cutters on a tungsten carbide insert bit.

tungsten carbide insert bit *n*: see *tungsten carbide bit*.

tunnage *n*: a tax imposed on ships that enter the United States; it is based on the tonnage of the ship. See *tonnage*.

tunnel *v*: in pipeline construction, to penetrate or pierce manually. Compare *bore*.

tunnel diode *n*: a diode with a special impurity that causes the diode to have negative resistance at low voltages in the forward bias direction and a short circuit in the negative bias direction. Tunnel diodes are used as oscillators and as microwave amplifiers. Also known as an Esaki tunnel diode.

turbidite *n*: a characteristic sedimentary deposit of the continental rise, formed by a turbidity current and composed of clay, silt, and gravel with the clay on top. See *turbidity current*.

turbidity current *n*: a dense mass of sediment-laden water that flows down the continental slope, typically through a submarine canyon. It carries clay, silt, and gravel quickly and turbulently downslope in a narrow tongue. When it reaches the more gradual slope of the continental rise, its energy is dissipated rapidly by friction, and the sediments settle out, largest particles first, to form a thick graded bed called a turbidite.

turbine *n*: a bladed rotor flowmeter component that turns at a speed that is proportional to the mean velocity of the stream and therefore to the volume rate of flow.

turbine flowmeter *n*: see *turbine meter*.

turbine fuel *n*: a fuel oil similar to diesel fuel that a turbine compressor burns. Some turbines do not require fuel as high in quality as a reciprocating engine, such as a diesel engine.

turbine meter *n*: a velocity-measuring device for fluids in which the flow is parallel to the rotor axis and the speed of rotation is proportional to the rate of flow. The volume of gas in gas measurement is determined by counting the revolutions of the rotor. In liquid turbine meter measurement, the meter and electronic instrumentation are combined to measure total flow and/or flow rate within the piping system.

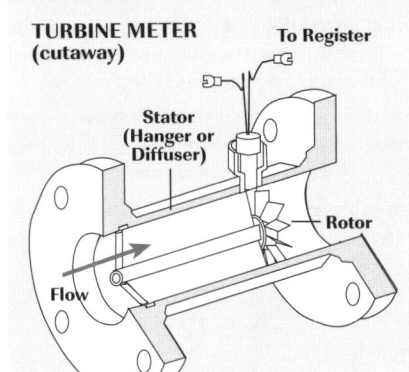

turbine motor *n*: see *turbodrill*.

turbocharge *v*: to supercharge an engine by use of a turbocharger.

turbocharger *n*: a centrifugal blower driven by exhaust gas turbines and used to supercharge an engine.

turbodrill *n*: a downhole motor that rotates a bit by the action of the drilling mud on turbine blades built into the tool. When a turbodrill is used, rotary motion is imparted only at the bit; therefore, it is unnecessary to rotate the drill stem. Although straight holes can be drilled with the tool, it is used most often in directional drilling.

turboexpander *n*: a device that converts the energy of a gas or vapor stream into mechanical work by expanding the gas or vapor through a turbine.

turbulent flow *n*: the erratic, nonlinear flow of a fluid, caused by high velocity. Characterized by random eddying flow patterns superimposed on the general flow progressing in a given direction.

turn *n*: a change in bearing (azimuth) of the hole. Usually spoken of as a right or left turn from the established hole drift.

turnaround *n*: 1. space that permits the turning around of vehicles, specifically on the drill site. 2. a period during which a plant is shut down completely for repairs, inspections, or modifications that cannot be made while the plant is operating.

turnaround season *n*: see *turnaround* (def. 2).

turnback *n*: see *crossover*.

turndown *n*: field terminology for rejecting a tank's contents on the basis of the gauger's evaluation and analysis.

turnkey contract *n*: a drilling contract that calls for the payment of a stipulated amount to the drilling contractor on completion of the well. In a turnkey contract, the contractor furnishes all material and labor and controls the entire drilling operation, independent of operator supervision. A turnkey contract does not, as a rule, include the completion of a well as a producer.

turns ratio *n*: in a transformer, a comparison (the ratio) between the number of turns in the primary winding and the number of turns in the secondary winding. See *transformer*.

turntable *n*: see *rotary table*.

turn to the right *v*: on a rotary rig, to rotate the drill stem clockwise. When drilling ahead, the expression "on bottom and turning to the right" indicates that drilling is proceeding normally.

turret mooring *n*: a system of mooring a drill ship on the drilling site, in which mooring lines are spooled onto winches mounted on a turret in the center of the vessel. Because all mooring lines are connected to the turret, the vessel is free to rotate around the turret axis and head into oncoming seas, regardless of direction.

TVD *abbr*: true vertical depth.

twin *n*: a well drilled on the same location as another well and closely offsetting it, but producing from a different zone.

twistoff *n*: a complete break in pipe caused by metal fatigue. *v*: to break something in two or to break apart, such as the head of a bolt or the drill stem.

two-cone bit *n*: a type of roller cone bit in which two cone-shaped cutting devices are mounted in such a way that they intermesh and rotate together as the bit drills.

The bit body may be fitted with nozzles, or jets, through which the drilling fluid is discharged. Compare *tricone bit*.

two-phase flow *n*: a vapor phase and a liquid phase flow in the same pipeline.

two-phase separator *n*: a separator that separates the well fluids into liquid and gas. The liquid (oil, emulsion, or water) goes out the bottom; the gas goes out the top. See *separator*.

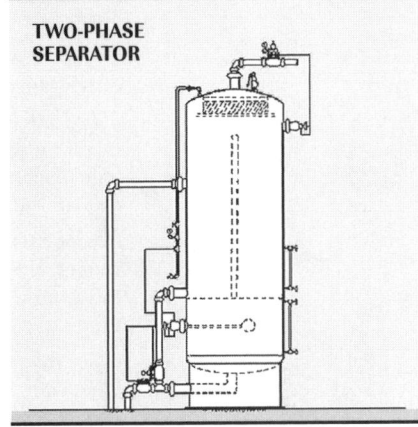

two-phase system *n*: a pipeline, plant, refinery, or other installation in which the fluid in the installation exists as a gas (vapor) and a liquid.

two-piece insert bowl *n*: see *insert bowl*.

two-step grooving system *n*: a pattern of drum spooling in which the wire rope is controlled by grooves to move parallel to drum flanges for half the circumference (180°) and then crosses over to start the next wrap. Also called counterbalance system.

two-stroke/cycle engine *n*: an engine in which the piston moves from top dead center to bottom dead center and then back to top dead center to complete a cycle. Thus, the crankshaft must turn one revolution, or 360°.

two-wire current transmitter *n*: in process control systems, a process transmitter that converts a process variable into a proportional 4-to-20 milliamp current signal that is sent to a two-wire control circuit.

twp *abbr*: township; used in drilling reports.

typhoon *n*: a tropical cyclone in the western North Pacific and most of the South Pacific.

UBD *abbr*: underbalanced drilling.

UBO *abbr*: underbalanced operations.

UEPS *abbr*: uninterruptible electrical power supply.

UIC *abbr*: Underground Injection Control.

UIPC *abbr*: Underground Injection Practices Council.

ULCC *abbr*: ultralarge crude carrier.

ullage *n*: the amount by which a tank or a vessel is short of being full, especially on ships. Ullage in a tank is necessary to allow space for the expansion of the oil in the tank when the temperature increases. Also called outage.

ullage gauge *n*: see *outage gauge*.

ullage hatch *n*: the opening in the top of a tanker or barge tank compartment through which ullage measurements and sampling operations are carried out.

ullage paste *n*: a paste that is applied to an ullage rule or dip tape and weight to indicate precisely the level at which the liquid meniscus cuts the graduated portion.

ullage reference point *n*: a point marked on the ullage hatch or on an attachment suitably located above or below the ullage hatch and situated at a distance above the bottom of a container greater than the maximum liquid depth in the container. Ullage measurements are taken from this reference point.

ullage rule *n*: a graduated rule attached to a dip tape to facilitate the measurement of ullage.

ullage tape-and-bob procedure *n*: see *outage tape-and-bob procedure*.

ultimate recovery *n*: total anticipated recovery of oil or gas from a well, lease, or pool.

ultimate strength *n*: the amount of tension, or tensile stress, applied to a piece of equipment that causes it to fail or continue to deform under a decreasing load.

ultra deep water *n*: water deeper than 3,000 feet (914 metres).

ultradeepwater *adj*: of or pertaining to operations in ultra deep water.

ultrahigh vacuum *n*: an extremely low pressure, usually only 10^{-10} millimetres of mercury or less.

ultralarge crude carrier (ULCC) *n*: a supertanker whose capacity is 500,000 deadweight tons or more. Compare *very large crude carrier*.

ultrasonic devices *n pl*: corrosion-monitoring devices that transmit ultrasonic waves through production structures to locate discontinuities in metal structure, which indicate corrosion damage.

ultrasonic meter *n*: flowmeter used (1) to measure transit times of an acoustic pressure wave with and against the flow (time-of-flight-type meter) to infer pipeline velocity or (2) to reflect sonic energy from scatterers in fluid back to a receiver (Doppler-type meter) to measure volumetric flow rate.

ultraviolet light *n*: light waves shorter than the visible blue violet waves of the spectrum. Crude oil, colored distillates, residuum, a few drilling fluid additives, and certain minerals and chemicals fluoresce in the presence of ultraviolet light. These substances, when present in mud, may cause the mud to fluoresce.

umbilical *n*: a line that supplies a diver or a diving bell with a lifeline, a breathing gas, communications, a pneumofathometer, and, if needed, a heat supply.

UMC *abbr*: underwater manifold center.

unaccounted-for crude oil *n*: the arithmetic difference between the indicated demand for crude oil and the total disposition of crude oil.

unaccounted-for gas *n*: the difference between gas taken into the distribution system and the known quantities of gas taken out of the system.

uncased crossing *n*: in pipeline construction, a road crossing bored without casing. The carrier pipe itself is pushed under the roadbed by the boring machine. To reduce potential damage to the coating, slick boring is frequently used. See *slick boring*.

uncased hole *n*: see *open hole*.

unclaimed property statute *n*: see *escheat*.

unconf *abbr*: unconformity; used in drilling reports.

unconformity *n*: 1. lack of continuity in deposition between rock strata in contact with one another, corresponding to a gap in the stratigraphic record. 2. the surface of contact between rock beds in which there is a discontinuity in the ages of the rocks. See *angular unconformity, disconformity*.

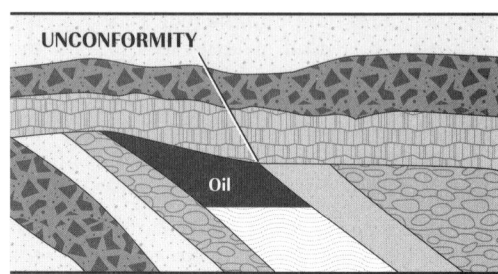

unconsolidated formation *n*: a loosely arranged, apparently unstratified section of rock.

unconsolidated sandstone *n*: a sand formation in which individual grains do not adhere to one another. If an unconsolidated sandstone produces oil or gas, it will produce sand as well if not controlled or corrected.

unconsolidated shales *n pl*: compacted beds of clays and silts that are soft.

uncontrolled blowout *n*: see *blowout*.

underbalanced *adj*: of or relating to a condition in which pressure in the wellbore is less than the pressure in the formation.

underbalanced choke *n*: a well-control choke that is specifically designed for underbalanced drilling. One model is a 3-in., rubber-lined, pressure-balanced choke.

underbalanced drilling (UBD) *v*: to carry on drilling operations with a mud whose density is such that it exerts less pressure on bottom than the pressure in the formation while maintaining a seal (usually with a rotating head) to prevent the well fluids from blowing out under the rig. Drilling under pressure is advantageous in that the rate of penetration is relatively fast; however, the technique requires extreme caution.

underbalanced liquid drilling *n*: see *flow drilling*.

underbalanced operations (UBO) *n pl*: processes on a drilling rig, a well servicing rig, or a coiled tubing unit that occur when the hydrostatic pressure of the fluid in the drill string or tubing string exerts less pressure than the pressure required to equal or exceed the pressure found in a porous and permeable formation.

underdelivery *n*: the amount by which the actual volume that has passed through the meter is less than the indicated volume registered by a meter. See *absolute error*.

underflow *n*: the lower discharge stream in a cone-shaped centrifuge. It moves downward and leaves by way of the apex.

undergauge bit *n*: a bit whose outside diameter is worn to the point at which it is smaller than it was when new. A hole drilled with an undergauge bit is said to be undergauge.

undergauge hole *n*: that portion of a borehole drilled with an undergauge bit.

underground blowout *n*: an uncontrolled flow of gas, salt water, or other fluid out of the wellbore and into another formation that the wellbore has penetrated.

Underground Injection Control (UIC) *n*: a program developed under SDWA to prevent injection well contamination of underground sources of drinking water. The UIC program divides injection wells into five classes (that range from common septic tanks to hazardous waste injection wells) and establishes specific requirements for construction, casing and cementing, plugging and abandonment, types of substances that can be injected, volume and pressures that can be injected, and mechanical integrity of the well.

Underground Injection Practices Council (UIPC) *n*: a committee of state regulators,

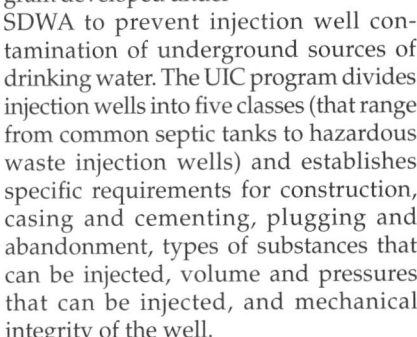

UNDERGAUGE HOLE

Environmental Protection Agency personnel, and industry representatives that works with the EPA in the formulation of new regulations.

underground sources of drinking water (USDW) *n pl*: fresh water sources located under the surface that are suitable for use as drinking water supplies.

underground storage area *n*: salt domes, depleted gas fields, or other suitable underground formations located in a company's major market areas into which gas is injected during low-demand summer months and from which it is withdrawn during high-demand winter months. May be 20 to 30 feet (6 to 9 metres) thick and can cover many square miles.

underground waste *n*: recoverable reserves lost as a result of damage to the reservoir.

underream *v*: to enlarge the wellbore below the casing.

underreamer *n*: a device used to underream. See *underream*.

underregistration *n*: the amount by which the indicated volume registered by a meter is less than the actual volume that has passed through the meter. Determined by means of a suitable standard device. Compare *overregistration*.

undertravel *n*: a phenomenon that occurs when the polished rod of a sucker rod pump is too short to allow the pump's plunger to travel the full length of its stroke.

underwater manifold center (UMC) *n*: this center allows the exploitation of smaller offshore fields. It is connected by pipeline to a platform more than 4 miles (6.4 kilometres) away. It collects fluids from the wells and sends them to the platform.

undivided agreement *n*: type of operating agreement that fixes the sharing of costs and benefits for the life of the unit.

undivided interest *n*: a fractional interest in minerals that, when conveyed, gives the new owner that fractional interest in the described tract. For example, a one-quarter undivided interest in an 80-acre (32-hectare) tract amounts to a one-quarter interest in the entire 80 acres, or 20 net undivided mineral acres (8 hectares). Compare *divided interest*.

unfocused log *n*: see *conventional electric log*.

uniformitarianism *n*: the geologic principle that the processes that are at work today on the earth are the same as, or very similar to, the processes that affected the earth in the past. The principle was formulated in 1785 by James Hutton, a

Scottish geologist. It became an accepted doctrine with the publication in 1830 of English geologist Charles Lyell's book *Principles of Geology*. Also called gradualism. Compare *catastrophism*.

uniform petroleum product *n*: product in which spot samples from top, upper, middle, lower, and outlet agree within the precision of the laboratory tests. Similarly, in pipeline transfers, spot samples taken at 1, 20, 50, 80, and 99 percent of the total volume agree within the precision of the laboratory tests.

uninterruptible electrical power supply (UEPS) *n*: an electrical system with built-in redundancy that supplies (powers) an entire electronic control system, and provides immediate changeover should one part of the system fail.

union *n*: a coupling device that allows pipes to be connected without being rotated. The mating surfaces are pulled together by a flanged, threaded collar on the union.

unipolar transistor *n*: a transistor that utilizes charge carriers of only one polarity, such as a field-effect transistor.

unit *n*: 1. a piece or several pieces of equipment performing a complete function (such as a beam pumping unit). 2. the joining of all or substantially all interests in a reservoir or field, rather than a single tract, to provide for development and operation without regard to separate property interests. 3. the area covered by a Unitization Agreement.

United States Army Corps of Engineers *n*: a federal agency involved in platform design. It also works to solve navigation problems. Address: 441 G Street NW; Washington, DC 20314-1000; 202-761-0660.

United States Coast Guard (USCG) *n*: an agency of the United States Department of Transportation that establishes regulations for offshore vessels and approves standards for plan approval, construction, operation, and inspection of the hull, machinery, electric systems, industrial systems, and safety and life-saving equipment of mobile offshore drilling units. The agency conducts rig stability tests, awards the original certification of the rig, and conducts annual inspections. It also establishes manning and personnel licensing requirements and investigates casualties. Address: 2100 Second Street SW; Washington, DC 20593; 202-267-2229; www.uscg.mil.

United States Code *n*: a series of volumes that is a published compilation of all the laws passed by Congress.

United States Fish and Wildlife Service (USFWS) *n*: a federal service that oversees commercial fishing investigations, fishing research, fishing statistics, sport fishing, and other areas related to the fishing industry. It also conducts wildlife research, enforces federal statutes and regulations applying to migratory birds, maintains the official list of threatened or endangered species, acquires habitats for threatened or endangered species, operates the national refuge system, manages predator and rodent control work, evaluates the impact of offshore development on fish and wildlife, develops programs to protect fish and wildlife from oil spills, coordinates individual and state efforts to rescue and rehabilitate oiled birds, and administers federal aid programs. Address: 4401 N. Fairfax Dr., Suite 634, Arlington, VA 22203; 703-358-1714.

United States Geological Survey (USGS) *n*: a federal agency within the Department of the Interior established in 1879 to conduct investigations of the geological structure, mineral resources, and products of the United States. Its activities include assessing onshore and offshore mineral resources; providing information that allows society to mitigate the impact of floods, earthquakes, landslides, volcanoes, and droughts; monitoring the nation's groundwater and surface water supplies and people's impact thereon; and providing mapped information on the nation's landscape and land use. Address: 12201 Sunrise Valley Drive; Reston, VA 20192; 703-648-4000.

unit fuel-injector assembly *n*: a device that combines an internal high-pressure fuel pump, an injector valve, and a spray nozzle, and that is used to inject diesel fuel into an engine's combustion chamber.

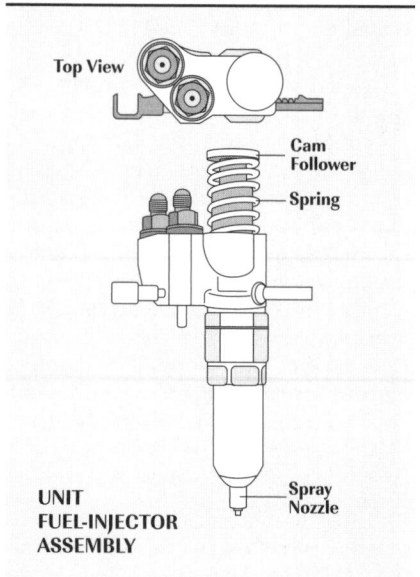

UNIT
FUEL-INJECTOR
ASSEMBLY

unitization *n*: the combining of leased tracts on a fieldwide or reservoirwide scale so that many tracts may be treated as one to facilitate operations like secondary recovery. Compare *pooling*.

unit of measurement *n*: the value of a quantity conventionally accepted as having a numerical value of 1. The unit of measurement of a quantity is fixed in order to make quantitative comparison possible between different values of this same quantity.

unit operator *n*: the oil company in charge of development and production in an oilfield in which several companies have joined to produce the field. Also called crew chief, rig operator.

units of volume *n pl*: the usual units of volume for petroleum measurement are the cubic metre, litre, imperial gallon, U.S. gallon, or the barrel (42 U.S. gallons, or 159 litres).

unity *n*: in mathematics, the number 1; a quantity regarded as 1.

univalent *n*: monovalent. See *valence*.

universal constant *n*: in an ideal gas, a number equal to the pressure of the gas times its molar volume divided by its temperature. See *molar volume, mole*.

"unless" clause *n*: see *drilling and delay rental clause*.

unloader *n*: see *circulation valve*.

unloading a well *n*: removing fluid from the tubing in a well, often by means of a swab, to lower the bottomhole pressure in the wellbore at the perforations and induce the well to flow.

unloading sub *n*: a special circulation valve that provides a way of equalizing tubing and annulus pressure. See *circulation valve*.

unproven area *n*: a wildcat area.

unrestricted ABS classification *n*: a classification for offshore rigs that indicates that the rig is constructed so that it can operate safely in offshore waters anywhere in the world.

unsaturated hydrocarbon *n*: a straight-chain compound of hydrogen and carbon whose total combining power has not yet been reached and to which other atoms or radicals can be added.

unslaked lime *n*: quicklime.

unstable *adj*: in process control, an undesirable oscillation or variation of a system's output at its final control element.

unstable air mass *n*: a large volume of air (an air mass) that is unsteady and has irregular motion. This instability can cause storms to form in the mass if enough moisture is also present.

unstable equilibrium *n*: a condition in which an offshore floating rig continues to incline after an external force caused an initial small angle.

unstable foam *n*: a fluid that occurs when a foam drilling fluid breaks into its components—that is, it is no longer gas emulsified in a liquid. Instead, it becomes separate drops of liquid in an airflow, or mist. Unstable foam cannot lift cuttings from the hole as well as foam. See *mist drilling*.

updip *adj*: higher on the formation dip angle than a particular point.

updip well *n*: a well located high on the structure.

upper kelly cock *n*: a valve installed above the kelly that can be closed manually to protect the rotary hose from high pressure that may exist in the drill stem.

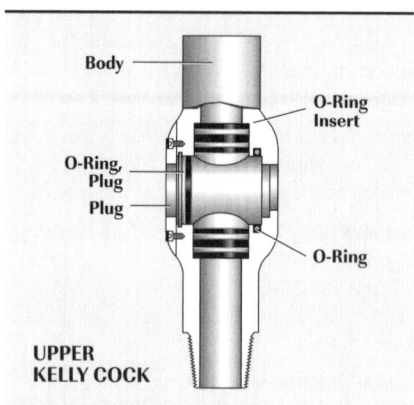

UPPER
KELLY COCK

upper range value (URV) *n*: the upper value of calibration of a process transmitter where its lower value is the lower range value (LRV).

upper sample *n*: in tank sampling, a spot sample taken at the mid-point of the upper third of the tank contents, that is, at one-sixth of the depth of liquid below the top surface.

upper string *n*: any part of the drill stem, tubing string, or casing string that lies in the upper part of the borehole.

upper tier *n*: a category of oil production for purposes of price control. Upper tier refers to oil that comes from reservoirs that began producing subsequent to 1972. It also refers to oil produced from wells having a production rate of 10 barrels (1,590 litres) per day or less for a continuous period of 12 months. Also called stripper oil.

upper troposphere *n*: the part of the earth's atmosphere that generally starts at an altitude of 18,000 to 19,000 feet (about 5.5 to 5.8 kilometres) and extends to about 50,000 feet (about 15 kilometres).

upset *n*: thickness forged to the end of a tubular (such as drill pipe) to give the end extra strength. *v*: to forge the ends of tubular products, such as drill pipe, so that the pipe wall acquires extra thickness and strength near the end. Upsetting is usually performed to strengthen the pipe so that threads, or threaded pieces, such as tool joints, can be added to the pipe.

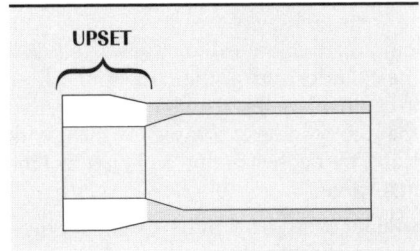

upset tubing *n*: tubing that is made with a thicker wall and larger outside diameter on both ends of a joint to compensate for cutting the threads.

upstream *n*: the point in a line or system situated opposite the direction of flow. *adv*: in the direction opposite the flow in a line.

upstream market *n*: a market for selling gas or petroleum before it is processed or refined. Compare *downstream market*.

upstructure *adj*: toward the upper portion of an updip formation.

urea *n*: a soluble, weakly basic, nitrogenous compound, $CO(NH_2)_2$, that is used in the manufacture of resins and plastics.

URV *abbr*: upper range value.

USCG *abbr*: United States Coast Guard.

USDW *abbr*: underground sources of drinking water.

useful volume *n*: in diesel engine fuel tanks, that part of the tank's capacity from which the fuel system can take fuel. Useful volume is less than the actual or total volume of the tank. Compare *total volume*.

use tests *n pl*: periodic equipment inspections to determine corrosion damage in fields that have been producing for several years.

USFWS *abbr*: United States Fish and Wildlife Service.

USGS *abbr*: United States Geological Survey.

utilization factor *n*: a ratio of the maximum demand of a system or part of a system to the rated capacity of the system or part of the system under consideration.

U-tube *n*: a U-shaped tube. *v*: to cause fluids to flow in a U-tube without adding an outside force. For example, heavy mud in the drill stem will force light mud in the annulus to flow.

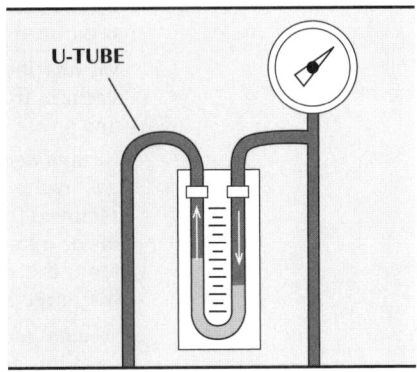

U-tubing *n*: the action of fluids flowing in a U-tube (as heavy mud forcing lighter mud down the drill stem and up the annulus).

V *sym*: volt.

VA *abbr*: volt-ampere.

VAC *abbr*: volts, alternating current.

vacuator valve *n*: on a dry air cleaner on an engine, a special valve that automatically opens to dump dust accumulated in the chambers of the air cleaner.

vacuum *n*: 1. a space that is theoretically devoid of all matter and that exerts zero pressure. 2. a condition that exists in a system when pressure is reduced below atmospheric pressure.

vacuum breaker *n*: a device used to prevent a vacuum from occurring in a tank, vessel, piping, and so on.

vacuum degasser *n*: a device in which entrained gas in the mud returning from the wellbore is removed from the mud by the action of a vacuum inside a tank. The gas-cut mud is pulled into the tank, the gas removed, and the gas-free mud discharged back into the mud pits.

vacuum gauge *n*: an instrument used on gas or gasoline engines to indicate the performance characteristics and load.

vacuum pressure *n*: any pressure below atmospheric pressure.

vacuum retort *n*: a device that measures the saturation, or the amount, of each fluid in a core sample by distilling the fluid from the core with heat at pressure below atmospheric. The vacuum retort is used to measure saturation on whole core samples.

vacuum tube *n*: see *electron tube*.

valence *n*: the tendency of elements to form compounds through a shift of electronic structure.

valence effect *n*: an effect that causes emulsions and colloids to separate; it is particularly prevalent in emulsions and colloids that contain a large number of high-valence ions. See *valence*.

valence electron *n*: a single electron or one of two or more electrons in the outer shell of an atom that is responsible for the chemical properties of the atom. Valence electrons of one atom combine with the valence electrons of other atoms to form many compounds. For example, when the valence electrons of sodium and chlorine combine, common table salt, or sodium chloride (NaCl), is formed.

valence number *n*: the number of electrons of an element involved in forming a compound.

valley line *n*: see *thalweg*.

valve *n*: 1. a device used to control the rate of flow in a line to open or shut off a line completely, or to serve as an automatic or semiautomatic safety device. Those used extensively include the check valve, gate valve, globe valve, needle valve, plug valve, and pressure relief valve. 2. British term for electron tube. See *electron tube*.

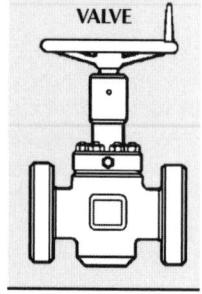

VALVE

valve arm *n*: in a two-diaphragm, four-chamber diaphragm displacement meter, the linkage that uses the motion from the crank to slide D-shaped valves over the three ported seats to fill and empty the four measurement chambers alternately.

valve overlap *n*: the phenomenon that occurs during the operating cycle of an engine in which both the intake and exhaust valves are open at the same time. Valve overlap occurs in a four-strokes-per-cycle engine at the end of the engine's exhaust stroke and the beginning of the intake stroke. Both valves being open allow exhaust gases to finish leaving the cylinder and, at the same time, allow inlet air to begin filling the cylinder. Valve overlap ensures that the cylinder is completely filled with fresh air; also, the cool incoming air helps cool the hot exhaust valve.

valve positioner *n*: an attachment that fits on a valve actuator, and which assures that the valve stem's movement follows the demands of a controller with great accuracy by acting as a force amplifier between the controller and the actuator. Valve positioners are used where friction occurs between valve stem and packing and when the force of flowing fluid on the valve plug causes unsatisfactory response.

valve pot *n*: the machined tapered opening on the fluid-end body of a mud pump that holds the valve seat.

valve seat *n*: a tapered metal ring that seals against the valve pot and holds a valve in a mud pump.

valve tray *n*: a tray installed in an absorber tower or fractionating column, similar to a bubble-cap tray except that the passageways permitting gas flow upward through the tray possess valves that reduce the size of the passages when the flow rate is reduced. Valve trays are more common than bubble-cap trays, because they are more efficient and less expensive.

vane anemometer *n*: a wind-speed measuring device that consists of flat blades (vanes) mounted in a circle so as to rotate under the action of the wind.

vapor *n*: a substance in the gaseous state that can be liquefied by compression or cooling.

vaporization *n*: 1. the act or process of converting a substance into the vapor phase. 2. the state of substances in the vapor phase.

vapor-liquid equilibrium ratio *n*: the partition coefficient, usually designated as K, which is equivalent to y/xm, where y is the mole fraction of a given component in the vapor phase that is in equilibrium with x, the mole fraction of the same component in the liquid phase. K is a function of temperature, pressure, and composition of the particular system.

vapor loss *n*: losses that occur to hydrocarbon liquids as the lighter components vaporize and leave the liquid.

vapor phase *n*: the existence of a substance in the gaseous state.

vapor point *n*: see *bubble point*.

vapor pressure n: the pressure exerted by the vapor of a substance when the substance and its vapor are in equilibrium. Equilibrium is established when the rate of evaporation of a substance is equal to the rate of condensation of its vapor.

vaporproof adj: not susceptible to or affected by vapors. For example, an electrical switch is made vaporproof so that a spark issuing from it will not cause an explosion in the presence of combustible gases (vapors).

vapor recovery n: a system or method by which vapors are retained and conserved.

vapor recovery unit (VRU) n: 1. in petroleum refining, a process unit consisting of a scrubber and a compressor. It is designed to recover petroleum from hydrocarbon vapors and safely to handle toxic gases produced from some wells. 2. in production operations, equipment designed to recover light ends that are released from oil in a stock or other tank.

vapor-tight tank n: a tank of conventional shape intended primarily for the storage of volatile liquids such as gasoline and so constructed that it can withstand pressures differing only slightly from atmospheric. Such tanks are equipped with special devices that permit gauging without opening the tank to the atmosphere.

vapor trail n: see *condensation trail*.

VAR abbr: volt-ampere reactive.

VAR$_C$ abbr: volt-ampere reactive in circuits with capacitors.

VAR$_H$ abbr: volt-ampere reactive in circuits with inductors. Also abbreviated VAR$_L$.

VAR$_L$ abbr: volt-ampere reactive in circuits with inductors. Also abbreviated VAR$_H$.

vara n: a Spanish unit of measurement, equal between 31 and 34 inches (78.74 and 86.36 centimetres). In Texas, it is equal to 33.33 inches (84.65 centimetres). Though an ancient unit of measurement, varas often show up on old survey reports in Texas and the southwestern U.S.

variable n: see *measurand*.

variable area flowmeter n: see *sample rotameter float*.

variable-area meter n: see *rotameter*.

variable bore ram n: a ram blowout preventer that contains blocks (rams) that can close and seal on a range of pipe sizes—for example, from 3½- to 5-inch (89- to 127-millimetre) pipe. Variable bore rams contain a large reserve of rubber in the ram block that specially designed antiextrusion plates force into sealing contact with smaller sizes of pipe. The antiextrusion plates also support the excess rubber when wellbore pressure is applied.

variable capacitor n: a capacitor that is constructed of stationary and movable plates. A shaft attached to the movable plate assembly is turned to rotate the movable plates in and out of the stationary plates. As the area of the movable plates increases or decreases within the stationary plates, the capacity varies. See *capacitor*.

variable clearance pocket n: a clearance pocket that can be adjusted to provide varying amounts of clearance space in a compressor cylinder.

variable costs n pl: the annual cost of operating and maintaining the system as well as fuel, gas loss, and unaccounted-for gas costs.

variable load n: on a floating offshore rig or ship, expendable items that increase or decrease the vessel's weight as they are used and replaced.

variable quantity n: in process control, the amount a process variable can change its chemical or physical composition because of changes in pressure, temperature, or volume.

variable ram blowout preventer n: a type of ram blowout preventer that has rams that can close on drill pipe of more than one size. For example, the ram preventer can be fitted with a set of variable rams that can properly close not only on 4-inch drill pipe, but also on 4½- and 5-inch drill pipe. If a drill string is made up of more than one size of pipe, variable rams eliminate the need for having more than one set of conventional pipe ram preventers available for the pipe sizes being run.

variable speed adj: of an electric motor when the motor's speed can be sped up or slowed down by varying the voltage or frequency of the electricity powering the motor.

variable wind n: a wind, generally light, or slow in speed, that frequently changes the direction in which it is blowing.

variation n: the angle by which a compass needle deviates from true north.

V-belt n: a belt with a trapezoidal cross section, made to run in sheaves, or pulleys, with grooves of corresponding shape.

VDC abbr: volts, direct current.

V-door n: an opening at floor level in a side of a derrick or mast. The V-door is opposite the drawworks and is used as an entry to bring in drill pipe, casing, and other tools from the pipe rack. The name comes from the fact that on the old standard derrick, the shape of the opening was an inverted V.

vector n: the representation of a quantity that gives the quantity both magnitude and direction and the components of which transform from one coordinate system to another in the same manner as the components of a displacement. In short, a vector is a quantity—for example force or velocity—made up of components of both direction and magnitude.

vector addition v: the adding together of the components that constitute a vector. See *vector*.

vee ring n: an elastomer seal energized by pressure. While it is circular in shape, the seal has a V-shaped cross section. See *O-ring*.

VEF abbr: vessel experience factor.

vegetation survey n: a method of detecting gas leaks in a line by noting the presence or absence of dead grass or other plants near the line. Natural gas kills plant roots over time.

vehicle n: in paint technology, the liquid phase of a paint that carries all the paint's constituents.

velocity n: 1. speed. 2. the timed rate of linear motion.

velocity head n: see *head*.

velocity meter n: a type of meter that measures the rate of fluid flow through it to determine the quantity of fluid flowing through it.

velocity of approach factor (E_v) n: a mathematical expression that relates the velocity of the fluid flowing in the meter tube upstream from the orifice to the velocity of the fluid flowing through the orifice.

velocity of flow equation n: in fluid measurement, a mathematical representation of fluid flow based on predetermined conditions.

velocity pressure n: the component of the pressure of a moving fluid that is due to the fluid's velocity and that is commonly equal to the difference between the impact pressure and the static pressure.

velocity profile n: the result of a velocity traverse. When fluid flows in a pipe, the fluid's velocity is not uniform owing to friction as the fluid moves past the pipe's walls. With some measuring devices, it is important that the device be placed properly within the flow to obtain accurate measurement. Running a velocity traverse to obtain the profile is necessary to properly place the measuring device. See *velocity traverse*.

velocity safety valve n: see *subsurface safety valve*.

velocity traverse n: the speed at which a fluid moves or passes through an object, related to a Pitot tube.

vena contracta *n*: squeezing of the gas flow.

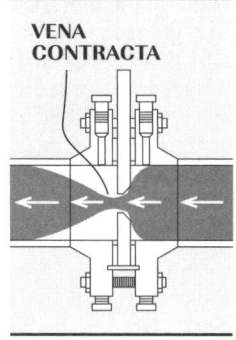

VENA CONTRACTA

vent *n*: 1. an opening in a vessel, line, or pump to permit the escape of air or gas. 2. a device installed on one end of that portion of a pipeline that crosses under a road. The vent marks the boundary of the highway right-of-way and provides an exit for any fluids should the pipeline develop a leak. It also aids in locating line breaks. *v*: to open a vessel or line so that air or gas can escape.

venturi effect *n*: the drop in pressure resulting from the increased velocity of a fluid as it flows through a constricted section of a pipe.

venturi nozzle *n*: a nozzle that, because of the venturi effect, increases the velocity of a fluid flowing through it.

venturi section *n*: in a fluid flow measuring device such as a venturi tube or a Dall tube, the part of the tube that contains the venturi, which is a constricted section that increases the flow's velocity. From the velocity increase, the quantity of flow can be inferred. See *venturi effect*.

venturi tube *n*: a short tube with a calibrated constriction that is used in instruments or devices such as jet hoppers. It was developed to take advantage of the principle that a fluid flowing through a constriction has increased velocity and reduced pressure.

venturi-tube meter *n*: a flowmeter used to determine the rate of flow and employing a venturi tube as the primary element for creating differential pressures in flowing gases or liquids. Compare *orifice meter*.

verification indicator *n*: an electrical measuring instrument that verifies the presence or absence of current or emf in a circuit. A compass needle and a dashboard ammeter are verification indicators.

vertical *n*: an imaginary line at right angles to the plane of the horizon. *adj*: of a wellbore, straight, not deviated.

vertical-cable survey *n*: a method of positioning hydrophones in the ocean.

vertical heater-treater *n*: a heater-treater whose cylinder stands upright, perpendicular to the ground. See *heater-treater*.

vertical hitch *n*: in crane operations, a method of attaching a sling to a load in which only one sling leg suspends the load.

vertical loop *n*: a change in the configuration of a pipeline whereby the pipe is curved from the horizontal to a 90° vertical run, then curved 90° to a horizontal run, curved again to a 90° vertical, and then curved 90° back to horizontal. The effect is to create turbulence in the fluid flow and distribute the S&W evenly in the flowing stream.

vertically integrated oil company *n*: a large oil company that is involved in all aspects of the oil industry, from exploration and production to transportation and refining.

vertically loaded anchor (VLA) *n*: an anchor type embedded into the seafloor using drag embedment techniques. After the flukes are tripped, the anchor achieves its holding capacity through soil reacting with the flukes. VLAs can withstand high vertical loads.

vertical permeability *n*: the permeability of a pay zone from the bottom to the top of the zone. Compare *horizontal permeability*.

vertical release *n*: a clause in an oil or gas lease that excludes, or releases, unproductive formations from the lease at the end of a specified period.

vertical separator *n*: a cylindrical separator standing upright, perpendicular to the ground.

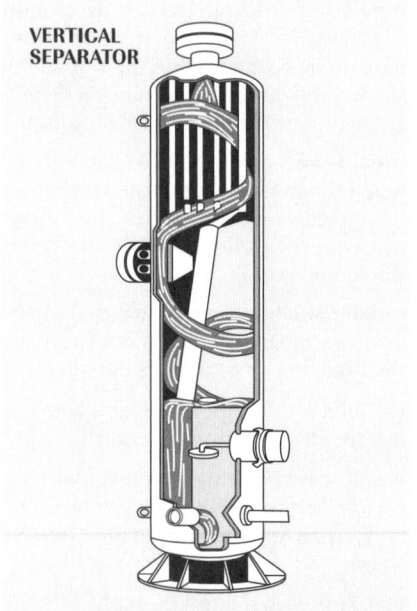

VERTICAL SEPARATOR

vertical station *n*: a preestablished location in the vertical plane along the tank shell corresponding to a given horizontal station. Compare *horizontal station*.

vertical stress *n*: in geology, the pressure created by the weight of all the liquid and rock above a given depth.

vertical support member (VSM) *n*: an H-shaped device that supports a pipeline above the ground. Vertical support members are used in parts of the world where frozen earth prevents normal burial of the line. See *lowering-up*.

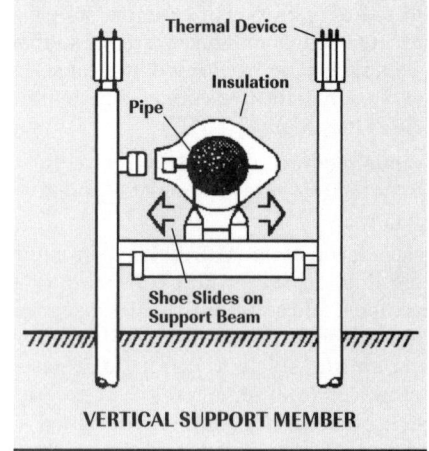

VERTICAL SUPPORT MEMBER

vertical tension (or component) *n*: vertical component of mooring line tension.

very large crude carrier (VLCC) *n*: a supertanker whose capacity is larger than 100,000 deadweight tons (90,720 deadweight tonnes) but less than 500,000 deadweight tons (453,600 deadweight tonnes). Compare *ultralarge crude carrier*.

vesicular *adj*: containing small cavities that are visible to the naked eye. A vesicular rock is a rock that contains small cavities.

vessel *n*: 1. a container used as a structural envelope to provide for functions that include storage, settling, separation, and filtration for liquid and/or vapor. Conditions of vessel use include atmospheric and elevated pressures and temperatures. 2. in nautical terms, virtually any watercraft that is larger than a rowboat, especially a ship or offshore drilling unit.

vessel experience factor (VEF) *n*: a calculated factor, based on the experience of recent voyages, that reflects the average difference between vessel and shore measurements in terms of percent.

vessel offset *n*: a phenomenon that occurs in drilling from floating offshore rigs (vessels) in which the vessel on the surface is not located directly above the wellhead on the seafloor. While anchoring and dynamic positioning techniques hold offset to a minimum, a certain amount occurs and subsea equipment must be designed to compensate for offset.

VFR *abbr*: visual flight rules.

V-G meter *n*: see *direct-indicating viscometer, Fann V-G™ meter*.

vibrating screen *n*: see *shale shaker*.

vibration dampener *n*: 1. a device, positioned in the drill stem between the bit and the drill collars, that absorbs impact loads and vibration from the up-and-down motion of the drill stem. Vibration dampeners are designed to transmit torque while absorbing reciprocative loads that decrease the efficiency of the drill bit. Also called a Shock Sub. 2. a device affixed to an engine crankshaft to minimize stresses that result from torsional vibration of the crankshaft.

vibration error *n*: the maximum change in operating output, at any measured value within a specific range, when vibration levels of specified amplitude and range or frequencies are applied to the transducer along specified axes.

vibrator hose *n*: short length of flexible hose that absorbs some of the vibrating pulsations that originate in a mud pump. Also called shock hose.

vintage *n*: used to indicate the period during which a gas sales contract was made or the well spud date.

visbreaking *n*: viscosity breaking, i.e., economically reducing the viscosity of the light end of a heavy fraction.

viscometer *n*: a device used to determine the viscosity of a substance. Also called a viscosimeter.

viscosimeter *n*: see *viscometer*.

viscosity *n*: a measure of the resistance of a fluid to flow. Resistance is brought about by the internal friction resulting from the combined effects of cohesion and adhesion. The viscosity of petroleum products is commonly expressed in terms of the time required for a specific volume of the liquid to flow through an orifice of a specific size at a given temperature.

viscosity index *n*: an index used to establish the tendency of an oil to become less viscous at increasing temperatures. Reference oils are a highly paraffinic Pennsylvania oil, rated 100, and a Gulf Coast naphthenic oil, rated 0.

viscous *adj*: having a high resistance to flow.

viscous flow *n*: see *laminar flow*.

visual flight rules (VFR) *n pl*: in aviation, rules established by the United States Federal Aviation Administration (FAA), that point out the weather conditions under which pilots may fly using only their ability to see outside their aircraft. Although the minimum visibility requirements vary in relation to the type of aircraft and the space in which the aircraft flies, in general, when the visibility drops below 3 miles or 5 kilometres—that is, when a pilot cannot see anything beyond

a point 3 miles or 5 kilometres from his or her aircraft, the pilot can no longer fly under VFR. Compare *instrument flight rules*.

viton *n*: a fluoroelastomer used in the seals installed in equipment exposed to H_2S. See *fluoroelastomer*.

vitrification *n*: a treatment in which electricity is applied to the soil to melt it.

vitriolic acid *n*: see *sulfuric acid*.

VIV *abbr*: vortex-induced vibration.

VLA *abbr*: vertically loaded anchor.

VLCC *abbr*: very large crude carrier.

VOC *abbr*: volatile organic compound.

voids *n pl*: cavities in a rock that do not contain solid material but may contain fluids.

volatile *adj*: readily vaporized.

volatile light ends *n pl*: those hydrocarbons that are readily vaporized.

volatile organic compound *n*: a readily vaporized compound composed of carbon-containing molecules derived from living organisms.

volatility *n*: the tendency of a liquid to assume the gaseous state.

volatilization *n*: a limited form of remediation; only the lighter fractions of the oil evaporate.

volcanology *n*: the scientific study of volcanoes and volcanic phenomena. Also called vulcanology.

volt (E or V) *n*: the unit of electric potential, voltage, or electromotive force in the metric system. See *electromotive force*.

voltage *n*: potential difference or electromotive force, measured in volts.

voltage-dependent resistor *n*: see *metallic-oxide varistor*.

voltage drop *n*: the voltage developed across an electrical component or conductor as current flows through component or conductor.

voltage gain *n*: in solid-state electronics, the increase in voltage produced by a transistor.

voltage regulator *n*: an electronic device, including certain solid-state diodes, that maintains a given voltage across a circuit within required limits despite variations in voltage or load.

voltage transmitter *n*: in process control, a four-wire process instrument that converts process variables such as temperature, pressure, level, and flow to a proportional signal of 1 to 5 volts DC. A voltage transmitter does not transmit data as accurately as a current transmitter because wire resistance and electrical interference affects its signal.

voltaic cell *n*: see *primary cell*.

volt-ampere (VA) *n*: an expression of apparent, or total, electric power in which power is the product of volts times amperes.

volt-ampere reactive (VAR) *n*: the reactive power in a circuit carrying alternating current when the product of the root-mean-square value of the voltage, expressed in volts, by the root-mean-square value of the current, expressed in amperes, and by the sine of the phase angle between voltage and current, equals 1. See *reactive power*.

voltmeter *n*: an instrument used to measure, in volts, the difference of potential in an electrical circuit.

volt-ohm-milliammeter (VOM) *n*: see *multimeter*.

volts, alternating current (VAC) *n*: electric current that flows through an electrical circuit in an alternating manner—that is, the current reverses direction in the circuit at regular intervals.

volts, direct current (VDC) *n*: electric current that flows through an electrical circuit in one direction only.

volume *n*: the amount of a substance that occupies a particular space.

volume booster relay *n*: in process control, a device attached to a valve in which pneumatic pressure in the relay improves (boosts) the valve's response time.

volume flow rate *n*: the amount, or quantity, of a fluid that passes a point for a given time.

volume meter *n*: see *positive-displacement meter*.

volume of displacement *n*: in offshore drilling from floating rigs, the volume of water the rig displaces as it floats in the water; the underwater volume of the rig. It is usually expressed in cubic metres or cubic feet.

volumeter *n*: see *positive-displacement meter*.

volumetric correction *n*: an equation that can be used to make calculating the maximum amount of casing pressure to expect when a gas kick is circulated out of a shut-in well more accurately reflect reality.

volumetric efficiency *n*: 1. actual volume of fluid put out by a pump, divided by the volume displaced by a piston or pistons (or other device) in the pump. Volumetric efficiency is usually expressed as a percentage. For example, if the pump pistons displace 300 cubic inches (4,916 cubic centimetres), but the pump puts out only 291 cubic inches (4,769 cubic centimetres) per stroke, then the volumetric efficiency of the pump is 97%. 2. for compressors, the ratio of the volume of gas actually delivered, corrected to suction temperature and pressure, to the piston displacement.

volumetric method *n*: a method of well control in which bottomhole pressure is kept constant when circulation is not possible and gas is migrating up the hole. Bottomhole pressure is maintained slightly higher than formation pressure while the gas is allowed to expand in a controlled manner as it moves to the surface.

voluntary pooling *n*: pooling of leased tracts willingly undertaken by all the parties involved, both working interest owners and royalty owners.

voluntary standard *n*: standards established generally by private sector bodies for use by any person or organization. Includes what are commonly referred to as industry standards and consensus standards, but does not include professional standards of personal conduct, institutional codes of ethics, private standards of individual firms, or standards mandated by law.

voluntary unitization *n*: unitization that is accomplished with the willing cooperation

of the affected parties, both working interest owners and royalty owners.

VOM *abbr*: volt-ohm-milliammeter.

vortex *n*: 1. a spiral motion of fluid within a limited area, especially a whirling mass of fluid that draws anything near it toward its center. 2. in gas and liquid measurement, a rotational-flow zone that forms on alternate sides of a nonstreamlined body placed in a high-velocity flowing stream.

vortex eliminator *n*: see *swirl pipe*.

vortex finder *n*: the short pipe in a cone-shaped separator that extends down into the cone body from the top and that forces the whirling stream of material to start downward toward the small end of the cone body.

vortex flowmeter *n*: in fluid-flow measurement, a device in a meter that uses a piezoelectric crystal to measure the rate of flow. As the fluid flows through the meter, vortices (spiral, or whirling, motions) occur, which the piezoelectric crystal detects. The crystal then produces AC voltage whose frequency is directly proportional to flow rate.

vortex-induced vibration (VIV) *n*: a shaking motion (a vibration) created in a marine riser system by the tendency of water to swirl (create a vortex) around the riser.

vortex shedding meter *n*: an instrument for detecting vortices induced in a fluid stream by a baffle in the fluid stream to measure instantaneous flow rate.

vortices *n pl*: rotational-flow zones that form on alternate sides of a nonstreamlined body placed in a high-velocity flowing stream.

VRU *abbr*: vapor recovery unit.

VSM *abbr*: vertical support member.

vug *n*: 1. a cavity in a rock. 2. a small cavern, larger than a pore but too small to contain a person. Typically found in limestone subject to groundwater leaching.

vuggy formation *n*: a rock formation that contains openings (vugs). See *vug*.

vuggy porosity *n*: secondary rock porosity formed by the dissolving of the more soluble portions of a rock in waters containing carbonic or other acids.

vugular formation *n*: see *vuggy formation*.

vulcanology *n*: see *volcanology*.

W *sym*: watt.

WACOG *abbr*: weighted average cost of gas.

wagon-wheel heater *n*: in pipeline construction, a multiheaded circular propane torch that is manually rotated inside the end of a pipe for preheating.

wait-and-weight method *n*: a well-killing method in which the well is shut in and the mud weight is raised the amount required to kill the well. The heavy mud is then circulated into the well while the kick fluids are circulated out. So called because one shuts the well in and waits for the mud to be weighted before circulation begins.

waiting on cement (WOC) *adj*: pertaining to the time when drilling or completion operations are suspended so that the cement in a well can harden sufficiently.

waiting on weather (WOW) *adj*: pertaining to the time when drilling or other operations are suspended because of bad weather.

walking beam *n*: the horizontal steel member of a beam pumping unit that has rocking or reciprocating motion.

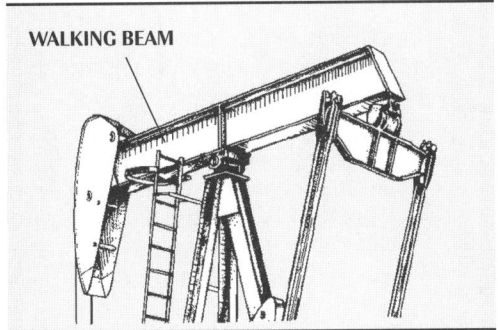

WALKING BEAM

wall-building ability *n*: the ability of a drilling mud to plaster the wall of the hole with solids from the drilling mud.

wall cake *n*: also called filter cake or mud cake. See *filter cake*.

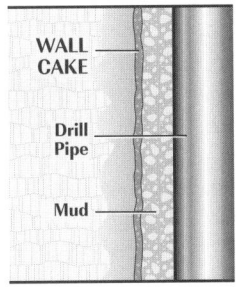

WALL CAKE

Drill Pipe

Mud

wall hook *n*: a device used in fishing for drill pipe. If the upper end of the lost pipe is leaning against the side of the wellbore, the wall hook centers it in the hole so that it may be recovered with an overshot, which is run on the fishing string and attached to the wall hook.

wall-hook guide *n*: see *wall hook*.

wall-hook packer *n*: see *hook-wall packer*.

wall sticking *n*: see *differential sticking*.

wall-stuck pipe *n*: see *differential sticking*.

wall thickness *n*: in tubular goods such as drill pipe, tubing, or casing, the width, or breadth, of the metal making up the sides (the wall) of the tubular. Wall thickness determines the inside diameter of the tubular and therefore affects the maximum size of a tool that can be run through the it.

wandering *n*: the tendency of the drill bit to deviate horizontally parallel to tilted strata.

War Emergency Pipelines (WEP) *n*: a government-financed, nonprofit corporation of eleven oil and pipeline companies established during World War II to build desperately needed pipelines such as Big Inch and Little Big Inch.

warm a connection *v*: see *heat a connection*.

warm front *n*: a front that is formed when warm air replaces cold air as the cold air retreats.

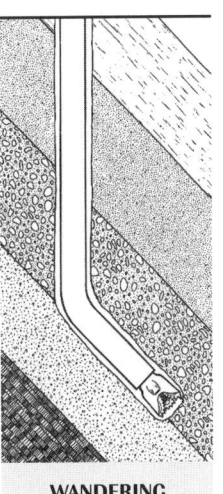

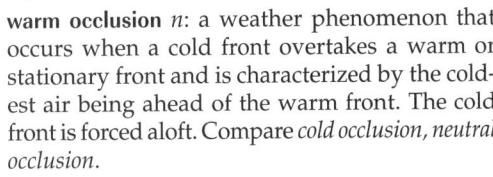

WANDERING

warm occlusion *n*: a weather phenomenon that occurs when a cold front overtakes a warm or stationary front and is characterized by the coldest air being ahead of the warm front. The cold front is forced aloft. Compare *cold occlusion, neutral occlusion*.

warps *n pl*: rock strata that are slightly tilted occur when broad areas of the crust rise or drop without fracturing.

warranty clause *n*: the clause in an oil and gas lease that assures title to the leased property by an express covenant to that effect.

warranty deed *n*: a deed in which the grantor stipulates by express covenant that the title to property is as it is represented to be and that the grantee's possession shall be undisturbed. Defects in title may include those that may have existed before the grantor obtained the title as well as any that have arisen during his or her ownership.

Warrington strand *n*: a wire-rope strand design that consists of a single, central wire surrounded by two outer layers of several wires. The wires in the inner layer are the same diameter as the inner wire. The wires in the outer layer are of two diameters, one smaller wire laying adjacent to one larger wire. A typical Warrington strand has 19 wires: 1 central wire, 6 inner wires, and 12 outer wires, 6 of which are smaller in diameter than the other 6.

wash *n*: a thin fluid that separates drilling mud from cement when cement is pumped downhole. A wash also removes mud from the walls of the hole by both turbulent and surfactant action.

washing *n*: 1. the high-pressure spraying of the crude oil cargo to dislodge or dissolve clingage and sediment from the walls, cross members, and lines in the compartments of a vessel during the unloading operation. 2. the use of a high-pressure water stream to dislodge clingage and sediment from the bulkheads, bottoms, and internal structures of a vessel's cargo tanks.

washing in *v*: displacing mud in the tubing by pumping salt water into it to start the flow of formation fluids.

wash oil *n*: see *absorption oil*.

washout *n*: 1. excessive wellbore enlargement caused by solvent and erosional action of the drilling fluid. 2. a fluid-cut opening caused by fluid leakage.

wash over *v*: to release pipe that is stuck in the hole by running washover pipe. The washover pipe must have an outside diameter small enough to fit into the borehole but an inside diameter large enough to fit over the outside diameter of the stuck pipe. A rotary shoe, which cuts away the formation, mud, or whatever is sticking the pipe, is made up on the bottom joint of the washover pipe, and the assembly is lowered into the hole. Rotation of the assembly frees the stuck pipe. Several washovers may have to be made if the stuck portion is very long.

washover *n*: the operation during which stuck drill stem or tubing is freed using washover pipe.

washover assembly *n*: see *washover pipe*.

washover back-off connector *n*: a fishing tool that is made up in a length of washover pipe connected to the top of the fish once the washover is completed, and then backed off the fish, thus enabling the washed-over part of the fish to be retrieved. The tool permits washover, back-off, and pulling to be carried out in one round trip.

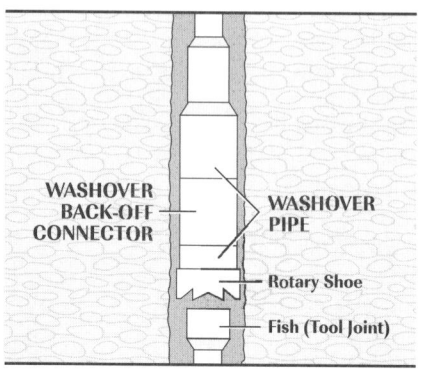

WASHOVER BACK-OFF CONNECTOR — WASHOVER PIPE
Rotary Shoe
Fish (Tool Joint)

washover pipe *n*: an accessory used in fishing operations to go over the outside of tubing or drill pipe stuck in the hole because of cuttings, mud, and so forth, that have collected in the annulus. The washover pipe cleans the annular space and permits recovery of the pipe. It is sometimes called washpipe.

washover shoe *n*: see *burn shoe*.

washover string *n*: the assembly of tools run into the hole during fishing to perform a washover. A typical washover string consists of a washover back-off connector, several joints of washover pipe, and a rotary shoe.

washpipe *n*: a short length of surface-hardened pipe that fits inside the swivel and serves as a conduit for drilling fluid through the swivel.

washpipe packing *n*: layers of dense flexible material that is stacked around the swivel's washpipe and the swivel's interior body seal. It helps seal between the static washpipe and the turning swivel body to ensure that high-pressure drilling mud flows through the swivel's stem and into the kelly.

wash tank *n*: a tank containing heated water through which crude-oil emulsion is forced to flow to remove water from the crude. Also called a gun barrel.

waste minimization *n*: the practice of reducing the amount of waste generated from a site through recycling programs and the use of nonhazardous chemical substitutes for hazardous chemicals.

watch circle *n*: circle of offset a rig will trace due to variables in environmental conditions.

water and sediment sample *n*: in tank sampling, a sample obtained with a thief to determine the amount of material at the bottom of the tank that cannot be sold.

water-back *v*: 1. to reduce the weight or density of a drilling mud by adding water. 2. to reduce the solids content of a mud by adding water.

water-base acid *n*: usually hydrochloric acid in water, a solution that is used to perform massive acid fractures in carbonate reservoirs. Acid concentrations of 22 percent to 28 percent provide maximum dissolving of carbonate rock. Small- to medium-sized sand can be added to the system to provide scouring of the fracture faces, along with corrosion inhibitors, demulsifiers, and clay stabilizing chemicals. Fresh water, in volumes of one to three times the acid volumes, is pumped between the various acid stages to dissolve sludge, which is a result of acid reacting with carbonate.

water-base gel *n*: a solution used in water-base fracturing fluids that causes the fracturing fluid to solidify, or gel. Water-base gel is used primarily in sensitive sandstone reservoirs, which, if a gel were not used, would absorb excessive amounts of water and decrease the effectiveness of the fracturing fluid.

water-base mud *n*: a drilling mud in which the continuous phase is water. In water-base muds, any additives are dispersed in the water. Compare *oil-base mud*.

water block *n*: a reduction in the permeability of a formation caused by the invasion of water into the pores.

water bottom *n*: water accumulated at (or sometimes added to) the bottom of the oil in a storage tank.

water-cement ratio *n*: the ratio of water to cement in a slurry. It is expressed as a percentage, indicating the number of pounds of water needed to mix 100 pounds (45.32 kilograms) of cement.

water coning *n*: the upward encroachment of water into a well caused by pressure drawdown from production.

water control *n*: additives used to prevent or minimize water loss into a formation during slurry placement.

watercourse *n*: a hole inside a bit through which drilling fluid from the drill stem is directed.

water cushion (W/C) *n*: water put into an empty string of pipe in a wellbore to prevent the pipe from being crushed by pressure in the annulus.

water cut *n*: the percentage of water in fluid produced from a well. *v*: to locate the oil-water interface and use the location to measure the volumes of oil and water in a shore tank or ship compartment.

water-cut index *n*: a measure of the amount of water in oil being produced by a well.

water-cut measurement *n*: 1. using the location of the oil-water interface to determine the volume of free water in a shore tank or vessel compartment. 2. the line of demarcation of the oil-water interface.

water-cut paste *n*: a material that changes color (usually to red) in water. The use of water-cut paste is one method by which the level of water in the bottom of an oil storage tank can be determined. The paste is applied to a plumb bob, which is lowered to the bottom of the tank and then retrieved. The water level is then measured off the bob by noting the depth of the red portion.

water dip *n*: the depth of free water in a container over and above the dip plate.

water distillation unit *n*: equipment used mostly on offshore or desert locations to convert salt water to potable fresh water.

water drive *n*: the reservoir drive mechanism in which oil is produced by the expansion of the underlying water and rock, which forces the oil into the wellbore. In general, there are two types of water drive: bottom-water drive, in which the oil is totally underlain by water; and edgewater drive, in which only a portion of the oil is in contact with the water.

water drive field *n*: an oilfield in which the primary reservoir drive mechanism is water.

watered-out *adj*: of a well, producing mostly water ("gone to water").

water encroachment *n*: the movement of water into a producing formation as production depletes the formation of oil and gas.

water finder *n*: a graduated rod, usually of metal, to which water-finding paper or paste can be applied. This paste or paper discolors on contact with water and thus affords a ready means of measuring the depth of water in a tank. Also called a water-finding rule.

water-finding paper *n*: a strip of paper coated with a chemical substance that changes color on immersion in water. When fastened to a water finder, it indicates the depth of free water in a container.

water-finding paste *n*: a paste containing a chemical that changes color in contact with water. When applied to a water finder, it indicates the level of free water in a container.

water-finding rule *n*: see *water finder*.

waterflooding *n*: a method of improved recovery in which water is injected into a reservoir to remove additional quantities of oil that have been left behind after primary recovery. Waterflooding usually involves the injection of water through wells specially set up for water injection and the removal of water and oil from production wells drilled adjacent to the injection wells.

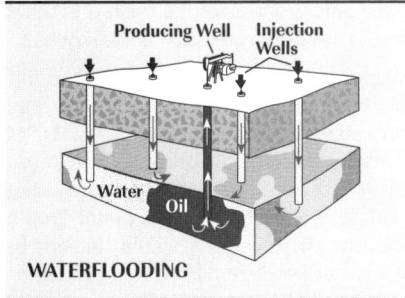

WATERFLOODING

waterflood kick *n*: the first indication of increased oil production as the result of a waterflood project.

water hammer *n*: a pressure concussion caused by suddenly stopping the flow of liquids in a closed container. See *fluid knock*.

water injection pump *n*: pump used in waterflooding or in water disposal. In waterflooding, the pump injects water into a well drilled in one part of a reservoir to force oil to a production well. In water disposal, the pump injects water produced from a well into a reservoir for disposal.

water injection well *n*: a well into which water is pumped to dispose of the water or to recover oil in a waterflood.

water-in-oil emulsion *n*: see *invert-emulsion mud*.

water jacket *n*: in the cooling system, a passageway inside the rim for circulating water.

water leg *n*: see *water outlet*.

waterline *n*: 1. the line on the hull of a ship or floating rig to which the surface of the water rises when the ship or rig is floating. 2. any of several lines parallel to this line, marked on the hull of a ship or rig, which indicates the depth to which the ship or rig submerges under various loads.

water loss *n*: see *fluid loss*.

water maker *n*: see *water distillation unit*.

water-oil ratio (WOR) *n*: a measure of the amount of produced water to the amount of produced oil in the total amount of fluids produced from a reservoir.

water outlet *n*: the part of a gun barrel that provides an outlet for the water that has separated from an emulsion and that regulates the amount of water held in the gun barrel.

water plane *n*: in naval architecture, a flat and level surface (a plane) parallel to the surface of the water and defined by its line of intersection with the vessel's hull.

water-plane area *n*: in naval architecture, the dimensions of the water plane created by a floating vessel; the area of a horizontal cross section of a floating vessel at the waterline for any given draft. See *water plane*.

water pollutant *n*: dredged spoil, solid waste, incinerator residue, filter back-wash, sewage, garbage, sewage sludge, munitions, chemical wastes, biological materials, radioactive materials, heat, wrecked or discarded equipment, rock, sand, cellar dirt, and industrial, municipal, and agricultural waste discharged into water.

water pressure *n*: in a floating object, the force exerted by the water on the bottom and sides of the floating object.

water-producing interval *n*: the portion of an oil or gas reservoir from which water or mainly water is produced.

water pump *n*: on an engine, a device, powered by the engine, that moves coolant (water) through openings in the engine crankcase, through the radiator or heat exchanger, and back into the crankcase.

water saturation *n*: the percentage of total pore space in a reservoir occupied by water. It is normally estimated by electric log calculations, although cores specially obtained may be analyzed for this value.

water slug *n*: a relatively small volume of water that is put into the drill string or tubing string on top of the drilling or workover fluid.

waterspout *n*: in general, the offshore equivalent of a tornado.

water string *n*: a string of casing used to shut off water above an oil sand.

water table *n*: 1. the structure at the top of the drilling derrick or mast that supports the crown block. 2. the underground level at which water is found.

water test *n*: final test on a pipeline in which the line is filled with water and pressured up to a higher-than-normal operating pressure. If no leaks occur, the line is put into service.

water-tight *adj*: not allowing water to pass in, out, or through.

water-tight door *n*: a door on ships or mobile offshore rigs that, when closed, blocks the passage of water and withstands its pressure.

water-tight integrity *n*: on a ship or offshore rig, the vessel's ability to prevent water from entering compartments on the vessel and thus maintain the vessel's ability to remain afloat in spite of several compartments being flooded. To provide watertight integrity, naval architects design vessels so that they are divided into many sections or compartments, each of which can be sealed closed to prevent water from entering—that is, each compartment can be made watertight by closing watertight doors or hatches.

water vapor *n*: water (H_2O) in its gaseous state (vapor) that can be liquefied by compression or cooling.

water well *n*: a well drilled to obtain a fresh water supply to support drilling and production operations or to obtain a water supply to be used in connection with an enhanced recovery program.

water-wet reservoir *n*: a hydrocarbon reservoir whose rock grains are coated with a film of water.

water-wet rock *n*: see *wettability*.

water zone *n*: the portion of an oil or gas reservoir occupied by water, usually the lowest zone in the reservoir.

watt (W) *n*: the unit of power in the metric system, equal to 1 joule per second, or 0.746 horsepower.

watt hour *n*: the application of 1 watt of power for 1 hour.

watthour meter *n*: a quantity recorder that measures power and integrates, or adds up, the amount of power used over a period of time. Also called an integrating wattmeter.

wattless power *n*: see *reactive power*.

wattmeter *n*: an instrument that measures the magnitude of the power in an electric circuit. It may read watts, kilowatts, or megawatts.

wave *n*: in electronics, a disturbance that begins at one point in a medium and goes to other points in the medium without permanently displacing the medium.

wave crest *n*: the top part of a wave.

wave cyclone *n*: a cyclone that develops and travels along a front. The circulation about the cyclone's center may give the front a wavelike shape. Also called wave depression.

wave depression *n*: see *wave cyclone*.

waveform *n*: in electronics, the pictorial representation of the shape (form) of a wave, which is obtained by plotting the displacement of the wave as a function of time, at a fixed point in space. See *wave*.

wave height *n*: the measurement of a wave from its trough to its crest.

wave height recorder *n*: a device that measures the heights of waves. Although many types are available, a commonly used one is firmly attached to the seabed or a stationary anchoring system at a depth of no more than about 65 feet or 20 metres. Devices in the instrument that transform pressure energy into height measurements (transducers) send signals to a recorder, which stores the information. Later, this information can be downloaded to a computer and analyzed.

wave interference *n*: the convergence of sea and swell or the convergence of two or more swells. See *sea, swell*.

wave length *n*: the horizontal distance between two successive wave troughs or crests.

wave period *n*: the time interval between the passage of two successive wave troughs or crests.

wave rose *n*: a diagram that depicts reported sea characteristics for a given area.

wave steepness *n*: the ratio between a wave's height and length.

wave trough *n*: the bottom part of a wave.

wave velocity *n*: the speed at which an ocean wave moves; it is usually measured in metres or feet per second.

wax *n*: see *paraffin*.

Wb *sym*: weber.

WC *abbr*: wildcat; used in drilling reports.

W/C *abbr*: water cushion; used in drilling reports.

wear bushing *n*: a fixed or removable cylindrical metal lining placed inside a fitting or piece of moving equipment that reduces friction and thus wear on the fitting.

wear elongation *n*: in a roller chain drive, an amount of stretching that causes the chain to engage improperly with the sprockets.

wear groove *n*: a groove in a metal part that when worn smooth indicates that the part should be replaced.

wear pad *n*: a flat metal protrusion of a downhole tool that is usually hardfaced and that protects the rotating tool from wear.

wear sleeve *n*: 1. a hollow cylindrical device installed around the swivel's rotating stem that absorbs the stem's rotation and helps form an oil seal between the stem and the oil-bath reservoirs inside the swivel. 2. in a riser adapter, a thin metal cylinder that fits inside the adapter and prevents premature wear on the inside bore of the adapter and drill stem components that rotate within it. See *riser adapter*.

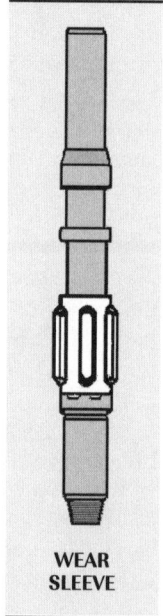

WEAR SLEEVE

weather *n*: the state of the atmosphere at a specific time with respect to heat or cold, wetness or dryness, calm or storm, and clearness or cloudiness. The term is also used to identify rain, storm, or other unfavorable atmospheric conditions. *v*: to allow an object or substance to age, usually under controlled conditions.

weathered crude *n*: crude oil that has lost an appreciable quantity of its entrained gas because of evaporation during storage.

weatherfax *n*: see *radiofax*.

weathering *n*: 1. the breakdown of large rock masses into smaller pieces by physical and chemical climatological processes. 2. the evaporation of liquid by exposing it to the conditions of atmospheric temperatures and pressure.

weathering test *n*: a Gas Processors Association test for LP gas for the determination of heavy components in a sample by evaporation of the sample as specified.

weather map *n*: a chart that indicates atmospheric conditions and circulation over an area at a given time. Also called surface chart.

weathervane *v*: of a storage tanker or shuttle vessel that is permanently moored, to make a 360° circle around the mooring as the weather changes.

weathervaning *n*: a vessel is allowed to change heading into the direction of the prevailing environment.

Webb-Wilson *n*: a brand name for one type of pipe tongs used on drilling rigs; sometimes used generically for all brands of tongs, even though other types of tongs exist.

weber (Wb) *n*: the metric unit of magnetic flux.

wedge *n*: 1. a metal or wooden piece that tapers to a thin edge and is used for raising heavy objects or for tightening by being driven into or between something. 2. something wedge-shaped, as a portion of liquid in the hold of a tanker ship.

wedge factor *n*: a formula for a trigonometric calculation to determine the oil remaining on board (ROB) when the vessel is out of trim and the ROB does not contact both forward and aft bulkheads. The formula assumes ROB to be a flowable liquid.

wedge formula *n*: a mathematical means of approximating the small quantities of liquid and solid cargo and free water on board before a vessel is loaded and after its cargo is discharged, based on cargo compartment dimensions and vessel trim. It is used only when a wedge exists and when the liquid does not touch all bulkheads.

Wedge Lock *n*: a device on Cameron ram blowout preventers that locks the rams on a subsea preventer closed so that even if hydraulic closing pressure is lost, the rams will remain closed and provide a seal. It is a wedge-shaped piston that is mounted in its own hydraulic housing on the back of each ram bonnet at 90 degrees to the operating piston's tail shaft. Once the rams are closed, the Wedge Lock pistons are hydraulically operated and they move behind the operating piston tail shaft and wedge it into place.

wedge socket *n*: wire rope fittings wherein the rope end is secured by a wedge.

wedge tables *n pl*: precalculated ship's tables based on the wedge formula and displayed in much the same way as the ship's innage/ullage tables. Used for small quantities when the liquid does not touch all bulkheads.

weephole *n*: a hole in the lantern ring of a liner packing in a mud pump. When the liner packing fails, fluid spurts out of the weephole with each stroke of the piston, indicating that the packing should be renewed. Also called tattletale hole, telltale hole.

weevil *n*: shortened form of boll weevil. See *boll weevil*.

weigh scale *n*: a device for determining either the mass or the weight of a body, depending on the apparatus and procedure employed.

weight *n*: 1. in mud terminology, refers to the density of a drilling fluid. 2. of a measurement, expresses degree of confidence in result of measurement of a certain quantity compared with result of another measurement of the same quantity. 3. the force with which a body is attracted to the earth or to another celestial body equal to the product of the object's mass and the acceleration of gravity.

weigh tank *n*: a tank used with a weigh scale for measurement of the liquid contents of the tank.

weight bar *n*: see *sinker bar*. Also called stem.

weight cut *n*: the amount by which drilling fluid density is reduced by entrained formation fluids or air.

weighted average cost of gas (WACOG) *n*: cost calculated as the total cost of all gas purchased during a base period divided by either the total quantity purchased (unit of production) or the system throughput (unit of sales) during the same period. This rate, plus any application of any pending surcharge adjustments, serves as the basis on which system tariff rates are computed and made effective.

weight indicator *n*: an instrument near the driller's position on a drilling rig that shows both the weight of the drill stem that is hanging from the hook (hook load) and the weight that is placed on the bit by the drill collars (weight on bit).

weighting material *n*: a material that has a high specific gravity and is used to increase the density of drilling fluids or cement slurries.

WEIGHT INDICATOR

weight-loaded regulator *n*: in process control, a valve containing a weight that controls the valve's opening and closing to control (regulate) the pressure in a vessel.

weight on bit (WOB) *n*: the amount of downward force placed on the bit by the weight of the drill collars.

weight-set packer *n*: a packer whose packing elements are activated when weight from the tubing string is applied. To set

some packers, rotation as well as weight is required. See *packer*.

weight up *v*: to increase the weight or density of drilling fluid by adding weighting material.

weir *n*: a metal plate installed in a separator, treater, or mud tank and used to regulate liquid level or, in the case of a metering separator, to measure flow.

welded tuff *n*: a pyroclastic deposit hardened by the action of heat, pressure from overlying material, and hot gases.

well *n*: the hole made by the drilling bit, which can be open, cased, or both. Also called borehole, hole, or wellbore.

wellbore *n*: a borehole; the hole drilled by the bit. A wellbore may have casing in it or it may be open (uncased); or part of it may be cased, and part of it may be open. Also called a borehole or hole.

wellbore cleanup *n*: see *wellbore soak*.

wellbore pressure *n*: 1. bottomhole pressure. 2. casing pressure.

wellbore soak *n*: an acidizing treatment in which the acid is placed in the wellbore and allowed to react by merely soaking. It is a relatively slow process, because very little of the acid actually comes in contact with the formation. Also called wellbore cleanup. See *matrix acidizing*. Compare *acid fracture*.

well completion *n*: 1. the activities and methods of preparing a well for the production of oil and gas or for other purposes, such as injection; the method by which one or more flow paths for hydrocarbons are established between the reservoir and the surface. 2. the system of tubulars, packers, and other tools installed beneath the wellhead in the production casing; that is, the tool assembly that provides the hydrocarbon flow path or paths.

well control *n*: the methods used to control a kick and prevent a well from blowing out. Such techniques include, but are not limited to, keeping the borehole completely filled with drilling mud of the proper weight or density during all operations, exercising reasonable care when tripping pipe out of the hole to prevent swabbing, and keeping careful track of the amount of mud put into the hole to replace the volume of pipe removed from the hole during a trip.

well-control equipment *n*: an assembly of several components, such as ram preventers, annular preventers, a choke and kill system, trip tanks, and mud-gas separators. On offshore floating rigs, well-control equipment also includes a marine riser system and a diverter system.

well density *n*: the ratio between the number of wells drilled in a field and the acreage. Under a 40-acre spacing pattern, the well density is one well per 40 acres.

well fluid *n*: the fluid, usually a combination of gas, oil, water, and suspended sediment, that comes out of a reservoir. Also called well stream.

wellhead *n*: the equipment installed at the top of the wellbore. A wellhead includes such equipment as the casinghead and tubing head. *adj*: pertaining to the wellhead (e.g., wellhead pressure).

wellhead connector *n*: on a floating drilling rig using a subsea blowout preventer stack, a tool that attaches the blowout preventer (BOP) stack to the wellhead. It is hydraulically operated from the surface and, when actuated, connects the BOP to the wellhead.

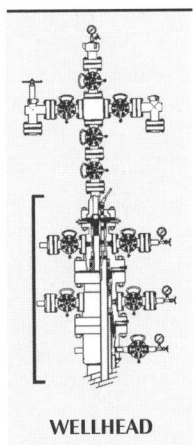

WELLHEAD

wellhead housing *n*: in offshore drilling, a tool that seats in a housing at the top of the conductor casing to support subsea equipment and subsequent casing strings and tubing hangers.

wellhead price *n*: the price received by the producer for sales at the well. See *crude oil average domestic first purchase price*.

well kick *n*: see *kick*.

well log *n*: see *log*.

well logging *n*: the recording of information about subsurface geologic formations, including records kept by the driller and records of mud and cutting analyses, core analysis, drill stem tests, and electric, acoustic, and radioactivity procedures. See *acoustic log, core analysis, driller's log, drill stem test, electric well log, mud analysis*, and *radioactivity log*.

well log library *n*: a private, or sometimes public, organization that maintains collections of oilfield data, particularly well logs. Users gain access to the information by paying membership dues or a user's fee.

well permit *n*: authorization, usually granted by a governmental conservation agency, to drill a well. A permit is sometimes also required for deepening or remedial work.

well platform *n*: an offshore structure with a platform above the surface of the water that supports the producing well's surface controls and flow piping.

well point *n*: in pipeline construction, a device installed in waterlogged soil to dry out areas along the ditch line. Functioning as a submersible pump, a well point is a hollow steel rod approximately 18 feet (5.5 metres) long driven into the ground. Its free end is connected by hose or tubing to a well-point pump that moves groundwater away from the ditch. Hundreds of well points may be required to stabilize an area.

well puller *n*: a member of a well-servicing crew.

well pusher *n*: see *toolpusher*.

well service and workover contractor *n*: a company that specializes in offshore well repair.

well servicing *n*: the maintenance work performed on an oil or gas well to improve or maintain the production from a formation already producing. It usually involves repairs to the pump, rods, gas-lift valves, tubing, packers, and so forth.

well-servicing rig *n*: a portable rig, truck-mounted, trailer-mounted, or a carrier rig, consisting of a hoist and engine with a self-erecting mast. See *carrier rig*. Compare *workover rig*.

well site *n*: see *location*.

well-site computer *n*: interprets information transmitted from the downhole sonde to make corrections for downhole conditions, make complex calculations, and print out various log formats. Also keeps track of the depth at which the logs are being obtained and warns of malfunctions.

well spacing *n*: the regulation of the number and location of wells over an oil or gas reservoir as a conservation measure. Compare *well density*.

well stimulation *n*: any of several operations used to increase the production of a well, such as acidizing or fracturing. See *acidize, formation fracturing*.

well stream *n*: see *well fluid*.

well surveying *n*: see *well logging*.

well test *n*: a test performed on a well to determine its production characteristics.

well test report *n*: the detailed results of a well test that present all test data, the calculations drawn from the data, and conclusions pertaining to the well's production potential.

well test surface package *n*: in well testing, an assembly of equipment that is attached to the wellhead of the well to be tested. The main component of a well test surface package is an oil and gas separator.

well velocity survey *n*: measures the time needed for a sound pulse generated at the surface to reach successive selected depths in the borehole. Used to correct problems in downhole acoustic logs caused by filtrate invasion of the formation and borehole irregularities. Also used to calibrate depths on seismic sections.

WEP *abbr*: War Emergency Pipelines.

West Texas intermediate *n*: crude oil produced from the Permian Basin of West Texas and eastern New Mexico. Its gravity falls between light and heavy crude oil. It is a "benchmark" crude in that its price is often quoted as a measure of general crude oil prices.

West Texas separator *n*: a type of separator often used on drilling or workover rigs to separate gas from mud and other liquids returning from the well. It is so called because the first types of this separator were fabricated in the oilfields of West Texas.

wet air cleaner *n*: see *oil-bath air cleaner*.

wet boring *n*: in pipeline construction, a boring technique similar to slick boring that is used for small-diameter pipe such as natural gas distribution lines. Water is the lubricant. Compare *slick boring*.

wet-bottom prover *n*: an open tank prover with a gauge glass on the bottom of the prover. Compare *dry-bottom prover*.

wet box *n*: see *mud box*.

wet-bulb thermometer *n*: a thermometer whose bulb is covered with a cloth, usually muslin or cambric, saturated with water. It is used to measure relative humidity.

wet cell *n*: a secondary cell in which the electrolyte is a dilute acid.

wet combustion *n*: the simultaneous or alternate injection of water and air into a formation during an in situ combustion operation. The water reduces heat loss by transmitting heat that would otherwise be left behind in the formation, reduces the amount of required air, and increases the rate of advance of the fire flood. Also called COFCAW (combination of forward combustion and waterflooding). Compare *dry combustion*.

wet film thickness *n*: in paint technology, the thickness of a coat of paint immediately after it is applied and before any solvents in the paint have evaporated.

wet foam *n*: foam in which the maximum amount of water or other liquid has been added. Compare *dry foam*. See *foam*.

wet gas *n*: 1. a gas containing water, or a gas that has not been dehydrated. 2. a rich gas.

wet glycol *n*: glycol that has absorbed water.

wet job *n*: said of a job in which tubing is pulled full of oil or water.

wetlands *n pl*: land or areas containing much soil moisture. Wetland areas include areas that support vegetation typically found in saturated soils; that are saturated, flooded, or ponded enough during the growing season to result in an absence of oxygen in their upper parts; or that are subject to wetland hydrology, which results in saturated soil conditions on at least a seasonal basis.

wet monsoon *n*: a summer monsoon that exhibits heavy rainfall.

wet natural gas *n*: a well stream containing mostly natural gas but with enough heavier hydrocarbons in it that it liquefies at surface temperature and pressure.

wet oil *n*: an oil that contains water, either as an emulsion or as free water.

wet suit *n*: a diving suit, usually made of neoprene material, designed to provide thermal insulation for a diver's body. A small amount of water enters the suit, is warmed by body heat, and protects divers for a short time.

wettability *n*: the relative affinity between individual grains of rock and each fluid that is present in the spaces between the grains. If oil and water are both present, the water is usually in contact with the surface of each grain, and the rock is called water-wet. If, however, the oil contacts the surface, the rock is called oil-wet.

wetting *n*: the adhesion of a liquid to the surface of a solid.

wetting agent *n*: see *surfactant*.

wetting phase *n*: the liquid in a two-liquid system that is attracted to the sides of the pore. See *nonwetting phase*.

wetting water *n*: a film of water that sticks to, or is adsorbed by, the solid rock material surrounding the pore spaces.

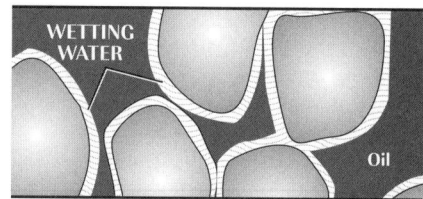

wet welding *n*: underwater welding performed without the use of a protective habitat.

wh *abbr*: white; used in drilling reports.

Wheatstone bridge *n*: in electronics, a special circuit that converts variable resistance from a sensing element into a proportional voltage. Output voltage can be calibrated in units of pressure, temperature, or other process variables;

or output voltage can be modified to produce a proportional current output for control or indication.

wheel ditcher *n*: a ditching machine for pipeline construction that has a large, rapidly rotating set of toothed scoops that lift dirt out of the ditch and feed it onto a conveyor mounted on the side of the machine. The wheel ditcher is used almost exclusively for ditching operations in stable soil.

wheel-type back-off wrench *n*: a wheel-shaped wrench that is attached to the sucker rod string at the surface and is manually turned to unscrew the string to allow it to be pulled from the well. Also called a back-off wheel or a circle wrench.

whelp *n*: (nautical) a sprocket tooth in a wildcat.

whip a connection *v*: see *heat a connection*.

whip line *n*: secondary hoist line.

whipping *n*: oscillation along the longitudinal axis of the drill stem.

whipstock *n*: a long steel casing that uses an inclined plane to cause the bit to deflect from the original borehole at a slight angle. Whipstocks are sometimes used in controlled directional drilling, in straightening crooked boreholes, and in side tracking to avoid unretrieved fish.

whipstock anchor packer *n*: a special-purpose packer placed in the casing to permit a sidetrack operation.

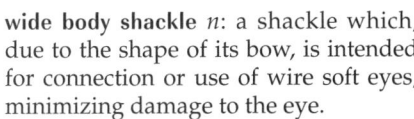

WHIPSTOCK

whirling *n*: see *bit whirl*.

whitecap *n*: a wave with a white, foamy top.

whole gale *n*: in the Beaufort wind scale, a wind whose speed is from 48 to 55 knots (55 to 63 miles per hour or 89 to 101 kilometres per hour).

whole mud *n*: all the components of the drilling mud, including reactive and nonreactive solids and the liquid phase (filtrate).

wickered *adj*: of or having wickers. Also called birdcaged.

wickers *n pl*: broken or frayed strands of the steel wire that makes up the outer wrapping of wire rope.

WICKERS

wide body shackle *n*: a shackle which, due to the shape of its bow, is intended for connection or use of wire soft eyes, minimizing damage to the eye.

wide-line chart *n*: an orifice chart where the differential and static pressure recordings indicate a wide range of pressure readings.

widowmaker (slang) *n*: on offshore drilling tender rigs, which feature a barge moored to a platform on which the wells are drilled, a narrow walkway placed between the barge and the platform that allows the crew to move between the two.

wildcat *n*: 1. a well drilled in an area where no oil or gas production exists. 2. (nautical) the geared sheave of a windlass used to pull anchor chain. *v*: to drill wildcat wells.

wildcatter *n*: one who drills wildcat wells.

wild well *n*: a well that has blown out of control and from which oil, water, or gas is escaping with great force to the surface. Also called a gusher.

willy-willy *n*: a tropical cyclone in Australia.

winch *n*: 1. a machine that pulls or hoists by winding a cable around a spool. 2. a device used to store, heave in, and pay out mooring wire on a rig.

wind *n*: a weather phenomenon created by unequal heating of the earth's surface and which is characterized by movement of the earth's atmosphere. Wind is observable by its effect on objects it contacts—for example, it causes tree branches to move and flags to wave.

wind arrow *n*: on a weather map, a symbol used to show the direction and wind speed at a specific location on the map. A dot at the base of the symbol indicates the location of the observation. One straight line (a shaft) goes out from the dot to indicate the wind's direction; short angular lines (feathers) and a blacked-in triangle (for the highest wind speeds) on the shaft indicate wind speed.

windbreak *n*: something that breaks the force of the wind. For example, canvas windbreaks installed around the outside of the rig floor on a drilling or workover rig afford the crew protection from strong, cold winds. Sometimes called a prefab.

windfall profit tax *n*: a federal excise tax on crude oil. It has a different rate for oil in a number of categories, for example, newly discovered oil, stripper oil, and stripper oil produced by independents. The tax rate is determined as a percentage—for instance, at 25%—of the difference between a base price and a market price. There are a

variety of exceptions and exemptions; interested parties should query the IRS or a tax accountant.

wind girder *n*: see *wind ring*.

wind guy line *n*: the wireline attached to ground anchors to provide lateral support for a mast or derrick. Compare *load guy line*.

winding *n*: a set of conductors installed to form the current-carrying element of a dynamo or a stationary transformer.

windlass *n*: 1. a device on an anchor-handling boat that propels the anchor chain to and from a chain locker, where it is stored. 2. an early hoist or winch which was a simple drum, or spool, sitting horizontally between two posts with one end of a rope attached.

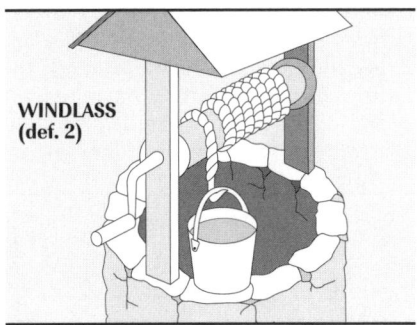

WINDLASS (def. 2)

wind-load rating *n*: a specification used to indicate the resistance of a derrick to the force of wind. The wind-load rating is calculated according to API specifications. Typical wind resistance in derricks is 75 miles per hour with pipe standing in the derrick and 115 miles per hour or higher without.

window *n*: 1. a slotted opening or a full section removed in the pipe lining (casing) of a well, usually made to permit sidetracking. 2. a period during which an operation or action can be carried out, such as the drilling window that occurs in offshore Arctic areas in summer or winter.

wind ring *n*: a horizontal stiffening and structural member installed near the top of a floating-roof tank to reinforce the tank wall against wind pressure. Also called a wind girder.

wind rose *n*: a diagram that plots wind direction and speed at a particular location.

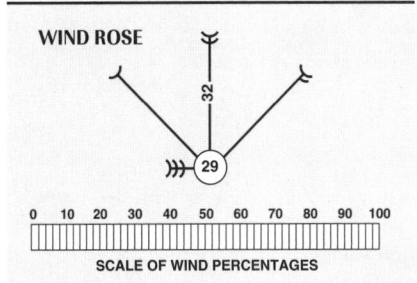

WIND ROSE

0   10   20   30   40   50   60   70   80   90   100

SCALE OF WIND PERCENTAGES

windward *n*: (nautical) upwind, i.e., the direction from which the wind is blowing. Also the side of a rig that is exposed to the effects of wind and waves (also known as upwind).

wind waves *n pl*: short-period waves formed by the contact of wind on the ocean surface.

wing *n*: a horizontal pipe that exits from one or two sides of a Christmas tree.

wing valve *n*: a valve on the wing of a Christmas tree. See *wing*.

wiper *n*: a circular rubber device with a split in its side that is put around drill pipe to wipe or clean drilling mud off the outside of the pipe as the pipe is pulled from the hole.

wiper plug *n*: a rubber-bodied, plastic- or aluminum-cored device used to separate cement and drilling fluid as they are being pumped down the inside of the casing during cementing operations. A wiper plug also removes drilling mud that adheres to the inside of the casing.

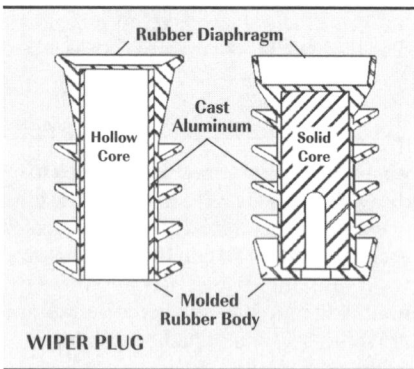

WIPER PLUG

wire *n*: in an anchoring system for semisubmersible rigs working in deep water, the strong cable (wire), one end of which is connected to a winch and the other end of which is attached to the end of the anchor's chain. By combining wire and chain, instead of using all chain, to create the length necessary to successfully anchor a semisubmersible in deep water, the anchoring system's weight and the storage space required for the anchoring system is reduced. (If only chain was used, the space needed to store the chain on board the rig while the anchors were not being used would be so large as to be impractical.)

wireline *n*: a slender, rodlike or threadlike piece of metal usually small in diameter, that is used for lowering special tools (such as logging sondes,

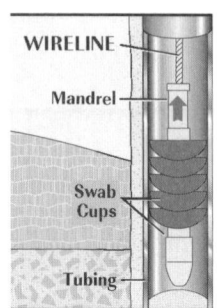

perforating guns, and so forth) into the well. Also called slick line. Compare *wire rope*.

wireline blowout preventer *n*: a device containing a valve that closes around the wireline to shut in the well if high pressure threatens a blowout. Also called a wireline valve.

wireline core barrel *n*: see *core barrel*.

wireline coring *n*: a coring method in which a core barrel assembly is lowered through the drill pipe on wireline without tripping out the drill pipe. After the core is drilled, the core and the core barrel are retrieved by an overshot assembly.

wireline cutting tool *n*: a device usually run on a solid wireline to cut another wireline stuck in a well.

wireline electric log *n*: see *electric logging*.

wireline entry guide *n*: a flared-end sub run on the end of the tubing string to permit easy access of wireline tools into the tubing ID.

wireline feeler *n*: a tool used to gouge and clean junk and debris from the casing in conjunction with a junk catcher.

wireline formation tester *n*: a formation fluid sampling device, actually run on conductor line rather than wireline, that also logs flow and shut-in pressure in rock near the borehole. A spring mechanism holds a pad firmly against the sidewall while a piston creates a vacuum in a test chamber. Formation fluids enter the test chamber through a valve in the pad. A recorder logs the rate at which the test chamber is filled. Fluids may also be drawn to fill a sampling chamber. Wireline formation tests may be done any number of times during one trip in the hole, so they are very useful in formation testing.

wireline grab *n*: a tool used to fish broken wireline out of tubing. Two or three barbed prongs extend along the inside of the U-shaped housing. Wireline becomes caught on these prongs and can be retrieved.

wireline log *n*: any log that is run on wireline.

wireline logging *n*: see *well logging*.

wireline operations *n pl*: the lowering of mechanical tools, such as valves and fishing tools, into the well for various purposes. Electric wireline operations, such as electric well logging and perforating, involve the use of conductor line, which in the oil patch is commonly but erroneously called wireline.

wireline pressure bomb *n*: a special container that holds a pressure-sensing device and a recorder that is run into and retrieved from a well on wireline to measure and record pressures in the well.

wireline preventer *n*: a manually operated ram preventer especially adapted for closure around a wireline.

wireline probe *n*: a diagnostic tool used to ascertain the position of a gas leak in the tubing of a gas-lift well.

wireline rig *n*: the smallest rig that raises and lowers a wireline or conductor line.

wireline service *n*: a general term used to refer to any servicing operation using a wireline.

wireline socket *n*: the connection used to attach wireline to the tool string. The wireline socket contains a rotating disk to allow the tool string to swivel horizontally and a small spring to cushion the line and to prevent the line from breaking during heavy operations. See *wireline tool string*.

wireline spear *n*: a special fishing tool fitted with prongs to catch and recover wireline that has broken off and been left in a well.

wireline stuffing box and lubricator *n*: a stuffing box and lubricator used with wireline during a bottomhole pressure test. See *stuffing box*.

wireline survey *n*: a general term often used to refer to any type of log being run in a well.

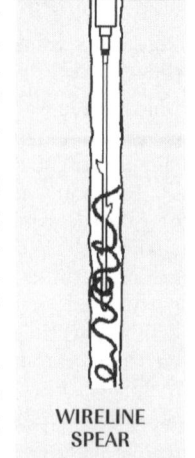

WIRELINE SPEAR

wireline tools *n pl*: special tools or equipment made to be lowered into and retrieved from the well on a wireline, e.g., packers, swabs, gas-lift valves, measuring devices.

wireline tool string *n*: the string of tools and equipment that is attached to conductive or nonconductive wireline and lowered downhole to perform wireline operations.

wireline truck *n*: a service vehicle or unit on which the spool of wireline is mounted for use in downhole wireline work.

wireline valve *n*: see *wireline blowout preventer*.

wireline well logging *n*: the recording of subsurface characteristics by wireline (actually conductor line) tools. Wireline well logs include acoustic logs, caliper logs, radioactivity logs, magnetic resonance logs, and resistivity logs.

wireline wiper *n*: a flexible rubber wiper used to scrape mud or oil from a wireline as it is pulled from a hole.

wireline workover *n*: a workover performed with wireline lowered into the well through the tubing string. A lubricator is rigged up over the wellhead and wireline tools are inserted through the lubricator under pressure. A variety of wireline tools is available.

wire rope *n*: a cable composed of steel wires twisted around a central core of fiber or steel wire to create a rope of great strength and considerable flexibility. Wire rope is used as drilling line (in rotary and cable-tool rigs), coring line, servicing line, winch line, and so on. It is often called cable or wireline; however, wireline is a single, slender metal rod, usually very flexible. Compare *wireline*.

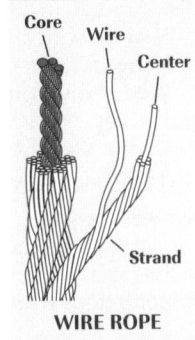

WIRE ROPE

wire-rope clip *n*: a device that secures the end of a bend in the wire rope to the body of the wire rope. A typical safety clamp consists of a U-shaped bolt that is fastened to the wire rope with a saddle-shaped base. Two nuts on the base thread onto the U-bolt to secure the clamp to the rope. Depending on the diameter of the wire rope, which determines the amount of rope that is looped back on itself to secure the rope to the tongs or other device, clamps can vary in number from one to three.

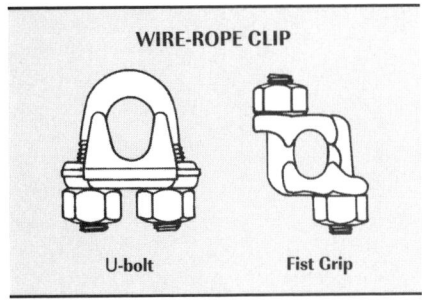

WIRE-ROPE CLIP

U-bolt   Fist Grip

wire rope core *n*: the center part of a wire rope construction which may be a bundle of wires or fiber.

wire rope drilling line *n*: see *drilling line*.

wire rope grip *n*: a special clamp that temporarily joins the end of two lengths of wire rope. See *swivel-type stringing grip*.

wire rope strand *n*: a bundle of wires that comprise part of a wire rope construction.

wire-wound resistor *n*: a resistor made from special wire, usually Nichrome, that is wound around a relatively small cylindrical core or tube. Wire-wound resistors can handle large amounts of power and maintain their ohmic value even under high temperatures.

wire-wrapped screen *n*: a relatively short length of pipe that has openings in its sides and a specially shaped wire wrapped around the pipe. It is used in wire-wrapped screen completions, usually in conjunction with a gravel pack. The screen and gravel pack block out sand and allow fluids to flow into the well through the openings in the screen. Also called a screen liner.

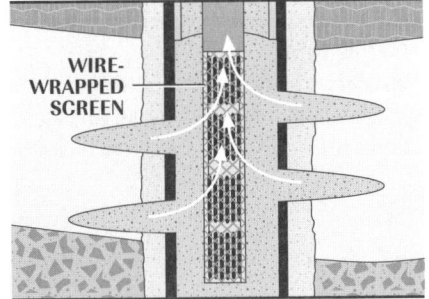

W-L; W/L *abbr*: wireline.

WMO *abbr*: World Meteorological Organization.

WOB *abbr*: weight on bit.

wobble *n*: movement between the mating surfaces of box and pin in a tool joint. *v*: to move in a rocking motion.

WOC *abbr*: waiting on cement; used in drilling reports.

WOE *abbr*: waiting on engineering.

WOG *abbr*: water-oil-gas.

WO/O *abbr*: waiting on orders; used in drilling reports.

wood alcohol *n*: see *methanol*.

WOR *abbr*: water-oil ratio.

work *n*: the acting of a force through distance, or the overcoming of a resistance to motion. No work is done unless motion is produced. Mathematically, work is force times distance.

work basket *n*: see *personnel basket*.

work boat *n*: a boat or self-propelled barge used to carry supplies, tools, and equipment to a job site offshore.

work hardening *n*: hardness developed in metals as a result of cold-working.

working barrel *n*: the outer shell of a downhole plunger pump. The pumping cycle starts with an upward stroke of the rods, which pulls the plunger up through the working barrel. On the upstroke, the traveling valve closes, the standing valve in the working barrel opens, fluid above the traveling valve is lifted out of the well, and a new charge is drawn into the pump. On the downstroke, the traveling valve opens, the standing valve closes, and the fluid is forced from the working

barrel through the traveling valve in the plunger and into the tubing. Repeated strokes bring the fluid to the surface.

working gas *n*: the volume of gas in an underground storage reservoir above the designed level of the base. It may or may not be completely withdrawn during any particular withdrawal season.

working interest *n*: the interest in oil or gas that includes responsibility for all drilling, developing, and operating costs. Also called leasehold interest, operating interest.

working-interest oil *n*: see *net production, working interest*.

working pressure *n*: 1. the maximum pressure at which an item is to be used at a specified temperature. 2. the current pressure functioning in a process.

working standard *n*: a standard that is calibrated against a reference standard and is intended to verify working measuring instruments of lower accuracy. Compare *primary standard, secondary standard*.

working string *n*: see *work string*.

working tape *n*: the measuring tape used to measure all of a tank's rings after it is calibrated to the standard tape. See *standard tape, strapping*.

work over *v*: to perform one or more of a variety of remedial operations on a producing oilwell to try to increase production. Examples of workover operations are deepening, plugging back, pulling and resetting liners, and squeeze cementing.

workover *n*: the performance of one or more of a variety of remedial operations on a producing oilwell to try to increase production. Examples of workover jobs are deepening, plugging back, pulling and resetting liners, and squeeze cementing. See *recompletion*.

workover fluid *n*: a special drilling mud used to keep a well under control while it is being worked over. A workover fluid is compounded carefully so that it will not cause formation damage.

workover rig *n*: a portable rig used for working over a well. See *production rig*.

work string *n*: 1. in drilling, the string of drill pipe or tubing suspended in a well to which is attached a special tool or device that is used to carry out a certain task, such as squeeze cementing or fishing. 2. in pipeline construction, the string of washpipe that replaces the pilot string in a directionally drilled river crossing. The work string remains in place under the river until the actual pipeline is made up and is ready to be pulled back across the river.

work unit *n*: a larger type of self-contained breathing gear that provides from 30 to 60 minutes of air. This apparatus differs from the emergency escape unit mainly in the size of the air bottle and, hence, in its capacity to store air. Compare *emergency escape unit.*

World Meteorological Organization (WMO) *n*: a United Nations agency that was formed to promote international cooperation in all aspects of meteorological activities. Address: Case Postale 2300, CH-1211, Geneva 2, Switzerland.

worm *n*: 1. a new and inexperienced worker on a drilling rig. 2. a short revolving screw with spiral-shaped threads.

worm gear *n*: the gear of a worm and a worm wheel working together.

worm's-eye map *n*: see *subgeologic map.*

worm wheel *n*: a toothed wheel gearing the thread of a worm.

worst-case spill *n*: the worst possible spill from a vessel or facility. Such a spill is determined by the maximum amount of discharge possible from the facility, or the total release of a full cargo from the firm's largest vessel, in the worst weather conditions.

WOW *abbr*: waiting on weather.

wrench flat *n*: a flat area on an otherwise round fitting to which a wrench can be applied (as on sucker rod coupling). Also called a wrench square.

wrench square *n*: see *wrench flat.*

wrist pin *n*: in an engine, the hard steel, hollow cylinder that attaches the piston to the piston rod. A circular opening in the piston is lined up with a corresponding circular opening in the rod and the wrist pin is pushed through the openings. Usually, special keys on each end of the pin lock the pin in the piston. Also called piston pin.

WTI *abbr*: West Texas intermediate.

wye connection *n*: a three- or four-wire connection, resembling the letter Y, used in three-phase circuits alone or in various combinations with delta connections.

X_C *sym*: capacitive reactance.

X-C polymer *n*: a biopolymer produced from a particular strain of bacteria on carbohydrates that produces large increases in apparent viscosity while maintaining fairly good control of filtration.

xln *abbr*: crystalline; used in drilling reports.

XOR *abbr*: exclusive or.

X-ray testing *n*: see *radiographic examination or testing*.

X-treme line casing *n*: brand name for streamlined casing, which features couplings threaded into the body of the casing itself; therefore, it does not require separate casing couplings to connect the joints. Streamlined casing is useful where casing of a given diameter must be used in a hole that is close to the same size as the casing.

xylene *n*: any of three flammable hydrocarbons, $C_6H_4(CH_3)_2$, similar to benzene. A commercial mixture is used as a solvent when oilfield emulsions are being tested.

yaw *n*: on a mobile offshore drilling rig or ship, the angular motion as the bow or stern moves from side to side. *v*: to move from side to side (as a ship).

Y-block *n*: see *Y-sub*.

Y connection *n*: see *wye connection*.

yd *abbr*: yard.

yd² *abbr*: square yard.

yd³ *abbr*: cubic yard.

yellow pod *n*: see *hydraulic control pod*.

yield *n*: the number of barrels of a liquid slurry of a given viscosity that can be made from a ton of clay. Clays are often classified as either high- or low-yield. A ton of high-yield clay yields more slurry of a given viscosity than a low-yield clay.

yield point *n*: the maximum stress that a solid can withstand without undergoing permanent deformation either by plastic flow or by rupture. See *tensile strength*.

yield strength *n*: in a joint of riser pipe, a measure of the amount of pressure inside or outside the riser that is required to permanently distort it.

yield value *n*: the resistance to initial flow, or the stress required to start fluid movement. This resistance is caused by electrical charges located on or near the surfaces of the particles. The values of the yield point and thixotropy, respectively, are measurements of the same fluid properties under dynamic and static states. The Bingham yield value, reported in pounds/100 square feet, is determined from a direct-indicating viscometer by subtracting the plastic viscosity from the 300-rpm reading. Also called yield point.

yoke *n*: a stationary piece of ferromagnetic material without windings, which permanently connects two or more magnet cores.

Young's modulus *n*: in mechanics, the ratio of a simple tension stress applied to a material to the resulting strain parallel to the tension.

Y-sub *n*: in multiple packer completion, a device, made up just below the uppermost packer, that switches tubular fluids to the annulus and annular fluids to the tubing above that packer. Also called a Y-block.

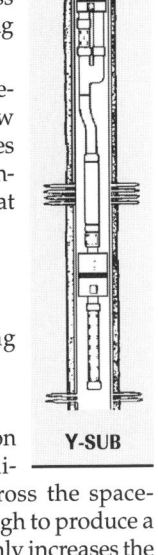

Y-SUB

Z *abbr*: distance of the righting arm.

Z *sym*: impedance.

Zener breakdown *n*: a phenomenon that occurs in semiconductor diodes when the electric field across the space-charge region becomes high enough to produce a form of field emission that suddenly increases the number of carriers in this region. This breakdown is typically several hundred to several thousand volts, depending on the semiconductor's construction, and is not destructive. Also called reverse breakdown, Zener effect.

Zener diode *n*: a solid-state diode that uses Zener breakdown to regulate voltage in a circuit in which it is installed. It operates on the principle that the voltage that occurs during breakdown is relatively constant and therefore can be used to regulate voltage. See *Zener breakdown*.

Zener effect *n*: see *Zener breakdown*.

zeolite *n*: hydrated silicates of aluminum with alkali metals.

zerk *n*: a special fitting on equipment that accommodates a similar fitting on a grease gun. The zerk allows grease to be injected, but forms a seal to prevent the entry of dirt into the equipment when the gun is removed.

zero *v*: to bring a differential pen to the zero point on an orifice chart.

zero adjustment *n*: in process control, a procedure in process transmitters that establishes the minimum, or zero, level of measurement in a process variable.

zero measurand output *n*: the output of a transducer with zero measurand applied.

zero point *n*: on an orifice chart, the point at which no differential pressure is exerted on the differential pen.

zero potential *n*: the actual potential of the surface of the earth taken as a point of reference. See *electric potential*.

zero-zero gel *n*: a condition wherein the drilling fluid fails to form measurable gels during a quiescent interval (usually 10 minutes).

zeta potential *n*: electrokinetic potential of a particle as determined by its electrophoretic mobility. This electric potential causes colloidal particles to repel each other and stay in suspension.

Ziegler-Nichols tuning *n*: in process control, a method used to establish the optimum performance of a closed-loop automatic control system. It uses an experimental procedure to establish system gain and time constants within proportional (P), proportional-plus-reset (PI), and proportional-plus-reset-plus-rate (PID) systems.

zinc chloride ($ZnCl_2$) *n*: a very soluble salt used to increase the density of water to points more than double that of plain water. Normally added to a system first saturated with calcium chloride.

Zodiac boat *n*: a relatively small inflatable boat, usually equipped with an outboard motor, which is mainly used for recreation and transporting marine scientists and divers. The boat is named after the Canadian firm Zodiak that makes them. Although the Zodiac brand is synonymous for these small inflatable boats, other brands, such as Eagle, are available. Also, Zodiak makes rescue boats, life rafts, and other inflatable products.

zone *n*: a rock stratum that is different from or distinguished from another stratum (e.g., a pay zone).

zone isolation *n*: the practice of separating producing formations from one another by means of casing, cement, and packers for the purposes of pressure control and maintenance, as well as the prevention of mixing of fluids from separate formations.

zone of lost circulation *n*: a formation that contains holes or cracks large enough to allow cement to flow into the formation instead of up along the annulus outside of the casing.

Abbreviations

A: ampere; area; cross-sectional area, in.^{2}; well spacing, acres

AAPG: American Association of Petroleum Geologists

AAPL: American Association of Petroleum Landmen

abd, abdn: abandoned

abs: absolute

ABS: American Bureau of Shipping

Ac: altocumulus

AC: alternating current

ACGIH: American Conference of Governmental and Industrial Hygienists

acre: not abbreviated

acre-ft: acre-foot

ACSR: aluminum cable steel-reinforced conductor

ACT: automatic custody transfer

ADS: atmospheric diving system

AESC: Association of Energy Service Companies

AFCI: arc fault circuit interrupter

AFE: authority for expenditure

AGA: American Gas Association

A-h: ampere-hour

AHTS: anchor-handling tug/supply vessel

AHV: anchor-handling vessel

AIME: American Institute of Mining, Metallurgical and Petroleum Engineers

A_i: cross-sectional area based on inside diameter of tubing, in.^{2}

AKO: adjustable kickoff tool

ALP: articulated loading platform

AM: amplitude modulation

θ: angle of deformation, radian

AMI: area of mutual interest

ANSI: American National Standards Institute

antilog: antilogarithm

ANWR: Arctic National Wildlife Refuge

A_o: cross-sectional area based on outside diameter of tubing, in.^{2}

AOF: absolute open flow

AOFP: absolute open flow potential

A_p: cross-sectional area through the packer seal, in.^{2}

A_s: cross-sectional area of steel in tubing, in.^{2}

API: American Petroleum Institute

API *MPMS*: API's *Manual of Petroleum Measurement Standards*

API **Spec 7**: *API Specification for Rotary Drilling Equipment*

APO: after payout

APR: trademark name for an annular pressure-responsive valve for a DST string

As: altostratus; arsenic

ASCII: American standard code for information interchange

ASME: American Society of Mechanical Engineers

ASSE: American Society of Safety Engineers

ASTM: American Society for Testing and Materials

atm: atmosphere

avg: average

AWG: American wire gauge

AWS: American War Standard

AWV: annulus wing valve

B: bottom of; center of buoyancy; formation volume factor; shear modulus, psi/radian

back emf: back electromotive force

BACT: best-available control technology

bbl: barrel

bbl/acre-ft: barrels per acre-foot

bbl/d, B/D, b/d, BPD, bpd: barrels per day

B_c: Bearden units of consistency-dimensionless

BCD: binary coded decimal

Bcf: billion cubic feet

Bcf/d, Bcf/D: billion cubic feet per day

BDC: bottom dead center

b_{dn}: slope of flowmeter velocity vs rps response curve (down runs)

BFPH: barrels of fluid per hour

BHA: bottomhole assembly

bhhp: bit hydraulic horsepower

bhp: brake horsepower

BHP: bottomhole pressure

BHT: bottomhole temperature

BII: business interruption insurance

bl: black

bld: bailed

BLM: Bureau of Land Management

BLPD: barrels of liquid per day

BO: barrels of oil

BOP: blowout preventer

BOPD: barrels of oil per day

BOPE: blowout preventer equipment

bpd, BPD: barrels per day

BPH: barrels per hour

BPO: before payout

brkn: broken

BS&W: basic sediment and water

Bscf/d, Bscf/D: billion standard cubic feet per day

Btu: British thermal unit

b_{up}: slope of flowmeter velocity vs rps response curve (up runs)

BW: barrels of water

BWPD: barrels of water per day

BWPH: barrels of water per hour

c: compressibility psi^{-1}

C: capacitance; carbon; Celsius; coefficient of expansion of steel per °F; coulomb; turbulence correction

°C: degrees Celsius

CAA: Clean Air Act

cal: calorie

CALM: catenary anchor leg mooring

CAO: computer assisted operations

CAODC: Canadian Association of Oilwell Drilling Contractors

CARE: Conservation Award for Respecting the Environment

CART: cam-actuated running tool

Cb: cumulonimbus

CBHT: circulating bottomhole temperature

CBL: cement bond log

Cc: cirrocumulus

cc: cubic centimetre

CCA: cold cranking ampere

CCL: casing collar log

cd: candela
CDU: conical drilling unit
CEQ: Council on Environmental Quality
CERCLA: Comprehensive Environmental Response, Compensation, and Liability Act of 1980
CFG: cubic feet of gas
CFR: Code of Federal Regulations; Coordinating Fuels and Equipment Research Committee
cgs: centimetre-gram-second
Δ: change from initial packer setting conditions
chk: choke
Ci: cirrus
Δt_c: compressional travel time, microsec/ft
CI: compression ignition
CIDS: concrete island drilling system
circ: circulated
CIV: chemical injection valve
Cl: chlorine
CLFP: choke-line friction pressure
cm: centimetre
CMC: carboxymethyl cellulose
cmil: circular mil
cm^2, sq cm: square centimetre
cm^3, cc: cubic centimetre
C_o: ultimate compressive strength of rock, psi
Co: cobalt
CO: carbon monoxide
CO_2: carbon dioxide
coax: coaxial cable
COH: change of heel
COFCAW: combination of forward combustion and waterflooding
comp: completed, completion
congl: conglomerate
cos: cosine
cosec: cosecant
COT: change of trim
cotan: cotangent
coul: coulomb
cp: centipoise
CP: casing pressure or casing point
CPC: computerized production control
C_{pl}: the correction for pressure on liquid
C_{ps}: the correction for pressure on steel
CPU: central processing unit
CRC: Coordinating Research Council, Inc.
crd: cored
crg: coring
CRT: cathode-ray tube
Cs: cirrostratus
CSM: catenary spread mooring
C_{tl}: the correction for temperature of liquid
C_{ts}: the correction for temperature of steel
cu: cubic
Cu: cumulus
CV: controlled variable
CWA: Clean Water Act
CZM: coastal zone management

d, D: day
d: inside diameter, in.

D: conduit diameter, in.; depth, ft; outside diameter, in.
ρ: density lb_m/cu ft; density, lb/gal; rock density, lb/ft³
$ρ_I$: density of available brine, lb/gal; density of fluid inside tubing, lb/cu in.
$ρ_o$: density of fluid outside tubing, lb/cu in.; density of oil
$ρ_t$: density of composite fluid
$ρ_w$: density of water
D&A: dry and abandoned
D&P: drilling and production
D, darcy: darcy
daN: decanewton
DC: direct current; drill collar; delivery capacity
DCFR: discounted cash flow rate of return
DCQ: daily contract quantity
DDC: deck decompression chamber
°API: degree API (American Petroleum Institute)
°F: degrees Fahrenheit
DEA: drag embedment anchor
DENLA: drag embedment normal load anchor
DF: derrick floor
D_I: inner pipe od, in.
dk: dark
dm: decimetre
dm³: cubic decimetre
dm³/s: cubic decimetres per second
D_o: outer pipe id or hole-size, in.
DOE: Department of Energy
DOI: Department of the Interior
dol, dolo: dolomite
DOT: Department of Transportation
DP: drill pipe
DPDT: double-pole, double-throw
DPST: double-pole, single-throw
DR: damage ratio
drlg: drilling
DST: drill stem test
DTL: dynamic tension limit
DWC: drilling with casing
DWF: deliberate well firing
dwt: deadweight ton

E: modulus of elasticity (30 × 10⁶ psi, for steel); volt; Young's modulus, psi
E&P: exploration and production
E_I: exponential integral
ECD: equivalent circulating density
eco-out: economic out
EED: energy, exploration, and development
EEEIPS: extra, extra, extra improved plough (plow) steel
EEIPS: extra, extra improved plough (plow) steel
EEPROM: electrically erasable programmable read-only memory
$\overline{σ}_h$: effective horizontal rock matrix stress, psi
$\overline{σ}_v$: effective vertical rock matrix stress, psi
efm: electronic flow measurement
Eh: oxidation-reduction potential

EHS: extremely hazardous substance
EIA: Electronic Industries Alliance; Energy Information Administration
EIPS: extra improved plough (plow) steel
EIS: environmental impact statement
elec log: electric log
elev: elevation
EMD: Electromotive Division of General Motors Corporation
emf: electromotive force
EMW: equivalent mud weight
EMWD: electromagnetic measurement while drilling
ENTELEC: Energy Telecommunications and Electrical Association
EOR: enhanced oil recovery
EPA: Environmental Protection Agency
EP mix: ethane-propane mix
EPCRA: Emergency Planning and Community Right-to-Know Act of 1986
epm: equivalents per million
EPROM: erasable programmable read-only memory
EPT: electromagnetic propagation tool
ERA: Economic Regulatory Administration
ERTS: Earth Resource Technology Satellite
ES: electric survey
ESA: Endangered Species Act
ESD: emergency shut down
est: estimated
et al: and others
ETBE: ethyl tertiary butyl ether
ETS: engine temperature switch
et ux: and wife
et vir: and husband
EUE: external upset end
XOR: exclusive or

$ρ_f$: fluid density desired, lb/gal
f: Fanning friction factor
F: center of flotation; Fahrenheit; farad; force, lb; force, psi
FAA: Federal Aviation Administration
FAR: federal aviation requirement
FCC: Federal Communications Commission
fcp: final circulating pressure
FE: flow efficiency
Fe: iron
FEA: Finite Element Analysis
FeO: ferrous oxide
FERC: Federal Energy Regulatory Commission
FeS: ferrous sulfide
FET: field-effect transistor
FIFRA: Federal Insecticide, Fungicide, and Rodenticide Act
FIH: fluid in hole
fl: flowed or flowing
flex joint: flexible joint
FLPMA: Federal Land Policy and Management Act of 1976
flocs: flocculates
fluor: fluorescence
fm: formation
FM: frequency modulation
FOB price: free on board price

FOGRMA: Federal Oil and Gas Royalty Management Act of 1982

FONSI: finding of no significant impact

FP: flowing pressure

FPC: Federal Power Commission

FPDSO: floating production, drilling, and system off-loader

FPSO: floating production and system off-loader

frac: fractured or fracturing

ft: foot, feet

ft-lb: foot-pound

ft/min, fpm: feet per minute

ft/s, fps: feet per second

ft^2, sq ft: square foot

ft^3, cu ft: cubic foot

ft^3/bbl, cu ft/bbl: cubic feet per barrel

ft^3/d, cu ft/d, cfd, cfD: cubic feet per day

ft^3/lb, cu ft/lb, cfp: cubic feet per pound

ft^3/min, cu ft/min, cfm: cubic feet per minute

ft^3/s, cu ft/s, cfs: cubic feet per second

FWKO: free-water knockout

g: gram

G: center of gravity; shear modulus, psi/radian

Ga: gallium

G&OCM: gas- and oil-cut mud

GAEOT: geometric average evaluation of tract

gal: gallon

gal/min: gallons per minute

GC: gas-cut

GCC: Gulf Cooperation Council

GCM: gas-cut mud

Ge: germanium

GFCI: ground fault circuit interrupter

GIN: generator identification number

GL: ground level

GLR: gas-liquid-ratio

GM: General Motors Corporation; metacentric height

Gohm, GΩ: gigohm, 1 billion ohms

GOR: gas-oil ratio

GOV: gross observed volume

GPA: Gas Processors Association

GPG: grains per gallon

gpm: gallons per minute

gr: gray

grn: green

GRN: gamma-ray-neutron

GRT: gross registered tonnage

GSV: gross standard volume

GSW: gross standard weight

GTO-SCR: gate-turnoff silicon-controlled rectifier

GVW: Grem vane wheel

GZ: the righting arm (or lever)

h: formation thickness; fracture height, ft; hour

H: hydrogen; henry; height of fluid column, ft; magnetizing force

HAP: hazardous airborne pollutant

HART: highway addressable remote transducer

HAZCOM: Hazard Communication Standard

HAZMAT: hazardous materials

HAZWOPER: Hazardous Waste Operations and Emergency Response Standard

HBP: held by production

HCN: hydrogen cyanide

HCS: Hazard Communication Standard

HGOR: high gas-oil ratio

hhp: hydraulic horsepower

hp: horsepower

hp-h, hp-hr: horsepower-hour

HLB: hydrophilic-lipophilic balance

HMI: human machine interface

HMPE: high modulus polyethylene

HOCM: heavily oil-cut mud

HP: hydrostatic pressure

HPNS: high-pressure nervous syndrome

HPU: hydraulic pressure unit

HSE: health, safety, and environment

H$_2$S: hydrogen sulfide

H$_2$SO$_3$: sulfurous acid

H$_2$SO$_4$: sulfuric acid

HWDP: heavy-walled drill pipe; heavy-weight drill pipe

Hz: hertz

I: current; moment of inertia, in.

IACS: International Association of Classification Societies

IADC: International Association of Drilling Contractors

IBOP: inside blowout preventer

ICC: Interstate Commerce Commission

ICP: initial circulating pressure

ICS: incident command system

ID: inside diameter

IDLH: immediately dangerous to life or health

I-ES: induction-electric survey

I-EUE: internal-external upset end

IFP: Institut Francais du Petrole

IFR: instrument flight rules

ig: igneous

IGBT: insulated-gate bipolar transistor

IHTS: immersion heater temperature switch

IMCO: Intergovernmental Maritime Consultative Organization

IMP: Instituto Mexicano del Petróleo (Mexican Petroleum Institute)

IMR: inspection, maintenance, and repair (used in diving)

in.: inch

in.^{2}, sq in.: square inch

in.^{3}, cu in.: cubic inch

in./s, ips: inches per second

in.lb: inch-pound

INGAA: Interstate Natural Gas Association of America

inside BOP: inside blowout preventer

IOGCC: Interstate Oil and Gas Compact Commission

IPAA: Independent Petroleum Association of America

IPR: inflow performance relationship

IPS: improved plough (plow) steel

ISA: Instrument Society of America

ITCZ: intertropical convergence zone

ITL: Information to Lessees and Operators

IUE: internal upset end

IWRC: independent wire rope center; independent wire rope core

J: joule; productivity index

J': modified productivity index for deliverability test

JEDEC: JEDEC Solid State Technology Association

JFET: junction-type field-effect transistor

JVCL: joint venture convention liability

k: permeability, md; millidarcys

K: bulk modulus, psi; permeability, darcys

K: dielectric constant; keel; Kelvin; potassium

K': consistency index, power law fluid; consistency index, lbf $\cdot$ sec^n/ft^2

°K: degrees Kelvin

KB: kelly bushing

K, C: constants for weighting material

kcmil: one thousand circular mils

k$_f$: fracture permeability; proppant permeability, md

kg: kilogram

kg/cm^3: kilograms per cubic centimetre

kg-m: kilogram-metre

kg/m^3: kilogram per cubic metre

KGS: known geologic structures

km: kilometre

kn, kt: knot

k$_o$: permeability to oil

KO: kicked off

kohm, kΩ: kilohm

KOP: kickoff point

kPa: kilopascal

k$_{rg}$: relative permeability to gas

k$_{ro}$: relative permeability to oil

k$_{rw}$: relative permeability to water

Ksi: one thousand pounds per square inch

kV, kv: kilovolt

kVA: kilovolt-ampere

kW, kw: kilowatt

kw-h, kwh, KWH: kilowatt-hour

L: fracture length from wellbore, ft; henry; inductance; length, in.; liter; litre

LACT: lease automatic custody transfer

LAER: lowest achievable emission rate

lb: pound

lb/bbl: pounds per barrel

lb/ft^3, pcf: pounds per cubic foot

LCD: liquid crystal display

LCM: lost circulation material

LDC: local distribution company

LED: light-emitting diode

LEPC: Local Emergency Planning Committee

LGS: low gravity solids

LIOGA: Louisiana Independent Oil and Gas Association

lm: lime

LMRP: lower marine-riser package

ln: logarithm base e

LNG: liquefied natural gas

LNGC: liquefied natural gas carrier

LOC: location

log: logarithm

LOPI: loss of production insurance

LOT: loaded on top

LPG: liquefied petroleum gas

LRG: liquefied refinery gas

LRV: lower range value

LS: long substrate

ls: limestone

lse: lease

LTL: Letter to Lessees and Operators

LVR: load vessel ratio

LWD: logging while drilling

m: slope of linear portion of transient pressure semilog plot, psi/cycle

m: metre

M: metacenter

m^2, sq m: square metre

m^3, cu m: cubic metre

m^3/d: cubic metres per day

mA, milliamp: milliampere

MAC: mobile arctic caisson

MACT: maximum achievable control technology

MARAD: Maritime Administration

MASP: maximum allowable surface pressure

Mbopd: thousands of barrels of oil per day

MBTA: Migratory Bird Treaty Act

MC: metal clad

Mcf: thousand cubic feet of gas

Mcf/d, Mcf/D: thousand cubic feet of gas per day

MCS: master control station

md: millidarcy

MD: measured depth

megohms, Mohms, $M\Omega$: 1,000,000 ohms

mer: maximum efficiency rate

METOC: Fleet Numerical Meteorology and Oceanography Detachment

mev: million electron volts

MER: maximum efficiency rate

μF, mF: microfarad

MFD: manufacturer's abbreviation for microfarad

MFE: trademark name for multiple formation evaluation

mg: milligram

mH: millihenry

μ: micron

microsec, ms: microsecond

MICT: moving in cable tools

MIL: military standard

mile: not abbreviated

millisec: millisecond

min: minute

MIR: moving in rig

MIT: mechanical integrity test

MJ: megajoule

ml: millilitre

MLP: maximum lawful price

mm: millimetre

MM bopd: millions of barrels of oil per day

MMBtu: one million Btus

MMcf: million cubic feet

mmf: magnetomotive force

MMI: man-machine interface

MMmcf/d, MMcf/D: million cubic feet per day

MMscf: million standard cubic feet

MMscf/d, Mmscf/D: million standard cubic feet per day

MMS: Minerals Management Service

mm^2: square millimetre

mm^3: cubic millimetre

MO: moving out

MODU: mobile offshore drilling unit

Mohm: megohm, or 1,000,000 ohms

mol: mole

MOP: margin of overpull

MOPU: mobile offshore production unit

MOSFET: metallic oxide semiconductor field-effect transistor

MOV: metallic oxide varistor

mp: melting point

MPa: megapascal

mph: miles per hour

MPI: magnetic particle inspection

MPL: multiposition lock

MPPRCA: Marine Plastic Pollution Research and Control Act of 1987

MROV: mean range of values

ms: microsecond

Mscf/bbl: thousand standard cubic feet per barrel

Mscf/d, Mscf/D: thousand standard cubic feet per day

MSDS: material safety data sheet

MSHA: Mine Safety and Health Administration

MSRC: Marine Spill Response Corporation

MTBE: methyl tertiary butyl ether

MUX: multiplex electronic control

mV, mv: millivolt

MWD: measurement while drilling

MWS: making well safe

n: power in productivity index formula

n': flow behavior index, power law fluid; flow behavior index, log slope of shear rate-stress plot, dimensionless

N: newton; normal; range extension factor for Fann viscometer (usually 1.0)

N_2: nitrogen

NAAQS: National Ambient Air Quality Standards

NACE: National Association of Corrosion Engineers

NAME: National Association of Maritime Educators

NAVOCEANO: Naval Oceanographic Office

NDT: nondestructive testing methods

NEC: National Electric Code

NEMA: National Electrical Manufacturers Association

NEPA: National Environmental Policy Act

NESHAP: National Emissions Standards for Hazardous Airborne Pollutants

nF: nanofarad

NFPA: National Fire Protection Agency

NGL: natural gas liquids

NGPA: Natural Gas Policy Act

Ni: nickel

NIOSH: National Institute for Occupational Safety and Health

NIST: National Institute of Standards and Technology

NLPGA: National LP-Gas Association

NO_2: nitrogen dioxide

NOAA: National Oceanic and Atmospheric Administration

NORM: naturally occurring radioactive materials

NOS: National Ocean Service

NPDES: National Pollutant Discharge Elimination System

NPS: National Park Service

N_{Re}: Reynolds Number

NRC: National Response Center; Nuclear Regulatory Commission

NRT: net registered tonnage

Ns: nimbostratus

NS: no show

NSFCC: National Strike Force Coordination Center

NSPS: New Source Performance Standards

NSV: net standard volume

NSW: net standard weight

NTL: Notice to Lessees and Operators

NWR: NOAA Weather Radio

NWS: National Weather Service

O: oxygen

O&G: oil and gas

O&GCM: oil- and gas-cut mud

O&SW: oil and salt water

OBQ: on-board quantity

OC: oil-cut

OCM: oil-cut mud

OCS: Outer Continental Shelf

OCSIP: OCS Oil and Gas Information Program

OCSLA: Outer Continental Shelf Lands Act

OD: outside diameter

OEE: operator's extra expense

OF: open flow

OH: open hole

Ω: ohm

ohms/mil-ft: ohms per mil-foot

OIM: offshore installation manager

ool: oolitic

OPA: Oil Pollution Act of 1990

op-amp: operational amplifier

OPEC: Organization of Petroleum Exporting Countries

OPS: Office of Pipeline Safety

ORP: oxidation-reduction potential

ORQ: oil rig quality

OSHA: Occupational Safety and Health Administration

OSH Act: Occupational Safety and Health Act of 1970

oz: ounce

μ: Poisson ratio

ΔP_f: frictional pressure drop, psi

ϕ: porosity, fraction

$\bar{p}$: average reservoir pressure, psi

p: pressure, psi

Pa: pascal

P&A: plug and abandon

PAC: polyanionic cellulose

Pb: lead

PB: plugged back

PBR: polished bore receptacle

PbS: galena

p_c: critical pressure, psia

PCB: polychlorinated biphenyl

PCC: permanent chain chaser

pcf, lb/ft³: pounds per cubic foot

PCV: positive crankcase ventilation

PDC: polycrystalline diamond compact

PDC log: perforating depth control log

PDI: paraffin-deposition interval

PDVSA: Petróleos de Venezuela S.A.

p_e: pressure at external radius, psi

Pe: photoelectric effect

PEL: permissible exposure limit

perf: perforated

pf: power factor

pF: picofarad

P_f: the phenolphthalein alkalinity of the filtrate

PFD: manufacturer's abbreviation for picofarad

p_{ff}: final flowing pressure on DST, psi

p_{fhm}: final hydrostatic mud column pressure on DST, psi

p_{fim}: initial hydrostatic mud column pressure on DST, psi

pH: an indicator of the acidity or alkalinity of a substance or solution

P_h: hydrostatic pressure, psi

p_i: pressure inside tubing at surface, psi

P_i: pressure inside tubing at the packer seal, psi

p_I: initial pressure, psi

PI: productivity index

PID: proportional, integral, and derivative

PIREPS: pilot reports

P.I.T.S.: Petroleum Industry Training Service (Canada)

pk: pink

pkr: packer

pl: pipeline

PLC: programmable logic controller

PLCA: Pipe Line Contractors Association

P_m: phenolphthalein alkalinity of mud

p_o: pressure outside tubing at surface, psi

P_o: pressure outside tubing above the packer seal, psi

POOH: pull-out-of-hole

$(P_I)_h$: borehole pressure to initiate horizontal fracture, psi

$(P_I)_v$: borehole pressure to initiate vertical fracture, psi

$p_{1\,hr}$: pressure on semilog straight line after 1 hr of transient test, psi

POP: percentage of proceeds; putting on the pump

por: pores; porosity

PPE: personal protective equipment

ppg, lb/gal: pounds per gallon

ppm: parts per million

Δp: pressure change, psi

p_r: formation pore pressure, psi

PR: production reduction insurance

δ: pressure drop in tubing due to flow, psi/in.

PSA: pressure setting assembly

PSD: prevention of significant deterioration

psi: pounds per square inch

psia: pounds per square inch absolute

psi/ft: pounds per square inch per foot

psig: pounds per square inch gauge

PTO: power takeoff

PV: plastic viscosity, Bingham plastic fluid; pore volume

PVD: photovoltaic diode

PVT: Pit Volume Totalizer; pressure, volume, and temperature

PVT analysis: pressure, volume, and temperature analysis

p_{wf}: flowing bottomhole pressure, psi

p_{ws}: shut-in bottomhole pressure, psi

q: flow rate stbd for liquid, Mcfd for gas

Q: flow rate, bbl/min

Q_b: pumping rate bbl/min

Q_{cf}: pumping rate cu ft/min

q_g: gas flow rate, Mcfd

q_o: oil flow rate, stbd

qt: quart

qtz: quartz

qtze: quartzite

q_o: flow rate of oil

q_t: flow rate of composite fluid

q_w: flow rate of water

r: radial clearance between concentric tubulars, in.; radius, ft

r_e: external radius or drainage radius, ft

r_w: well bore radius, ft

R: Rankine; ratio of od to id of tubular; rel; resistance

°R: degrees Rankine

RAD: radioactive densitometer

RAM: random-access memory; recirculating average mixer

RBOP™: trademarked abbreviation for rotating blowout preventer

r_c: radius of closed fracture near wellbore, ft

RC: resistance-capacitance

RCM: recirculating cement mixer

RCP: reduced circulating pressure

RCRA: Resource Conservation and Recovery Act

r_e: drainage radius, ft

rec: recovered

RFG: reformulated gasoline

RIH: run-in-hole

RMA: Radio Manufacturers Association

rmg: reaming

RMS: robot maintenance system; root-mean-square

RMS value: root-mean-square value

ROB: remaining on board

ROD: rich-oil demethanizer

ROM: read-only memory

ROP: rate of penetration

ROV: remotely operated vehicle

ROW: right-of-way

RP: rock pressure

rpm: revolutions per minute

RQ: reportable quantity

RSPA: Research and Special Projects Administration

RSTC: rig safety and training coordinator

RT: register ton; rotary table

RTD: resistance temperature detector

RTTS™: trademark for a retrievable squeeze tool

RTU: remote terminal unit

RUCT: rigging up cable tools

RUR: rigging up rotary rig

r_w: wellbore radius, ft

Δt_s: shear wave travel time, microsec/ft

s, sec: second

S: sulfur; van Everdingen-Hurst skin factor

S_g: gas saturation

S_h: horizontal tensile strength of rock, psi

S_v: vertical tensile strength of rock, psi

SAFE: Safety Award for Excellence program

SALM: single anchor leg mooring

S&W: sediment and water

SAPP: sodium acid pyrophosphate

sat: saturated or saturation

SAWRS: Supplemental Aviation Weather Reporting Station

SBGD: seabed gas diverter

SBHT: static bottomhole temperature

SBM: synthetic-based mud

Sc: stratocumulus

SCBA: positive-pressure self-contained breathing apparatus

scf: standard cubic feet

scf/d, scf/D: standard cubic feet per day

SCR: silicon-controlled rectifier

SCSSV: surface-controlled subsurface safety valve

sd: sand or sandstone

SDO: shut down for orders

SDWA: Safe Drinking Water Act

sdy: sandy

SEA: suction embedment anchor

sec: secant; second; section

sed: sediment

SEG: Society of Exploration Geophysicists

seis: seismograph

SEM: subsea electronic module

SEPLA: suction embedment plate anchor

SERC: State Emergency Response Commission

S_g: gas saturation, fraction

SG: show of gas

SGA: Southern Gas Association

sh: shale

S_i: stress in inside fibre of tubular, psi

Si: silicon

SI: shut in; International System of units, or metric system; spark ignition

SIBHP: shut-in bottomhole pressure

SIC: standard industrial classification

SICP: shut-in casing pressure

SIDPP: shut-in drill pipe pressure

sin: sine

SIP: shut-in pressure

SLAR: side-looking airborne radar

v_s: slippage velocity-difference between heavier and lighter fluid

S/m: siemens per metre

SMGC: surface marine gridded climatology data

SNG: synthetic or substitute natural gas

S_o: oil saturation, fraction; stress in outside fibre of tubular, psi

SO: show of oil

SO&G: show of oil and gas

SOLAS: safety of life at sea

SO_2: sulfur dioxide

SP: self-potential; spontaneous potential

SPBM: single-point buoy mooring

SPCC: Spill Prevention, Countermeasures and Control

spd: spudded

SPDT: single-pole, double-throw

SPE: Society of Petroleum Engineers

sp gr: specific gravity

spm: strokes per minute

SPM: subsea pilot manipulated

SPR: Strategic Petroleum Reserve

SPS: submerged production system

SPST: single-pole, single-throw

S_r: shear rate, sec^{-1}

sq: square

sq ft, ft^2: square feet

sq in., $in.^2$: square inches

squ: squeeze

S_S: shear stress, lb/ft^2

ss: sand or sandstone

SSO: slight show of oil

SSTT: subsea test tree

SSU: Saybolt Second Universal

SSV: surface safety valve

St: stratus

S/T: sample tops

STB: stock-tank barrel

STB/d, STB/D: stock-tank barrels per day

STDW: Standards of Training, Certification, and Watchkeeping

std: standard

stds: stands

stn: stain

ε: strain, in./in.

σ: stress, psi

strks: streaks

sul wtr: sulfur water

sur: survey

SW: salt water

swbd: swabbed

swbg: swabbing

sx: sacks

sync amplifier: synchronization amplifier

sync control: synchronization control

t: time (hours); tonne

t': time, minutes

T: temperature, °F; tesla; ton; top of

TA: temporarily abandoned

tach: tachometer

tan: tangent

TAPS: Trans-Alaska Pipeline System

Tcf: trillion cubic feet

Tcf/d, Tcf/D: trillion cubic feet per day

TCP: tubing conveyed perforator

TD: total depth

TDC: top dead center

TEFC: totally enclosed fan-cooled motor

TEG: triethylene glycol

TENV: totally enclosed nonventilated motor

TEOR: thermal enhanced oil recovery

TFL: through-the-flow-line

THROT: tubing hanger running and orientation tool

TLM: taut leg mooring

TLV: threshold limit value

t_p: time well on production before shut-in, hours

t'_p: time well on production before shut-in, minutes

σ_h: total horizontal stress, psi

σ_v: total vertical stress, psi

TOV: total observed volume

TP: tubing pressure

TPC: tonnes-per-centimetre

TPI: tons-per-inch

TPQ: threshold planning quantities

tram: trammel

t_s: stabilization time, hours

TSCA: Toxic Substance Control Act

TSD: treatment, storage, and disposal

TSP: thermally stable polycrystalline

tstg: testing

TVD: true vertical depth

tsp: township

twp: township

Δt: shut-in time, hours

$\Delta t'$: shut-in time, minutes

UBD: underbalanced drilling

UBO: underbalanced operations

UEPS: uninterruptible electrical power supply

UIC: Underground Injection Control

UIPC: Underground Injection Practices Council

UKCS: United Kingdom Continental Shelf

ULCC: ultralarge crude carrier

UMC: underwater manifold center

URV: upper range value

USDW: underground sources of drinking water

USCG: United States Coast Guard

USFWS: United States Fish and Wildlife Service

USGS: United States Geological Survey

ΔV_i: volume increase, bbl per bbl initial fluid

μ: viscosity, cp

μ_a: apparent viscosity, cp

μ_o: viscosity oil, cp

μ_g: viscosity gas, cp

μ_w: viscosity water, cp

v: Poisson's ratio; rock volume, $in.^3$

$\bar{v}$: average flow velocity

V, v: volt

V: flow velocity, ft/sec; volume, bbls

VA: volt-ampere

VAC: volts, alternating current

VAR: volt-ampere reactive

VAR_C: volt-ampere reactive in circuits with capacitors

VAR_H: volt-ampere reactive in circuits with inductors. Also abbreviated VAR_L.

VAR_L: volt-ampere reactive in circuits with inductors. Also abbreviated VAR_H.

V_c: critical velocity, ft/sec

VDC: volts, direct current

VEF: vessel experience factor

VFR: visual flight rules

VIV: vortex-induced vibration

VLA: vertically loaded anchor

VLCC: very large crude carrier

VOC: volatile organic compound

VOM: volt-ohm-milliammeter

V_p: compressional wave velocity

V_s: shear wave velocity

VRU: vapor recovery unit

VSM: vertical support member

w: fracture width, in.

w_i: weight of fluid contained inside tubing, lb/in.

w_o: weight of annulus fluid displaced by bulk volume of tubing, lb/in.

w_s: weight of tubing, lb/in.

W, w: watt

W: weighting material needed, lb/bbl of initial fluid

WACOG: weighted average cost of gas

Wb: weber

WC: wildcat

W/C: water cushion

WEP: War Emergency Pipelines

wh: white

W-L, W/L: wireline

WMO: World Meteorological Organization

WOB: weight on bit

WOC: waiting on cement

WOE: waiting on engineering

WOG: water-oil-gas

WO/O: waiting on orders

WOR: water-oil ratio

WOW: waiting on weather

WTI: West Texas intermediate

X_c: capacitive reactance

XOR: exclusive or

yd: yard

yd^2: square yard

yd^3: cubic yard

YP: yield point, Bingham plastic fluid

y_w: water holdup

Z: distance of the righting arm; gas deviation factor; impedance

$ZnCl_2$: zinc chloride

SI Units

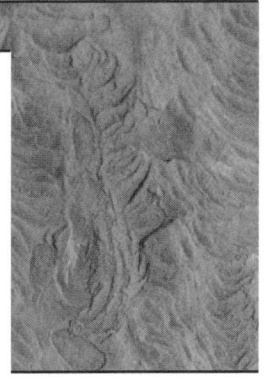

Quantity	Unit Name	Symbol	Formula
Base Units			
Length	metre	m	
Mass	kilogram	kg	
Time	second	s	
Electric current	ampere	A	
Temperature	kelvin	K	
Amount of substance	mole	mol	
Luminous intensity	candela	cd	
Supplementary Units			
Plane angle	radian	rad	
Solid angle	steradian	sr	
Derived Units			
Area	square metre		m^2
Volume	cubic metre		m^3
Speed, velocity	metre per second		m/s
Acceleration	metre per second squared		m/s^2
Density	kilogram per cubic metre		kg/m^3
Concentration	mole per cubic metre		mol/m^3
Specific volume	cubic metre per kilogram		m^3/kg
Luminance	candela per square metre		cd/m^2
Moment of force	newton metre		$N \bullet m$
Derived Units With Special Names			
Frequency	hertz	Hz	$1/s$
Force	newton	N	$kg \bullet m/s^2$
Pressure, stress	pascal	Pa	N/m^2
Energy, work, quantity of heat	joule	J	$N \bullet m$
Power	watt	W	J/s
Electric charge	coulomb	C	$A \bullet s$
Electric potential	volt	V	W/A
Electric resistance	ohm	Ω	V/A
Electric conductance	siemens	S	A/V
Electric capacitance	farad	F	C/V
Magnetic flux	weber	Wb	$V \bullet s$
Inductance	henry	H	Wb/A
Magnetic flux density	tesla	T	Wb/m^2

Quantity	Unit Name	Symbol	Formula
Luminous flux	lumen	lm	cd•sr
Illuminance	lux	lx	lm/m^2
Activity of radionuclides	becquerel	Bq	s^{-1}
Absorbed dose of ionizing radiation	gray	Gy	J/kg

Non-SI units allowable with SI

Quantity	Unit Name	Symbol	Formula
Time	minute	min	1 min = 60s
	hour	h	1 h = 3 600s
	day	d	1 d = 86 400s
	year	a	
Plane angle	degree	°	$1° = \pi/180$ rad
	minute	′	$1′ = \pi/10\ 800$ rad
	second	″	$1″ = \pi/648\ 000$ rad
Capacity or volume	litre	L	$1\ L = 1\ dm^3$
Temperature	degree Celsius	°C	interval of 1°C=1K
Mass	tonne	t	1 t = 1 000 kg
Revolution	revolution	r	1 r = 2 rad
Marine and aerial distance	nautical mile		1 nautical mile = 1 852 m
Marine and aerial velocity	knot	kn	1 nautical mile per hour = (1 852 600)m/s
Land area	hectare	ha	$1\ ha = 10\ 000\ m^2$
Pressure	standard atmosphere	atm	1 atm = 101.325 kPa

SI Units for Drilling

Quantity or Property	Conventional Units	SI Unit	Symbol	Multiply by
Depth	feet	metres	m	0.3048
Hole and pipe diameters	inches	millimetres	mm	25.4
Bit size				
Weight on bit	pounds	decanewtons	daN	0.445
Nozzle size	32ds inch	millimetres	mm	0.794
Drill rate	feet/hour	metres/hour	m/h	0.3048
Volume	barrels	cubic metres	m³	0.1590
	U.S. gals/stroke	cubic metres per stroke	m³/stroke	0.00378
Pump output and flow rate	U.S. gpm	cubic metres per minute	m³/min	0.00378
	bbl/stroke	cubic metres per stroke	m³/stroke	0.159
	bbl/min	cubic metres per minute	m³/min	0.159
Annular velocity Slip velocity	feet/min	metres per minute	m/min	0.3048
Liner length and diameter	inches	millimetres	mm	25.4
Pressure	psi	kilopascals megapascals	kPa MPa	6.895 0.006895
Bentonite yield	bbl/ton	cubic metres per tonne	m³/t	0.175
Particle size	microns	micrometres	Mm	1
Temperature	°Fahrenheit	°Celsius	°C	(°F - 32)/1.8
Mud density	ppg (U.S.)	kilograms per cubic metre	kg/m³	119.82
Mud gradient	psi/foot	kilopascals per metre	kPa/m	22.621
Funnel viscosity	s/quart (U.S.)	seconds per litre	s/L	1.057
Apparent and plastic viscosity	centipoise	millipascal seconds	mPa•s	1

315

Quantity or Property	Conventional Units	SI Unit	Symbol	Multiply by
Yield point Gel strength and stress	$lb_f/100\ ft^2$	pascals	Pa	0.4788 (0.5 for field use)
Cake thickness	32ds inch	millimetres	mm	0.794
Filter loss	millimetres or cubic centimetres	cubic centimetres	cm^3	1
MBT (bentonite equivalent)	lb/bbl	kilograms per cubic metre	kg/m^3	2.85
Material concentration	lb/bbl	kilograms per cubic metre	kg/m^3	2.85
Shear rate	reciprocal seconds	reciprocal seconds	s^1	1
Torque	foot-pounds	newton metres	N•m	1.3558
Table speed	revolutions per minute	revolutions per minute	r/min	1
Ionic concentration in water	equivalents per million	moles per cubic metre	mol/m^3	1
Corrosion rates	$lb/ft^2/year$	grams per square metre per day	g/m²•d	13.377
	mils per year	millimetres per year	mm/a	0.0254

SI Equivalents

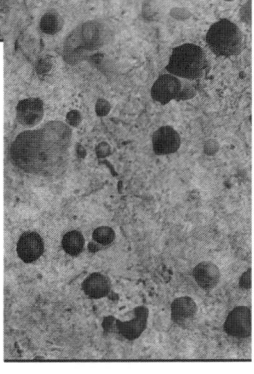

Length

1 millimetre = 0.04 inch

1 centimetre = 0.39 inch

1 metre = 39.37 inches = 1.09 yards

1 inch = 2.54 centimetres

1 foot = 3.05 decimetres

1 yard = 0.91 metre

1 mile = 1.61 kilometres

1 foot = .305 metre

Area

1 square centimetre = 0.15 square inch

1 square decimetre = 0.11 square foot

1 square metre = 1.20 square yards

1 hectare = 2.47 acres

1 square kilometre = 0.39 square mile

1 square inch = 6.45 square centimetres

1 square foot = 9.29 square decimetres

1 square yard = 0.83 square metre

1 acre = 0.40 hectare

1 square mile = 2.59 square kilometres

Pressure

1 kilopascal = 0.145 pound per square inch

1 kilopascal per metre = 0.044 pound per square
  inch per foot

1 pound per square inch = 6.894 kilopascals

1 pound per square inch per foot = 22.62 kilopascals per metre

Volume

1 cubic centimetre = 0.06 inch

1 cubic metre (stere) = 1.31 cubic yards

1 cubic inch = 16.39 cubic centimetres

1 cubic foot = 0.28 cubic decimetre

1 cubic yard = 0.75 cubic metre

Capacity

1 millilitre = 0.06 cubic inch

1 litre = 61.02 cubic inches = 1.507 liquid quarts

1 decalitre = 0.35 cubic foot = 2.64 liquid gallons

1 fluid ounce = 29.57 millilitres

1 U.S. gallon = 3.785 litres

1 barrel (oil) = 159 litres

Weight

1 gram = 0.04 ounce

1 kilogram = 2.20 pounds

1 metric ton (tonne) = 0.98 English ton

1 ounce = 28.35 grams

1 pound = 0.45 kilogram

1 English ton = 1.02 metric tons

Density

1 kilogram per litre = 8.34 pounds per gallon

1 kilogram per litre = 62.5 pounds per cubic foot

1 pound per gallon = 0.119 kilogram per litre

1 pound per cubic foot = 0.016 kilogram per litre

SI PREFIXES

Value	Prefix	Symbol	Value	Prefix	Symbol
10^{18}	exa	E	10^{-1}	deci	d
10^{15}	peta	P	10^{-2}	centi	c
10^{12}	tera	T	10^{-3}	milli	m
10^{9}	giga	G	10^{-6}	micro	M
10^{6}	mega	M	10^{-9}	nano	n
10^{3}	kilo	k	10^{-12}	pico	p
10^{2}	hecto	h	10^{-15}	femto	f
10^{1}	deca	da	10^{-18}	atto	a